William R. Hendee (Ed.)

Biomedical Uses of Radiation

William R. Hendee (Ed.)

Biomedical Uses of Radiation

Part A – Diagnostic Applications

Weinheim · New York · Chichester · Brisbane · Singapore · Toronto

Editor:
Dr. William R. Hendee
Medical College of Wisconsin
Office of Research, Technology, and Information
8701 Watertown Plank Road
Milwaukee, WI 53226
USA

Library of Congress Card No. applied for.

A catalogue record for this book is available from the British Library.

Deutsche Bibliothek Cataloguing-in-Publication Data:
Biomedical Uses of Radiation / William R. Hendee (Ed.). – Weinheim ; New York ; Chichester ; Brisbane ; Singapore ; Toronto : Wiley-VCH
ISBN 3-527-29668-9
Part A. Diagnostic Applications. – 1999
Part B. Diagnostic Applications. – 1999

Printed on acid-free and chlorine-free paper

Composition: kühn & weyh Software GmbH, D-79110 Freiburg
Printing: Strauss Offsetdruck GmbH, D-69509 Mörlenbach
Bookbinding: J. Schäffer GmbH & Co. KG, D-67269 Grünstadt
Printed in the Federal Republic of Germany

Preface

In 1995 the world celebrated the centennial anniversary of the use of radiation for both diagnostic and therapeutic purposes in medicine. This celebration emphasized the many contributions of radiation to the diagnosis and treatment of disease and to improvements in the health and well being of people around the world. It also drew attention to the ongoing progress in these contributions that continues even today. In fact, never before have changes and improvements in the medical uses of radiation occurred as rapidly as they are happening now. Computers, digital radiation detectors, image reconstruction from projections, information networks, and functional and metabolic imaging are contributing to an ongoing revolution in medical imaging with radiation. In therapeutic medicine, revolutionary new treatment methods with radiation sources both outside and inside the body are now possible through advances in nanoelectronics and computer technologies. These advances reflect the efforts of scientists worldwide working in collaboration with physicians and technologists to advance the applications of radiation to improve health and relieve pain and suffering.

The effective use of radiation in medical diagnosis and therapy requires a sophisticated understanding of physics principles and engineering concepts and their applications to anatomic and physiologic challenges intrinsic to human health and disease. Exposition of these principles, concepts and applications is the primary objective of this two volume series entitled Biomedical Uses of Radiation. The books are intended for readers who are interested primarily in the scientific and technical features of radiation medicine rather than mainly in the clinical applications. Hence, the books are directed toward an audience with an understanding of physics and mathematics at the beginning graduate level. Each of the chapters has been written by a recognized expert in the field, and is carefully laid out to guide the reader efficiently to a clear understanding of the technical features of the subject under consideration.

Working on this two-volume series has been a distinct pleasure for me. Each of the authors has been diligent with regard to deadlines and formatting, and dedicated to the belief that the books are needed and will be highly useful to physicists and engineers working in the field as well as to students just entering the discipline. One individual to whom we are all indebted is Ms. Terri Komar, editorial assistant in my office. She has kept us all on track and focused on our shared objectives, and has done so with great efficiency and good humor. My wife Jeannie also deserves special mention; her support and enthusiasm gives me constant sustenance.

William R. Hendee
Milwaukee, Wisconsin USA

Contents

Volume A – Diagnostic Applications

Volume B – Therapeutic Applications

List of Contributors

Benjamin R. Archer
Baylor College of Medicine
Department of Radiology
Onc. Baylor Plaza
Houston, TX 77030
USA

Peter A. Bandettini
Biophysics Research Institute
Medical College of Wisconsin
8701 Watertown Plank Road
Milwaukee, WI 53226
USA

Rasmus M. Birn
Biophysics Research Institute
Medical College of Wisconsin
8701 Watertown Plank Road
Milwaukee, WI 53226
USA

George T. Y. Chen
Department of Radiation and Cellular Oncology
Univcrsity of Chicago
8541 S. Maryland Ave MC 0085
Chicago, IL 60637
USA

William T. Chu
Lawrence Berkeley National Laboratory
University of California
Berkeley, CA 94720
USA

Paul M. DeLuca
Department of Medical Physics
University of Wisconsin
1300 University Avenue
1530 MSC
Madison, WI 53706-1532
USA

Larry A. DeWerd
Department of Medical Physics
University of Wisconsin
1300 University Avenue
1530 MSC
Madison, WI 53706-1532
USA

Robert L. Dixon
Bowman Gray School of Medicine
Wake Forest University
Winston-Salem, NC 27157-1088
USA

Kathleen M. Donahue
Biophysics Research Institute
Medical College of Wisconsin
8701 Watertown Plank Road
Milwaukee, WI 53226
USA

F. Marc Edwards
Radiation Oncology Associates
Johnson County Radiation Therapy Facility
12000 W 110th St
Overland Park, KS 66210
USA

Eric G. Hendee
Department of Medical Physics & Human Oncology
University of Wisconsin
K4/B100 Clinical Science Center
Madison, WI 53792-0600
USA

William R. Hendee
Medical College of Wisconsin
8701 Watertown Plank Road
Milwaukee, WI 63226
USA

Donald E. Herbert
University of South Alabama
Department of Physics and Statistics
269 CSAB
Mobile, AL 36688
USA

David E. Hintenlang
Department of Nuclear Engineering Science
University of Florida
P.O. Box 100385
Gainesville, FL 32610-0385
USA

Russell, K. Hobbie
Department of Physics and Astronomy
University of Minnesota
148 Physics Building
116 Church St. SE
Minneapolis, MI 55455
USA

Geoffry S. Ibbott
University of Kentucky Medical Center
Department of Radiation Medicine
800 Rose Street
Lexington, KY 40536-0084
USA

Roger H. Johnson
Marquette University
Department of Biomedical Engineering
Olin Engineering Center
Milwaukee, WI 53201-1881
USA

Marie Foley Kijewski
Brigham and Women's Hospital
Harvard Medical School
Department of Radiology
75 Francis Street
Boston, MA 02115
USA

Carolyn M. Kimme-Smith
UCLA School of Medicine
Department of Radiological Sciences
Los Angeles, CA 90024
USA

Haakil Lee
Department of Radiology and Radiological Sciences
R-1311 MCN
Vanderbilt University Medical Center
Nashville TN 37232-2675
USA

Azam Niroomand-Rad
Department of Radiation Medicine
Georgetown University
3800 Reservoir Road, NW
Washington, DC 20007
USA

Jatinder, R. Palta
Department of Radiation Oncology
University of Florida
P.O. Box 100385
Gainesville, FL 32610-0385
USA

Charles A. Pelizzari
Department of Radiation and Cellular Oncology
University of Chicago
5841 S. Maryland Ave MC 0085
Chicago, IL 60637
USA

Satish C. Prasad
SUNY Health Science Center
Department of Radiation/Radiation Physics
750 E. Adams St.
Syracuse, NY 13210
USA

William R. Riddle
Department of Radiology and Radiological Science
R-1311 MCN
Vanderbilt University Medical Center
Nashville, TN 37232-2675
USA

E. Russell Retenour
Department of Radiology, Box 292
University of Minnesota
420 Delaware, St. SE
Minneapolis, MI 55455
USA

John C. Roeske
Department of Radiation and Cellular Oncology
University of Chicago
5841 S. Maryland Ave MC 0085
Chicago, IL 60637
USA

John A. Rowlands
Reichmann Research Building
Sunnybrook Health Science Center
2075 Vayview Avenue
Toronto ON M4N 3M5
Canada

Thaddeus V. Samulski
Duke University Medical Center
Department of Radiation Oncology
Box 3085
Durham,NC 27710
USA

Douglas J. Simpkin
St. Luke's Medical Center
2900 West Oklahoma Avenue
Milwaukee, WI 53201
USA

Bruce R. Thomadsen
Department of Medical Physics & Human Oncology
University of Wisconsin
K4/B100 Clinical Science Center
Madison, WI 53792-0600
USA

Benjamin M. W. Tsui
University of North Carolina
Biomedical Engineering & Radiology
CB #7525
152 McNider Hall
Chapel Hill, NC 27599-7575
USA

1 Production and Interaction of X rays Radiation Measurement Quantities and Units

Roger H. Johnson

Biomedical Engineering, Marquette University

1.1 Introduction

This chapter deals with the production of X rays and their mechanisms of interaction with materials. The emphasis is placed on X rays useful for diagnostic imaging, that is in the energy range between 10 and 140 keV. X-ray interactions are discussed primarily from the standpoint of imaging, as opposed to diffraction or spectroscopy. Brief coverage of important developments not yet clinically utilized, such as synchrotron and plasma X-ray sources, is also included. Because these sources produce radiation over a wide spectral range, including intense soft X rays useful for biological X-ray microscopy, microtomography and spectroscopy, production and interaction of low-energy X rays are discussed, though in considerably less depth. Though electron-impact sources are still used to produce most X rays used in clinical applications and for biomedical research, the unique properties of some of the newer devices, most notably laser-produced plasma and synchrotron X-ray sources, have made possible discoveries which never would have been made in their absence. As these sources become ever cheaper and more efficient, they are certain to have an important impact in basic research, and possibly in clinical practice as well.

1.2 Production of X rays

This section describes the physics involved in and a few of the devices used for the production of X rays. Most types of artificially-produced ionizing photon beams arise from one of two basic mechanisms: Bremsstrahlung generation or characteristic emission. A description of these mechanisms will be followed by sections on the three most important classes of devices used to produce X rays for biomedical imaging: electron-impact X-ray tubes, synchrotrons and laser-produced plasma X-ray sources. Less common and recently proposed X-ray sources like special-purpose microfocal tubes, z-pinch plasmas, and X-ray lasers will be discussed in less depth.

1.2.1 Bremsstrahlung and Characteristic Radiation

When high-energy electrons interact with matter, their kinetic energy is lost to heat and radiative processes. Though nearly all the incident electron energy is converted to heat, the radiative processes are significant enough to be the source of most X rays used for imaging. The necessary parts of an X-ray source of the type most commonly employed in the clinic are a cathode (source of electrons) and an anode, housed in an evacuated enclosure to facilitate the passage of electrons between them, and a high-voltage generator to supply the potential difference between the cathode and anode which imparts to the electrons their kinetic energy. X rays were in fact discovered when Roentgen observed evidence of Bremsstrahlung emission upon discharging high-voltage sparks in a low-pressure glass tube.

When energetic electrons, or other "swift" charged particles, interact with matter, there are four types of interaction which result eventually in the loss of all their kinetic energy[1]:

1) Elastic collision with atomic electrons in the target. Significant only for very-low-energy electrons (<100 eV), these interactions may be thought of as taking place with the target atom as a whole. The incident electron is deflected by the field of the atomic electrons, but the amount of energy transferred to the target atom is less than the lowest ionization potential.
2) Elastic collision with a target nucleus. This interaction, which has a relatively high probability of occurring, causes deflection of the incident electron, but is not accompanied by excitation of the target nucleus, nor by radiation, resulting only in the loss of a sufficient (small) amount of kinetic energy to conserve momentum between the interacting particles. The first two types of interaction do not produce energetic photons and will not be considered further.
3) Inelastic collision with bound atomic electrons in the target. Most of the incident electron kinetic energy is lost by this mechanism. Bound electrons may either be excited or ejected from the target atom (ionization). If the energy of a colliding incident electron exceeds their binding energy and inner-shell (K and L) electrons are ejected from target atoms, these vacancies are filled by downward cascades from outer orbitals as shown in Figure 1.1[2], an event accompanied by emission of characteristic photons which may contribute a significant though small fraction (10–25% for tungsten in the diagnostic energy range) of the flux available for imaging.
4) Inelastic collision with a target nucleus. If incident electrons pass close to a target nucleus but escape capture, they will be deflected in their path by the electric field of the nucleus and may lose energy either to nuclear excitation or, more commonly, to radiation as shown in Figure 1.2. It is this radiation, termed "Bremsstrahlung" (German for "braking radiation"), which constitutes the primary emission from clinical X-ray tubes. Although highly unlikely at low incident electron energies, and relatively rare at diagnostic energies, at extremely high energies (>10 MeV) the likelihood of these inelastic nuclear collisions may surpass that of ionization. At all energies the ratio of these radiative to ionizational energy losses is greater for high Z than for low. In the diagnostic range, even for high-Z elements, it is less than 0.1. If an incident electron of kinetic energy T experiences a direct hit with a target atom nucleus, losing all its energy in a single interaction, a photon of energy $h\nu = T$ will be produced. Such photons are very rare, and possess the highest energy possible for electron-target interactions.

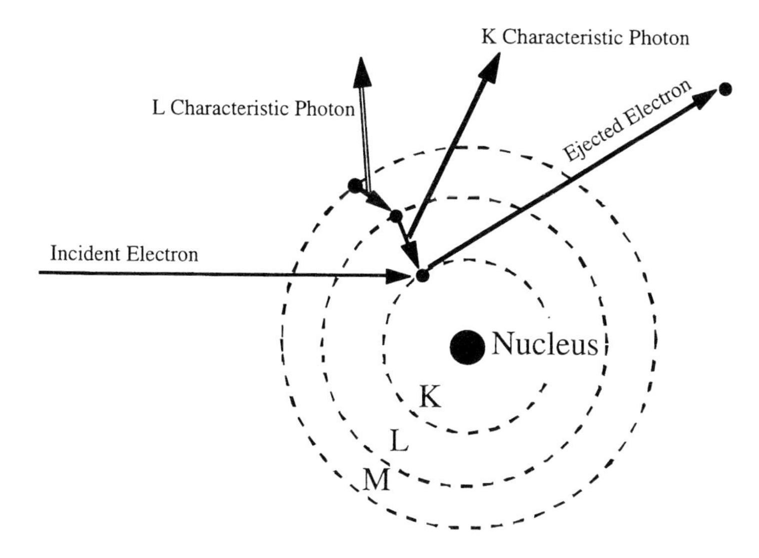

Figure 1.1. Fluorescent X-ray emission. An inner-shell electron is ejected and the downward cascade to fill the vacancy results in emission of photons of characteristic energies. (Adapted from Bushberg *et al.*, 1994[2].)

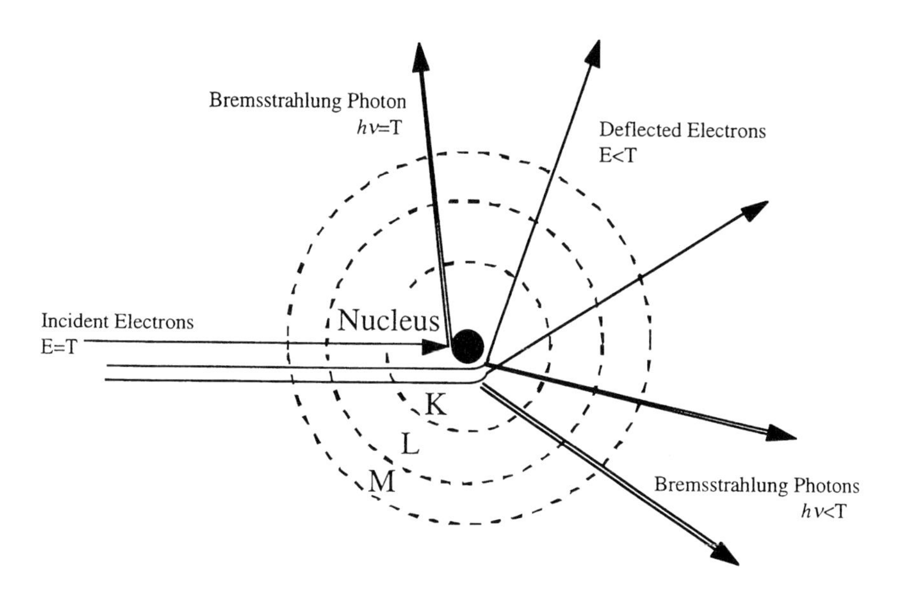

Figure 1.2. Bremsstrahlung emission. Incident electrons are accelerated in the nuclear field emitting energetic photons to conserve energy and momentum. (Adapted from Bushberg *et al.*, 1994[2].)

In a thick target, all of the kinetic energy of the incident electrons will eventually be lost by a combination of these four types of interaction. Each electron may undergo a large number of interactions before being stopped. After inelastic interactions involving atomic electrons, the ejected electrons may themselves have substantial kinetic energy. These δ rays are fast charged particles which will also be stopped by a combination of the four interactions above. A 100-keV electron may undergo 1000 such interactions before being stopped [3].

From classical theory, if it is assumed that all target electrons are free electrons (justified in the case where the target electron's binding energy is negligible compared with the kinetic energy it receives), interaction with atomic electrons is reduced to the simple case of an elastic coulomb collision. For such collisions, the probablility of transferring an energy Q to the target electron is inversely proportional to Q^2, heavily favoring interactions resulting in small energy losses over large-loss interactions. In fact, about half the kinetic energy of incident electrons is lost in a large number of low-loss ionizational collisions and the other half to a much smaller number of higher-loss interactions in which δ rays or Bremsstrahlung are produced. Limitations of the classical theory for inelastic scattering of electrons by atomic elcctrons are detailed by Evans[1].

As derived by Evans from Bethe's original retardation law, for a non-relativistic electron of velocity V, rest mass $m_0 \approx 10^{-27}$ g, and kinetic energy $T = ½\, m_0 V^2$ interacting with a target of atomic number Z, the energy loss per unit pathlength due to ionizational collisions is approximately given by[1]:

$$\left(\frac{\mathrm{d}T}{\mathrm{d}s}\right)_{ion} = 1.257 \times 10^{-6} \frac{e^4}{m_0 V^2} NZ \ln \frac{m_0 V^2}{I\sqrt{2}} \tag{1.1}$$
$$= 6.280 \times 10^{-7} \frac{e^4}{T} NZ \ln \frac{T\sqrt{2}}{I} \quad \text{J/cm}$$

where $e = 4.80 \times 10^{-10}$ esu is the charge on the electron, N is Avogadro's number (atoms/cm^3) and I is the geometric mean of the ionization and excitation potentials of the target atom. I is a nebulous quantity, difficult to determine theoretically or experimentally, but it has been approximated by

$$I \approx kZ \tag{1.2}$$

where k is a constant decreasing from 18 eV for H to 10 eV for Pb[1]. The inverse relationship between stopping power and electron energy is evident from Eq. (1.1). In the diagnostic energy range, while most of the radiation is produced by Bremsstrahlung, most of the incident electron energy is lost to ionizational collisions. A different form for the above equation for electron retardation by ionization and more complex relationships between I and Z are reviewed by Rao-Sahib and Wittry[4], who also show that reasonable agreement with experiment is obtained when k is set to the popular value of 11.5 eV[5].

For inelastic nuclear interactions, classical theory holds that radiation will be produced whenever charged particles are accelerated, and that the amplitude of the emitted electromagnetic radiation will be proportional to the acceleration undergone by the charge. The acceleration of a particle with charge ze and mass M by a nucleus of charge Ze is proportional to $\frac{Zze^{2}}{}$[1].

Since intensity is proportional to the square of the product of the charge and the amplitude, the intensity of emitted radiation is proportional to $\frac{Z^{2}z^{4}e^{6}}{}$. It is seen that the Bremsstrahlung intensity varies as the square of the atomic number of the target, a prediction confirmed for thin (but not thick, see below) targets by experiment, and inversely as the square of the mass of the incident particle, explaining why electrons produce about one million times more Bremsstrahlung than protons or alpha particles of the same velocity.

Most of the quantum mechanical theory for Bremsstrahlung is posed in terms of, for example, loss of kinetic energy per unit pathlength of the incident radiation. Since it is practically impossible to know anything about pathlengths of electrons in solid targets, many experiments, which have been used to successfully validate quantum mechanical theoretical findings, have been carried out on extremely thin target foils in which, on average, one or zero interactions would take place, making the foil thickness a reasonable approximation for the pathlength traversed by emerging quanta. A plane wave representing the incident electron enters the nuclear field of a target atom and has a small probability of emitting a photon after being scattered. In the diagnostic energy range, for nuclei of charge Ze and incident electrons of kinetic energy T and total energy $T + m_0c^2$, the differential cross section for emission of a photon of energy between $h\nu$ and $h\nu + \mathrm{d}(h\nu)$ is:

$$\mathrm{d}\sigma_{\mathrm{rad}} = \sigma_0 B Z^2 \frac{T + m_0c^2}{T}\frac{\mathrm{d}(h\upsilon)}{h\upsilon}\ \mathrm{cm}^2/\mathrm{nucleus}\,, \tag{1.3}$$

$$\text{where } \sigma_0 = \frac{1}{137}\left(\frac{e^2}{m_0c^2}\right)^2 = 5.80\times10^{-28}\ \mathrm{cm}^2/\mathrm{nucleus}\,, \tag{1.4}$$

and B is a somewhat controversial coefficient which varies slowly with Z and T. Results originally due to the nonrelativistic quantum theoretical approach of Sommerfeld[1,6] suggest that, over the diagnostic energy range, B decreases gradually from 10 to about 7 as the photon energy increases from zero to T. Compared to classical theory, which predicts a large number of low-loss interactions, quantum mechanical theory predicts a smaller number of large-loss events. Total energy-loss predictions are about equal, but the expected emitted photon spectrum differs markedly between the two models. Experiments confirm the correctness of the quantum mechanical approach.

1.2.2 Electron-Impact X-ray Sources

By far the most common type of X-ray source is the electron-impact X-ray tube. All clinical X-ray sources, including those used for standard radiography, CT scanning and mammography, are of this type. A typical high-flux X-ray tube for diagnostic radiology is shown in Figure 1.3[7]. Electrons emitted from a heated filament are accelerated through a large potential and made to impinge upon a small area of a solid metallic target or anode. The "technique" or "technique factors" employed in making an X-ray exposure refer to the voltage (kVp for "kilovolts, peak") applied to the tube and the electron current (mA) passing through it. The kVp primarily influences the energy of photons in the beam, and therefore their penetrating capacity, while the mA is linearly related to the total photon energy (flux) in the beam. Most of the radiation produced in these tubes is the broadband Bremsstrahlung emitted when the electrons are slowed to rest in the anode material. If the electron accelerating voltage is high enough to impart to the electrons adequate energy to eject inner-shell electrons from atoms in the target, a small fraction of the emitted flux may consist of photons possessing energies characteristic of electron shell transitions. This section describes the circuits and hardware required for routine production of X rays by tubes, the design of tubes for clinical imaging, the spectra they emit and methods of shaping them, and specialized sources for mammography and research applications.

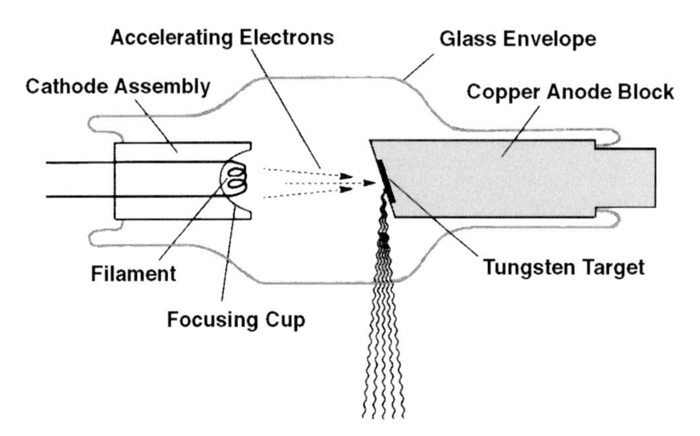

Figure 1.3. Schematic of a stationary-anode, clinical X-ray tube. (Reprinted with permission from Wolbarst, 1993[7].)

1.2.2.1 Electrical Circuits for X-ray Tubes

All electron-impact X-ray tubes require a stable high voltage to be applied between the cathode, from which the electrons are emitted, and the positive anode upon which they impinge. The magnitude of this applied voltage determines the penetrating capacity of the X-ray beam produced by the tube. Since there is an optimum applied voltage for any particular imaging task, it is important that the power supply provide a stable, constant direct current (DC) voltage. Here we briefly consider the circuit elements required to achieve this objective, and how they are combined in practical high-voltage generators for clinical X-ray sources.

Transformers are used to increase or decrease the magnitude of the available line voltage. In order to acquire enough energy to produce X rays useful for clinical imaging, electrons have to be accelerated through very high electric field gradients produced by applied potential differences of between 10 000 and 140 000 volts. Since the alternating current (AC) power supplied by electrical utilities to consumer hospitals is usually either 110 or 220 volts, a device is needed to step these voltages up many fold. Transformers in X-ray generators change the amplitude or voltage of an alternating current by the process of mutual electromagnetic induction[8]. In the simplest air core transformer, two adjacent, insultated coils of wire each set up in the other an induced electromotive force. The input coil is called the primary, and the output is referred to as the secondary. Alternating current flowing through the primary causes a magnetic field to be set up in its vicinity. The oscillating magnetic flux links with the secondary coil and induces an AC current (I) to flow through it. The voltage (V) in each coil is proportional to the number of turns (N), or windings:

$$\frac{V_p}{N_p} = \frac{V_s}{N_s}, \tag{1.5}$$

where the subscripts refer to the primary and secondary. Ignoring the slight loss of energy, the output and input powers are equal:

$$V_p I_p = V_s I_s . \tag{1.6}$$

If there were 500 turns on a 220-volt, 55-kilowatt primary, the output of a 250 000-turn secondary would be 110 000 volts at 500 milliamps. Step-up transformers have more turns on the secondary and increase the voltage at the output relative to the input, while step-down transformers are used to decrease the output voltage.

Real transformers are not perfectly efficient, and energy losses – due to resistance of the windings, eddy currents, and magnetic domain hysteresis – are ultimately manifested as heat. Proper design should keep these losses to less than five percent. Thicker copper windings, particularly on the high-current primary, minimize resistance losses. The efficiency of the simple insulated coils of the air core transformer can be improved by insertion of iron cores into the windings as in the open core transformer depicted in Figure 1.4[8]. Magnetization of the core markedly increases the intensity of the magnetic flux set up by the alternating current through the coil, but significant losses persist due to leakage at the ends of the cores. Leakage is much lower in closed core transformers, shown in Figure 1.5, since the magnetic flux passes through a continuous core from one coil to the other. The most efficient and popular design is the shell-type transformer shown in Figure 1.6. Both heavily-insulated coils are wound around the same central pole in the core. The cores of closed core and shell type transformers are made up of laminated steel layers. Insulation between the layers minimizes losses due to eddy currents induced by the AC in the windings.

An autotransformer is a special type of transformer used to regulate the voltage input to the primary of the high-voltage step-up transformer. In an X-ray tube, the autotransformer provides the ability to select the kVp applied to the tube. Although the same objective could be accomplished with a variable resistor, autotransformers are selected for kVp control of X-ray tubes, since energy losses are much less than those associated with resistors. As shown in Figure 1.7, an autotransformer consists of a single winding. The taps on the primary side span a fixed number of turns, usually somewhat less than the total. One of the taps on the secondary side is variable, allowing the output to span a selectable number of turns. The ratio of the autotransformer output to the input voltage is determined by the number of turns spanned at the output

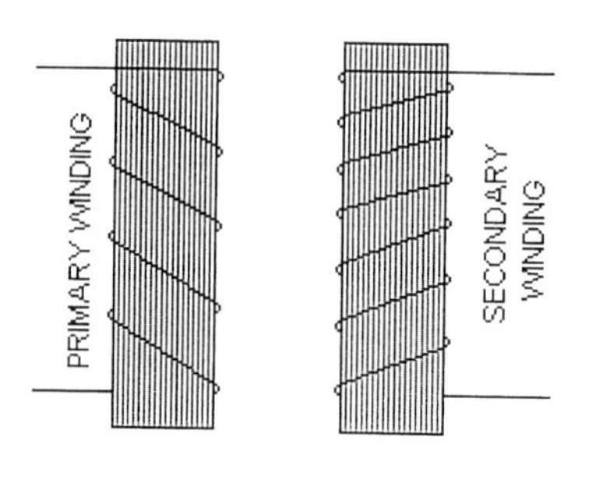

Figure 1.4. Open core transformer for X-ray generator. Iron cores within the windings improve efficiency of magnetic flux generation. (Adapted from Selman, 1994[8].)

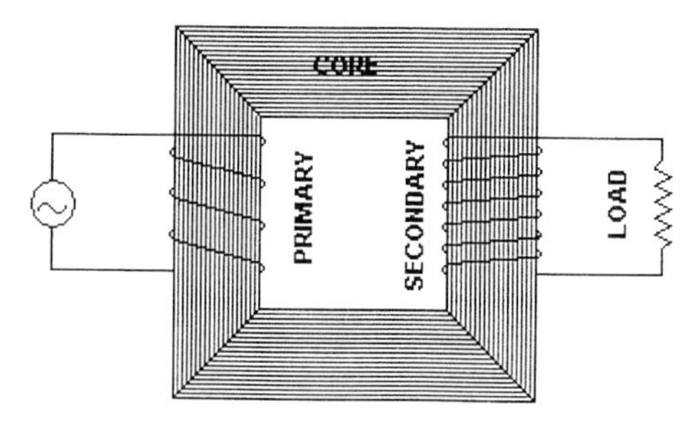

Figure 1.5. Closed core transformer. Leakage is reduced and efficiency improved by continuous conducting path between windings. (Adapted from Selman, 1994[8].)

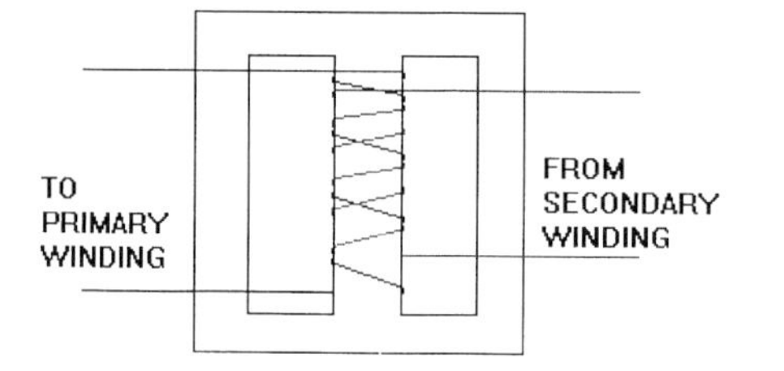

Figure 1.6. Shell-type transformer. Primary and secondary windings share the central member of the core for maximum efficiency. (Adapted from Selman, 1994[8].)

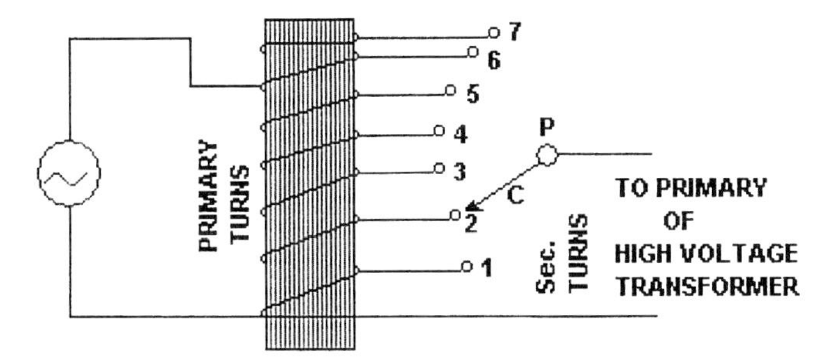

Figure 1.7. Schematic of autotransformer for kVp slection. (Adapted from Selman, 1994[8].)

relative to the number of turns spanned at the input. Autotransformers are suitable for selection of output voltages which differ from the input voltage by less than an order of magnitude.

Rectifiers permit the flow of electrons through X-ray tubes in one direction only: from the cathode to the anode. If a high-voltage alternating current were applied to the cathode-anode system of an X-ray tube, current would only flow through the tube for that half of the AC cycle during which the cathode was negative relative to the anode. Rectification, defined as changing alternating to direct current, requires construction of a circuit through which current can only flow in one direction. Since the heated filament is an efficient souce of electrons separated by vacuum from the anode, normally a very inefficient source of electrons, application of an alternating current to the tube would result in generation of a pulsating X-ray beam which would be on during only half of the cycle period: The tube would self-rectify. This is undesirable since not only is the tube only producing X rays half the time, but most of the time the potential applied to the tube will not be equal to the optimal kVp for the imaging task at hand. Further, a self-half-rectified tube will be able to withstand less heat loading than the optimal design because the anode can never be allowed to reach temperatures so high that they promote electron emission from the target material. Such electrons would flow to the filament during the half-cycle it was positively biased with respect to the "anode", damaging the filament and shortening its life. Anode damage (melting, pitting and cracking) is minimized by bombarding the surface with the steadiest possible stream of electrons over the required exposure time: another motivation for stable, DC tube excitation.

An important design goal for the X-ray generator is to produce a stable constant voltage for application to the tube. Rectification may be accomplished by the use of solid state diodes, which have largely replaced the vacuum diode tubes of older-generation X-ray apparatus. Diodes permit the flow of current in only one direction. If a battery and load are connected in series with a p-n junction semiconductor diode, the electrons from the n-type half of the diode can only cross the "potential hill" at the junction when the negative battery terminal is connected to the n-type side (the diode is in forward bias):

If the connections are reversed (reverse bias), no current will flow. A large number of solid state diodes (about 150) need to be packaged in series to produce a single rectifier (still referred to as a "diode") for an X-ray tube because a single semiconductor (doped silicon) diode element can only withstand a reverse bias of about 1000 volts or less before breakdown occurs by the avalanche or Zener effect [9].

For half-wave ("single-pulse") rectification, a single diode in series between the transformer secondary and anode, or two diodes, one on the anode side and the other on the cathode side, will serve to allow passage of electrons through the diode, and hence the tube, only during the half cycle when the diode is forward biased. The tube will never be allowed to enter reverse bias, but X rays will only be generated over less than half of the AC cycle (Figure 1.8, top panel) [2]. Full-wave (two-pulse) rectification is achieved by arranging four diodes in a bridge circuit as shown in Figure 1.9[7]. During each AC half cycle, electrons will flow through two parallel diodes in the direction opposite to the arrow of the diode symbol. The tube will never enter reverse bias and current will flow through the tube during the entire cycle. While more efficient than half-wave rectification and sometimes used in practice, especially on older systems, a single-phase, full-wave-rectified waveform is not optimal for application to the X-ray tube since the kVp will still differ from the optimal value during a significant portion of the cycle (Figure 1.8, second panel).

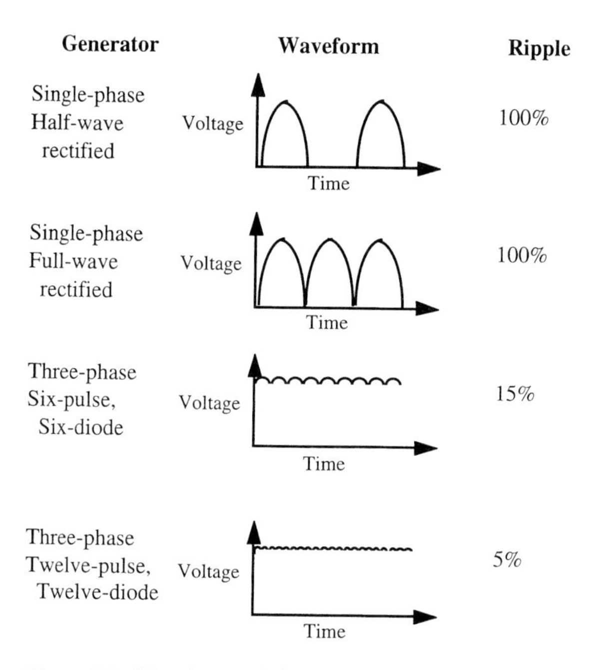

Figure 1.8. Waveforms of the potential applied across the cathode-anode gap by various types of X-ray generator. (Adapted from Bushberg *et al.*, 1994[2].)

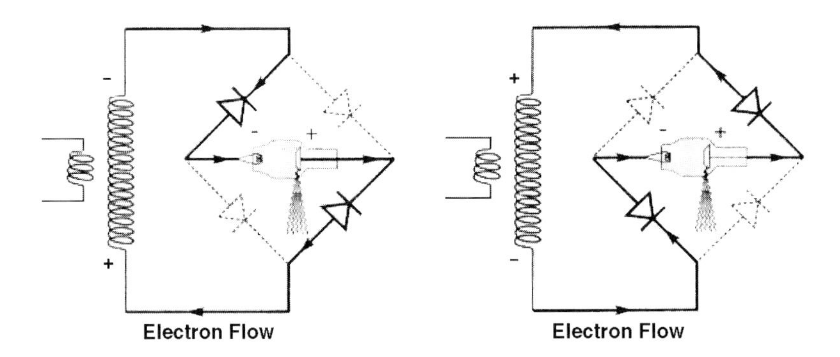

Figure 1.9. The four-diode bridge circuit for full-wave rectification. (Reprinted with permission from Wolbarst, 1993[7].)

A three-phase generator serves to maintain the potential across the tube at a more constant value than a single-phase unit. The power supply to a three-phase generator consists of three, single-phase power supplies which are 120 degrees out of phase with each other. Three transformers are required to step up the voltage. The three primaries are connected in a "delta" or triangular configuration; the secondaries may be configured either as a delta or a star (or "wye") as shown in Figure 1.10 and 1.11[7,8]. The "six-pulse, six-diode" configuration of Figure 1.10 supplies six pulses to the tube every 1/60th second. A slight modification provides "12-pulse, 12-diode" excitation with the circuit of Figure 1.11. The waveforms corresponding to these circuits are shown in the third and fourth panels of Figure 1.8.

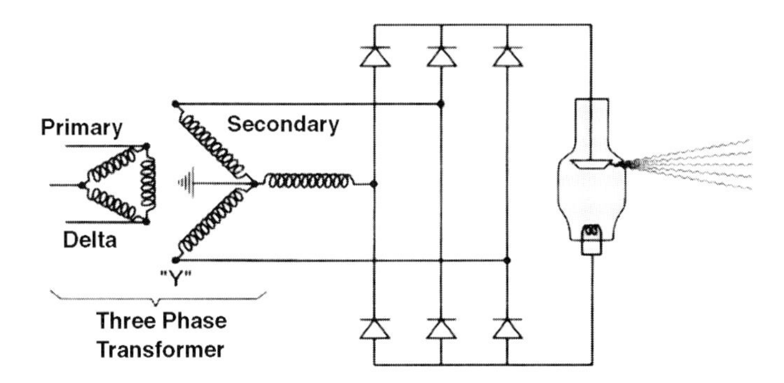

Figure 1.10. The delta, wye, three-phase transformer for six-pulse, six-diode high-voltage excitation. (Reprinted with permission from Wolbarst, 1993[7].)

Medium- and high-frequency X-ray generators represent the state of the art in X-ray tube electronics. These devices incorporate circuits which full-wave rectify, then smooth the (single- or three-phase) line voltage using capacitive filtering. An inverter "chops" the nearly-constant DC voltage at between 5

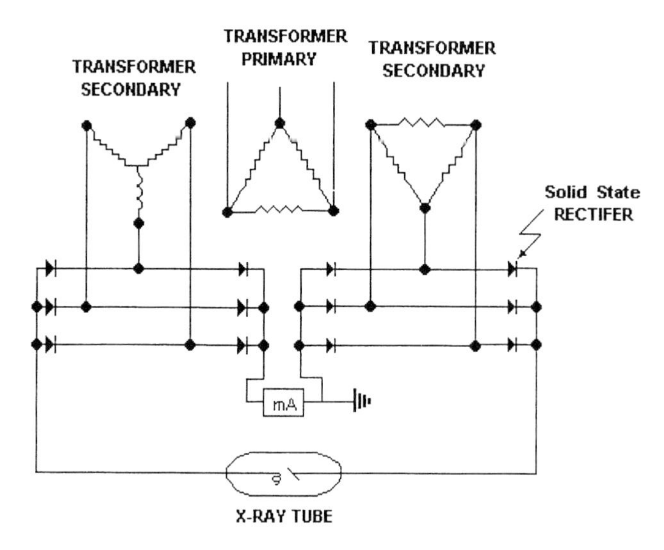

Figure 1.11. Simplified schematic of the generator circuit for twelve-pulse, twelve-diode tube excitation. (Adapted from Selman, 1994[8].)

and 100 kHz, producing a high-frequency, low-voltage AC waveform which is easily step-up transformed and stabilized. This high-frequency, high-voltage AC is full-wave rectified and smoothed a second time, providing a nearly-constant potential to the tube. In addition, transformer efficiency improves with AC frequency, permitting smaller, lighter generator designs. High-frequency generators (20 or 100 kHz) are frequently employed for mammography due to dose reductions of approximately 25% [8].

Voltage ripple, expressed as a percent, is defined as:

$$\%\text{Ripple} = \left[\frac{V_{max} - V_{min}}{V_{max}}\right] \times 100 . \tag{1.7}$$

Half-wave and full-wave rectification both produce waveforms with 100% ripple; three-phase, six-pulse and three-phase, twelve-pulse generators exhibit about 15% and 5% ripple, respectively; and medium- and high-frequency generators produce about the same voltage ripple as the best three-phase units: 5% or less.

1.2.2.2 Cathodes and Anodes

The electrons accelerated to produce X rays in the tube are emitted from a filament, similar to that in a light bulb. The filament is usually a fine, coiled tungsten wire. A low-voltage (6–12 volts) current (3–5 amps) is passed through the filament, heating it, lowering the work function of the metal, and facilitating the release of electrons from the wire's surface by thermionic emission. At white heat, electrons may be literally "boiled off" the surface of the wire. The filament current is supplied by a low-voltage circuit. A rheostat (variable resistor), saturable reactor, or high-frequency filament control circuit allows the current flowing through the filament, and therefore the tube current (mA) to be varied. Small increases in temperature can lower the work function of the filament material significantly and cause large increases in tube current. Older equipment usually had an ammeter in series with the filament circuit by which the filament current could be monitored. Modern X-ray tubes are equipped with a high-frequency filament controller or a space charge compensator which uses a feedback mechanism during exposure to match the filament current with the chosen mA (tube current). A small change in filament current causes a large change in the tube mA and therefore the X-ray flux emitted from the tube. The filament controller corrects for temporal fluctuations in the supply line voltage so that changes of +/–10% in line voltage result in only fractional percent changes in filament current. Both ends of the filament wire are connected to the low-voltage filament circuit, and one of them is also connected to the negative terminal of the high-voltage power supply across the tube.

The filament may be contained in a "cathode block" which includes a cup-shaped surface, facing the anode, negatively biased with respect to the filament. This focusing cup repels the emitted electrons and serves to confine them to a more tightly-focused beam. In the absence of an applied potential between the cathode and the anode, electrons emitted from the filament cluster in its vicinity. Their mutual repulsion limits the emission of additional electrons from the wire, and actually forces some of the electrons back into the filament. At equilibrium, a "space charge" or cloud of electrons, forms at the filament tip. When a high voltage is applied to the tube, making the anode positive with respect to the filament, electrons are drawn from the space charge and accelerated toward the anode. The excited volume at the anode surface in which the electrons are stopped is called the focal spot. Bremsstrahlung and possibly characteristic X rays are emitted in all directions from the focal spot. Some tubes are provided with two filaments: a larger coil of coarser wire for short, high-mA exposures (to minimize image degradation due to patient motion, for example), and a shorter, finer coil for longer exposures used for exams in which perception of fine image detail is important and motion is less of a problem. The smaller filament produces a smaller focal spot lessening image degradation due to penumbral blurring.

The essential requirements of the anode, or tube target, are that it emit X rays efficiently in response to electron bombardment and that it withstand without damage the high thermal loads imposed by the rapid deposition of large amounts of energy. The first requirement favors a high-Z material; the second, high melting point, good thermal conductivity, and toughness. A denser target possesses the additional, lesser, advantage that electrons are stopped in a smaller excitation volume from which the X rays are emitted. A number of candidate anode materials and their important physical properties are shown in Table 1.1. Most of them are not commonly used in conventional tubes for clinical imaging but may be employed in the special purpose research and microfocal instrumentation described in Sections 1.2.2.8 and 1.2.4.2. Tungsten (W), with its very high melting point and reasonable thermal conductivity and density provides a near-ideal combination of properties for use as an anode material. Tungsten is the choice for almost all X-ray tube anodes. (Its high melting point also explains its utilization as the filament material.) A tungsten-rhenium alloy which is tougher, and therefore resists cracking and tearing in response to thermal stresses set up by steep thermal gradients, is employed in some tubes.

Table 1.1.

Target Material	Atomic Number	Atomic Weight	Thermal Conductivity (W/cm·K)	MP (C)	Density
Be	4	9	2.18–1.68	1273	1.85
Al	13	27	2.36–2.40	660	2.70
Ti	22	48	0.224–0.207	1660	4.54
Cr	24	52	0.965–0.921	1857	7.18
Fe	26	56	0.865–0.720	1535	7.87
Co	27	59	1.05–0.890	1495	8.90
Ni	28	59	0.941–0.827	1453	8.90
Cu	29	64	4.03–3.95	1083	8.96
Mo	42	96	1.39–1.35	2617	10.22
Pd	46	106	0.716–0.730	1554	12.02
Ag	47	108	4.29–4.26	962	10.50
Sn	50	119	0.759–0.704 (Orientat.-depend.)	232	5.7–7.3
Ta	73	181	0.574–0.577	2996	16.65
W	74	184	1.77–1.63	3410	19.30
Pt	78	195	0.717–0.717	1772	21.45
Au	79	197	3.19–3.13	1064	18.88

The separation from the filament to the anode is generally on the order of a few centimeters. The polished surface of the anode which is bombarded with electrons is inclined at a slight angle, generally in the 7–20° range, with respect to a plane perpendicular to the filament-anode axis, as shown in Figure 1.12[8]. Due to the line-focus principle, the apparent focal spot "seen" by the film or detector is smaller than the bombarded area of the anode, a phenomenon known as foreshortening. This allows the thermal load on the target to be distributed over a larger area for a given effective focal spot size. The width of the focal spot (in the plane perpendicular to the page in Figure 1.12) is the same as that of the excited volume on the target, but the apparent height of the focal spot (H_a), at the center of the field, is related to the real height of the bombarded area (H_b) by $H_a = H_b \sin\Theta$, where Θ is the angle of the anode surface with respect to the vertical. The apparent height of the focal spot toward the anode side of the tube is less than this value (as evident from Figure 1.12), providing higher spatial resolution, while the focal spot dimension toward the cathode side of the tube is larger.

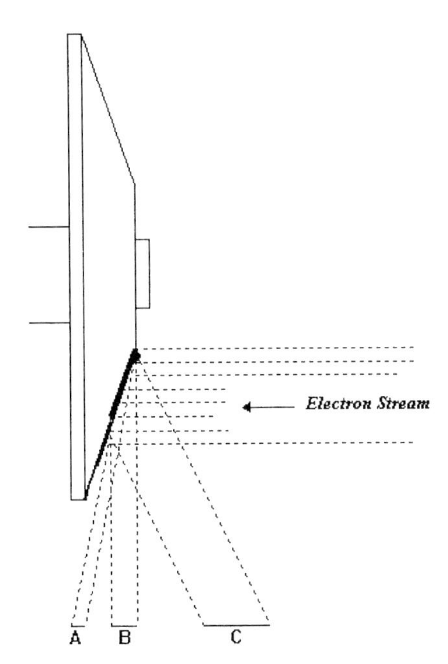

Figure 1.12. An angled anode distributes the thermal load over a larger area for a given focal spot size seen by the film. Foreshortening causes the effective focal spot height to vary over the film as shown. (Adapted from Selman, 1994[8].)

An undesirable consequence of an angled anode is known as the "heel effect". Even though the range of electrons in a tungsten target at diagnostic energies is on the order of microns, X rays emerging toward the anode side of the focal spot have to escape, on average, from deeper depths in the target, and undergo considerably more self absorption in the anode on their path toward the film, than rays on the filament side of the beam. This increased filtration preferentially removes lower-energy photons from the beam, causing the average and equivalent energies of the beam to be greater toward the anode side of the exposed field.

1.2.2.3 Sources for Diagnostic Radiology

In stationary anode tubes, the fixed anode assembly is often composed of a copper bulk material with a polished tungsten button embedded into the surface forming the actual target. The superior heat conduction properties of copper aid in dissipating the extreme thermal loads encountered during operation of the tube. In rotating anode tubes, a tungsten-rhenium-coated molybdenum anode disk is mounted on the shaft of a high-speed motor. During exposures, the anode rotates and the energetic electrons impinge upon an angled, annular strip around the anode called the focal track. This increases the bombarded target area by several hundred fold compared to stationary anode devices, and provides dramatically increased heat loading capacity. Molybdenum, with its lower thermal conductivity, prevents transmission of excessive heat to the motor bearings. Rotating anodes are generally either three or five inches in diameter, and rotate at speeds between 3 000 and 10 000 rpm. Larger diameters and higher rotation speeds are associated with higher-heat-load equipment like helical CT scanners which have adequate heat loading capacity to provide continuous, multi-second exposures. While general-purpose stationary anode tubes have focal spots in the 2–4 mm range, rotating anode, dual filament tubes may have focal spots between 0.3 and 2 mm.

A complete circuit diagram of a generator for a full-wave rectified X-ray tube is shown in Figure 1.13[8]. The primary, or low-voltage circuit comprises all the components upstream of the high-voltage, step-up transformer. The kVp meter across the primary circuit is calibrated at the factory to establish the precise relationship between the AC voltage at the secondary of the auto-transformer and the DC potential across the tube. In general, the center of the secondary of the step-up transformer is grounded: a large, positive potential is applied to the anode, while a negative potential of equal magnitude is applied to the cathode. A milliammeter in series with the secondary circuit indicates the tube current.

In operation, for a given filament current and no applied tube potential, a space charge of magnitude proportional to the filament current will form on the filament side of the cathode-anode gap. As the kVp across the tube is increased from zero, at low kVp more and more electrons will be drawn from the space charge and accelerated toward the anode, causing a rapid increase in the tube current as shown in Figure 1.14[2]. Above a threshold kVp, all electrons emitted from the filament will immediately be accelerated toward the anode: The space charge will have been entirely depleted, and the tube current will have reached the maximum value achievable with that particular filament current. In the low-kVp regime, the tube current and therefore the emitted X-ray flux is said to be space-charge-limited. The tube current depends on the rate at which the applied potential can pull electrons out of the space charge. After the space charge is depleted, tube output is "emission limited", meaning that the filament temperature imposes the limit on the rate of electron release. In the emission-limited regime, where X-ray tubes should be operated, tube current can be controlled independently from accelerating

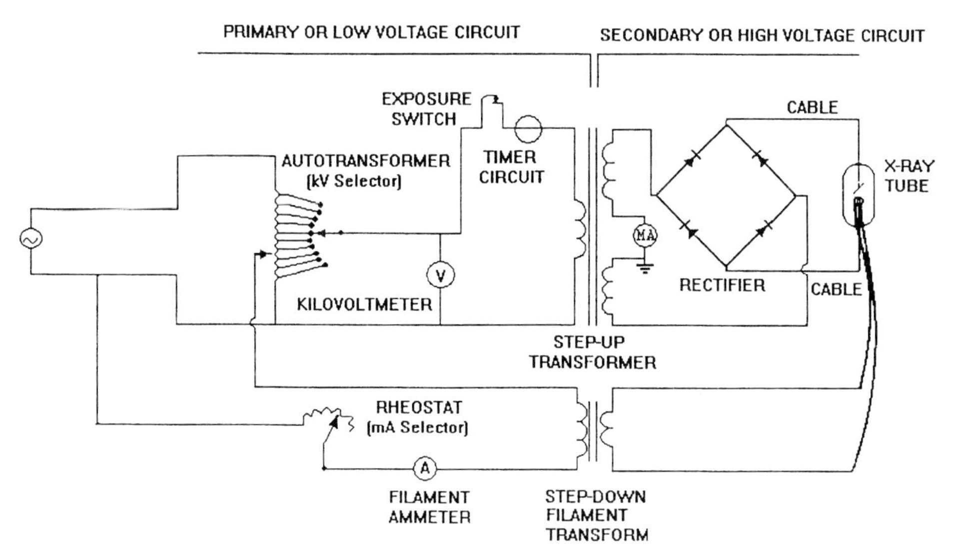

Figure 1.13. Circuit diagram of a generator for a full-wave rectified X-ray tube. (Adapted from Selman, 1994[8].)

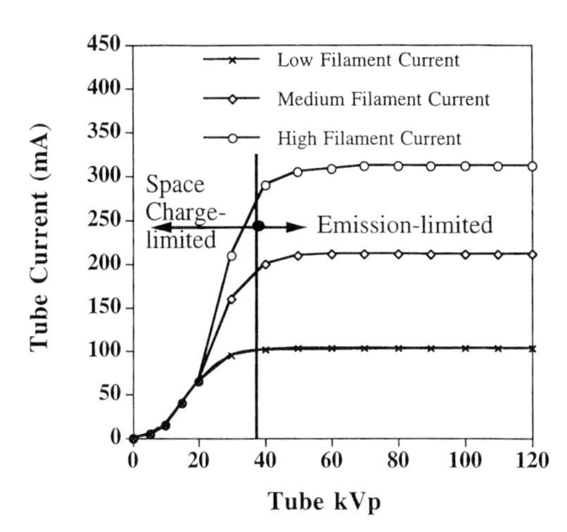

Figure 1.14. Tube current as a function of kVp for various filament temperatures. Operation in the emission-limited output regime permits independent control of kVp and mA. (Adapted from Bushberg *et al.*, 1994[2].)

potential by varying the filament temperature. The space charge compensator automatically controls the filament temperature to allow the kVp to be raised while the selected mA is maintained.

The field-emission X-ray source is a specialized type of electron impact source in which the filament is not heated. Field emission is an alternative to thermionic emission which usually (in electron microscopes) exploits the fact that a single crystal of tungsten or other metal, when correctly oriented with respect to the applied field, has a much lower work function than rolled or drawn material, allowing electrons to tunnel through the potential barrier at the surface under the influence of a strong electric field in the absence of heating. The technology was originally reduced to routine practice in electron microscopy[10,11] because the very high current densities at the cathode (10^3–10^7 A/cm^2 compared to 10 A/cm^2 for thermally-assisted tungsten) allowed the formation of nm-order electron probes orders of magnitude brighter than available in conventional instruments.

Use of the field-emission principle in microfocal X-ray tubes was advocated in 1957 by Pattee[12] and possibly earlier by others[13]. Subsequently, use of field emission sources for diagnostic radiology has been suggested[14], though they have never come into routine use. In the larger-focal-spot, high-flux design proposed for clinical use, the anode was a tungsten cone with axis coincident with the beam axis. An array of sharp needle assemblies surrounding the cone served as the cathode. Though unclear from the literature, it is likely that polycrystalline cathode needles were used, increasing the average work function at the surface due to random grain orientation and compromising the available beam current. Pulsing the target briefly to voltages as high as 350 kV resulted in the emission of a high-current stream of electrons from the needle tips which produced a flash of X rays. A sequence of pulses at 1000 Hz was used to obtain the exposures required clinically. The advantages of the design included the circular symmetry of the focal spot which avoided the deleterious effects on image quality and exposure uniformity due to the heel effect, the source size was independent of tube current, the spatial resolution at the image plane could be improved by a factor of two or more, and the tubes could be very compact. The primary disadvantages were the difficulty in obtaining adequate flux for imaging the torso and the very high vacuum ($\sim 10^{-9}$ torr) required to avoid contamination of the cathode surface. Even a few atoms on the crystal surface can raise the work function dramatically, seriously reducing the available flux.

1.2.2.4 X-ray Tube Ratings

X-ray tubes are generally rated by manufacturers as to the permissible power loading of the focal spot, the heat storage characteristics and cooling rate of the anode, and the cooling rate of the housing. These ratings are supplied as charts, which must be specific for each tube and combination of operating conditions: generator type, focal spot size and anode rotation speed. For example, a stationary-anode tube is far more susceptible to thermal damage from excessive power loading than is a rotating-anode tube, which dissipates heat over a larger surface area on the target. Similarly, three-phase units tolerate heat considerably better than single-phase tubes. Variations in focal spot size and anode rotation speed are also important considerations.

Power loading limitations determine the maximum exposure time that can be tolerated by the anode without damage for a single exposure at a particular combination of tube voltage and current. A typical single-exposure rating chart is shown in Figure 1.15. Either kVp or mA is plotted vs. exposure time with each of the several curves corresponding to a discrete mA or kVp

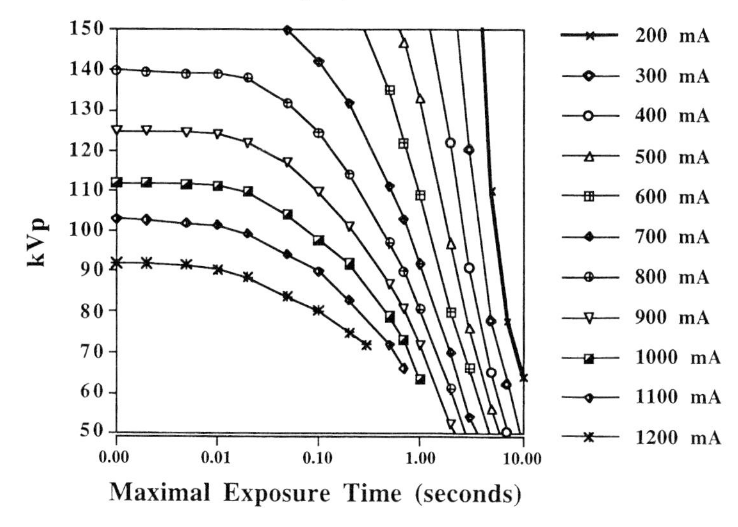

Figure 1.15. Rating charts for an X-ray tube under various operating conditions. (Adapted from Bushberg *et al.*, 1994[2].)

value, respectively. The safe operating regions are below and to the left of the curves. For a given exposure time, the product of kVp and mA on each of the several curves will always be approximately constant and equal to the tube's power rating in kW.

Heating and cooling of the anode and tube housing is quantified in heat units (HU). This convention evolved as a convenient way to calculate energy deposition into the device from the parameters readily available to the operator: kVp, mA and exposure time. For a single-phase, full-wave rectified unit:

$$1\,\text{HU} = 1\,\text{kVp} \times 1\,\text{mA} \times 1\,\text{second} \approx 0.75\,\text{J}, \tag{1.8}$$

since the average voltage is about 75% of the nominal setting[3]. For three-phase generators (nearly constant voltage):

$$1\,\text{kVp} \times 1\,\text{mA} \times 1\,\text{second} = 1\,\text{J} \approx 1.35\,\text{HU}. \tag{1.9}$$

A typical chart depicting anode thermal characteristics is shown in the left panel of Figure 1.16[3]. For the rising curves, if a tube is operated continuously (fluoroscopic mode) for the length of time on the x axis at the heat input rate associated with the curve, the number of heat units on the y axis will have been deposited in the anode. The curves are nonlinear because as the anode temperature rises heat escapes faster. At longer times the curves tend to level out as the rate of heat dissipation almost equals the deposition rate. The descending curve quantifies the cooling rate after operation is discontinued and

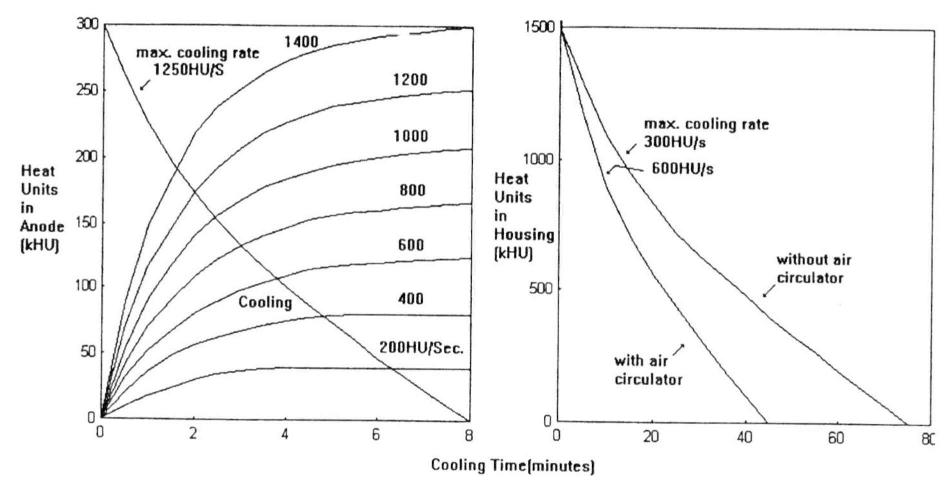

Figure 1.16. Left panel shows anode heating characteristics for various heat input rates versus operation duration and the time required for cooling. Right panel shows housing cooling curves with and without air circulation. (Adapted from Johns and Cunningham, 1983[3].)

is useful for determining the waiting time required before another continuous exposure can be initiated. The right panel in Figure 1.16 shows a housing cooling curve. The advantage of active cooling of the housing is evident.

1.2.2.5 X-ray Tube Spectra

In day-to-day clinical practice, it is generally not possible to accurately characterize or measure the spectrum of photon energies present in the X-ray beam. For research applications, for example to develop quantitative, multispectral imaging techniques and verify their performance, it is highly desirable to know the X-ray spectrum accurately. Attempts have therefore been made to measure and calculate it as discussed below. For routine purposes, however, it generally suffices to know the values of a small number of parameters which affect the number of photons emitted from the tube and govern the shape of the spectrum of photon energies.

The spectral distribution of the emitted photons is primarily influenced by the kVp, the target material, and by beam filtration. The potential applied to the tube governs the maximum energy of photons in the beam: higher kVp produces higher-energy radiation. The energies at which characteristic emissions occur are a function of the binding energies of anode atomic electrons causing higher-Z targets to produce higher-energy characteristic peaks in the emitted spectrum. Inherent filtration (attenuation by the glass envelope of the tube and any other material layers through which the beam must pass) and added filtration preferentially remove low-energy photons from the beam (mostly by photoelectric absorption, as described in Section 1.3.3.1), shifting the average energy of the beam to higher values.

The important characteristics of the shape of the emitted spectrum are embodied in the "quality" of the beam, as measured by the half-value layer. "Harder" beams have "higher quality", which means that they contain a higher proportion of high-energy photons, possess higher average and equivalent energies, and can therefore penetrate greater material thicknesses. "Lower quality" beams are beams which contain relatively more "soft", or low-energy photons. The half-value layer, or HVL, is defined as the thickness of any material required to reduce the exposure rate of an X-ray beam to half its incident value[3,8,15]. Since X-ray attenuation is exponential (as discussed in Section 1.3):

$$\frac{I}{I_0} = \frac{1}{2} = e^{-\mu \cdot \mathrm{HVL}} \tag{1.10}$$

$$\ln\left(\frac{1}{2}\right) = -\mu \cdot \mathrm{HVL} \tag{1.11}$$

$$\mathrm{HVL} = \frac{0.693}{\mu} \tag{1.12}$$

where μ is defined as the linear attenuation coefficient of the material used to attenuate the beam. Definitions of HVL found in the literature are often ambiguous as to whether the quantity being halved is exposure rate or intensity as Eq. (1.10) would imply[2,16–21]. For monoenergetic beams, defining HVL in terms of intensity or exposure would be equivalent, but for polyenergetic beams this is not the case because exposure is defined in terms of liberated charge per mass of air (see Section 1.4.4). Because the mass energy absorption coefficient of air exhibits a strong inverse energy dependence, the HVL of a soft polychromatic beam would be found to be thinner using exposure measurements than using intensity measurements. Energy-integrating detectors suitable for intensity measurements do not lend themselves to routine use in radiology, primarily due to count rate limitations. Since exposure is nearly always the quantity measured in practice it seems reasonable to define HVL in these terms, in spite of the apparent conflict with Eq. (1.10) and the very definition of the linear attenuation coefficient (both of which apply strictly for monoenergetic beams).

For a polyenergetic beam such as those employed in diagnostic imaging, attenuation of the beam by any material will preferentially remove lower-energy photons from the beam. The emergent beam will therefore be "hardened" or shifted to higher energies. If it is desired to determine the second HVL of the beam, defined as the thickness of the material to again halve the exposure, it will be necessary to insert more attenuating material to stop the higher-energy photons: the second HVL will be thicker than the first. It is possible to approximately reconstruct the spectrum by analysis of attenuation data as described by Delgado and Ortiz[22] and references 1–16 therein.

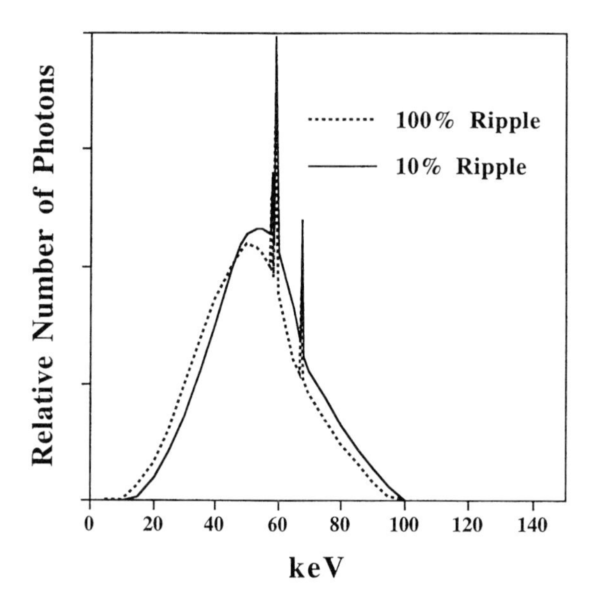

Figure 1.17. Effect of tube potential waveform on output spectrum. (Adapted from Bushberg *et al.*, 1994[2].)

The voltage waveform applied to the tube can also have an effect on beam quality. Any drop of the applied voltage due to ripple will result in a reduction of the maximum photon energy in the beam, causing a spectral shift toward lower energies as shown in Figure 1.17, and reducing beam intensity and penetrating power[2]. In general this will result in higher dose to the patient and greater heat loading of the anode.

The "equivalent energy" of the beam is defined as the energy of a monochromatic beam which would produce the same results in terms of its penetrating capacity[3]. A broadband X-ray beam having an equivalent energy $h\nu$ has the same HVL as a monochromatic beam of energy $h\nu$. After measuring the HVL in a practical situation, one can calculate the linear attenuation coefficient according to $\mu = 0.693/\mathrm{HVL}$, then refer to a table of linear attenuation coefficients as a function of energy for the material used in the HVL determination to find the equivalent energy of the beam.

Intensity, with units of energy per unit time per unit area (e. g. $\mathrm{keV/sec \cdot cm^2}$), is a measure of the energy in the beam. The beam intensity varies[16] linearly with the product of the tube current and exposure time (mAs) and with $(\mathrm{kVp})^2$. As described earlier, classical theory predicts that the efficiency of Bremsstrahlung production should vary as the square of the target atomic number, a relationship which holds reasonably accurately for intensity emitted from thin targets[1]. But quantum mechanics predicts, and most texts state[2,3,18], that for thick targets the intensity variation with Z is linear:

$$\text{Intensity} \propto \text{mAs} \times \text{kVp}^2 \times Z_{\text{anode}} \tag{1.13}$$

Rao-Sahib and Wittry measured the X-ray intensity produced at discrete kVp settings corresponding to characteristic emission lines using a curved crystal, wavelength-dispersive spectrometer[4]. They investigated 23 elements from $Z = 6$ to $Z = 92$ using polished flat targets in an X-ray microprobe with a take-off angle of 53°. The experimental results did not confirm a linear variation with Z, but indicated that intensity depended on Z^n, where n varied from 1.38 for aluminum ($Z = 13$) to 1.19 for gold ($Z = 79$). In many cases quantum mechanical theory serves to provide correction factors or supplemental terms to the classical results, inclusion of which often provides reasonable agreement with experiment, especially for non-relativistic electrons.

In research settings, when it is required to have precise knowledge of the X-ray spectrum, a histogram of photon energies can be obtained using a solid-state, energy-dispersive X-ray detector such as a high-purity germanium (HPGe) or a lithium-drifted silicon (Si:Li) crystal interfaced to a multichannel analyzer[15]. If the count rate is low enough, these detectors can register (detect and count) individual photons and determine their energies to sub-keV accuracy by pulse height analysis. In practice, the count rates (photon fluence) from clinical X-ray tubes are much too high for this type of detector and, if nothing were done to reduce them, pulse pile-up would result. Multiple low-energy depositions in the crystal might be erroneously summed to yield spurious high-energy counts and inaccurate spectral measurement. To remedy this, either a pinhole aperture may be placed in front of the detector or the detector positioned at a great distance from the source[23]. In the former case, the spectral intensity would have to be normalized (scaled up) using a separate measurement, perhaps with an ionization chamber, since the true energy fluence at the detector would not be known; in the latter, air absorption becomes a factor. Detector systems of this type normally require cooling of the crystal to cryogenic temperatures, are rather expensive, and are not readily available to many imaging researchers. Less expensive, room-temperature detectors based on silicon photodiodes have been used to characterize diagnostic X-ray spectra[24], but the inferior energy resolution of these devices and the necessity of correcting the data for photoelectron escape and scattered counts make these methods of spectral measurement unreliable.

Most direct measurements of X-ray spectra have been obtained for studies of electron-probe X-ray microanalysis, but the results can be compared to the same theories relevant for diagnostic X-ray tubes. Green and Cosslett used a crystal wavelength-dispersive spectrometer in conjunction with a flow proportional counter to measure the X-ray spectrum from a germanium target and to estimate the efficiencies of characteristic X-ray production from ten elemental targets spanning the periodic table from carbon to gold[25,26]. X rays were detected at a 45° takeoff angle from polished flat solid targets mounted in a special-purpose microfocal X-ray tube. For all elements the efficiencies ranged between 10^{-6} and 10^{-3} characteristic X rays per steradian per incident electron, depending primarily upon the ratio of the incident electron energy to the relevant shell ionization energy. At a fixed kVp, the characteristic production efficiency was found to be a rapidly decreasing function of Z.

These experimental results showed good agreement with their theoretical prediction of the efficiencies for K, L and M shell characteristic X-ray production, accounting for ionization by both electron bombardment and photoelectric absorption of the Bremsstrahlung (see Section 1.3)[27]. The higher fraction of characteristic radiation produced by low-Z (compared to the ubiquitous tungsten) targets explains their suitability for studies benefitting from quasimonochromatic, low-energy x-radiation.

Because of the difficulties associated with accurate spectral measurements, a number of investigators have attempted to derive analytical methods to predict the spectral output of X-ray sources. The earliest analytical result for the intensity emitted from a thick target may be due to Kulenkampff and Kramers[28,29]:

$$I_\upsilon = kZ(E_0 - h\upsilon)\,, \tag{1.14}$$

where I_υ is the emitted intensity of photons of energy $h\upsilon$, k is a constant, Z is the atomic number of the target, and E_0 is the energy of the incident electrons. If I, E and $h\upsilon$ are in keV and the energy interval unit is 1 keV, then the constant k is equal to[4] 2.2×10^{-6}. The units of Kramers's result as presented are:

$$\frac{\text{keV}}{\text{electron} \cdot 4\pi\,\text{steradians} \cdot \text{keV interval}}\,.$$

Conversion using:

$$\frac{I_\text{b} \times 6.2422 \times 10^{18}\,\dfrac{\text{electrons}}{\text{sec}} \times \dfrac{1-\cos\alpha}{2}}{\text{keV}/\text{photon}}\,,$$

where I_b is the tube current in amps, yields the photon flux per keV interval onto a circular detector subtending a cone of half-angle α.

Kramers's equation gives the triangular intensity distribution predicted by quantum mechanics and provides reasonably accurate, if fortuitiously so, estimates of X-ray intensity production in thick targets. Fortuitious, some have claimed, because two simplifying assumptions in Kramers's approach may have cancelled each other out[6]: He assumed that the coefficient B in the equation for the differential radiative cross section of the target nucleus (Eq. 1.3) is constant, when in fact it is a slowly decreasing function of energy, and he neglected self-absorption in the target. Assuming B to be constant would cause the predicted spectrum to be too hard, but neglecting the beam-hardening effect of self-filtration in the anode compensated in part for this error.

Soole[6], Birch and Marshall[30], and Tucker, Barnes and Chakraborty[31] all used similar approaches to derive analytical expressions for the spectra emitted from thick metallic targets bombarded by electrons of constant kinetic energy. Tucker, *et al.* have shown that the number of photons with energy between $h\upsilon$ and $h\upsilon$ + d$h\upsilon$ produced per incident electron (before self-filtration by the target material) can be derived from Eq. (1.3) to be:

$$N(h\upsilon)\mathrm{d}h\upsilon = \frac{\sigma_0 Z^2}{A}\frac{\mathrm{d}h\upsilon}{h\upsilon}\int_{h\upsilon}^{T_0}\frac{B(T+m_0c^2)}{T}\left(\frac{1}{\rho}\frac{\mathrm{d}T}{\mathrm{d}x}\right)^{-1}\mathrm{d}T, \tag{1.15}$$

where the symbols have the same meaning as in Eq. (1.3), and ρ and A are the target density and atomic mass. Taking attenuation by the target material into account, this becomes:

$$N(h\upsilon)\mathrm{d}h\upsilon = \frac{\sigma_0 Z^2}{A}\frac{\mathrm{d}h\upsilon}{h\upsilon}\int_{h\upsilon}^{T_0}\frac{B(T+m_0c^2)}{T}\mathrm{F}(h\upsilon,T)\left(\frac{1}{\rho}\frac{\mathrm{d}T}{\mathrm{d}x}\right)^{-1}\mathrm{d}T, \tag{1.16}$$

where

$$\mathrm{F}(h\upsilon,T)=\exp\left[\frac{-\mu_{\text{target}}(h\upsilon)\left(T_0^2-T^2\right)}{\rho c\sin\theta}\right] \tag{1.17}$$

gives the attenuation due to filtration by the anode according to the Thomson-Whiddington relation[1]. The depth, x, in the anode at which X rays are produced, and the distance, d, through which the exiting beam must penetrate are related by $d = x/\sin\theta$, where θ is the anode angle, and the depth at which electrons with initial energy T_0 have residual energy T is given by:

$$x = \frac{T_0^2 - T^2}{\rho c}, \tag{1.18}$$

where c is the Thomson-Whiddington (T-W) “constant”. Variability between the results for calculated Bremsstrahlung spectra obtained by various researchers is due in part to the values, shown in Table 1.2, assumed for the T-W constant, which increases slowly with electron energy. Soole used values derived from the works of Green and Cosslett[27] and Tothill, while Birch and Marshall and Tucker, *et al.* used values derived from Bichsel and the empirical relationship of Katz and Penfold[30].

Because of multiple scattering in the target (large number of small energy losses), the depth to which electrons penetrate in the anode will be considerably smaller than their pathlength. A number of practical difficulties, including straggling, make it difficult to experimentally determine the range of electrons in metals. Katz and Penfold gave an early review of the electron energy-range data and proposed a formula for the range (R_0) of electrons with incident energies (E in MeV) between .01 and 3.0 MeV[1]:

$$R_0\left(\mathrm{mg}/\mathrm{cm}^2\right) = 412E^{(1.265-0.0954\ln E)}. \tag{1.19}$$

Table 1.2.

Incident Electron Energy (keV)	T-W constant (10^5keV^{2}m^2/kg) Soole	Birch and Marshall
20	.29	–
25	–	.39
50	.44	.54
75	–	.625
100	.54	.70
150	.57	.84
200	.59	1.0
250	.61	–

This relationship yields shallower depths of penetration than Eq. (1.18) above. For example, for 100-keV electrons in tungsten, the Katz-Penfold equation gives a range of 6.9 microns, while the T-W equation using Birch and Marshall's value for the constant predicts that electrons will have lost all their energy at a depth of 7.4 microns, and Soole's assumed value of the constant yields 9.6 microns.

An important contribution of Tucker's group was to parameterize B as[31]:

$$B=\begin{cases}\left[A_0+A_1T_0\right]\left[1+B_1\left(\frac{h\upsilon}{T}\right)+B_2\left(\frac{h\upsilon}{T}\right)^2+B_3\left(\frac{h\upsilon}{T}\right)^3+B_4\left(\frac{h\upsilon}{T}\right)^4\right] & ,\ h\upsilon\le T\\ 0 & ,\ h\upsilon>T\end{cases}, \quad (1.20)$$

where $h\upsilon$ is X-ray photon energy, T_0 is the incident electron energy, T is the electron energy, and the coefficients A and B are parameters determined by least-squares fitting of the model's results to a large number of spectra determined experimentally by another group[32]. The coefficients A determine radiation output and were found to be $A_0 = 3.685 \times 10^{-2}$ photons/electron and $A_1 = 2.9 \times 10^{-5}$ photons/(electron · keV). The coefficients B govern the spectral shape and were –5.049, 10.847, –10.516 and 3.842.

For spectral modeling purposes, the characteristic contribution can simply be superimposed on the Bremsstrahlung component. The intensity of production of characteristic radiation has been found by many investigators to follow the relationship[26,33]:

$$I_K = C\left(\frac{T_0}{E_k}-1\right)^{1.63}, \quad (1.21)$$

where C is a constant, T_0 is the incident electron energy, and E_k is the K-shell binding energy. Most experiments have placed the value of the exponent within the range 1.0–1.7, depending upon the value of C. For tungsten, a value between 1.63 and 1.65 is often used. C, and therefore efficiency of characteristic X-ray production for a given T_0, increases rapidly with decreasing target atomic number. Because of this, attempts to produce quasimonochromatic beams from electron-bombarded targets for biomedical applications have often employed copper, aluminum and silver anodes. For fixed ratios of T_0/E_k (>1), C increases with Z. Characteristic X-ray production initially increases with depth into the target, then decreases as the electrons lose energy. Tucker modeled the production efficiency with depth as a parabolic function which reaches zero at the depth where the average electron energy equals the K-shell binding energy[31]:

$$N(h\upsilon_i) = A_k\left(\frac{T_0}{E_k}-1\right)^{n_k} f(h\upsilon_i)\int_0^R P\left(\frac{x}{R}\right)\exp\left(\frac{-\mu(h\upsilon_i)x}{\sin\theta}\right)dx\,, \tag{1.22}$$

where $N(h\upsilon_i)$ is the number of $h\upsilon_i$ characteristic X rays per incident electron emerging from the target, A_k and n_k are parameters adjusted to fit model results to experimental data, $f(h\upsilon_i)$ is the fractional emission for the various characteristic X rays (K_{α_1}, K_{α_2}, K_β, etc.), R is the distance at which the average electron kinetic energy equals the binding energy, and x is depth within the target. The parabolic production efficiency function used was:

$$P\left(\frac{x}{R}\right)=\begin{cases}\left(\frac{3}{2}\right)\left[1-\left(\frac{x}{R}\right)^2\right] & , x\le R\\ 0 & , x>R\end{cases} \tag{1.23}$$

For 90% tungsten/10% rhenium targets, very good fits to the data were obtained with $A_k = 1.349 \times 10^{-3}$ photons/electron and $n_k = 1.648$.

In general, probably because of the large numbers of adjustable parameters (in particular the parameterization of B), Tucker's results seem to provide the best fit to experimental data, while Kramers's result produces a spectrum of the same shape shifted about 5 keV toward lower energies and lacking the characteristic contribution. Birch and Marshall's publication shows a similar comparison to Kramers's model (though Tucker's implementation of Birch and Marshall produced a badly-skewed spectrum). Considering its extreme simplicity, the accuracy of Kramers's result remains remarkable.

1.2.2.6 Beam Filtration

The X-ray spectrum produced in the anode has the triangular energy distribution predicted by Kramers's rule. There is a preponderance of low-energy photons and very few photons with energies close to T_0. In general the beam must pass through several layers of material before emerging from the X-ray tube: the target material itself, the glass envelope surrounding the tube and/

or a vacuum window, usually composed of aluminum or beryllium. In addition it may pass through thin layers of oil and plastic. These unavoidable attenuating layers are referred to as inherent filtration, and serve to remove almost all the X rays below about 10 or 15 keV from the beam. Inherent filtration is usually specified in terms of equivalent millimeters of aluminum. Clinical X-ray tubes will typically have several millimeters inherent aluminum filtration (with a minimum imposed by regulations designed to reduce patient dose), while special purpose microfocal tubes may have a few hundred microns of beryllium ($Z = 4$) or less. As the tube ages and the anode and filament materials vaporize in the imperfect vacuum, the inherent filtration increases due to deposition on the output window. For most purposes, even further filtration of the beam is warranted, since low-energy photons have a reduced likelihood of reaching the image receptor and contributing to the image signal but do increase the dose to the patient. The most common practice is to add additional aluminum filtration. Some manufacturers provide the choice of using either a thin copper or aluminum filter.

For research purposes in which narrow-band or quasimonochromatic ratiation is useful, it is possible to employ anode/filter material combinations which enhance the production and transmission to the specimen of a selected energy band of radiation. Figure 1.18 shows the K-absorption edges of elements which occur between about 15 and 52 keV[34]. Since the fluorescent X rays are emitted with energy equal to the difference in binding energy between the two shells involved (usually the K and L shells), they will have

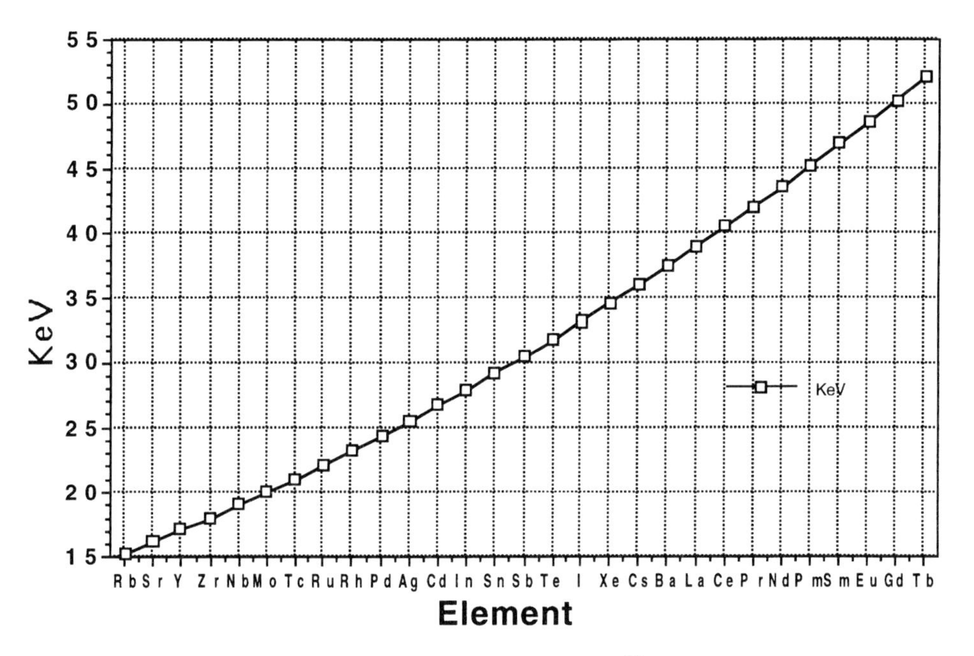

Figure 1.18. Energies of K absorption edges. (Data from Weast[34].)

an energy slightly below the K-edge and the target material itself will be relatively transparent to its own characteristic radiation. For example[34], a platinum target, with its K-edge at about 78.4 keV, emits K_{α_1} and K_{α_2} fluorescent X rays at about 66.8 and 65.1 keV, respectively, with a $K_{\alpha_2}/K_{\alpha_1}$ emission probability ratio of 0.583. Pt itself could be used as a filter to provide transmission of the radiation between 65.1 and 78.4 keV, while providing increased filtration of lower-energy photons. Enhanced removal of flux just below the characteristic emissions could be achieved by use of an ytterbium filter (K-edge at 61.3 keV), and narrow-band illumination achieved with the combination of either ytterbium and tungsten (K-edge = 69.5 keV) or ytterbium and tantalum (K-edge = 67.4 keV). A serious problem with absorption filtering to tailor the spectrum is that flux is severely reduced, particularly if filters with K-edges below the band of interest (like the ytterbium in this example) are used since, with increasing energy, absorption decreases exponentially below a K-edge, makes a step function up at the edge, then decreases exponentially again. Ytterbium would still absorb quite strongly between 65 and 67 keV.

Another application of beam filters is to equalize the exposure at the image receptor. Equalization or compensation filters present varied thicknesses over the image field to compensate for large variations in body thickness, such as occur between the neck and shoulders or the lungs and mediastinum for chest imaging, or for varied ray path lengths across the field[2]. Tapered wedge filters or hollowed-out trough filters are used for beam equalization in these applications. There is currently research underway to develop equalization methods for mammography where large density variations and the requirements to image the more attenuating areas in the center and and near the chest wall are in conflict with the desirability of visualizing the skin line, which tends to get overexposed. Bowtie filters are used in computed tomography to compensate for the ray path lengths at the periphery of the field being shorter than those in the center.

1.2.2.7 Sources for Mammography

Mammography is currently an area of intensive research for several reasons, including the challenging nature of the imaging task, the increased emphasis on women's health and early detection of cancer, the availability of funding, and the imposition of new regulations intended to improve the consistency of quality control nationwide. Being composed entirely of soft tissue – fat, glandular tissue and parenchyma – there is very little object contrast available for transmission X-ray imaging. The difficulty lies in maximizing image contrast while keeping dose to the acceptable, low level required of any screening procedure. In general, dose and contrast both decrease with increasing kVp.

Since subject contrast (see Section 1.3.6 for a discussion of object, subject and image contrast) in this instance is improved by use of low-energy X rays, mammography tubes are optimized to produce radiation in the 17- to 25-keV range. Though efforts are underway to produce monochromatic beams for breast imaging with electron impact tubes using graphite mosaic crystal monochromators[35,36] and the possibility of mammography using synchrotron[37,38] and Compton backscattered (see Section 1.2.5) radiation is being investigated[39,40], it is currently impractical to obtain narrow-band radiation in the desired range for mammography. Breast imaging is therefore the most common application of beam filtering to tailor the spectrum from an electron-impact tube. A number of modeling studies have been carried out to determine the optimal spectrum for mammography[41,42]. Most mammography tubes are operated between 22 and 30 kVp and employ molybdenum anodes and filters. Mo has the K-edge at 20.01 keV, K_{α_1}, K_{α_2} and K_{β_1} lines at 17.48, 17.37 and 19.61 keV, respectively, and $K_{\alpha_2}/K_{\alpha_1}$ and K_{β}/K_{α} emission probability ratios of 0.525 and 0.197. The production and transmission of radiation with energies between 17 and 20 keV is enhanced with a Mo-Mo anode-filter combination, which increases the photoelectric-to-Compton interaction ratio and improves image contrast for the soft tissues of the breast. Recently some mammography tubes have been provided with both molybdenum and rhodium anodes and filters. Rh has the K-edge at 23.26 keV and K_{α_1}, K_{α_2} and K_{β_1} lines at 20.22, 20.07 and 22.72 keV, respectively. Mo-Mo, Mo-Rh, and Rh-Rh target-filter combinations are possible. (One manufacturer also supplies an aluminum filter.) Studies have indicated that Mo-Rh and Rh-Rh combinations provide some dose reduction relative to Mo-Mo for all breasts at the expense of contrast in the case of small or fatty breasts, both effects due to the higher average energy of the rhodium-filtered beam[43]. For thick or dense breasts, however, the dose reduction is not accompanied by contrast degradation, suggesting that perhaps both target-filter combinations have their place, depending on patient characteristics.

One requirement of a mammography tube is that it produce a small focal spot for full-breast studies (approximately 300 microns) and even smaller spots for spot magnification studies (about 100 microns). This is achieved by use of a single fine wire filament and a focusing cup which is activated (negatively biased with respect to the filament) for the small-spot mode. The anode is always of the rotating variety with either a molybdenum track or selectable molybdenum and rhodium tracks. The exit window must be thinner than in conventional tubes, and in modern equipment is made of beryllium instead of aluminum or glass, in order to transmit the low-energy radiation. The anode angle is usually smaller than in conventional tubes, which is acceptable because the smaller resulting focal spot is achieved without the problem of restricting the field size. The adverse consequences of the heel effect are minimized by positioning the filament nearer the patient's head than the target so the more intense side of the beam is toward the chest wall.

The Mammography Quality Standards Act (MQSA) of 1992 (interim guidelines currently in effect; final rule expected by 1998) attempts to improve the quality of mammographic imaging nationwide by regulating the facilities performing imaging studies[44]. Among the requirements are annual surveys of the equipment by qualified medical physicists. The focal spot size may be measured by either of the two methods described in the Mammography Quality Control Manual of the American College of Radiology. The older method involves characterizing the focal spot size using a fine slit to measure the line spread function of the system[45] if a preliminary star pattern measurement falls out of specification. The newer method characterizes the high-contrast spatial resolution of the system in line pairs per millimeter by use of a bar (line pair) pattern[46]. A complete description of modern mammography equipment, including the tube and generator, and the performance characteristics to be satisfied can be found in the recommended specification by Jaffe *et al.*[47].

1.2.2.8 Specialized Sources for Research

Besides tubes for clinical use, a wide variety of electron-impact sources have been developed for research and other purposes[48]. One of the most common reasons for this is that standard tubes are designed to produce optimal images from a single type of object: the human body. The attenuation characteristics of individuals in the population and of the various intact body tissues all fall within a relatively narrow range requiring X rays between about 20 and 140 keV. In the research setting it is common to study objects either much more attenuating (usually in nondestructive testing) or much less attenuating, as is usually the case in basic biological research, than the human body. These studies require more versatile instrumentation. For biological research the most common requirement is for low voltage. X-ray apparatus for low-voltage applications must have very thin aluminum or beryllium windows to transmit the low-energy photons. Another parameter of electron-impact sources which has been optimized for high-resolution imaging and chemical analysis applications is the focal spot size.

The extreme case is X-ray microscopy of thin or ultrathin soft tissue sections. Today, most of this type of research is carried out on synchrotrons or with laser-produced plasma X-ray sources, to be described briefly later, but in the 1930's through the 1960's there was a great deal of activity in X-ray microscopy using electron-impact instrumentation. The interest started with Uspenski in 1914[49] who was the first to appreciate the potential of point-projection magnification, and Ardenne in 1939[50], who may have been the first to construct a functional instrument for the purpose. In the 1950's, Cosslett and Nixon revitalized the field by producing highly-functional instruments and valuable new results[51–57]. Several manufacturers produced commercial instruments in the 1950's and 1960's, but interest dwindled thereafter, and no dedicated versions are available today.

The two common ways to obtain high-resolution images of biological objects are contact microradiography and point-projection microscopy. For contact imaging, the specimen is placed in intimate contact with a film or photoresist for exposure. The size of the focal spot is not important, since resolution is limited by the film or resist. The developed image is then magnified in the light or electron microscope for viewing. Point-projection magnification is achieved by placing the specimen very close to the source and the detector some distance away. In this geometry, the focal spot size limits the resolution. Most of the older X-ray microscopes employed sources of the semi-thin, transmission anode variety, which means that the electron beam was focused onto the vacuum side of a thin metal foil target (which either served to terminate the vacuum or was adhered to a beryllium window), through which the X-ray beam emerged into the specimen chamber, usually at atmospheric pressure.

Henke describes an instrument of this type built for contact imaging with large focal spots and very large (up to 1000 mA) tube currents[58]. Excitation voltages and target foils were chosen to provide illumination suitable for chemically-specific imaging of biological constituents. For example the aluminum K-line provided 1.49-keV, the copper L-line 0.93-keV, the iron L-line 0.70-keV, the chromium L-line 0.57-keV, the titanium L-line 0.45-keV and the carbon K-line 0.28-keV radiation. Filters of the same material as the anode were used to attenuate the lower energy photons. Accelerating voltages between 0.3 and a few keV were used which necessitated keeping the target free from all carbon and tungsten contamination from the vacuum system and filament, otherwise the low-voltage electron beam would be prevented from reaching the target and exciting the fluorescent radiation. This was accomplished in part by an inverted design in which the large, coiled tungsten filament was positioned below the anode.

Cosslett and Nixon's instrument shared this inverted design feature, as did commercial versions by General Electric and Philips. Saunders used a Cosslett-Nixon projection X-ray microscope, which produced focal spots as small as 0.1 micron by the use of sophisticated electron optics[56], to produce beautiful micrographs of a wide variety of biological specimens[59–62]. Bellman used a hot-cathode, solid-anode diffraction tube with thin beryllium windows to perform what he called microarteriography on 120- to 450-micron sections[63]. The use of capillary optics to focus broadband X-ray beams to small spots has opened up still another method to improve the flux available in a given size spot[64–77]. Imaging and X-ray fluorescence analysis instruments employing either single, conical focusing capillaries or arrays thereof have started to be developed and applied to biomedical research.

A number of groups have modified scanning electron microscopes to produce micron-order focal spots for biological investigations[78–84]. Preferably the specimen chamber is removed and replaced with one having a hole in the roof to which the target foil is mounted on a beryllium window; alternatively, the target and specimen are both accommodated in the vacuum. The lens systems on these instruments are designed to focus the electron beam to tens of nanometers, and very low tube currents are obtained (fractions of a microamp) even with the largest apertures and lowest lens currents, necessitating exposure times of tens of seconds or even minutes.

There are two or three manufacturers that currently market microfocal (1- to 20-micron focal spots) X-ray sources, primarily to the nondestructive testing and microelectronics communities. In a few instances, these instruments, which incorporate sophisticated electron lenses and either solid or transmission-type anodes, have been employed for biomedical research[85–91]. A system currently in use for microtomography of pulmonary microvasculature employs a commercial solid-anode tube which operates over the 5–100 keV range with tens of microamps beam current[92]. The anode is cooled by circulating fluid, allowing operation at 300 watts. The tube is demountable with turbo and backing pumps continually replenishing the vacuum, allowing periodic polishing of the anode and rapid interchange of target materials, and simplifying filament replacement. The anode is a 6-mm diameter rod which can be rotated under vacuum to present fresh surface to the electron beam. Focal spots as small as three microns are available; the beam emerges through a 500-micron beryllium window.

1.2.3 Synchrotron X-ray Sources

For many imaging and spectroscopy applications in biomedical research, it would be desirable to have a source of X rays which provides far higher flux than conventional electron-impact tubes do. When narrow spectral bands of radiation provide the desired signal and information, attempts to filter or tailor the beam from broadband Bremsstrahlung sources using thin-foil filters generally result in reduction of the available flux to such low levels that data collection times become unacceptably long or the acquired photon statistics become unfavorable with respect to signal-to-noise. Because of the kVp^2 dependence of flux available from electron-impact tubes, they are not well-suited for generation of X rays with energies below about 5 keV bright enough to image even very small soft-tissue samples. Synchrotrons are currently the best available sources for fulfilling requirements of high flux and tunable monochromaticity over a spectral range which extends from the infrared to hard X rays. Synchrotron radiation is also highly polarized and collimated, pulsed, and partially coherent, which can be advantageous properties for certain applications. Though exhorbitant in cost and inaccessible to the majority of biomedical researchers, synchrotrons fill an important niche in X-ray imaging and spectroscopy science, and have facilitated discoveries which never would have been possible in their absence.

A number of excellent reviews and comprehensive works devoted to the application of synchrotron radiation to biomedical research exist[93–97], as do introductory and advanced texts and papers in the field[98–107]. This section discusses the salient features of synchrotrons as sources of radiation for biomedical research and summarizes some of the important applications and discoveries these sources have made possible.

1.2.3.1 Storage Rings

Synchrotron radiation is emitted when charged particles are accelerated. In particular, when positrons or electrons traveling at near relativistic velocities are curved in their trajectories, they emit radiation in a direction tangent to their paths as a result of the centripetal acceleration imposed by applied magnetic fields. (Storage rings can be used to accelerate either positrons or electrons. In this section "electron" is used to mean either positive or negative electrons.) By application of an rf potential, a synchrotron accelerates electrons around a circular path. The electrons are held to the fixed path by application of a time-varying magnetic field. First described in 1944 by Ivanenko and Pomeranchuk[103], who noted that betatron operation could be halted due to high energy losses by electrons to radiation, synchrotron radiation was first observed experimentally by Elder, Gurewitch, Langmuir and Pollock at GE's Schenectady Research Laboratory, where a 70-MeV cyclic electron accelerator was in operation[108,109]. This device, called an electro-synchrotron, gave the radiation its name. (Technically, a synchrotron accelerates charged particles to high energies for a short time span, whereas a storage ring maintains a beam of charged particles circulating for hours or days, but radiation emitted from storage rings has also come to be known as "synchrotron radiation".)

From the 1940's through most of the 1970's, large synchrotrons were built for high-energy physics research, for example as electron-positron colliders, and scientists interested in exploiting the unique properties of the radiation produced were relegated to the role of "parasitic users". A 1965 National Academy of Sciences report on uses of synchrotron radiation prompted Ednor Rowe and Fred Mills to convert the 240-MeV storage ring Tantalus, at the Synchrotron Radiation Center (SRC) at the University of Wisconsin in Madison, into the first dedicated synchrotron light source, coming on-line in 1968. By the late '70's, the broad array of applications for synchrotron radiation had assumed sufficient importance to merit the design and construction of storage rings, like the 1-GeV Aladdin at the Wisconsin SRC, dedicated to beam production. Since then, important light sources including the Photon Factory in Tsukuba and Spring-8 in Nishi-Harima, Japan, the National Synchrotron Light Source (NSLS) at Brookhaven National Laboratory on Long Island, the European Synchrotron Radiation Facility (ESRF) in Grenoble, the Advanced Light Source in Berkeley and the Advanced Photon Source at Argonne, have come on-line and fostered rapid progress in synchrotron radiation science by thousands of facility and visiting scientists. Smaller, "compact synchrotron" sources, costing a few instead of hundreds of millions of dollars, are now being routinely delivered for X-ray lithography[104,110–112], and can also be used for imaging and other research.

As shown in Figure 1.19[99], a storage ring consists of an evacuated pipe through which the electrons are made to orbit by means of two basic types of magnet which comprise the "magnet lattice" of the ring. Focusing sextupole and quadrupole magnets act as lenses, confining the electrons to a tight pencil beam by application of a nonuniform magnetic field, whereas bending magnets usually deflect electrons in their trajectory by application of a uniform magnetic field perpendicular to the electrons' trajectory. The storage ring consists of curved sections of pipe, around which the bending magnets are arranged, and straight sections, capable of accommodating insertion devices (wiggler and undulator in Figure 1.19, see Section 1.2.3.2), which connect the circular bending arcs. Beamlines are straight sections of evacuated pipe emer-

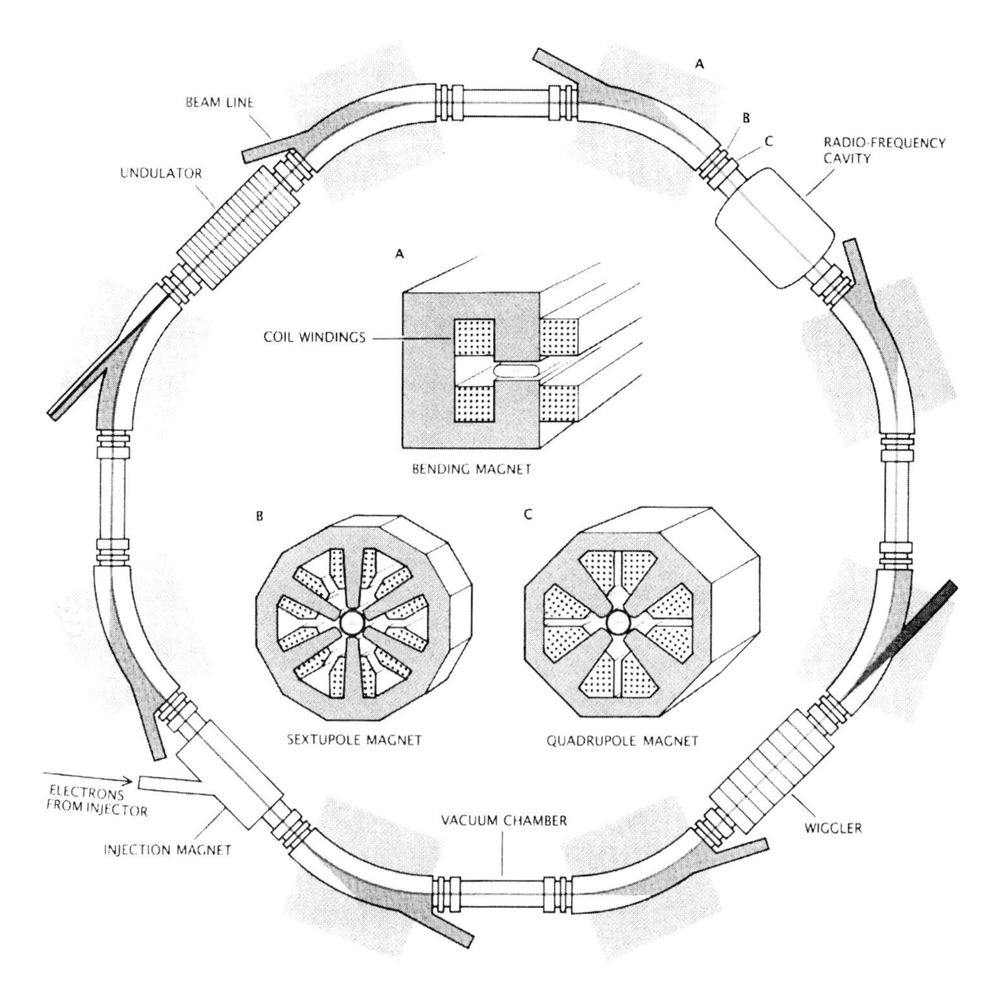

Figure 1.19. Schematic illustrating the important components of a synchrotron light source. (Reprinted with permission from Winick, 1987[99].)

ging tangentially from the storage ring, which serve as conduits to channel radiation from the electrons to experiment stations arranged around the ring. A large synchrotron with kilometer-order circumference can accommodate tens of beamlines. The beams exit through vacuum-tight, radiolucent windows and are generally incident upon a variety of optical elements which divert the radiation and tailor its properties for particular experiments. Crystal (focusing and non-focusing) monochromators, grazing-incidence and multilayer mirrors, diffraction gratings and zone plates are common examples of synchrotron radiation optical elements[113–115].

For the nonrelativistic case ($\beta = v/c \ll 1$, where v is particle velocity and c is the speed of light), the power P radiated by an accelerating particle of charge e was given by Larmor as[104,116,117]:

$$P = \frac{2e^2}{3c^3}\left(\frac{\mathrm{d}v}{\mathrm{d}t}\right)^2 = \frac{2e^2}{3m^2c^3}\left(\frac{\mathrm{d}p}{\mathrm{d}t}\right)^2, \tag{1.24}$$

where p is the charged particle's momentum. For the particular case of an electron traversing a circular orbit of radius R with constant velocity:

$$P = \frac{2e^2v^4}{3c^3R^2}. \tag{1.25}$$

Under these conditions, the radiation emission is in a dipole pattern, with the maximum intensity in a plane perpendicular to the orbit plane as shown in Figure 1.20[99], top panel.

In the relativistic case ($\beta \approx 1$), the radiated power is:

$$P = \frac{2e^2c}{3R^2}\beta^4\left(\frac{E}{mc^2}\right)^4. \tag{1.26}$$

The ratio of the total to the rest energy of the particle, E/mc^2, is denoted by γ and, for an electron, is equal to $1957E$, where E is in GeV. It can be seen that the radiated power for a proton would be about 2000^4, or 10^{13}, less than for an electron, which explains why electrons and positrons are used in storage rings, and protons are not. In the relativistic case, the radiation emission pattern is markedly peaked in the forward direction of electron motion (tangent to the orbital trajectory) as shown in Figure 1.20, second panel.

The energy lost by an electron in the course of a single orbit is:

$$\Delta E = \frac{4\pi e^2\beta^3\gamma^4}{3R}, \tag{1.27}$$

which, in practical units, amounts to:

$$\Delta E = \frac{88.47E^4}{R}, \text{ and } P = \frac{88.47E^4I}{R} = 26.54BE^3I, \text{ since } R \quad \frac{3\ 33E}{B}, \tag{1.28}$$

where ΔE is in keV, E is the accelerator energy in GeV, R is the storage ring radius in meters, P is the radiated power in kilowatts, I is the beam current in amps, and B is the field strength in Tesla (1 T= 10^4 Gauss). Considering modest ring parameters like 1 GeV, 500 mA, and 1 T, the enormity of the power radiated from synchrotrons becomes evident (the Spring-8 is an 8-GeV facility). Appropriate adjustments to the foregoing equation provide the power available per unit horizontal angle. The energy lost to radiation has to be replaced if the particles are to be kept in orbit, which is the function of rf cavities inserted into the storage ring (Figure 1.19).

The instantaneous power radiated per unit wavelength λ, and per unit (vertical angle ψ) radian is:

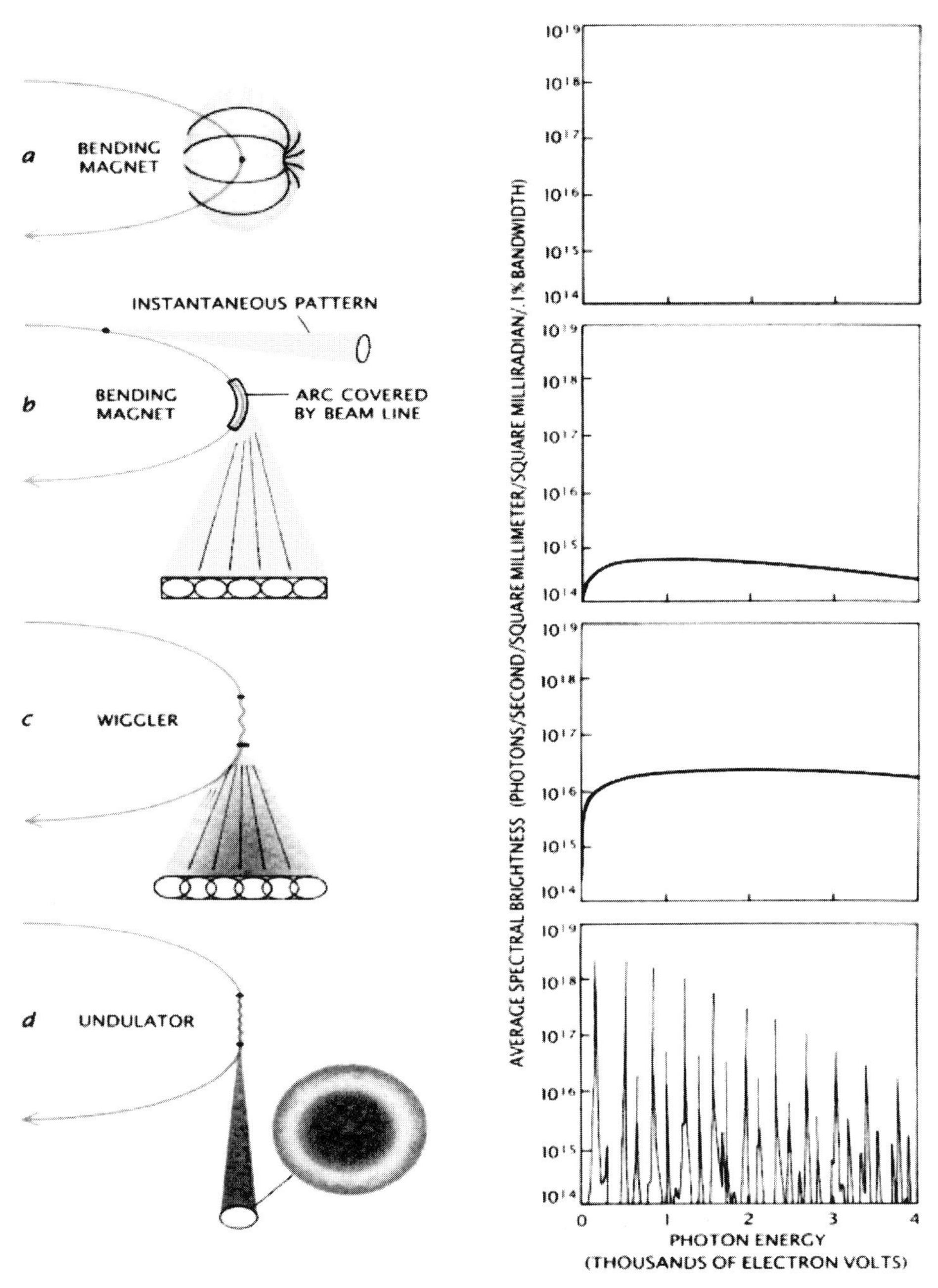

Figure 1.20. Emission envelopes and spectra for various synchrotron components. (Reprinted with permission from Winick, 1987[99].)

$$P(\lambda,\psi,t)=(2.72\times10^{-9})\frac{e^2c}{R^3}\left(\frac{\lambda_c}{\lambda}\right)^4\gamma^8\left[1+(\gamma\psi)^2\right]^2$$

$$\left[K_{2/3}^2(\xi)+\frac{(\gamma\psi)^2}{1+(\gamma\psi)^2}K_{1/3}^2(\xi)\right]\frac{\mathrm{J}}{\mathrm{sec}\cdot\mathrm{rad}\cdot\mathrm{cm}}, \tag{1.29}$$

where $\xi=\left(\frac{\lambda_c}{2\lambda}\right)\left[1+(\gamma\psi)^2\right]^{3/2}$, $K_{1/3}$ and $K_{2/3}$ are modified Bessel functions of the second kind, and λ_c is the critical wavelength:

$$\lambda_c=\frac{4\pi R}{3\gamma^3}, \tag{1.30}$$

or the critical energy is:

$$\varepsilon_c=\frac{3\hbar c\gamma^3}{2R}. \tag{1.31}$$

In practical units:

$$\lambda_c=\frac{5.59R}{E^3}=\frac{18.64}{BE^2}\text{ and }\varepsilon_c=\frac{2.218E^3}{R}=0.67E^2B, \tag{1.32}$$

where λ_c is in Ångstroms and ε_c is in keV. Half the total power is radiated at energies above, and half at energies below, the critical energy. In a typical first-generation 1-GeV storage ring with 1-T bending magnets, the critical energy is 0.665 keV. High-field insertion devices, such as wigglers described later, are used to raise the critical energy to considerably higher levels. Since the critical energy varies with γ^3, storage rings operating at higher accelerating energies (up to 8 GeV) also produce much higher-energy radiation, even from bending magnets: well into the diagnostic energy range (10 to 100+ keV).

Integrated over all emission angles, the radiated power spectral distribution is:

$$P(\lambda,t)=(9.87\times10^{-9})\frac{e^2c}{R^3}\left(\frac{\lambda_c}{\lambda}\right)^3\gamma^7\int_{\lambda_c/\lambda}^{\infty}K_{5/3}(\xi)\,\mathrm{d}\xi\,\frac{\mathrm{J}}{\mathrm{sec}\cdot\mathrm{cm}}, \tag{1.33}$$

while integrated over all energies (wavelengths), the radiated power angular distribution is:

$$P(\psi,t)=(4.375\times10^{-8})\frac{e^2c}{R^2}\gamma^5\left[1+(\gamma\psi)^2\right]^{-5/2}\left[1+\frac{5}{7}\frac{(\gamma\psi)^2}{1+(\gamma\psi)^2}\right]\frac{\mathrm{J}}{\mathrm{sec}\cdot\mathrm{rad}}. \tag{1.34}$$

Winick[116] reproduces a number of useful equations for distribution functions originally derived by Green[118]. Particularly handy for calculating the available flux in a given spectral band are the notations:

$$G_0(y) = \int_y^\infty K_{5/3}(\xi)\mathrm{d}\xi\,, \tag{1.35}$$

$$G_\mathrm{i}(y) = y^\mathrm{i} G_0(y)\,, \tag{1.36}$$

and the equation:

$$N_\mathrm{k}(\lambda) = 1.256\times10^{10} k\gamma G_\mathrm{i}(y) \frac{\text{photons}}{(k\gamma)\cdot\text{sec}\cdot\text{mA}\cdot(\text{mrad}\,\theta)}\,, \quad \text{for all}\,\psi\,, \tag{1.37}$$

where y $\frac{\lambda_C}{\lambda}$ $\frac{\varepsilon}{\varepsilon_C}$,

which gives the flux, per second, per mA, per mrad horizontal angle, integrated over vertical angular emission range ψ and within the fractional bandwidth $k = \Delta\lambda/\lambda$. For example, using these equations and the appropriate table of Bessel functions and integrals, the flux available in a 10% bandwidth at the critical wavelength can be shown to be[104,116]:

$$N_{0.1}(\lambda_c) = 1.601\times10^{12} E \frac{\text{photons}}{\text{sec}\cdot\text{mA}\cdot(\text{mrad}\,\theta)}\,, \quad \text{for all}\,\psi\,. \tag{1.38}$$

For photon energies corresponding to wavelengths above and below λ_C, the flux, integrated over the vertical angular emission range ψ, can be approximated by:

$$N(\lambda) \approx (9.35\times10^{16}) I \left(\frac{R}{\lambda_c}\right)^{1/3} \left(\frac{\Delta\lambda}{\lambda}\right) \frac{\text{photons}}{\text{sec}\cdot\text{mrad}\theta}\,, \quad \text{for}\ \lambda >> \lambda_c\,, \tag{1.39}$$

and by: $N(\lambda) \approx (3.08\times10^{16}) IE \left(\frac{\lambda_c}{\lambda}\right)^{1/2} \mathrm{e}^{-\lambda_c/\lambda} \left(\frac{\Delta\lambda}{\lambda}\right) \frac{\text{photons}}{\text{sec}\cdot\text{mrad}\,\theta}$, for $\lambda << \lambda_c$, (1.40)

where I is the beam current in amps, R is the ring radius in meters, and E is the beam energy in GeV. The flux peaks at a wavelength about three times λ_C, falls off very rapidly at energies above the critical energy due to the factor $\mathrm{e}^{-\lambda_c/\lambda}$, and falls off more slowly at lower energies.

For energies close to the critical energy, the vertical emission angle ψ varies with γ^{-1} and, for accelerating energies in excess of 1 GeV, is less than about 1 mrad. While this remarkable collimation accounts for the extremely high available fluxes, it can cause practical problems with alignment of experimental components, particularly if the beam is unstable and moves around. Sophisticated beam monitoring equipment is therefore essential for synchrotron radiation experiments, as is flexibility in experiment station positioning. Beam monitors can be incorporated into closed-loop systems which perturb the electron orbit to keep the beam position stable. At energies other than the critical energy:

$$\psi = \frac{1}{\gamma}\sqrt[3]{\lambda / \lambda_c}\ , \text{ for } \lambda >> \lambda_c\ , \tag{1.41}$$

and $$\psi = \frac{1}{\gamma}\sqrt[2]{\lambda / \lambda_c}\ , \text{ for } \lambda << \lambda_c\ . \tag{1.42}$$

In general, synchrotron radiation is polarized, with the electric vector parallel to the acceleration vector. In the electron's instantaneous direction of motion, polarization is complete, with the electric vector in the orbital plane. The degree to which the radiation is polarized depends on the emission angle and wavelength: polarization decreases as these increase. Polarization is also decreased due to incoherent betatron oscillations of the many electrons producing the radiation at any instant.

Synchrotron radiation is emitted in very short, high-frequency pulses because electrons are injected into and accelerated by rf cavities around the ring in bunches or buckets[116]. The frequency of radiation emission pulses depends on the orbital frequency of the electrons and the rf frequency, which must be an integral multiple, or harmonic, of the orbital frequency. Typical rf frequencies are in the 50- to 500-MHz range. For example, if the orbital frequency were 1 MHz, an rf freqency of 300 MHz would represent the 300-th harmonic of the orbital frequency, and a maximum of 300 bunches of electrons could be accelerated if all the buckets were full. The total beam current is directly related to the number of filled buckets. Often, only a few buckets are full: only one in the case of collider experiments, in which case the beam current is very low. The duration of each radiation pulse is usually between a few and a few hundred nanoseconds, but may be in the picosecond range. The pulsed nature of the radiation emission is beneficial for some experiments and detrimental to others.

While the radiation intensity produced by synchrotron sources surpasses that of almost all point sources of X rays, their brightness advantage is far greater still due to the low emittance, or angular divergence, of the beam. Low emittance of the electron beam is a requirement for optimal undulator performance. Beam emittance is a tradeoff between betatron oscillations caused when electrons recoil as a result of emitting radiation in direction transverse to their motion, and the damping action of rf cavities which impart momentum to the electrons only in their direction of motion. The emittance of the electron beam can be controlled by varying the field strength of the focusing quadrupole magnets of the storage ring. For transmission imaging of small objects and for X-ray diffraction the extremely high brightness of synchrotron X-ray beams is an important advantage of these sources. On the other hand, the extremely low beam divergence, particularly in the vertical direction, can be a drawback for transmission imaging and tomography of larger specimens. For such applications, either the beam must be spread in the vertical direction or in both directions by the use of optics such as curved crystals, or the specimen may be scanned vertically through the beam with the transmitted intensity recorded as a succession of linear projections which are then combined in software to form a two-dimensional image. It is difficult to achieve uniform illumination intensity using beam-spreading optics, which also impose a significant flux penalty.

1.2.3.2 Insertion Devices

The properties of the radiation emitted from storage rings (bending-magnet radiation) can be improved for many biomedical applications by the use of insertion devices. These are electromagnetic devices inserted into the straight sections of the storage ring. There are three types of insertion device commonly used to produce radiation beams with the desired qualities: wavelength shifters, wigglers and undulators. All three types impart to the orbiting electrons or positrons additional motions, or deflections from their circular path. These deflections are produced by the periodic magnetic structure of the device: A periodic, linear array of alternating north-south pairs of magnets are arranged outside the evacuated tube of the storage ring causing the orbital electrons to oscillate in their paths. The accelerations accompanying these perturbations imposed on the circular particle trajectory cause more electromagnetic radiation to be given off. Figure 1.21 shows the brightness emitted by various X-ray sources, including tubes, bending magnets, wigglers and undulators[99]. Vertical magnetic fields cause electron oscillations in the horizontal (orbital) plane. Most insertion devices are of the vertical-field type, since the horizontal acceptance aperture of the beam is larger than the vertical[119]. The three types of insertion device differ in terms of the spectrum and

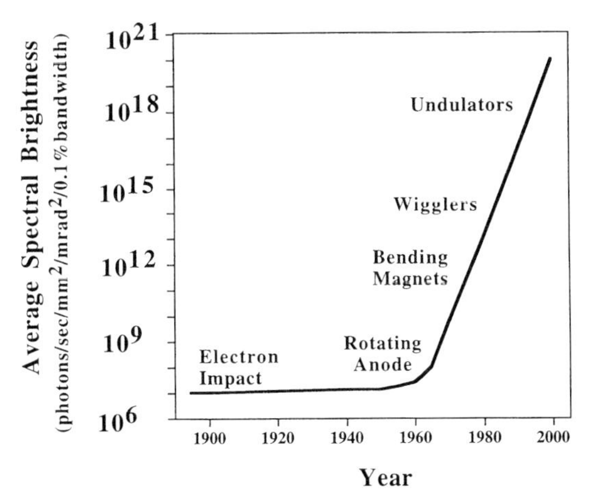

Figure 1.21. Brightness of synchrotrons bending magnets and insertion devices compared to electron-impact sources. (Adapted from Winick, 1987[99].)

intensity of the radiation they produce. Ideally, all three types of insertion device are designed in such a way that the trajectories of electrons emerging from the device are unaffected: They are the same as they would have been in the absence of the insertion device, and the particles continue on along their approximately circular orbit in the storage ring. The three types of device exert their various effects on the emitted spectrum according to the strength of the imposed magnetic fields and the number of poles or pairs of magnets.

The simplest type of insertion device, the wavelength shifter, has three poles: a strong central pole and entrance and exit poles with half the field strength and reverse polarity of the central pole. (A wavelength shifter is, in fact, just the simplest possible wiggler.) The sum of the fields in the longitudinal orbital direction is zero, with the peripheral poles serving to compensate for the deflection of the strong central pole, allowing the net electron trajectory to remain unaffected. The spectrum produced by a wavelength shifter is qualitatively very similar to that of a bending magnet with a field strength equal to the central pole. Since the field of a wavelength shifter can be made much higher than that of large-radius bending magnets, the main effect of a wavelength shifter is to up-shift the emitted energy spectrum.

Wigglers and undulators are the most common insertion devices. Wigglers[119–122] have a relatively smaller number of relatively higher-field poles compared to undulators, and therefore cause a smaller number of larger deflections in the electrons' trajectories. The spectrum emitted by a wiggler or undulator depends on the amount of deflection and on the ratio of the deflection angle α to the radiation cone opening angle $1/\gamma$, as shown in Figure 1.22[104]. This ratio is called the deflection parameter, $k = \alpha\gamma$. In a wiggler, the deflection angle is large compared to the radiation emission angle, and the emitted spectrum is similar to that from a bending magnet of equivalent field strength, but with much higher flux as shown in Figure 1.20, third panel. Higher wiggler field strengths may be used to extend the range of available

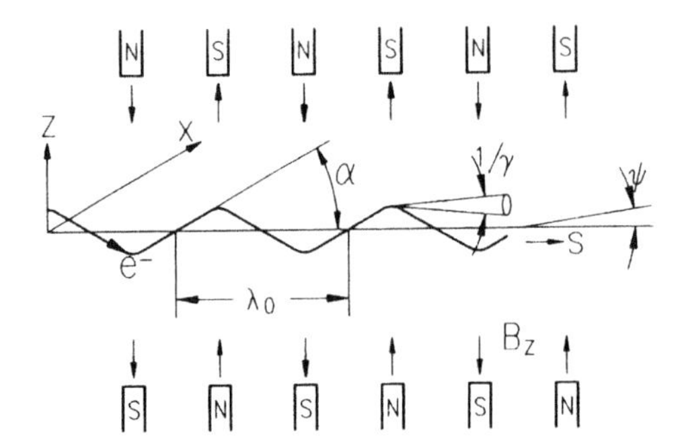

Figure 1.22. Electron trajectory through a multipole magnet. Definitions of the deflection angle α and the radiation emission cone opening angle $1/\gamma$. (Reprinted with permission from Weihreter[104].)

photon energies into the tens of keV. The factor by which the flux provided by a wiggler is increased relative to a (large-radius) bending magnet of comparable field strength is proportional to the number of poles in the wiggler. Undulators generally provide high-flux radiation in the UV or soft-X-ray spectral region[123–126]. An undulator produces a large number of small oscillations in the electrons' trajectories. The deflection angle is small compared to the emission angle (or of the same order), and the flux gain compared to bending magnets is proportional to the square of the number of poles. Undulator radiation from multiple electrons can constructively or destructively interfere, producing an energy spectrum characterized by a large number of narrow-band peaks as shown in Figure 1.20, bottom panel. The gap between the poles may be varied to tune the spectral peaks to the desired energy.

1.2.3.3 Applications of Synchrotron Sources

Synchrotron X-ray sources have been exploited for a broad range of applications in biomedical research including X-ray microscopy[125,127–147], microtomography[148–162], protein crystallography[94,95,163–171], extended X-ray absorbtion fine structure analysis (EXAFS)[95,172], holography[173–178], angiography[168,179–200] and mammography[35,37,38,201]. The tunable monochromaticity and high flux available were the motivations to utilize synchrotron radiation for almost all of these applications. Coherence and polarizability of the radiation were essential for others.

High flux at low energy is particularly important for X-ray microscopy of thin specimens, one of the earliest applications of synchroton radiation to biomedical research[127–129,202–206]. X-ray microscopy in the "water window" presents attractive possibilities for studies of biological specimens with nanometer resolution. In this spectral region, between the oxygen K-edge at about 23 Å and the carbon K-edge at 44 Å, water, which makes up the bulk of cellular matter, is almost transparent and nucleic acids and proteins absorb strongly. This offers the opportunity to image unstained, wet tissues at atmospheric pressure with a radiation which, though lethal, should cause less radiation damage than electron microscopy in the overlapping portion of the resolution regimes of the techniques[204–206].

X-ray microscopy can be performed using four basic modes or imaging geometries: contact, scanning, imaging and holographic[207]. Early studies utilized contact microscopy, in which the specimen is placed in intimate contact with the detector, usually a high-resolution X-ray photoresist of the type used in X-ray lithography[133,175,202,207–216]. The chemical structure of the resist is altered by absorption of x radiation transmitted through the specimen. After exposure the specimen is removed from the surface and the resist is chemically developed, yielding a relief map in which the height is either proportional to (positive resist) or inversely proportional to (negative resist) X-ray exposure. The theoretical resolution of the resist, usually polymethylmethacrylate (PMMA) or a copolymer thereof, is on the order of ten nanometers for soft X rays, and resolution as high as thirty nanometers has been obtained.

Until recently, the relief map in the developed resist had to be replicated by deposition of a metal or polymer film. After dissolving away the resist, the replica was viewed in the scanning or transmission electron microscope. The necessity to produce a replica contributed to the tedious nature of this technique. With the advent of scanned-probe microscopes in the 1980's, an improved, direct method for viewing the developed resist became available[217].

Scanning X-ray microscopy is another possibility using synchrotron radiation[127,131,218–223]. In this geometry, the beam is focused to a small pencil or point, using a zone plate or reflective optical element (usually augmented by an aperture upstream from the sample), the specimen is scanned through the X-ray beam in a raster or boustrophedonic manner, and the image acquired point-by-point by a single-element detector. The diameter of the beam at the specimen and the accuracy of the scanning mechanism determine the spatial resolution, which may be on the order of nanometers, the highest obtained by any mode of X-ray imaging. Though the apparatus is rather expensive and the possibility of "flash" or realtime imaging is precluded by long imaging times, scanning has been the most productive mode of synchrotron X-ray microscopy due to the high spatial resolution and relatively modest optical requirements.

Imaging microscopy utilizes a condenser zone plate to simultaneously illuminate the entire area of the thin specimen to be imaged and a zone plate objective to form the transmitted image on the detector. Although the state of the art is advancing, it remains challenging to fabricate the zone plates which limit the performance of this method[142,144,224–227]. The resolution attainable is roughly equal to the width of the outermost annular ring, generally limited to several tens of nanometers.

X-ray holography has long been proposed as a possible method to obtain ultra-high-resolution, three-dimensional information from biological objects[173–178,228–230], but has remained an ellusive goal. This is in part due to the lack of adequate technology to present the image formed by the recombined reference and sample beams to the human viewer in an acceptable or even useful way. Until practical X-ray lasers in the appropriate wavelength range are developed (see Section 1.2.5), coherent beams from undulators present the best candidate sources for holographic application.

Synchrotron X-ray microtomography has been applied or suggested for application to a number of biological investigations[148–150,152,154,158,160,162,231–236]. Compared to microscopy of thin sections, microtomography of real specimens provides more modest spatial resolution in the tens of microns range, though the possibility of higher spatial resolution has been demonstrated on test objects[237]. The most impressive biological results to date have been obtained on cancellous bone[161]. One of the problems with synchrotrons as sources for 3D microtomography is the almost planar nature of the beam, necessitating vertical translation of the specimen through the beam and confining volumetric imaging to the stacking of serially-reconstructed slices in software. There has recently been an interesting demonstration of the feasibility of phase contrast[238] synchrotron X-ray tomography[160,162,239] which exploits the larger differential in "phase retardation cross sections" (vs. attenuation cross sections) of various healthy and malignant soft tissues and the partial coherence of synchrotron radiation.

In the 1980's, there was a flurry of activity in synchrotron coronary angiography, which was even advocated as a mass-screening technique[179–184, 186–188,190,191,240]. It was the markedly-improved signal-to-noise ratio, relative to broad-band illumination, at a given (preferably low) dose afforded by the tunable monochromatic beam applied to iodine-enhanced subtraction angiography that prompted these efforts. It is unlikely that a screening method requiring such elaborate instrumentation as synchrotron sources will ever become economically viable, but some recent studies have shown excellent potential for similar, much higher resolution, techniques in a basic-research context[198,199]. With the current emphasis on women's health issues in general and breast cancer detection in particular, it has been suggested that synchrotron X-ray sources be applied to mammography[37,38,201]. Again, it seems unlikely that sychrotron mammography will become accessible to the general population, but characterization of the X-ray absorption properties of benign and diseased breast tissues of various types and precancerous and malignant stages could be a valuable contribution to the practice of X-ray mammographic cancer screening.

1.2.4 Plasma X-ray Sources

Plasmas, which have been referred to as the fourth state of matter, are described by the physics of highly-ionized matter. While a plasma is a collection of electrons, ions, neutral atoms and molecules which, taken as a whole, is electrically neutral, the degree of ionization is very high[115]. The interparticle spacing within a plasma is relatively high, and the particles' motion is dominated by interactions of their electric and magnetic fields. High-temperature plasmas can emit both Bremsstrahlung and characteristic line radiation. Typically, the radiation is emitted in short bursts, ranging from microseconds for some glow-discharge plasmas to femtoseconds for modern laser-produced sources. Plasmas are characterized by the temperature, density, and oscillation frequency of their electrons, which dominate in the X-ray emission processes. A diagram showing the ranges of these parameters for various plasmas is shown in Figure 1.23[115]. The parameter ranges characteristic of pinch and laser plasmas useful for X-ray production fall in the upper right corner of the diagram.

There are a number of methods to produce plasmas energetic enough to emit hard and soft X rays, including focusing an electron or ion beam onto a plasma[241,242], producing a micropinch of high-Z electrode material by vacuum spark discharge[243], imploding thin wire arrays by passage of tremendous electrical currents[244], tokamaks[245,246], and focusing an intense optical laser

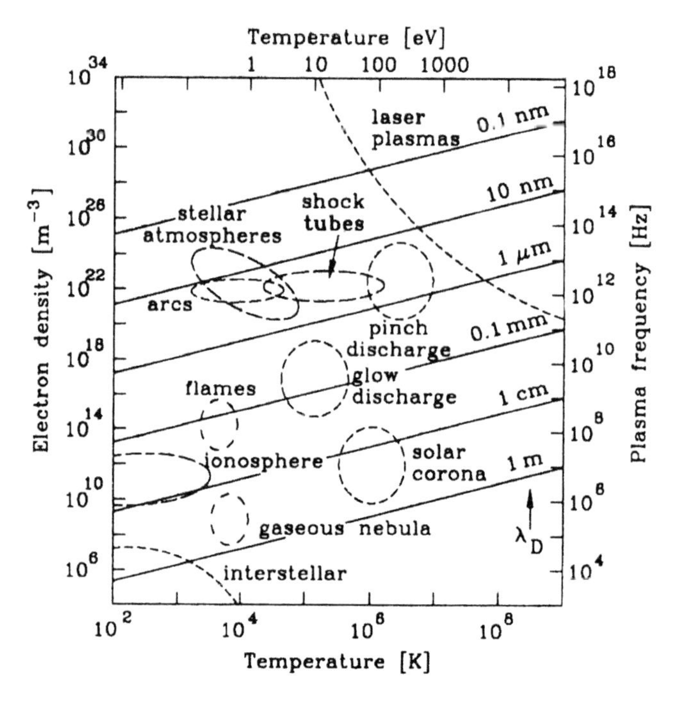

Figure 1.23. Electron density, temperature and oscillation frequency for various types of plasma. (Reprinted with permission from Michette and Buckley[115].)

pulse onto a solid or liquid target. Z-pinch and laser-produced plasmas (LPP's) are the most popular for laboratory production of X rays for imaging, spectroscopy and lithography.

This section briefly describes gas and laser-produced plasma X-ray sources. Both types have found increasing application in microscopy and imaging in recent years, and the laser-produced variety seems to hold realistic promise for offering a viable, high-flux, compact alternative to the electron-impact X-ray source for clinical use[247,248]. While electron-impact sources have evolved to a highly-mature technological state in which their properties and emission spectra are quite well understood, this is not true of plasma sources. Efforts to understand the basic physics describing the emission processes, to improve the efficiency of energy conversion and reliability, and to characterize the time course and spectral content of the complex emissions, especially from laser-produced plasmas, constitute an area of intensely feverish research in the 1990's. This introduction is intended to convey the basic principles on which these devices operate, to present a few of the imaging applications to which the sources have been applied, and to serve as a gateway into the vast literature documenting the work and progress in the field. Detailed discussion, derivations and equations relevant to the physics of plasma X-ray sources are available in a number of sources[115,249–264].

1.2.4.1 Gas Plasma Sources

In a gas jet plasma or Z-pinch source, a high current is rapidly passed through a cylindrical annular gas jet, for example by discharging a bank of capacitors[265–268]. Typically, atoms of an inert gas such as argon, krypton or xenon are stripped into a highly-ionized charge state. Nitrogen has also served as the working gas[269]. The high self-magnetic fields generated by the current confine the plasma and compress it "into a pinch", causing implosion of the cylinder and producing the high electron temperatures required for emission of a burst of X rays. Typically, a few hundred joules of X-ray energy are emitted in a 50- to several-hundred-nanosecond FWHM pulse[265, 266], though the Proto-II diode system at Sandia has been reported to produce many kilojoules of radiation[267] and even more energetic plasmas undergoing development at Sandia threaten to rival the National Ignition Facility[244]. Several tens of millijoules per square cm may be available at the specimen surface.

The spectrum consists of characteristic lines superimposed on a white, Bremsstrahlung background, usually spanning the VUV to soft-X-ray range (20 to several hundred angstrom wavelengths), but possibly extending down into the sub-angstrom range[267]. If the wavelengths above 44 Å are not desired, they may be attenuated, for example by a thin aluminum layer at the specimen or by another gas puffed into the beamline. For water-window microscopy, carbon dioxide or nitrogen used as the working gas produce intense emission lines in the 33- to 41-Å and 24- to 29-Å ranges, respectively. It is possible to use one "discharge gas", such as argon or helium, optimized for initiation of plasma discharge by the high current, and a separate "emitting gas", such as nitrogen or carbon dioxide, optimized with respect to its spectral emission properties[270]. In X-ray microscopes which use zone plate objectives, $\Delta\lambda/\lambda$ must be very small for efficient focusing, and narrow spectral lines are isolated from the continuum with condenser zone plate/monochromator combinations[269,271–273]. A commercial z-pinch source was manufactured in the 1980's, and incorporated into several X-ray microscopes for biological research[211,212].

1.2.4.2 Laser-produced Plasma Sources

When an optical laser beam is focused onto the surface of a solid target, plasma formation can occur if the breakdown threshold is exceeded. The breakdown threshold, defined as the lowest optical power density required to ionize atoms in the target medium and produce a plasma, varies for different targets, but is easily exceeded with many modern lasers. If the focused beam intensity is in excess of 10^{16} W/cm^2, formation of a highly-ionized, solid-density plasma results. Rapid recombination and other processes in the high-temperature plasma give rise to emissions spanning the spectrum from visible to hard X-ray. High-intensity irradiation of target atoms ionizes them by multiphoton excitation: Many photons are absorbed simultaneously allowing photons of energy much lower than electrons' binding energy to strip electrons from their atomic orbits. Irradiation of a target atom with intense laser light causes it to develop a time-dependent dipole moment and produce emission of photons at odd multiples of the laser light frequency[274]. This generation of odd harmonics of the incident frequency is called optical harmonic generation. Shortening the pulse length and increasing the intensity has made possible the observation of harmonics up to the 109th for 811-nm irradiation and the 143rd for 1053-nm laser light[251]. These harmonic emissions and other line emissions resulting from electron transitions (bound-bound transitions in ions) are superimposed on the continuum emission from radiative recombination (free-bound transitions of electrons and ions) and Bremsstrahlung (free-free transitions of electrons)[252]. The total conversion efficiency of light to X-ray energy is less than 1% at 10^{16} W/cm^2, but could exceed 10%[252] at intensities over 10^{18} W/cm^2. For low-Z targets like carbon, mylar tape, or liquid droplets the plasma consists mostly of hydrogen-like and helium-like ions giving a relatively small continuum background and strong K-shell emission in the water window[275–277]. For medium-Z targets, the continuum contribution is increased overlaid with a denser line spectrum from L-shell emission from neon-like and other ions. High-Z targets produce the largest continuum.

Since the invention of the laser in 1960, optical power output has increased by twelve orders of magnitude[278]. Q-switching in the 1960's followed by mode locking around 1970 increased the available power to about 10^6 and 10^9 watts, respectively, providing intensities (also called power densities or focused irradiance) of 10^{12} and 10^{16} watts/cm^2. But well into the 1980's, rather large and very expensive laser systems, like the VULCAN and SPRITE at Rutherford Appleton Labs[275,279–284], the GDL at Rochester's Laboratory for Laser Energetics[285], or the NOVA at Lawrence Livermore[286,287], were required to obtain the very high intensities in what were relatively large focal spots (100 to 300 microns compared to the 40 to 100 microns which are commonplace today) required for production of adequate X-ray flux for single-shot imaging. The invention of chirped-pulse amplification in the late 1980's made feasible relatively-inexpensive ($100 000 to $160 000), compact, bench-top terawatt (10^{12} watt; "t^3", or "t-cubed", for table-top terawatt[278]) lasers, capable of delivering a joule per pulse in sub-picosecond pulses and providing intensities[249,251] of 10^{18} to 10^{19} W/cm^2.

Chirped-pulse amplification achieves this high irradiance in compact devices by shortening the pulse length after amplification, avoiding the nonlinear optical responses (Kerr effect) which degrade the beam in almost all high-gain, broadband materials at powers in the terawatt range[251]. An ultrashort (femtosecond), low-power (nanojoule) pulse is first chirped, or lengthened by factors as large as 10^4, by one optical grating, amplified by factors between 10^6 and 10^{11} while in the dispersed, low-power state, then condensed to its original duration again, or reconstructed, by a second optical grating. A chirped pulse possesses a time-dependent frequency because the pulse stretcher has a frequency-dependent phase function. The pulse compressor must then possess the conjugate phase function to perfectly reconstruct the pulse. Probably the most popular amplification systems are Nd:glass[288,289] and Ti:sapphire[262–264,290–294]. A Ti:sapphire system can produce pulses under 10 femtoseconds in duration.

There are several advantages to high-power, short pulse systems: hydrodynamic motion in a plasma is almost neglibible for a 100 fs pulse; very high energy densities can be achieved; the electric field due to the light pulse is many times stronger than the Coulomb field binding even the innermost electrons of high-Z elements; and very high oscillatory energies (also called quiver or ponderomotive energy) are achieved by the stripped free electrons. These attributes of short pulses from t-cubed lasers make them attractive for plasma generation in LPP X-ray sources. The high energy densities result in high output X-ray flux which scales as laser intensity raised to powers between 1.9 and 2.8, depending on the target material[295]. The highly-ionizing nature of the irradiation broadens the array of suitable target materials and enhances plasma electron temperature and X-ray output. The high quiver energies of the free electrons is accompanied by the emission of harder X rays. This is because the quiver energy is rapidly thermalized after short-pulse irradiation, producing quasi-Maxwellian plasmas with many hot electrons which emit high-energy X rays (and electrons): MeV X rays are emitted[251] from plasmas generated by irradiation with 10^{18} W/cm^2. Intensities in the 10^{14} to 10^{17} watts/cm^2 range are associated with high-brightness, soft-X-ray sources (low fluxes of hard X rays are also produced at these intensities), whereas intensities approaching the petawatt range, on the order of 10^{17} to 10^{20} watts/cm^2 are required to produce hard X-ray beams with flux high enough to be useful for clinical imaging. The properties of the plasma and of the X-ray emission can be controlled in large measure by the intensity, polarization, and wavelength of the incident laser pulse and by choice of target material[254–257,278,296]. It has also been demonstrated that X-ray emission from solid targets can be enhanced by use of a relatively long, less-intense prepulse before the high-intensity main pulse (e.g. 30- to 200-ps, 10^{13} to 10^{14} W/cm^2 prepulse followed by 1-ps main pulse)[254,255,297–300]. In this case, the main laser pulse is interacting not with a solid but with a pre-plasma. The prepulse may be either of the same, or a different wavelength than the main pulse.

A slight variant on the LPP X-ray source theme, which apparently has the potential to produce even brighter beams of coherent, multi-keV radiation, is being proposed and investigated by Rhodes and colleagues[301–303]. "Hollow atoms", of xenon, for example, are produced when clusters of atoms from a pulsed supersonic gas jet are exposed to 300-fs pulses of UV laser light at 10^{18} W/cm^2. The quiver energy of stripped, outer-shell electrons is transferred through energetic collisions to inner-shell electrons, ejecting them from the atom. The orbit-filling cascade results in intense X-ray emission. One proposed use for the radiation is Fourier transform holographic microimaging of biological specimens[178].

The radiation in the 1- to 10-keV range from LPP's, though in general incoherent, can be orders of magnitude brighter than that available from synchrotron sources[278]. The source size, though larger than the best microfocal electron-impact tubes and larger than synchrotron beams at the source, is typically on the order of 40 to 100 microns, allowing for high-magnification imaging. Each pulse from an LPP is much longer than an individual pulse from a synchrotron; the repetition rate of LPP's is far lower than that of synchrotrons[251], ranging from about 10^{-3} to 10^3 Hz. Thus, the peak power of LPP X-ray sources is higher than synchrotrons and much higher than conventional tubes, but the average power is much lower. Efforts are underway to increase the repetition rate of chirped-pulse amplified lasers, which is often only a few Hz or less, since imaging of thick objects, and certainly clinical imaging, currently would take hundreds or thousands or pulses[247,248]. For 100-fs pulses at a typical repetition rate of 10 Hz, a continuous imaging time of minutes would result in a true exposure time of less than a nanosecond. For this reason, the best argument for LPP X-ray imaging that can be made at present is for small specimens which can be imaged in a single shot.

1.2.4.3 Applications of Plasma Sources

Intensive research on short-pulse LPP's is underway, partly because of their potential as X-ray sources and as the gain medium for X-ray lasers, but mostly because of their potential use for indirect approaches to inertial confinement fusion[251,252,286,298]. LPP's as X-ray sources have been used for time-resolved absorption experiments and X-ray diffraction, for X-ray lithography at the 13- to 14-nm wavelength for which high-efficiency multilayer reflective optics are available[265,304–316], and for X-ray microscopy[132,143,211,212,217,266,275,289,304,317–330]. Compared to synchrotron sources, LPP X-ray sources have the advantages of low cost, not requiring an ultra-high vacuum, emitting very high brightness beams in short bursts, high spatial stability and reproducibility, and small source size (smaller than z-pinch and plasma focus sources).

The simplest arrangement for X-ray imaging with plasma sources is shown in Figure 1.24[331]. A short focal-length lens focuses the laser beam through a vacuum window onto a solid (or tape, or droplet) target mounted in the imaging chamber. X rays are emitted into 2π, passing through the vacuum to the specimen and film. The minimum source-to-specimen distance, which is decreased in the interest of increasing the flux on the sample, is limited by physical constraints to at least a few millimeters[321]. Another problem with the arrangement, and with LPP sources in general, is the production of debris, including "hot rocks" up to several microns in size, which can be ablated from the target surface. This leads to fouling of the specimen chamber as well as contamination of the image. Often, silicon nitride windows (Si_3N_4) are placed in front of the specimen to protect it from debris and allow it to be kept at atmospheric pressure. An environmental sample cell for resist-based imaging is shown in Figure 1.25[332].

Normal-incidence, reflective optical systems have been proposed and implemented to shield the specimen from debris and to increase the flux on the specimen/detector assembly. Examples are the multilayer spherical mirror

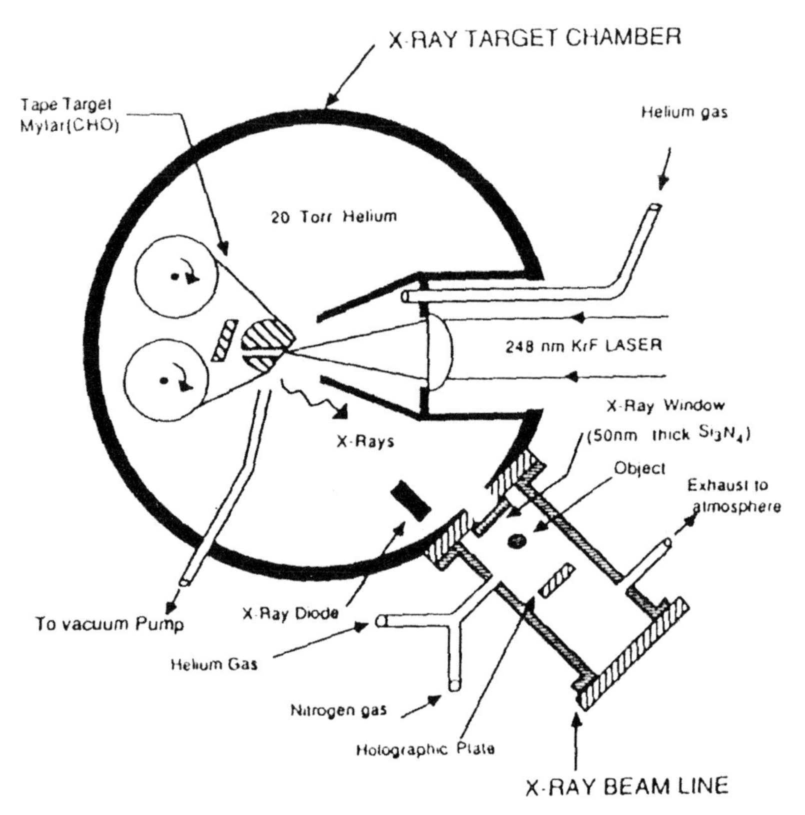

Figure 1.24. X-ray microscope based on a laser-produced plasma source. (Reprinted with permission from Turcu *et al.*, *J. Appl. Phys.* 73, 8081 (1993)[331]. Copyright 1993 American Institute of Physics.)

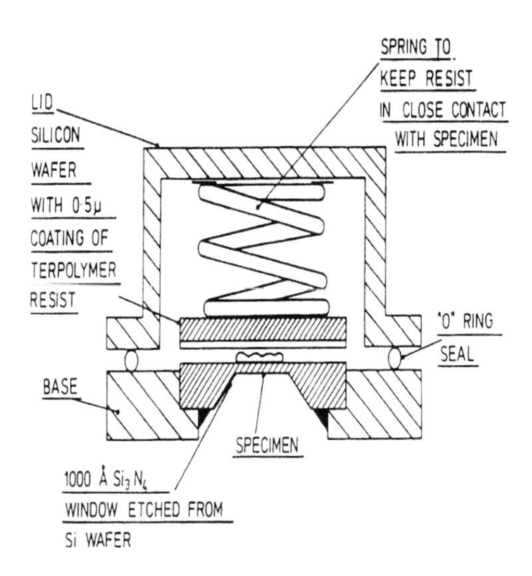

Figure 1.25. An environmental sample cell for resist-based X-ray microscopy using a laser-produced plasma source. (Reprinted with permission from Rosser *et al.*[332].)

and the Schwarzschild objective, used in conjunction with a thin, transmissive target, shown in Figure 1.26[322]. Historically, the problem with optics has been low efficiency, though reflectivities as high as 60% have been reported for multilayer mirrors optimized for particular (multinanometer) spectral bands[333]. Any conceivable optical scheme is likely to suffer from debris deposition, and fast shutters or gas backing pressures are difficult to implement or only partial solutions. The most promising approaches to avoid debris contamination appear to be reducing its production by use of thin targets (e.g. mylar tape[275,312,319,331], or metal-doped glass targets[316]) with the specimen and any optical components on the opposite side from the plasma, or using droplet (alcohol or water) targets produced by a vibrating ink-jet nozzle[315,334] or commercial vibrating orifice droplet generator[335]. With droplet targets, the source size can be carefully controlled and exactly known, since it is equal to the droplet dimension for small droplets, and debris depostion is reduced by more than three orders of magnitude compared to thin tape targets.

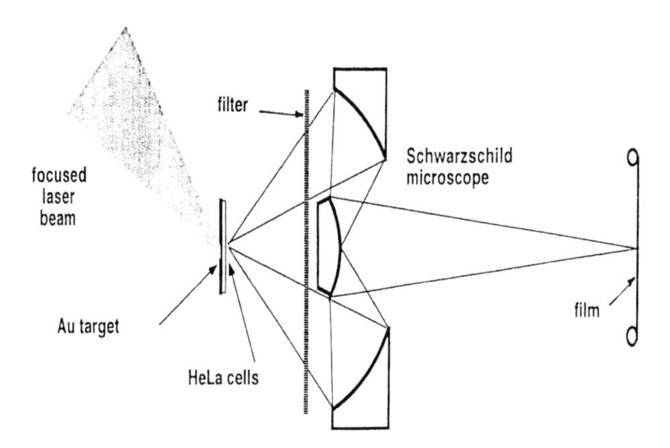

Figure 1.26. Laser-produced plasma based X-ray microscope incorporating Schwartzschild objective for magnification imaging. (Reprinted with permission from Richardson *et al.*[322].)

In 1979, researchers at Rochester's Laboratory for Laser Energetics reported on nanosecond biological X-ray diffraction using an LPP[285]. They showed diffraction patterns of cholesterol and rat spinal roots. Panessa used a pulsed-plasma source and resists to record contact micrographs of proteoglycans[336] and myosin filaments[337], while Feder et al. imaged red blood cells with a z-pinch source as early as 1982[265]. Widespread attention was focused on the potential for high-resolution X-ray imaging of biological specimens using plasma X-ray sources when Feder and colleagues published micrographs, obtained using a gas jet plasma and photoresist detector, of live human platelets on the cover of *Science* in 1985[317]. Other groups were contemplating the possibility of biological imaging using LPP sources at about the same time[279,280,318].

In the 1980's, Michette imaged epidermal hairs, as well as flagellates and human fibroblasts, using very high-energy (30 to 80 joule) pulses from the VULCAN and SPRITE lasers focussed onto mylar tape targets[275]. More recently, his group has been involved in development of scanning X-ray microscopes using LPP sources. Either the specimen[319] or the source[320] may be scanned. Scanning permits the use of a single-element silicon diode detector, but incurs other practical problems, particularly when the source is scanned. Rosser, *et al.*, used a 2-J laboratory Nd:glass-amplified laser beam focused onto gold targets to produce water-window radiation for imaging cultured fibroblasts and other objects[304]. A group at Stanford used an LPP source, consisting of an Nd:YAG focused onto a solid copper target, and the Schwarzschild arrangement of Figure 1.26 to obtain pictures of test patterns in 1989[338,339]. Artyukov built a similar optical setup and used it with an Nd:glass laser and rhenium target to image mesh standards at micron-order resolution[340]. Horikawa used an Nd:YAG laser, iron-target LPP source, grazing incidence condenser, multilayer Schwarzschild objective, and scintillator/microchannel plate detector to image unstained myofibrils at sub-micron resolution[329,341].

Tomie and a team at the Electrotechnical Laboratory in Tsukuba showed detailed images (up to 40-nm resolution) of sea urchin and barnacle sperm, algae, and chinese hamster cells, recorded on resist using water-window LPP source radiation, mostly from yttruim targets[217,321,330]. They state that the requirement for X-ray intensity increases as the fourth power of resolving power and that a few tens of millijoules per square centimeter are required when the resist detection efficiency is 10%. To approach the practical resolution limit of 30 nm imposed by the resist, a dose of 5×10^5 Gray is required, about 10^5 higher than the threshold for biological damage. If manifestation of morphological damage is to be avoided in the recorded image, the exposure time has to be less than 1 ns, a compelling argument for development of LPP sources. Solem has calculated that 30-ps exposures are required for 10-nm resolution[342], the theoretical resolution limit for X-ray photoresists[205,343].

Shinohara and Kinjo have imaged human chromosomes using very soft, second-harmonic (527-nm) radiation from gold targets irradiated with 300-ps, 26-joule pulses focused into 100-micron focal spots (10^{15} W/cm^2)[136,289,322,327,344–349]. They estimated that the flux at the specimen (2 cm from the source, protected by a SiN window) was about 1.4×10^{15} photons/cm^2, with photon energies of 250–1250 eV. This would translate to approximately 0.17 J/cm^2, in agreement with the exposure requirements calculated by others[175]. Stead and colleagues investigated the effects of electron microscopy fixatives on biological materials using X-ray microscopy of plant epidermal hair cells[326]. The X-ray images, demonstrating 150-nm resolution on weakly-attenuating carbonaceous material, were acquired on photoresists using a Rutherford Appleton Nd:glass laser and thin-foil gold targets. Resists were read out with scanning and transmission electron and atomic force microscopes. This group had previously demonstrated 50-nm resolution on dense, particulate and diatom structures[350].

Although still technically challenging to perform and scant on results, X-ray microscopy with LPP sources appears to hold promise for very high-resolution, one-shot imaging of biological objects. Whether this type of source will evolve to compete in the clinic with rotating-anode, electron-impact tubes[247,248] depends upon whether the debris-deposition problem can be solved, either by the use of appropriate, efficient optics or innovative target materials and geometries, and whether the repetition rates (and thereby average power output) can be increased to the level required for sub-second total exposure times.

1.2.5 X-ray Lasers and Free Electron Lasers

Most lasers produce bright, coherent, visible or UV light by amplification occuring over multiple passes through a Fabry-Perot reflecting cavity or resonator containing the pumped gain medium[351]. There currently exist no suitable, efficient mirrors for X-ray beams which can be used to produce multiple passes through the gain medium (energetic plasma)[352], but because of widespread interest in the desirable properties X-ray laser light would possess for a variety of applications including biological microscopy and holography, research on X-ray lasers has persisted. In fact, efforts to exploit, in a practical and cost-effective way, the various schemes proposed for the production of X-ray lasers have increased since the first demonstrations of X-ray lasing in 1985[353,354]. Since no suitable X-ray mirrors exist (and even if they did they would likely be destroyed by energy deposition from optical and x-radiation of the proximal plasma gain medium), amplification must be achieved on a single pass of photons through the gain medium, usually a plasma formed by an intense optical laser focused onto a line along which lasing occurs. When gain is achieved in an energetic plasma, the lasing is said to occur by amplification of spontaneous emission or ASE. Until recently, X-ray lasers remained exotic and expensive propositions, primarily because of the requirement for extremely high-intensity visible lasers to produce the plasmas used to pump X-ray lasers. But with the emergence of chirped-pulse amplified[249,355] and other less expensive optical lasers with intensities in the terawatt to petawatt range, and with the demonstration of X-ray lasing in capillary discharge plasmas[356,357], it appears that low-cost, "personal"[358] X-ray lasers may soon become a reality[359]. Partly because of their complexity and partly because the shortest-wavelength lasing demonstrated to date is at the long-wavelenth end of the spectrum useful for biological imaging research, a thorough discussion of X-ray lasers is beyond the scope of this chapter. The interested reader should consult other excellent sources[206,282,283,303,352–354,358,360–386].

Both of the successful approaches to lasing at X-ray wavelengths use high-power optical lasers to create the high-temperature plasmas required as pump sources. In their most-investigated incarnations, these two alternative methods use electron-collisional recombination pumping[352] (also called recombination pumping) with H-like or Li-like ions[354,358,372,385,387–389], and electron-collisional excitation pumping (often called collisional pumping) with Ne-like or Ni-like ions[353,368,378–382,386]. (Other pumping mechanisms, including photoionization, photoexcitation and charge transfer, have been proposed on theoretical grounds, but have not proven themselves experimentally[352].) Although a number of other laboratories are contributing to the rapid developments in the X ray laser field, perhaps mainly because of their places of discovery these two approaches to X-ray lasing are often identified with groups at Princeton and Lawrence Livermore (LLNL), respectively.

In the recombination scheme, a laser-produced plasma containing highly-stripped ions is the gain medium. After a brief pulse of the optical laser, the plasma is cooled rapidly, undergoing rapid three-body combination and producing population density inversion, a necessary condition for gain. An important advantage of the recombination scheme is that it scales rapidly to shorter wavelengths with increasing plasma ion charge (and therefore target Z), though with decreasing axial extent of the plasma column in which lasing occurs. But it is difficult to obtain and maintain a long, uniform plasma column using recombination pumping. The Princeton group uses a strong magnetic field to confine the plasma column, which helps in this regard. The recombination X-ray laser should exhibit a much higher pumping efficiency than the collisionally-pumped laser, but saturated output has not been approached[390]. One of the earliest applications of X-ray lasers to biological microscopy involved recombination-type lasers using a 1 kJ CO_2 optical laser to produce lasing at 18 (carbon target) and 4 nm (lead target). The X-ray laser was used to image cervical cancer cells and other specimens on a resist detector[388], and it is also proposed to perform biological microscopy at several hundred times magnification with a microscope employing a Schwarzschild objective and CCD-based detector[389].

In the collisional excitation scheme for Ne-like selenium, gain is achieved mainly by collisionally exciting ions to the 3p level, setting up a population inversion between the 3p and 3s levels because of very rapid decay from the latter to the ground state. The Ni-like tantalum X-ray laser, which produced the first significant gain at wavelengths below the carbon K-edge, works in a similar way, with population inversion between the 4d and 4p levels. The collisional excitation scheme has the advantage that it is self-replenishing in the sense that the initial state from which pumping occurs and the ground state to which decay occurs after lasing are the same. Also, the lower state is relatively translucent to the laser wavelength allowing plasma column lengths of centimeters. The length of the plasma column over which lasing occurs is important because amplification has to occur in a single pass. The most important parameter with regard to laser output intensity is the gain length (GL), calculated as the product of the gain (cm^{-1}) and the plasma column length. Initial experience with collisional excitation lasers came out of the inertial confinement fusion facilities at LLNL, including the NOVA and Novette lasers. The first demonstrations[378] of this scheme involved exploding thin selenium foils using intensities of 7×10^{13} W/cm^2 focused on 1.8-cm lengths of foil and producing GL of 8 at 206.4 and 209.8 Å, since raised to 16 and

15.2, respectively [358]. Nickel-like lasers, created by exploding thin tantalum, tungsten, or gold foils, are very similar in principle to their Neon-like, collisional excitation analogs but require less optical laser power at a given X-ray wavelength and have produced the shortest wavelength X-ray lasers to date: 35.6 Å for Ni-like Au[378]. An excellent summary of the gain lengths achieved at wavelengths between 35.6 and 326 Å is presented by Skinner[358]. A GL = 7 has been demonstrated for Ni-like tungsten at 43.18 Å, the first lasing inside the water window, and a GL = 8 for Ni-like tantalum at 44.83 Å, just above the carbon K-edge, at a wavelength ideal for biological X-ray holography [391]. The highest possible output power (saturation) is achieved at GL's between about 15 and 20 which, for collisionally-excited lasers, requires about 1 kJ of driver laser energy at X-ray wavelengths around 20 nm, but over 10 kJ in the water window[392]. It will be necessary to substantially reduce this power requirement, using pre-plasma-forming prepulses or other methods, if table-top lasers are to be used successfully to drive high-output X-ray lasers.

Da Silva and the group and LLNL have used the NOVA-driven, Ni-like tantalum X-ray laser to image a variety of specimens, including rat sperm nuclei, at 44.83 Å[379,380,382]. The optical system consisted of a near-normal-incidence sperical WC/C multilayer mirror condenser with 5% efficiency and narrow bandpass at 44.83 Å, a 500-annulus zone-plate with 450-Å outer-zone width as objective lens, and a microchannel plate (MCP) intensified detector. The MCP limited the spatial resolution to over 500 Å, but correlation with TEM and AFM images provided useful information about the chemical content of imaged volumes in the nucleus. These investigators suggest that cheaper, shorter-wavelength, higher-power X-ray lasers, coupled with improved, short-wavelength multilayer mirrors and improved zone plates and detectors, will bring X-ray laser microscopy into broader usage.

Another interesting, if exotic, tunable, high-flux X-ray source currently under investigation utilizes a phenomenon known as Compton backscattering[39,40,201,393]. When an intense visible or infrared laser beam collides head-on with a near-relativistic electron beam photons interacting with free electrons change their energy and direction to conserve energy and momentum. The group at the Vanderbilt free electron laser (FEL)[394,395] is proposing to turn the infrared laser light beam produced by the FEL wiggler back on the electron beam so that the two collide head on. Two-micron infrared photons backscattered from 43-MeV electrons in the FEL beam will have a maximum energy of 17.6 keV, in the range thought optimal for breast imaging. The Compton backscattered beam is to be diverted using graphite mosaic optics and focused and collimated with a Kumakhov optic[76], to produce a monochromatic, parallel beam accessible for clinical imaging. Another group at Duke is proposing a similar FEL-based approach to produce 12.2 MeV photons for gamma-ray spectroscopy and therapy[396].

If a short-pulse, infrared laser beam is normally incident on an MeV electron beam, the Thomson scattering process produces what has been called the laser synchrotron X-ray source[397]. A group at Lawrence Berkeley Laboratory has produced 300 femtosecond pulses of 30 keV x-radiation using terawatt laser pulses incident at 90° on a relativistic accelerator electron beam[398]. The energy of the highly-directional X-ray beam can be controlled by varying the energy of the electron and laser beams and the angle at which they interact. Although very low fluxes were produced in the first experiments, the researchers are optimistic that these can be increased to useful levels.

While electron-impact sources still produce most of the X-ray beams used in clinical and even research applications, it appears that new types of source will prove useful in the future, particularly when high average power is not a requirement and short, ultra-intense pulses can be used to advantage.

1.3 Interaction of X rays with Matter

Viewed from the broadest standpoint, electromagnetic waves interact with matter when their particles possess energies corresponding to energy-state differences in atoms or molecules of the medium. For example, if a radio wave has a frequency equal to the precession frequency of nuclear spins aligned in a strong magnetic field, it may couple to the spins and excite some of them from the lower-energy parallel to the antiparallel state, a phenomenon which forms the basis for the nuclear magnetic resonance (NMR) experiment. A microwave may interact with an electron in a strong magnetic field, inducing a change in energy states which can be exploited using electron spin resonance (ESR) methods.

For infrared, visible and UV radiation, one can express the total energy of a multiatomic molecule as[399]:

$$E = E_{\text{electronic}} + E_{\text{vibrational}} + E_{\text{rotational}} \,.$$

Rotational energies are about two orders of magnitude lower than vibrational, which are about two orders of magnitude lower than electronic. If a microwave photon has energy corresponding to the difference between rotational states of a molecule, it may excite the molecule to the higher-energy state. Thus, absorbtion of microwave radiation by a gas can give rise to rotational absorption spectra. A higher-energy infrared photon may induce changes between molecular vibrational states. Visible and UV radiation can excite outer electronic transitions. Molecular absorption spectra from these interactions can be very complex, since each electronic state may be associated with a number of vibrational and rotational states. In all cases in which interactions occur, frequencies in the beam of electromagnetic radiation can be said to resonate with particular energy-state transitions in the interaction medium. These resonance phenomena may give rise to discrete, measurable changes in the transmitted frequency spectrum if instrumentation of adequate sensitivity and energy resolution is available.

Similarly, higher energy photons in the soft- and diagnostic X-ray ranges may interact with and give up energy and momentum to atomic electrons in the absorbing medium. Very high-energy x and gamma rays can undergo interactions with bound electrons, nucleons, and the nuclear fields of absorber atoms. These interactions produce changes in the energy content of the beam, cause the emission of electrons or nucleons, or transiently excite atoms in the interaction medium. Detection of these energy changes, emitted subatomic particles, or photons forms the basis of a number of imaging and spectroscopic techniques from which insight into biological structure and function can be gained. For clarity, a brief summary of the various X-ray energy ranges, names they are often referred to by, and the interactions of most importance in each range are given in Table 1.3[115,400].

Table 1.3.

Energy Range (keV)	Common Names	Relevant Interactions
< 0.120	vacuum ultraviolet	absorption, elastic scatter
0.120–0.500	ultrasoft X rays, water-window X rays (.280–.560 keV)	photoelectric absorption coherent scatter
.100–20	soft X rays Grenz rays	photoelectric absorption coherent scatter
20–140	diagnostic X rays gamma rays	Compton scatter photoelectric absorption
120–300	orthovoltage X rays gamma rays	Compton scatter
300–1000	intermediate-energy X rays gamma rays	Compton scatter
>1000	megavoltage X rays therapeutic x and gamma rays	pair production photonuclear reactions Compton scatter

When related to X-ray beam propagation and interactions, the word "attenuation" refers to any process which causes a decrease in the number of photons passing through a unit area perpendicular to the beam's propagation direction per unit time interval. Attenuation occurs by three primary mechanisms: geometry, scattering and absorption. The decrease due to simple geometry in photon flux from a point source is often called the inverse square law. The photon flux is inversely related to the square of the distance from the source at which the flux measurement is made:

$$\phi_2 \quad \frac{\phi_1 d_1^2}{d_2^2},$$

where d_1 and d_2 are the distances from the source at which fluxes ϕ_1 and ϕ_2 are measured. For electron-impact and other Lambertian-emitting point sources of X rays, the flux intercepted by the sample may be dramatically increased by placing the sample close to the source. If a large-format detector is placed some (greater) distance away, a magnified image is produced, which improves the detector system's spatial resolution referred to object space. This can be an important advantage for digital detectors which generally have pixel sizes between about 10 microns (for the smallest-pixel CCD sensors) and 100 to 200 microns (for photostimulable phosphor plates and amorphous silicon detector panels). In practice, especially for tomographic applications, these advantages have to be traded off with the problems created by the large cone angle subtended by the specimen when it is imaged in close proximtiy to the source.

From the standpoint of their physical interactions with matter, there is no distinction between X rays and gamma rays of the same energy: they differ only in their site of origin. Gamma rays are of nuclear and X rays of extranuclear origin. X rays are produced by the acceleration of free electrons or other charged particles (Bremsstrahlung) or by bound electron transitions (characteristic radiation), while gamma rays are produced by nuclear transitions. In this section we consider the interaction of X rays with matter. The energy range of primary interest here is between about 10 and 150 keV since clinical diagnostic imaging utilizes photons of these energies. The descriptions of the interactions of diagnostic X rays which follow apply also to the attenuation of the gamma rays used for SPECT and PET imaging as described in Volume 1, Chapter 3.

Interactions of photons in the 100-eV to 10-keV range relevant to X-ray microscopy and holography will be described briefly, but is treated more fully in other texts[114,115,401]. In these energy regimes, interaction may involve inner and outer orbital electrons. Interactions of higher-energy X rays with matter, which may include nuclear processes, are discussed in Volume 2, Chapter 1.

1.3.1 Indirect versus Direct Ionization

A radiation beam is an agent of energy transfer[20]. When a beam encounters biological or other matter, it may interact with atoms in the medium and cause changes in the nature of the beam and of the material. Energy transfer processes may include total absorption and various types of scatter, all of which may be of interest for different types of biomedical imaging experiments. It is helpful to divide electromagnetic radiations into different categories. The infrared, visible and UV interactions discussed above were non-ionizing interactions, characteristic of relatively low-energy exchange events: Vibrational and rotational states could be influenced and outer orbital electrons could be excited to higher energy states, but separation of charge was not a possibility.

Ionizing radiations are energetic enough to entirely remove atomic electrons or protons from their atoms[3,20]. In the case of x radiation, photons in the beam must possess energies higher than the binding energies of the ejected orbital electrons for ionization to occur. Directly ionizing radiation accomplishes the charge separation directly by interaction between the coulomb forces of the involved particles. Directly ionizing particles include electrons, protons, and other heavy charged particles. Indirectly ionizing radiations, including photons and neutrons, are uncharged particles which may partake in interactions which release energetic charged particles such as electrons and protons from atoms in the medium. Photons, which are the indirectly ionizing radiation of broadest interest in biomedical research, eject electrons, and neutrons eject protons, from atoms in the medium. Although the initial ejection of the charged particle occurs by direct interaction with the incident particle, these radiations are still called "indirectly ionizing" because the bulk of the ionization is due to subsequent interactions of the released, energetic charged particle.

The initial interaction with the X-ray photon may result in some scattered radiation and sets in motion a fast electron. This high-speed electron may cause electronic excitations and ionizations, and may break molecular bonds. If an electron interacts with a nucleus, it may produce additional Bremsstrahlung which, along with any scattered radiation, may contribute to additional indirectly ionizing events. So a complex series of interactions may take place before all the photon energy is lost to electronic motions.

The minimum energy required for ionization cannot be precisely fixed, because it differs from molecule to molecule. A photon may be capable of ionizing an atom when it is bound in one molecule, but not when it is bound in another. Some texts give the cutoff energy for ionizing radiation as 12.6 eV, since this is the ionization potential for hydrogen in water, an important constituent of biological tissue. In tissues, photons must have energy greater than about 10 eV to be capable of ionization.

1.3.2 Scattering

X photons can interact with atomic electrons by elastic or inelastic scattering. In the case of elastic scattering, no kinetic energy is given up by the photon to the attenuating medium, while inelastic scattering involves the loss of some of the incident photon's energy. As the name implies, scattered photons are not stopped or completely removed from the (broad) beam, but are caused to deviate from their original, straight-line path from the source, with no loss of energy for elastically scattered photons, or with diminished energy for inelastically scattered photons. Both types of scatter play a role in image formation because the direction and energy of scattered photons depends on the properties of the scattering material as well as the energies of the incident photons.

1.3.2.1 Elastic Scattering

When photons are elastically scattered, no momentum is transferred to the scattering medium and the wavelengths of the incident and scattered photons are equal. It was found shortly after their discovery that X rays could not be efficiently reflected from polished surfaces but instead were scattered in all directions. J.J. Thomson explained this as the result of the classical interaction of the incident electromagnetic wave with individual atomic electrons which, in his approach, can be regarded as free[1,3,20,115]. In the classical presentation of elastic scattering, valid for photon energies large compared to atomic electrons' binding energies but small campared to m_0c^2 = 511 keV, individual atomic electrons are set to oscillating by the force of the sinusoidal electric field of the incident wave. The vibrating charge radiates at a frequency equal to that of the incident beam, causing no decrease in energy. This type of elastic scattering is called Thomson or classical scattering.

When the (diagnostic-energy: $\beta/c \ll 1$) photon wavelength is similar to the diameter of atoms in the attenuating medium, all atomic electrons, including tightly-bound inner-shell electrons, oscillate and reradiate in phase. This elastic scattering by atomic electrons as a group is sometimes referred to as Rayleigh scattering or coherent scattering, since it is a cooperative phenomenon. Since the distinctions between Thomson and Rayleigh scattering theory can be confusing and are of doubtful practical significance, it can be helpful to think of elastic scattering as being due to a coherent scattering process caused by atomic electrons as a group and characterized by a differential cross-section per unit angle given by[3,20]:

$$\frac{d\sigma_{coh}}{d\Theta} = \frac{r_0^2}{2}\left(1+\cos^2\Theta\right)\left[F(x,Z)\right]^2 2\pi \sin\Theta, \tag{1.43}$$

where the momentum transfer variable $x = \dfrac{\sin\Theta/2}{\lambda}$.

This equation gives the fraction of the incident energy scattered into the cone between angle Θ and $\Theta + d\Theta$ which subtends the solid angle:

$$d\Omega = 2\pi \sin\Theta d\Theta. \tag{1.44}$$

The expression:

$$\frac{d\sigma_0}{d\Omega} = \frac{r_0^2}{2}\left(1+\cos^2\Theta\right) \tag{1.45}$$

is called the Thomson coefficient and correctly predicts the fraction of energy scattered per electron per unit solid angle for photons of zero (negligible) energy. Quantum mechanical corrections to Thomson's derivations were required to accurately describe scattering at higher energies as described in the next section.

The atomic form factor $F(x,Z)$ in Eq. (1.43) approaches zero for large angles Θ and Z for small Θ and falls off dramatically as x (proportional to photon energy) increases[402]. These two effects imply that coherent elastic scattering is significant for low energy X-ray interactions with high-Z elements. The relative contributions of coherent and Compton (incoherent) scatter and photoelectric absorption to attenuation in carbon and gold are shown in Figure 1.27[115]. For body tissues in the diagnostic energy range, coherent scattering accounts for less than 5% of total scatter[3,402]. In addition, the angular distribution is markedly peaked in the forward direction, with half the coherently-scattered photons being contained in cone angles of 38°, 29°, 23°, 19°, 15°, 12° and 9° for 10-, 20-, 30-, 40-, 60-, 80- and 100-keV photons, respectively. Total coherent cross sections σ_{coh} are are obtained by integrating Eq. (1.43) from 0 to 180 degrees and are tabulated for each element as functions of energy[402,403]. The smooth angular distribution of elastically-scattered radiation given by Eq. (1.43) is accurate for randomly ordered aggregates of atoms. If a narrow-energy-band photon beam interacts with an ordered array of atoms like a crystalline lattice (or grating) and the lattice spacing is of the order of the photon wavelength, positive reinforcement of the coherent scatter from crystal planes can occur, causing sharp peaks in the angular distribution of scattered radiation emission. This phenomenon, called Bragg scattering, forms the basis for X-ray crystallography.
In the diagnostic energy range, coherent scattering plays a small role and can often be neglected. For energies below 1 keV, at which most X-ray microscopic imaging is performed, absorption dominates and coherent scattering plays an important role. At soft X-ray energies, the interaction with an atom is described by the conplex amplitude[115]:

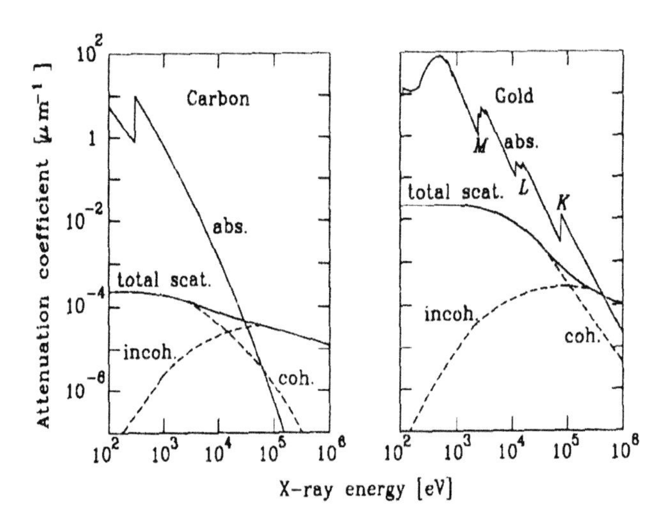

Figure 1.27. Scatter and absorption contributions to the linear attenuation coefficients of carbon and gold as functions of X-ray energy. (Reprinted with permission from Michette and Buckley[115].)

$$A(\Theta,h\nu) = A_T(\Theta)\left[f_1(h\nu) + if_2(h\nu)\right], \tag{1.46}$$

where A_T is the amplitude given by the Thomson coefficient, Θ is the scattering angle and $f_1 + if_2$ is the complex number of electrons in the atom[404]. $f_1(h\nu)$ and $f_2(h\nu)$ are the atomic scattering factors which describe scattering and absorption, respectively, by the atom. They are independent of Θ at low energy. The scattering intensity for a single electron predicted by the Thomson coefficient of Eq. (1.45) must be multiplied by the square of the number of electrons involved to obtain the coherent scattering cross section for a multi-electron atom[115,404]:

$$\frac{\mathrm{d}\sigma_{coh}}{\mathrm{d}\Omega} = \frac{r_0^2}{2}\left(1+\cos^2\Theta\right)\left|f_1 + if_2\right|^2, \tag{1.47}$$

which, integrated over all angles gives:

$$\sigma_{coh} = \frac{8\pi r_0^2}{3}\left|f_1 + if_2\right|^2. \tag{1.48}$$

The refractive index, a key parameter for soft X-ray microscopy, is given by:

$$n = 1-\delta - i\beta = 1-K\left(f_1 + if_2\right), \tag{1.49}$$

where $$\delta = \frac{r_0\lambda^2}{2\pi} n_a f_1 \tag{1.50}$$

is the refractive index decrement,

$$\beta = \frac{r_0\lambda^2}{2\pi} n_a f_2 \tag{1.51}$$

is the absorption index,

and $$K = \frac{n_a r_0 \lambda^2}{2\pi}, \tag{1.52}$$

in which $n_a = \frac{\rho N_0}{A}$, ρ is the density, N_0 Avogadro's number and A the atomic weight. After traveling a distance x through an attenuating medium, the incident wave amplitude A_0 is reduced to:

$$A = A_0 \exp\left(\frac{-2\pi\beta x}{\lambda}\right)\exp\left(\frac{i2\pi\delta x}{\lambda}\right), \tag{1.53}$$

and the intensity is given by:

$$I = |A|^2 = I_0 \exp\left(\frac{-4\pi\beta x}{\lambda}\right). \tag{1.54}$$

The phase change is seen to be $2\pi\delta x/\lambda$ and the attenuation is $e^{-\mu x}$ where the attenuation coefficient

$$\mu = \frac{4\pi\beta}{\lambda} = 2 n_a r_0 \lambda f_2 . \tag{1.55}$$

f_1 and f_2 are thus associated with the phase shift and absorption components of the refractive index, respectively. While the phase of visible light waves is retarded by interaction with a material, that of X-ray waves is usually advanced relative to the incident wave. The tabulated optical constants of Henke[405] and others are usually obtained by absorption measurements of f_2 using thin foils, followed by calculation of f_1 from the Kramers-Kronig relationships[115,404]. It is noted that the ratio f_1/f_2 increases rapidly with X-ray energy which forms the basis of the argument for recent experiments using relatively hard X-ray phase imaging for soft tissues[160,162,238,239].

1.3.2.2 Inelastic Compton Scattering

Compton scattering, which dominates among the competing attenuation mechanisms at higher diagnostic energies, involves the transfer of kinetic energy from incident photons to atomic electrons with which they collide[1,3,15,20,115]. For the chance of a Compton interaction to be high, the incident photon energy must be large compared to the binding energy of the electron involved, and inelastic scattering usually involves outer-shell, loosely-bound atomic electrons. These electrons are ejected from the atom: Compton scattering is an ionizing event which contributes to exposure and dose.

When the photon energy is not negligible compared to m_0c^2 its momentum can no longer be neglected and the momentum it loses after being deviated through non-zero scattering angles must be conserved by transfer to the ejected Compton electron. Considering the involved electron to be at rest and free (reasonable at the photon energies for which inelastic scattering is important), a Compton interaction is analogous to a collision between a (lighter) queue ball and a (heavier) billiard ball. The plane of the interaction (containing the balls' centers) is defined by the plane containing the photon's incident and scattered trajectories. Since there is no momentum out of this plane, the path of the fast electron must also lie in this plane, and the interaction can be pictured as shown in Figure 1.28[3]. The energies of the incident photon, scattered photon and receding electron are represented by $h\nu$, $h\nu'$ and E respectively. The scattering angle for the photon and recoil angle for the electron are Θ and Φ, both with respect to the incident photon trajectory.

Using relativistic expressions for kinetic energy and momentum and conservation of energy and momentum, it can be shown that the reduced energy of the scattered photon is given by:

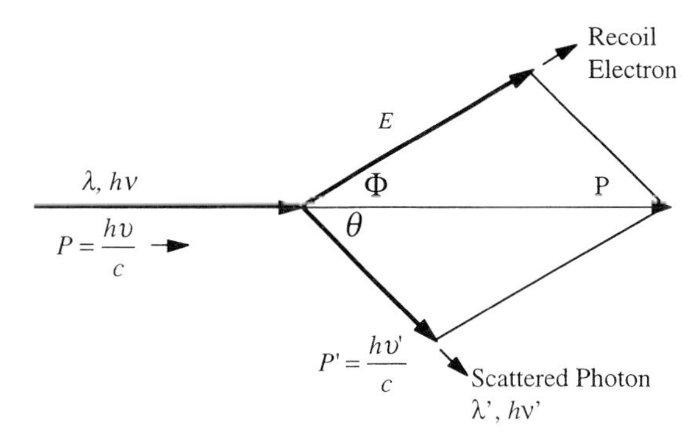

Figure 1.28. Compton scattering geometry. (Adapted from Johns and Cunningham[3].)

$$h\nu' = h\nu\left[\frac{1}{1+\alpha(1-\cos\Theta)}\right], \tag{1.56}$$

and the kinetic energy of the recoil electron is:

$$E = h\nu\left[\frac{\alpha(1-\cos\Theta)}{1+\alpha(1-\cos\Theta)}\right], \tag{1.57}$$

where dimensionless $\alpha = \frac{h\nu}{m_0c^2} = \frac{h\nu}{511\text{keV}}$, with $h\nu$ in keV, or, in terms of wavelength:

$$\lambda' = \lambda + \frac{h}{m_0c}(1-\cos\Theta) = \lambda + 0.0243(1-\cos\Theta). \tag{1.58}$$

The quantity $\frac{h}{m_0c}(1-\cos\Theta)$ is known as the Compton shift and gives the increase in wavelength of the scattered photon[1]. It is interesting to note that the Compton shift in wavelength at a particular scattering angle is independent of incident photon energy, but the Compton shift in terms of energy depends very strongly on energy. For example, at 90° for a 10-keV photon the Compton shift in energy is 0.2 keV (2%) whereas for a 10-MeV photon it is 9.51 MeV (>95%). True to the billiard-ball analogy, the above equations confirm that if the electron takes a direct hit, the photon will be backscattered through 180° imparting to the electron the maximum possible kinetic energy:

$$E_{\text{max}} = h\nu\left[\frac{2\alpha}{1+2\alpha}\right], \tag{1.59}$$

while if the photon barely ticks the electron the photon will be minimally deviated in its path, $\cos\Theta = 1$, the photon's energy after scattering will be almost equal to its incident energy, and the electron will be perturbed in a direction almost perpendicular to the incident photon trajectory.

The probability of a Compton interaction was first calculated by Klein and Nishina in 1928[1,406]. The Klein-Nishina equation, based on Dirac's relativistic theory, has been repeatedly shown to correctly predict experimental results by taking the recoil of the electron into account. The differential cross section for Compton scattering is given by the product of the Thomson coefficient, Eq. (1.45) above, and a correction term designated f_{KN}:

$$\frac{d\sigma_{Compton}}{d\Omega} = \frac{r_0^2}{2}\left(1+\cos^2\Theta\right)\left(f_{KN}\right), \tag{1.60}$$

$$\text{where } f_{KN} = \left[\frac{1}{1+\alpha(1-\cos\Theta)}\right]^2 \left\{1+\frac{\alpha^2(1-\cos\Theta)^2}{[1+\alpha(1-\cos\Theta)]\left(1+\cos^2\Theta\right)}\right\}. \tag{1.61}$$

At diagnostic energies, the Compton scattering cross section is slightly less at a scattering angle Θ of 180° than at 0°, with a minimum at 90°, whereas for higher-energy photons scattering becomes increasingly peaked in the forward direction. The effect of incorporating f_{KN} into the equation for the Compton cross section is to decrease the amount of scatter predicted by Thomson's equation, particularly for high-energy photons and large scattering angles.

1.3.3 Absorption

As an alternative to scattering, a photon may be completely absorbed and removed from the beam, a process more likely than scattering at low energies as shown in Figure 1.27 and also at extremely high energies. Photons may be completely removed from the beam by photoelectric absorption for soft- and diagnostic X rays, or by pair production or photodisintegration at very high energies.

1.3.3.1 Photoelectric Absorption

The photoelectric effect, for the discovery of which Einstein received the Nobel Prize in physics in 1921, involves a collision of the incident photon with a bound atomic electron[1,3,15,20]. The electron is ejected from the atom with kinetic energy equal to the difference between the photon's energy and the binding energy of the orbital electron:

$$KE_{PE} = h\nu - BE. \tag{1.62}$$

Electrons may be ejected from any shell, with the required energy decreasing rapidly from the K to the L to the *M* and to orbitals still more distant from the nucleus. Electrons ejected by low-energy photons are ejected in a direction approximately perpendicular to the incident photon trajectory and increasingly toward the forward direction as photon energy is raised. The photoelectric cross section varies with the fourth power of the absorber's atomic number, linearly with absorber density and inversely with absorber atomic weight and, in general, inversely with the third power of photon energy:

$$\sigma_{PE} \propto \frac{\rho Z^4}{A(h\nu)^3} . \tag{1.63}$$

Photoelectric absorption is most likely at energies just higher than the various electron binding energies causing sharp discontinuities in the curve relating the photoelectric cross section to photon energy. These jumps in the photoelectric cross section are referred to as the K-, L- and M- edges and so forth, according to the shell from which the electron is ejected. Photoelectric absorption is most likely to involve an electron whose binding energy is just below the photon energy: Photons with sufficient energy are most likely to interact with K-shell electrons. The minimum kinetic energy required for a photoelectron to escape from the surface of a material is called the work function Φ_0 of the material[34] and is on the order of 2 to 6 eV.

After ejection of an inner-shell electron by the photoelectric effect, the absorber atom is left in an excited state. It can return to stability either by emission of characteristic radiation or by ejection of an Auger electron. Both characteristic radiation and Auger electron ejection occur as a consequence of the inner-shell vacancy being filled by an electron from an outer shell "falling in" to take the place of an ejected photoelectron. An Auger electron is an electron ejected from a shell more distant from the nucleus than that in which the vacancy occured. The fluorescent yield, defined as the ratio of characteristic photons to Auger electrons emitted after photoelectric absorption, increases from a few percent in soft tissue to 95% in lead[3]. The energy of a characteristic photon is equal to the difference in binding energies (BE) between the inner shell from which the photoelectron was ejected and the outer shell from which the electron filling the vacancy originated:

$$h\nu_{characteristic} = BE_{inner} - BE_{outer} . \tag{1.64}$$

The kinetic energy of an Auger electron is equal to the difference in binding energies of the two shells involved decremented by the energy required to eject the Auger electron:

$$KE_{Auger} = BE_{inner} - \left(BE_{outer} + BE_{Auger}\right) . \tag{1.65}$$

Since Auger electrons are usually ejected from the same shell as that from which the electron making the downward transition to fill a vacancy originated:

$$KE_{Auger} = BE_{inner} - 2BE_{outer} . \tag{1.66}$$

Due to their charge, Auger electrons are absorbed in surrounding tissues. Characteristic X rays caused by photoelectric absorption are few enough in number and low enough in energy that their contribution to transmitted X-ray images can be ignored.

1.3.3.2 Pair Production

Pair production, defined as the conversion of a photon to an electron-positron pair, may occur above a threshold of 1.02 MeV. A highly-energetic photon may interact with the strong field of the nucleus and be eliminated from the beam. In its place a positive and a negative electron appear. The magnitude of the threshold energy for pair production is determined by the energy equivalent of the mass of these two electrons. Above the threshold, the cross section for pair production increases rapidly with photon energy. Pair production does not occur in the diagnostic energy range, but is an important interaction mechanism in radiation therapy, discussed in Volume 2.

1.3.3.3 Photodisintegration

Photonuclear interactions involve the absorption of a photon by the nucleus followed by the ejection of a nucleon. Very high threshold photon energies (2.22 MeV for hydrogen, 18.7 MeV for carbon and 15.7 MeV for oxygen[20]) are required to overcome the nuclear binding energies of protons and neutrons, and photodisintegration is not relevant in diagnostic applications of ionizing radiation.

1.3.4 Attenuation Cross Sections and the Linear Attenuation Coefficient

As alluded to earlier, attenuation involves all the processes causing a reduction in the number of photons in an X-ray beam as it traverses an interaction medium or absorber. In diagnostic and other imaging applications of ionizing radiation, ignoring the inverse square law (see the beginning of Section 1.3), the total attenuation cross section, σ_{tot}, includes contributions from all the involved processes described above: elastic and inelastic scattering and photoelectric absorption:

$$\sigma_{tot} = \sigma_{coh} + \sigma_{Compton} + \sigma_{PE}. \tag{1.67}$$

The cross sections σ are a measure of the likelihood that a particular interaction will occur for an individual target or atom in the absorber. Cross sections are tabulated[403] with units of area (e.g. Barnes: 1 Barn = 10^{-24} $cm^2/atom$).

For a narrow monoenergetic beam of energy $h\nu$, the number of photons remaining in the beam after traversing a thickness x of absorber is given by[1,3,15,20]:

$$N = N_0 e^{-\mu x} \tag{1.68}$$

where N_0 is the number of photons incident on the absorber and μ is defined as the linear attenuation coefficient of the absorber material at energy $h\nu$. In terms of intensity:

$$I = I_0 e^{-\mu x}. \tag{1.69}$$

The linear attenuation coefficient has units of reciprocal distance (e.g. cm^{-1}) and can be calculated as[20]:

$$\mu = n_v \sigma_{tot} = n_m \rho \sigma_{tot}, \tag{1.70}$$

where n_v and n_m are the number of atoms per unit volume or per unit mass, respectively, in the attenuating medium of infinitessimal thickness dx, σ_{tot} is the total cross section, and ρ is the density of the medium (e.g. grams per cubic centimeter). The linear attenuation coefficient is a function of the incident photon energy and the atomic number and density of the absorber. A related quantity, the mass attenuation coefficient is defined as:

$$\frac{\mu}{\rho} = n_m \sigma_{tot} \tag{1.71}$$

and has units of area per unit mass (e.g. cm^2/gram). The mass attenuation coefficient is independent of absorber density. For dose and radiation damage considerations, it is often necessary to calculate the energy actually absorbed in the object using the energy absorption coefficient:

$$\mu_{ab} = \mu \left(\frac{E_{ab}}{h\nu} \right), \tag{1.72}$$

where E_{ab} is the average energy absorbed per interaction.

Eq.s 1.68 and 1.69 will only be valid for experiments involving monoenergetic photons where the measurement of transmitted intensity is made in the narrow beam geometry illustrated in Figure 1.29[20]. As shown, this geometry can be achieved by placing the absorber close to the collimated radiation source and the small, single-element detector some distance away. This arrangement excludes most of the scattered radiation from the measurement, since, except for very small scattering angles, it will diverge from the narrow beam and escape detection. In practice, for broad-band Bremsstrahlung spectra, the transmitted intensity must be calculated as an integral over photon energy, $h\nu$:

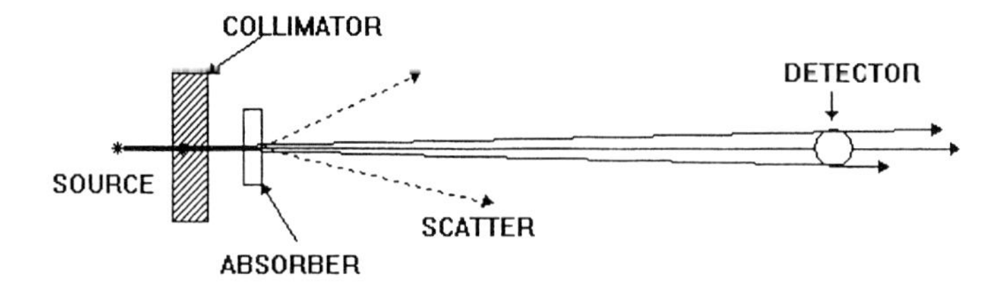

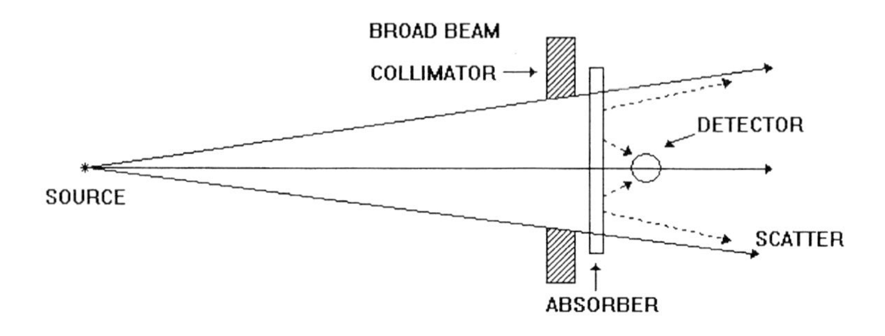

Figure 1.29. Narrow and broad beam geometry for measuring attenuation. (Adapted from Anderson[20].)

$$I = \int_0^{h\nu_{\max}} I_0(h\nu) e^{-\mu(h\nu)x} \mathrm{d}h\nu. \tag{1.73}$$

For modeling and other purposes, it may be desirable to fit the tabulated cross-section data, since the energy increments are generally rather course (e.g. 5, 10 or 20 keV). This has often been done successfully using an equation of the form[407,408]:

$$\mu(h\nu) = a_1 f_1(h\nu) + a_2 f_2(h\nu) = a_1 \frac{1}{(h\nu)^3} + a_2 f_{KN}(h\nu), \tag{1.74}$$

where a_1 and a_2 are functions of physical properties of the absorber and $f_{KN}(h\nu)$ is the appropriate form of the Klein-Nishina equation. The first term is associated with absorption due to the photoelectric effect and the second accounts for Compton scattering. The functions are given by:

$$a_1 = \frac{K_1 \rho Z^4}{A} \tag{1.75}$$

$$\text{and } a_2 = \frac{K_2 \rho Z}{A}, \tag{1.76}$$

where the K's are constants, ρ is the material density, Z the atomic number and A the atomic weight of the absorber. The appropriate form of the Klein-Nishina formula is[3]:

$$f_{KN}(h\nu) = \frac{3}{4}\sigma_0 \left\{ \left(\frac{1+\alpha}{\alpha^2} \right) \left[\frac{2(1+\alpha)}{1+2\alpha} - \frac{\ln(1+2\alpha)}{\alpha} \right] + \frac{\ln(1+2\alpha)}{2\alpha} - \frac{1+3\alpha}{(1+2\alpha)^2} \right\}, \tag{1.77}$$

where $\alpha = \frac{h\nu}{m_0 c^2} = \frac{h\nu}{511 keV}$,

and the total cross section σ_0 (the Thomson classical scattering coefficient) is given by:

$$\sigma_0 = \frac{8}{3}\pi r_0^2 = 66.525 \times 10^{-26}\ cm^2, \tag{1.78}$$

in which the classical electron radius is:

$$r_0 = \frac{ke^2}{m_0 c^2} = 2.81794 \times 10^{-13}\ cm,$$

and the constant

$k = 8.9875 \times 10^{13} \frac{Ncm^2}{C^2}$ arises from Coulomb's law.

For practical purposes the 3/4 σ_0 of Eq. (1.77) can be lumped with K_2 in Eq. (1.76), and $f_{KN}(h\nu)$ taken as the expression within the braces. Eq. (1.74) can be fit to the tabulated data (using minimum least squared error or other appropriate criterion) over the energy ranges of interest by finding a set of coefficients K_1 and K_2 for each element of interest. In general it will be necessary to calculate different sets of K's on each side of absorption edges spanned by the energy range at hand. Attenuation by compounds can be estimated according to the rule of mixtures [3,409–411]:

$$\bar{Z} = \sqrt[m]{a_1 Z_1^m + a_2 Z_2^m + \ldots + a_n Z_n^m + \ldots}, \tag{1.79}$$

where $\bar{Z}$ is the effective atomic number of a mixture of materials of atomic number Z_1 to Z_n and a_1 to a_n are the fractional numbers of electrons per gram of the various materials.

1.3.5 Interaction of X rays with Detector Materials

A very important class of interactions of X rays with matter involves their absorption and conversion in imaging and other X-ray detectors. Often, it is the electrons liberated by photoelectric and Compton interactions in the active detector material which ultimately produce the signal which is sensed. Alternatively, visible light emitted when an atom excited by one of these mechanisms returns to stability may be detected by an electronic detector or form latent image centers in the silver bromide grains of a film emulsion. In order to detect X rays efficiently and minimize the required exposure to the patient, a detector material should be dense and of high atomic number to maximize the number of interactions described by the foregoing equations. There is almost always a tradeoff between detector thickness and spatial resolution: A thick detector will stop and convert more X rays, but scatter of X rays and possibly other quanta within the detector itself degrades the spatial resolution of the acquired image, obscuring fine detail.

Recent developments promise to yield electronic detector arrays capable of detecting X rays directly; that is, converting X-ray energy directly to a measurable quantity such as voltage or charge using very dense, novel solid state materials such as CdZnTe and HgCdTe [412,413]. Another approach uses amorphous selenium for direct X-ray detection[414,415]. But most available detectors, including photostimulable phosphor plates[416–428] and the new amorphous silicon arrays which theoretically could eventually also be used in a direct mode[429–433], first convert the X rays to visible light which is then detected, either by film or by electronic imaging detectors including photomultiplier tubes (PMT's), charge-coupled devices (CCD's) and tube-based cameras such as Vidicons and Plumbicons [434–440]. There are two primary reasons for this. The first is that most X-ray sensitive materials are too penetrable (either their mass density is too low or they have to be fabricated thin in order to preserve spatial resolution) to stop X rays efficiently. The second reason, which applies to CCD's and most silicon-based solid state devices, is that the active sensing material is not radiation hard: With continued exposure key properties like dark noise and charge transfer efficiency are degraded to unacceptable levels[441–443].

The scintillating materials used to convert the transmitted X-ray pattern to light in the visible spectrum fall into two major categories: granular phosphors[444–455] and solid crystals[456,457]. It is interesting to note that the first scintillation detector was used by Roentgen in his discovery of X rays when he observed the glowing of a barium platinocyanide-covered paper screen in a darkened room while experimenting with electrical discharges in a partially-evacuated tube[33,458–461]. Historically, the most common type of imaging X-ray detector has been the film-screen system. Thin granular phosphor screens (between about fifty and two hundred microns thick) on one or both sides of the film absorb X rays and convert them with some relatively low efficiency (between about five and twenty percent) to light, generally in the green portion of the visible spectrum to which silver bromide film emulsion is most sensitive. The phosphor particles, usually with densities between five and eleven grams per cubic centimeter and diameters in the five- to twenty-micron range, are held together by a plastic binder matrix. A reflective front-side coating increases the light emitted in the direction of the film. Besides conversion efficiency, the main problem is that light scatters laterally within the screen material, degrading the spatial resolution of the image at the film plane. Opacifiers are often added to the binder matrix to decrease this spreading of light at the cost of reduced efficiency, but the spatial resolution is generally no better than the screen thickness.

The advantage of solid crystalline scintillators like cesium iodide (CsI) is that the material can be grown in sheets of close-packed single-crystal columns up to several hundred microns thick. Absorption efficiency is therefore as good or better than that of the thickest phosphor screens. Light emitted as a result of X-ray absorption in a columnar crystal does not spread appreciably laterally into adjacent crystals due to the refractive index mismatch at the boundary, but is channeled down the crystal to the film or other detector stage. Spatial resolution is therefore limited by the diameter of the individual crystal columns. The problems with crystal scintillators include their expense and susceptibility to hygroscopic or other degradation. Their conversion efficiency is also somewhat lower than the more efficient granular phosphors.

Early detectors for computed tomography (CT scanners) were based on ionization of pressurized gases such as xenon[462–464]. These detectors were very linear and stable, and some scanner manufacturers still employ them, but they are rather inefficient due to the low density of the gas. Cadmium tungstate scintillators coupled to photodiodes have found broad application in CT scanners; others such as bismuth germanate, CsI and sodium iodide have been tried; and a number of novel scintillators, including ceramic materials, have been developed for the purpose[465]. Problems with monocrystalline materials include afterglow, low mechanical strength, hysteresis (response depends on irradiation history), and hygroscopic attack. Promotion of transparency of ceramic materials can be problematic.

1.3.6 Object, Subject and Image Contrast

While there is no universally-accepted set of terms to describe the contrast at the various stages of the imaging procedure, it is important at least to recognize the factors involved at each stage, facilitating clarity of explanation and understanding. If one considers an object or small inclusion, represented by subscript 1, to be imaged against a background designated by subscript 0, the object contrast may be defined in terms of their linear attenuation coefficients as:

$$C_{object} = \frac{|\mu_1 - \mu_0|}{\mu_0}. \qquad (1.80)$$

This quantity embodies the physical differences in density and atomic number available to potentially allow discrimination of the object from the background in a displayed image. Object contrast is sometimes called intrinsic or patient contrast. Subject contrast may be defined as:

$$C_{subject} = \frac{|\Phi_1 - \Phi_0|}{\Phi_0}, \tag{1.81}$$

where Φ is the photon fluence emerging from the patient after passing through the inclusion and the background. Since detectors' responses are usually linear with energy, it might be more appropriate to define subject contrast as:

$$C_{subject} = \frac{|\Psi_1 - \Psi_0|}{\Psi_0}, \tag{1.82}$$

where Ψ is the energy fluence emerging on the detector side of the patient. In fact, though the results are not exactly the same, it makes little difference if photon fluence, energy fluence or exposure are used in the calculation of subject contrast[3]. Image contrast has to do with the sum total result of the whole system: density differences in the patient, X-ray spectrum utilized, efficiency and response characteristics of the detector and any other components used to present the image to the observer, etc. It may be defined as:

$$C_{image} = \frac{|I_1 - I_0|}{I_0}, \tag{1.83}$$

where I indicates the flux of visible light energy presented to the observer's eye.

Object contrast is due to intrinsic differences in the imaged patient's tissues, and can only be improved by the addition of some contrast-enhancing agent, such as iodine injected into the vascular compartment. Subject contrast is affected by the X-ray spectrum used to produce the image. In general, the differences between attenuation coefficients of various tissues and materials are larger at low energies than at high, so low-energy beams will produce higher subject contrast. The low-energy limits are set by the penetrating capacity of X rays. In addition, illumination preferentially accentuating energies just above an absorption edge in one of the tissues of interest can produce enhanced subject contrast. The most important determinant of conspicuity of features or discontinuities in the patient is image contrast. The contrast-producing characteristics of the detector system can be influenced to a larger degree than is usually the case for contrast agents or technique selection. Films and film-screen systems are available with a wide variety of contrast characteristics, and the development process can also be tailored to increase or decrease image contrast. The offset (dark level) and gain of electronic detectors can be adjusted to emphasize contrast over particular ranges of intensity (subject contrast). Window width and level of electronic display devices are usually under realtime control of the operator and have made a large impact in improving the quality of the images presented to the observer by, on the one extreme, accentuating very small differences in the detected X-ray intensity or, on the other, allowing simultaneous display of a very wide dynamic range at comfortable levels of illumination.

1.4 Quantities and Units

The purpose of this section is to present some of the fundamental concepts involved in the measurement and quantification of radiation intensity, exposure and dose and to summarize, in tabular form, the units used in this book. Discussion of some special units adopted historically for use in the radiation science community but now being phased out of use is included, though the standard units will be used throughout this book.

Not long after their discovery, evidence started to mount of the potentially harmful effects that X and gamma rays could have on living organisms[466], though debate over the root cause of these effects continued into the 1920's. The formation of an international radiation protection committee was discussed at the first meeting of the International Congress of Radiology in 1925, and the International Committee on X-ray and Radium Protection (ICRP) was formed in 1928 during the Congress's second meeting in Stockholm[467]. Since 1931 the ICRP has issued a series of reports in which they made recommendations as to quantities and units to be used to describe radiation and related entities, apparatus and methods to make these measurements in the clinic, and requirements for adequate shielding for personnel and patients. Another body whose purpose it is to compile reports and formulate recommendations regarding the measurement of ionizing radiation and standardization of the units used to describe the measured quantities is the International Commission on Radiation Units (ICRU)[468–470]. The ICRU has issued a number of reports over the years, which specialized in the definition and dissemination of units to be used in the radiological sciences and in radiation protection. The Commission defined some units specific to radiological quantities. Today, all involved bodies are advocating adoption of the units of the Systeme International (SI units), recommended by the Comite International des Poids et Mesures (CIPM). In 1975, the General Conference on Weights and Measures (CGPM) accepted two SI units, the gray and the bequerel, specific to the radiological sciences[3].

1.4.1 Fundamental Constants, Quantities and Units

The fundamental units of mass, length, time and current, from which most other units are derived, are, of course, the kilogram, the meter, the second and the ampere. A number of standard physical and electrical units and quantities are defined in Table 1.4 (adapted from Johns and Cunningham[3] and Smith[9]).

Table 1.4.

Quantity	Symbol	Meaning	SI Unit/ abbreviation	Conversions or fundamental units
mass	m		kilogram / kg	
length	l		meter / m	
time	t		second / s	
current	I		ampere / A	
velocity	v	v = l/t	m/s	
acceleration	a	v/t	m/s^2	
force	F	ma	newton / N	$kg \cdot m/s^2$
energy	E	Fl	joule / J	$kg \cdot m^2/s^2$ 1 eV = 1.602×10^{-19}J
power	P	E/t	watt / W	J/s
frequency	f, ν	number/length or time	hertz / Hz	s^{-1}
charge	Q	It	coulomb / C	As
potential	V	E/Q	volt / V	J/C
capacitance	C	Q/V	farad / F	C/V
resistance	R	V/I	ohm / Ω	V/A
field strength	ε	force/ unit charge	*	V/m
magnetic flux density	B	force/ unit charge momentum	Tesla / T	1T = 10^4 Gauss Wb/m^2
magnetic flux	ϕ	integral of magnetic flux density	Weber / Wb	$T \cdot m^2$

A number of radiological units and quantities are defined in Table 1.5 (compiled from ICRU Report 51[470], Johns and Cunningham[3], and Attix[400], and discussed in the following sections).

Table 1.5.

Quantity	Symbol	Meaning	SI Unit/ abbreviation	Fundamental units or Conversions
exposure	X	charge liberated per unit mass	C/kg	Q/m 1 Roentgen = 2.58×10^{-4} C/kg
energy imparted	ε	net energy absorbed in matter (see text)	*	J
absorbed dose	D	energy imparted per unit mass	gray / Gy	J/kg 1 Gy = 100 rad 1 Gy = 10^4 erg/g (kerma has same definition)
absorbed dose rate	$\dot{D}$	D/t	*	J/kg.s
quality factor	Q	weighting factor for varying biological effectiveness of radiation	*	
dose equivalent	H	QD	sievert / Sv	1 Sv = 100 rem
dose equivalent rate	$\dot{H}$			
activity	A	disintigrations/ second	becquerel / Bq	s^{-1} 1 Ci = 3.7×10^{10} Bq
fluence	Φ	particles/unit area	*	m^{-2}
fluence rate or flux density	ϕ	fluence/unit time	*	$m^{-2} \cdot s^{-1}$
energy fluence	Ψ	energy/unit area	*	J/m^2
intensity or energy flux	ψ	energy fluence/unit time	*	J/m^2s
linear energy transfer	LET	energy lost by particle per unit pathlength	*	J/m
lineal energy	y	energy imparted per mean chord length through volume	*	J/m

* No SI unit applies

1.4.2 Particle Fluence and Flux

We saw earlier that radiation beams are agents of energy transfer. There are a number of quantities used, either colloquially or in technical communication, to describe the passage or travel of energy through space or matter. Fluence, with units of number per unit area, refers to the number of quanta, particles or photons N passing through a unit area perpendicular to the beam:

$$\Phi = \frac{N}{A}\left(\frac{N}{m^2}\right). \tag{1.84}$$

Fluence rate (also called flux density), with units of number per unit area per unit time, is the fluence per unit time:

$$\phi = \frac{N}{A \cdot t}\left(\frac{N}{m^2 \cdot s}\right). \tag{1.85}$$

1.4.3 Energy Fluence and Flux

Energy fluence, with units of energy per unit area, describes the total energy of the particles traversing a unit area:

$$\Psi \quad \frac{NE}{A} \quad \frac{J}{m^2} \quad . \tag{1.86}$$

For monoenergetic photons, for example, one can determine the total energy by multiplying the particle fluence by the energy of each photon. Energy-integrating detectors, such as solid state or crystal-based systems, may be used to estimate the energy fluence in practice, though this is not done routinely in the clinic. Mathematically, the total energy in a polychromatic beam incident on an area can be calculated from:

$$\Psi = \sum_{i=0}^{E_{max}} n_i E_i , \tag{1.87}$$

where n_i is the number of particles with energy E_i, and the sum is over energies where i is incremented from zero up to the energy possessed by the most energetic particles in the beam (up to the value of kVp, for example). Energy flux, colloquially called intensity and equivalent to energy fluence rate, is the energy fluence per unit time:

$$\psi = \frac{NE}{A \cdot t}\left(\frac{J}{m^2 \cdot s}\right). \tag{1.88}$$

1.4.4 Exposure: the Roentgen

In general, it is practically difficult to measure intensity or fluence, and also not as useful (from a radiation protection standpoint, at any rate) as measuring a quantity more closely related to the potentially-harmful effects of ionizing radiation. Radiation damage from indirectly ionizing beams is not caused by the photons (or neutrons) directly, but by the fast electrons ejected from atoms in the absorbing medium. It is true that the initial interaction event is caused by the photons themselves, either in photoelectric or Compton interactions, but the number of these primary events is far smaller than that of the subsequent interactions of the ejected electrons with tissue atoms before they are brought to rest. Exposure is the quantity that was defined early on to quantify the potential of an X-ray beam to cause separation of charge in matter. The Roentgen is a unit of exposure originally defined as the amount of exposure required produce one electrostatic unit per cubic centimeter of air at standard temperature and pressure:

$$1R = \frac{1\,\text{ESU}}{\text{cm}^3\ \text{of air at STP}} = \frac{1\,\text{ESU}}{0.001293\,\text{g air}}\,. \tag{1.89}$$

In the SI system, the unit of exposure is coulombs per kilogram with:

$$1\ \text{Roentgen} = 2.58\times10^{-4}\ \text{coulomb/kg in air}\,. \tag{1.90}$$

1.4.5 Absorbed Dose: The Gray, the Rad and the Kerma

Biological effects depend on the amount of energy absorbed in tissue as a result of exposure to ionizing radiation, not on the ability of radiation to ionize air. The ICRU defines energy imparted to matter in a volume as[470]:

$$\varepsilon = R_{in} - R_{out} + \sum Q\ (J)\,, \tag{1.91}$$

where R_{in} is the radiant energy incident on the volume which means the sum of the energies of all charged and uncharged particles that enter the volume, R_{out} is the radiant energy emerging from the volume, and ΣQ is the sum of all changes of mass into energy or vice versa (e.g. annihilation of a positron or electron-positron formation in pair production) which occur in the volume. Radiant energy is defined as the energy of particles emitted, transferred or received out of or into the volume. Absorbed dose is then defined as:

$$D = \frac{\varepsilon}{m}\ \left(\frac{J}{kg} = Gy\right), \tag{1.92}$$

and absorbed dose rate by:

$$\dot{D} = \frac{D}{t}\left(\frac{Gy}{s}\right). \tag{1.93}$$

Quantification of absorbed dose has often been in terms of the older units, ergs/g and rads, but these are being phased out in favor of the gray. Kerma, with units of gray, J/kg, rad, or erg/g, is a previously-recommended term[471] denoting absorbed dose which still finds use, particularly in the radiation therapy field.

Since it requires 33.85 eV to produce one ion pair in air and each ion pair has a charge of 1.6×10^{-19} coulomb, 33.85 joules are absorbed from an exposure of 1 C/kg in air. Thus an exposure of 1 C/kg in air results in an absorbed dose of 33.85 Gray. In general, the absorbed dose may be calculated using:

$$\text{energy absorbed (Gray)} = \Psi\left(\frac{J}{m^2}\right)\frac{\mu_{ab}}{\rho}\left(\frac{m^2}{kg}\right), \tag{1.94}$$

where $\frac{\mu_{ab}}{\rho}$ is defined as the energy absorption coefficient[3] or the mass energy absorption coefficient[18]. The energy absorption coefficient is defined as the product of the mass absorption coefficient and the fraction of photon energy absorbed in a particular medium per interaction:

$$\frac{\mu_{ab}}{\rho} = \frac{\mu}{\rho} \cdot \left(\frac{\overline{E}_{ab}}{h\nu}\right). \tag{1.95}$$

Tables are available giving the energy absorption coefficient as a function of photon energy[472].

1.4.6 Dose Equivalent: LET, Q, the Sievert and the Rem

Alpha particles and other charged species cause far more biological damage per quanta than do photons of the same energy. One reason for this is that they deposit their energy in the tissue over far shorter distances. Linear energy transfer (LET in J/m or, traditionally, in keV/micron) is a measure of the rate at which an ionizing particle deposits energy along its path in tissue. Alpha particles may deposit energy at rates up to 1000 times higher than electrons[3]. High-LET particles produce stronger biological effects than do low-LET species. The quality factor, Q, is introduced to take the varying "biological effectiveness", or the potential of different types of radiation to cause adverse biological effects, into account. Q varies with LET, and is quantified for a particular radiation by comparing the dose of the radiation required to produce a particular biological effect with the dose of a reference radiation required to elicit the same effect. The reference radiation may be medium-energy (e.g. 200-keV) X rays or ^{60}Co radiation[18,473]. Dose equivalent, H, expresses the damaging effects of different radiations on a common scale by incorporating the quality factor:

$$H = DQ\ (Sv). \qquad (1.96)$$

The Sievert is the SI unit for dose equivalent. An older unit still in common use is the rem. Quality factors for several types of radiation are listed in Table 1.6[474]. In the field of radiation biology, a similar modifying factor, the relative biological effectiveness or RBE, is used to compare the potential for damage of various types of radiation on a common scale.

Table 1.6.

Radiation	Q
X rays, gamma rays, beta particles	1
thermal neutrons	5
neutrons and protons	20
alpha-particles	20
heavy recoil nuclei	20

1.5 References

1. R. D. Evans. *The Atomic Nucleus.* (Robert Krieger, Malabar, Florida, 1955).
2. J. T. Bushberg, J. A. Seibert, E. M. Leidholdt, J. M. Boone. *The Essential Physics of Medical Imaging.* (Williams and Wilkins, Baltimore, 1994).
3. H. E. Johns, J. R. Cunningham. *The Physics of Radiology.* (Charles Thomas, Springfield, Illinois, 1983).
4. T. S. Rao-Sahib, D. B. Wittry, *J. Appl. Phys.* **45**, 5060 (1974).
5. J. A. Wheeler, R. Ladenburg, *Phys. Rev.* **60**, 754 (1941).
6. B. W. Soole, *J. Phys.* **B5**, 1583 (1972).
7. A. B. Wolbarst. *Physics of Radiology* . (Appleton and Lange, Norwalk, Connecticut, 1993).
8. J. Selman. *The Fundamentals of X-ray and Radium Physics.* (Charles Thomas, Springfield, Illinois, 8th edition, 1994).
9. R. J. Smith. *Electronics: Circuits and Devices.* (Wiley, New York, 3rd edition, 1987).
10. A. V. Crewe, D. N. Eggenberger, J. Wall, L. M. Welter, *Rev. Sci. Instrum.* **39**, 576 (1968).
11. J. I. Goldstein, et al., Scanning Electron Microscopy and Microanalysis (Plenum, New York, 1981).
12. H. H. Pattee, in *X-ray Microscopy and Microradiography* ,V. E. Cosslett, *et al.*, Eds., (Academic, New York, 1957), p. 278.
13. W. P. Dyke, J. K. Trolan, W. W. Dolan, F. J. Grundhauser, *J. Appl. Phys.* **25**, 106 (1954).

14. F. M. Charbonnier, J. P. Barbour, W. P. Dyke, *Radiology* **117**, 165 (1975).
15. J. E. Turner. *Atoms, Radiation, and Radiation Protection*. (Wiley, New York, 1995).
16. M. M. Ter-Pogossian. *The Physical Aspects of Diagnostic Radiology*. (Harper and Row, New York, 1969).
17. G. L. Clark. *Applied X rays*. (McGraw Hill, New York, 1955).
18. W. R. Hendee, E. R. Ritenour. *Medical Imaging Physics*. (Mosby, St. Louis, 1992).
19. P. Sprawls. *Physical Principles of Medical Imaging*. (Medical Physics Publishing, Madison, Wisconsin, 1993).
20. D. W. Anderson. *Absorption of Ionizing Radiation*. (University Park Press, Baltimore, 1984).
21. W. Huda, R. Slone. *Review of Radiologic Physics*. (Williams and Wilkins, Baltimore, 1995).
22. V. Delgado, P. Ortiz, *Med. Phys.* **24**, 1089 (1997).
23. J. Rheinlander, L. Gerward, A. Lindegaard-Andersen, J. Larsen, *J. X-ray Sci. Technol.* **3**, 166 (1992).
24. K. Aoki, M. Koyama, *Med. Phys.* **16**, 529 (1989).
25. M. Green, in *X-ray Optics and X-ray Microanalysis*, H. H. Pattee, V. E. Cosslett, A. Engstrom, Eds., (Academic, New York, 1963), p. 185.
26. M. Green, V. E. Cosslett, *Brit. J. Appl. Phys. (J. Phys. D series 2)* **1**, 425 (1968).
27. M. Green, V. E. Cosslett, *Proc. Phys. Soc.* **78**, 1206 (1961).
28. H. Kulenkampff, *Annalen der Physik Leipzig* **69**, 548 (1922).
29. H. A. Kramers, *Phil. Mag.* **46** (series 6), 836 (1923).
30. R. Birch, M. Marshall, *Phys. Med. Biol.* **24**, 505 (1979).
31. D. M. Tucker, G. T. Barnes, D. P. Chakraborty, *Med. Phys.* **18**, 211 (1991).
32. T. Fewell, R. E. Shuping, K. R. Hawkins. *Handbook of Computed Tomography X-ray Spectra. HHS (FDA) Report No. 81–8162.* (U.S. Government Printing Office, Washington, 1981).
33. A. H. Compton, S. K. Allison. *X rays in Theory and Experiment*. (Van Nostrand, Princeton, 1967).
34. R. C. Weast, M. J. Astle, W. H. Beyer, Eds. *CRC Handbook of Chemistry and Physics*. (CRC Press, Boca Raton, Florida, 1983).
35. R. Beccherle, *et al.*, in *Digital Mammography*, A. G. Gale, S. M. Astley, D. R. Dance, A. Y. Cairns, Eds., (Elsevier, Amsterdam, 1994), p. 153.
36. M. Gambaccini, A. Taibi, A. Del Guerra, F. Frontera, M. Marziani, *Nuc. Inst. Meth. Phys. Res.* **A365**, 248 (1995).
37. R. E. Johnston, *et al.*, *Radiology* **200**, 659 (1996).
38. E. Burattini, *et al.*, *Radiology* **195**, 239 (1995).
39. F. E. Carroll, *Lasers in Surgery and Medicine* **11**, 72 (1991).
40. P. A. Tompkins, C. C. Abreu, F. E. Carroll, Q.-F. Xiao, C. A. MacDonald, *Med. Phys.* 21, 1777 (1994).
41. R. Fahrig, M. J. Yaffe, *Med. Phys.* **21**, 1473 (1994).
42. R. Fahrig, M. J. Yaffe, *Med. Phys.* **21**, 1463 (1994).
43. E. L. Gingold, X. Wu, G. T. Barnes, *Radiology* **195**, 639 (1995).
44. 102nd Congress. *Mammography Quality Standards Act of 1992*. (U.S. Government Printing Office, Washington, 1992).
45. J. Law, *Brit. J. Radiol.* **66**, 44 (1993).
46. M. M. Goodsitt, H.-P. Chan, B. Liu, *Med. Phys.* **24**, 11 (1997).
47. M. J. Yaffe, *et al.*, *Radiology* **197**, 19 (1995).
48. M. Yoshimatsu, S. Kozaki, in *X-ray Optics Applications to Solids*, H. J. Queisser, Ed., (Springer, Berlin, 1977), p. 9.
49. N. Uspenski, *Physik. Zeitschr.* **15**, 717 (1914).
50. M. von Ardenne, *Die Naturwissenschaften* **27**, 485 (1939).
51. V. E. Cosslett, W. C. Nixon, *Nature* **168**, 24 (1951).
52. V. E. Cosslett, W. C. Nixon, *Nature* **170**, 436 (1952).
53. V. E. Cosslett, W. C. Nixon, *J. Appl. Phys.* **24**, 616 (1953).

54. V. E. Cosslett, W. C. Nixon, *X-ray Microscopy.* (Cambridge University Press, London, 1960).
55. V. E. Cosslett, W. C. Nixon, and H. E. Pearson, in *X-ray Microscopy and Microradiography,* V. E. Cosslett, *et al.*, Eds., (Academic, New York, 1957), p. 96.
56. W. C. Nixon, in *X-ray Microscopy and Microradiography,* V. E. Cosslett, *et al.*, Eds., (Academic, New York, 1957), p. 34.
57. W. C. Nixon, *Nature* **175**, 1078 (1955).
58. B. L. Henke, in *The Encyclopedia of Microscopy,* G. L. Clark, Ed., (Reinhold, New York, 1961), p. 675.
59. R. L. D. C. H. Saunders, in *X-ray Microscopy and X-ray Microanalysis,* A. Engstrom, V. Cosslett, H. Pattee, Eds., (Elsevier, Amsterdam, 1960), p. 244.
60. R. L. D. C. H. Saunders and L. Van der Zwan, in *X-ray Microscopy and X-ray Microanalysis,* A. Engstrom, V. Cosslett, H. Pattee, Eds., (Elsevier, Amsterdam, 1960), p. 293.
61. R. L. D. C. H. Saunders, in *Vth International Congress on X-ray Optics and Microanalysis*, G. Mollenstedt, K. H. Gaukler, Eds., (Springer, Berlin, 1969), p. 550.
62. R. L. D. C. H. Saunders, M. A. Bell, V. R. Carvalho, in *Vth International Congress on X-ray Optics and Microanalysis*, G. Mollenstedt, K. H. Gaukler, Eds., (Springer, Berlin, 1969), p. 569.
63. S. Bellman, H. A. Frank, P. B. Lambert, B. Oden, J. A. Williams, in *X-ray Microscopy and X-ray Microanalysis,* A. Engstrom, V. Cosslett, H. Pattee, Eds., (Elsevier, Amsterdam, 1960) p. 259.
64. D. A. Carpenter, M. A. Taylor, *50th Annual Meeting of the Electron Microscopy Society of America* (San Francisco Press, 1992), p. 1758.
65. W. M. Gibson, M. A. Kumakhov, *50th Annual Meeting of the Electron Microscopy Society of America* (San Francisco Press, 1992), p. 1726.
66. B. R. York, *50th Annual Meeting of the Electron Microscopy Society of America* (San Francisco Press, 1992), p. 1764.
67. D. Kruger, C. Abreu, W. Peppler, C. MacDonald, C. Mistretta, *Med. Phys.* **21**, 875 (1994).
68. A. Attaelmanan, S. Larsson, A. Rindby, P. Voglis, A. Kuczumow, *Rev. Sci. Instrum.* **65**, 7 (1994).
69. H. N. Chapman, K. A. Nugent, S. W. Wilkins, A. V. Rode, in *SPIE Volume 1741 Soft X-ray Microscopy*, (SPIE, Bellingham, Washington, 1992), p. 40.
70. K. Furuta, *et al.*, *Rev. Sci. Instrum.* **64**, 135 (1993).
71. N. Yamamoto, Y. Hosokawa, *Jpn. J. Appl. Phys.* **27**, L2203 (1988).
72. H. N. Chapman, *et al.*, in *X-ray Microscopy III,* A. G. Michette, G. R. Morrison, C. J. Buckley, Eds., (Springer, Berlin, 1992), p. 131.
73. K. Lewotsky, *Laser Focus World*, (September, 1994), p. 30.
74. H. N. Chapman, A. Rode, K. A. Nugent, S. W. Wilkins, *Applied Optics* **32**, 6333 (1993).
75. D. J. Thiel, D. H. Bilderback, A. Lewis, *Rev. Sci. Instrum.* **64**, 2872 (1993).
76. S. B. Dabagov, *et al.*, *Nuc. Inst. Meth. Phys. Res.* **B103**, 99 (1995).
77. A. Attaelmanan, P. Voglis, A. Rindby, S. Larsson, P. Engstrom, *Rev. Sci. Instrum.* **66**, 24 (1995).
78. D. J. Pugh, P. D. West, *J. Microsc.* **103**, 227 (1974).
79. G. Fuhrmann, H. Halling, R. Moller, J. Vell, V. Z. Shuang-Ren, in *IEEE Nuclear Science Symposium and Medical Imaging Conference*, (IEEE, New York, 1991), p. 2012.
80. R. H. Johnson, A. C. Nelson, D. H. Burns, in *12th International Congress for Electron Microscopy*, (San Francisco Press, 1990), p. 518.
81. P. C. Cheng, S. P. Newberry, H. G. Kim, I. S. Hwang, in *Modern Microscopies: Techniques and Applications*, P. J. Duke, A. G. Michette, Eds., (Plenum, New York, 1990), p. 87.
82. P. C. Cheng, *et al.*, in *X-ray Microscopy III*, A. G. Michette, G. R. Morrison, C. J. Buckley, Eds., (Springer, Berlin, 1992), p. 184.

83. A. Y. Sasov, *J. Microsc.* **147**, 169 (1987).
84. A. Y. Sasov, *J. Microsc.*. **156**, 91 (1989).
85. L. A. Feldkamp, S. A. Goldstein, A. M. Parfitt, G. Jesion, M. Kleerekoper, *J. Bone Min. Research* **4**, 3 (1989).
86. J. L. Kuhn, S. A. Goldstein, L. A. Feldkamp, R. W. Goulet, G. Jesion, *J. Orthop. Res.* **8**, 833 (1990).
87. S. M. Hames, M. J. Flynn, D. A. Reimann, in *IEEE Nuclear Science Symposium and Medical Imaging Conference*, (IEEE, New York, 1992), p. 1331.
88. J. T. De Assis, R. T. Lopes, J. L. Rodrigues, in *IEEE Nuclear Science Symposium and Medical Imaging Conference*, (IEEE, New York, 1993), p. 1731.
89. F. W. Prior, *et al.*, in *Digital Mammography*, A. G. Gale, S. M. Astley, D. R. Dance, A. Y. Cairns, Eds., (Elsevier, Amsterdam, 1994), p. 143.
90. A. V. Clough, J. H. Linehan, C. A. Dawson, *Am. J. Physiol.* **272**, (Heart Circ. Physiol. 41), H1537 (1997).
91. D. R. Harder, M. L. Schulte, A. V. Clough, C. A. Dawson, *Am. J. Physiol.* **263**, H1616 (1992).
92. R. H. Johnson, H. Hu, S. T. Haworth, P. S. Cho, C. A. Dawson, J. H. Linehan, *Phys. Med. Biol.* **43**, 929 (1998).
93. E. Burattini, A. Balerna, Eds. *Biomedical Applications of Synchrotron Radiation*. (IOS Press, Amsterdam, 1996).
94. H. B. Stuhrmann, Ed. *Uses of Synchrotron Radiation in Biology*. (Academic, London, 1982).
95. B. Chance, *et al.*, Eds. *Synchrotron Radiation in the Biosciences*. (Oxford University Press, London, 1994).
96. J. C. Giacomini, H. J. Gordon, in *Handbook on Synchrotron Radiation, Volume* **4**, S. Ebashi, *et al.*, Eds., (North-Holland, Amsterdam, 1991), p. 349.
97. E. Rubenstein, *Nuc. Inst. Meth. Phys. Res.* **222**, 302 (1984).
98. H. Winick, S. Doniach, Eds. *Synchrotron Radiation Research*. (Plenum, New York, 1980).
99. H. Winick, *Scientific American* **257**, 88 (1987).
100. H. Winick, Ed. *Synchrotron Radiation Sources A Primer Volume 1*. (World Scientific, Singapore, 1994).
101. S. Ebashi, M. Koch, E. Rubenstein, Eds. *Handbook on Synchrotron Radiation Volume 4*. (North-Holland, Amsterdam, 1991).
102. G. Margaritondo. *Introduction to Synchrotron Radiation*. (Oxford University Press, New York, 1988).
103. A. A. Sokolov, I. M. Ternov. *Radiation from Relativistic Electrons*. (American Institute of Physics, New York, 1986).
104. E. Weihreter. *Compact Synchrotron Light Sources*. (World Scientific, Singapore, 1996).
105. A. Craievich, Ed. *Synchrotron Light: Applications and Related Instrumentation*. (World Scientific, Singapore, 1988).
106. E. B. Hughes, *et al., Acta Radiologica* **S365**, 43 (1983).
107. C. Kunz, Ed. *Synchrotron Radiation*. (Springer, Berlin, 1979).
108. F. R. Elder, A. M. Gurewitsch, R. V. Langmuir, H. C. Pollock, *Phys. Rev.* **71**, 829 (1947).
109. F. R. Elder, R. V. Langmuir, H. C. Pollock, *Phys. Rev.* **74**, 52 (1948).
110. C. Archie, *IBM J. Res. Develop.* **37**, 373 (1993).
111. M. N. Wilson, *et al., IBM J. Res. Develop* **37**, 351 (1993).
112. H. Takada, K. Furukawa, T. Tomimasu, *Opt. Eng.* **27**, 550 (1988).
113. A. I. Erko, V. V. Aristov, B. Vidal. *Diffraction X-ray Optics*. (Institute of Physics, Bristol, 1996).
114. A. G. Michette. *Optical Systems for Soft X rays*. (Plenum, New York, 1986).
115. A. G. Michette, C. J. Buckley, Eds. *X-ray Science and Technology*. (Institute of Physics, Bristol, 1993).

116. H. Winick, in *Synchrotron Radiation Research,* H. Winick, S. Doniach, Eds., (Plenum, New York, 1980), p. 11.
117. E. Burattini, in *Biomedical Applications of Synchrotron Radiation,* E. Burattini, A. Balerna, Eds., (IOS Press, Amsterdam, 1996), p. 3.
118. G. Green, *BNL Report No.50522* (Department of Energy, 1977).
119. H. Winick, G. Brown, K. Halbach, J. Harris, *Physics Today,* (May, 1981), p. 50.
120. G. Brown, K. Halbach, J. Harris, H. Winick, *Nuc. Inst. Meth.* **208**, 65 (1983).
121. J. E. Spencer, H. Winick, in *Synchrotron Radiation Research*, H. Winick, S. Doniach, Eds., (Plenum, New York, 1980), p. 663.
122. W. Thomlinson, D. Chapman, N. Gmur, N. Lazarz, *Nuc. Inst. Meth. Phys. Res.* **A266**, 226 (1988).
123. D. Attwood, *et al., Applied Optics* **32**, 7022 (1993).
124. G. S. Brown, in *Synchrotron Radiation Research Volume 2 Advances in Surface and Interface Science*, R. Z. Bachrach, Ed., (Plenum, New York, 1992), p. 317.
125. C. Capasso, *et al., J. Vac. Sci. Technol.* **A9**, 1248 (1991).
126. H. Maezawa, Y. Suzuki, H. Kitamura, T. Sasaki, *Applied Optics* **25**, 3260 (1986).
127. P. Horowitz, J. A. Howell, *Science* **178**, 608 (1972).
128. P. Horowitz, *Ann. New York Acad. Sci.* **306**, 203 (1978).
129. E. Spiller, *et al., Science* **191**, 1172 (1976).
130. J. Kirz, D. Sayre, in *Synchrotron Radiation Research,* H. Winick, S. Doniach, Eds., (Plenum, New York, 1980), p. 277.
131. M. Howells, J. Kirz, D. Sayre, G. Schmahl, *Physics Today,* (August, 1985), p. 22.
132. R. Feder, *Physics Today* (January, 1986), p. S12.
133. S. Richards, A. D. Rush, D. T. Clarke, W. J. Myring, *J. Microsc.* **142**, 1 (1986).
134. C. Jacobsen, *et al., Phys. Med. Biol.* **32**, 431 (1987).
135. C. Jacobsen, S. Lindaas, S. Williams, X. Zhang, *J. Microsc.* **172**, 121 (1993).
136. K. Shinohara, H. Nakano, Y. Kinjo, M. Watanabe, *J. Microsc.* **158**, 335 (1990).
137. D. L. Shealy, R. B. Hoover, T. W. Barbee, A. B. C. Walker, *Opt. Eng.* **29**, 721 (1990).
138. R. B. Hoover, U. S. Patent No. 5,107,526 (April 21, 1992).
139. B. W. Loo, S. Williams, S. Meizel, S. S. Rothman, *J. Microsc.* **166**, RP5 (1992).
140. G. R. Morrison, *SPIE Volume 1741 Soft X-ray Microscopy,* (SPIE, Bellingham, Washington, 1992), p. 186.
141. G. R. Morrison, M. T. Browne, *Rev. Sci. Instrum.* **63**, 611 (1992).
142. R. E. Burge, C. J. Buckley, G. F. Foster, A. Miller, T. Wess, *J. X-ray Sci. Technol.* **3**, 311 (1992).
143. P. Guttmann, *et al.*, in *SPIE Volume 1741 Soft X-ray Microscopy,* (SPIE, Bellingham, Washington, 1992), p. 52.
144. M. Hayashida, *et al.*, U. S. Patent No. 5,204,887 (April 20, 1993).
145. J. Stohr, *et al., Science* **259**, 658 (1993).
146. S. Williams, *et al., J. Microsc.* **170**, 155 (1993).
147. B. La Fontaine, *et al., Appl. Phys. Lett.* **66**, 282 (1995).
148. A. C. Thompson, *et al., Nuc. Inst. Meth. Phys. Res.* **A222**, 319 (1984).
149. U. Bonse, *et al., Nuc. Inst. Meth. Phys. Res.* **A246**, 644 (1986).
150. P. Spanne, M. L. Rivers, *Nuc. Inst. Meth. Phys. Res.* **B24/25**, 1063 (1987).
151. J. H. Kinney, *et al., Rev. Sci. Instrum.* **59**, 196 (1988).
152. K. Engelke, M. Lohmann, W. R. Dix, W. Graeff, *Rev. Sci. Instrum.* **60**, 2486 (1989).
153. J. H. Dunsmuir, H. W. Deckman, B. P. Flannery, K. L. D'Amico, S. R. Ferguson, in *Industrial Computerized Tomography,* (ASNT, Columbus, Ohio, 1989), p. 135.
154. W. Graeff, K. Engelke, in *Handbook on Synchrotron Radiation Volume 4,* S. Ebashi, *et al.*, Eds., (North-Holland, Amsterdam, 1991), p. 361.
155. J. H. Kinney, M. C. Nichols, *Annu. Rev. Mater. Sci.* **22**, 121 (1992).
156. J. V. Smith, in *X-ray Optics and Microanalysis 1992*, B.P. Kenway, *et al.*, Eds., (Institute of Physics, Bristol, 1992), p. 605.
157. M. D. Silver, in *SPIE Volume 2516 X-ray Microbeam Technology and Applications,* (SPIE, Bellingham, Washington, 1995).
158. T. Takeda, *et al., Rev. Sci. Instrum.* **66**, 1471 (1995).

159. T. Takeda, *et al.*, in *SPIE Volume 2708 Physics of Medical Imaging*, (SPIE, Bellingham, Washington, 1996), p. 685.
160. A. Momose, T. Takeda, Y. Itai, K. Hirano, in *SPIE Volume 2708 Physics of Medical Imaging*, (SPIE, Bellingham, Washington, 1996), p. 674.
161. M. Pateyron, *et al.*, in *SPIE Volume 2708 Physics of Medical Imaging*, (SPIE, Bellingham, Washington, 1996), p. 417.
162. A. Momose, T. Takeda, Y. Itai, K. Hirano, *Nature Medicine* **2**, 473 (1996).
163. H. D. Bartunik, R. Fourme, C. Phillips, in *Uses of Synchrotron Radiation in Biology*, H. B. Stuhrmann, Ed., (Academic, London, 1982), p. 145.
164. H. D. Bartunik, in *Handbook on Synchrotron Radiation Volume 4*, S. Ebashi, M. Koch, E. Rubenstein, Eds., (North-Holland, Amsterdam, 1991), p. 147.
165. M. G. Strauss, I. Naday, I. S. Sherman, *IEEE Trans. Nuc. Sci.* **NS-34**, 389 (1987).
166. M. G. Strauss, *et al.*, in *SPIE Volume 1447 Charge-Coupled Devices and Solid State Optical Sensors II,* (SPIE, Bellingham, Washington, 1991), p. 12.
167. I. Naday, M. G. Strauss, I. S. Sherman, M. R. Kraimer, E. M. Westbrook, *Opt. Eng.* **26**, 788 (1987).
168. D. Cline, A. Garren, J. Kolonko, in *Nuclear Science Symposium and Medical Imaging Conference*, (IEEE, New York, 1993), p. 1799.
169. R. Fourme, in *Biomedical Applications of Synchrotron Radiation,* E. Burattini, Ed., (IOS Press, Amsterdam, 1996), p. 77.
170. R. Fourme, in *Biomedical Applications of Synchrotron Radiation,* E. Burattini, Ed., (IOS Press, Amsterdam, 1996), p. 209.
171. W. A. Eaton, E. R. Henry, J. Hofrichter, *Science* **274**, 1631 (1997).
172. J. Bordas, in *Uses of Synchrotron Radiation in Biology*, H. B. Stuhrmann, Ed., (Academic, London, 1982), p. 107.
173. J. C. Solem, G. C. Baldwin, *Science* **218**, 1172 (1982).
174. J. C. Solem, G. F. Chapline, *Opt. Eng.* **23**, 193 (1984).
175. C. Jacobsen, M. Howells, J. Kirz, S. Rothman, *J. Opt. Soc. Am.* **A7**, 1847 (1990).
176. I. McNulty, *et al.*, *Science* **256**, 1009 (1992).
177. I. McNulty, *et al.*, in *SPIE Volume 1741 Soft X-ray Microscopy*, (SPIE, Bellingham, Washington, 1992), p. 78.
178. K. Boyer, J. C. Solem, J. W. Longworth, A. B. Borisov, C. K. Rhodes, *Nature Medicine* **2**, 939 (1996).
179. H. D. Zeman, *et al.*, *IEEE Trans. Nuc. Sci.* **NS-29**, 442 (1982).
180. E. B. Hughes, *et al.*, *Nuc. Inst. Meth.* **208**, 665 (1983).
181. E. B. Hughes, *et al.*, *Acta Radiol.* **S365**, 43 (1983).
182. E. B. Hughes, *et al.*, *Nuc. Inst. Meth. Phys. Res.* **B10/11**, 323 (1985).
183. E. B. Hughes, *et al.*, *Nuc. Inst. Meth. Phys. Res.* **A246**, 719 (1986).
184. J. N. Otis, *et al.*, *IEEE Trans. Nuc. Sci.* **NS-31**, 581 (1984).
185. A. Akisada, *et al.*, *Nuc. Inst. Meth. Phys. Res.* **A246**, 713 (1986).
186. W. R. Dix, *et al.*, *Nuc. Inst. Meth. Phys. Res.* **A246**, 702 (1986).
187. E. Rubenstein, *et al.*, *Proc. Natl. Acad. Sci. USA* **83**, 9724 (1986).
188. E. Rubenstein, *Ann. Rev. Biophys. Biophys. Chem.* **16**, 161 (1987).
189. R. K. Smither, E. M. Westbrook, *Nuc. Inst. Meth. Phys. Res.* **A266**, 260 (1988).
190. A. C. Thompson, *et al.*, *Nuc. Inst. Meth. Phys. Res.* **A266**, 252 (1988).
191. A. C. Thompson, *et al.*, *Rev. Sci. Instrum.* **60**, 1674 (1989).
192. S. A. Audet, *Rev. Sci. Instrum.* **60**, p. 2276 (1989).
193. K. Nishimura, *et al.*, *Rev. Sci. Instrum.* **60**, p. 2260 (1989).
194. K. Ueda, *et al.*, *Rev. Sci. Instrum.* **60**, p. 2272 (1989).
195. J. C. Giacomini, H. J. Gordon, in *Handbook on Synchrotron Radiation Volume 4,* S. Ebashi, M. Koch, E. Rubenstein, Eds., (North-Holland, Amsterdam, 1991), p. 349.
196. K. Hyodo, K. Nishimura, M. Ando, in *Handbook on Synchrotron Radiation Volume 4,* S. Ebashi, M. Koch, E. Rubenstein, Eds., (North-Holland, Amsterdam, 1991), p. 55.
197. T. Takeda, *et al.*, *Med. Biol. Eng. Comput.* **32**, 462 (1994).
198. H. Mori, *et al.*, *Circulation* **89**, 863 (1994).

199. H. Mori, *et al.*, *Circ. Res.* **76**, 1088 (1995).
200. H. Mori, *et al.*, *Radiology* **201**, 173 (1996).
201. F. E. Carroll, *et al.*, *Invest. Radiol.* **29**, 266 (1994).
202. R. Feder, D. Sayre, E. Spiller, J. Topalian, *J. Appl. Phys.* 47, 1192 (1976).
203. R. Feder, *et al.*, *Science* **197**, 259 (1977).
204. D. Sayre, J. Kirz, R. Feder, D. M. Kim, E. Spiller, *Ultramicroscopy* **2**, 337 (1977).
205. D. Sayre, R. Feder, D. M. Kim, E. Spiller, *Science* **196**, 1330 (1977).
206. D. Sayre, J. Kirz, R. Feder, D. M. Kim, E. Spiller, *Ann. New York Acad. Sci.* **306**, 286 (1978).
207. M. R. Howells, J. Kirz, D. Sayre, *Scientific American*, (February, 1991), p. 88.
208. M. Hatzakis, *J. Electrochem. Soc.* **116**, 1033 (1969).
209. D. L. Spears, H. I. Smith, *Electronic Letters* **8**, 102 (1972).
210. R. Feder, E. Spiller, J. Topalian, *Polymer Engineering and Science* **17**, 385 (1977).
211. R. Feder, *et al.*, in *X-ray Microscopy*, G. Schmahl, D. Rudolph, Eds., (Springer, Berlin, 1984), p. 279.
212. R. Feder, V. Mayne-Banton, in *Examining the Submicron World*, R. Feder, J. W. McGowan, D. M. Shinozaki, Eds., (Plenum, New York, 1986), p. 277.
213. I. Haller, R. Feder, M. Hatzakis, E. Spiller, *J. Electrochem. Soc.* **126**, 154 (1979).
214. W. Meyer-Ilse, in *SPIE Volume 1140 X-ray Instrumentation in Medicine and Biology*, (SPIE, Bellingham, Washington, 1989), p. 226.
215. G. D. Kubiak, R. Q. Hwang, M. T. Schulberg, D. A. Tichenor, K. Early, *Applied Optics* **32**, 7036 (1993).
216. K. Early, *et al.*, *Applied Optics* **32**, 7044 (1993).
217. T. Tomie, *et al.*, *Science* **252**, 691 (1991).
218. Y. I. Borodin, *et al.*, *Nuc. Instrum. Meth. Phys. Res.* **A246**, 649 (1986).
219. F. Cinotti, *et al.*, in *X-ray Microscopy: Instrumentation and Biological Applications,* P. C. Cheng and G.J. Jan, Eds., (Springer, Berlin, 1987), p. 311.
220. H. Rarback, *et al.*, *Rev. Sci. Instrum.* **59**, 52 (1988).
221. C. J. Buckley, H. Rarback, in *Modern Microscopies: Techniques and Applications* P. J. Duke, A. G. Michette, Eds., (Plenum, New York, 1990), p. 69.
222. D. Morris, *et al.*, *Scanning* **13**, 7 (1991).
223. W. Meyer-Ilse, M. Moronne, C. Magowan, P. Selvin, *Scanning* **15**, Suppl.III, 35 (1993).
224. G. Schmahl, D. Rudolph, P. Guttmann, O. Christ, in *X-ray Microscopy*, G. Schmahl, D. Rudolph, Eds., (Springer, Berlin, 1984), p. 63.
225. V. Bogli, *et al.*, *Opt. Eng.* **27**, 143 (1988).
226. B. Lai, *et al.*, *Appl. Phys. Lett.* **61**, 1877 (1992).
227. A. A. Krasnoperova, *et al.*, *J. Vac. Sci. Technol.* **B11**, 2588 (1993).
228. E. Spiller, in *X-ray Microscopy: Instrumentation and Biological Applications,* P. C. Cheng and G.J. Jan, Eds., (Springer, Berlin, 1987), p. 224.
229. M. Howells, *et al.*, *Science* **238**, 514 (1987).
230. D. Joyeux, F. Polack, R. Mercier, in *SPIE Volume 1140 X-ray Instrumentation in Medicine and Biology,* (SPIE, Bellingham, Washington, 1989), p. 399.
231. P. Spanne, *Phys. Med. Biol.* **34**, 679 (1989).
232. J. C. Elliott, D. K. Bowen, S. D. Dover, S. T. Davies, *Biological Trace Element Research* **13**, 219 (1987).
233. W. Graeff, K. Engelke, *Microradiography and Microtomography*, HASYLAB/DESY Report No. F41-9004, July, 1990, p. 1.
234. F. A. Dilmanian, *et al.*, in *IEEE Nuclear Science Symposium and Medical Imaging Conference*, (IEEE, New York, 1991), p. 1831.
235. K. W. Jones, *et al.*, in *X-ray Microscopy III*, A. G. Michette, G. R. Morrison, C. J. Buckley, Eds., (Springer, Berlin, 1992), p. 431.
236. K. Engelke, W. Graeff, L. Meiss, M. Hahn, G. Delling, *Invest. Radiol.* **28**, 341 (1993).
237. W. S. Haddad, *et al.*, *Science* **266**, 1213 (1994).
238. P. Cloetens, R. Barrett, J. Baruchel, J. P. Guigay, M. Schlenker, *J. Phys. D: Appl. Phys.* **29**, 133 (1996).

239. A. Momose, J. Fukuda, *Med. Phys.* **22**, 375 (1996).
240. E. N. Dementiev, *et al.*, *Rev. Sci. Instrum.* **60**, 2264 (1989).
241. R. A. McCorkle, *Ann. New York Acad. Sci.* **342**, 53 (1980).
242. S. Bowyer, *Applied Optics* **32**, 6930 (1993).
243. A. Schulz, R. Burhenn, F. B. Rosmej, H. J. Kunze, *J. Phys. D: Appl. Phys.* **22**, 659 (1989).
244. J. Goldstein, *Science* **277**, 306 (1997).
245. J. H. Dave, *et al.*, *J. Opt. Soc. Am.* **B4**, 635 (1987).
246. M. Sato, *Rev. Sci. Instrum.* **58**, 481 (1987).
247. C. Tillman, A. Persson, C.-G. Wahlstrom, S. Svanberg, K. Herrlin, *Appl. Phys.* **B61**, 333 (1995).
248. K. Herrlin, *et al.*, *Radiology* **189**, 65 (1993).
249. D. Strickland, G. Mourou, *Opt. Commun.* **56**, 219 (1985).
250. M. D. Perry, A. Szoke, O. L. Landen, E. M. Campbell, *Phys. Rev. Lett.* **60**, 1270 (1988).
251. M. D. Perry, G. Mourou, *Science* **264**, 917 (1994).
252. M. M. Murnane, H. C. Kapteyn, R. W. Falcone, *Phys. Rev. Lett.* **62**, 155 (1989).
253. M. M. Murnane, H. C. Kapteyn, M. D. Rosen, R. W. Falcone, *Science* **251**, 531 (1991).
254. J. C. Kieffer, *et al.*, *Phys. Rev. Lett.* **62**, 760 (1989).
255. J. C. Kieffer, *ibid.*, **68**, 480 (1992).
256. J. C. Kieffer, *et al.*, *Phys. Fluids* **B5**, 2676 (1993).
257. J. C. Kieffer, *et al.*, *Applied Optics* **32**, 4247 (1993).
258. S. Augst, D. Strickland, D. D. Meyerhofer, S. L. Chin, J. H. Eberly, *Phys. Rev. Lett.* **63**, 2212 (1989).
259. M. Pessot, J. Squier, G. Mourou, D. J. Harter, *Opt. Lett.* **14**, 797 (1989).
260. H. C. Kapteyn, M. M. Murnane, A. Szoke, A. Hawryluk, R. W. Falcone, in *Ultrafast Phenomena VII,* C. P. Harris, E. P. Ippen, G. A. Mourou, A. H. Zewail, Eds., (Springer, Berlin, 1990), p. 122.
261. M. Chaker, *et al.*, *Phys. Fluids* **B3**, 167 (1991).
262. J. Zhou, C.-P. Huang, C. Shi, M. M. Murnane, H. C. Kapteyn, *Opt. Lett.* **19**, 126 (1994).
263. J. Zhou, C.-P. Huang, M. M. Murnane, H. C. Kapteyn, *Opt. Lett.* **20**, 64 (1995).
264. I. P. Christov, H. C. Kapteyn, M. M. Murnane, C.-P. Huang, J. Zhou, *Opt. Lett.* **20**, 309 (1995).
265. J. Bailey, Y. Ettinger, A. Fisher, R. Feder, *Appl. Phys. Lett.* **40**, 33 (1982).
266. R. Feder, J. S. Pearlman, J. C. Riordan, J. L. Costa, *J. Microsc.* **135**, 347 (1984).
267. R. B. Spielman, *et al.*, *J. Appl. Phys.* **57**, 830 (1985).
268. R. E. Stewart, D. D. Dietrich, R. J. Fortner, R. Dukart, *J. Opt. Soc. Am.* **B4**, 396 (1987).
269. B. Niemann, *et al.*, *Optik* **84**, 35 (1990).
270. W. Neff, R. Holz, R. Lebert, F. Richter, U. S. Patent No. 5,023,897, June 11, 1991.
271. D. Rudolph, B. Niemann, G. Schmahl, J. Thieme, in *X-ray Microscopy II*, D. Sayre, M. Howells, J. Kirz, H. Rarback, Eds., (Springer, Berlin, 1988), p. 216.
272. W. Neff, J. Eberle, R. Holz, R. Richter, R. Lebert, in *X-ray Microscopy II*, D. Sayre, M. Howells, J. Kirz, H. Rarback, Eds., (Springer, Berlin, 1988), p. 22.
273. D. Rothweiler, W. Neff, R. Lebert, F. Richter, M. Diehl, in *X-ray Optics and Microanalysis 1992*, B.P. Kenway, *et al.*, Eds., (Institute of Physics, Bristol, 1992), p. 479.
274. J. L. Krause, K. J. Schafer, K. C. Kulander, *Phys. Rev. Lett.* **68**, 3535 (1992).
275. A. G. Michette, *et al.*, *J. Phys. D. Appl. Phys.* **19**, 363 (1986).
276. G. Zeng, *et al.*, *J. Appl. Phys.* **67**, 3597 (1990).
277. G. Zeng, *et al.*, *J. Appl. Phys.* **69**, 7460 (1991).
278. C. J. Joshi, P. B. Corkum, *Physics Today*, (January, 1995), p. 36.
279. P. T. Rumsby, *J. Microsc.* **138**, 245 (1985).
280. R. W. Eason *et al.*, *Optica Acta* **33**, 501 (1986).
281. A. D. Stead, T. W. Ford, *Annals of Botany* **64**, 713 (1989).

282. R. E. Burge, *et al.*, in *X-ray Optics and Microanalysis 1992*, B.P. Kenway, *et al.*, Eds., (Institute of Physics, Bristol, 1992), p. 487.
283. R. E. Burge, M. T. Browne, P. Charalambous, G. Slark, P. Smith, in *SPIE Volume 1741 Soft X-ray Microscopy*, (SPIE, Bellingham, Washington, 1992), p. 170.
284. H. He, *et al.*, *Rev. Sci. Instrum.* **64**, 26 (1993).
285. R. D. Frankel, J. M. Forsyth, *Science* **204**, 622 (1979).
286. N. M. Ceglio, *Ann. New York Acad. Sci.* **342**, 65 (1978).
287. F. Ze, *et al.*, *J. Appl. Phys.* **66**, 1935 (1989).
288. P. Maine, G. Mourou, *Opt. Lett.* **13**, 467 (1988).
289. Y. Kinjo, *et al.*, *J. Microsc.* **176**, 63 (1994).
290. N. Sarakura, Y. Ishida, H. Nakano, Y. Yamamoto, *Appl. Phys. Lett.* **56**, 814 (1990).
291. J. D. Kmetic, J. J. Macklin, J. F. Young, *Opt. Lett.* **16**, 1001 (1991).
292. D. E. Spence, P. N. Kean, W. Sibbet, *Opt. Lett.* **16**, 42 (1991)
293. J. L. A. Chilla, O. E. Martinez, *J. Opt. Soc. Am.* **B10**, 638 (1993).
294. C. P. J. Barty, C. L. Gordon, B. E. Lemoff, *Opt. Lett.* **19**, 1442 (1994).
295. P. D. Gupta, P. A. Naik, H. C. Pant, *J. Appl. Phys.* **56**, 1371 (1984).
296. L. A. Gizzi, *et al.*, *Phys. Rev. Lett.* **76**, 2278 (1996).
297. D. G. Stearns, O. L. Landen, E. M. Cambell, J. H. Scofield, *Phys. Rev.* **A37**, 1684 (1988).
298. K. A. Tanaka, *et al.*, *J. Appl. Phys.* **63**, 1787 (1988).
299. J. A. Cobble, *et al.*, *J. Appl. Phys.* **69**, 3369 (1991).
300. R. Bobkowski, J. N. Broughton, R. Fedosejevs, R. J. Willis, M. R. Cervenan, *J. Appl. Phys.* **76**, 5047 (1994).
301. A. McPherson, B. D. Thompson, A. B. Borisov, K. Boyer, C. K. Rhodes, *Nature* **370**, 631 (1994).
302. A. B. Borisov, A. McPherson, K. Boyer, C. K. Rhodes, in *X-ray Lasers*, D. C. Eder, D. L. Matthews, Eds., (American Institute of Physics, New York, 1994), p. 134.
303. A. B. Borisov, A. McPherson, B. D. Thompson, K. Boyer, C. K. Rhodes, *J. Phys. B: At. Mol. Opt. Phys.* **B28**, 2143 (1995).
304. R. J. Rosser, *et al.*, *Applied Optics* **26**, 4313 (1987).
305. D. A. Tichenor, *et al.*, *Opt. Lett.* **16**, 1557 (1991).
306. F. Bijkerk, in *X-ray Optics and Microanalysis 1992*, B.P. Kenway, *et al.*, Eds., (Institute of Physics, Bristol, 1992), p. 471.
307. F. Bijkerk, E. Louis, G. E. van Dorssen, A. P. Shevelko, A. A. Vasilyev, *Applied Optics* **33**, 82 (1994).
308. E. Louis, et al., *Microelectronic Engineering* **21**, 67 (1993).
309. R. L. Kauffman, D. W. Phillion, R. C. Spitzer, *Applied Optics* **32**, 6897 (1993).
310. M. Richardson, *et al.*, *Applied Optics* **32**, 6901 (1993).
311. L. A. Hackel, R. J. Beach, C. B. Dane, L. E. Zapata, *Applied Optics* **32**, 6914 (1993).
312. S. J. Haney, K. W. Berger, G. D. Kubiak, P. D. Rockett, J. Hunter, *Applied Optics* **32**, 6934 (1993).
313. G. E. Sommargren, L. G. Seppala, *Applied Optics* **32**, 6938 (1993).
314. C. Cerjan, *Applied Optics* **32**, 6911 (1993).
315. L. Malmqvist, L. Rymell, H. M. Hertz, *Appl. Phys. Lett.* **68**, 2627 (1996).
316. H. Nakano, T. Nishikawa, N. Uesugi, *Appl. Phys. Lett.* **70**, 16 (1997).
317. R. Feder, *et al.*, *Science* **227**, 63 (1985).
318. R. J. Rosser, R. Feder, A. Ng, P. Celliers, *J. Microsc.* **140**, RP1 (1985).
319. A. G. Michette, *et al.*, *Rev. Sci. Instrum.* **64**, 1478 (1993).
320. A. G. Michette, R. Fedosejevs, S. J. Pfauntsch, R. Bobkowski, *Measurement Science and Technology* **5**, 555 (1994).
321. T. Tomie, *et al.*, in *SPIE Volume 1741 Soft X-ray Microscopy*, (SPIE, Bellingham, Washington, 1992), p. 118.
322. M. Richardson, *et al.*, in *SPIE Volume 1741 Soft X-ray Microscopy*, (SPIE, Bellingham, Washington, 1992), p. 133.
323. J. Fletcher, R. Cotton, C. Webb, in *SPIE Volume 1741 Soft X-ray Microscopy*, (SPIE, Bellingham, Washington, 1992), p. 142.

324. R. A. Cotton, M. D. Dooley, J. H. Fletcher, A. D. Stead, T. W. Ford, in *SPIE Volume 1741 Soft X-ray Microscopy*, (SPIE, Bellingham, Washington, 1992), p. 204.
325. H. K. Pew, G. L. Stradling, in *X-ray Microscopy III*, A. G. Michette, G. R. Morrison, C. J. Buckley, Eds., (Springer, Berlin, 1992), p. 206.
326. A. D. Stead, R. A. Cotton, A. M. Page, M. D. Dooley, T. W. Ford, in *SPIE Volume 1741 Soft X-ray Microscopy*, (SPIE, Bellingham, Washington, 1992), p. 351.
327. K. Shinohara, *et al.*, in *SPIE Volume 1741 Soft X-ray Microscopy*, (SPIE, Bellingham, Washington, 1992), p. 386.
328. H. Aritome, *et al.*, in *SPIE Volume 1741 Soft X-ray Microscopy*, (SPIE, Bellingham, Washington, 1992), p. 129.
329. Y. Horikawa, K. Nagal, S. Mochimaru, Y. Iketaki, *J. Microsc.* **172**, 189 (1993).
330. H. Kondo, T. Tomie, *J. Appl. Phys.* **75**, 3798 (1994).
331. I. C. E. Turcu, *et al.*, *J. Appl. Phys.* **73**, 8081 (1993).
332. J. Rosser, *et al.*, *J. Microsc.* **138**, 311 (1985).
333. S. P. Vernon, D. G. Stearns, R. S. Rosen, *Opt. Lett.* **18**, 672 (1993).
334. L. Rymell, H. M. Hertz, *Opt. Commun.* **103**, 105 (1993).
335. K.-D. Song, D. R. Alexander, *J. Appl. Phys.* **76**, 3302 (1994).
336. B. J. Panessa, R. A. McCorkle, P. Hoffman, J. B. Warren, G. Coleman, *Ultramicroscopy* **6**, 139 (1981).
337. B. J. Panessa-Warren, in *X-ray Microscopy*, G. Schmahl, D. Rudolf, Eds., (Springer, Berlin, 1984), p. 268.
338. J. A. Trail, R. L. Byer, J. B. Kortright, in *X-ray Microscopy II*, D. Sayre, M. Howells, J. Kirz, H. Rarback, Eds., (Springer, Berlin, 1988), p. 310.
339. J. A. Trail, R. L. Byer, *Opt. Lett.* **14**, 539 (1989).
340. I. A. Artyukov, A. I. Fedorenko, V. V. Kondratenko, S. A. Yulin, A. V. Vinogradov, *Opt. Commun.* **102**, 401 (1993).
341. M. Yasugaki, Y. Horikawa, U. S. Patent No. 5,131,023, July 14, 1992.
342. J. C. Solem, *J. Opt. Soc. Am.* **B3**, 1551 (1986).
343. E. Spiller, R. Feder, in *X-ray Optics Applications to Solids*, H. J. Queisser, Ed., (Springer, Berlin, 1977), p. 35.
344. K. Shinohara, *et al.*, *Photochemistry and Photobiology* **44**, 401 (1986).
345. K. Shinohara, Y. Kinjo, M. C. Richardson, *et al.*, in *SPIE Volume 1741 Soft X-ray Microscopy*, (SPIE, Bellingham, Washington, 1992), p. 386.
346. K. Shinohara, A. Ito, *J. Microsc.* **161**, 463 (1991).
347. K. Shinohara, *et al.*, in *X-ray Microscopy III*, A. G. Michette, G. R. Morrison, C. J. Buckley, Eds., (Springer, Berlin, 1992), p. 347.
348. A. Ito, K. Shinohara, *Cell Structure and Function* **17**, 209 (1992).
349. A. Ito, K. Shinohara, H. Nakano, T. Matsumura, K. Kinoshita, *J. Microsc.* **181**, 54 (1996).
350. T. W. Ford, A. D. Stead, C. P. B. Hills, R. J. Rosser, N. Rizvi, *J. X-ray Sci. Technol.* **1**, 207 (1989).
351. D. C. O'Shea, W. R. Callen, W. T. Rhodes. *Introduction to Lasers and their Applications.* (Addison-Wesley, Reading, Massachusetts, 1977).
352. R. C. Elton. *X-ray Lasers.* (Academic, Boston, 1990).
353. D. L. Matthews, *et al.*, *Phys. Rev. Lett.* **54**, 110 (1985).
354. S. Suckewer, C. H. Skinner, H. Milchberg, C. Keane, D. Voorhees, *Phys. Rev. Lett.* **55**, 1753 (1985).
355. M. Pessot, P. Maine, G. Mourou, *Opt. Commun.* **62**, 419 (1987).
356. J. J. Rocca, *et al.*, in *X-ray Lasers,* D. C. Eder, D. L. Matthews, Eds., (American Institute of Physics, New York, 1994), p. 359.
357. G. P. Collins, *Physics Today*, (October, 1994), p. 19.
358. C. H. Skinner, *Phys. Fluids* **B3**, 2420 (1991).
359. Y. Nagata, *et al.*, *Phys. Rev. Lett.* **71**, 3774 (1993).
360. L. I. Gudzenko, L. A. Shelepin, *Sov. Phys.-Dok.* **10**, 147 (1965).
361. L. I. Gudzenko, L. A. Shelepin, S. I. Yakovlenko, *Sov. Phys.-Usp.* **17**, 848 (1975).
362. J. Peyraud, N. Peyraud, *J. Appl. Phys.* **43**, 2993 (1972).

363. J. C. Weisheit, *J. Phys. B: Atom. Molec. Phys.* **8**, 1556 (1975).
364. J. C. Weisheit, C. B. Tarter, J. H. Scofield, L. M. Richards, *J. Quant. Spectrosc. Radiat. Transfer* **16**, 659 (1976).
365. J. M. Green, W. T. Silfvast, *Appl. Phys. Lett.* **28**, 253 (1976).
366. W. W. Jones, A. W. Ali, *J. Appl. Phys.* **48**, 3118 (1977).
367. W. W. Jones, A. W. Ali, *J. Phys. B: Atom. Molec. Phys.* **11**, L87 (1978).
368. D. Matthews, *et al.*, *J. Opt. Soc. Am* **B4**, 575 (1987).
369. A. K. Davé, G. J. Pert, *J. Phys. B: Atom. Mol. Phys.* **18**, 1027 (1985).
370. W. T. Silvast, O. R. Wood, *J. Opt. Soc. Am.* **B4**, 609 (1987).
371. J. E. Trebes, *et al.*, *Science* **238**, 517 (1987).
372. C. H. Skinner, D. Kim, A. Wouters, D. Voorhees, S. Suckewer, in *X-ray Microscopy II*, D. Sayre, M. Howells, J. Kirz, H. Rarback, Eds., (Springer, Berlin, 1988), p. 36.
373. N. H. Burnett, P. B. Corkum, *J. Opt. Soc. Am.* **B6**, 1195 (1989).
374. N. H. Burnett, G. D. Enright, *IEEE J. Quant. Elect.* **26**, 1797 (1990).
375. D. C. Eder, *Phys. Fluids* **B2**, 3086 (1990).
376. D. C. Eder, D. L. Matthews, Eds. *X-ray Lasers.* (American Institute of Physics, New York, 1994).
377. P. Amendt, D. C. Eder, S. C. Wilks, *Phys. Rev. Lett.* **66**, 2589 (1991).
378. B. J. MacGowan, in *SPIE Volume 1741 Soft X-ray Microscopy*, (SPIE, Bellingham, Washington, 1992), p. 2.
379. L. B. Da Silva, *et al.*, *Science* **258**, 269 (1992).
380. L. B. Da Silva, *Opt. Lett.* **17**, 754 (1992).
381. L. B. Da Silva, in *SPIE Volume 1741 Soft X-ray Microscopy*, (SPIE, Bellingham, Washington, 1992), p. 154.
382. R. Balhorn, *et al.*, *in SPIE Volume 1741 Soft X-ray Microscopy*, (SPIE, Bellingham, Washington, 1992), p. 374.
383. D. I. Chiu, in *X-ray Optics and Microanalysis 1992*, B.P. Kenway, *et al.*, Eds., (Institute of Physics, Bristol, 1992), p. 511.
384. D. Attwood, *Physics Today*, (August, 1992), p. 24.
385. D. S. DiCicco, D. Kim, R. Rosser, S. Suckewer, *Opt. Lett.* **17**, 157 (1992).
386. R. Cauble, *et al.*, *Science* **273**, 1093 (1996).
387. S. Suckewer, C. H. Skinner, R. Rosser, U. S. Patent No. 5,177,774, January 5, 1993.
388. C. H. Skinner, *et al.*, *J. Microsc.* **159**, 51 (1990).
389. D. S. DiCicco, D. Kim, L. Polonsky, C. H. Skinner, S. Suckewer, in *SPIE Volume 1741 Soft X-ray Microscopy*, (SPIE, Bellingham, Washington, 1992), p. 160.
390. S.-S. Han, *et al.*, in *X-ray Lasers,* D. C. Eder, D. L. Matthews, Eds., (American Institute of Physics, New York, 1994), p. 235.
391. R. A. London, M. D. Rosen, J. E. Trebes, *Appl. Opt.* **28**, 3397 (1989).
392. M. H. Key, C. G. Smith, in *X-ray Lasers,* D. C. Eder, D. L. Matthews, Eds., (American Institute of Physics, New York, 1994), p. 423.
393. F. E. Carroll, *et al.*, *Invest. Radiol.* **25**, 465 (1990).
394. C. A. Brau, *Free Electron Lasers* (Academic Press, Cambridge, 1990).
395. C. Pellegrini, *et al.*, *Nuc. Instrum. Meth. Phys. Res.* **A331**, 223 (1993).
396. K. J. Weeks, V. N. Litvinenko, J. M. J. Madey, *Med. Phys.* **24**, 417 (1997).
397. P. Eisenberger, S. Suckewer, *Science* **274**, 201 (1996).
398. R. W. Schoenlein, *et al.*, *Science* **274**, 236 (1996)
399. D. A. Skoog, D. M. West. *Fundamentals of Analytical Chemistry 3rd Edition.* (Holt, Rinehart and Winston, New York, 1976).
400. F. H. Attix. *Introduction to Radiological Physics and Radiation Dosimetry.* (Wiley, New York, 1986).
401. J. D. Jackson. *Classical Electrodynamics 2nd Edition.* (Wiley, New York, 1975).
402. J. H. Hubbell, *et al.*, *J. Phys. Chem. Ref. Data* **4**, 471 (1975).
403. E. F. Plechaty, D. E. Cullen, R. J. Howerton, *Report UCRL-50400* (Lawrence Livermore National Laboratory, University of California, 1975).
404. C. Jacobsen, in *Biomedical Applications of Synchrotron Radiation*, E. Burattini, A. Balerna, Eds., (IOS Press, Amsterdam, 1996), p. 59.

405. B. L. Henke, E. M. Gullikson, J. C. Davis, *At. Data Nucl. Data Tables* **54**, 181 (1993).
406. O. Klein, Y. Nishina, *Zeitschrift Physik* **52**, 853 (1929).
407. R. E. Alvarez, A. Macovski, *Phys. Med. Biol.* **21**, 733 (1976).
408. H. K. Huang, C. K. Wong, F. L. Roder, in *IEEE International Workshop on Physics and Engineering in Medical Imaging*, (IEEE, New York, 1982), p. 122.
409. D. R. White, R. J. Martin, R. Darlison, *Br. J. Radiol.* **50**, 814 (1977).
410. D. R. White, M. Fitzgerald, *Health Physics* **33**, 73 (1977).
411. D. R. White, *Med. Phys.* **5**, 467 (1978).
412. W. J. Hamilton, *et al.*, in *Nuclear Science Symposium and Medical Imaging Conference*, (IEEE, New York, 1993), p. 232.
413. R. C. Schirato, R. M. Polichar, J. H. Reed, S. T. Smith, in *SPIE Volume 2009 X-ray Detector Physics and Applications II*, (SPIE, Bellingham, Washington, 1993), p. 48.
414. W. Zhao, *et al.*, in *SPIE Volume 2708 Physics of Medical Imaging*, (SPIE, Bellingham, Washington, 1996), p. 523.
415. D. L. Lee, L. K. Cheung, E. F. Palecki, L. S. Jeromin, in *SPIE Volume 2708 Physics of Medical Imaging*, (SPIE, Bellingham, Washington, 1996), p. 511.
416. D. M. Korn, A. R. Lubinsky, J. F. Owen, in *SPIE Volume 626 Medicine XIV/PACS IV*, (SPIE, Bellingham, Washington, 1986), p. 108.
417. A. Lubinsky, Owen, JF and Korn, DM, in *SPIE Volume 626 Medicine XIV/PACS IV*, (SPIE, Bellingham, Washington, 1986), p. 120.
418. J. Miyahara, K. Takahashi, Y. Amemiya, N. Kamiya, Y. Satow, *Nuc. Instrum. Meth. Phys. Res.* **A246**, 572 (1986).
419. B. R. Whiting, J. F. Owen, B. H. Rubin, *Nuc. Instrum. Meth. Phys. Res.* **A266**, 628 (1988).
420. Y. Amemiya, *et al.*, *Nuc. Instrum. Meth. Phys. Res.* **A266**, 645 (1988).
421. Y. Amemiya, S. Kishimoto, T. Matsushita, Y. Satow, M. Ando, *Rev. Sci. Instrum.* **60**, 1552 (1989).
422. C. E. Floyd, H. G. Chotas, J. T. Dobbins, C. E. Ravin, *Med. Phys.* **17**, 454 (1990).
423. R. H. Templer, *Nuc. Instrum. Meth. Phys. Res.* **A300**, 357 (1991).
424. F. K. Koschnick, J. M. Spaeth, R. S. Eachus, W. G. McDugle, R. H. D. Nuttall, *Phys. Rev. Lett.* **67**, 3571 (1991).
425. F. K. Koschnick, J. M. Spaeth, R. S. Eachus, *J. Phys.: Condensed Matter* **4**, 8919 (1992).
426. H. von Seggern, *Nuc. Instrum. Meth. Phys. Res.* **A322**, 467 (1992).
427. J. M. Spaeth, F. K. Koschnick, R. S. Eachus, W. G. McDugle, R. H. D. Nuttall, *Nucl. Tracks Radiat. Meas.* **21**, 73 (1993).
428. G. V. Semisotnov, *et al.*, *J. Molec. Biol.* **262**, 559 (1996).
429. L. E. Antonuk, *et al.*, *IEEE Trans. Med. Imaging* **13**, 482 (1994).
430. L. E. Antonuk, *et al.*, *Med. Phys.* **19**, 1455 (1992).
431. L. E. Antonuk, *et al.*, *Med. Phys.* **21**, 942 (1994).
432. J. H. Siewerdsen, *et al.*, *Med. Phys.* **24**, 71 (1997).
433. L. E. Antonuk, *et al.*, *Med. Phys.* **24**, 51 (1997).
434. S. M. Gruner, J. R. Milch, G. T. Reynolds, *Nuc. Instrum. Meth.* **195**, 287 (1982).
435. S. M. Gruner, *Rev. Sci. Instrum.* **60**, 1545 (1989).
436. A. R. Cowen, A. Workman, *Phys. Med. Biol.* **37**, 325 (1992).
437. V. G. M. Althof, *et al.*, *Med. Phys.* **23**, 1845 (1996).
438. J. M. Boone, T. Yu, J. A. Seibert, *Med. Phys.* **23**, 1955 (1996).
439. T. Yu, J. M. Boone, *Med. Phys.* **24**, 565 (1997).
440. S. Hejazi, D. P. Trauernicht, *Med. Phys.* **24**, 287 (1997).
441. P. F. Schmidt, D. V. McCaughan, R. A. Kushner, *Proc. IEEE* **62**, 1220 (1974).
442. V. A. J. van Lint, Proc. *IEEE* **62**, 1190 (1974).
443. J. Janesick, T. Elliott, F. Pool, *IEEE Trans. Nuc. Sci.* **NS-36**, 572 (1989).
444. G. W. Ludwig, J. D. Kingsley, *J. Electrochem. Soc.* **117**, 348 (1970).
445. G. W. Ludwig, J. S. Prener, *IEEE Trans. Nuc. Sci.* **NS-19**, 3 (1972).
446. R. K. Swank, *J. Appl. Phys.* **44**, 4199 (1973).
447. J. H. Chappell, S. S. Murray, *Nuc. Instrum. Meth. Phys. Res.* **221**, 159 (1984).

448. D. M. de Leeuw, T. Kovats, S. Herko, *J. Electrochem. Soc.* **134**, 491 (1987).
449. E. Sluzky, K. Hesse, *J. Electrochem. Soc.* **135**, 2893 (1988).
450. E. Sluzky, K. Hesse, *J. Electrochem. Soc.* **136**, 2724 (1989).
451. D. J. Mickish, J. Beutel, in *SPIE Volume 1231 Image Formation*, (SPIE, Bellingham, Washington, 1990), p. 327.
452. G. Blasse, *IEEE Trans. Nuc. Sci.* **NS-38**, 30 (1991).
453. G. E. Giakoumakis, *et al.*, *Med. Phys.* **20**, 79 (1993).
454. S. M. Gruner, S. L. Barna, M. E. Wall, M. W. Tate, E. F. Eikenberry, in *SPIE Volume 2009 X-ray Detector Physics and Applications II*, (SPIE, Bellingham, Washington, 1993), p. 98.
455. D. Cavouras, I. Kandarakis, G. S. Panayiotakis, E. K. Evangelou, C. D. Nomicos, *Med. Phys.* **23**, 1965 (1996).
456. K. Oba, M. Ito, M. Yamaguchi, M. Tanaka, *Adv. Elect. Electron. Phys.* **74**, 247 (1988).
457. T. S. Curry, J. E. Dowdey, R. C. Murry. *Christensen's Introduction to the Physics of Diagnostic Radiology.* (Lea and Febiger, Philadelphia, 1984).
458. W. C. Rontgen, *Nature* **53**, 274 (1896).
459. W. K. Rontgen, *CA A Cancer Journal for Clinicians* **22**, 153 (1972).
460. O. Glasser. *Dr. W.C. Rontgen.* (Charles C Thomas, Springfield, Illinois, 1945).
461. O. Glasser. *Wilhelm Conrad Rontgen and the Early History of the Roentgen Rays.* (Charles C Thomas, Springfield, Illinois, 1934).
462. M. Yaffe, A. Fenster, H. E. Johns, *J. Comp. Assist. Tomog.* **1**, 419 (1977).
463. A. Fenster, *J. Comp. Assist. Tomog.* **2**, 243 (1978).
464. H. N. Cardinal, A. Fenster, *Med. Phys.* **15**, 167 (1988).
465. D. A. Cusano, C. D. Greskovich, F. A. DiBianca, U. S. Patent No. 4,421,671, December 20, 1983.
466. E. Thomson, *Boston Med. Surg.* **135**, 610 (1896).
467. L. S. Taylor. *Radiation Protection Standards.* (CRC Press, Cleveland, 1971).
468. ICRU Report No. 30. *Quantitative Concepts and Dosimetry in Radiobiology.* (International Commission on Radiation Units and Measurements, Washington, 1979).
469. ICRU Report No. 33. *Radiation Quantities and Units.* (ICRU, Washington, 1980).
470. ICRU Report No. 51. Quantities and Units in Radiation Protection Dosimetry. (ICRU, Washington, 1993).
471. ICRU Report No. 10. *Radiation Quantities and Units.* (ICRU, Washington, 1962).
472. NBS. *National Bureau of Standards Handbook 85.* (U.S. Government Printing Office, Washington, 1964).
473. I. A. Cunningham, A. Fenster, *Med. Phys.* **11**, 303 (1984).
474. NCRP. *NCRP Report Number 91.* (National Council on Radiation Protection, Washington, 1987).

2 X-ray Imaging: Radiography, Fluoroscopy, Computed Tomography

John A. Rowlands

Sunnybrook Health Science Center, University of Toronto

2.1 Basics of X-ray Imaging

X-ray images are based on transmission of quanta through the body, with the image contrast resulting from variations in thickness and composition of the internal anatomy. Currently, most X-ray images involve the use of film (film-screen combinations), but digitization of X-ray images has begun and the concept of CT imaging, a digital reconstruction of a cross-section of the body, is well established.

2.1.1 Introduction

X-ray imaging was the earliest and still is the most important method of imaging the interior of the human body. It differs from all the other imaging modalities in that it requires irradiating the patient with ionizing radiation to produce an image. The justification for using ionizing radiation on the human body is that the benefits are much greater than the potential damage. For example, one benefit of X-ray imaging is the high resolution possible. In fact, X-ray images are shadows of the internal parts of the body and the contrast in X-ray images is generated by the absorption or scattering of X rays. Thus, scattered radiation is inherent within the contrast mechanism as is the primary interaction of radiation with the body in the diagnostic range. It can fall on the image receptor and by reducing contrast corrupt the direct shadow image. Thus appropriate methods to reduce scatter are important.

2.1.2 Spectrum

The only practical method currently available for producing X rays for diagnostic purposes is the X-ray tube. The bremsstrahlung spectrum produced by a conventional X-ray tube is spread over a large range of energies. The highest energy (expressed in keV) of the spectrum corresponds to the potential (kVp) applied across the tube. Without the inherent filtration in the tube, X rays of close to zero energy would be present. Self-filtration due to absorption in the X-ray tube target, the glass envelope of the tube, the insulating oil, and window of the housing substantially reduce the number of low energy X rays in the spectra. The low energy part of the X-ray spectrum is of limited value in imaging because it is strongly absorbed within the body. Therefore such low energies contribute mostly to the patient exposure and little to the final image. Thus, to improve the form of X-ray spectra in the low energy region, filtration is added. The X-ray spectrum for the most common tube target material (tungsten W) consists primarily of bremsstrahlung radiation. Although W K-fluorescence lines (characteristic radiation) will be present for kVp above the K-absorption edge, the fractional amount of energy in these lines compared to bremsstrahlung is negligible. However, for lower atomic number Z targets than W (*e.g.* Mo) the fraction of energy in the characteristic K-lines can be much larger. Thus a typical mammographic spectrum obtained using a Mo target contains a considerable amount of energy in the characteristic spectral lines which results in a narrower (better) spectrum than from bremsstrahlung alone. The spectrum of a mammographic tube is normally shaped by use of Mo filters, although other elements are possible. For tungsten tubes in the normal diagnostic spectral range, aluminum filtration is used. An example spectrum for a W target tube is shown in Figure 2.1.

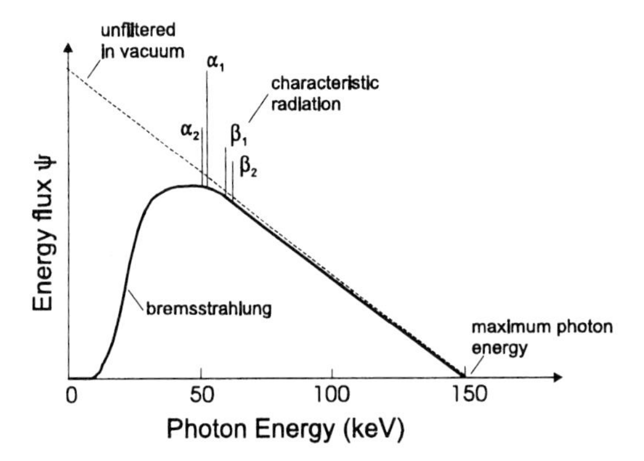

Figure 2.1. X-ray spectrum from a tungsten target tube. (Reproduced from Reference 13, by permission.)

2.1.3 X-ray Attenuation

The initial image acquisition operation is identical in all X-ray detectors. In order to produce a signal, the X-ray quanta must interact with the detector material. The probability of interaction or *quantum detection efficiency* η for quanta of energy E is given by:

$$\eta = 1 - e^{-\mu(E)T} \tag{2.1}$$

where μ is the linear attenuation coefficient of the detector material and T is the thickness of the detector. The quantum efficiency can be increased by making the detector thicker or by using materials which have a higher atomic number or density. The quantum efficiency will, in general, be highest at low energies, gradually decreasing with increasing energy. However, if the material has an atomic absorption edge in the energy region of interest, then quantum efficiency increases dramatically above this energy, causing a local minimum in η. Example curves of η plotted as a function of energy for materials useful as radiological detectors are shown in Figure 2.2.

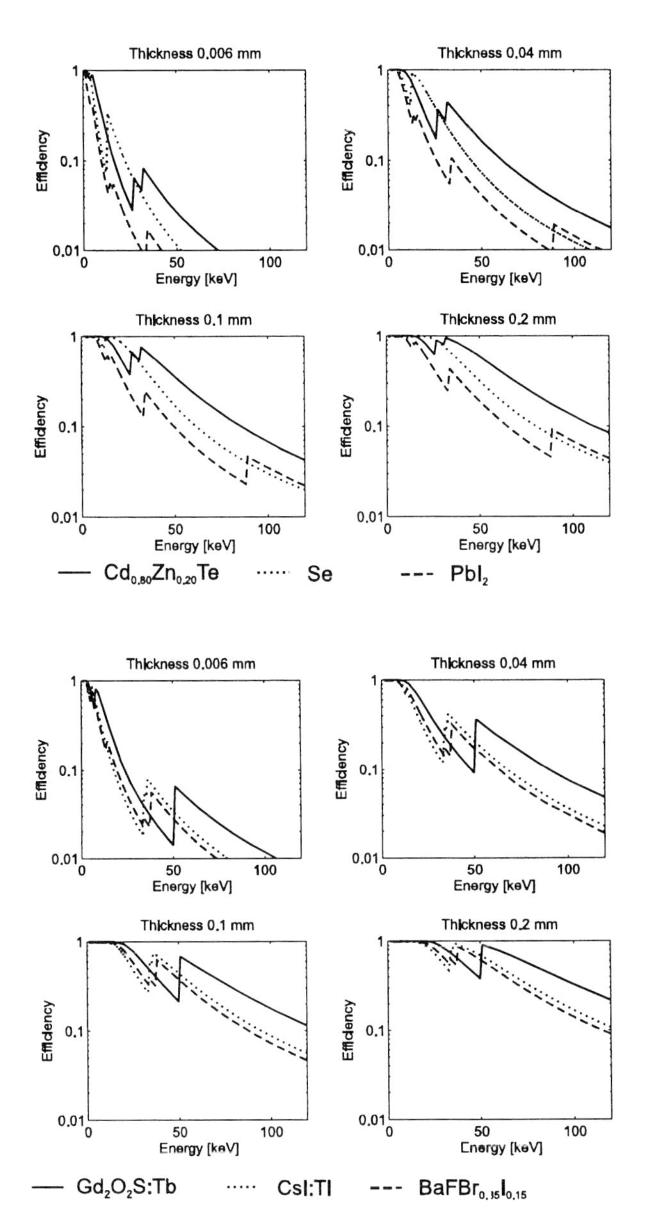

Figure 2.2. Attenuation as a function of energy for detectors of radiological interest. (Images courtesy of J. Mainprize.)

2.1.4 X-ray Noise

X-ray noise is inherent in all X-ray images. Noise in an image can always be reduced by increasing the radiation exposure and hence the dose to the patient. The converse is true – noise increases if exposure is reduced. The amount of noise will be increased by improper detector design or inappropriate choice of beam quality for the particular X-ray application. The statistics of X-ray noise generation can be understood with reference to standard theories of Poisson noise distribution[1]. Factors relating to the propagation of noise through the complete X-ray imaging system include: formation of the image by interaction with the patient's body; the collection of this image by the detector; and the corrupting effects of scattered radiation and electronic noise.

Interaction with the detector can be represented as a binomial process with probability of success, η, and it has been shown that the distribution of interacting quanta is still Poisson with standard deviation:

$$\sigma = (N_0 \eta)^{1/2}. \tag{2.2}$$

Therefore the signal-to-noise ratio is:

$$\mathrm{SNR} = S/\sigma = \frac{N_0 \eta}{(N_0 \eta)^{1/2}} = (N_0 \eta)^{1/2}. \tag{2.3}$$

Thus decrease in η results in a reduction of SNR. A full analysis of signal and noise propagation in a detector system must take into account the spatial frequency dependence of both signal and noise. Signal transfer can be characterized in terms of the modulation transfer function $MTF(f)$. Noise is described by the noise power or Wiener spectrum $W(f)$. A useful quantity for characterizing the overall performance of imaging detectors is their spatial frequency f dependent detective quantum efficiency DQE(f). This describes the efficiency of the detector in transferring the signal-to-noise ratio contained in the incident X-ray pattern to its output. Ideally, DQE$(f) = \eta$ for all f. However, additional noise sources will reduce this value and cause the DQE to decrease with increasing spatial frequency. A 100% DQE means that the detected X-ray image quality is exactly equivalent to that present in the aerial X-ray image. If DQE is 10%, then the resulting image noise would correspond to an ideal image using 1/10 the number of X rays actually incident on the detector. But what effect does reduced DQE have on the appearance of the image? For Poisson processes, the noise variance is given by the number of X rays detected per unit area. Thus, a DQE of 10% causes the noise variance to increase by a factor of 10, *i.e.* the standard deviation of noise goes up by a factor of $\sqrt{10}$ compared to a DQE of 100%. This is illustrated in Figure 2.3 which shows simulated X-ray images of the same objects obtained with varying amounts of radiation.

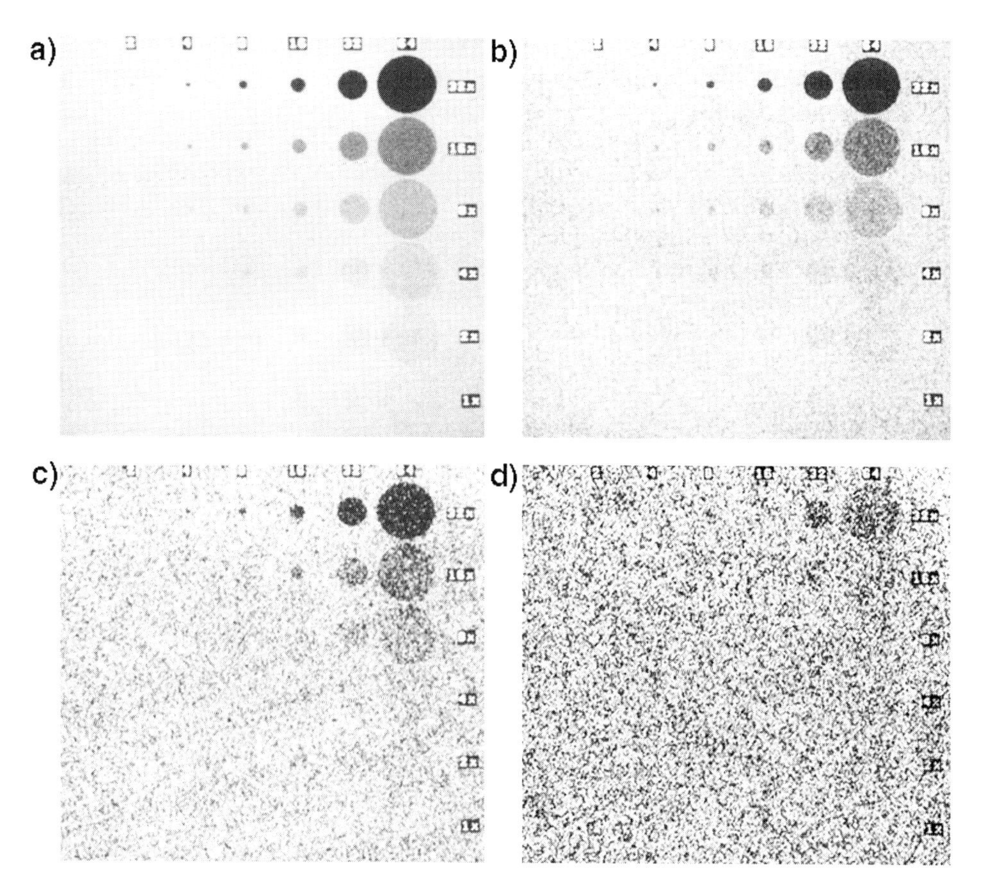

Figure 2.3. Image obtained with various numbers of X rays per pixel: (a) 1000 X rays/pixel, (b) 100 X rays/pixel, (c) 10 X rays/pixel, and (d) 1 X-ray/pixel. For comparison with X-ray images, fluoroscopic images are taken with ~1μR/frame which corresponds to ~6 photons/pixel/frame and radiographs ~100 μR, which is ~100 photons/frame. (Images courtesy of M.S. Rzeszotarski.)

2.1.5 Resolution

The resolution of an image is the smallest visible interval. The better the resolution the greater the ability to view small details. Spatial resolution in radiography is determined by factors unrelated to the detector and by the detector characteristics. The former includes unsharpness arising from geometrical factors such as the effective focal spot size of the X-ray source, the magnification between the anatomical structure of interest, and the plane of the image detector and relative motion between the X-ray source, patient and image receptor during the exposure. Detector-related factors include its effective aperture size, sampling interval and any lateral signal spreading effects within the detector or readout. There are two general ways resolution can be measured: first, using a lead bar pattern; and second, a more formal method using the MTF. The MTF represents the modulation at f normalized to the modulation at $f = 0$, both measured at the output of the imaging system. In Figure 2.4, images of sinusoidal patterns are shown before and after detection by detectors with various MTFs. The use of MTF is convenient as it readily permits the computation of the overall MTF of a system from its components using the formula:

$$MTF_T(f) = MTF_A(f)\, MTF_B(f) \ldots \tag{2.4}$$

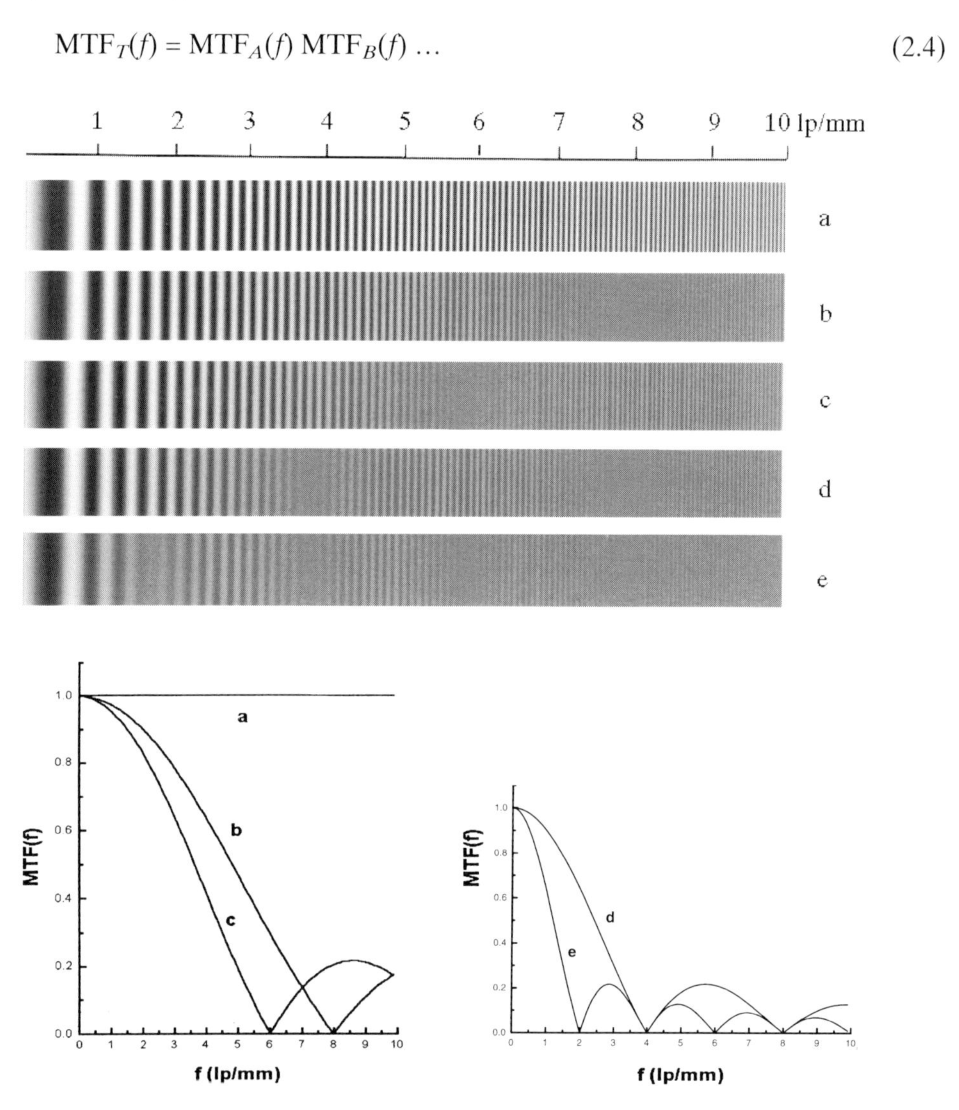

Figure 2.4. An X-ray sinusoidal bar pattern with spatial frequency varying up to 10 lp/mm (line pairs per millimetre) reproduced at approximately five times the actual size: (a) the original pattern before detection, (b) after detection by an imaging system with MTF as shown – a sinc function with first zero at 8 lp/mm, (c) 6 lp/mm, (d) 4 lp/mm, and (e) 2 lp/mm. (Images courtesy of W. Ji.)

The unsharpness introduced by the focal spot depends on the radiographic magnification of the image. The concept of radiographic magnification is shown in Figure 2.5 where the image size of an infinitesimal object is shown projected from the focal spot. This spatial blur can be related to the MTF by taking the Fourier transform[2]. Where resolution loss is due to motion, the velocity and the exposure time of the given image are important. For the heart, very fast exposure times are required as heart muscle may contract rapidly. For bony tissue, longer exposure times can be used since this tissue is relatively stable and there is little chance of motion blurring. By obtaining a pinhole image in the presence of motion, a MTF of motion can be obtained. The components of MTF described can then be readily combined to obtain the complete system MTF using the concept of cascaded systems (Eq. 2.4).

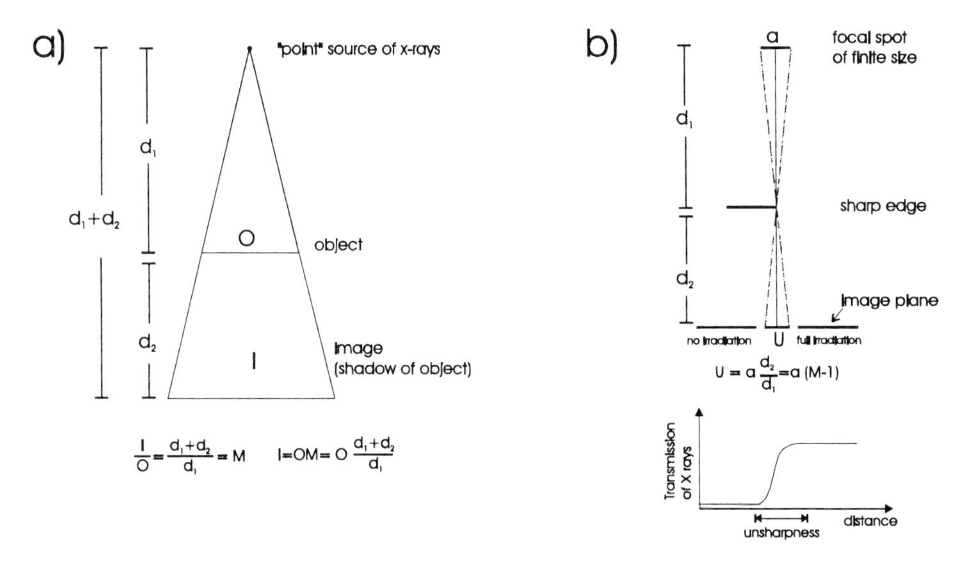

Figure 2.5. Radiographic magnification: (a) the geometrical increase in object size *O*, and (b) the focal spot blurring at image plane.

2.1.6 Contrast

Contrast in images of the body is created by the differences in attenuation along paths through the body. The X-ray attenuation follows the exponential attenuation expression or Beer's law[3]:

$$N = N_0 e^{-\mu(E)T} \tag{2.5}$$

where N is the transmitted X-ray fluence for an incident fluence N_0. If the scattered radiation is eliminated, contrast is a combination of attenuation processes due to Compton scattering and the photoelectric effect. Since the body mainly consists of low atomic number materials, the dominant mechanism for contrast is Compton scattering over most of the diagnostic X-ray energy range. However, for very low energy X rays, typically those used in mammography, *i.e.* the range of 10 to 30 keV, the photoelectric effect is also important. The two largest contrast generating materials in the human body are bony tissue (radiologically dense due to its calcium content) and air pockets (*e.g.* gas in bowel, chest cavity) which may displace soft tissue to provide large contrast differentials.

The image contrast C can be defined in terms of the fluence measured through the tissue of interest N_2 to a reference path N_1 through adjacent tissue, then:

$$C = \frac{N_2 - N_1}{N_1} \tag{2.6}$$

From the linear attenuation coefficients of these various tissues the contrast of typical objects may be calculated as shown in Table 2.1 for a 1 cm block of tissue.

Table 2.1. Contrast arising from a 1 cm block of tissue compared to a 1 cm block of muscle.

Tissue	μ (cm^{-1})	N/N_0 ($x = 1$ cm)	C wrt muscle (%)
Muscle	0.180	0.835	0
Air	0	1	20
Blood	0.178	0.837	0.2
Bone	0.480	0.619	–26

Only bone and air have much contrast in the aerial X-ray image when compared to the human vison threshold of ~4%. Artificial contrast can be generated by the use of ingested or injected contrast materials. For example, if some blood in a vessel is displaced by a solution of heavy elements, such as iodine, a visible contrast difference can be generated. Another example is air contrast when a liquid is replaced by air. The attenuation of all materials is strongly dependent on the X-ray energy as well as the atomic number of the contrast material. Therefore, the contrast of X-ray images can be modified by changing the X-ray energy used in the beam. This is achieved by changing the kilovoltage and filtration.

2.1.7 Scatter

X-ray scatter is always a factor in X-ray imaging. The amount of scatter present in an image is related to the thickness of the body part as well as the size of the field of view, see Figure 2.6.

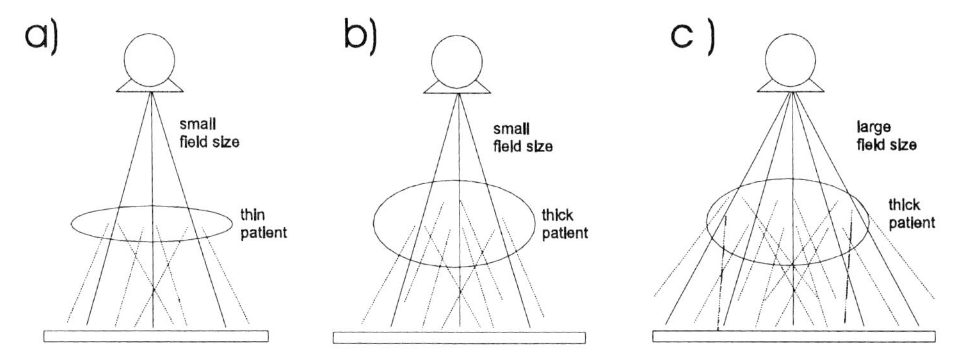

Figure 2.6. Amount of scatter depending on irradiation conditions: (a) Small field-of-view – small body thickness. Scatter is primarily directed backwards but only the forward scatter is relevant to X-ray imaging. A small body part creates little scatter and a small field-of-view ensures most of that scattered radiation does not reach the detector. (b) Thicker body part with small field-of-view. Scatter radiation still primarily backwards, although forward scatter is increasing. Attenuation combined with attenuation of the primary radiation means the ratio of scatter at the primary is increasing and can be the equivalent of primary or larger. (c) Large field-of-view with a thick body part. There is significant scatter to the primary ratio and it is very difficult to avoid all the scattered radiation impinging on the sensor.

The effect of scatter radiation can be greatly reduced, although never eliminated, by using an anti-scatter grid which by use of oriented lead plates selectively reduces the scatter compared to primary radiation. More effective methods for reducing scatter are based on reducing the field of view. Where systems reduce the field of view to a single pencil beam, the scatter in the image is entirely eliminated. Distancing the detector from the body part can also allow scatter to escape effectively – this is called an air gap.

2.1.8 Digital Imaging

At some stage in the digital imaging system, the aerial X-ray image is sampled both in the spatial and intensity dimensions. In the spatial dimension, samples are obtained as averages of the intensity over picture elements or *pixels*. These are ideally square, and are spaced at equal intervals throughout the plane of the image. In the intensity dimension, the signal is digitized into one of a finite number of levels or *bits*. To avoid degradation of image quality, it is important that the pixel size and the bit depth are appropriate for the requirements of the imaging task.

2.1.9 Summary – Approach

X-ray images are obtained from a system[4]. To understand the system, its components and even sub-components must first be understood. The sub-components are common to many components and are thus examined first. Then the components are assembled. Finally, complete systems are discussed in terms of both their parts and the clinical requirements.

2.2 Sub-components

Sub-components include film, lenses and phosphors, the basis of most existing conventional and some digital X-ray systems. The concept of electrostatic X-ray transducers, used in some advanced sensor designs, will then be introduced. Different X-ray scanning approaches as well as laser and electron beam scanning will be examined. Finally, self-scanned readout structures such as charge-coupled devices and large area active matrix arrays will be discussed.

2.2.1 Photographic Film

Photographic film is a unique material that is sensitive to a very few quanta of light. It can create a latent optical image in a fraction of a second exposure, maintain this latent image for months, and eventually be developed without significant loss of information. It is also used as a display and archiving medium. The only method of producing photographic film is the silver process[2].

The photographic process uses a thin layer of silver halide crystals suspended in gelatin and supported on a transparent film base. The key feature which gives unexposed film its long shelf life is that more than one light photon must impinge on an individual crystal of silver halide in order to create a stable latent image. A single light photon creates an electron that is stable for a short time (about a second). If another few electrons (the exact number depends on various factors) are released in the same crystal within this time, the electrons stabilize each other and the latent image is established. This multi-electron process is key to understanding, not only the long shelf life and the H&D (Hurter and Driffield) curve which is the characteristic response of film to light, but also problems such as reciprocity failure and latent image fading.

2.2.1.1 Processing

After exposure of the film to light, the latent image is formed as the excitation centers on the individual film crystals. Processing converts the latent image to a viewable permanent image and is made possible by the suspension of the crystals in a thin layer of water permeable gelatin supported on a transparent plastic substrate. Chemicals are transported to the crystals without disturbing the their positions when the film is dipped into chemical solutions. The development process turns a sensitized transparent crystal of silver halide into a grain of metallic silver that absorbs light and therefore is opaque. (Since these grains are very small, they appear black in the same way that any finely powdered metal appears black, not with a characteristic silvery color that one might otherwise expect.) After the latent image has been developed, there are unexposed and therefore undeveloped transparent silver halide crystals within the gelatin. During the fixing stage, these undeveloped silver halide crystals are removed chemically. Next is the water wash where the processing chemicals are completely removed, leaving only the insoluble silver grains embedded in pure gelatin. Drying removes the excess water solvent from the gelatin and results in a completely permanent archival material known as photographic film.

2.2.1.2 H&D Curve, Reciprocity Failure, and Latent Image Fading

The characteristic curve of photographic film, the H&D curve, is shown schematically in Figure 2.7. The optical density [OD = $\log_{10}(I_I / I_T)$, where I_I is the intensity of readout light incident on a developed film and I_T is the transmitted intensity] is plotted against the logarithm of optical exposure to the undeveloped film. The most important part of the curve is the straight line region with slope γ. This slope cannot continue indefinitely and saturation occurs at high exposures, resulting in the shoulder of the curve. At low exposures there is some film density, known as *base plus fog*, even in the absence of any exposure. This results in the curvature at low exposures known as the *toe*. The origin of the H&D curve can be understood with respect to the basic

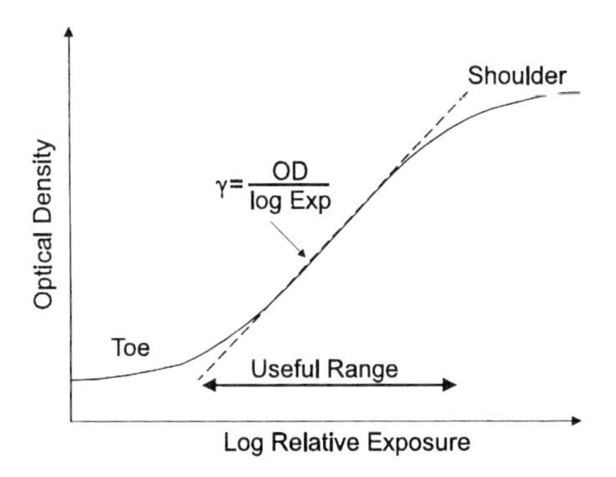

Figure 2.7. The characteristic curve of photographic film, the H&D curve.

photographic process as follows. A small amount of light will have difficulty in creating any kind of developable image due to the instability of single electron excitations. Thus, small amounts of light cause little darkening of the film. When enough light is incident, the film starts to develop (the toe), then it responds more rapidly and the straight line part of the curve emerges. Once there is sufficient light to develop most of the grains, then saturation begins, giving rise to the shoulder of the curve and a flattening at high exposure. These characteristics can be modified in a vast range of ways – by changing the slope of the curve and/or changing the sensitivity of the curve which allows the adaptation of the film to different applications. Thus the factors affecting the H&D curve are important.

Reciprocity failure refers to the inability of the film to respond in a uniform manner to X-ray exposures that are arbitrarily long or short. To understand the meaning of reciprocity failure, the meaning of *reciprocity* must first be understood. If for example, the photon flux is I and the exposure time is t, then an integrated photon flux proportional to It would fall on the film. If the I was doubled (*i.e.* doubling the amount of light falling per second) and the t halved, the same amount of light It would reach the film. If the film showed *reciprocity*, the same OD would be obtained in either case. Film has remarkably good reciprocity in the range of exposures normally used in photographic cameras, *i.e.* from a thousandth to a fraction (1/10 or 1/5) of a second. This covers most exposure times encountered in radiography and therefore reciprocity failure is not important. However for the long exposure times of 1–5 s used in tomography and mammography, reciprocity failure is important. The darkening of the film will then be reduced. The reason for reciprocity failure is in the photographic process – the photographic grain is designed not to be activated unless sufficient light falls on it in a short time to stabilize the image. Although this saves it from fogging in the dark, the by-product is reciprocity failure at long exposure times.

Latent image fading is another radiographic problem arising from the photographic process. The latent image, once formed, is not completely stable. For example, a crystal in which ten light photons were absorbed, would most probably be unconditionally stable, but one in which only three light photons were absorbed might only be stable for a few minutes, after which it could cease to be sensitized. In order that image densities be kept consistent, equal lengths of time should lapse between the exposure of the film and its development. Finally, it should be noted that the instability of film processing systems is the single most problematic aspect of a modern radiology department. Maintaining a match in characteristics between films processed from one day to the next, from morning to afternoon and from one film processor to another is an essential chore called *film quality control*.

2.2.2 Lenses – Electron and Optical

It is helpful to unify our understanding of change of scale in radiography by comparing and contrasting the operation of electron and optical lenses. The differences between electron and optical lenses are significant in radiography and perhaps best illustrated by looking at how images can be formed using electrons or light photons. In X-ray imaging, the use of a primary image sensor comparable to the size of the body part being imaged and then transferring it to a smaller sensor, is a common procedure. This change of scale, sometimes called *demagnification,* is desirable as it is often more practical to manufacture a small sensor than a large one. On the path to the small image, the optical lenses are constrained by the laws of geometric optics, *i.e.* the path of light in a uniform medium is a straight line. Thus, as shown in Figure 2.8, when imaging from a large object to a small optical image, invariably much of the light is lost. However, in the case of electron lenses, the path of the particles need not be a straight line allowing collection of all the information. This principle is used in X-ray image intensifiers (XRII). It is also possible to use

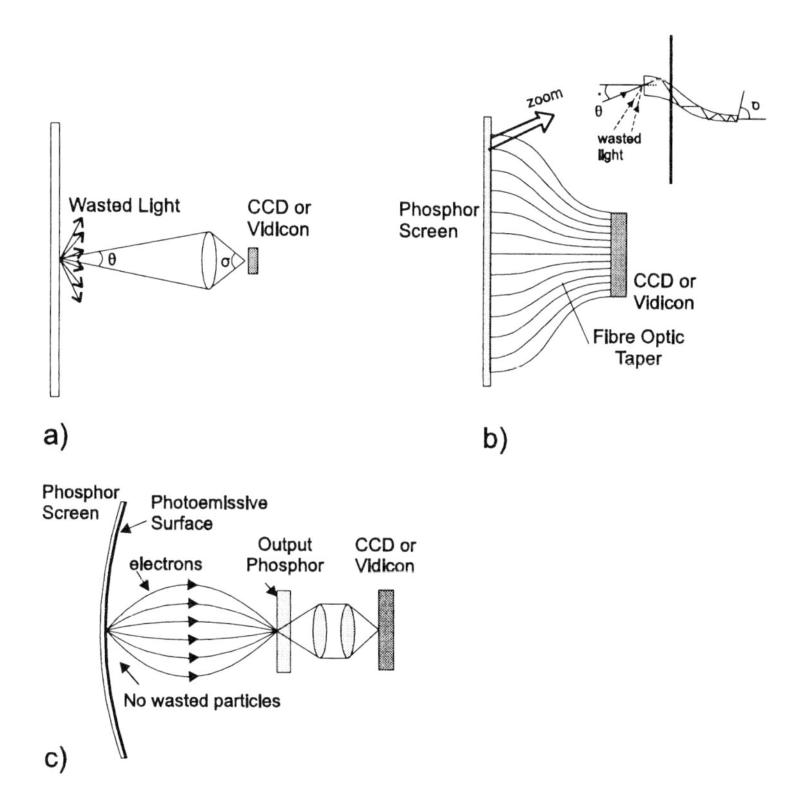

Figure 2.8. (a) Optical lenses, (b) fibre optic tapers, and (c) electron lenses used to demagnify a large object with a correspondingly large aerial X-ray image to a small optical image. (Redrawn from reference 22, with permission.)

fibre optic bundles or tapers to effect the optical coupling and required change of scale. These bundles consist of an array of optical fibers of constant diameter fused to form a light guide. The fibers form an orderly array so that there is a one-to-one correspondence between the elements of the X-ray image and an optical imaging device. To accomplish the required change in scale, the fibreoptic bundle is tapered by heating and softening the glass and drawing it. While facilitating the construction of a detector to cover the required anatomy in the patient, increasing demagnification by tapering also further reduces coupling efficiency by limiting the acceptance angle at the fibre optic input. Overall, fibre optics are more efficient than lenses at high demagnification, but introduce phenomena unique to themselves such as image shear and chickenwire artifacts.

2.2.3 Phosphors

Most current X-ray imaging detectors employ a phosphor in the initial stage to absorb the X rays and produce light. As shown in Figure 2.9, phosphors work by exciting electrons from the valence band to the conduction band where they are free to move a small distance within the phosphor. Some of these electrons will decay back to the valence band without giving off any radiant energy, but in an efficient phosphor, most of the electrons will return to the valence band through a local state (created by small amounts of impurities called activators) and in the process emit light. Thus phosphors can be relatively efficient converters of the large incident energy of the X-ray to light photons. Since light photons each carry only a few eV of energy, many light photons are created from the absorption of a single X-ray.

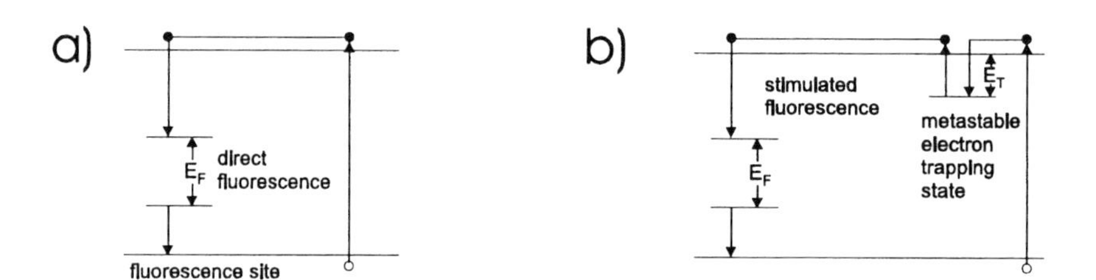

Figure 2.9. Energy band diagrams of crystals used in conventional phosphors (left) and photostimulable phosphors (right). (Reproduced from reference 22, with permission.)

This *quantum amplification* is the *conversion gain* g_I of the phosphor. For example, if Gd_2O_2S phosphor absorbs an X-ray photon of energy $\varepsilon = 60$ keV, this corresponds to the same energy as 25 000 photons of green light ($E = 2.4$ eV). However, because of competing energy loss processes, the *conversion efficiency* of Gd_2O_2S is only about 15%, so that on average only 4500 light quanta are released per interacting X-ray quantum.

By contrast, in photostimulable phosphors the immediate light output is not of interest. Instead, the energy stored in traps is subsequently stimulated in the readout process. The efficiency of the storage function can be improved by increasing the probability of electron trapping. On the other hand, when these electrons are released by the stimulating light during readout, the probability of their being retrapped instead of producing signal would then be higher, thus reducing the efficiency of readout. The optimum balance occurs where the probabilities of an excited electron being retrapped or stimulating fluorescence are equal. In addition, the decay characteristics of the emission must be fast enough that the image can be read in a conveniently short time. The energy levels in the phosphor are critical to effective operation. The energy difference between the trap level and the conduction band edge ε_T must be small enough so that stimulation with laser light is possible, yet large enough to prevent thermal excitation of the electron from the trap. Finally, the wavelength of the emitted light has to be adequately separated from that of the stimulating light quanta to avoid contaminating the measured signal. Photostimulable phosphors are commonly in the barium fluorohalide family, typically $BaFBr:Eu^{2+}$, where the atomic energy levels of the Europium activator determines the characteristics of light emission[5].

2.2.3.1 X-ray Screens

To effectively create an X-ray image, a transparent phosphor would be ineffective since light could move large distances within the phosphor and cause blurring. Instead X-ray screens are made highly scattering or *turbid*. These screens consist of a layer of phosphor, in the form of very fine powder, incorporated within a non-radiative but optically transparent binder supported on the surface of a plastic substrate. X-rays cause the phosphor to emit light near the point of incidence on the screen. After their formation, the light quanta must escape the phosphor as efficiently and as close to their point of formation as possible.

Figure 2.10(a), (b) and (c) illustrate the effect of phosphor thickness and the depth of X-ray interaction on spatial resolution of a phosphor detector and the complete loss of resolution for transparent screens. In settled phosphor screens, the phosphor grains are optically highly scattering due to the large refractive index of phosphors compared to the plastic binder. The scattering is sufficiently intense that the flow of photons is considered diffusive and the layer turbid. This results in a limit to the lateral spreading of the light relative to the thickness of the layer. Other optical effects are used to change the imaging properties of the screen, *e.g.* a reflective backing helps increase the amount of light escaping the front of the screen but at the cost of increased blurring and hence reduced resolution. Typically, without the backing, fewer than one-half of the created light quanta escape the phosphor and are potentially available to be recorded. The addition of the reflective layer

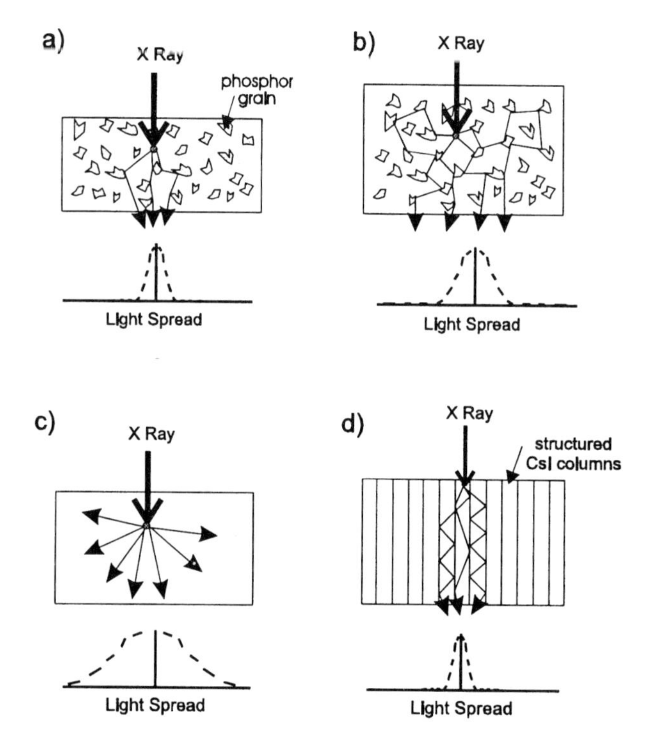

Figure 2.10. Phosphor screen resolution and the effect of screen thickness on blurring: (a) thin settled phosphor powder screen, (b) thicker settled phosphor screen, (c) transparent phosphor screen, and (d) structured phosphor screen, *e.g.* CsI.

increases the escape fraction to almost 100%. Light absorbing dye can also be added to the screen to enhance the resolution more than elimination of the reflective backing, again at the cost of signal loss.

The disadvantages of powdered phosphor screens can be overcome by using structured phosphors, such as CsI, which can be created in the form of a fibre optic plate, as shown in Figure 2.10 (d). The structured phosphor guides light created within the bulk of the material to the surface without much blurring. However, separation between fibers is created by cracking and as a result the channeling of light is not perfect.

Screens are most commonly used in conjunction with film sandwiched in a cassette. The film may be either between two screens or on the surface of a single screen such that the light from the screen impinges on the part of the film closest to it, thereby forming a sharp image. Screens can also be used, coupled optically, either with a fibre optic or a lens to other optical imaging devices, *e.g.* CCDs or video camera tubes.

2.2.3.2 Pulse Height Spectra of Phosphors

The behavior of phosphor screens and devices containing phosphors can be better understood by examination of pulse height spectra. These are obtained using a photomultiplier coupled to the phosphor as shown in Figure 2.11(a).

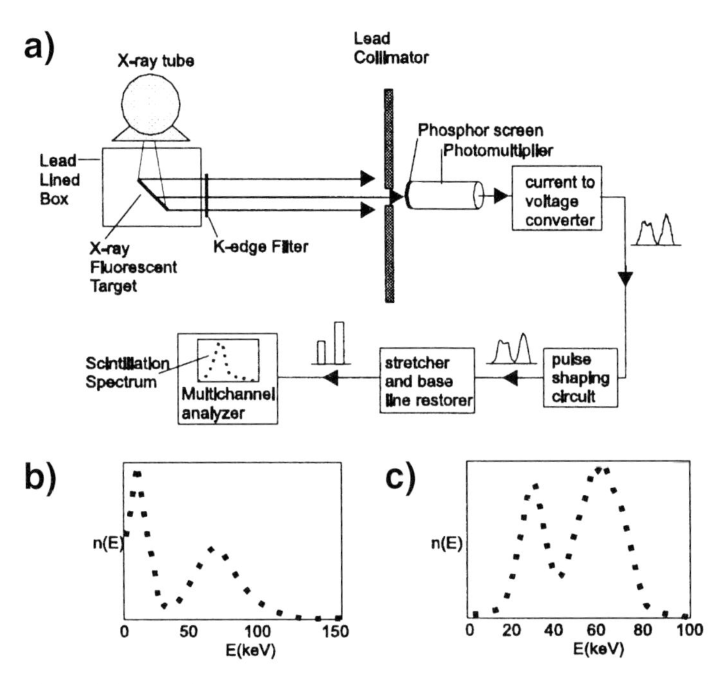

Figure 2.11. (a) Apparatus for obtaining a pulse height spectrum PHS, (b) PHS from a radiographic screen with Gd_2O_2S screen, and (c) PHS from an X-ray image intensifier with a CsI screen.

X rays impinge on the phosphor one at a time. Multiple light photons are given off from the phosphor for each absorbed X-ray. The light is absorbed by the photocathode of the photomultiplier and amplified essentially noiselessly. By subsequent electronic processing called *shaping,* a pulse is created whose height corresponds to the number of light photons emitted in that individual X-ray event. A multi-channel analyzer sorts the pulses according to their height and thereby creates the spectrum. A perfect phosphor would, assuming monoenergetic X rays are impinging on it, produce pulses all of equal height given by the conversion gain g_1. The width of the spectrum would show the deviation from perfect operation. The presence of multiple peaks as shown in Figure 2.11 (b) and (c) indicates the presence of K-escape and other physical effects. The pulse height spectrum can be used to understand degradation in detective quantum efficiency due to an imperfect spectrum. Swank[6] described this effect and the "Swank factor" A_s characterizes this additional noise source as $\mathrm{DQE}(0) = \eta A_s$. The Swank factor is calculated in terms of the moments of the pulse height distribution of g_1 as:

$$A_s = \frac{M_1^2}{M_0 \, M_2} \tag{2.7}$$

where M_i indicates the ith moment of the distribution.

The actual number of quanta produced by an interacting X-ray will depend both on its incident energy and the mechanism of interaction with the phosphor crystal. The most likely type of interaction, the photoelectric effect, will result in both an energetic photoelectron and either a second (Auger) electron or a fluorescent X-ray quantum. The energy of fluorescence depends on the shell in which the photoelectric interaction occurred. For example, K-shell interactions for Gd_2O_2S have a threshold of 50.2 keV and produce the most intense fluorescence just below 43 keV. However, for a CsI phosphor as in an X-ray image intensifier, the fluorescence is at 30–33 keV. The fluorescent quanta are either reabsorbed in the phosphor or escape. In the latter case, the energy deposited in the phosphor from the X-ray quantum is reduced giving rise to a second peak in the distribution with a lower value of g_I. The effect of fluorescence loss is to broaden the distribution of g_I, causing a decrease in A_S. Further increase in the spectral widths and hence decrease in the A_S can be attributed to the optical effects used to enhance resolution such as added dye or omission of a reflective backing.

2.2.3.3 Photostimulable Phosphor Screens

A photostimulable phosphor screen is very similar to a conventional X-ray screen except it uses a phosphor that contains traps for excited electrons as shown in the band structure diagram, Figure 2.9 (b). Thus, after excitation by X rays, the photostimulable phosphor contains a latent image consisting of the trapped electrons. This latent image can be read using photoluminescence. By stimulating the screen point-by-point with red laser light, electrons are excited from the traps. This allows them to reach luminescence centers and emit blue light which is then detected by a photomultiplier. The amount of light given off at each point corresponds to the number of charges stored in the latent image, *i.e.* the number of X rays that impinged on that particular point.

Currently there is only one kind of stimulable phosphor (the BaFBr family) in common use. This is because conventional phosphors only have to give off light, but the requirements of storing the image, exciting with red light and giving off blue light has only been possible with BaFBr. It is desirable to find materials with higher X-ray absorption than this material, however nothing has yet been developed. Would it be possible to improve the resolution of photostimulable phosphor systems by use of a transparent phosphor? It has been shown earlier that a transparent phosphor used with film would have very poor resolution. However, the situation for a stimulable phosphor is quite different from that of film-screen. For a stimulable phosphor, the resolution arises from the precision with which the stimulating laser light is positioned. Thus a transparent medium could give a higher resolution image than a turbid phosphor due to the elimination of scattering of the readout laser beam.

2.2.3.4 Output Phosphors

Output phosphors are used in XRII and cathode ray tubes (CRTs). They emit light when energetic (20–30 keV) electrons travel through a vacuum and impinge on them. The range of such electrons is only a few microns so all the energy can be absorbed by using a thin phosphor layer permitting high resolution. The phosphor surface on which electrons are incident is usually covered with a very thin layer of aluminum to establish an electrical potential and prevent light created in the phosphor layer from returning into the device, scattering and reducing image contrast. Since light has to be transmitted outside the vacuum envelope, the phosphor is built on a glass plate through which it is viewed. Unfortunately light is given off in all directions and some will be trapped within the glass by total internal reflection. This causes halation (*i.e.* halos arise around bright objects resulting in a loss of contrast as trapped light re-enters the phosphor and is then re-emitted making it viewable) as shown in Figure 2.12. Several methods for the suppression

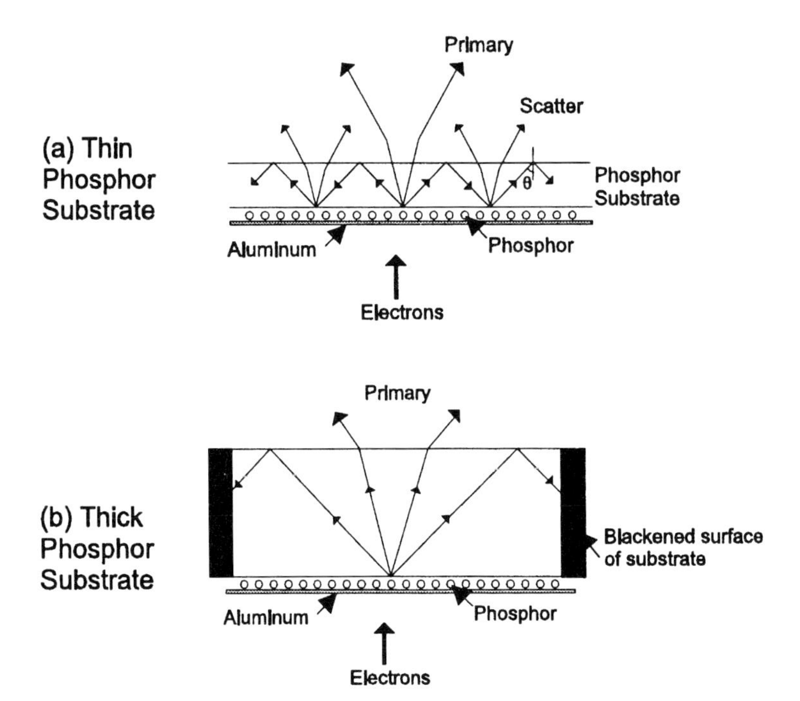

Figure 2.12. Origin of veiling glare or contrast loss in output phophor by halation: (a) multiple reflections within a thin substrate, and (b) elimination of halation by use of a thick substrate or *light trap*.

of halation are in common use. These are: use of grey (*i.e.* light absorbing) glass which preferentially attenuates the unwanted light since it has on average a longer path length; use of fibre optic substrate; and substantial increase in thickness of the substrate so that the totally reflected light is absorbed at the blackened edge of the substrate. This last approach is called a *light trap* and is the best current method for XRII output phosphors. However it requires specially designed lenses to accurately focus a plane image through such a large thickness of glass.

2.2.4 Electrostatic X-ray Transducers

X rays interacting with an electrostatic X-ray transducer release positive and negative ions (in gas or liquid) or electrons and holes (in a solid state detector). In each case, the released carriers are charged particles. In a suitable material, a large number of carriers will be released for each X-ray that interacts due to the large energy of diagnostic X rays. The amount of energy that has to be absorbed from the X-ray in order to release a pair of carriers is called the *excitation energy* W and is related to the conversion gain g_I and X-ray energy E by $g_I = E/W$. X rays interacting in the sensor release carriers, because they are charged, can be guided directly to the surfaces of the plate by the applied electric field as shown in Figure 2.13. Therefore, the latent charge image on the surface is not as blurred as in a phosphor layer of comparable thickness.

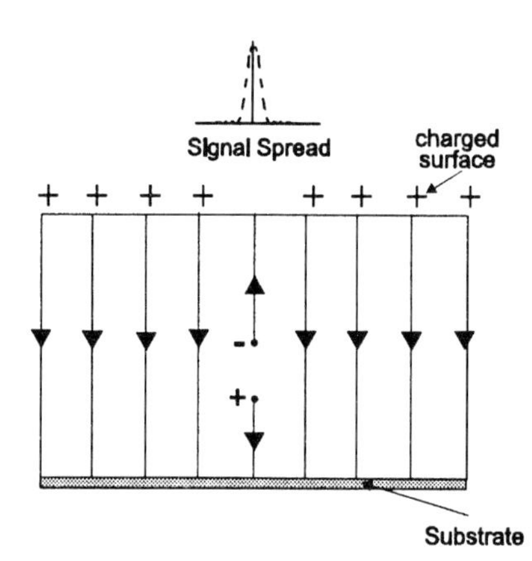

Figure 2.13. Resolution in an electrostatic transducer.

2.2.4.1 Ionization Chamber

Radiation passing through air causes ionization. In the presence of an applied field, these ions can be separated and collected on electrodes. This is the basis of the ionization chamber used in radiation dosimetry[3]. The fundamentals of the ionization process and the particular value of ionization energy for air and other gases are well established. Portable ion chambers have air equivalent walls which as far as possible mimic the effect of the free air chamber shown in Figure 2.14. Imaging devices based on ionization in gases such as xenon are the basis of detectors used in computed tomography.

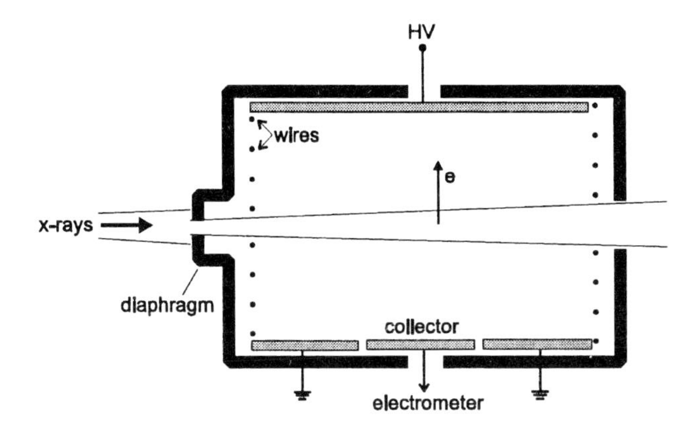

Figure 2.14. Free air ion chamber.

Radiation entering the ion chamber causes ionization, *i.e.* positive ions move in the direction of the electric field and negative ions in the direction antiparallel to the electric field. As soon as ions start to move toward the electrodes on the surface of the chamber, a current flows in the external circuit and continues until all ions are cleared from the chamber.

2.2.4.2 Photoconductors and Semiconductors

Detectors based on photoconductors or semiconductors are solid state analogs of the gaseous ionization chamber[7]. Radiation detectors must not have any free carriers in the absence of radiation, otherwise a current would flow continuously. The free carriers generated by the action of ionizing radiation are used to indicate the presence of radiation. In a metal, there are already so many carriers that the additional few released by radiation cannot be detected.

Therefore the state of matter required in a solid state detector is one in which there are few carriers naturally occurring, *i.e.* semiconductors or insulators. In a semiconductor, very few free carriers are present and in insulators, there are essentially none at room temperature. Another requirement is that those carriers released by radiation have a long enough lifetime that they can reach the electrodes. Such insulators are called *photoconductors*. Figure 2.15 shows the band structure of photoconductors and semiconductors.

Comparing Figure 2.15 with Figure 2.9, the band structure of phosphors, there is very little difference. In fact, phosphors are insulators with very large forbidden band gaps. The band structure for semiconductors and photocon-

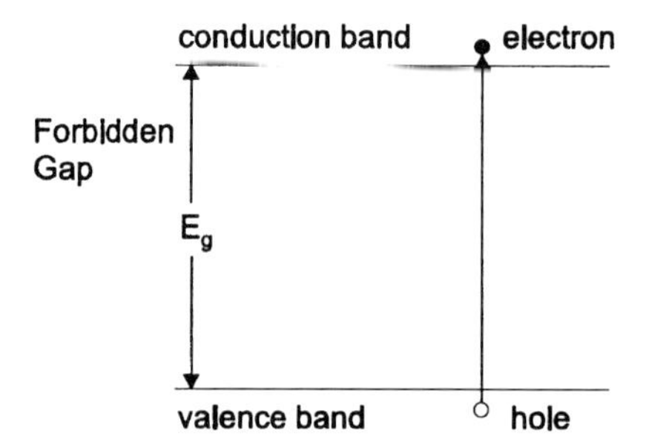

Figure 2.15. Band structure of a photoconductor or semiconductor depending on the value of the forbidden energy gap E_g of the material.

ductors differs only in the band gap energy E_g. When an X-ray impinges on a solid state material, ionization occurs from the valance band to the conduction band. The minimum possible energy to cause this ionization is E_g. An optical photon with energy E_g can excite an electron from the valance band to the conduction band. Diagnostic X rays have energy thousands of times larger than E_g. The first interaction of diagnostic X rays in a detector would usually, because detector materials have high Z, be through the photoelectric effect in which a very energetic photoelectron is released. This electron passing through the solid state material causes further ionization. Under these cir-

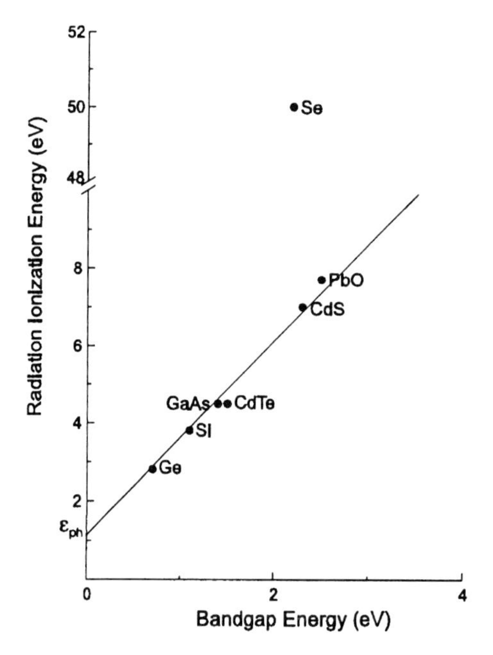

Figure 2.16. A plot of the energy W required to release an electron-hole pair in a solid state material plotted as a function of the forbidden band gap energy E_g of the material.

cumstances, the amount of energy necessary to create an electron-hole pair is not simply E_g since there are other competing mechanisms. This was studied by Klein[8] and shown in Figure 2.16 is a plot of W as a function of E_g for several semiconductors and insulators used as radiation detectors or spectrometers.

The data fit a straight line with a small intercept. The slope of the curve is approximately three which indicates that on average an energy of 3 E_g has to be absorbed to release an electron-hole pair. The intercept indicates a parasitic loss due to phonons. Almost all photoconductors and semiconductors fit Klein's curve provided the electric field is large enough and the material is pure enough that all the freed charge carriers are collected. Important exceptions to this curve include amorphous selenium, the most commonly used solid state photoconductor in X-ray imaging. It is a photoconductor with E_g = 2.2 eV. Its great advantage is that it is uniform in imaging properties to a very fine scale (an amorphous material is entirely free from granularity) and can be easily and inexpensively made in large areas. The first medical application of amorphous selenium was in xeroradiography, a technical and commercial success in its day, where a latent charge image on the surface of an amorphous selenium plate was readout with toner, *i.e.* the photocopier process. Xeroradiography is no longer competitive, probably because of the toner readout method, not the underlying properties of amorphous selenium[9]. Amorphous selenium is still the only practical large area semiconductor or photoconductor suitable for medical imaging. Investigation of other possible materials such as CdZnTe, PbI_2 and TlBr are ongoing to overcome the disadvantages of amorphous selenium – its high W, high activation potential and relatively low Z.

2.2.5 Laser Beam Scanning

Scanning is an essential function of many radiological systems both for readout and display. One approach is to scan a laser beam. Lasers have many unique features such as coherence, monochromaticity, Gaussian beam shape and very high collimation. In laser scanning the last two features are the most important. Figure 2.17 shows a generic laser beam scanner. The beam emerging from a gas laser can be well collimated and has a Gaussian beam profile if the laser is operated in the fundamental transverse mode (TEM_{00}). For solid state lasers, special aspheric lenses are required to make the beam have a Gaussian profile. The diameter and the divergence of a beam are intimately related in Gaussian laser optics. The larger the beam diameter, the smaller its divergence.

The laser beam is expanded or contracted to an appropriate diameter for further processing. The deflection of the laser is accomplished using either a galvanometer with a mirror attached or a rotating polygonal mirror. Subsequently, the laser beam passes into the entrance pupil of an F/θ lens whose purpose is three fold. First, the laser beam is focused to a sharp point at the

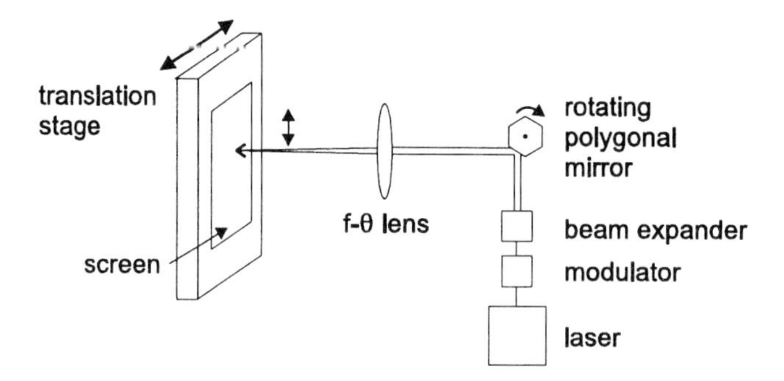

Figure 2.17. Laser beam scanner showing principal components: laser, intensity modulator laser beam expander, scanning mirror, and F/θ lens.

image plane. The greater the beam diameter at the entrance of the lens, the smaller the focal spot size (limited by diffraction). Second, constant angular diversion of the laser beam must correspond to a constant linear displacement of the laser spot irrespective of the deflection angle. Third, the entrance pupil of the lens must be large enough and far enough away from the lens to permit the scanning operation at all necessary angles. The modulation of the laser beam intensity is accomplished with either an electro-optic or acoustic-optic modulator in the laser beam. Modulation can also be achieved with solid state lasers by changing the laser drive current. The most commonly used lasers are the helium neon laser – a gas laser producing red light; the argon ion laser – a gas laser producing blue light; and the solid state laser that produces infrared or red light. Further developments of solid state lasers to allow them to create blue light are ongoing. The laser beam scanner is used as the basis for laser cameras to produce large area films from digital input and laser readout of the stimulable phosphor systems.

2.2.6 Electron Beam Scanning

Electron beam scanning systems are all vacuum tube devices. The source of the electrons is thermionic emission from a heated cathode and the resulting electrons generated are focused into a beam by an electron gun. The gun focuses these electrons onto a distant screen. The deflection of the beam is accomplished by other elements such as magnetic yoke or electrostatic plates. A generic electron beam scanning system is shown in Figure 2.18. Electron beam scanning is a principal component in vidicon tubes which are optical sensor tubes; in large area X-ray sensitive vidicons; and in the cathode ray tubes used to display electronic images.

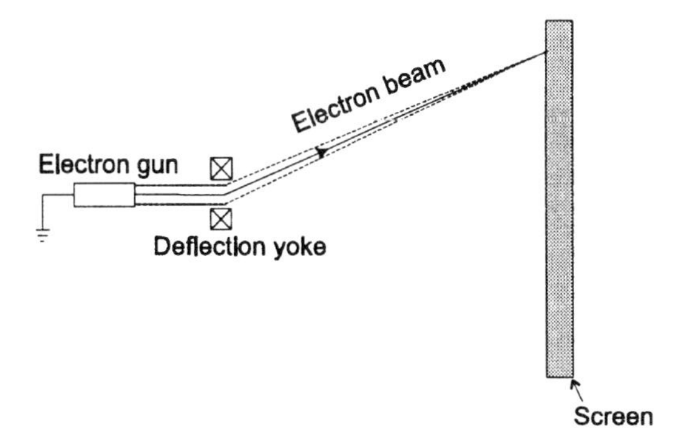

Figure 2.18. Electron beam scanning system showing: (left) electron gun, (centre) scanning elements such as magnetic field coil or electrostatic electrodes, and (right) screen where the electrons land.

2.2.7 Vidicons, Great and Small

The simplicity of the vidicon has made it the preeminent optical imaging sensor amongst vacuum tube devices for videofluoroscopy – its only challenge is from solid state devices. The characteristics of the tube can be modified by changing the photoconductor in the target. The construction of the input target of the different kinds of vidicons is illustrated in Figure 2.19.
The complete vidicon tube is shown in Figure 2.20. This is the most common type with a triode gun, beam focusing and deflection achieved by magnetic focus and scan coils (not shown) around the tube. The source of the beam is the heated cathode. All potentials in a vidicon tube are as a convention referenced to the cathode. The potential on a cylindrical electrode G1, with an axial hole controls the current emitted from the cathode. G2 (typically +400 V) accelerates the beam and defines a cross-over point between G1 and G2. G3 is the focusing electrode and is used in conjunction with a solenoid around the tube to establish focus of the electron beam at the target plane. Where the beam exits G3 there is a very fine mesh called G4 designed to ensure that the potential across the exit is uniform and thus maintains focus across the entire field-of-view. Large area vidicons, directly sensitive to X rays, can also be made. Several differences in addition to the increased scale are required. The most important is the increased thickness of photoconductor to efficiently absorb X rays.

2.2.8 Self-scanned Readout Structures

A self-scanned readout structure may be defined as one in which the image created in a certain plane is readout in that same plane. The advantage of such structures is that they are compact. The two major types of self-scanned devices used in radiography are charge-coupled devices, known as CCDs, and

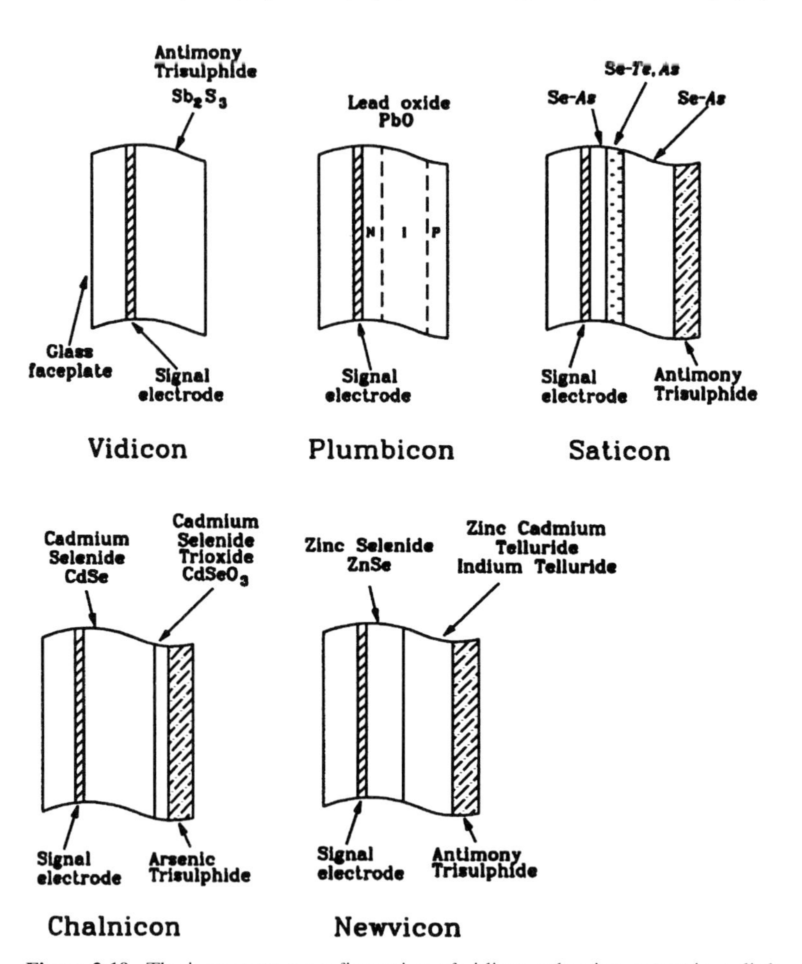

Figure 2.19. The input target configuration of vidicon tubes important in radiology. (Reproduced from AAPM Medical Physics Monograph Number 20 "Specification, acceptance testing and quality control of diagnostic x-ray imaging equipment", AIP, 1994, with permission).

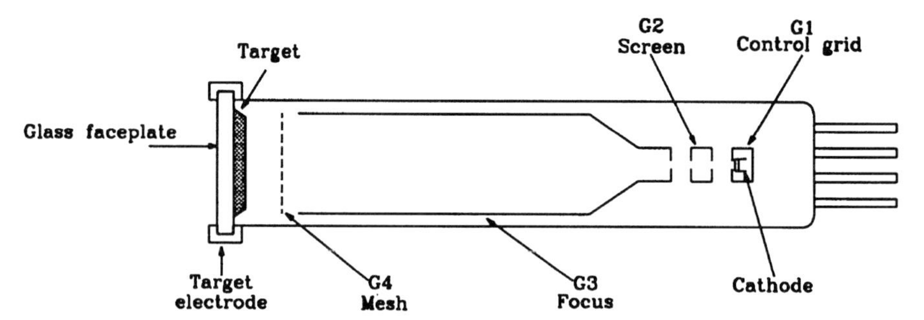

Figure 2.20. Electrode arrangement of a triode gun vidicon tube. (Reproduced from AAPM Medical Physics Monograph Number 20 "Specification, acceptance testing and quality control of diagnostic x-ray imaging equipment", AIP, 1994, with permission).

large area thin film active matrix arrays that are being proposed for large area direct detectors of radiation as well as for liquid crystal displays. Both of these systems depend on an enabling technology called photolithography.

2.2.8.1 Photolithography

Photolithography is the most important technique for constructing high resolution circuits layer-by-layer onto a flat substrate, thus allowing circuits of arbitrary complexity. Semiconductor devices such as transistors, capacitors and resistors can be insulated from each other and connected by appropriate placement of patterned layers of insulator, semiconductor or metal. In Figure 2.21, the principle of production of one single layer, similar to the ten or twenty layers used in modern integrated circuits or *chips,* is shown.

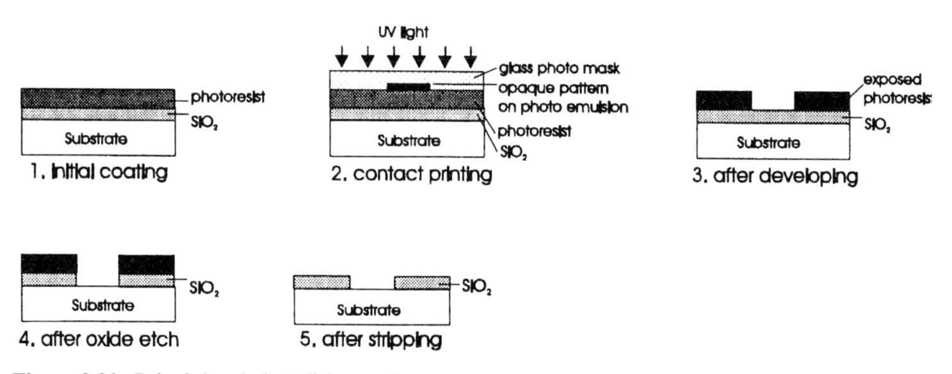

Figure 2.21. Principle of photolithography.

The photolithographic process depends on a material called photoresist[10]. This is not the silver halide material discussed in conjunction with film, but a less sensitive material which is exposed using UV (*i.e.* ultraviolet light). First the layer to be patterned is evaporated onto the substrate. This layer could be a semiconductor to make active devices; a metal layer to connect layers; or an insulator to isolate previously made layers. The substrate is then placed on a high speed *spinner* and a small quantity of the photoresist is dispensed onto the centre of the slowly spinning material. The rotational speed is increased and by a combination of centrifugal force and surface tension, a uniform layer of photoresist about a micron thick is produced. The spinner is stopped abruptly, leaving the photoresist with a uniformly flat surface. This is dried in an oven before the next stage, in which a latent image is formed in the photoresist by projection of UV light through a photomask, which is a glass or quartz substrate with a pattern formed in chrome. Where light falls on the photoresist, polymerization occurs, chemically changing the photoresist. The non polymerized material is removed either chemically (dipping it in a solvent bath) or by ion etching leaving the photoresist pattern. A uniform layer (*e.g.* semiconductor, metal or insulator) is etched through the patterned photoresist, and finally, the photoresist pattern is removed. This sequence, from a bare substrate to one layer of patterned material, can be repeated with different materials as many times as necessary to create a fully functional circuit.

2.2.8.2 Charge Coupled Devices

Charge coupled devices (CCDs) are simple and elegant sensor and readout systems made by photolithography[11]. They are normally made using high purity single crystal silicon. To understand the mechanics of the CCD principle, an analogy is helpful. If a team of people pass buckets from hand to hand, the level of water in each bucket could represent the amount of charge and therefore the signal sent by the device. Moving the buckets from hand to hand keeps the signal from each pixel isolated while it is transferred. This is similar to the operation of a CCD built on a layer of silicon, as illustrated in Figure 2.22, where the buckets are potential wells. Charge is transferred from one well to the next in the channel, without mixing, by appropriate synchronous operation of the electrodes above the channel. A CCD is an integrated circuit formed by depositing a layer of oxide insulator followed by a series of electrodes, called *gates* on a semiconductor substrate to form an array of metal-oxide-semiconductor (MOS) capacitors. By applying potentials to the gates, the material below is depleted to form charge storage *wells*. These store charge injected into the CCD or generated within the semiconductor by the photoelectric absorption of optical quanta. When the potentials over adjacent

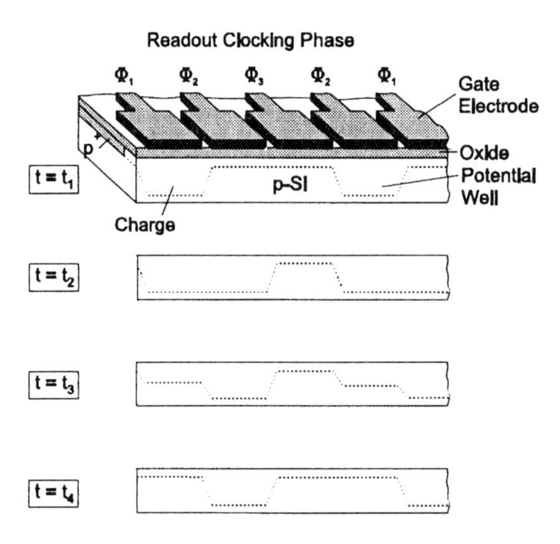

Figure 2.22. Structure of a charge-coupled device array illustrating motion of stored charge in one direction as the potential wells are adjusted under control of the gate electrode potentials. (Reproduced from Reference 22, with permission).

gates are varied appropriately, the charge is transferred from well to well. If in making an optical sensor from a CCD, the electrodes are made transparent to light, then silicon above the wells can function as a photodetector. Charge generated by light is deposited in the well closest to the point of incidence of light. After the image has been formed, the charge coupling principle can transfer the image charge from its point of creation to a single readout preamplifier. In Figure 2.22, a linear array is shown but area arrays can be similarly made by connecting vertical charged coupled columns. In one type of complete optical detector, a frame transfer CCD, an entire image is transferred across the array and under a light shield before readout as shown in Figure 2.23(a). Shown is an array with 6 × 4 elements but in practice 500 × 500 or 1000 × 1000 elements or more may be used. Alternatively, *interline readout* CCDs, Figure 23(b), have a line of optically-shielded storage pixels adjacent to each column of detector elements.

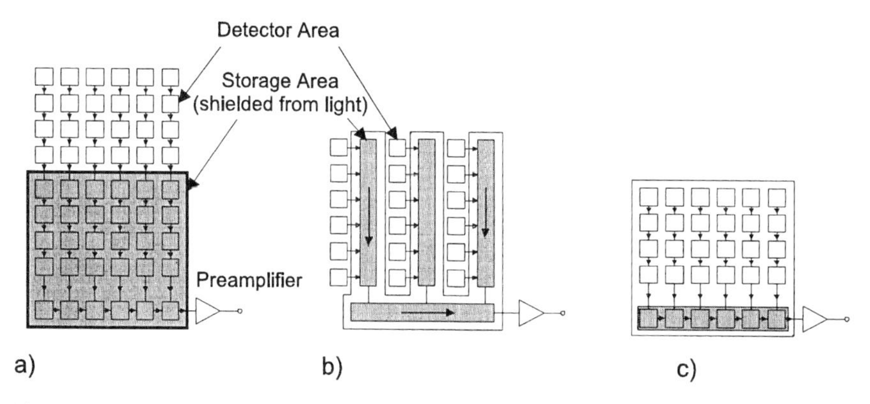

Figure 2.23. Typical readout configurations of CCDs: (a) frame transfer, and (b) interline transfer, both for staring applications; and (c) time domain integration TDI for scanning applications.

In any CCD, the charge is transferred over many adjacent elements before it reaches the preamplifier. It is therefore critical that the efficiency of each transfer be extremely high. Poor transfer efficiency can cause a serious loss of spatial resolution in the detector. For example, if the image charge must be shifted across n elements and the efficiency per transfer is ϵ, then the overall charge transfer efficiency is ϵ^n. Even if ϵ is as large as 0.999, the charge transfer efficiency falls to 90% over 100 transfers and 37% over 1000 transfers. In commercial CCDs, values of ϵ as high as 0.999999 are achievable, thus maintaining integrity of the charge packets and eliminating resolution loss.

The well storage capacity of the device is very important in radiological applications to maintain high dynamic range. Depending on pixel size, capacities of 100 000 to several million electrons are possible. For scanning systems, it is often more practical to operate the CCD in time delay integration (TDI) mode, as shown in Figure 2.23(c). Here a storage section is not required as the charge is simultaneously integrated and shifted down the CCD detector columns towards the horizontal readout register. The image on the CCD must move in synchrony with the shifted charges representing the image to prevent motion blurring. Thus CCDs are precision sub-components within the imaging system.

2.2.8.3 Large Area, Thin Film Active Matrix Arrays

In theory, large area CCDs could be made. Practically, there is a problem related to the production yield. When very large area (*e.g.* 30 cm × 30 cm) devices are required, a different approach with simpler, more modest circuitry to ensure reliable functioning is used, and a production yield that can be maintained at a practical, economic level. Thus cheaper substrates made of glass are preferred to the high purity single crystal silicon used in CCDs and computer chips. However, similar concepts are used in the photolithographic processes, although the line widths will be generally larger since the scale of devices will be greater. The reason large area devices are required is to allow humans to directly interact with them. For example, some displays have a large size requirement and some X-ray sensors need to be as large as the body part being imaged. The use of glass substrates mandates that the processing temperatures be kept below glass softening temperature thus limiting the kind of processing available and the materials used.

The most common semiconductor used in thin film active matrix arrays is amorphous silicon, developed in the last twenty years following Spear's discovery that the otherwise intractable pure material could be made useful by inactivating dangling bonds. This was achieved by incorporating appropriate amounts of hydrogen, yielding *a*-Si:H hydrogenated amorphous silicon. A related material made by recrystalizing *a*-Si:H is polycrystalline silicon (*poly* Si) which has a higher carrier mobility than *a*-Si:H and is thus preferred in more advanced devices. Only a relatively small repertoire of devices is possible using active matrix technology, *e.g.* small (~1–2 pF) capacitors made by overlaying electrodes separated by insulating layers; optically sensitive photodiodes; and finally active devices called thin film transistors or TFTs, which are electrically activated switches and are the thin film equivalent of a MOSFET (metal oxide field effect transistor). The principle of operation of a TFT is shown in Figure 2.24.

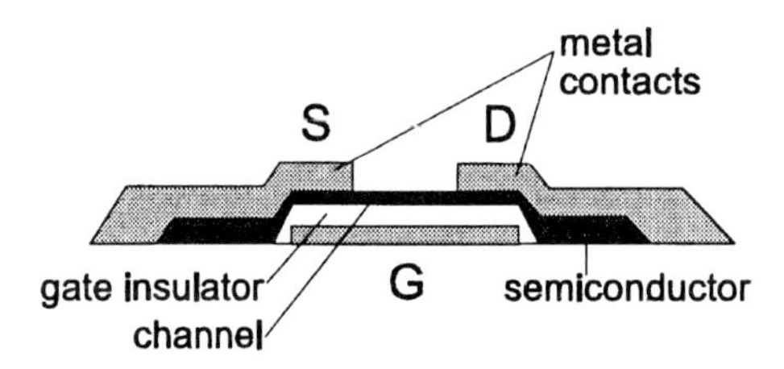

Figure 2.24. Schematic diagram of a thin film transistor or TFT.

The control element of the TFT is the gate. With the gate negative, the TFT is off and there is no conduction between the drain and the source of the TFT. Turning the gate potential positive creates a conductive channel beneath the gate. This in turn connects the drain and source, effectively turning the transistor on. The elements of a large area thin film active matrix array are shown in Figure 2.25. Currently, the peripheral functions (scanning control, charge amplification, multiplexing and display driver) are performed by single crystal silicon integrated circuits made separately from the active matrix array and connected by wirebonding. More advanced concepts under development would integrate all these functions onto the active matrix substrate, thus achieving a complete imaging system on glass.

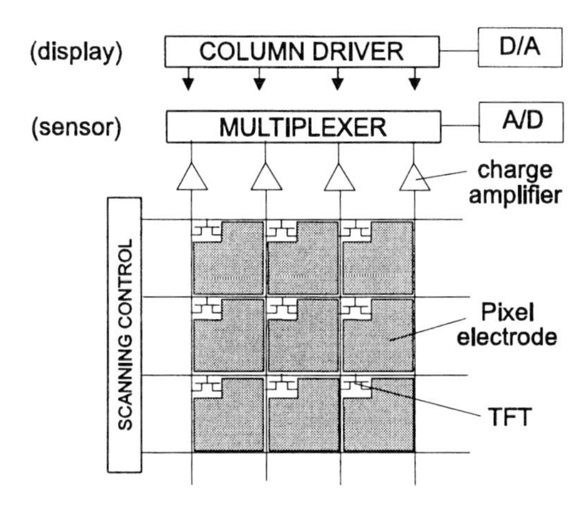

Figure 2.25. Overview of active matrix array that can be used either as a sensor or display component depending upon the nature of the peripheral column electronics.

Such a large area array could be used as a display or a sensor. In either case, scanning operation is accomplished by activating the common gate line for a row, thereby turning on one row of TFTs at a time. Charge can either be introduced (display application) or readout (sensor application) from a row using the data lines connected to each column. Once that row is completed, the gate line turns off all the TFTs within it. This continues in sequence throughout the entire array. For use as a display, the pixel electrodes control the polarization state of a liquid crystal layer, thus forming an electrically controlled spatial optical modulator. For use as an X-ray sensor, a layer of photoconductor is deposited so that each electrode reads out the latent image charge created by the action of X-rays either *directly* (in the photoconductor) or *indirectly* using a phosphor layer first to change X-ray energy to light that is subsequently detected in the photoconductor layer or individual photodiodes.

This completes the major sub-components of an imaging system. Each must be chosen with care and understanding since their physical and material properties ultimately define and limit the complete system.

2.3 Components

Components, built from the previously discussed sub-components, may be complete devices that are unique to X-ray imaging, or more commonly devices, which have been adapted from components used in non-medical applications.

2.3.1 Fundamentals

2.3.1.1 Geometry of Irradiation – Pencil, Slit, Slot or Area and Scatter

"What X-ray geometry should be used?" is the first fundamental decision facing the designer of an X-ray imaging system. Conventionally, producing an X-ray image involves exposing the entire area of interest simultaneously and detecting it with an area sensor as in film-screen radiography.

Other approaches are possible, but the simplest is to obtain a pencil beam of radiation (accomplished by collimating the broad area flux from the X-ray tube by means a lead blocker with a small hole in it) and scanning it over the patient, one point at a time. A single sensor aligned with the pencil beam creates an image of the patient. Other variations between a pencil and area beams can be seen in Figure 2.26.

Slit irradiation is obtained with a fan beam of radiation and an aligned single line detector which is scanned perpendicularly to the line across the patient. Both pencil beam and slit beam scanning are extremely inefficient in the utilization of X rays. Most of the X rays are removed by the collimator and a full scan imposes an enormous heat load on the tube. It is possible to improve the efficiency of such systems by employing a multi-line or "slot" detector where the X-ray beam extends across the full image field in one dimension and is several lines wide in the other. There are two types of slot detectors: first, a slot is moved discontinuously across the width, a single exposure made and the multiple lines readout. This process is repeated until the entire area is covered. Second, in time domain integration, TDI, the slot beam detector moves continuously and the image is readout one line at a time.

Why use these complicated scanning methods to produce images when it appears that irradiating a static area detector is much simpler? Several concepts must be balanced. Each method has advantages and disadvantages but the most important consideration is scattered radiation. A reduced area

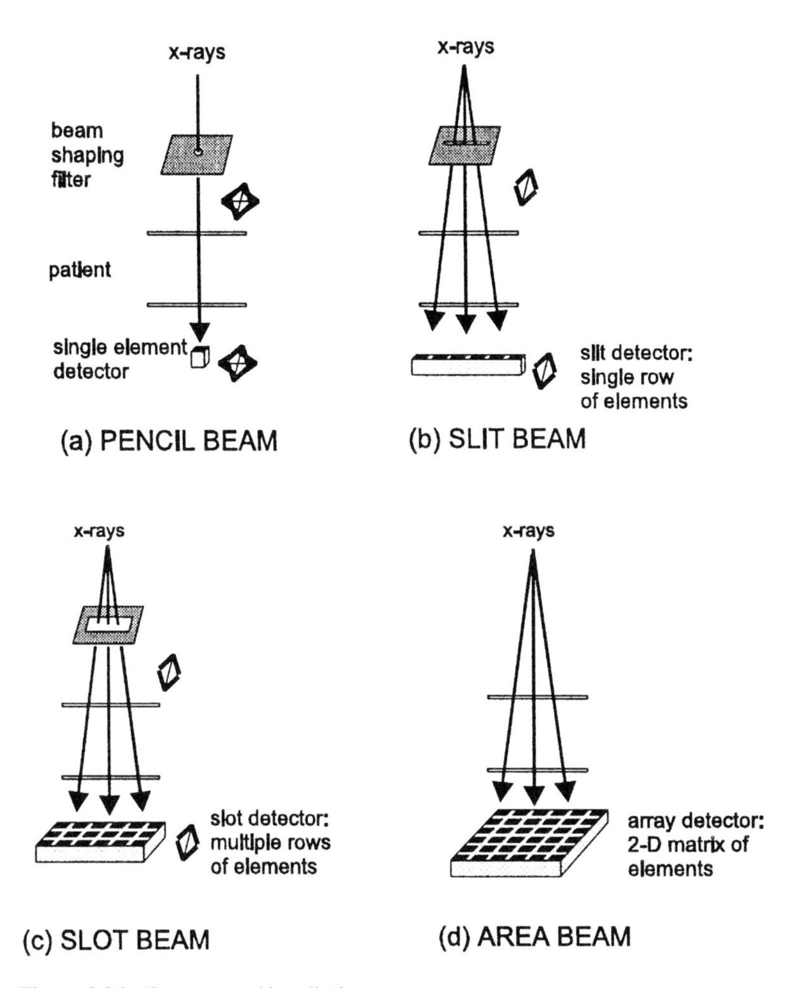

Figure 2.26. Geometry of irradiation.

detector can, by use of blockers, be much more efficient than area detectors at eliminating scatter. Also, reduced dimensionality detectors are easier to make than area detectors. However, the shortest exposure and the least loading of an X-ray tube is possible with the area detector. It will be seen later during discussion of applications that all geometries have a place in radiology.

2.3.1.2 Primary and Secondary Quantum Sink and Noise

The primary quantum sink of a detector refers to the efficiency with which the detector absorbs and utilizes X rays. If the primary quantum sink successfully absorbs most of the incident X rays, the final image may still be of poor quality because other events may occur and corrupt the image quality. This concept, called a *secondary quantum sink*, can be understood with reference

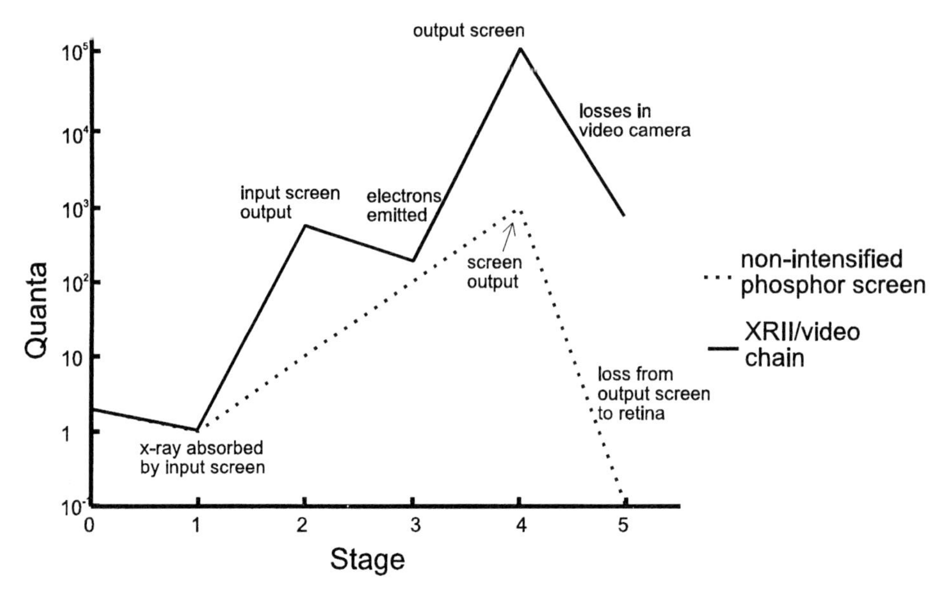

Figure 2.27. Quantum accounting diagram for non intensified (direct viewing of a phosphor screen) and intensified fluoroscopy (using an XRII/video chain).

to Figure 2.27, a quantum accounting diagram. Fluoroscopy with and without the use of an XRII were chosen as examples to illustrate the operation without and with a secondary quantum sink respectively. The propagation of signal through the various energy conversion stages of the imaging system is illustrated. Thus, in the diagram, N_0 quanta are incident on a specified area of the detector surface (Stage 0). A fraction of these, given by the quantum detection efficiency η, interact with the detector (Stage I). In a perfect imaging system η would be equal to 1.0. The mean number N_1 of quanta interacting represents the "primary quantum sink" of the detector. The fluctuation about N_1 is $\sigma_{N_1} = (N_1)^{1/2}$. This defines the signal-to-noise ratio SNR, of the imaging system which increases as the square root of the number of quanta interacting with the detector. Regardless of the value of η, the maximum SNR of the imaging system occurs at this stage. If the SNR of the imaging system is essentially determined there, the system is said to be *X-ray quantum limited* in its performance. Generally, the SNR is reduced as the signal passes through the imaging system because of losses and additional sources of fluctuation.

To avoid losses at subsequent stages, the detector must provide adequate quantum gain g_1 directly following the initial X-ray interaction. Stages II and III illustrate respectively the processes of creation of many light photons from a single interacting X-ray (referred to as conversion gain g_1) and the escape of quanta from the phosphor with mean probability g_2. Light absorption, scattering and reflection processes are important. Further losses occur in the coupling of the light by the eye lens that transfers light to the retina

(Stage IV) and in the spectral sensitivity and optical quantum efficiency of the retina (Stage V). Since the conversion gain of the phosphor is insufficient to overcome these losses, and the number of retinal impulses falls below that at the primary quantum sink, a *secondary quantum sink* is formed. The statistical fluctuations at this stage become an important additional noise source. This is especially important when the spatial-frequency-dependent DQE is analyzed. Figure 2.27, shows that if an XRII is used and by electron optical amplification the brightness of the image is maintained, then the number of quanta representing the image at all stages of the imaging chain exceeds the number of quanta at the primary detection stage. Thus for non-intensified fluoroscopy, the appearance of noise in the image would be much larger than the intensified system despite the same primary quantum sink.

2.3.1.3 Resolution

The resolution of an image depends on the resolution of the sensor, the X-ray tube focal spot size and magnification, and effects due to motion. In a digital imaging system, there may be further loss of resolution due to inadequate sampling and aliasing. For linear systems, the loss of resolution from each component can be combined using the concept of modulation transfer function MTF.

The final MTF(f) is the net result of all the blurrings shown in Figure 2.28 for an XRII.

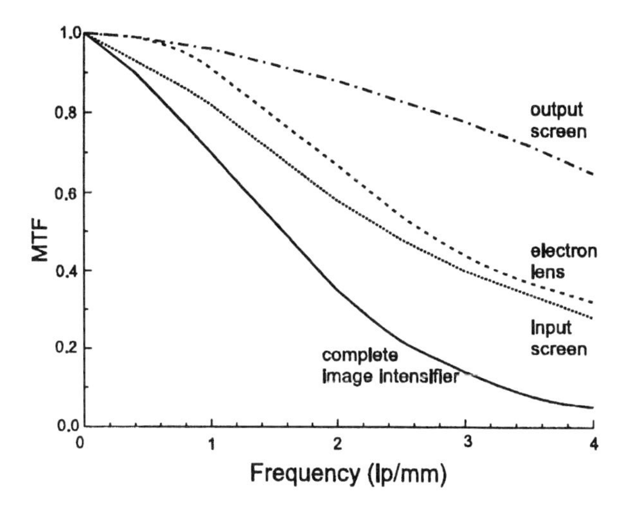

Figure 2.28. Combination of many MTFs due to different resolution loss effects within an XRII.

Multiplying each component MTF at a given frequency yields the system MTF. In contrast, obtaining resolution in the spatial domain, rather than spatial frequency domain, requires a convolution of the individual components. This is a much more difficult mathematical operation and is the reason that the frequency domain description is preferred.

2.3.1.4 Motion

Motion in X-ray imaging is a problem because the finite response time of a sensor and the duration of the X-ray exposure can cause blurring. The amount of blurring that can be tolerated in a system varies as does the amount of motion expected in different parts of the body. The motion MTF(f) in the spatial domain can be related to the time response $R(v)$ of the sensor in the temporal frequency domain by considering that a sine grating of spatial frequency f moving at velocity V will create a temporal modulation of frequency v at a point:

$$v = f\,V \tag{2.8}$$

Therefore:

$$\mathrm{MTF}_M(f) = R(v) \tag{2.9}$$

2.3.1.5 Dynamic Range

Dynamic range is often misunderstood as it has many uses and meanings. Every image contains points of different intensities, otherwise it would not be an image. The dynamic range within an image is the range of intensities to be expressed. Thus, dynamic range is a property of the imaging task rather than the sensor. However, the sensor must correctly represent the available image information. To do this, the image must be correctly exposed, *i.e.* the maximum and minimum of the image must be matched to those of the sensor. The dynamic range of a sensor is limited at low exposure by irreducible external noise; at high exposure image sensors are limited by saturation whereby further response is paralyzed.

In practice, the required dynamic range for an imaging task has two parts. The first describes the ratio between the X-ray attenuation of the most radiolucent and most radio-opaque paths through the patient to be included on the same image. The second is the precision of the X-ray signal to be measured in the image part representing the most radio-opaque anatomy. If there was a factor of 50 in attenuation across the image field and 1% precision in measuring the signal in the most attenuating region was needed, then the dynamic range requirement would be 5000. Thus, the dynamic range requirements for certain applications can often exceed the capabilities of available detectors.

Although the requirements for dynamic range differ between imaging tasks, some general principles for requirements of each modality may be established. First it is important to recognize that the X rays are attenuated exponentially. Thus an added half-value layer thickness of tissue will attenuate the beam to ½ of its original value, while the same half value thickness omitted will increase the X-ray intensity by a factor of 2. A mean exposure value X_{mean} for a system is established by irradiating a uniform phantom and factors above and below this mean value are of interest. For example, in fluoroscopy generally a range of 1–100 is useful which corresponds to a minimum of 1/10 X_{mean} and a maximum of 10 X_{mean}.

2.3.2 X-ray Tube and Generator

In Chapter 1, alternatives to conventional X-ray tubes were discussed. In medical radiography, the X-ray tube is the only device currently used and its parts relevant to the imaging task will be discussed[13]. The focal spot size is important for system resolution. Also related to the X-ray tube and the generator is the blurring due to motion through the inability of the tube to output radiation at an arbitrarily high rate. The track on the anode of the tube limits the instantaneous power that the tube can absorb and hence controls the instantaneous X-ray output. The bremsstrahlung process is so inefficient that most of the power delivered to the tube is wasted as heat that then has to be absorbed. Three limits to the X-ray output of a tube can be identified. First, the instantaneous output is limited by the ability of the target track to absorb power without vaporizing. This power limit is enhanced by rotating the anode during the X-ray exposure and by angling the target. The focus can then be physically long, yet from the direction of use, be foreshortened and appear symmetrical. The second limitation is the amount of energy that can be absorbed by the anode without melting. The material tungsten is usually used and is capable of being heated to incandescence without damage. The anode is cooled by radiating heat to the tube housing. The housing heat capacity provides the third limitation to the output of the tube. These three limitations demand consideration in the daily use of the tube. The design of tubes are optimized for particular tasks and improvements are possible. The heat capacity of the anode can be increased by adding graphite, a material with a very high heat capacity to mass combined with the ability to be heated to very high temperature. The heat absorbing ability of the housing can be improved by cooling, using circulating oil or forced air.

The generator must provide power to the X-ray tube in a co-ordinated manner to satisfy the requirements of imaging. Three separate electrical inputs are required by the tube: current to the filament; high potential to the anode; and alternating current to drive the rotor connected to the anode. The tube requires an appropriate amount of current to the filament so that the temperature is sufficient to provide an adequate electron tube current by thermionic emission. Very large potentials (up to 150 kV) are required across diagnostic X-ray tubes. However by use of a grounded centre-tapped transformer the anode and cathode of the tube do not have to exceed 75 kV from ground, thus reducing by two the insulation requirements of the shockproof cables. The cathode cable also supplies the filament current.

The generator is calibrated to the tubes connected to it so that an individual exposure beyond the capabilities of the tube is not permitted. For systems with the capability of a series of exposures, the same applies. However, there are often no generator limitations to prevent individual exposures adding up to a damaging level. There is an obligation on the part of the operator to be aware of the tube limitations and avoid over stressing it. Such limitations are often reached in situations where testing is in progress and may damage or destroy a tube. The generator also co-ordinates the timing and operation of other imaging components in the X-ray examination room. For example, when the X-ray exposure control is first depressed, a *preparation* signal is sent to the generator which starts the rotation of the anode and increases the filament current until its temperature is appropriate to provide the required space charge limited tube current. After the high voltage is applied to the tube, which occurs after the X-ray exposure control is depressed to the exposure level, the exposure will be made unless it is blocked by one of many further interlocks that are present in most X-ray rooms. Examples are the Bucky grid set in motion by the preparation signal that must be in exactly the right position before the exposure is triggered; or when the system is coupled to a cine camera that must be in a particular state before the exposure is allowed. After the exposure is started, it may be terminated by releasing the hand trigger by the operation of a phototimer or by the operation of a simple timer.

2.3.3 Grids

Grids and air gaps are the only practical means to reduce scatter in area images. The most common type of grid uses a solid material spacing as shown in Figure 2.29. In using a grid, the primary X-ray radiation must have a known geometrical relationship between the focal spot and the detector for every point on the image. Scattered radiation generated within the body does not have such a relationship. It is therefore possible to create a grid structure that preferentially allows primary radiation to reach the detector while simultaneously cutting off scattered or secondary radiation. Conventional grids are linear, *i.e.* made up of straight flat strips of lead separated by aluminum so that they are unselective in the direction parallel to the strips. The selectivity S of a grid is defined as the ratio $S=T_P/T_S$ (where T_P is the primary transmission and T_S is the scatter transmission) and should be as high as possible. T_P is limited by the geometric space between the absorbing septa and the absorption of the aluminum inter-spacing. The value of T_S is related to the *grid ratio* R_G which is the ratio between the height to the separation of the lead septa. Typical values are 2–12. The *contrast improvement factor* K is the ratio of the contrast with and without the grid and values are in the range

1.5–3.5. The *Bucky factor* B_F is the ratio of patient exposure without and with the grid *in order to obtain the same film density*. Typical values are in the range of 2–3. Thus, a grid is not a perfect device.

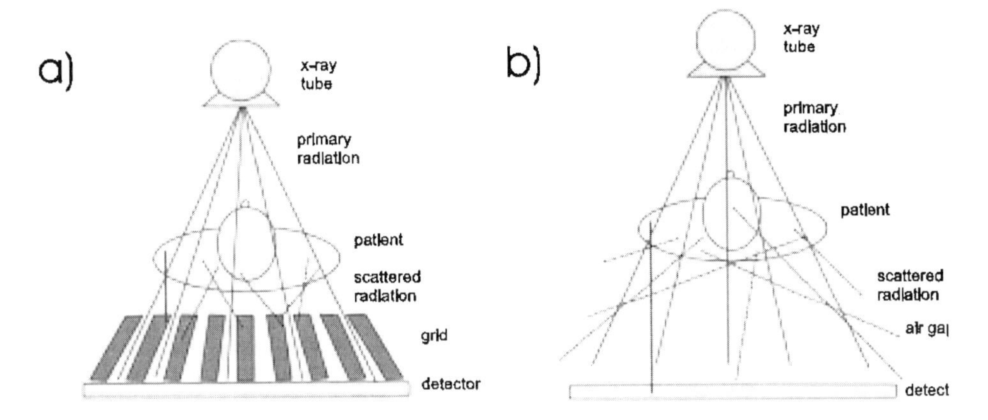

Figure 2.29. Scatter reduction methods: (a) conventional grid, and (b) air gap.

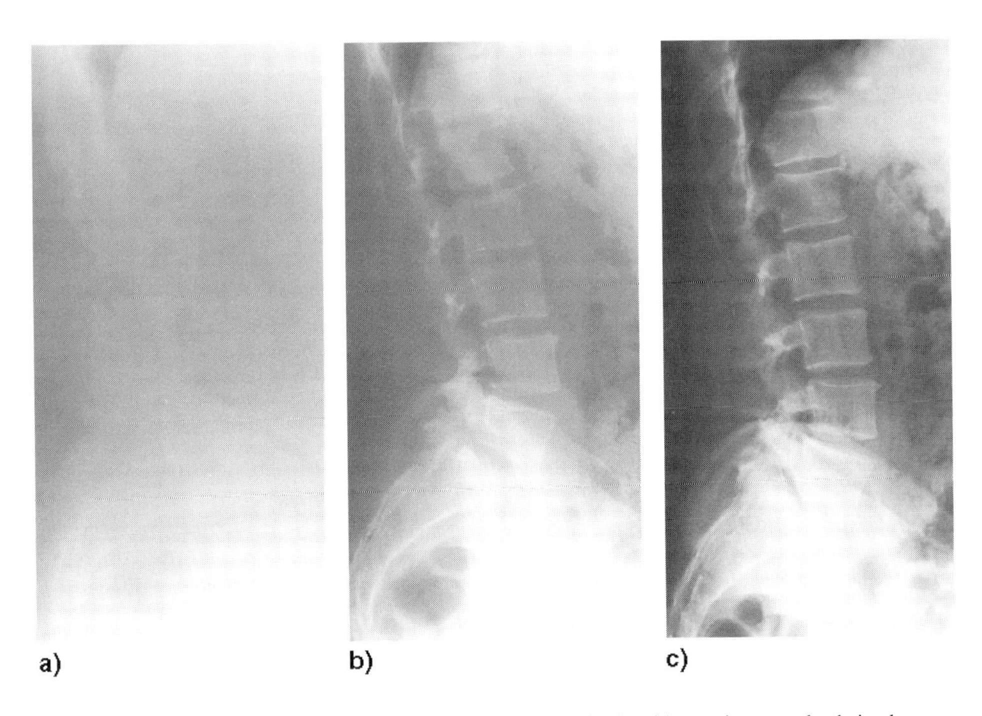

Figure 2.30. Images of a lateral view of the lower spine obtained by various methods in the presence of scatter and after the removal of scatter: (a) image obtained without any scatter reducing device in place, (b) scatter reduced using a conventional grid, and (c) scatter practically eliminated using an air-spaced scanning grid. (Images courtesy of G.T. Barnes.)

Figure 2.30 shows scatter corrupting a clinical image and the comparative success using a conventional and a more advanced air interspaced scanning grid. This demonstrates that a grid is essential to visualize useful information when body parts as thick as the spine are imaged radiographically. Also it shows that conventional grids are far from ideal. Although the air spaced grid is superior, it is not generally used. For all grids, there is a need to eliminate visualizing the grid structure within the image. Motion of the grid perpendicular to the grid lines during the X-ray exposure is the method usually used to accomplish this.

2.3.4 Film-Screen Combinations

The most commonly used image sensor in radiography is the film-screen combination consisting of a cassette with phosphor screens. The cassette can be opened to allow a film to be inserted.

When closed, the film is kept in close contact with the screen, or more commonly a pair of screens, facing toward the film as shown in Figure 2.31.

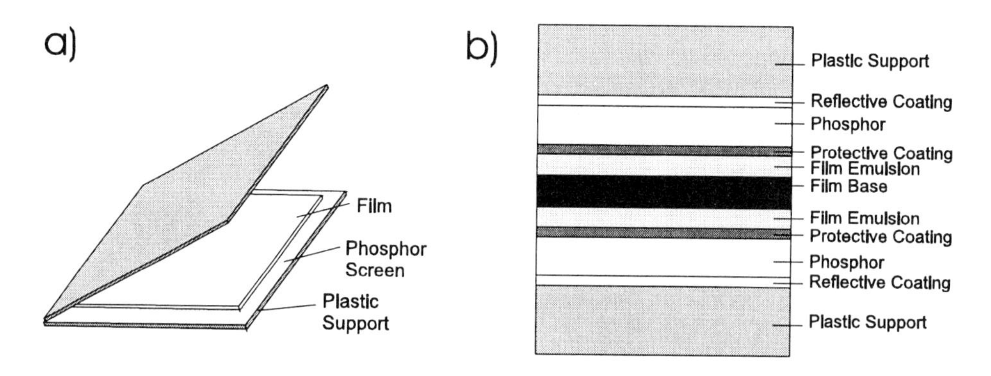

Figure 2.31. Film-screen: (a) opened cassette to show placement of film and position of screens, and (b) cross-sectional view through a dual screen system used in general purpose radiography.

X rays are incident from above and must pass through the front of the cassette before reaching the screens. When the X rays impinge on a screen, light is emitted and is transferred to the screen surface where it enters the film. In Figure 2.31 (a), a cassette and film are shown in the open condition, a single screen is in contact with a single emulsion on one side of the film, or as shown in Figure 2.31 (b), two screens face the film which has two emulsions, one on either side of the film base. In advanced film design, an anti-halation layer is placed between the two emulsions. During X-ray exposure, the anti-halation layer is opaque and prevents light crossing over from one emulsion to the other, thus reducing blurring. The anti-halation layer is removed rendering the film transparent for viewing during the film development. Key factors in the success of a film-screen cassette are: first, excellent contact between the screen and the film to prevent blurring or loss of light, and second, the front surface of the cassette must be easily penetrated by X rays and yet not cause scatter. Often at the back of the cassette is a lead layer to prevent X-ray back-scatter. Film-screen combinations are not very efficient at absorbing X rays because there is a trade-off between resolution and efficiency. Only if it is acceptable to have a relatively blurred image is it possible to use a screen thick enough to be efficient in absorbing X rays. High resolution film-screen combinations absorb no more than 10 or 20% of the X rays whereas high efficiency general purpose screens may absorb ~ 30–40% of the radiation incident. Since the X rays pass through the film on their way to the screen, some exposure will be developed due to direct interactions of X rays with the film emulsion. However the attenuation of films compared to screens is so small that this can usually be ignored.

2.3.5 Computed Radiography

Computed radiography CR is a digital X-ray system using photostimuable phosphor screens. As shown in Figure 2.32, CR screens are exposed in a cassette, similar to a film-screen cassette, but with only one screen in the cassette at a time. This reduces the absorption efficiency compared to a dual screen combination in a film-screen cassette and is a disadvantage of CR. In practice, depending on the laser intensity, the readout of a stimulable phosphor plate yields only a fraction of the stored signal. This is a disadvantage with respect

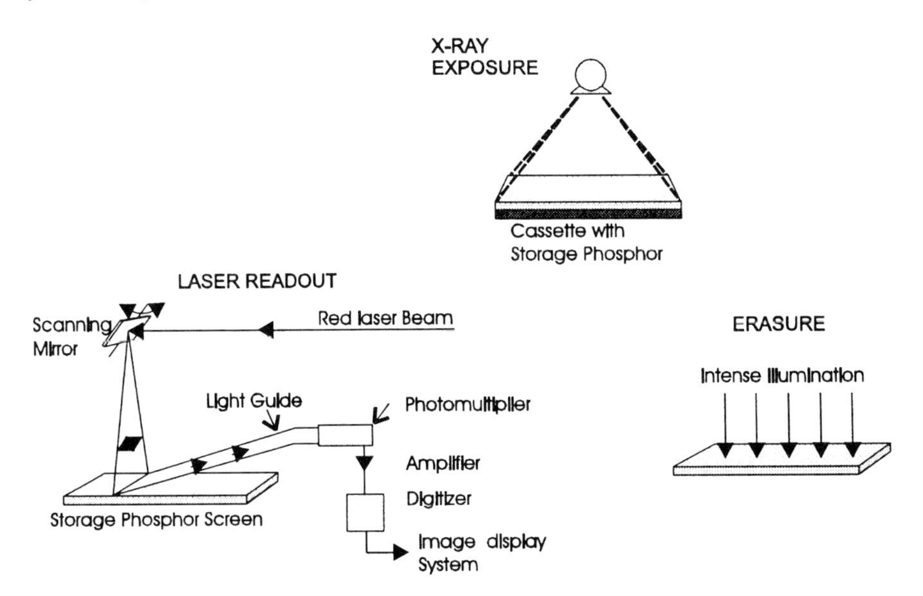

Figure 2.32. Computed radiography system based on the use of reusable photostimulable phosphor plates housed in cassettes.

to sensitivity and readout noise. However, it can be helpful by allowing the plate to be *pre-scanned* – *i.e.* read out with only a small part of the stored signal, to allow automatic optimization of the sensitivity of the electronic circuitry for the main readout.

The readout system for photostimulable phosphor plates uses a red laser beam scanning system to stimulate the screen on a point-by-point basis, exciting blue light from the screen. The blue light is collected by a light guide that is a critically important component in the avoidance of a secondary quantum sink, *i.e.* the light is funneled to a photomultiplier tube that detects and amplifies the signal. The advantages of stimulable phosphor systems are that they are digital systems with a very high dynamic range enhanced by pre-scanning.

2.3.6 X-ray Image Intensifier

The XRII is a crucially important component of real time X-ray imaging systems (*i.e.* fluoroscopic) and digital subtraction angiography systems, and a complete XRII is shown in Figure 2.33. From the diagram it can be seen that the XRII is a vacuum tube device[14,15]. X-rays are converted to fluorescent light in the large input phosphor screen usually of 5″ to 16″ in diameter. The fluorescence illuminates a photocathode (evaporated directly onto the phosphor) and frees electrons into the vacuum space. The electrons are accelerated through a potential difference of ~25 kV and electrostatically focused by the electrodes onto a small (~2.5 cm diameter) output phosphor. The light produced by the output phosphor is then optically coupled to a video camera or other optical devices such as cine cameras or small format (photofluorographic) cameras.

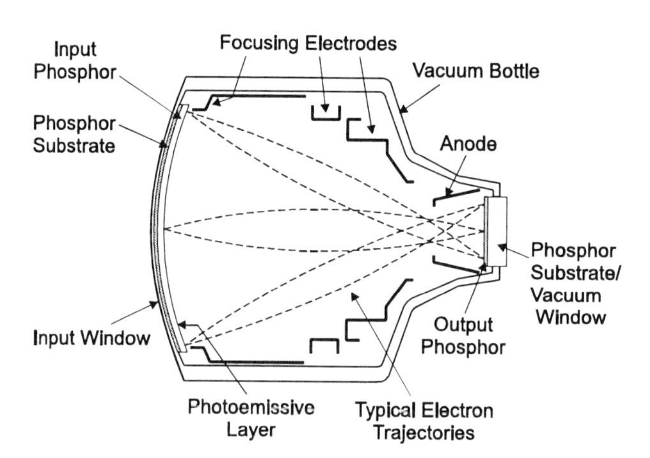

Figure 2.33. X-ray image intensifier. (Reproduced from Reference 22, with permission).

The input window is necessary to preserve the vacuum within the XRII, but causes two problems: it absorbs X rays, thus reducing the quantum efficiency and scatters X rays causing loss of contrast due to *veiling glare*. The function of the input phosphor is to provide a high value of η and g_I and to convey the light in as sharp an image as possible to the photocathode. The phosphor CsI(Na) is universally used for this purpose – its relatively high Z and high packing density lead to high η in the diagnostic energy range. The unique advantage of CsI is that it can be prepared (by evaporation and subsequent heat treatment) in the form of fibre optic light guide. The purpose of the photocathode is to efficiently convert light photons to electrons. The greatest efficiency is obtained when the spectral sensitivity of the photocathode is matched to the phosphor spectrum. Thus, CsI(Na) which has an output spectrum centered on 420 nm is always used. The photocathode is extremely sensitive to contamination so it has to be made *in situ*, within the otherwise completed XRII after a very high vacuum has been established. The vacuum has to be maintained continuously thereafter. An important factor in achieving gain within an XRII is the 25 keV of energy each electron released from the photocathode receives from the electrostatic field before striking the output phosphor. An additional gain process arises from the electron-optical demagnification, *i.e.* reduction in the size of the output image with respect to the input. The magnification or *zoom mode* of any XRII can be changed by appropriate modification of the electrode potentials. The function of the output phosphor is to convert the incident electron image to a visible light image. It should do this with the greatest possible efficiency and with the least blurring.

There are great advantages inherent in the design of the XRII. First, the intimate coupling of the phosphor and photocathode provides much higher collection efficiency of the light than with lens or fibre optic methods. Secondly, the collection and focusing of emitted electrons by the electrostatic field is also very efficient and these two factors outweigh the inefficiency of the photocathode. Thirdly, the acceleration of the electrons within the tube provides a high gain which more than compensates for subsequent losses in the imaging system. Finally, the demagnification of the image in the tube allows efficient lens coupling to the next stage. The optical image formed on the output phosphor can then be coupled to other optical devices and to understand that, the optical distributor must be examined.

2.3.7 Optical Distributor

The principle used in the optical distributor is that if light is focused to a parallel beam, then the path length between one lens and another can be made of arbitrary length. Thus the first component of the optical distributor is the collimator lens on the output of the image intensifier that takes the output image and collimates it, *i.e.* makes it into a parallel beam, so that it can be directed by mirrors into another camera lens focused to infinity. By interchanging mirrors the image can be directed from one optical device to another. By use of a partially-silvered mirror, it is possible to guide light to two separate imaging systems simultaneously. For example, if the majority of the light is sent to a small film camera, the remainder can be diverted to a video camera to verify that the image was correctly acquired.

2.3.8 Video Camera – CCD vs. Vidicon

A video camera changes a visible light scene into an electrical version of the same scene. In medical fluoroscopy, the video camera is optically coupled to the output of an XRII. Characteristic of all videofluoroscopic images is the dominance of X-ray quantum noise made possible by the large gain of the XRII. The development of all video camera tubes and the emergence of the dominant vidicon tube is discussed by McGee[16]. Solid state sensors (CCDs) made using high purity silicon integrated circuits[11] are starting to be used in limited applications. It can be expected that CCDs will become more important in the future. All modern video sensors store charge representing the image at each pixel *continually* and over the whole active area *simultaneously* (*i.e.* the pixels sense light in parallel). Parallel *detection* is essential to a sensitive camera, but, a practical system is usually *scanned* in pixel serial form.

The pattern with which the camera target is scanned is called the *raster*. Four different features of the raster format can be identified: aspect ratio; number of active lines on the screen; update or frame rate (*i.e.* how often the image is completely rewritten); and interlace type (interlaced 2:1 or 1:1, also called non-interlaced or progressively scanned). If video systems were specifically designed for videofluoroscopy then a square image format or *aspect ratio* would surely have been used to best fit the inherently circular form of the image from the XRII. However, broadcast standards and perhaps just the availability of monitors has dictated the essentially universal acceptance of the broadcast standard of 4:3 (4 wide by 3 high) aspect ratio in fluoroscopy. The number of lines used to create the image is important to maintaining resolution and also in maintaining a believable image, free from horizontal lines. The number of lines quoted for any video standard is the total number including some which are purposely blank so as to allow time for the beam to get back to the beginning or *retrace*. For example in a 525 line system, there are ~483 *active* or displayed lines.

To produce an image sequence that is fluid in motion (*i.e.* free from jerkiness) requires a frame rate greater than ~24 frames per second. However, a higher refresh rate is necessary to avoid image flicker, to which the human eye is very sensitive. It was established early in the history of television, that refreshing the image at > 50 Hz avoided visible flicker. To maintain a reasonably low video bandwidth, a compromise was made that was to only refresh the entire image every 1/30 s (*i.e.* 30 frames/s). Thus to avoid flicker the image is displayed in two fields, each taking 1/60 s (*i.e.* 60 fields/s). One field consists of all the odd lines with the other field displaying the even lines. This 2:1

interlace standard has been universally adopted in videofluoroscopy and the two fields combined are shown in Figure 2.34. It has been found that static images place more demanding requirements on the refresh rates than moving images. Fortunately, the deficiencies of these historical decisions can be overcome using digital systems that can store the previous field and rapidly redisplay it in conjunction with the next field. This procedure is called *de-interlacing*. The number of lines can also be increased by digital interpolation, a procedure called *up-scanning*.

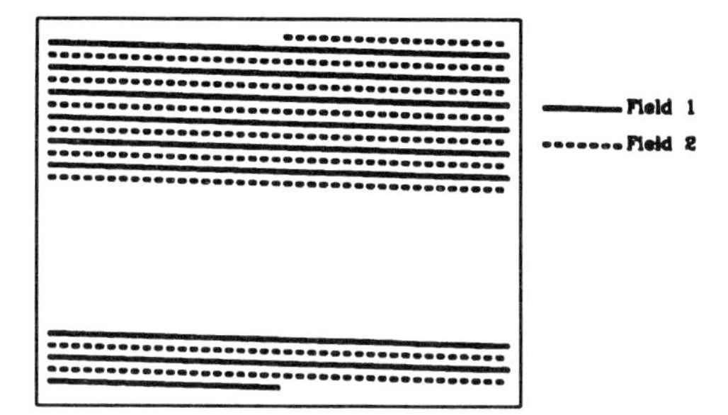

Figure 2.34. Two separate fields of an interlaced raster combined into a complete image. (Reproduced from AAPM Medical Physics Monograph Number 20 "Specification, acceptance testing and quality control of diagnostic x-ray imaging equipment", AIP, 1994, with permission).

Figure 2.35 illustrates the basis of operation of a vidicon tube camera. Vidicons are vacuum tube devices with an electron beam arranged to scan a target and continually return the free surface to ground. The other side of the target is maintained at a potential V_a through a transparent electrode. Thus V_a appears across the target layer giving rise to the electric field E. The target is photoconductive: *i.e.* an excellent insulator in the dark, but readily allows the passage of charge carriers (electrons and holes) freed by the action of light. The electric field E, within the photoconductor causes electrons to drift to the

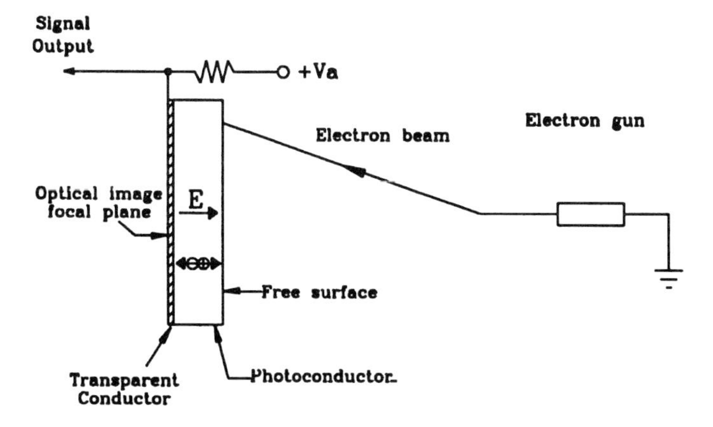

Figure 2.35. Basis of operation of the optical vidicon camera tube. (Reproduced from AAPM Medical Physics Monograph Number 20 "Specification, acceptance testing and quality control of diagnostic x-ray imaging equipment", AIP, 1994, with permission).

transparent electrode while holes are drawn to the free surface. Thus a latent charge image forms on the free surface of the illuminated target. Readout of the latent image is performed by the scanning beam as it supplies electrons to return the surface of the target to ground. During the time the beam dwells on a particular pixel, a charge equal to the latent image charge at that pixel flows from the beam to the target. A preamplifier connected to the target forms the video signal from this current. If the scanning is not absolutely uniform, then geometrical distortion and signal distortion (shading) can result. In a well-designed vidicon camera the only added noise is amplifier noise, most of which arises in the first stage of the preamplifier. Amplifier noise is *triangular* or peaked at higher spatial frequencies. Thus X-ray noise is dominant at low spatial frequencies, while the amplifier noise is dominant at high frequencies.

A block diagram of the electronics[17] controlling a vidicon video camera is shown in Figure 2.36.

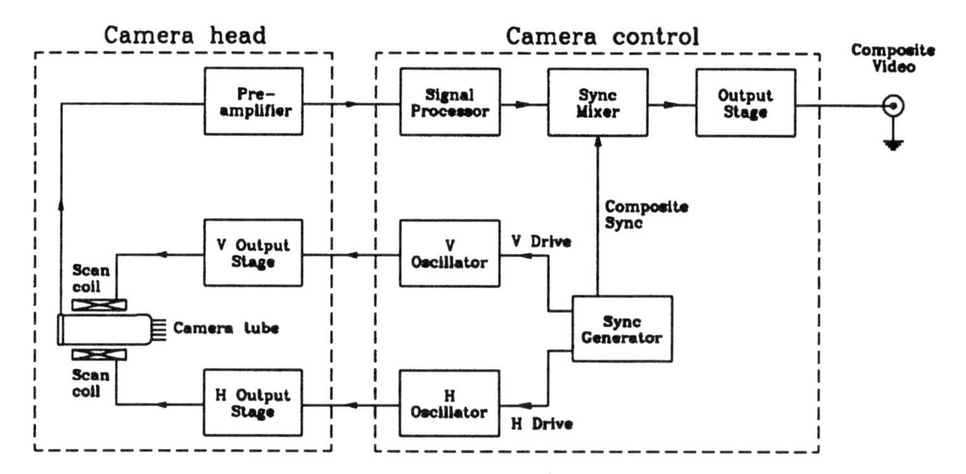

Figure 2.36. Block diagram of a typical camera used in fluoroscopy. (Reproduced from AAPM Medical Physics Monograph Number 20 "Specification, acceptance testing and quality control of diagnostic x-ray imaging equipment", AIP, 1994, with permission).

It performs two major functions. The first is achieved by analog circuits that carry the signal current from the camera tube target. The second is performed by circuits that support the operation of the tube (*e.g.* the potentials and currents necessary to generate and deflect the electron beam as well as the digital synchronization signals to co-ordinate the image scanning).

The operation of the CCD camera is similar to that of the vidicon except the readout is accomplished across the target surface rather than through a vacuum space. The cameras are also similar in terms of components handling analog signals co-ordinated by a synchronization circuit. However, rather than analog scan circuits, the CCD camera has shift registers and clocks working at the pixel frequency. Solid state cameras can therefore be much more compact than those for vacuum tube vidicons.

2.3.9 Video Monitor

The video monitor converts information from electronic into optical form. To analyze the monitor, a means of converting the optical form back into an electronic form is needed. Such test equipment is generally lacking in radiological situations. This is unfortunate since the monitor is possibly the single most important link in the imaging chain. Figure 2.37 shows a conceptual view of the key component of the video monitor, the CRT[17], which is a vacuum tube device. An electron beam is formed in the electron gun and under the control of magnetic scanning coils scans an image in a raster. The intensity of the beam impinging on the output phosphor is controlled by the grid potential. The intensity of light given off is dependent on the current landing on the screen. The relationship between the light intensity I and the video potential applied to the grid V is called the *characteristic curve.* Characteristic curves for various settings of the brightness control for a typical monitor are shown in Figure 2.38. The contrast control changes I for a given V, *i.e.* it acts as a gain control. The brightness control adjusts the I when V=0. Which settings are optimum? The straight line curve, obtained at low brightness, is the characteristic used for broadcast television. The slope γ on a log-log characteristic curve (*i.e.* $I = V^{\gamma}$) is often used to characterize the curve. This is the γ of the CRT and varies between 2 and 3 depending on the design of the CRT. However monitors are not used in this manner in fluoroscopy. The characteristic shown as medium brightness is preferred – a dynamic range of 100 to 1 in the input X-ray image coincides with the approximately 100 to 1 optical output possible with the monitor that is within the range of the human eye's acceptability at a given accomodation level. At higher brightness, this mapping can no longer be achieved. The maximum available light level from a monitor is ~ 120 cd/m^2. This compares with ~ 1 000 cd/m^2 from an ordinary light box and the more than 10 000 cd/m^2 from a *bright light.* The ambient

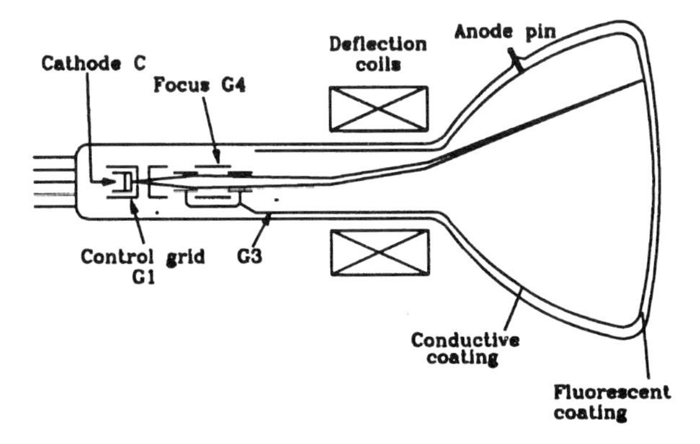

Figure 2.37. The operation of a cathode ray tube or CRT, the principal component of a video monitor. (Reproduced from AAPM Medical Physics Monograph Number 20 "Specification, acceptance testing and quality control of diagnostic x-ray imaging equipment", AIP, 1994, with permission).

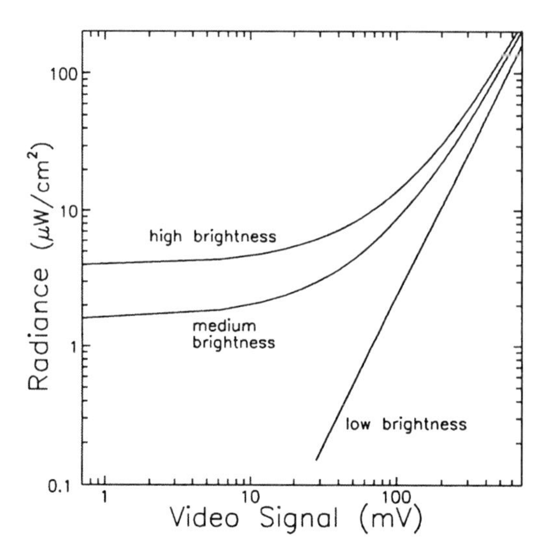

Figure 2.38. Characteristic curves of monitor depending on setting of brightness control. (Reproduced from AAPM Medical Physics Monograph Number 20 "Specification, acceptance testing and quality control of diagnostic x-ray imaging equipment", AIP, 1994, with permission).

light adds a bias and reduces the output dynamic range of the monitor as well as modifying the contrast scale. The presence of ambient light on the face of the monitor is superficially similar to increasing the brightness control on the monitor. However, it differs in detail and in Figure 2.39 the characteristic curve of a monitor in the presence of ambient light is plotted. With ambient light, the monitor characteristic knee is more abrupt, reducing the effective dynamic range of the monitor compared to the best case in Figure 2.38. Therefore subdued ambient illumination is required for maximum visibility of details over the required X-ray dynamic range.

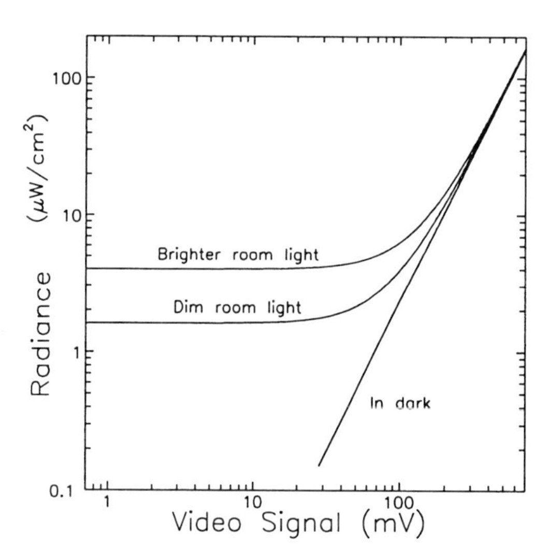

Figure 2.39. Characteristic curves of monitor depending on ambient light level. (Reproduced from AAPM Medical Physics Monograph Number 20 "Specification, acceptance testing and quality control of diagnostic x-ray imaging equipment", AIP, 1994, with permission).

2.3.10 Multiformat Camera

A multiformat camera is used to obtain a hard copy of an image appearing on a video monitor. Photographing the output of an image on a video monitor is difficult because the electron beam is scanning continually and operation of the shutter at an inappropriate time causes artifacts on the film. The multiformat camera solves this problem by synchronizing the opening and closing of the shutter to the retrace period between frames. At the same time it juxtaposes many images on a single piece of film. Often 14 ″ × 17″ film is used to conveniently fit on the film view boxes ubiquitous in X-ray departments. The quality of images with multiformat cameras is quite high, but not as high as with laser cameras. Since the image does not have to be in digital form – it simply has to be available to be viewed on a monitor – multiformat cameras are often preferable. Video signals can be looped through as many monitors as is necessary. One critical point is that termination of the video cable should be done once and only once to avoid reflections in the cable to avoid ghosts appearing on the images.

2.3.11 Laser Camera

A laser camera enables digital image data to be output to photographic film. In this way, it serves a function for digital data similar to that which a multiformat camera does for analog data. In construction, the laser camera consists of a mechanism to bring the film into the focal plane of a laser scanner. The laser scanner was shown earlier. The disadvantage of a laser camera is that it must store all the images to be printed before it can start to print. Thus it is inherently a more complicated device than a multiformat camera.

2.3.12 Dual-energy Decomposition Algorithms

The human body consists of tissues which attenuate through the photoelectric effect and Compton effects. In principle, a separation of these components is possible because the energy dependence of these two processes are different[4]. Once this separation is established, then the data can be recombined to produce images containing weighted averages of the two processes to represent, *e.g.* the difference between bone tissue and soft tissue or between air and soft tissue. In order to make these separations, X-ray images at different energies need to be obtained. Contrast materials such as iodine with K-edges in the diagnostic range have markedly different attenuation properties below than above the K-edge at 33 keV. In principle, a third component can be separated. However, dual energy decomposition procedure is the norm.

Every sample has two components of attenuation: that due to photoelectric absorption and that due to Compton scattering. Measurements of the attenuation μ of the sample is made at two different energies, *i.e.* $A_T(E_1)$ and $A_T(E_2)$. This can be decomposed into component attenuations from thickness d_{PE} of a pure photoelectric and d_C of a pure Compton attenuators, *i.e.*:

$$A_T(E_1) = \mu_{PE}(E_1) d_{PE} + \mu_C(E_1) d_C \quad (2.10)$$

$$A_T(E_2) = \mu_{PE}(E_2) d_{PE} + \mu_C(E_2) d_C \quad (2.11)$$

If μ_{PE} and μ_C are known as a function of energy, then all factors are known except d_{PE} and d_c and can be solved from these simultaneous equations. However, it is impractical to use monoenergetic beams and thus beam hardening will occur. Fortunately, beam hardening is relatively easy to correct. The decomposition can only be achieved precisely if the data is entirely free of scattered radiation. Using dual energy decomposition a considerable amount of noise is unavoidable due to the great similarity between the energy dependence of PE and Compton effects.

Once the pure PE and Compton scattering components are known, then reconstruction into biologically meaningful components can be performed. For example, a dual energy radiograph through the thorax consists of bone and soft tissue. The amount of each can be established provided the photoelectric and Compton attenuation coefficients are known at the appropriate energies for these tissues. If d_B is thickness of bone (expressed in cm^2/g) and similarly d_{ST} is the thickness of soft tissue then:

$$d_{PE} \mu_{PE} = d_B \mu_{PE}^{B} + d_{PE} \mu_{PE}^{ST} \quad (2.12)$$

$$d_C \mu_C = d_B \mu_C^{B} + d_{ST} \mu_C^{ST} \quad (2.13)$$

This manipulation permits us to create both a bone and a soft tissue image when dual energy decomposition is used. The same concept can be used to calibrate the system using, for example, a highly PE absorber such as Al and a highly Compton absorber such as Plexiglass as the component materials. This is useful in constructing phantoms and is more practical than using real soft tissues and bone.

2.3.13 Tomographic Reconstruction Algorithms

It is often convenient to measure the attenuation coefficient at every point in the human body rather than a projection as is normal in X-ray imaging. In Figure 2.40(a), the first step is to obtain accurate attenuation measurements along many paths through a slice of the body part to be imaged, *i.e.* the basis for computed tomography CT[18]. To simplify this, examine the case of a square object consisting of four elements. If measurements are made from

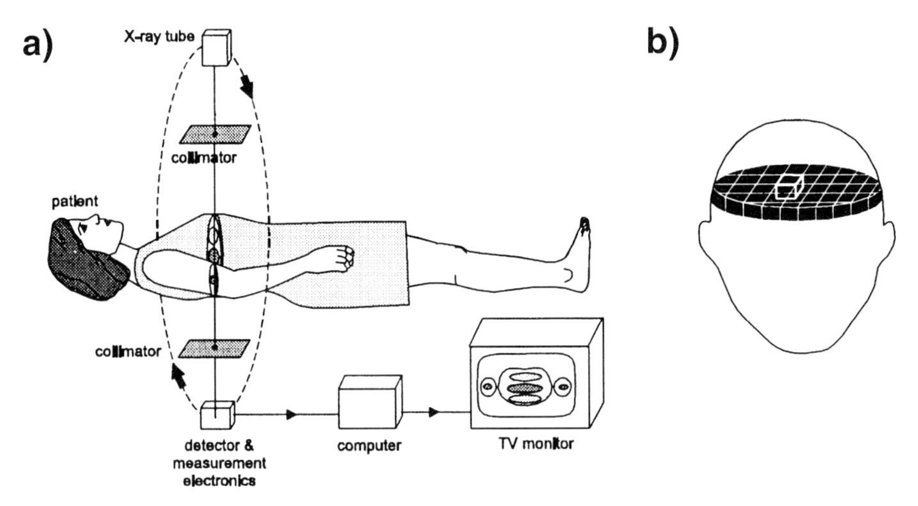

Figure 2.40. Concept of tomographic reconstruction. (Reproduced from Reference 12, with permission).

top to bottom and side to side for the two beams yielding four different measurements, then there are four unknowns – the attenuations of each element (Figure 2.41). A set of four simultaneous equations permits unambiguous isolation of the attenuations. This concept can be readily extended to many more pixels in a slice (or *voxels*) than these four. Many more directions (or *views*) through the body part must be taken to ensure sufficient independent data to permit the reconstruction.

In practice, such algebraic reconstruction methods are not used except in special circumstances, since they are computationally inefficient and require several iterations for adequate accuracy. Thus, the direct convolution back projection method and its variants are almost universally used.

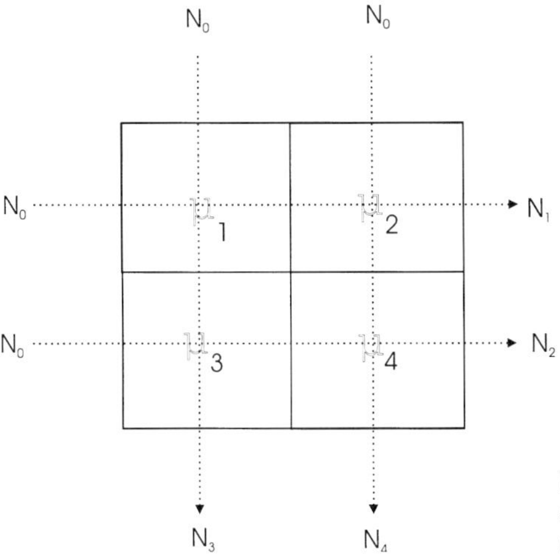

Figure 2.41. Measurement and reconstruction attenuation coefficients of a two by two block of tissue.

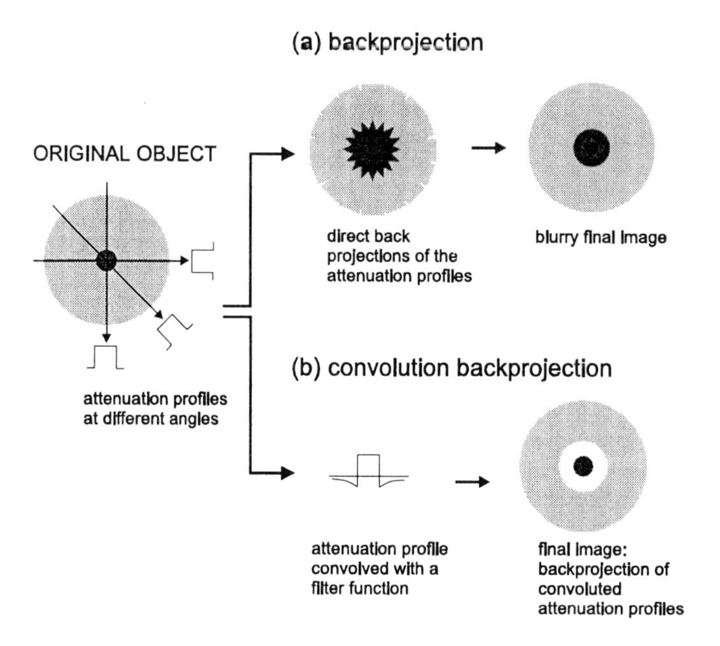

Figure 2.42. Concept of back projection and convolution back projection for computed tomographic CT reconstruction.

The concept of back projection is shown in Figure 2.42. Views obtained at a large number of angles (typically 256 or 512 views for 512 × 512 voxel reconstruction) are projected backward in the computer, *i.e.* the measured attenuation data is smeared back along its path. This process is repeated for all views. The resulting image is disappointing, an accurate but incredibly blurred version of the original object. The point spread function is $1/r$ where r is the radial distance from the object point. In principle, deconvolution of this blurred point spread function should recover the original object. This is achieved by convolving the measurement in the direction perpendicular to the ray as shown in Figure 2.42. This simple procedure results in a computationally robust and unique solution to the reconstruction.

The attenuation value in CT uses a normalized scale based on a reference material (almost universally water). The attenuation value is defined in Hounsfield Units (HU) by the equation:

$$\mu_{rel} = 1000\frac{\mu_{object} - \mu_{water}}{\mu_{water}}. \tag{2.14}$$

That is, the CT numbers for a tissue type are the fractional difference of the linear attenuation coefficient with respect to water, measured in units of 0.1% of the attenuation of water. Thus the CT numbers for many tissues in the body are close to each other. However, the rms noise in the measurement can be quite low (~1–5 HU), thus permitting delineation of different tissue types. This is quite different from projection radiography where the small differences in μ between soft tissue types are impossible to image. In contrast, digital manipulation of the CT image (windowing and leveling) permits an arbitrary selection of CT numbers to be mapped onto the CRT display. On this scale, soft tissues are near zero (*e.g.* blood, liver tumors ~+40 HU, breast and fat –100 to –200 HU, lung –200 to –800 HU, air –1000 HU, and dense bone in the range of +2000 to +3000 HU).

Several corrections to the measured attenuation data are required before an accurate reconstruction, free of artifact, can be obtained. Normalization is a correction of the response of each detector in relationship to the output of the tube. Beam hardening, which arises from the preferential absorption of softer radiation from a polyenergetic beam, causes deviations from the ideal exponential (Beer's) law. This can be corrected by measuring the attenuation of known thicknesses of materials and using a *look-up* table that converts measured transmission to thickness for a particular spectrum without relying upon Beer's law.

2.4 Systems

Having examined the components of complete imaging systems, we can now address the clinical needs for imaging various parts of the body. The single most important X-ray imaging modality is of the chest and will be dealt with first.

2.4.1 Chest Radiography

Radiography of the chest is unique since it is the only X-ray imaging modality where the patient is in the upright posture as shown in Figure 2.43. It is also unusual in using a very large Source (tube) to Detector distance (SID). The convention must be maintained as the appearance of images are greatly modified by change in SID due to relative magnification differences. A possible reason for the large SID is to permit an increase in air-gap to reduce scatter without greatly increasing the image size by radiographic magnification.

2.4.1.1 Film-Screen

The conventional method of imaging the chest is with film-screen radiography and film-screen chest radiography is probably still the single most common kind of medical image. The primary problem of chest imaging is to adequately visualize the lungs, which consist of low density tissue suspended in an air cavity enclosed by the rib cage. Simultaneously other structures appear in the image including the spine and heart (the mediastinum) which have

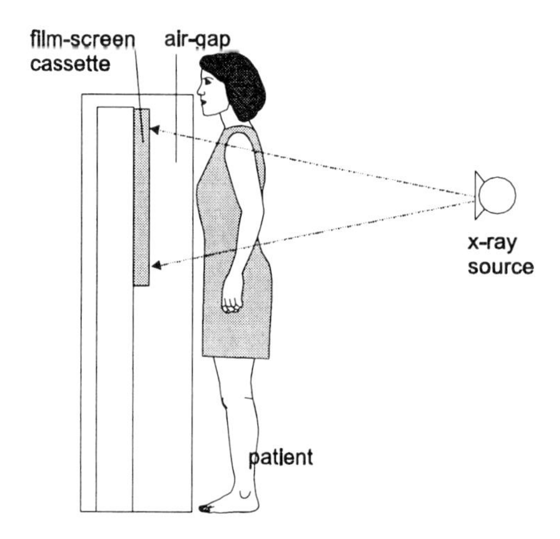

Figure 2.43. Chest radiography geometry.

much greater thickness and therefore much greater X-ray attenuation than the lungs. The dynamic range of the image is therefore higher than can be conveniently imaged on a film-screen combination where the optimum part of the H&D curve only extends over about a factor of ten in X-ray exposure. The scatter in a chest image is a problem. In the lung fields, the scatter to primary ratio is of the order of 2 which can be greatly improved by the use of a grid or air gap; but in the mediastinum, the scatter to primary ratio can be ten or more. Until the 1960's, there was no attempt to image the mediastinum and very low energy X rays were used (60–80 kVp) providing very good contrast in the lung fields but little contrast in the mediastinum. Later, it was realized that the mediastinum could be seen if higher kVp was used. Now it is usual to use 120–150 kVp, which degrades the lung fields but enhances the mediastinum. Grid and air gap application is difficult to optimize as the scatter in the mediastinum, although causing a loss of contrast can actually be beneficial in bringing the film off the toe of the H&D curve where no perceptible blackening occurs. The overall compromise of these factors has not yet succumbed to detailed analysis and thus remains a clinical decision.

A recent innovation that can be used to extend the useful dynamic range of film-screen is to make an asymmetric film-screen pair. For chest imaging, the design concept uses the front screen and film optimum for the lung field, *i.e.* a relatively slow film and a thin, high resolution screen. The back screen and film for the mediastinum uses a much faster film and a thick, blurry screen resulting in a low resolution image of the poorly exposed mediastinum. The use of an anti-halation layer between the film emulsions is then essential to optically isolate the films.

2.4.1.2 Equalization

The purpose of scanned equalization radiography SER is to radiographically reduce the dynamic range of the aerial X-ray image before it is incident on the film-screen X-ray receptor. The contrast of the image is enhanced by ensuring that almost all information is presented at the optimal (steepest) part of the H&D curve. The cost is a loss of broad area contrast, making the image appear rather alien and requiring considerable physician re-education.

The general concept of SER is to modulate the intensity of the beam entering the patient by monitoring and controlling the exit exposure. There is a detector behind the patient which provides feedback to the generator to keep the exposure at that point constant. The appearance of the image is critically dependent on the size of the X-ray aperture that defines the beam. If the aperture is very small, then the equalization of the output intensity results in an image with no contrast whatsoever. A larger aperture results in a region of equalization equivalent to a spatial high pass filter, reducing or eliminating large area variations of transmitted intensity. This is the requirement for chest radiography where the large variation of transmission of the lungs compared to the spine and heart (mediastinum) often overwhelms the dynamic range of film.

The principle of operation of a single beam approach[19,20] to SER is shown in Figure 2.44. The fore collimator defines a rectangular beam which scans the patient; the aft collimator prevents scattered radiation reaching the shielded parts of the film-screen. The X-ray detector, which scans with the aft collimator, measures the radiation transmitted through the patient and feeds back an appropriate signal to the X-ray generator to control X-ray intensity. This single beam approach places a heavy load on the X-ray tube and also requires a relatively long total exposure time. These deleterious factors can

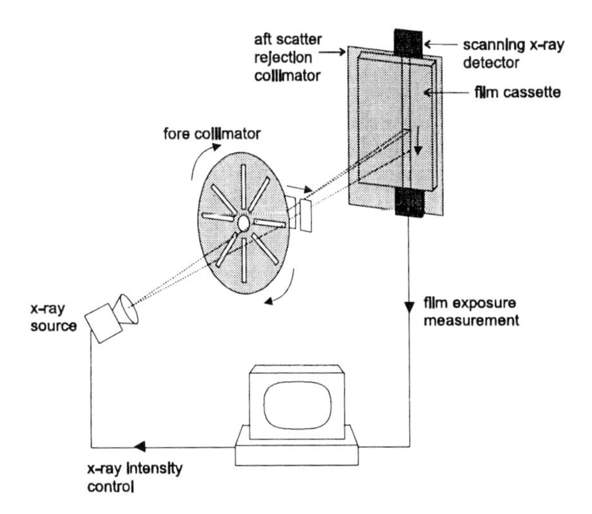

Figure 2.44. Concept of radiographic equalization.

be alleviated using multiple beam approaches where the tube output in a wide slit is modulated by means of movable filters. The disadvantage of this approach is the smaller dynamic range.

2.4.1.3 Computed Radiography

Systems for CR can be of two general types, captive or reusable cassettes. In the former, the receptor and its readout are integrated into the X-ray machine. While this requires a specially-designed machine with higher capital cost, it also eliminates the need for loading, unloading and carrying the cassettes to a separate reader thereby reducing labor costs. The use of a single or a limited number of receptors simplifies the task of correction for non-uniformities of the receptors. A reusable cassette system shown in Figure 2.32 may be advantageous where a high degree of portability is required, such as in intensive care situations or operating rooms, with the added advantage of being compatible with existing radiographic units. Both systems use photostimulable phosphor screens. In the cassette based system, photostimulable screens store a latent image during the exposure phase that is readout with a laser scanner. In practice, these systems are widely used for emergency and bedside radiography where the variable readout sensitivity compensates for under and overexposure problems often experienced with film-screen radiography because automatic exposure control is not feasible in these applications. The newer implementation of stimulated phosphor technology is to eliminate cassettes and integrate all the imaging functions (*i.e.* X-ray exposure, readout and erasure). This is achieved by enclosing the screen within a darkened enclosure as shown in Figure 2.45.

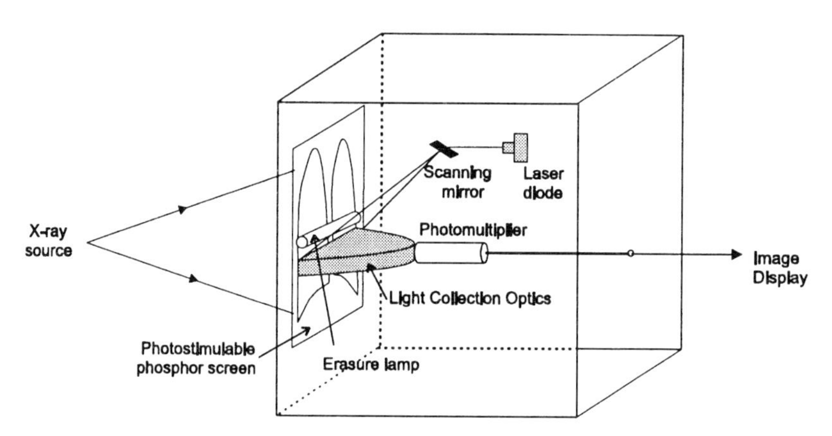

Figure 2.45. Integrated (cassetteless) stimulated phosphor readout system.

2.4.1.4 Other Digital Chest Imaging Methods

There have been recent reviews of digital imaging[21,22]. A recently introduced chest radiography system[23] uses electrometers to readout a latent image charge formed on the surface of a charged amorphous selenium drum shown in Figure 2.46.

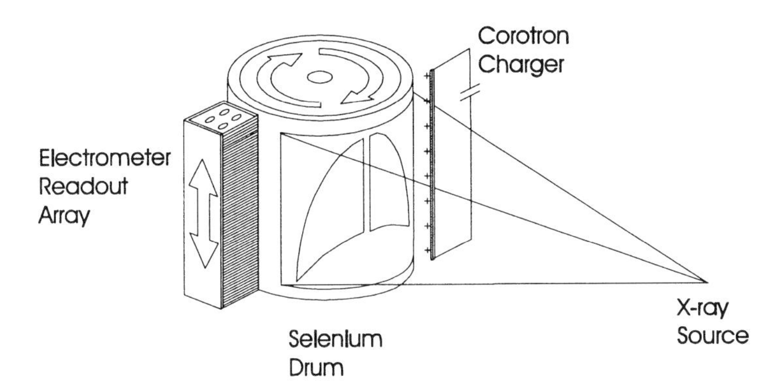

Figure 2.46. Digital chest imager using large selenium drum.

A large metal drum has a thin (0.5 mm) layer of amorphous selenium evaporated onto its surface. The selenium is activated (*i.e.* made sensitive to X rays) by charging its surface to a high potential (~ 1500 V) by rotating the drum under a corona charging system or *corotron*. The drum is then stopped and the X-ray exposure made. A latent image of charge is thus formed on the drum surface. The drum is then rotated under an array of electrometers which scan the image and read it into a computer. A high quality image is available on a computer monitor within a few seconds of the X-ray exposure. This approach satisfies the ideal requirements – high image quality and instant readout but is still bulky, much larger than a cassette. A major image quality problem arises from the interaction of the electrometer separation from the drum surface and image resolution. This requires incredibly fine mechanical tolerances to maintain close separation over the whole area of the image.

A new system was developed in the nineties with the maturing of *active matrix* technology. It promises to solve the bulkiness problem by being compact enough to fit within a cassette and further improve image resolution. With the *direct* system[24], resolution loss on readout can be entirely eliminated by defining pixel electrodes directly onto the amorphous selenium surface. This is achieved by evaporating an amorphous selenium layer on a large area readout structures based on the use of active matrix arrays as shown in Figure 2.47. The active matrix array consists of millions of individual pixel electrodes connected by TFTs (one for each pixel) to electrodes passing over the entire array to subsidiary electronics on the periphery. The TFTs act as switches to control the readout of image charge a line at a time. Other photoconductor materials are under consideration to replace the amorphous selenium layer.

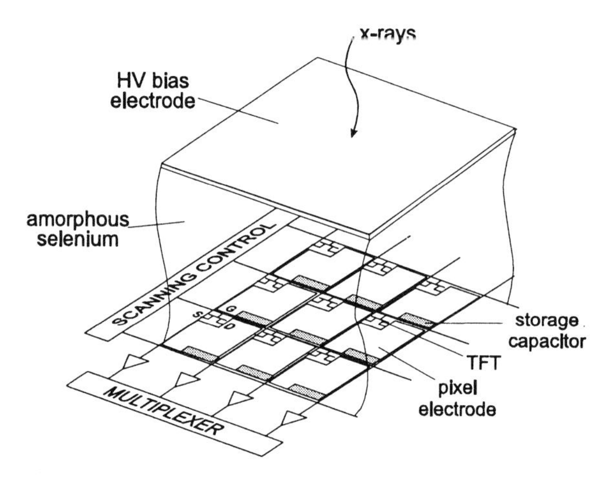

Figure 2.47. Large area flat panel system for digital radiography and fluoroscopy based on a *direct* conversion active matrix readout system using a photoconductor such as amorphous selenium.

In the *indirect method*[25, 26] shown in Figure 2.48 a phosphor layer is used to absorb X rays and the resultant light photons are detected by a large area photo-diode array readout with TFTs integrated onto the plate at each pixel. The individual photodiodes are usually made with amorphous silicon, onto which a conventional X-ray absorbing phosphor, such as Gd_2O_2S, is placed or thallium-doped cesium iodide (CsI:Tl) is grown. The detector pixels are configured as photodiodes which convert the optical signal from the phosphor to charge and store it on the pixel capacitance. The signal is read out similarly to the direct method by activation of control lines connected to the gates of the TFTs located on each detector pixel.

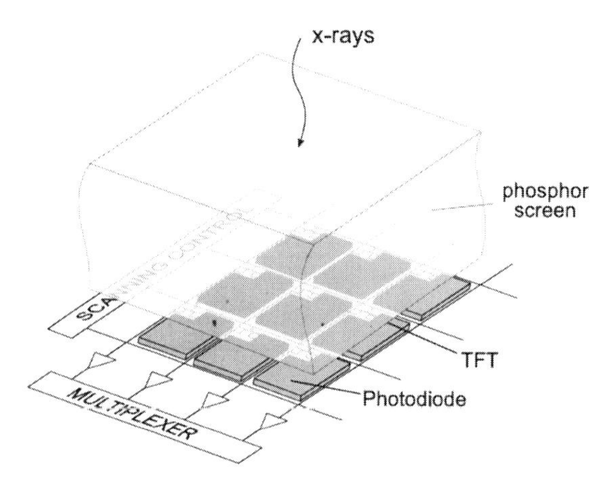

Figure 2.48. Large area flat panel system for digital radiography and fluoroscopy based on an *indirect* conversion active matrix readout system with a phosphor layer used to convert X-ray energy to light that is subsequently readout using an active matrix array of photodiodes and TFT switches.

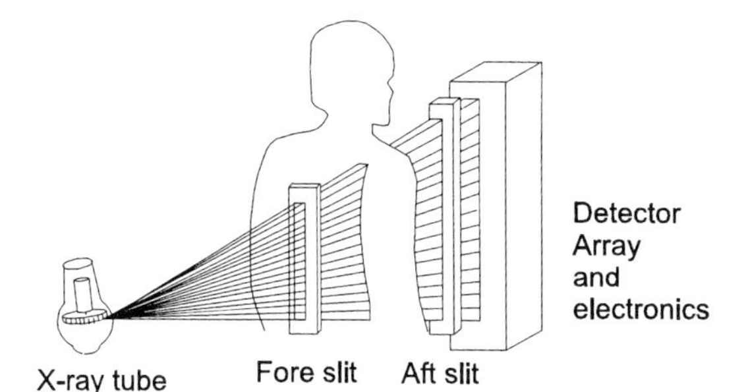

Figure 2.49. Scanning linear detector used for digital chest imaging, including dual energy.

Another approach to digital chest radiography used a scanning linear array[27] as shown in Figure 2.49. This approach, because of the scanning geometry, is very free of scattered radiation and so proves to be an excellent basis for dual energy chest imaging. A dual diode array in which a front detector sensed the lower energy X-ray while the back detector sampled the higher energy X rays. Optimization studies showed that the increase in separation of these two spectral sensitivities was enhanced by introducing a copper filter between the front and back detector. This results in a loss of overall quantum efficiency but modeling confirmed this as the overall optimum. Example chest images obtained using this dual-energy approach are shown in Figure 2.50. These images demonstrate the huge increase of conspicuity possible when the distraction of the structure of the ribs is eliminated from the lung fields.

2.4.2 Fluoroscopy

The imaging chain used in XRII fluoroscopy is shown in Figure 2.51. The XRII absorbs the incident X-ray image, amplifies it in an essentially noise free manner and outputs it as an optical image. The optical image is then *distributed* by lenses and mirrors to one of several possible optical imaging systems, depending on the clinical application. The optical imaging systems of interest are video cameras (used for real-time viewing of X-ray image sequences or *videofluoroscopy*) and small film cameras (used for static imaging on photographic film which is also called indirect radiography or *photofluorography*).

Very high doses can accumulate due to the length of interventional fluoroscopic procedures. This requires that a low radiation exposure rate be used. Nevertheless the image quality must be adequate for visualization. Therefore the system must be X-ray quantum limited. The XRII is the key component that provides sufficient gain to permit imaging limited by X-ray quantum statistics at the extremely low exposure rates typical of fluoroscopy. The resulting real-time images are usually displayed using a video system (conventional or CCD) optically coupled to the XRII. The need for radiographs or stored image sequences during fluoroscopic procedures has previously been satisfied

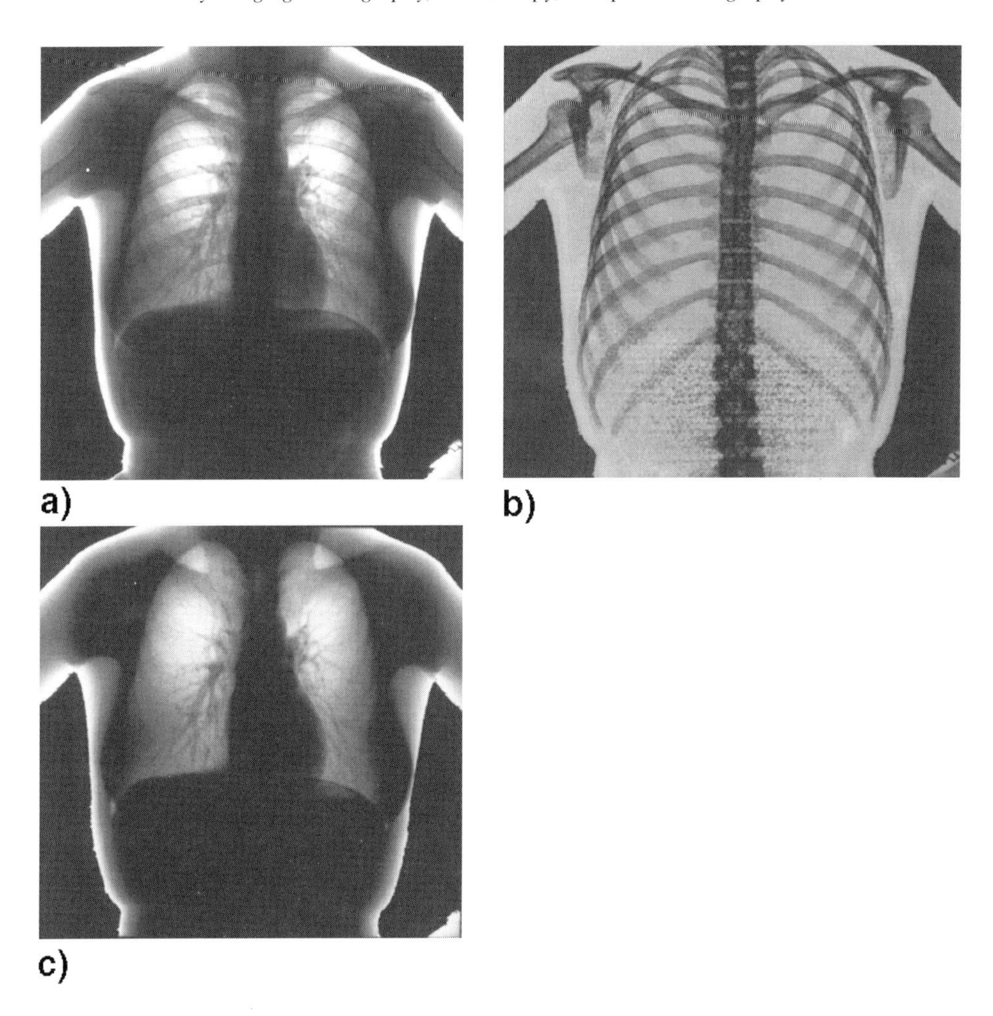

Figure 2.50. Dual energy X-ray images of the chest presented as: (a) original image, (b) bone only, and (c) soft tissue only images. (Images courtesy of G.T. Barnes.)

with optical attachments to the XRII such as small format (or photofluorographic) and movie (ciné) cameras. Recently, instant radiography and instant ciné have been made possible by digitization of the video signal. Although similar to the imaging chain in fluoroscopy, the difference is that the video camera used in digital radiography is capable of integrating the charge from a complete radiographic exposure and digitizing [28,29].

There are several image quality parameters to consider including: noise, lag, MTF, spatial distortion, and the contrast transfer function or characteristic curve. There is no single optimum value for lag in fluoroscopic video systems. Usually the dominant component of lag in such a system is the camera tube. The beneficial action of lag, reducing quantum noise, has to be balanced

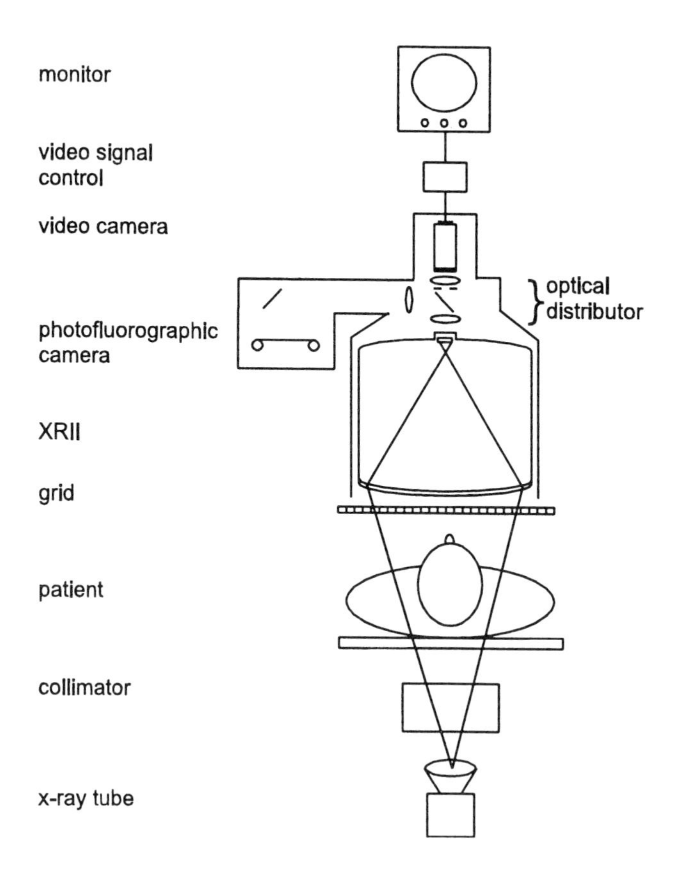

Figure 2.51. Overview of fluoroscopic/fluorographic clinical imaging chain.

against the detrimental effect, image blurring for moving objects. Thus the optimum depends upon the degree of motion in the particular clinical problem. In cardiac work, the least possible degree of lag would be optimum, so that a Plumbicon, Saticon tube or CCD with lag ~ 1 field would be chosen. In bone imaging, a long lag would be beneficial by reducing noise and not introducing blurring as there is essentially no motion of the bones. Thus the Sb_2S_3 vidicon with lag of the order of 8 fields may be suitable. The noise of a camera is triangular – *i.e.* peaked at high frequency. Thus a tremendous reduction in *rms* noise can be achieved by reducing the bandwidth. Visually, high frequency noise is not disturbing, but a small amount of low frequency noise may be extremely irritating. The measurement of *rms* noise without knowledge of bandwidth therefore is not very useful. The slope of the characteristic curve of vidicons, in the straight line region, is γ. The γ for most vidicon types used in videofluoroscopy is unity (the exception is the Sb_2S_3 vidicon with $\gamma \sim 0.7$). If the XRII is in good focus visualization of 1.4 lp/mm in the 9″ mode and 2 lp/mm in the 6″ mode in the horizontal (scan) direction is possible. Non-uniformity of response commonly causes problems. The most common is reduced intensity at the edges of the image or *shading*. Some cameras have circuits built-in to the signal processor to correct shading by modifying gain as a function of scanning beam position. This circuit may also correct similar problems originating in other parts of the imaging chain such as the XRII (shading) or the coupling optics (optical vignetting).

Setting the optical aperture size for the video camera to correctly balance conflicting requirements is an important matter. The optimum setting of the aperture for clinical applications is to attain the widest range around the mean exposure rate E'_{mean} possible within these limitations. If the dynamic range is defined as the ratio of maximum video signal I_{max} to amplifier noise N_A, then for gastro-intestinal videofluoroscopic rooms the mean signal I_{mean} is of the order of $10\ N_A$ and $I_{max} \sim 10\ I_{mean}$ (*i.e.* a dynamic range of 100). Thus when a camera is used in dual modalities, *e.g.* fluoroscopy and digital radiography, the aperture has to be changed when the modality changes.

2.4.2.1 Automatic Exposure Control

Two kinds of systems are in general use in videofluoroscopy to accommodate changes in patient attenuation. In the first, no attempt to change the patient exposure rate ($E'_{Pat.}$) is made, only changes in the gain of the video system are made. In this situation, a $Sb_2\ S_3$ vidicon tube is usual because its optical sensitivity can be varied by adjusting the target potential. This violates the As Low As Reasonably Available (ALARA) principle as a small, easily penetrated body part is accorded the same radiation exposure as a thick, difficult to penetrate part. In the second system, these deficiencies are overcome by use of an automatic exposure control or AEC. Here the incident exposure rate (and in some cases the beam quality) is adjusted according to the X-ray attenuation of the patient. The radiation transmitted through the patient and incident on the XRII – $E'_{meas.}$ – is measured and kept constant by using feedback to the generator.

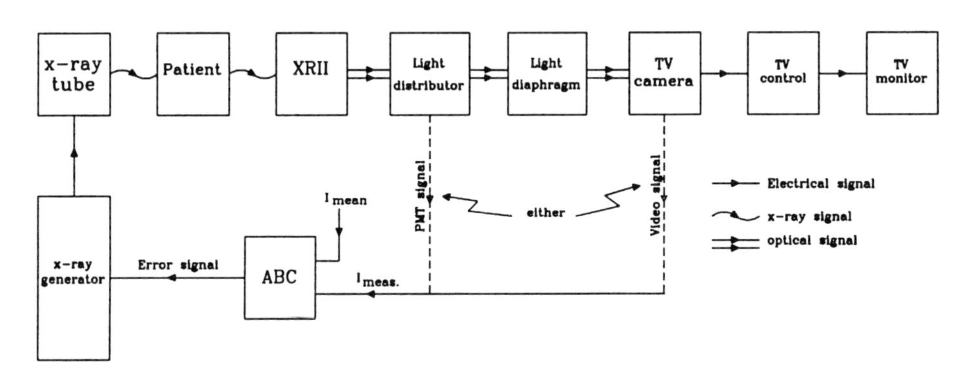

Figure 2.52. Schematic block diagram of the components of an XRII/video chain showing the feedback operation of the Automatic Exposure Control (AEC) in videofluoroscopy. (Reproduced from AAPM Medical Physics Monograph Number 20 "Specification, acceptance testing and quality control of diagnostic x-ray imaging equipment", AIP, 1994, with permission).

In Figure 2.52, the X-ray control circuitry is shown. Both the fluoroscopic X-ray exposure rate and the exposure for the photospot film are controlled by servo mechanisms – the Automatic Exposure Control or AEC for video-fluoroscopy and the phototimer for photofluorography. The AEC attempts to maintain a constant optical flux rate at the output of the XRII by adjusting the X-ray factors (kVp and/or mA) with changes in attenuation due to patient thickness variation. The time dependence change in radiation level as the patient attenuation (*e.g.* patient movement or swallowing of barium contrast material) is changed is shown in Figure 2.53.

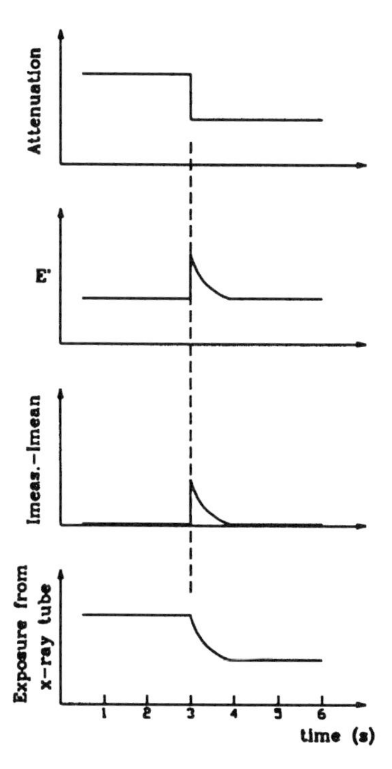

Figure 2.53. Timing diagrams illustrating operation of the AEC in fluoroscopy when attenuation of patient has a step function change as shown. (Reproduced from AAPM Medical Physics Monograph Number 20 "Specification, acceptance testing and quality control of diagnostic x-ray imaging equipment", AIP, 1994, with permission).

2.4.2.2 Real-time Digital Image Processing

In principle, image quality can be improved using real-time digital processing[30] by: first, increasing the signal-to-noise ratio by averaging in either the spatial or temporal domain; and second, modifying the display contrast to better represent the information in the image. Due to the speed of operation required, specialized hardware is necessary to implement real-time image processing. Improvement of image quality could be used: first, to improve the diagnostic efficacy of fluoroscopy at current exposure rates, or second, to permit reduction of exposure rate to the patient while maintaining image quality. Motion detection methods are used in combination with averaging to avoid image smearing in the presence of motion. The algorithm detects motion by comparing a stored image to the new data on a pixel-by-pixel basis. If the new image is more than a threshold value different from the old, motion is deemed to have occurred. The threshold levels are not easily identified and may vary for bright and dark areas. Noise reduction by temporal averaging using digital recursive filtration in this manner has been known for many years and such equipment is commercially available. However the clinical acceptance of the technique has been disappointing, due presumably to these threshold and contrast problems.

2.4.2.3 Dose Reduction in Fluoroscopy – Region of Interest (ROI)

More efficient use of radiation in fluoroscopy can be achieved by giving normal exposures to regions most vital to the procedure while simultaneously reducing exposure in the periphery which mainly gives spatial perspective and context. Such a region of interest (ROI) approach has been demonstrated using a mask subtraction method, in which the mask is either pre-computed or calculated on-the-fly on the basis of simplifying assumptions such as the ROI is a circle[31,32]. Another proposed method is that the dynamic response to imaging parameters, such as zoom mode or kVp change, is possible without *a priori* information about the exposure profile except the profile changes gently. The concept is to use digital processing to equalize the image, *i.e.* eliminate low frequency components of the image including the ROI filter used to reduce exposure. Using any of these methods the area exposure product can be shown to be reduced by a factor of the order of four while maintaining the image quality in the central region approximating 1/3 of the full diameter of the XRII at the cost of increased image noise in the peripheral region. The ROI method has not yet come into general usage but nevertheless presents a promising method for reducing dose during flurosocopy.

2.4.2.4 Photofluorography vs. Digital Fluorography

Photofluorographs, commonly called small format or 100 mm films, are obtained by exposing film to the optical image produced by an XRII. This technique has many advantages compared to film-screen. First, unlike film-screen, the size of the film can be chosen independently from the size of the body part to be imaged. Secondly, due to the gain of the XRII, the patient exposure can be reduced and exposure times shortened so as to reduce motion blurring. Thirdly, due to the smaller form of the film, they appear sharper and noise is subjectively less important.

For digital fluorography, the video camera (vidicon or CCD) is operated in pulse progressive readout (PPR) mode. In PPR, prior to the X-ray exposure pulse, previous images are eliminated by *scrubbing*, *i.e.* the camera sensor is cleaned by continuous scanning. When a signal is received from the generator because the operator has requested an X-ray exposure (the preparation signal), the camera is *blanked* (*i.e.* the scanning is stopped at the end of the next full frame of scrubbing). The camera then signals the generator and a phototimed X-ray exposure is made. After completion of the exposure, this image is *progressively* read from the video camera sensor by restoring the beam current (or for CCDs, scanning pulses) at the beginning of the video frame and digitizing at a rate and bandwidth compatible with the required image quality.

The amount of radiation used per frame can be controlled by varying the efficiency of optical coupling between the intensifier and camera sensor using an optical diaphragm. Opening the diaphragm permits reduction of the exposure per frame which increases the quantum noise and *vice versa*. Most PPR systems are quantum noise limited in the range 10–100 μR/frame (25 cm mode of XRII) and 20–200 μR/frame (15 cm mode). Usually exposures at the higher end of these ranges are used to provide the lowest noise images.

By using a video camera in PPR mode, it is possible to efficiently capture images over a wide range of X-ray exposure times, (1024 × 1024 pixel video systems coupled to intensifiers of 15 or 25 cm input diameter have been shown capable of clinical image quality equivalent to 100 mm photofluorography[28,29]). Digital correction of shading and structural non-uniformities of the imaging system as well as automatic image enhancement algorithms can further improve the acceptability of PPR digital images. PPR systems with 1024 lines are suitable for gastrointestinal studies which currently use 100 mm photofluorography with a 25 cm field of view. Larger intensifier fields are needed for general radiography that demand increased matrix sizes to maintain resolution.

The phototimer system for fluorography requires a much faster response than the AEC for fluoroscopy to permit the short exposures that will freeze patient motion. During the actual X-ray exposure it is only practical to vary exposure time. A relatively small range of exposure times is reasonable: it must be larger than the fastest switching time of the generator, yet shorter than 100 ms to efficiently stop motion. Prior to x ray exposure, the choice of the X-ray kVp and current has to be made according to the fluoroscopic measurement of the attenuation of the patient. The phototimer integrates the current from the radiation detector until it reaches a preset threshold Q_{mean}. At this point, a logical command is sent to the generator to terminate the exposure.

An alternative to the XRII/video system is a directly X-ray sensitive, large area video camera[33] shown in Figure 2.54. In principle, the whole fluoroscopic imaging chain (XRII, optical distributor and multiple optical devices) could be replaced by a large area vidicon. It is a single stage device containing a layer of photoconductor (*e.g.* amorphous selenium or PbI_2) as the X-ray transducer. Due to the reduced number of stages compared to an XRII/video system, it has the potential for higher resolution. It is inherently a flat field device so distortion and shading are less serious than for an XRII. X-ray sensitive vidicons have been investigated previously but none were satisfactory for medical application. In the past, an X-ray vidicon lacked a useful system for: first, storing cine sequences of video images; or seond, storing single high quality small format radiographs produced by the 100 mm photofluoro-

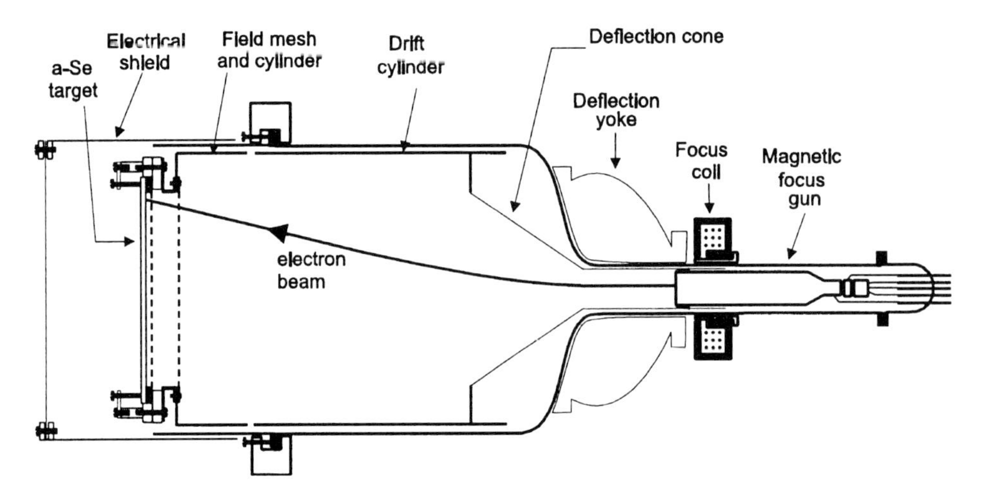

Figure 2.54. Large area X-ray sensitive vidicon.

graphic camera. Today video tape recorders and digital frame stores enable a large area vidicon to be reconsidered. It is however still just as bulky as the XRII system. In the ideal system, there would be a digital, flat-panel able to perform all current radiological modalities at reduced dose. It would, immediately after the patient's X-ray exposure, provide a high quality radiograph on a video monitor. It would also be usable for real-time imaging (*e.g.* digital cine loops and fluoroscopy). The physical form would be similar to a film-screen cassette (except for the addition of an umbilical cord) that could easily fit into the Bucky tray of existing X-ray rooms. This ideal is so far best realized by the use of flat panel systems based on active matrix arrays. The major remaining problem is how to reduce the noise of flat panel systems to that achievable with an XRII or large area vidicons so as to permit quantum noise limited operation at fluoroscopic exposure rates (*i.e.* 20 μR/s).

2.4.3 Gastrointestinal Imaging

Gastro-intestinal (GI) fluoroscopy is the most common fluoroscopic procedure. Most GI fluoroscopic examinations are now performed using *double-contrast*. In this technique, a patient drinks contrast material, usually a suspension of a powdered compound of barium, and then swallows a quantity of gas pills. The stomach acid activates the pills generating gaseous CO_2. The gas inflates the GI tract and allows the GI walls to be separated and uniformly coated by the contrast material. Through the two-dimensional X-ray projection of the contrast coated GI tract, the outline of the GI walls and the internal folds and crevices are made visible in the fluoroscopic image.

Motion is the essence of fluoroscopic GI studies. During the examination, the radiologist needs to be able to view the motion of relevant organs. There is a problem due to the temporal lag of the XRII/video system. Combined with the motion present in the GI examination, the temporal lag causes image blurring, the degree of which depends on the velocity of motion. However, the temporal lag reduces the image noise by effectively averaging successive frames. This leads to a controversy between the use of the very *laggy* Sb_2S_3 vidicon camera tube and the essentially lag-free Plumbicon camera tube. Sb_2S_3 vidicon tubes typically average 8 video frames resulting in smoother (less noisy) images but with excessive smearing or tailing due to motion blurring. On the other hand, Plumbicon (PbO) tubes do not significantly average frames and have annoying levels of noise which conceal fine detail and low contrast objects but almost no image smearing. If digital temporal averaging (averaging consecutive images in the image sequence) uses a large lag parameter to reduce noise appreciably, it causes unacceptable motion blurring in those regions of the image where motion is occurring. However, the ability of motion detection circuits to detect motion in low contrast regions is limited. In GI fluoroscopy, velocities on the order of 10–30 mm/s are commonly seen. A clinically useful balance between motion blurring and noise reduction is very difficult. Nevertheless, a compromise between the amount of noise reduction required, the amount of motion in the individual patient and how much motion blurring can be tolerated must be made.

2.4.4 Digital Subtraction Angiography (DSA)

Digital systems for subtraction angiography are now in widespread clinical use[34]. The principle used in DSA is that iodine contrast media partially replaces blood in the vessels, resulting in considerable extra attenuation and making it possible to image the circulating blood. Initially dual and even triple energy schemes based on the presence of a K absorption edge at ~33 keV were used but were found to have problems maintaining low enough noise and freedom from artifacts. Thus, it is more practical and satisfactory to make images before and after the injection of iodine and *subtract* away the first image, ideally leaving an image of just the iodinated blood. If a path through the body has length x, then blood is replaced with iodine/blood mixture in a vessel of thickness d. Then:

$$I_1 = I_0 e^{-\mu_T x}, \quad I_2 = I_0 e^{-\mu_T x} e^{-\mu_I d} \tag{2.15}$$

where it is assumed that the iodine takes up negligible volume. To perform the separation of the iodine image, a log subtraction is performed (*i.e.* $\log I_1 - \log I_2$). This results in :

$$[\log I_0 - \mu_T x] - [\log I_0 - \mu_T x - \mu_I d] = \mu_I d. \tag{2.16}$$

Thus a linear image in d is obtained. It should be noted that scatter corrupts the noise level in the image as this, adding in quadrature, is not subtracted. Furthermore, the presence of scatter destroys the quantitative nature of the subtraction. For these reasons it is desirable to reduce scatter in DSA procedures to the greatest degree possible. Grids must be used as it is impossible to use scanning systems as the demands of consistent images free of motion to permit an accurate subtraction could not be attained.

Subtraction angiography was, prior to the invention of DSA, performed by photographic subtraction. However, this technique was never very popular and no longer in use because the DSA method offers greater convenience and permits immediate diagnosis.

2.4.5 Mammography, an Exception to Every Rule

Mammography is interesting, challenging and unique. It is the only projection X-ray imaging modality that attempts to visualize soft tissue contrast[36]. The photoelectric effect is used to achieve soft tissue contrast and this requires the use of very low kVp. The geometrical situation in mammography is also unusual, only one half of the field of the X-ray tube is used as shown in Figure 2.55. The purpose of this arrangement is to ensure that the central ray grazes the chest wall, eliminating the possibility of missing breast tissue close to the chest wall.

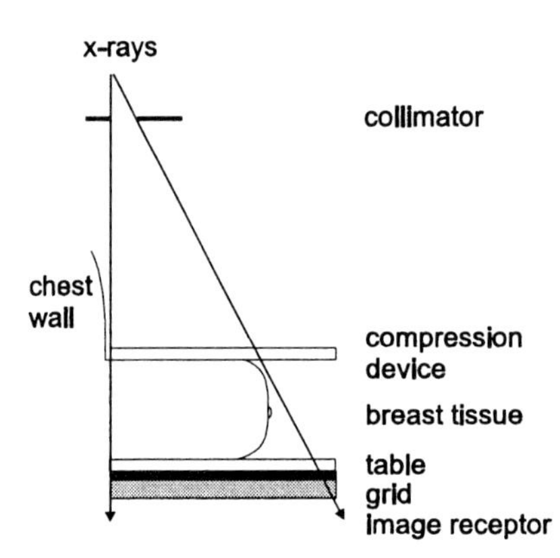

Figure 2.55. Geometrical arrangement in mammography.

Originally, mammography used non-screen film which produced excellent images albeit with very large radiation doses. This was alleviated by the introduction of *xeromammography*, a close technical relative of electrophotography – *i.e.* the photocopier process. Xeromammography[37]– the first large scale medical application of amorphous selenium – used only minor modifications of the photocopier process. It was a technical and commercial success in its day; it significantly reduced dose compared to non-screen film and also had a huge dynamic range as the toner development process suppressed large area contrast. However, with the introduction of mammographic film-screen systems it is no longer competitive. Film-screen has a much smaller dynamic range than non-screen film or xeromammography and therefore requires an extreme amount of breast compression to equalize the X-ray path length to the point that the whole breast can be visualized. The standard procedure is film-screen with a single screen with the film facing the radiation so that radiation first impinges on the part of the screen closest to the film. For the very highly absorbed radiation used in mammography, this is preferentially absorbed close to the screen surface. Digital radiography is starting to be important to clinical practice, but is still undergoing development. Small field of view digital systems are available and full field of view systems are undergoing clinical trials. The first full field systems are scanning systems and area systems made of tiled arrays of CCDs. Active matrix systems are in the preliminary stages of development and the challenge is to make pixels small enough and at an affordable cost. The current view is that 100 μm pixels will not be small enough but 50 μm will be more than adequate. Clinical and scientific studies are necessary to resolve this resolution problem. The intrinsically high resolution of amorphous selenium combined with the simplicity of the active matrix panel makes this approach for mammography very promising. The availability of such digital systems will facilitate the clinical application of computer aided diagnosis[38,39].

2.4.6 Motion Tomography

Tomography is an X-ray technique that blurs out the shadows of superimposed structures to allow examination of the structures of interest. Tomography requires an X-ray tube, an X-ray film and a rigid connecting rod that rotates about a fixed fulcrum. When the tube moves in one direction, the film moves in the opposite direction. The film is in a tray under the X-ray table so that it is free to move without disturbing the position of the patient. The fulcrum or pivot is the only point in the system that remains stationary. The amplitude of tube travel is called the tomographic angle. The plane of interest within the patient is positioned by the operator to lie at the fulcrum, and is the only plane that remains in sharp focus. All points above and below this focal plane are blurred as illustrated in Figure 2.56).

2.4.6.1 Linear, Circular and All Sort of Motions

In the simplest systems the tube path is a straight line, consequently *linear tomography*. Other paths are possible such as circular, elliptical, hypocycloidal, Figure-8 and trispiral. The more the tomographic motion differs from the shape of the object being examined, the less likely it is to produce artifacts. Thus tomographic units have been designed to operate with a wide variety of curvilinear motions, the simplest non-linear motion being circular. The different motions are desirable because sharp points in the image away from the focal plane will be blurred in the form of the motion. The more complicated the motion, the less likely a blurred object will be confused with a real lesion. The width of the blurring out of the plane is controlled by: tomographic angle; the distance of the object from the focal plane; and the distance from the film. Thus, the effective section thickness is reduced as θ increases. If θ is large (> 60°) then it is called wide-angle tomography and is intended to isolate a single narrow section close to the fulcrum. In contrast, narrow-angle tomography (θ < ~20°), zonography, is only intended to exclude distant planes.

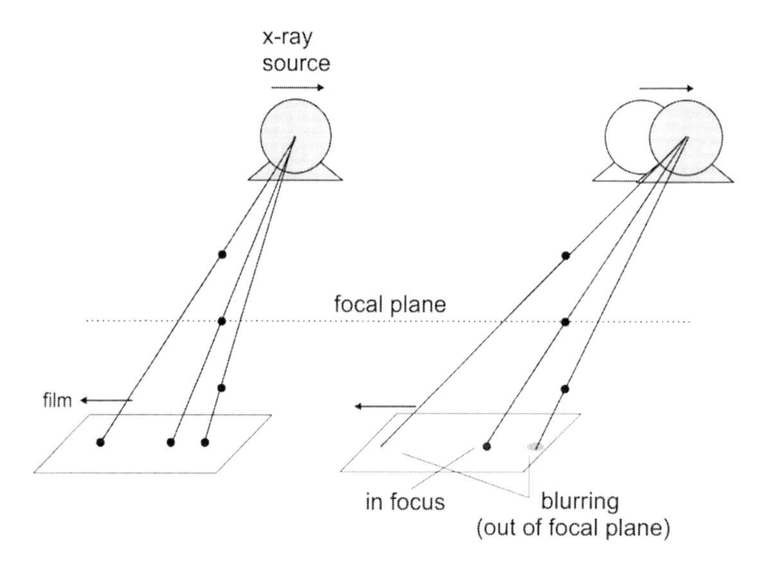

Figure 2.56. Principle of tomographic blurring

2.4.6.2 Motion Tomosynthesis

In conventional motion tomography both the X-ray tube and sensor move about the pivot point. If however a stationary digital sensor is used, capable of being readout many times while the tube moves (note that exposure per subimage is much less than normal, so total exposure is equivalent to conventional film-screen), then tomograms can be synthesized at any desired level within the patient by post-processing and creating a fulcrum at any arbitrary level. This approach, while theoretically possible, will only become practicable with flat panel detectors of sufficient geometrical accuracy and readout rate. Preliminary feasibility has been demonstrated with XRII but the inherent distortion and need for numerous corrections has inhibited its general application. The advantage of this approach is that it reduces patient dose as multiple tomograms can be obtained at the same total dose as a single conventional tomogram.

2.4.7 Computed Tomography

CT was immediately accepted because of the obvious benefits of true transverse tomography and the ability of CT to display subtle differences in tissue attenuation[18]. These outweighed the desire for high spatial resolution which could not be achieved with the coarse detectors and limited computer capacity available at the time, but which could be achieved with standard radiographic projection imaging. To this day, the resolution of CT is worse than that of projection imaging, but it is continuing to undergo improvement.

2.4.7.1 First and Second, Third and Fourth Generation CT

The first practical CT scanners, later called first generation scanners, are shown in Figure 2.57. In first generation machines, a pencil beam of radiation and associated detector were set at a particular angle and then physically moved across the slice until an adequate number of samples were obtained. The entire detector/X-ray tube collimator arrangement was rotated through a small angle and the movement and measurements perpendicular to the ray repeated. This rotation and sliding are repeated until an adequate number of views and angles are sampled. These measurements were even more laborious to perform than to describe and hence the improvements represented by the second and third generations were made. More detectors are used so reducing (second) and eliminating (third) the transverse sliding motion. The third generation machine is the most commonly used variant as it permits accurate collimation and the minimum number of detectors to eliminate the sliding transverse motion. The fourth generation machine uses a fixed ring of detectors which eliminates the need to rotate the detectors, but requires for a given resolution many more detectors. The fourth generation arrangement complicates the positioning of scatter eliminating septa normally placed between detectors. However, the fourth generation machine is much less susceptible to reconstruction artifacts than the third.

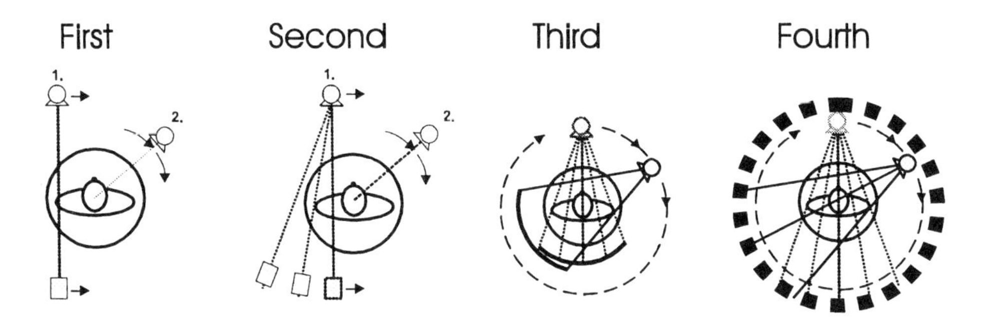

Figure 2.57. The CT generation gap.

The speed of acquisition and reconstruction of CTs has been rapidly increasing. The limitations are computer power to maintain the reconstruction rate; the heat loading on the tube; and the inertial forces on the X-ray tube bearings. Now most CT scanners use slip rings to permit continuous rotation of the tube rather than the ~1½ rotations limited by rolling up of high voltage cables. The possibility of rapid continuous rotation has led to a further development – *spiral scanning*.

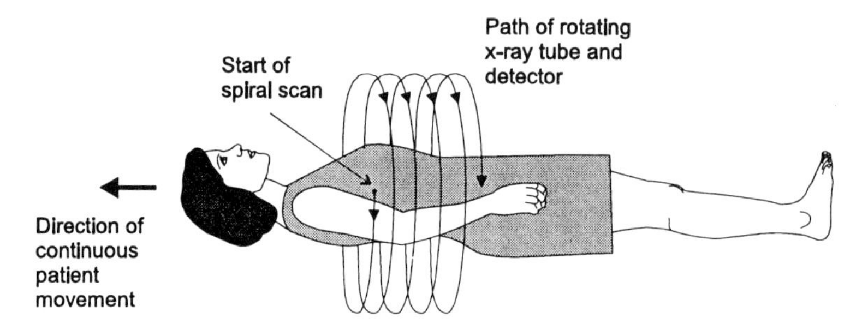

Figure 2.58. Spiral scanning concept.

2.4.7.2 Spiral Scanning CT

Spiral scanning (Figure 2.58) shows that the X-ray tube rotates continuously while the patient couch moves linearly. Thus with respect to the patient, the focal spot traces a spiral around the patient, hence the name. This approach permits volumetric scanning of a large section of the patient within a single breath hold (*e.g.* 20 s) so avoiding patient motion. This ensures much more consistent data from slice to slice than is possible with a conventional scanner and thus facilitates true 3-D reconstruction.

2.4.7.3 Cardiac/Electron Beam Source CT

The rapid motion of the heart precludes acquisition of real-time data sufficient to reconstruct a fully consistent image of a single heart beat. Instead data from conventional CT scanners are stored according to the phase of the heartbeat measured by a cardiac monitor. Thus an image of the moving heart, *averaged* over many heartbeats, is created. In imaging the diseased heart, it may not be a reasonable assumption that each heartbeat is the same. There is interest in a CT scanner fast enough to obtain all the views necessary to reconstruct an image in much less than a heartbeat (~1/s). The resolution time needed (1/30 s) is clearly impossible for a massive rotating element such as an X-ray tube. The only practical solution is based on a unique scanning electron beam X-ray tube shown in Figure 2.59. Several slices of data are obtained simultaneously by use of several X-ray target rings and several rows of X-ray detectors. Such devices are commercially available and are used at specialist cardiac facilities.

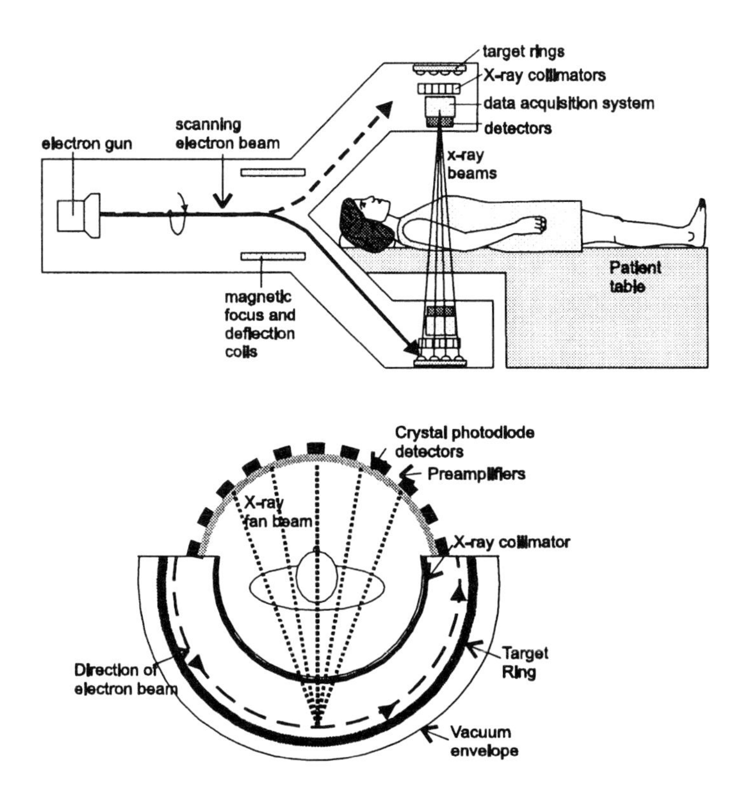

Figure 2.59. Cardiac CT scanning concept.

2.4.7.4 General CT facts and figures

To determine the position of the patient before cross-sectional images are made, a conventional projection image is acquired. This requires the X-ray tube in a fixed position and the patient table is translated linearly as data is acquired.

The noise in CT images is related to the patient dose and the reconstruction kernel used. However noise is not significantly affected by the size of the patient. Generally the same exposure is used for all patient sizes, except children, so no AEC is necessary. For most situations the dose is quite considerable – in the range 1–5 Rad to the imaged body part. Unlike in projection radiography (where dose is greatest at the plane of entry of X rays and least at the sensor), in CT, due to the highly penetrating X rays used combined with rotation around the body and the application of bow tie filters to equalize the exposure to all detectors, dose deposition is almost uniform. Compared to the cross-sectional slices, the dose delivered by the digital projection image can be assumed negligible.

All CT scanners are prone to developing annoying artifacts in the cross-sectional images. Efforts have to be made by the manufacturers to reduce these to manageable levels, but drifts and faults can cause them to reappear.

For example, slight errors in a single detector gives rise to rings in the image. Streaks are also common artifacts arising from sharp edges of highly attenuating materials. The sensors in CT scanners are often xenon ionization chambers. Although a very old technology, they are still desirable because they can be made consistent and free of ring artifacts. Furthermore they do not have lag that can cause inconsistencies. The more advanced types of detectors usually use phosphors in conjunction with photodiodes. However, lag is a continual problem with such detectors and each manufacturer struggles with specialized proprietary (usually undisclosed) solutions.

2.5 Summary – Does X-ray Imaging have a Future?

Over 100 years ago during the epic three weeks that Roentgen discovered the existence of X rays and even, serendipitously, viewed the bones of his finger while holding up a small fluorescent screen, he doubted his sanity. How could he dare to believe he was seeing inside the live human body? Since then and for most of this century, X rays have been the only non-invasive imaging method. Now many more methods have made their appearance and one by one, X-ray methods are being replaced. Can we anticipate that in the next millennium X rays will cease to be used for medical imaging?

What are the advantages of X rays? In general terms, the usefulness of X rays is to visualize morphological changes and differences with very high spatial and temporal resolution and low noise. The chest X-ray perhaps best epitomizes the strengths of imaging. With a relatively low exposure, it provides a window into the thorax and is used everyday to diagnose hundreds of different diseases and pathologies with only a glance.

The film-screen combination is still the dominant imaging modality. It has a DQE of less than 30% at zero spatial frequency and drops rapidly – within the region of important spatial frequencies – to less than 1%. The promise of flat panel systems is that they will be able to improve the DQE to > 30% throughout the spatial frequency range of interest. Probably the single, most important concept for the modernization of conventional radiography and fluoroscopy is the flat panel electronic sensor which, within the size limitations of a conventional film-screen cassette, will perform the acquisition and digitization of an X-ray image and communicate this image immediately to a video monitor.

Similarly, the CT scanner appears to be on the brink of important new technical improvements beyond those possible today. By use of multiple rings of detectors, greater resolution in the scan direction and a reduced heat loading on the tube will be possible, eventually leading to reduced cost.

The answer to whether these new X-ray technologies will continue to grow and prosper and continue to offer their aid in diagnosis and treatment will depend on better understanding the risks of low levels of radiation. Our present knowledge is based on linear extrapolation from much higher exposures (*e.g.* atom bomb survivors) and may be too pessimistic or even too optimistic. Similarly, our understanding of the risks of other modalities, currently regarded as harmless, could change. There are strengths (benefits) and weaknesses (risks) in all medical procedures and a reasonable balance is desired. Some factors are structural (the equipment), some are technical (contrast media), and some are accidental (falling off the examination table). All are subject to change and improvement as our understanding increases.

2.6 Acknowledgments

My work has been supported by three Canadian granting agencies: the Medical Research Council, the National Cancer Institute, and the Natural Sciences and Engineering Research Council. I am also indebted to Litton Systems Canada, Inc. and Noranda Inc. for financial support over the years and grateful to those who have provided useful comments and aided me to better understand this subject.

2.7 References

1. H. Barrett and W. Swindell, *Radiological Imaging*, (Academic Press, New York, 1981).
2. J.C. Dainty and C. Shaw, *Imaging Science*, (Academic Press, London, 1974).
3. H.E. Johns and J.R. Cunningham, *The Physics of Radiology*, (Thomas, Springfield, IL, 1983).
4. A. Macovski, *Medical Imaging Systems*, (Prentice-Hall, New Jersey, 1983).
5. K. Kato, in *Specification, Acceptance Testing and Quality Control of Diagnostic X-ray Imaging Equipment*, J.A. Seibert, G.T. Barnes and R.G. Gould, Eds. (American Institute of Physics, Woodbury, NY, 1994), pp. 731–769.
6. R.K. Swank, *Journal of Applied Physics*, **44**, 4199–4203 (1973).
7. G.F. Knoll, *Radiation Detection and Measurement*, (Wiley, Toronto, 1989).
8. C.A. Klein, *Journal of Applied Physics*, **39**, 3476–3478 (1968).
9. I. Brodie and R.A. Gutcheck, *Medical Physics*, **12**, 362–367 (1985).
10. I. Brodie and J.J. Murray, *The Physics of Micro/Nano-Fabrication*, (Plenum Press, New York, 1992).
11. M.F. Thompsett, in *Electronic Imaging*, T.P. McLean and P. Schagen, Eds. (Academic Press, London, 1979), pp. 55–101.
12. R.A. Street, *Hydrogenated Amorphous Silicon*, (Cambridge University Press, Cambridge, 1991).
13. E. Krestel, *Imaging Systems for Medical Diagnostics*, (Siemens, Berlin, 1990).
14. K.G. Vosburg, R.K. Swank and J.M. Houston, *Advances in Electronics and Electron Physics*, **43**, 205–244 (1977).
15. P. DeGroot, in *AAPM Summer School on Specification, Acceptance Testing and Quality Control of Diagnostic X-ray Imaging Equipment*, J.A. Siebert, G.T. Barnes

and R.G. Gould, Eds., (American Institute of Physics, Woodbury, NY, 1992), pp. 477–510.
16. J.D. McGee, in *Electronic Imaging*, T.P. McLean and P. Schagen, Eds. (Academic Press, London, 1979), pp. 11–54).
17. G.P. McGinty, *Video Cameras: Operation and Servicing*, (Howard Sams and Co., Indianapolis, 1984).
18. L.W. Goldman and J.B. Fowlkes, Eds., *AAPM 1995 Summer School on Medical CT and Ultrasound: Current Technology and Applications*, (American Institute of Physics, Woodbury, NY, 1995).
19. D.B. Plewes, *Medical Physics*, **10**, 646–654 (1983).
20. D.B. Plewes and E. Vogelstein, *Medical Physics*, 10, 655–663 (1983).
21. H. Rougeot, in *AAPM 1993 Summer School on Digital Imaging*, W.R. Hendee and J.H. Trueblood, Eds. (American Institute of Physics, Woodbury, NY, 1993), pp. 49–96.
22. M.J. Yaffe and J.A. Rowlands, *Physics in Medicine and Biology*, **42**, 1–39 (1997).
23. U. Neitzel, I. Maack and S. Guenther-Kohlfahl, *Medical Physics*, **21**, 509–516 (1994).
24. W. Zhao and J.A. Rowlands, *Medical Physics*, **22**, 1595–1604 (1995).
25. L.E. Antonuk, J. Boudry, W. Wang, D. McShan, E.J. Morton, J. Yorkston and R.A. Street, *Medical Physics*, **19**, 1455–1466 (1992).
26. U.W. Schiebel, N. Conrads, N. Jung, M. Weilbrecht, H. Wieczorek, T. Zaengel, M.J. Powell, I.D. French and C. Glasse, *Proceedings of SPIE*, **2163**, 129–140 (1994).
27. M.D. Tesic, R.A. Mattson, G.T. Barnes, R.A. Sones and J.B. Stickney, *Radiology*, **148**, 259–2264 (1983).
28. D.M. Hynes, J.A. Rowlands and E.W. Edmonds, *Journal of Canadian Association of Radiologists*, **40**, 262–265 (1989).
29. J.A. Rowlands, D.M. Hynes and E.W. Edmonds, *Medical Physics*, 16, 553–560 (1989).
30. J.A. Rowlands, *Proceedings of SPIE,* **1652**, 294–303 (1992).
31. S. Rudin, D.R.Bednarek, *Medical Physics*, **19**, 1183–1189 (1993).
32. M.S. Labbe, M.Y. Chiu, M.S. Rzeszotarski, A.R. Bani-Hashemi and D.L. Wilson, *Medical Physics*, **21**, 471–481 (1994).
33. R. Luhta and J.A. Rowlands, *Proceedings of SPIE*, **1896**, 38–49 (1993).
34. C.A. Mistretta, in *Digital Radiography*, **314**, 18–23 (1981).
35. S.J. Riederer and R.A. Kruger, *Radiology*, **147**, 633–638 (1983).
36. A. Haus and M.J. Yaffe, Eds., *Syllabus of Categorical Course on Technical Aspects of Mammography*,(Radiological Society of North America, Oak Brook, IL, 1994).
37. J.W. Boag, *Physics in Medicine and Biology*, **18**, 3–37 (1973).
38. H.P. Chan, K. Doi, S. Galhotra, C.J. Vborny, H. MacMahon and P.M. Jokich, *Medical Physics*, **14**, 538–548 (1987).
39. M.L. Giger, N. Ahn, K. Doi, H. MacMahon and C.E. Metz, *Medical Physics*, **17**, 861–865 (1990).

3 Radioactivity, Nuclear Medicine Imaging and Emission Computed Tomography

Benjamin M. W. Tsui

Department of Biomedical Engineering and Department of Radiology, The University of North Carolina at Chapel Hill

3.1 Introduction

Radioactivity was observed first in 1896 by Antoine H. Becquerel[1] and in the following year by the wife and husband research team Marie and Pierre Curie[2]. Another significant milestone was the discovery of artificially produced radioisotopes by Irene Curie (daughter of Marie and Pierre Curie) and her husband Frederick Joliot in 1933[3]. In 1935, Georg Hevesy[4] first used radiophosphorus for metabolic studies in rats. Since then, tracer amounts of radioisotopes have been used in studies of physiological function and in treatment of diseases in animals and humans. Clinical nuclear medicine imaging received major boosts with the development of the scintillation camera in the late 1950s by Hal O. Anger[5] and the discovery of ^{99m}Tc by Powell Richards in 1960[6] followed by its application in humans[7].

In this chapter, the properties of radioactivity and the biomedical uses of radioisotopes in medical imaging are discussed. In nuclear medicine, pharmaceuticals are labeled with specific radioisotopes to form radiopharmaceuticals that are administered to patients. Depending on their biokinetic properties, the radiopharmaceuticals are distributed and taken up by various organs and/or tissue types. The uptake of a specific radiopharmaceutical depends on the physiological function and the status of different organs and tissue types. By using radiation detectors that detect the radiation emitted by radioisotopes, the distribution and uptake of the pharmaceutical can be determined. Conventional nuclear medicine (NM) imaging techniques provide two-dimensional (2D) images of the *in vivo* radioactivity distribution. Emission computed tomography (ECT) imaging techniques, including positron emission tomography (PET) and single photon emission computed tomography (SPECT), provide three-dimensional (3D) representations of *in vivo* radioactivity distributions. From these images, the physiological status of the target organ or tissue type can be determined. This information is used by physicians to differentiate normal from disease states in patients and for diagnosis of the cause of disease states.

3.2 Fundamentals of Nuclear Physics, Radioisotopes and Radioactivity

3.2.1 Nuclear Physics

3.2.1.1 Composition of Atom and Nucleus

In 1897, J.J. Thomson[8] suggested that the atom consists of fundamental particles and established the e/m ratio of cathode rays, which were later identified as *electrons*. He proposed an early model of the atom in which a number of electrons were distributed randomly within a sphere consisting of an equal number of positively charged particles. From a series of scattering experiments, Ernst Rutherford[9] concluded in 1911 that the atom consists of a small central nucleus with radius of $\sim 10^{-15}$ m and contains positively charged *protons*. The nucleus is surrounded by electrons in a cloud about 10^{-10} m in radius. However, Rutherford's model was problematic because, according to the model, stationary electrons would be pulled into the nucleus.

The classical model of the atom was established by Niels Bohr[10] in 1913 based on three fundamental postulates. They are:

1. Electrons revolve about a stationary central nucleus only in orbits.
2. Atomic electrons can only exist in orbits with discrete energy levels.
3. Energy is required or released only when an electron changes from one orbit to another.

The Bohr model suggests that energy is released as photons when an electron undergoes a transition from one orbit to another near the nucleus. Excellent agreement between predictions from the Bohr model and experimental spectroscopic data was found in the case of hydrogen but not for multielectron atoms. For these atoms, Bohr's model was extended to include elliptical-shaped orbits and wave properties of the electrons. To account for all the fine structures of the observed spectroscopic data, the energy state of each orbit electron is uniquely represented by a set of four quantum numbers.

The composition of the nucleus was thought to consist of protons until 1932, when Chadwick[11] discovered a new particle with a mass very nearly equal to that of the proton but having no charge. The new particle was called the *neutron*. Table 3.1 shows the properties of the fundamental components of the atom and nucleus.

Table 3.1. Properties of the fundamental components of atom and nucleus

Particle	Symbol	Charge	Mass (MeV/c^2)	Spin (h)
electron	e^-	–1	0.511	½
proton	p^+	+1	938.211	½
neutron	n	0	939.505	½

The most recent theory suggests that electrons and nucleons (proton and neutrons) are members of a family of elementary particles. They can be classified into three main categories. Leptons are extranuclear particles and include electrons, neutrinos and muons. Hadrons are nuclear particles and include baryons, such as protons and neutrons, and hyperons, antibaryons and mesons. The third category consists of particles that mediate forces and includes photons, gravitons, gluons and intermediate vector bosons.

3.2.1.2 Nuclear Forces

Elementary particles exert forces on each other and are constantly created and annihilated. Creation, annihilation and force are, in fact, related phenomena that are collectively referred to as interactions. There are four types of interactions, i.e. nuclear (strong), electromagnetic, weak and gravitational, with relative strengths of 1, 10^{-2}, 10^{-13} and 10^{-38} respectively.

The gravitational interaction of matter is important on a large scale, although it is the weakest of the elementary particle interactions. Electromagnetic interactions are responsible for binding electrons to nuclei in atoms and molecules. However, if only electromagnetic interactions existed in the nucleus, the nucleons would fly apart because of the repulsion of the positively charged protons.

There are two main nuclear interactions. The weaker one is the so-called *weak interaction* that governs the radioactive decay of atomic nuclei. The stronger one of the two is the attractive nuclear interaction, called the *exchange force,* between two nucleons. This force is effective only over a very short distance and is responsible for binding of protons and neutrons to form nuclei.

Before the mid-nineteenth century, interactions, or forces, were commonly believed to act at a distance. Michael Faraday initiated the idea that an interaction is transmitted from one body to another through a field. The field theory has been extended to describe nuclear interactions. The attempt to unify the four interactions into a single conceptual whole was initiated by Albert Einstein before 1920, an effort that is still continuing.

3.2.2 Radioisotopes

Radioisotopes are unstable nuclides that spontaneously transform to stable nuclides through nuclear transitions with the release of radiation. Unstable nuclides include heavy atoms, which transform into lighter and more stable atoms, proton-rich nuclides which transform a proton into an electron with the release of positrons, and neutron-rich nuclides which transform a neutron into a proton with the release of electrons. Nuclear transitions can be divided into the following categories.

3.2.2.1 Alpha ($^4_2$He) Decay

Alpha decay can be described by the reaction

$$
{}^{A}_{Z}V \rightarrow {}^{A-4}_{Z-2}W^{(*)} + {}^{4}_{2}\mathrm{He} + Q
$$
$$
\quad \hookrightarrow {}^{A-4}_{Z-2}W + \gamma \tag{3.1}
$$

where Z is the atomic number, A is the mass number, γ represents the photon emission, Q is the energy released [kinetic energy (KE) + recoil energy] and $^{(*)}$ indicates nuclear excitation. An example is the transition of radium-226 to radon-222 through alpha decay, as shown in Figure 3.1 (a). Because of their short range in tissues, alpha particles are not used for imaging purposes. However, alpha emitters, such as astatine-211 and bismuth-212, have been used in radiotherapy to kill cancerous cells at close range.

3.2.2.2 Beta$^-$ Decay (e^-)

Beta$^-$ decay can be described by the reaction

$$
{}^{A}_{Z}V \rightarrow {}^{A}_{Z+1}W^{(*)+} + e^- + \bar{v} + Q
$$
$$
\quad \hookrightarrow + e^- \longrightarrow {}^{A}_{Z+1}W^{(*)} \tag{3.2}
$$

where e^- and $\bar{v}$represent the beta$^-$ particle or electron and the antineutrino, respectively, and Q is the energy released (KE + recoil energy). An example is the transition of xenon-133 to cesium-133 through beta$^-$ decay, as shown in Figure 3.1 (b). The distribution of kinetic energy of the beta particles is shown in Figure 3.2. It indicates that most of the emitted beta particles receive less than half of the maximum kinetic energy. Typically, the average energy $\bar{E}$ is equal to about one-third of the maximum energy, E_{MAX}.

Beta$^-$ decay occurs in high-N/Z or neutron-rich nuclei and involves the transition

$$
n \rightarrow p + e^- + \bar{v} \tag{3.3}
$$

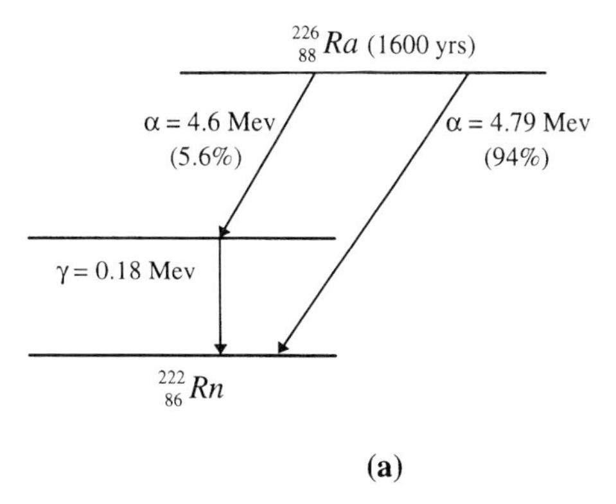

(a)

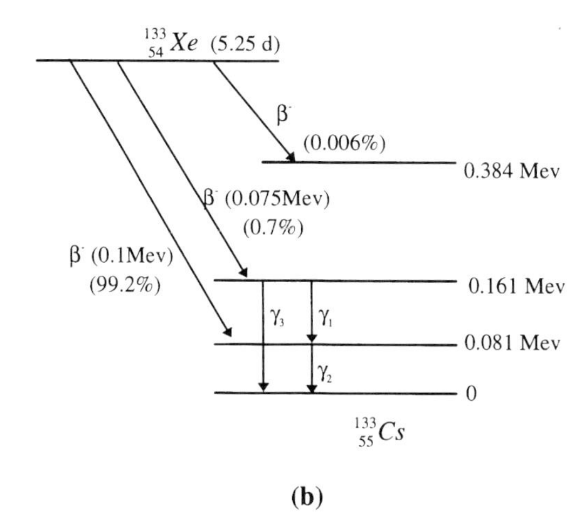

(b)

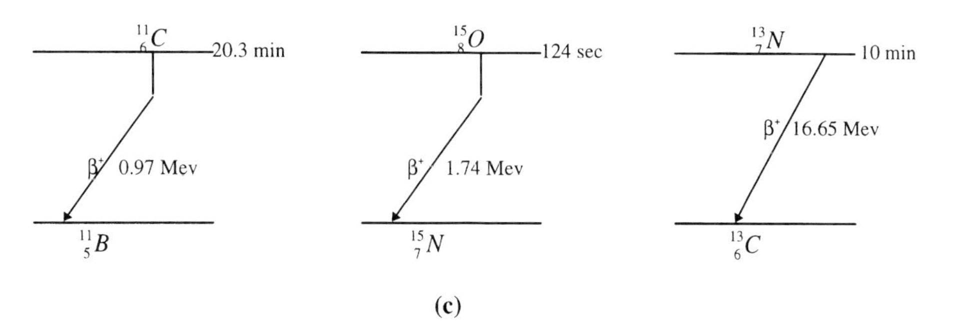

(c)

Figure 3.1. Examples of atomic and nuclear transitions. (a) Alpha (or α) decay. (b) Beta⁻ (or β^-) decay. (c) Beta⁺ (β^+ or positron) decay.

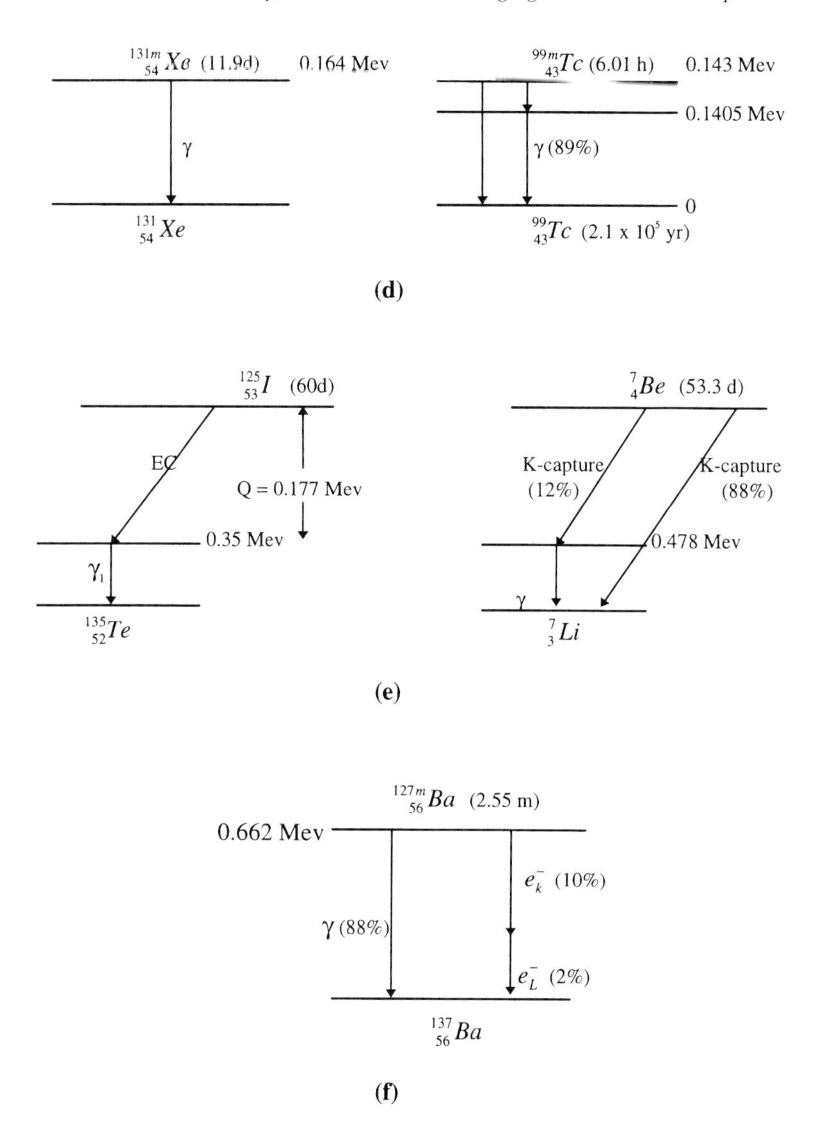

Figure 3.1. Examples of atomic and nuclear transitions. (d) Isomer decay. (e) Electron capture. (f) Internal conversion.

The existence of the antineutrino (and neutrino) was hypothesized by Pauli in 1927 to satisfy the conservation of energy, angular momentum and lepton in the β^- decay process. The antineutrino $\bar{\nu}$ was discovered in 1953 by Reines and Cowan, 26 years after its existence was hypothesized.

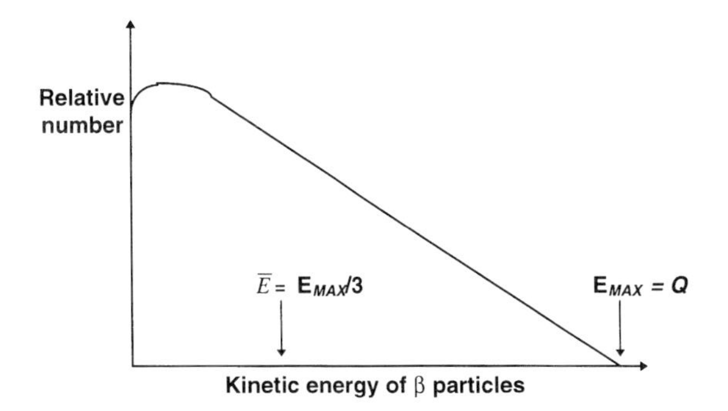

Figure 3.2. The distribution of kinetic energy of β particles from beta⁻ and beta⁺ decay transitions. The maximum kinetic energy of the beta particles is equal to the energy release, Q, from the nuclear transition.

3.2.2.3 Beta⁺ or Positron Decay (e^+)

Beta⁺ or *positron decay* can be described by the reaction

$$^{A}_{Z}V \rightarrow {}^{A}_{Z-1}W^{(*)-} + e^+ + \nu + Q$$
$$\quad \hookrightarrow {}^{A}_{Z-1}W^{(*)} + e^- \text{ (released)} \tag{3.4}$$

where Q is the energy released (K.E. + recoil energy) and ν represents a neutrino. Examples of beta⁺ or positron decay are shown in Figure 3.1 (c).

Beta⁺ decay occurs in low-N/Z or proton-rich nuclei and involves the transition

$$p \rightarrow n + e^+ + \nu \tag{3.5}$$

The positron that is released from the process will undergo annihilation with an electron, with two 511 keV photons emitted 180° apart. Beta⁺ or positron decay is important in positron emission tomography (PET), which is based on coincidence detection of the two 511 keV annihilation photons from positron-emitting radionuclides.

Beta⁺ decay is energetically possible only if

$$M_V \geq M_W + 2m_e \tag{3.6}$$

where M_V and M_W are the atomic masses of the radionuclides before and after beta⁺ decay. For proton-rich nuclei that do not meet the energy requirement for beta⁺ decay, an alternative decay scheme to a stable nucleus is electron capture, which is described in section 3.2.2.5.

3.2.2.4 Isomer Decay

Isomer decay can be described by the reaction

$$ {}_{Z}^{A}V^{\mathrm{m}} \rightarrow {}_{Z}^{A}V + \gamma \tag{3.7}$$

The excited radionuclide ${}_{Z}^{A}V^{\mathrm{m}}$ is often the product of decay from a radioactive parent and has a long half-life. The long-lived *metastable* or *isomeric* state may last from seconds to days before decay into the stable nuclei ${}_{Z}^{A}V$ accompanied by gamma ray emission. Examples of isomer decay are shown in Figure 3.1 (d).

3.2.2.5 Electron Capture (K-capture)

Electron capture involves both atomic and nuclear transitions. It can be described by the reaction

$$ {}_{Z}^{A}V + e^{-}_{K,L} \rightarrow {}_{Z-1}^{A}W^{*(*)} + \nu + Q \tag{3.8}$$

where $^{(*)}$ indicates nuclear excitation and * indicates atomic excitation. Examples of electron capture are shown in Figure 3.1 (e).

Electron capture involves the transition

$$ p + e^{-}_{K,L} \rightarrow n + \nu \tag{3.9}$$

where an orbital electron is "captured" by the nucleus and combines with a proton to form a neutron. Usually, the captured electron is from the K or L shell because of the proximity. The EC process is usually accompanied by emission of characteristic X rays. It is more frequent among heavier elements because the orbital electrons are closer.

Electron capture may occur in lieu of β^{+} decay for positron-rich nuclides when

$$ 0 < M_{\mathrm{V}} - M_{\mathrm{W}} < 2m_{\mathrm{e}} \tag{3.10}$$

The energy required for the electron capture transition to occur is

$$ M_{\mathrm{V}}c^{2} - M_{\mathrm{W}}c^{2} > E_{\mathrm{K\text{-}shell}} + I \tag{3.11}$$

where $E_{\mathrm{K\text{-}shell}}$ is the binding energy of the K-shell electrons and I the energy released is small.

3.2.2.6 Internal Conversion

In an isomeric transition, instead of emission of a gamma photon, an alternative transition called *internal conversion* may occur. Like electron capture, internal conversion involves both atomic and nuclear transitions and can be described by

$$
\begin{array}{l}
{}^{A}_{Z}V^{*} \rightarrow {}^{A}_{Z}V^{*} + e^{-}_{K,L(ejected)} \\
\qquad\quad \hookrightarrow {}^{A}_{Z}V + \gamma_{\text{x-ray}} \text{ or Auger } e^{-}
\end{array}
\tag{3.12}
$$

where the excited nucleus ${}^{A}_{Z}V^{*}$ is de-excited by ejecting a K (or L) electron, called a *conversion electron*. Internal conversion and γ–ray emission compete with each other. The ratio between internal conversion and γ–ray emission is given by the internal conversion coefficient α, defined as

$$\alpha = \frac{Ne}{N\gamma} \tag{3.13}$$

where Ne is the number of conversion electrons emitted and $N\gamma$ is the number of γ rays emitted.

The total internal conversion coefficient is given by

$$\alpha = \alpha_{\text{K}} + \alpha_{\text{L}} + \alpha_{\text{M}} + \ldots\ldots\ldots\ldots \tag{3.14}$$

where α_{i} are partial internal conversion coefficients from the ith shell. The positive ion ${}^{A}_{Z}V^{*+}$ will promptly fill its electron shell vacancy with an electron from an outer shell, with release of energy in the form of *characteristic X rays* or by transferring the energy to another orbital electron. The electron that receives the energy may be ejected from the atom and is called an *Auger electron*.

The *fluorescent yield* is the number of "fluorescent" X-ray photons emitted for each K-shell vacancy.

$$\omega = \frac{\#\ of\ K\ X-ray\ quanta\ emitted}{\#\ of\ vacancies\ in\ K\ shell} \tag{3.15}$$

3.2.3 Radioactivity

Radioactive decay occurs when an unstable nucleus undergoes transformation(s) into another more stable nucleus accompanied by radiation or particle emission. It is a spontaneous process and can only be described in terms of probabilities and average rates.

3.2.3.1 Decay Equations

In a sample containing N radioactive atoms of a certain radionuclide, the average rate at which the atoms are decaying is the *activity* A and is proportional to N, i.e.

$$A = \frac{dN}{dt} = -\lambda N \tag{3.16}$$

where the minus sign indicates that the number of atoms is decreasing and λ is the decay constant.

The basic unit of activity

$$A = \lambda N \tag{3.17}$$

is the Curie (Ci), which is defined as 3.7×10^{10} disintegrations per second. In tracer studies, the amount of radioactivity used is of the order of μCi or 10^{-6} Ci. In nuclear medicine imaging, amounts of radioactivity used are generally of the order of mCi or10^{-3} Ci. In SI (System International) units, the unit of radioactivity is the Bq (bequerel), which is equal to one disintegration per second.

Integrating Eq. (3.16) gives

$$N(t) = N_o e^{-\lambda t} \text{ or } A(t) = A_o e^{-\lambda t} \tag{3.18}$$

where N_o and A_o are the number of atoms and the activity present at time t = 0, respectively, and $N(t)$ and $A(t)$ are the number of atoms and the activity present at time t. The *decay factor* $e^{-\lambda t}$ represents the decay of a constant fraction of atoms or activity per unit time. Figure 3.3 shows plots of the activity or number of atoms as a function of time.

3.2.3.2 Half-life

Half-life is defined as the time required for N to equal $N_o/2$, i.e. for half of the original amount of radioactive atoms to decay. It can be shown that

$$T_{1/2} = 0.693/\lambda \tag{3.19}$$

3.2.3.3 Average (Mean) Life τ

The *average (mean) life* τ is the average time the atoms remain in the sample before decaying

$$N_o \tau = \int_0^\infty N_o e^{-\lambda t}\, dt \tag{3.20}$$

It can be shown that

$$\tau = 1/\lambda \text{ or } \tau = 1.44\, T_{1/2} \tag{3.21}$$

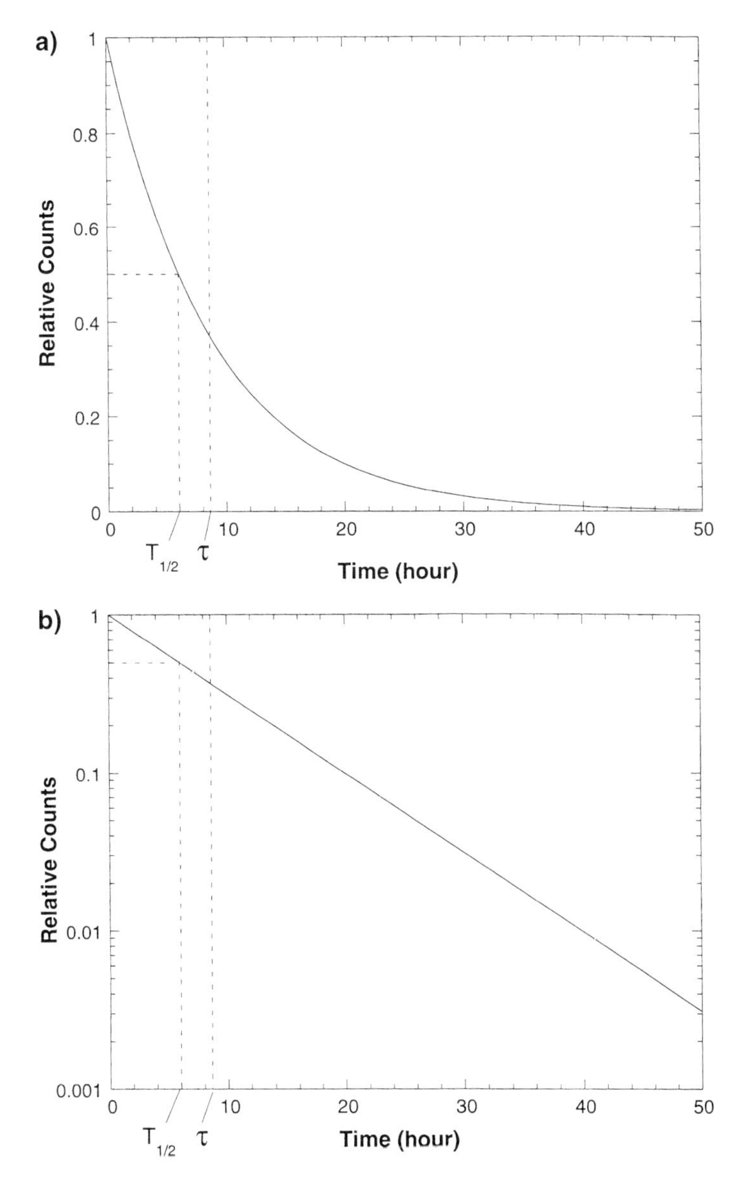

Figure 3.3. Plot of the decay factor $e^{-\lambda t} = e^{-0.693/T_{1/2}}$ as a function of time for Tc-99m with a half-life $T_{1/2} = 6$ hours in (a) linear and (b) semi-log scale.

3.2.3.4 Decay of Mixed and Unrelated Radionuclides

In many applications, it is necessary to determine the type and amount of a variety of radionuclides in a mixture. An example is in the identification and determination of radioactive contamination. Suppose a sample consists of a mixture of n unrelated radionuclides and the decay equation of the ith radionuclide is given by

$$N_i(t) = N_i^o \mathrm{e}^{-\lambda_i t} \tag{3.22}$$

where λ_i is the decay constant and N_i^o and $N_i(t)$ are the number of the ith radionuclide at time 0 and t respectively. The decay equation of the entire sample is given by

$$N_i(t) = \sum_{i=1}^{n} N_i^o \mathrm{e}^{-\lambda_i t} \tag{3.23}$$

where $N(t)$ is the total number of radioactive atoms in the sample.

3.2.3.5 Radioactive Series Growth and Decay

The decay of a radionuclide may result in a series of n intermediate radioactive nuclides before reaching a stable nuclide. As shown in Figure 3.4, P, D_1, D_2, ... and D_n denote the number of parent, daughter, granddaughter, ... and the nth radioactive granddaughter atoms. At time $t = 0$, let

$$\begin{aligned} & P(0) = \mathrm{P}_0 \\ \text{and} \quad & D_1(0) = D_2(0) = D_3(0) = \ldots\ldots\ldots\ldots\ldots = D_n(0) = 0 \end{aligned} \tag{3.24}$$

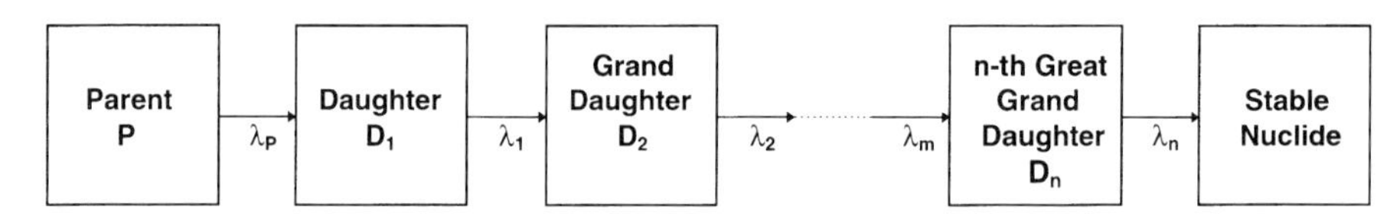

Figure 3.4. Radioactive series decay from parent with decay constant λ_P to n radioactive daughters and grand daughters with decay constants λ_i where $i = 1,2,\ldots,n$.

At time t, the amount of the nth radioactive daughter in the series is given by

$$\frac{\mathrm{d}D_\mathrm{n}}{\mathrm{d}t} = \lambda_\mathrm{m} D_\mathrm{m} - \lambda_\mathrm{n} D_\mathrm{n} \tag{3.25}$$

where λ_m and λ_n denote the decay constant of the immediate parent and the nth daughter.

By solving a total of N equations with $i = 1, 2, \ldots\ldots, N$ in Eq. (3.25), we find

$$D_\mathrm{n}(t) = P_\mathrm{o}\,(h_P \mathrm{e}^{-\lambda_P t} + h_1 \mathrm{e}^{-\lambda_1 t} + h_2 \mathrm{e}^{-\lambda_2 t} + \ldots\ldots + h_n \mathrm{e}^{-\lambda_n t}) \tag{3.26}$$

where

$$\begin{aligned} h_\mathrm{P} &= \frac{\lambda_P}{\lambda_n - \lambda_P} \frac{\lambda_1}{\lambda_1 - \lambda_P} \frac{\lambda_2}{\lambda_2 - \lambda_P} \ldots\ldots\ldots\ldots \frac{\lambda_m}{\lambda_m - \lambda_P} \\ h_1 &= \frac{\lambda_P}{\lambda_1 - \lambda_1} \frac{\lambda_1}{\lambda_n - \lambda_1} \frac{\lambda_2}{\lambda_2 - \lambda_1} \ldots\ldots\ldots\ldots \frac{\lambda_m}{\lambda_m - \lambda_1} \end{aligned} \tag{3.27}$$

$$\vdots \qquad \vdots \quad \vdots \qquad \vdots \qquad\qquad \vdots$$

$$h_n = \frac{\lambda_P}{\lambda_P - \lambda_n} \frac{\lambda_1}{\lambda_1 - \lambda_p} \frac{\lambda_2}{\lambda_2 - \lambda_n} \cdots\cdots\cdots \frac{\lambda_m}{\lambda_m - \lambda_n}$$

and $h_P + h_1 + h_2 + \ldots\ldots\ldots + h_n = 0$ (3.28)

The above equations are called the *Bateman Equations*[12].

3.2.3.6 Parent–Daughter Decay

An important radioactive series involves the simple parent–daughter decay into a stable nuclide. It is found in most generators for producing radioactive nuclides. From Eq. (3.26), the amount of daughter radionuclide is given by

$$D(t) = P_o\,(h_P e^{-\lambda_P t} + h_D e^{-\lambda_D t}) \tag{3.29}$$

Substituting the expressions for h_P and h_D in the Bateman Equation shown in Eq. (3.27), we find

$$D(t) = P_o \frac{\lambda_P}{\lambda_D - \lambda_P} (e^{-\lambda_P t} - e^{-\lambda_D t}) \tag{3.30}$$

The amounts of the parent and daughter depend on the relative values of the decay constants of the parent and daughter radionuclides. Specifically, when $\lambda_P << \lambda_D$ or $(T_{1/2})_P >> (T_{1/2})_D$, Eq. (3.30) becomes

$$D(t) \approx P_o \frac{\lambda_P}{\lambda_D}(1-e^{-\lambda_D t}) \tag{3.31}$$

Equation (3.31) shows that the activity of the daughter, $\lambda_D D(t)$, builds up as a function of time and approaches that of the parent, $\lambda_P P_o$, at time $t \to \infty$. When that occurs, the parent and daughter are in *secular equilibrium*.

When $\lambda_P < \lambda_D$, Eq. (3.30) cannot be simplified as in the above. The activity of the daughter builds up as a function of time, eventually exceeds that of the parent, reaches a maximum and decreases in the same manner as the decay of the parent. As a result, when $t >> 1/(\lambda_D - \lambda_P)$, the parent and daughter are in *transient equilibrium,* and we have

$$\frac{\lambda_D D(t)}{\lambda_P P(t)} \approx \frac{\lambda_D}{\lambda_D - \lambda_P} = \frac{(T_{1/2})_P}{(T_{1/2})_P - (T_{1/2})_D} > 1 \tag{3.32}$$

An important example of transient equilibrium is the decay of ^{99}Mo with a 66-hour half-life to ^{99m}Tc with a 6-hour half-life. This parent–daughter decay is the basis of the ^{99}Mo – ^{99m}Tc radionuclide generator that is widely used in nuclear medicine clinics.

When $\lambda_P \approx \lambda_D$, Eq. (3.20) cannot be simplified. The ratio of the parent and daughter activity will increase linearly with time.

When $\lambda_P > \lambda_D$, there is also no equilibrium between parent and daughter activities. Here, the daughter activity builds up, reaches a maximum and decreases. When the parent activity becomes insignificant, the daughter activity will decay according to its decay constant.

3.2.4 Production of Radionuclides

Most naturally occurring radionuclides are heavy elements (e.g. uranium and radium) or have very long half-lives (e.g. ^{40}K with $T_{1/2} \sim 10^9$ years). As a result, they are not suitable for use in nuclear medicine imaging purposes. All radionuclides used in nuclear medicine are man-made or "artificial" (see Table 3.2). The production of radionuclides from stable nuclides is an important issue. There are several methods of producing radionuclides.

Table 3.2. Radionuclides used in nuclear medicine and emission single-photon computed tomography imaging.

Radionuclide	Half-life	Energy of Primary Photons (keV)	Abundance of Primary Photons
Technetium-99m	6 h	140	89%
Thallium-201	73 h	72 – 84	99%
Gallium-67	3.3 d	~88, 185, 300	68%, 21%, 16.6%
Indium-111	2.8 d	172, 245	91%, 100%
Iodine-131	8.04 d	365	81%
Iodine-123	13.2 h	159	84%
Xenon-127	36.4 d	172, 203, 375	25.5%, 68.3%, 17.2%
Xenon-133	5.25 d	81, 161	43%, 0.06%

3.2.4.1 Neutron Activation (Reactor-produced)

When a stable nuclide is struck by a neutron, its nucleus may capture the neutron and become neutron rich and, as described in section 3.2.2.2, unstable. The unstable neutron-rich nuclide will tend to decay through β^- decay to a stable nuclide. Neutron activations are usually achieved in a nuclear reactor because of its abundance of neutrons. The most common type of *neutron activation* is the (n, γ) reaction with the release of a γ photon, which can be represented by

$$^{A}_{Z}X\ (\mathrm{n}, \gamma)\ ^{A+1}_{Z}X \tag{3.33}$$

Here, the target and product nuclides are different isotopes of the same element. The product is not carrier free, i.e. free of stable atoms of the same element as the radioactive nuclide), and may have very low specific activity. Another common type of neutron activation is the (n, p) reaction, represented by

$${}^{A}_{Z}X\ (\mathrm{n, p})\ {}^{A}_{Z-1}Y \tag{3.34}$$

Here, a proton is ejected resulting from absorption of a neutron into the nucleus, and the target and product are different elements.

3.2.4.2 Charged Particle Activation (Cyclotron-produced)

In charged-particle activation, positively charged particles, such as protons, deuterons (${}^{2}_{1}H$ nuclei), tritrons (${}^{3}_{1}H$ nuclei), and α particles (${}^{4}_{2}H_{e}$ nuclei), are accelerated to a very high speed and directed onto a target nuclide. Electrons are not used because they have a large relativistic mass increase, even at relatively low energy (≤ 100 keV). The incident particles must have enough energy to penetrate the repulsive Coulomb force exerted by the nucleus. Two common types of charged particle activation reactions are:

$${}^{A}_{Z}X\ (\mathrm{p, n})\ {}^{A}_{Z+1}Y \tag{3.35}$$

and

$${}^{A}_{Z}X\ (\mathrm{d, n})\ {}^{A+1}_{Z+1}Y \tag{3.36}$$

In both cases, the product nuclei are proton rich and tend to undergo β^+ decay or electron capture transitions to stable nuclei. The target and product are different elements, with different atomic numbers, and as a result the product is carrier free. The reaction usually has a small activation cross-section, or probability of activation, and the incident particle beam has a low intensity, resulting in a relatively high cost of production. β^+ emitters, such as ^{11}C (20 min), ^{13}N (10 min) and ^{15}O (2 min), can be used to label a wide variety of physiological tracers and are useful in PET.

3.2.4.3 Photonuclear Activation

Another possible method of radioisotope production is *photonuclear activation,* which can be represented by

$${}^{A}_{Z}X\ (\gamma, \mathrm{n})\ {}^{A-1}_{Z}X \tag{3.37}$$

and

$${}^{A}_{Z}X\ (\gamma, \mathrm{p})\ {}^{A-1}_{Z-1}X \tag{3.38}$$

Photonuclear activation requires high-energy photons (≥ 10 MeV) and is not a common method of radionuclide production.

3.2.4.4 Radionuclide Generator

A typical generator consists of a parent–daughter radionuclide pair in which the daughter can be separated and extracted from the parent. The most important generator used in nuclear medicine is the $^{99}Mo - ^{99m}Tc$ generator system, with the half-lives of the parent ^{99}Mo and daughter ^{99m}Tc being 66 hours and 6 hours respectively. The decay process of the $^{99}Mo - ^{99m}Tc$ generator satisfies the condition of transient equilibrium described in section 3.2.3.6.

3.2.4.5 Activation Rate

The activation rate of a target bombarded by a beam of high-energy particles is given by

$$\left(\frac{dN}{dt}\right)_{act.} = \sigma\phi Nx[\text{activations/sec}] \tag{3.39}$$

where ϕ is the flux density (particles/cm^2 sec) of the incident beam, σ is the activation cross-section, N is the number of target nuclei/cm^3 and x is the thickness of the target.

The activation rate per unit mass is given by

$$\frac{1}{m}\left(\frac{dN}{dt}\right)_{act.} = \frac{(dN/dt)}{(Nx)A/N_o} = \frac{6.023\times 10^{23}\sigma\phi}{A} \tag{3.40}$$

where A is the sample weight and N_o is the Avogadro's number, i.e. 6.023×10^{23}.

We can consider the irradiation beam to act as an inexhaustible supply of long-lived "parent" nuclei generating "daughter" nuclei at a constant rate with $\lambda_{parent} << \lambda_{daughter}$. Let $D(t)$ be the number of daughter nuclei at time t; then, from section 3.2.3.6,

$$D(t) = \frac{\lambda_{parent} N(t)}{\lambda_{daughter}}(1 - e^{\lambda_{daughter} t}). \tag{3.41}$$

As $dN/dt = \sigma\phi Nx = \text{constant} = \lambda_{parent}N(t)$, we have

$$D(t)\lambda_{daughter} = \sigma\phi Nx(1 - e^{\lambda_{daughter} t}). \tag{3.42}$$

The *saturation activity* is given by

$$A(t \to \infty) = D(t \to \infty)\,\lambda_{daughter} = \sigma\phi Nx \tag{3.43}$$

The specific activity at time t_2 after a period of time t_1 bombardment is given by

$$A(t_1 + t_2) = \sigma\phi Nx(1 - e^{\lambda_{\mathrm{daughter}} t_1})\,(1 - e^{\lambda_{\mathrm{daughter}} t_2{}^2}) \tag{3.44}$$

3.3 Principles of Conventional Nuclear Medicine Imaging

3.3.1 Introduction

In conventional nuclear medicine imaging[13], a tracer amount of a given radiopharmaceutical, i.e. a pharmaceutical labeled with a radioisotope, is administered to the patient. The pharmaceutical carries the radioisotope to different organs or tissues depending on its biokinetic properties. For example, ^{99m}Tc-labeled sulfur colloid is trapped in normal liver tissue but not cancerous liver tissue and is used to detect liver metastases. Thallium-201 acts as a potassium analog and can be used as a perfusion agent to evaluate the perfusion characteristic of the myocardium. The isotope most widely used in nuclear medicine is ^{99m}Tc. Its popularity is the result of its almost ideal half-life of 6 hours and the fact that it decays with the emission of 140 keV photons. Other isotopes used in nuclear medicine include ^{123}I, ^{131}I, ^{121}In and ^{67}Ga. Unlike other medical imaging modalities, nuclear medicine provides functional and physiological information about the patient through the biokinetic properties of the administered radiopharmaceuticals.

All the radioisotopes used in nuclear medicine imaging emit gamma rays, which travel through the patient's body and are detected by a position-sensitive radiation detector placed outside the body to form an image. In conventional planar nuclear medicine imaging, the 3D distribution of radioactivity is recorded by the radiation detector in a 2D plane representing a projection of the 3D information. In emission computed tomography (ECT), cross-sectional images are reconstructed from multiple-projection images obtained from around the patient.

The goal of nuclear medicine imaging is to produce images that accurately represent the radioactivity distribution *in vivo*. However, the instrumentation used, as well as physical factors, such as attenuation and scatter, and other factors related to the biokinetics of the radiopharmaceutical and to the patients cause undesirable degradation of the acquired data and subsequent images. Research and development have been devoted to improving the instrumentation, image processing and reconstruction methods in order that nuclear medicine images with the highest quality and quantitative accuracy can be obtained.

3.3.2 Instrumentation

Most of the instruments used in nuclear medicine imaging include a positionally sensitive radiation detector system, a collimator, and computer and image display systems. The position-sensitive radiation detector system and its associated collimator have major effects on the nuclear medicine images, as described below.

3.3.2.1 Radiation Detectors

Radiation detectors are used to detect and evaluate incident photons for various applications, ranging from radioactivity assays, radiation monitoring and dosimetry measurements to imaging[14]. The radiation detection process involves transforming the energy of the photon into an electronic signal for subsequent processing. There are various types of radiation detectors. Gas-filled detectors include ionization chambers, proportional counters and Geiger–Müller (GM) tubes. Although these detectors have found wide applications as dosimeters, dose calibrators and gas-flow counters, they are seldom used in nuclear imaging, primarily because of their low detection efficiency. Liquid scintillation detectors, which are useful in detecting β-particles and low-energy X- and γ-rays, are also not used in nuclear medicine imaging. Semiconductor detectors have the advantage of superb energy resolution and allow an exceptional degree of scatter elimination. However, the popular Si and Ge detectors must be operated at a low temperature (e.g. liquid nitrogen for Ge) to avoid undesirable noise currents. Another problem with semiconductor detectors is caused by the presence of impurities. To avoid this problem, either a very pure crystal (e.g. high-purity Ge) is used or compensating impurities, e.g. lithium (Li), are needed. Either of these approaches increases the cost of the detector. Finally, it is difficult to fabricate semiconductors with a thickness greater than about 1 cm. These disadvantages limit the use of semiconductor detectors in nuclear medicine. However, the recent development of 'room-temperature' semiconductor materials, e.g. zinc cadmium telluride (ZnCdTe), has renewed interest in the potential use of semiconductor detectors in nuclear medicine.

Most radiation detectors in nuclear medicine are based on crystalline solids of inorganic scintillators. Table 3.3 shows the properties of four scintillation crystals commonly used in nuclear medicine. Among them, the thallium-activated sodium iodine, NaI(Tl), crystal is by far the most widely used as a result of its excellent radiation detector properties, availability and low cost. Bismuth germanate, $Bi_3Ge_4O_{12}$ or BGO, cesium fluoride, CsF, and barium fluoride, BaF, are used in positron emission tomography (PET) systems.

When a γ-ray emitted from an radioisotope interacts with a scintillation crystal, ionization or excitation of atoms of the crystal occurs. Most of the energy released during these ionization and excitation processes is dissipated as heat. Some of the energy is released in the form of visible light photons, i.e.

Table 3.3. Properties of scintillators used in nuclear medicine and emission computed tomography systems.

Scintillator	NaI(Tl)	BGO[1]	CsF	BaF
Effective A	50	74	53	52
Density (gm/cm^3)	3.67	7.13	4.61	4.89
Index of Refraction	1.78	2.15	1.48	1.56
Scintillation Photon Yield (per KeV)	40	4.8	2.5	2.0
Decay Time	230	300	2.5	0.8, 620[2]
Wavelength of scintillations (Å)	4150	4800	3900	2250, 3100[2]
Hydroscopic	Yes	No	Very	No

[1] $Bi_3Ge_4O_{12}$
[2] Fast and slow components, respectively

scintillation emissions. The amount of light produced by a single γ-ray is very small. A major advance in scintillation detector technology is the development of photomultiplier tubes (PMTs), which amplify the small light scintillations into electronic signals that can be accurately detected and measured.

Figure 3.5 shows a typical scintillation detector consisting of a NaI(Tl) scintillator and a PMT. The PMT consists of a photocathode coated with a photoemissive surface operating at a negative potential. The photocathode absorbs the scintillation light photons with the ejection of electrons. The ejected electrons are directed towards stages of dynodes with increasingly higher positive potentials. When an accelerated electron hits a dynode, several secondary electrons are ejected. The total number of secondary photons generated depends on the energy of the electrons that strike the dynode (or the potential difference between adjacent stages of the dynodes) and the number of dynodes. A typical PMT consists of about 10 dynodes with a total amplification factor of the order of ~10^7. This large multiplication factor provides output signals that have sufficient magnitude for further processing.

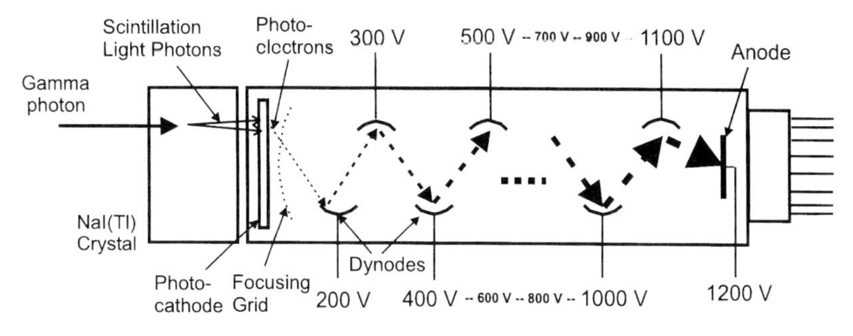

Figure 3.5. Schematic diagram of a typical scintillation detector that consists of a NaI(Tl) scintillator and a photomultiplier tube (PMT).

In addition, the magnitude of the output signal of a scintillation detector is proportional to the energy of the incident photon. However, there are several factors that introduce variations in this relationship. They include fluctuations, due to electrical noise, and statistical variations in the number of light photons produced per incident γ-photon, the number of photoelectrons released from the photocathode and the multiplication factor of each of the dynodes. Among these factors, the most important is the variation in the number of photoelectrons from the photocathode.

Figure 3.6 shows a typical energy spectrum of ^{99m}Tc detected using a NaI(Tl) detector. The photopeak in the energy spectrum is approximately Gaussian in shape. The *energy resolution*, R, is often defined as

$$R(E) = \frac{FWHM(E)}{E} \times 100\% \tag{3.45}$$

where E is the gamma ray energy at which the energy resolution is defined and FWHM (full width at half maximum) is the width of the Gaussian-shaped photopeak at half the peak height. The energy resolution $R(E)$ of a scintillation detector is inversely proportional to the energy of the incident photon:

$$R(E) = K\frac{\sqrt{E}}{E} = \frac{K}{\sqrt{E}} \tag{3.46}$$

where K is a constant. Equation (3.46) shows that the energy resolution is inversely proportional to the square root of the gamma ray energy.

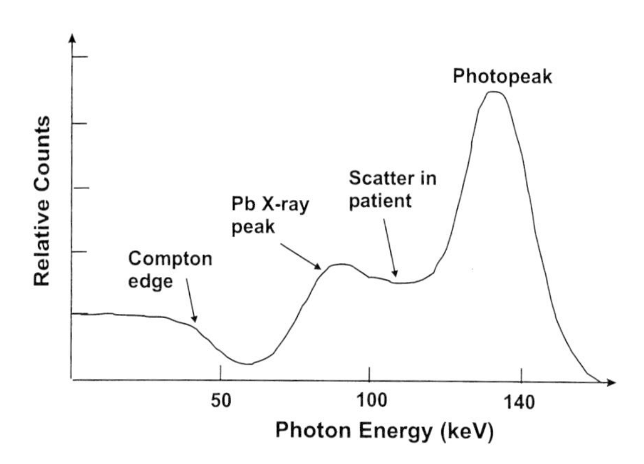

Figure 3.6. Typical energy spectrum obtained from a patient injected with ^{99m}Tc-labeled radiopharmaceutical using a scintillation camera with a lead collimator. The lead X-ray peak is a result of interaction of the 140 keV photons from ^{99m}Tc with the lead septa of the collimator. The Compton edge represents the maximum energy deposited in the scintillation crystal and corresponds to the energy of the recoil electron for 180° Compton scattering events.

3.3.2.2 Scintillation Cameras Used in Nuclear Medicine Imaging

To provide position information for the detected photons during formation of an image, a small radiation detector can be scanned over the area to be imaged. An image can be formed by recording the magnitude of the detected signal at each of many different scanning positions. Early planar nuclear medicine imaging systems consisted of a single or an array of small radiation detectors. These systems required long scanning times to complete the acquisition of data for an image.

A large-area stationary detector permits faster acquisition of a nuclear medicine image because it eliminates scanning motion. The most popular nuclear medicine imaging system is the scintillation camera developed by Anger [15], which is often called the *Anger* camera. Figure 3.7 shows a diagram of a modern scintillation camera. It consists of a large-area circular (e.g. 40 cm in diameter) or rectangular (e.g. 40 × 50 cm) NaI(Tl) crystal with a typical thickness of just less than 1 cm for detection of low-energy photons (e.g. 140 keV r rays from ^{99m}Tc).

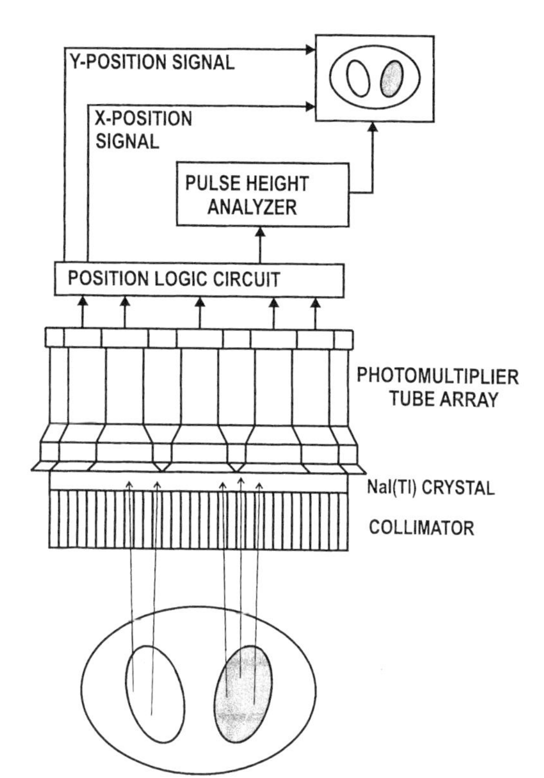

Figure 3.7. Schematic diagram of the components of a typical conventional scintillation camera system. [From: B.M.W. Tsui, *Physics of SPECT. Radio-Graphics* **16(1)**,173–183 (1996)]

Behind the scintillation crystal is an array of PMTs. When a photon is absorbed at a particular location in the crystal, the resulting scintillation light photons are detected by the PMTs. The number of light photons intercepted by a particular PMT, and its resulting output signal, depend on the distance of the particular PMT from the point of photon absorption. From the output signals of the array of PMTs, the location of the interaction of any photon in the crystal can be determined. In addition, the sum of the output of the PMTs is proportional to the energy absorbed in the crystal for the incident photon. A typically modern scintillation camera has an energy resolution of about 10%.

Because of the finite energy resolution of the scintillation detector, absorbed energies within an *energy window* are accepted. A wider energy window accepts more primary photons at the expense of an increased number of scatter photons detected in the same energy window. The choice of an optimum energy window is important to ensure good image quality.

3.3.2.3 Collimators

A collimator used in nuclear medicine imaging can be regarded as being the analog of a lens used in optical imaging. It is placed in front of the crystal of the scintillation camera, as shown in Figure 3.8. The most commonly used camera collimator consists of a large array of small holes with parallel walls that are made of lead. The collimator hole and septa restrict the direction of the photons that are detected by the crystal. By using an appropriate septal thickness, photons that travel in the directions that intercept the septa will be stopped and unable to reach the crystal for detection. The septa must be sufficiently thick to stop most of the intercepting photons to minimize the amount of septal penetration. On the other hand, overly thick septa will lower the detection efficiency.

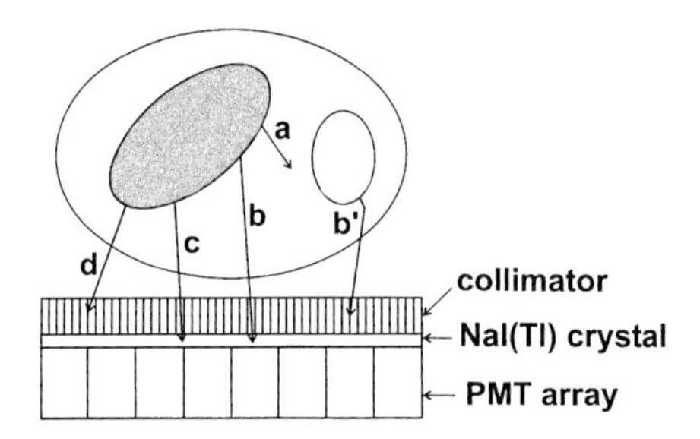

Figure 3.8. Schematic diagram of photon interactions inside a patient (ellipse) and collimation of the exiting photons before detection. Photon path *a* indicates a photoelectric interaction in which the photon is totally absorbed by an atom. Photon path *b* and *b′* indicate a Compton scatter interaction in which the photon changes its direction of travel, with partial loss of its original energy. *b*, A scattered photon, and *c*, a primary, unscattered photon, pass through holes of the collimator and are detected by the radiation detection. *b′* and *d* are photons that are blocked by the collimator hole septa and are not detected by the scintillation camera. [From: B.M.W. Tsui, *Physics of SPECT. RadioGraphics* **16(1)**,173–183 (1996)]

The ratio of the hole size to the hole length determines the acceptance angle of photons and the spatial resolution of the collimator. The geometric efficiency of a multihole collimator for a scintillation camera is approximately proportional to the area of the hole aperture or the square of the spatial resolution.

The collimator is the major factor that determines both the resolution and the efficiency of the scintillation camera system. The use of an appropriate collimator design is important to ensure the best image quality obtained from a nuclear medicine imaging system. A typical scintillation camera system includes several different collimators for different clinical applications and imaging conditions. Examples are high-sensitivity, general-purpose and high-resolution collimators with different trade-offs of spatial resolution and detection efficiency; also included are low-energy, medium-energy and high-energy collimators for use with radioisotopes that emit photons with different energies.

To improve the trade-off between spatial resolution and detection efficiency over the conventional parallel-hole design, other collimator designs have been developed. Figure 3.9 shows different collimator design geometries. The most commonly used parallel-hole collimator design has an object to image size ratio of one. The converging-hole collimator design (e.g. fan-beam and cone-beam) provides higher detection efficiencies (~ 1.5 times for fan-beam and ~ two times for cone-beam) compared with the parallel-hole design for the same spatial resolution at a cost of smaller field-of-view. Converging-hole collimators are useful for imaging organs or body regions that are smaller than the size of the scintillation camera. The pinhole collimator provides a substantial increase in detection efficiency when the object is close to the pinhole aperture. However, the much smaller field-of-view allows imaging of only small organs, such as the thyroid. The pinhole collimator is not suitable for imaging large objects because of the rapid decrease in detection efficiency as a function of distance. The diverging-hole collimator provides a larger field-of-view than the parallel-hole collimator at the expense of lower detection efficiency.

3.3.3 Quantitative Nuclear Medicine Imaging

The ultimate goal of nuclear medicine imaging is the accurate determination of radioactivity *in vivo* so that accurate clinical diagnoses can be made. However, there are a number of factors that interfere with the achievement of the goal. These factors can be divided into physical factors and factors related to the radiopharmaceutical, patient, imaging system and the image reconstruction methods.

Physical factors include noise fluctuations in the detected counts and the attenuation and scatter of photons in patients. Photons with energies less than 1 MeV have two possible ways of interacting with tissues in the patient, namely photoelectric interaction and Compton scattering. During these inter-

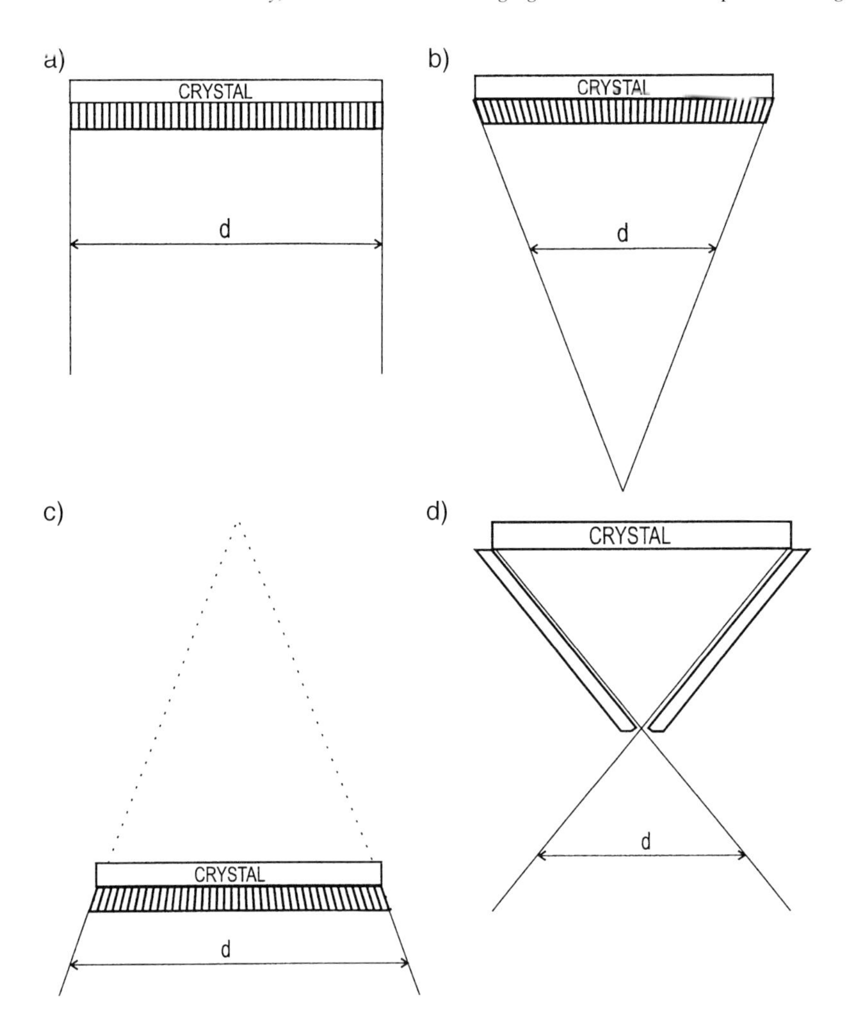

Figure 3.9. Common types of collimator designs for scintillation camera. (a) Parallel-hole collimator. (b) Converging-hole collimator. (c) Diverging-hole collimator. (d) Pinhole collimator. The useful field of view is indicated by D. 4. [*From: Tsui BMW. Collimator Design, Properties and Characteristics. (In) SNM The Scintillation Camera, Simmons GH, ed. (Society of Nuclear Medicine), Chapter 2, pp. 17–45, 1988*]

actions, the emitted photons may be absorbed in the patient's body or scattered from their original paths with loss of energy. These photons will not be intercepted by the detector and are considered to have been attenuated. The degree of attenuation depends on the energy of the photons, the attenuation properties of the tissues along the path of the photons and the thickness of the tissue they have to travel through before they reach the detector. When a photon beam with intensity I_o, is incident on a slab of material with thickness d, the intensity of the transmitted photon beam, I, is given by

$$I = I_0 e^{-\mu d} \tag{3.47}$$

where μ is the linear attenuation coefficient of the material at the energy of the incident photons. For 140 keV photons, for example, the half-value layer (HVL) (the thickness of the soft tissue that will stop half of the incident photons) is about 4.5 cm. Attenuation is the major influence on the quantitative accuracy of nuclear medicine images.

As shown in Figure 3.8, when a photon experiences a Compton scattering interaction, it changes its direction of travel and loses some of its energy. Because of the finite energy resolution of the camera, scattered photons originating from some distance away may be detected within the energy window used. Detection of scattered photons results in reduced image contrast and a degradation in image quality.

3.4 Principles of Single-photon Emission Computed Tomography (SPECT)

3.4.1 Introduction

Single-photon emission computed tomography (SPECT) is a medical imaging modality that combines conventional nuclear medicine imaging techniques with methods for image reconstruction from projections. In SPECT, planar nuclear medicine images (projection images) are acquired from different orientations around the patient. Image reconstruction methods are then applied to the projection images to form cross-sectional (i.e. tomographic) images of the patients. Under ideal situations, when the projection data are proportional to simple sums of the radioactivity distribution along the projection rays, reconstructed images provide accurate representations of radioactivity distributions without the effects of overlapping structures in planar images.

3.4.2 Instrumentation

SPECT instrumentation uses a radiation detector that is similar to those used in conventional nuclear medicine systems. In addition, the instrumentation includes a special gantry that moves the radiation detector so that projection data can be acquired around the patient. To reduce the acquisition time to acquire projection data, SPECT systems with multiple detector arrays, and single and multiple cameras have been developed and are commercially available[16].

3.4.2.1 Multiple Detector-based SPECT Systems

Figure 3.10 illustrates three examples of multiple-detector-based SPECT systems. They are the MARK IV system[17], the Headtome-II[18] and the Cleon brain SPECT system[19]. These systems use an array of small radiation detectors for acquiring projection data at specific projection angles. By placing a ring of detectors around the patient, the time required to acquire a complete set of projection data can be reduced compared with the time required to rotate a single camera around the patient. Another advantage of a multiple-detector-based SPECT system is its ability to measure high count rates, required for first-pass cardiac imaging when the bolus of radiopharmaceutical is followed right after injection. However, multiple-detector systems are more expensive and difficult to maintain than camera-based SPECT systems. With advances in high-count-rate digital cameras and multiple-camera-based SPECT systems, the advantages of multiple-detector-based SPECT systems are diminishing. Recently, new multiple-detector-based SPECT systems have been developed that allow simultaneous acquisition of multiple-image slices.

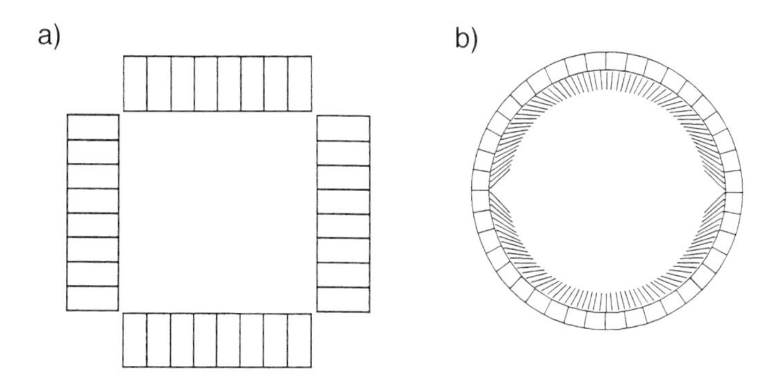

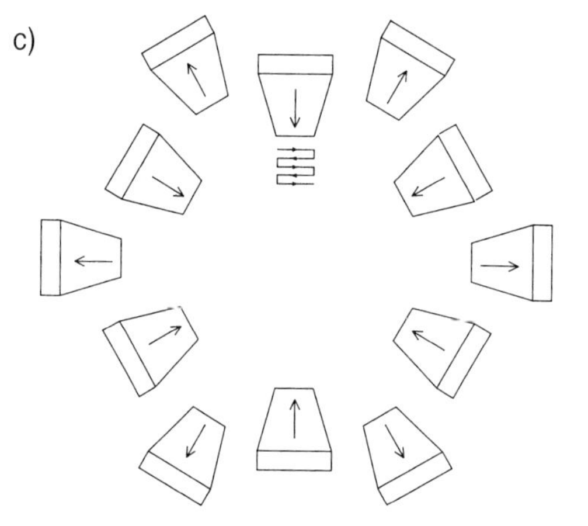

Figure 3.10. Examples of multi-detector based SPECT systems. (a) The MARK IV system consists of four arrays of eight individual NaI(Tl) detectors arranged in a square configuration. (b) The Headtome-II system consists of a circular ring of detectors. A set of collimator vanes that swings in front of the discrete detector is used to collect projection data from different views. (c) A unique Cleon brain SPECT system consists of 12 detectors that scan both radially and tangentially. [*From: Tsui BMW. SPECT (Single-Photon Emission Computed Tomography). (In) Biomedical Engineering Handbook, J. Bronzino, Ed., (CRC Press, Inc., Boca Raton, FL), pp. 1055–1076, 1995*]

3.4.2.2 Rotating Camera-based SPECT Systems

Most commercial SPECT systems consist of a single or of multiple scintillation cameras mounted on a rotating gantry, as shown in Figure 3.11. Rotating camera-based SPECT systems allow the use of state-of-the-art camera technology. The total detection efficiency of the system is proportional to the number of cameras used. A dual-camera SPECT system has two cameras positioned 180° apart as shown in Figure 3.11 (b). The system can also be used for dual-view whole-body scanning. When they are configured at 90°, the system can be used for fast acquisition of 180° data in cardiac SPECT. Dual-camera SPECT systems that provide a variable angle between the two cameras are increasing in popularity. The triple-camera SPECT system design [Figure 3.11(c)] consists of three cameras configured such that they completely surround the object to be imaged. This design provides the best compromise between high detection efficiency and system cost. It is ideal for brain and whole-body SPECT imaging when fast acquisition of 360° projection data is required. Currently, the dual-camera SPECT system is most popular, principally because of the large demand for cardiac and whole-body SPECT imaging.

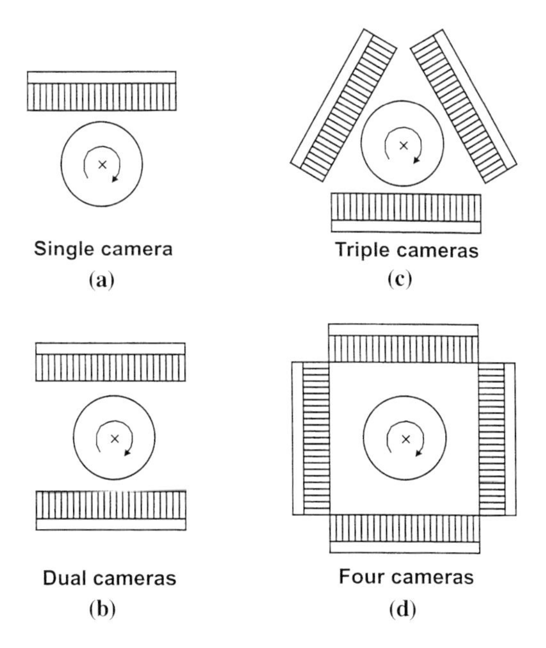

Figure 3.11. Schematic diagrams show typical configurations of commercial SPECT imaging systems that are based on single or multiple rotating scintillation cameras. The increased number of scintillation cameras around the patient results in increased detection efficiency or improved spatial resolution. [*From: Tsui, BMW, Physics of SPECT. RadioGraphics* **16(1)**,173–183, 1996]

3.4.3 Image Reconstruction Techniques

3.4.3.1 Analytical Methods

The basic method for 2D image reconstruction from projections was developed for X-ray CT. As shown in Figure 3.12, the projections are in the form of a one-dimensional (1D) data array. Assume that the projection data $p_\theta(t)$ at projection angle θ are given by the sum or integral of the radioactivity distribution f(x,y)

$$p_\theta(t) = \Lambda \int_0^\infty f(x,y)\,\mathrm{d}s \tag{3.48}$$

where θ is the angle between the projection array and the x-axis, s is the distance along a perpendicular to the projection data and Λ is the gain factor. Equation (3.48) is called the Radon transform[20]. The reconstruction problem is to seek a solution of the inverse of the Radon transform[21].

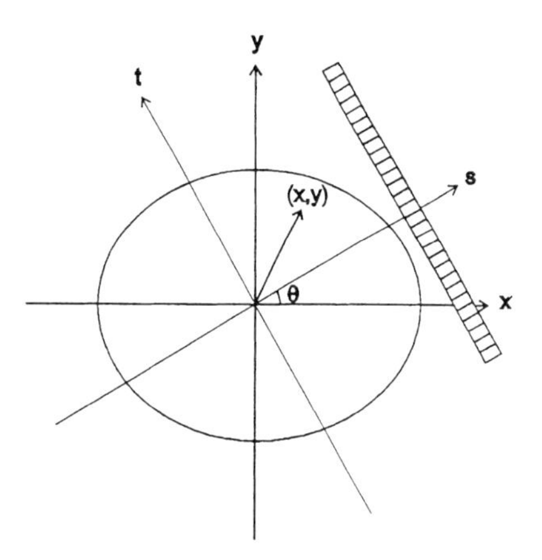

Figure 3.12. Schematic diagram of the two-dimensional image reconstruction problem. The projection data are line integrals of the object distribution along rays that are perpendicular to the detector. A source point (x,y) is projected onto a point (t,θ), where t is a position along the projection and θ is the projection angle. [*From: Tsui BMW. SPECT (Single-Photon Emission Computed Tomography). (In) Biomedical Engineering Handbook, J. Bronzino, Ed., (CRC Press, Inc., Boca Raton, FL), pp. 1055–1076, 1995]*

From the central slice theorem, the cross-section of the Fourier transform of the reconstructed image $F(\nu_x, \nu_y)$ along the θ direction where $\tan\theta = y/x$ is equal to the Fourier transform of the projection data along the same direction $p_\theta(s)$, i.e.

$$F(\nu_x, \nu_y) = P_\theta(\nu_t) \tag{3.49}$$

where $P_\theta(\nu_t)$ is the one-dimensional Fourier transform of $p_\theta(t)$: $P_\theta(\nu_t) = FT^{-1}[p_\theta(t)]$.

Equation (3.49) can be used directly in image reconstruction. However, a more common form of the reconstruction algorithm is the backprojection of the filtered projection in the spatial frequency domain. The reconstructed image estimate $\hat{f}(x,y)$ is given by

$$\hat{f}(x,y) = \boldsymbol{B}\{FT^{-1}[W(\nu_t)P_\theta(\nu_t)]\} \quad (3.50)$$

where $\boldsymbol{B}$ is the backprojection operator

$$B[P_\theta(t)] = \int_0^{2\pi} P_\theta(\mathrm{t})\mathrm{d}\theta \quad (3.51)$$

and FT^{-1} is the 1D inverse Fourier transform operator. The filter $W(\nu)$ is a 1D ramp filter.

Equivalently, in the spatial domain, the reconstructed image estimate can be obtained from the convolution backprojection algorithm

$$\hat{f}(x,y) = \int_0^{2\pi} w(s-s')P_\theta(t)\mathrm{d}\theta \quad (3.52)$$

where $w(s)$ is the Fourier transform of the 1D ramp function $W(\nu)$.

These analytical image reconstruction algorithms are efficient, robust and relatively easy to implement. Consequently, they are the most popular reconstruction method found in clinical SPECT systems. An alternative approach is the use of iterative reconstruction algorithms, which have found particular use in quantitative SPECT image reconstruction.

The analytical reconstruction methods described above provide an exact solution to the inverse Radon transform problem only under ideal conditions. In SPECT, this means that the projection data are exact sums of the radioactivity along the projection rays. In practice, the projection data are affected by a variety of image-degrading factors, such as those described in section 3.3.3. Direct application of the analytical reconstruction methods to measured projection data leads to inaccuracies in the reconstructed images and often a degradation in image quality.

3.4.3.2 Iterative Reconstruction Algorithms

In a typical iterative algorithm applied to SPECT data, an initial estimate (typically a uniformly emitting distribution) for the radioactivity source distribution is first assumed. A set of projections is calculated from the initial source distribution using a known, or assumed, *transition matrix*. The (i, j) element of the transition matrix is equal to the probability that a gamma ray emitted from the jth voxel in the radioactivity distribution is detected in the ith bin of the projection data set. The transition matrix can be used to include available knowledge about the attenuation and scatter properties of the object and the detection system used in acquiring the projection data. The calculated projection data are compared with the measured projection data and, using an appropriate updating scheme, a correction is applied to the estimated source distribution. This process is repeated until the current estimate of the radioactivity satisfies a preselected criterion.

The correction applied to update the estimated source distribution is based on certain criterion. One possible criterion is the minimum mean squares error (MMSE), which minimizes the discrepancies between the measured and the calculated projections[22]. Other criteria include the maximization of various conditional functions, such as the maximum entropy (ME)[23] or maximum likelihood (ML)[24], and Bayesian methods, including maximum *a posteriori* approaches[25]. The ML approach has generated the most interest in recent years. The Bayesian approach is more general and may provide improved reconstructions, as *a priori* source information is explicitly included in the method.

Although iterative reconstruction approaches were first evaluated in the late 1960s and early 1970s as solutions to the reconstruction problem[26, 27], they have not been used in commercial CT imaging systems. The wide acceptance of the alternative analytical reconstruction methods is related to its high computational efficiency. A major drawback of clinical implementation of iterative reconstruction methods has been the extensive computational requirement.

More recently, the use of iterative reconstruction methods in SPECT has attracted renewed interest. This interest has resulted from the idea that characteristics of the imaging system, as well as physical processes such as attenuation and scatter, can be directly incorporated into iterative algorithms via the transition matrix. This allows accurate compensation for these image-degrading factors, which is impossible to achieve with analytical reconstruction methods. Recently, advances in computer hardware and custom-dedicated processors have significantly reduced the processing time required by the iterative algorithm. In addition, fast iterative algorithms, such as the ordered-subset (OS) expectation-maximization (EM) algorithm,[28] have shown much faster converging rates. The combination of these advances with the optimization of software implementation could well allow the application of iterative reconstruction to clinical use in the near future.

3.4.4 Quantitative SPECT Methods

Like conventional nuclear medicine imaging (section 3.3.3), projection data acquired in SPECT are affected by various image degrading factors[29]. Recently, much of the research in SPECT has been devoted to the development of quantitative SPECT methods, with the goal of compensating for degrading factors so that SPECT images with the highest quality and quantitative accuracy can be achieved.

3.4.4.1 Compensation for Attenuation

Attenuation is primarily caused by absorption of photons by photoelectric interactions and scattering of photons during Compton interactions. Attenuation is a major factor affecting the quantitative accuracy of SPECT data. Hence, attenuation compensation is essential if SPECT quantitation is to be improved. Most attenuation compensation methods are based on various assumptions concerning the nature of the source activity and the attenuating medium. These methods can be classified as (1) preprocessing compensation; (2) intrinsic compensation; and (3) post-processing compensation.

Preprocessing methods[30] attempt to correct measured projection data before image reconstruction. They are based on the arithmetic or geometric mean of projection data acquired on the opposite side of the patient. Preprocessing methods provide only approximate compensation, particularly for multiple-source distributions. The intrinsic compensation method is based on a direct solution of the inverse Radon transform that includes the attenuation effect. However, an exact solution is found only when uniform attenuation is assumed[31]. In post-processing approaches[32], a reconstructed image is first obtained using the conventional analytical reconstruction method without any attenuation compensation. Attenuation compensation is achieved by multiplying the uncompensated image by a correction matrix whose element is equal to the reciprocal of an estimated average attenuation along all projection rays that span between the element and the boundary of the attenuation medium.

The attenuation compensation methods described above are easy to implement. Because of the underlying assumptions however, they provide only approximate attenuation compensation. In clinical situations that involve complex radioactivity source and non-uniform attenuation distributions, these methods may not be adequate. An example is cardiac SPECT: the chest region consists of lung and soft tissues that have different attenuation coefficients. Anatomical variations, such as the size of the body, the shape of the diaphragm and the size and shape of breasts in female patients, produce different artifacts in the reconstructed image that may affect clinical diagnosis[33]. To compensate for a non-uniform attenuation, iterative reconstruction methods have been developed that incorporate the attenuation coefficient distribution in the transition matrix[34]. The methods involve acquisition of transmission CT data using a radionuclide source. The attenuation coefficient distribution is obtained by reconstructing the transmission CT data. Preliminary clinical data indicate that compensation for non-uniform attenuation has the potential to improve clinical diagnosis[35].

3.4.4.2 Compensation for Compton Scattered Photons

The detection of Compton scattered photons leads to a degradation in quantitative accuracy and image quality, especially a reduction in image contrast. In typical clinical SPECT imaging situations using ^{99m}Tc, the ratios of scattered to primary photons is about 20–30% for brain and may be as much as 40% for body imaging. Therefore, compensation for scatter is important to the improvement of both the quality and the quantitative accuracy of SPECT images. A large effort has been devoted to this problem, and a number of approaches have been developed to compensate for scatter.

One approach is based on the convolution–subtraction technique[36]. In this technique, scatter is modeled as a convolution of the measured projection data with an empirically derived function. The resulting estimate of the scatter contribution is then subtracted from the original projections to provide compensated projections.

A simple and popular scatter compensation method is based on data obtained from an energy window that is placed directly over the photopeak and another energy window that is placed below the photopeak window. The method assumes that the scatter image acquired from the lower energy window is a close approximation to a constant multiplied by the true scatter component of the image acquired from the photopeak window[37]. Hence, scatter compensation is achieved by subtracting an empirically determined fraction of the image reconstructed, using events acquired within the lower energy window, from the image reconstructed from events acquired from the photopeak window. However, the approach requires a scatter-multiplier that depends on the specific patient to be imaged, the imaging system characteristics and the imaging parameters used[38].

The scatter compensation methods described above and other similar methods assume that the scatter response function is either stationary and/or symmetrical in shape. For some SPECT imaging situations, these first-order approximate scatter compensations may be sufficient. In reality, however, the scatter response function varies as a function of depth and with distance from the edge of the scatter medium[39]. To compensate accurately for the effects of scatter, this information must be taken into consideration.

An exact model of scatter can be incorporated into iterative reconstruction algorithms to permit accurate compensation for scatter[40]. The approach does not yield higher image noise, as in the subtraction method. Because of the intensive calculations involved, practical implementation of the exact scatter compensation method requires accurate parameterization of the scatter response function[41] and high-speed computational hardware.

3.5 Principles of Positron Emission Tomography (PET)

3.5.1 Introduction

Like SPECT, positron emission tomography (PET) combines radiation detection techniques with image reconstruction from projections[42]. The major difference between the two ECT techniques is the use of radioisotopes that emit positrons. In particular, the positron emitters carbon-11, nitrogen-13 and oxygen-15 can be used to label physiologically active compounds that are involved in biochemical and metabolic activities in the body. Fluorine-18, perhaps the most commonly used radioisotope in PET, can form a strong carbon–fluorine bond that is similar to the hydrogen–carbon bond and can exhibit steric effects similar to hydrogen. The ability of PET to image these compounds provides a unique opportunity to assess patient physiological function that is impossible to achieve with other medical imaging techniques.

Imaging of positron emitters involves detection of the two 511 keV photons released during annihilation of the positron with an electron. Table 3.4 lists the important radioisotopes that are used in PET. The have short half-lives and often require on-site production. As a result, PET imaging requires special instrumentation designs, and image processing and reconstruction methods.

Table 3.4. Properties of positron emitters used in PET.

Positron Emitters	Half Life (min)	Maximum Positron Kinetic Energy (MeV)	Positron Range (mm)	Production Reaction	Means of Production
C-11	20.3	0.97	2.06	$^{14}N(p,\alpha)^{11}C$	Cyclotron
N-13	9.96	1.19	3	$^{16}O(p,\alpha)^{13}N$ $^{13}C(p,n)^{13}N$	Cyclotron
O-15	2.07	1.7	4.5	$^{14}N(d,n)^{15}O$ $^{15}N(p,n)^{15}O$ $^{16}O(p,pn)^{15}O$	Cyclotron
F-18	109.8	0.635	1.4	$^{18}O(p,n)^{18}F$ $^{20}Ne(d,\alpha)^{18}F$	Cyclotron
Rb-82	1.27	3.15	13.8	$^{82}Sr \rightarrow {}^{82}Rb$	Generator
Ga-68	68.3	1.88	5.4	$^{68}Ge \rightarrow {}^{68}Ga$	Generator

3.5.2 Positron Emission and Coincidence Detection

As described in section 3.2.2.3, proton-rich nuclei belong to a class of unstable nuclides that decay by either electron capture or emission of a positron β^+. Depending on its kinetic energy, the positron will travel a certain range before annihilation with an electron, yielding two 511 keV photons. The two annihilation photons travel in directions that are $180° \pm 0.4°$ apart. The basic principle of PET imaging involves the simultaneous detection of the two 511 keV photons, a technique known as coincidence counting. As shown in Figure 3.13, coincidence detection requires two detectors directly facing each other and connected by a coincidence circuit. An event is registered when both detectors detect a 511 keV photon within the resolving time of the coincidence circuit.

Because the two 511 keV photons are emitted in opposite directions, a detected coincidence event indicates that an annihilation event has occurred within the region common to the field-of-views of both detectors. This approach to collimation is called *electronic collimation*. The electronic collimation provides much higher detection efficiency than the lead collimator and contributes to the 10-fold increase in detection efficiency for PET compared with SPECT. Also, a line connecting the opposing coincidence detectors can be used to define a projection line in the PET reconstruction.

As shown in Figure 3.13, there are several types of photon interactions and detected events that occur in coincidence imaging. True coincidence events are registered when the two 511 keV photons from the same annihilation travel within the volume subtended by the pair of detectors and when the timing and detected energies are within the resolving time and the energy window used. An attenuated event occurs when one of the 511 keV photons is attenuated. A scatter event is recorded when one or both of the 511 keV photons are scattered before detection. An accidental or random event occurs when two photons from unrelated annihilations are detected as a coincidence event. The detection of attenuated, scatter and random events affects the quantitative accuracy and quality of PET images. Like SPECT, compensation for these image degradation factors is important to the development of PET imaging.

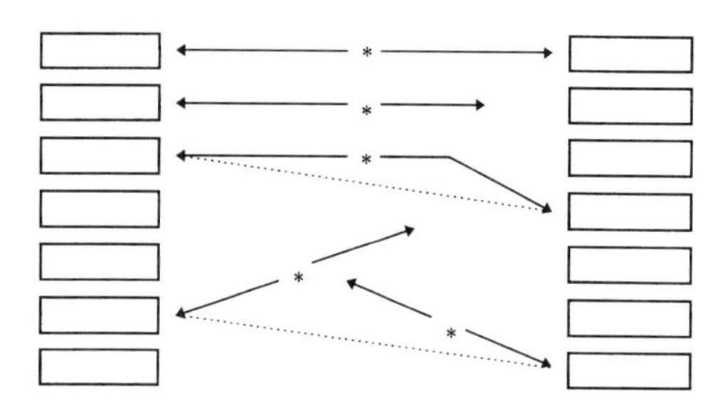

Figure 3.13. Different types of photon interactions and detected events that occur in coincidence imaging. From top to bottom: a true coincidence (one of the 511 keV photons is attenuated), an attenuated event a detected scattered event and an accidental or random event in which two photons from unrelated annihilations are detected as a coincidence event.

3.5.3 Basic PET System Designs

A PET system consists of multiple detectors surrounding the patient in the geometry of a ring or hexagon. Each detector is in coincidence with several detectors on the opposite side of the system. The multiple coincidence lines form the projections necessary for image reconstruction.

The detection efficiency of a PET system depends on the scintillation material used. Table 3.3 lists the properties of several scintillators that have been used in PET system designs. The NaI(Tl) scintillator was used in early PET systems because of its availability and relatively low cost. However, thick NaI(Tl) crystals are required to adequately stop the high-energy 511 keV photons leading to a loss in intrinsic resolution. In most modern PET systems, the scintillator bismuth germanate (BGO) is used because of its high stopping power for 511 keV photons. For example, a 3-cm-thick BGO will stop ~90% of the incoming 511 keV photons. In coincidence detection, this translates into an overall detection efficiency of $(0.9)^2$, or 0.81. However, BGO has a relatively long scintillation light decay time compared with other scintillators, resulting in a lower count rate capacity. Time-of-flight detection (in which the relative arrival times of the two 511 keV photons are used to determine the position of annihilation) has been proposed in PET system design to provide improved image quality. The CsF scintillator with a very short decay time was used in the early system designs. A major drawback of the CsF crystal is its extremely high hydroscopicity. Recently, the BaF scintillator, which is much less hydroscopic, has replaced CsF for use in time-of-flight detection. Time-of-flight detection has not been adopted in commercial PET system designs, principally because of its higher cost.

The spatial resolution of a PET system improves with smaller detector size, larger number of detectors and smaller ring size. The larger number of detectors also increases the counting capability and lower the dead time of the PET system. However, the smaller detector size requires a larger number of detectors and a high system cost. The optimal diameter of a PET system is of the order of 80 cm.

The early development of PET was limited by the size of the detectors and the PMTs required to be attached to each detector. Increasing the number of crystals, PMTs and associated electronics required to achieve improved spatial resolution became prohibitive. Also, the size of the PMTs imposes a limit on the smallest size for the detector and the spatial resolution of the system. Modern PET systems use large detector modules that are slotted into different depths to form a surface with an array of small crystals. The back of the crystal is viewed by a small number of PMTs less than the number of the smaller slotted crystals. Typically, a detector module consists of a 10×10 cm crystal whose surface is slotted into an array of 8×8 smaller crystals. There is an array of four PMTs at the back of the crystal. The outputs of the four PMTs allow determination of the position at which the photon hits the crystal.

A typical clinical PET system consists of multiple rings of BGO block detectors to provide a spatial resolution on the order of 4–5 mm. In 2D data acquisition mode, in which septa are used to define separate image slices, the total system sensitivity is on the order of 177 kcps/μCi/cc for a uniform cylindrical source 20 cm in diameter. In 3D data acquisition mode, in which the septa are removed, the total system sensitivity increases to about 1460 kcps/μCi/cc. These system characteristics provide much improved PET image quality compared with that obtained from earlier systems.

3.6 Referenccs

1. Becquerel, *Compte Rendu de l'Academie des Sciences* **122**, 420, 559, 1086 (1896).
2. Curie, S. Curie, *CR Acad Sci (Paris)* **127**, 173 (1898).
3. Curie, F. Joliot, *Nature* **133**, 201–204 (1934).
4. Hevesy. Adventures in Radioisotope Research: The Collected Papers of George Hevesy. (Pergamon Press, New York, 1962).
5. Anger, *Rev Sci Instrum* **29**, 23 (1958).
6. Richards, A survey of the production at the Brookhaven National Laboratory of radioisotopes for medical research, *5 Congresso Nucleare Rome.*, Rome (Comitato Nacionale Ricerche Nucleari, 1960).
7. Harper, K.A. Lathrop, W. Siemens, et al., *Radiat. Res.* **16**, 593 (1962).
8. Thompson, *Phil. Mag.* **44**, 293 (1897).
9. Rutherford, *Phil. Mag.* **21**, 669 (1911).
10. Bohr, *Phil. Mag.* **26**, 1, 476 (1913).
11. Chadwick, *Proc. Roy. Soc.* **A136**, 692 (1932).
12. Evans. *The Atomic Nucleus.* (McGraw-Hill, New York, 1955).
13. Wagner, Z. Szabo, J.W. Buchanan, Eds., *Principles of Nuclear Medicine* (W.B. Saunders Company, Philadephia, 1995).
14. Knoll. *Radiation detection and measurement.* (John Wiley & Sons, New York, ed. Second Edition, 1986).
15. Anger, *Journal of Nuclear Medicine* **5**, 515–531 (1964).
16. Jaszczak, B.M.W. Tsui, in *Principles of Nuclear Medicine,* H. N. Wagner, Z. Szabo, J. W. Buchanan, Eds. (W.B. Saunders, Philadelphia, 1995) pp. 317–341.
17. Kuhl, R.Q. Edwards, *Radiology* **80**, 653 (1963).
18. Hirose, Y. Ikeda, Y. Higashi, et al., *IEEE Trans. Nucl. Sci.* **Ns-29**, 520 (1982).
19. Stoddart, H.A. Stoddart, *IEEE Trans. Nucl. Sci.* **NS-26**, 2710 (1979).
20. Radon, *Ber Verh Sachs Akad Wiss* **69**, 262–277 (1917).
21. Brooks, G. Di Chiro, *Phys. Med. Biol.* **21**, 689–732 (1976).
22. Huesman, G.T. Gullberg, W.L. Greenberg, et al. *User Manual, Donner Algorithms for Reconstruction Tomography.* (University of California, Lawrence Berkeley Laboratory, 1977).
23. Minerbo, *Comput Graph Image Process* **10**, 48–68 (1979).
24. Rockmore, A. Macovski, *IEEE Transactions on Nuclear Science* **NS-23**, 1428–1432 (1976).
25. Levitan, G.T. Herman, *IEEE Transactions on Medical Imaging* **MI-6**, 185–192 (1987).
26. Kuhl, R.Q. Edwards, *Radiology* **91**, 975–983 (1968).
27. Muehllehner, *J. Nucl. Med.* **9**, 337 (1968).
28. Hudson, R.S. Larkin, submitted to IEEE Trans. Med. Imag. (1993).
29. Jaszczak, R.E. Coleman, F.R. Whitehead, *IEEE Trans. Nucl. Sci.* **NS-28**, 69–80 (1981).
30. Budinger, G.T. Gullberg, in *Reconstruction Tomography in Diagnostic Radiology and Nuclear Medicine,* M. M. Ter-Pogossian, Ed. (University Park Press, Baltimore, 1977) pp. pp. 315–342.
31. Tretiak, C.E. Metz, *SIAM Journal of Applied Mathematics* **39**, 341–354 (1980).
32. -T. Chang, *IEEE Trans. Nucl. Sci.* **NS-25**, 638–643 (1978).
33. DePuey, E.V. Garcia, *Journal of Nuclear Medicine* **30**, 441–449 (1989).
34. Tsui, G.T. Gullberg, E.R. Edgerton, et al., *J.Nucl.Med.* **30**, 497–507 (1989).
35. Ficaro, J.A. Fessler, P.D. Shreve, ct al., *Circulation* **93**, 463–473 (1996).
36. Axelsson, P. Msaki, A. Israelsson, *Journal of Nuclear Medicine* **25**, 290–294 (1984).
37. Jaszczak, K.L. Greer, C.E. Floyd, Jr., et al., *J. Nucl. Med.* **25**, 893–900 (1984).
38. Koral, F.M. Swailem, S. Buchbinder, et al., *J. Nucl. Med* **31**, 90–98 (1990).
39. Frey, B.M.W. Tsui, *IEEE Transactions on Nuclear Science* **38**, 789–794 (1991).
40. Floyd, R.J. Jaszczak, R.E. Coleman, *IEEE Transactions on Nuclear Science* **NS-32**, 779–785 (1985).

41. Frey, Z.-W. Ju, B.M.W. Tsui, A fast projector-backprojector pair modeling the asymmetric, spatially varying scatter response function for scatter compensation in SPECT imaging, *Conference Record of the 1992 IEEE Nuclear Science Symposium and Medical Imaging Conference*, Orlando, FL (1992).
42. Mandelkern, M.E. Phelps, in *Diagnostic Nuclear Medicine,* A. Gottschalk, P. B. Hoffer, E. J. Potchen, Eds. (Williams & Wilkins, Baltimore, 1976), Vol. I, pp. 128–149.

4 Effects of Biomedical Ionizing Radiation and Risk Estimation

F. Marc Edwards

Johnson County Radiation Therapy Facility

4.1 Introduction

The term "biomedical ionizing radiation" encompasses many types of radiation with a wide range of energies and intensities. Electromagnetic radiation is used for diagnosis and therapy, with energies ranging from 20 keV characteristic X rays used in mammography imaging to several MeV x- and γ-rays used in cancer treatment. Particulate radiation, primarily electrons, but including protons, neutrons, and heavy ions, is also used in cancer treatment. The intensity of biomedical radiation may vary from negligible (at or below normal background levels) to many orders of magnitude above background. All biomedically useful radiation interacts with living matter, resulting in the deposition of energy. The nature of this interaction and the amount of energy deposited generally determine the biological effect on the organism. The bioeffect may be desired, as in the case of radiation therapy treatment for cancer, or may be an unavoidable consequence of the use of radiation, such as the possible induction of cancer as a result of a diagnostic X-ray procedure. This chapter will present an overview of the bioeffects of ionizing radiation and risks associated with their use.

4.1.1 Elements of Bioeffects Description

The elements of bioeffect description are similar for all uses of medical radiation. First, there must be a method to quantify and measure the amount of biomedical radiation. Systems of quantities and units, as well as standardized methods for their measurement, have been established. The fundamental units of ionizing radiation are based upon the amount of energy absorbed per unit mass of absorber. Units used for quantifying biological effects also may take into account the linear energy transfer of the radiation and the homogeneity of dose across various organs. Second, the possible types of interactions of the biomedical radiation must be known. This includes information about how an interaction takes place and what type of interaction products are produced and with what probability. Understanding of energy deposition at the atomic and molecular level, as well as understanding of the chemical byproducts of radiation interactions, is essential for the study of bioeffects. Third, the mechanisms by which an interaction ultimately leads to biologic damage should be understood as thoroughly as possible. This step is often the weakest link in an examination of hazardous agents. For example, it is well-known that X rays interact to produce ion pairs and free radicals and that these interaction products cause DNA damage. However, the subsequent mechanisms leading from damage at the molecular level to the ultimate expression of that damage at the organic or even tissue level are less well known.

Fourth, it is necessary to know the possible types of ultimate expression of damage. In general these are classified as somatic, genetic, or fetal to denote the affected tissue. A somatic effect is any damage ultimately occurring to the individual originally exposed to the harmful agent. Somatic effects resulting from high levels of exposure are termed deterministic (also called non-stochastic effects). Deterministic effects are caused by cell killing, and the severity of the effect is proportional to the level of exposure. Deterministic effects often occur relatively soon (hours to weeks) after irradiation. For example, sterility in males can be induced by exposure of the testes to a high (2–3 Gy) dose of ionizing radiation. Non-deterministic effects (also called stochastic effects) are thought to result from DNA mutation. The probability of a non-deterministic effect is proportional to exposure but the severity of the effect is not. Stochastic effects may result from low level exposure, and may occur years after irradiation. The archetypal non-deterministic somatic effect is cancer. A genetic effect is any damage occurring in future generations as a result of exposure of parent germ cells to the harmful agent. Depending on the type of damage to the germ cell, genetic effects can range from mild, such as eye color changes, to severe, such as mental retardation. A fetal effect is any damage occurring as a result of exposure to the harmful agent during the embryonal or fetal stage of development. This type of damage is highly dependent on timing of the exposure with respect to gestational age and can range from mild or severe birth defects to childhood malignancies.

Fifth, damage resulting from exposure to the biomedical radiation should be distinguished from damage resulting from other agents. This distinction is rarely an easy task, particularly for low level exposure in uncontrolled circumstances. If biomedical radiation produced damage having unique symptoms, for example in a manner analogous to anthrax bacillus causing characteristic black lesions, the identification of damage would be simple, even at very low incidence rates. Unfortunately, with very few exceptions, the types of damage resulting from exposure to low levels of ionizing radiation can also be caused by a multitude of other agents, many of which occur naturally. Hence, these non-unique ionizing radiation effects can be distinguished only as a higher-than expected incidence rate of normally occurring conditions. Both the identification of an effect and quantification of its magnitude are critically dependent upon the availability of proper unexposed control groups and careful interpretation of results.

Sixth, it is necessary to establish a quantitative relationship between the magnitude of exposure and the magnitude of its effect. For meaningful risk estimation it is insufficient to know only that a particular effect is possible. The dose-response relationship, which expresses the probability of a particular response or effect as a function of dose of harmful agent, also must be known. Elucidation of dose-response relationships is less problematic for high dose, deterministic effects. However, significant problems are encountered with dose-response relationships for stochastic low dose effects. In this case, it is often experimentally difficult to obtain statistically significant data in dose region of most interest. Because a dose-response model must then be used to extrapolate from high levels of dose to lower levels of dose, the quantitative estimate of low-level risk is limited by the accuracy of the dose-response relationship and the extrapolation procedure.

Finally, it is desirable to have epidemiological studies on human populations in order to avoid problems inherent in trying to use animal data to quantify human risk. While the types of effects and even the shape of dose-response relationships may be similar for many species, the quantitative magnitude of the effect is unique for each organism. Because ethical considerations prohibit the type of human experimentation required for optimal information, inadvertent or unavoidable exposure is usually the sole source of human data. Unfortunately, these data are often characterized by uncertain dosimetry and improper controls, thereby further complicating quantitative risk estimation.

4.2 Common Elements of Effects of High and Low Level Ionizing Radiation

Ionizing radiation is a generic term encompassing a wide range of electromagnetic radiation and elementary particles with the common ability to produce ion pairs during their interaction with matter. High-level ionizing radiation is used in medicine for the treatment of malignant disease, while low-level ionizing radiation is used in imaging procedures. Radiation oncology utilizes X rays, γ-rays, electrons (and much less frequently neutrons and other elementary particles) to kill tumor cells within the body. In the US in 1990 it was estimated that over 1300 treatment facilities treated approximately 490 000 new patients and delivered about 15 million patient-treatments[1]. In diagnostic radiology, X rays (for transmission imaging) and γ-rays (for emission imaging) are the most prevalent agents used for imaging. Of the 294 million imaging procedures (not including dental imaging) performed in the US in 1990, approximately 90 per cent utilized ionizing radiation[1]. In developed nations, the average exposure to the entire population from medical sources is approximately equal to that from natural sources. This is three to four orders of magnitude higher than the average exposure from other man made sources such as nuclear power generation. Medical radiation is by a wide margin the largest single man-made contributor to the total population exposure.

4.2.1 Units of Ionizing Radiation

Units of ionizing radiation are fully discussed in Vol. 1 Chapter 1, consequently, only a brief review is given here. In early investigations of ionizing radiation, radiation "exposure" was defined in terms of the quantity of x or γ-

rays that produced a specific amount of ionization in air. The unit of radiation exposure was the roentgen, abbreviated R. This quantity suffered the disadvantages of being applicable only to a particular type of radiation (x- or γ -rays) and a particular medium (air). The quantity "exposure" and its special unit (R) are no longer used for scientific communication. For many instances in which exposure was used to specify the amount of ionizing radiation, the quantity air kerma is now used. For purposes of biologic effects it is more appropriate to focus on the amount of energy absorbed by the medium exposed to radiation. Absorbed dose is defined to be the quantity of energy absorbed per unit mass of a medium. The SI (Systeme International[2]) unit for absorbed dose is the gray, where 1 Gy = 1 joule/kg. The previous unit for absorbed dose was the rad, an acronym for radiation absorbed dose. Since one rad was defined to be 0.01 joule/kg, it is easily shown that 1 Gy equals 100 rad. The size of the rad was originally chosen such that exposure of soft tissue to one R would result in approximately one rad of absorbed dose. While much of the early research in radiation bioeffects was reported using units of roentgen or rad, these older units have been supplanted by SI units.

The energy imparted as a consequence of one gray absorbed dose is comparatively small. For instance, it is known that the whole-body lethal dose of X rays is approximately 5 Gy. A 70-kg person receiving a dose of 5 Gy absorbs a total of 350 joules of energy, which is approximately equal to 84 calories. If this energy were eventually converted entirely to heat, the temperature of the body would rise only 0.001°C. Even in radiation oncology treatments, in which a patient typically receives approximately 2 Gy to a localized region of the body, it is impossible for the patient to "feel" the radiation as an increase in temperature. Diagnostic imaging utilizes orders of magnitude less dose to localized regions, typically on the order of 0. 1 to 10 milligray. As will be seen below, the bioeffects of ionizing radiation result from the pattern of energy deposition on microscopic level rather than the average total energy deposition.

The absorbed dose from any type of ionizing radiation (X rays, γ-rays, electrons, α-particles, etc.) can be quantified in units of grays. However, one gray of X rays and one gray of α-particles have significantly different biological effects, owing to differences in the manner in which the energy is absorbed. The "equivalent dose" is a special unit that reflects the biological effectiveness of the radiation[3]. The SI unit of equivalent dose is the sievert (Sv). The older unit of equivalent dose, the rem (acronym for rad equivalent mammal), is no longer in use. Equivalent dose is the product of absorbed dose and a "radiation weighting factor". The radiation weighting factor, which depends upon the linear energy transfer (LET) of the radiation, describes the relative biological effectiveness of the radiation. Ionizing radiation may be crudely divided into low LET (< 1 keV/μ) and high LET (>10 keV/μ) radiation. The radiation weighting factor of low LET radiation, such as X rays, γ-rays and electrons, is approximately equal unity, while for high LET radiation, such as neutrons and heavy charged particles, the factor ranges from 2 to 20. A high LET radiation having a radiation weighting factor of 10 would cause approximately ten times as much biological damage as low LET radiation. Most biomedical uses of ionizing radiation involve low LET radiation, hence the Gy and Sv are of equal magnitude. Important exceptions are the use of neutrons and heavy charged particles for cancer treatment and environmental exposure to the alpha emitter radon. In addition, the dosimetry of the A-bomb survivor population, which was exposed to ionizing radiation of a wide range of LET (due to the neutron component of the fission weapon), must be described using equivalent dose.

In many situations it is necessary to estimate risk from non-uniform or partial body low level radiation. Special units have been developed for this purpose. The "effective dose" has associated with it the same probability of the occurrence of cancer and genetic effects whether received by the whole body via uniform irradiation or by partial body or individual organ irradiation. Effective dose is calculated as the sum of the products of the equivalent dose of each tissue or organ with its "tissue weighting factor." Tissue weighting factors, derived from individual organ risk data, vary from 0.01 to 0.20. The SI units of equivalent dose and effective dose are the same (Sv). However, the magnitude of equivalent dose and effective dose are identical only for uniform whole body irradiation. Effective dose should never be used for determination of possible effects of high level irradiation.

4.2.2 Interactions and Mechanisms of Damage

A comparatively small amount of energy in the form of ionizing radiation can cause significant biological damage because of the manner in which ionizing radiation interacts with matter. Charged particles, such as beams of electrons from a linear accelerator or alpha particles from radioactive decay, directly ionize matter. As discussed in Vol. 1 Chapter 1, electromagnetic radiation can produce secondary charged particles in the form of energetic electrons as a result of photoelectric, Compton, and pair production interactions. Depending on their charge, mass and initial energy, charged particles will travel from their site of origin before they are absorbed. As the charged particles pass close to and interact with bound atomic or molecular electrons in the medium, they cause some of the bound electrons to be ejected from their atoms, thereby producing ion pairs. The distribution of primary energy loss events along a charged particle track reflects both the properties of the charge particle and the medium. The average rate of energy loss and range of charged particles is described by the Bethe-Block equation, however, more detailed descriptions may be obtained by incorporating the properties of the medium in terms of its dipole oscillator strength distribution[4]. Because photon and electron interactions are probabilistic, energy deposition is spatially and temporally non-uniform. Although the average energy deposited over a large volume is quite small, the local energy deposition is comparatively high for those microscopic volumes that happen to lie at or near an interaction site.

There are two consequences of the local deposition of energy that can lead to biological damage. First, the chemical bonds of an essential biomolecule can be broken as a direct result of ionization. This type of event is termed "direct action". Second, reactive chemical species can be produced that subsequently react with essential biomolecules to produce molecular damage. This effect is termed "indirect action". Interaction of ionizing radiation with water produces excited and ionized water molecules and free electrons. These products quickly interact to form various chemical species, predominately H^3O^+, $OH^{\bullet}$, $H^{\bullet}$, H_2 and e^-_{eq}. Since the hydroxyl free radical ($OH^{\bullet}$) has an unpaired electron, it is chemically very reactive and is known to play a crucial role in indirect radiation effects. The hydroxyl radical has been estimated to account for 70% of damage caused by reactive species[5]. The presence of dissolved oxygen in the aqueous environment of living cells also plays a role chemistry of radiation damage. Molecular oxygen interacts with radiation hydrolysis products to form hyperoxyl radicals and hydrogen peroxide, both of which are chemically reactive. Molecular oxygen also acts to "fix" the otherwise repairable damage caused by free radicals. Hence the presence of oxygen acts to enhance the biological effects of ionizing radiation, a phenomenon that is clinically important in the treatment of tumors.

It is well known from selective irradiation of cell components that DNA is the only significant "target" molecule for biologic damage. The nature of DNA damage depends not only on the spatial distribution of ionization and chemical species, but also upon the structure of the DNA molecule. DNA in the nucleus exists as two helical strands, each composed of a linear sequence of nucleotide bases. A nucleotide consists of a nitrogen base (cytosine, thymidine, adenine, or guanine), a deoxyribose sugar, and a phosphate. Sugar-phosphate bonds connect the bases in sequence, while hydrogen bonding between bases pairs (cytosine and guanine, or thymidine and adenine) is responsible for the complementary nature of the two strands. The structure of the DNA molecule at various levels of magnification is illustrated in Figure 4.1. The DNA molecule does not normally exist in a simple linear form, but is highly looped and folded. DNA is wound around histone proteins like "beads on a string" to form nucleosomes. The nucleosomes coil upon themselves to form solenoids, while the solenoids, in turn, loop upon themselves many times to form chromatin. During mitosis, the chromatin becomes condensed into chromosomes. The several hundred million base pairs in a single chromosome, which, if extended, would be about 6 cm long, are packed into a structure of a few microns in size. Thus, a single ionization track could intersect the DNA molecule multiple times.

Types of DNA damage include damage to the sugar moieties resulting in single- and double-strand breakage, damage or deletion of a base, chemical cross-linking of two strands, and combinations of damage to a local area, termed "locally multiply damaged sites"[6]. The steps leading from X-ray interaction to DNA damage are illustrated in Figure 4.2. In the left-most section of Figure 4.2, an electron interacts with water to form an $OH^{\bullet}$ radical which damages a single base. In the middle section, the track of one electron produces sufficient

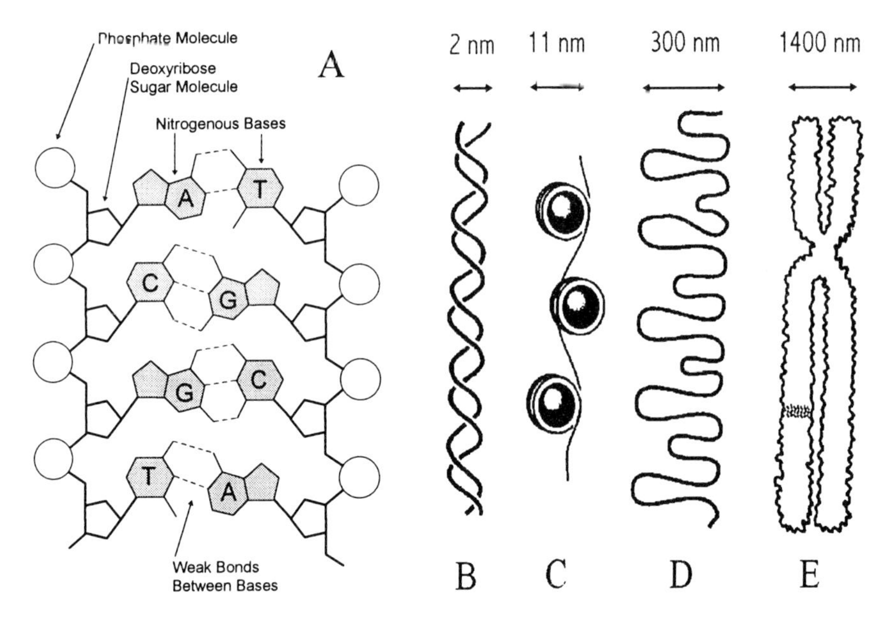

Figure 4.1. Illustration of the structure of DNA. In 4.1A, four base pairs are illustrated showing the chemical structure of the phosphate, sugar and base molecules as described in the text. In 4.1B, a short region of the DNA molecule is shown illustrating the diameter of the molecule. DNA does not normally exist in the cell is the simple linear form of 4.1B, but is coiled, looped and folded. In 4.1C, a nucleosome is formed by DNA wound on histone proteins. Nucleosomes coil upon themselves to form solenoids, which then loop upon themselves to form chromatin. During mitosis, the chromatin becomes condensed into chromosomes. Figure 4.1D illustrates a section of a chromosome in extended form, while Figure 4.1E shows an entire chromosome in metaphase. The entire human genome consists of 3 billion base pairs contained in 23 chromosomes. Figures 4.1B–E modified from Hall[15].

ionization to directly ionize and break both sugar molecules, forming a double strand break. In the right-most section, two independent electrons cause two single-strand breaks that randomly fall close enough together to form a double-strand break. Gel electrophoresis assay techniques[7] allow quantitative analysis of radiation induced single-strand breaks (SSBs) and double-strand breaks (DSBs) in plasmid or small viral DNA. SSBs and DSBs may be caused either by direct or indirect action. On average, SSBs are produced four times more frequently than base damages, 20 times more frequently than DSBs and 30 or more times more frequently than cross-links[8]. The presence of free radical scavengers also has a dramatic effect on the ratio of SSBs to DSBs.

The number of strand breaks formed depends on absorbed dose. SSBs exhibits a linear dependence with dose. This is because only one event or "hit" is required to break a single strand even though the hit may be due to direct ionization or the formation of hydroxyl radicals. The formation of DSBs is more complex. DSBs due to direct ionization of both strands follow a linear dose dependence, since a single event breaks both strands. It also may be possible for a single hydroxyl-induced event to produce sufficient radicals to

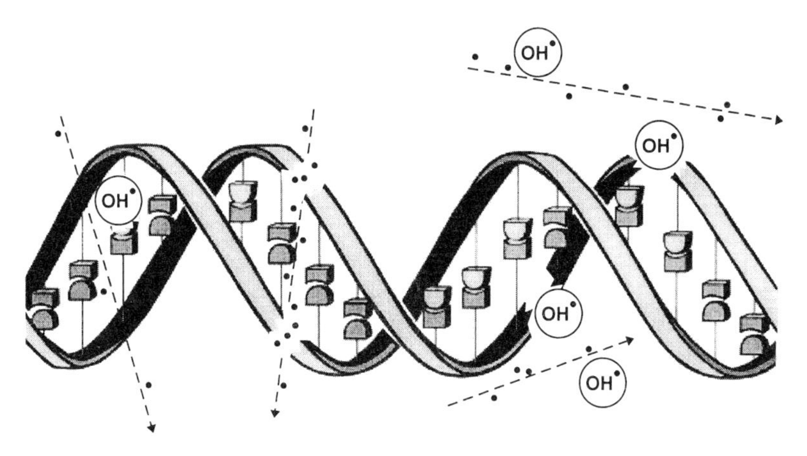

Figure 4.2. Schematic illustration of the mechanisms leading to DNA damage. An incident X-ray undergoes Compton scattering, producing a scattered X-ray and a fast (energetic) electron. This electron creates several hundred ion pairs as it loses energy. In some cases ionization occurs in DNA atoms, directly damaging the DNA. In other cases the radiation hydrolysis of cellular water molecules creates hydroxyl free radicals ($OH^{\bullet}$), which damage DNA through chemical reactions. Types of damage from both mechanisms include single- and double-strand breakage, single-base deletion, and base cross-linking.

break both strands. However, two random SSBs also may occur with sufficient spatially correlation to induce a double strand break. Since random accumulation of two hits is required, the dose dependence is quadratic. Hence the formation of DSBs from all processes follows a linear-quadratic dose dependence. The relative contribution of direct and indirect effects on the formation of SSBs and DSBs has been studied by Le Sech et. al.[9] using nearly monoenergetic low-LET synchrotron radiation that could be tuned to phosphorus K-shell resonant absorption of 2.152 keV. Their results, showing linear and linear-quadratic dose dependence of strand breaks, is shown in Figure 4.3. By comparing SSB and DSB production at photon energies off the resonant absorption peak, to SSB and DSB production observed when the photon energy is equal to the K-shell absorption energy, Le Sech et. al. were able to separate direct from indirect effects. The results showed that, on resonance, the SSBs exceeded DSBs by a factor of 14 and that direct effects accounted for about 47% of SSBs and 58% of DSBs. For high-LET radiation the contribution of indirect effects becomes less important, and both SSB and DSB production shows a linear dependence on dose.

Cells have considerable capacity for repair of DNA damage. An accumulating body of evidence suggests that SSBs play a limited role in cell lethality, and perhaps even in malignant transformation[10]. Studies have shown that doses which cause an average of one lethal event per cell, induce 20–80 DSBs. A lethal event may involve two or more DSBs in close proximity (within 10 nm), that is incorrectly repaired or is not able to be repaired. Such an event could result in loss of part of the DNA within a chromosome or exchange of DNA between chromosomes. Since DNA damage becomes

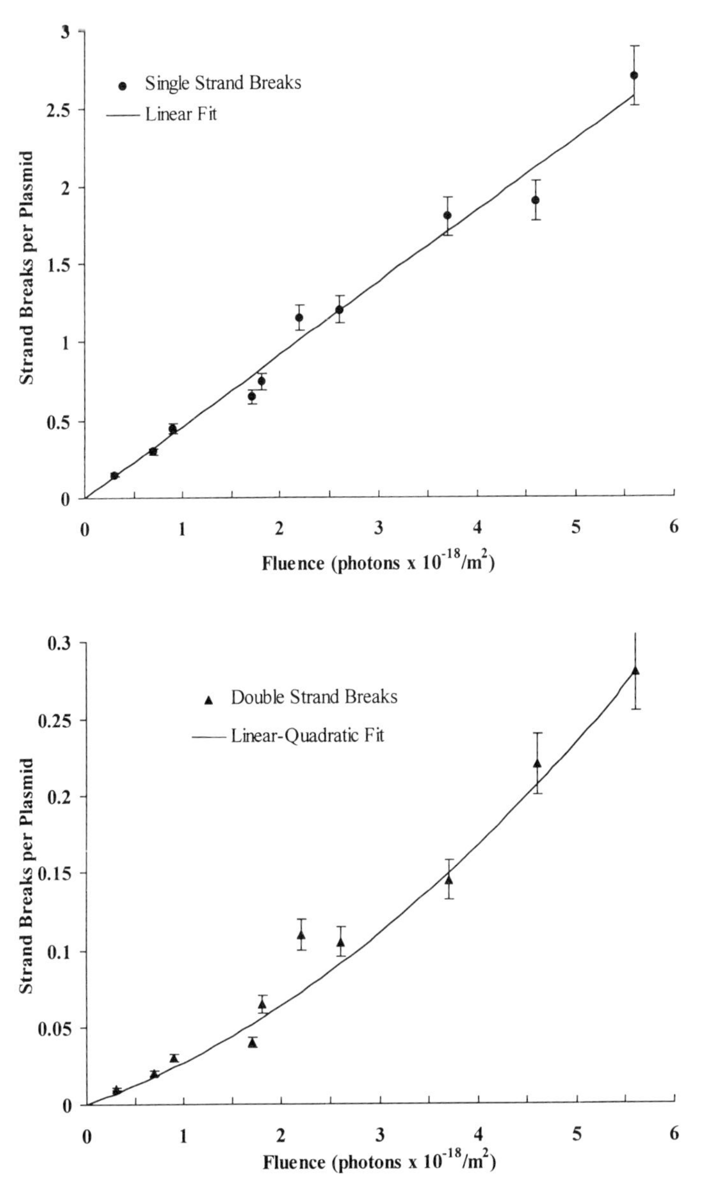

Figure 4.3. Dose dependence of single strand breaks and double strand breaks observed in plasmid DNA irradiated with monoenergetic low LET photons. Data re-plotted from Le Sech et. al.[9] with permission.

apparent at replication, rapidly proliferating cells have a greater chance of expressing damage than do slowly proliferating ones. If the damage makes replication impossible, cell death occurs. If the damage is not lethal, but alters the genetic sequence or structure, mutations can occur.

4.3 Effects of High Level Ionizing Radiation

The term "high level effects," usually means those bioeffects of ionizing radiation due to significant cell killing. High level effects are often "acute effects," in which the bioeffect is observed within a short period of time (hours to weeks) after exposure. High level effects are deterministic. A minimum level of exposure (i.e. a threshold dose) is required to induce the effect, and the level of the effect is proportional to the level of exposure. The magnitude of "high" depends upon the biological endpoint. In the context of human whole body exposure, the lowest dose associated with any acutely observable effect is about 0.1 to 0.5 Gy, the whole body dose above which chromosomal aberrations in blood lymphocytes are observed. Since acute effects of cells or organs often involve even higher dose, an informal dividing line of 0.1 Gy (received in a short period of time) is often use to separate high level from low level effects. Effects in radiation therapy patients, in a which small volume of tissue receives 20 to 70 Gy of a period of weeks, obviously are high level effects. With the possible exception of some interventional procedures, patients receiving diagnostic imaging studies, in which partial volumes of the body are exposed to effective doses of the order of 1 mGy or less, are not subject to high level effects. Similarly, radiation workers, whose typical annual exposure is less than 10 mSv, are not subject to high level effects.

4.3.1 Mechanisms of Damage from High Level Ionizing Radiation

As discussed above, the fundamental mechanism of radiation damage is the local deposition of energy resulting in DNA damage from either direct ionization or production of reactive chemical species. The initial damage of most consequence to cell survival is double strand breaks. For cell death to occur, fairly severe DNA damage, probably involving two or more double strand breaks or several local multiply damaged sites, must be produced. Such damage may be impossible for the DNA to repair, or the repair process may result in the loss or inactivation of important genes.

High level radiation effects are quantified in terms of the survival of cells, tissues or organisms after irradiation. Death is usually assessed by determining the reproductive capacity of irradiated cells. Hence, the DNA of cells may be so damaged that they are unable to undergo a single successful mitosis after irradiation, or they may continue to divide for a few generations but ultimately be unable to sustain a colony of cells. Two mechanisms of cell death are caused by ionizing radiation; necrosis and apoptosis. Necrosis is degenerative in nature and always the outcome a severe injury. Necrotic cells are characterized by the loss of membrane integrity before DNA degradation, and DNA fragments are random is size. Apoptosis, or programmed cell death, is a process occurring during normal development of cells and tissues. During apoptosis the DNA fragments in a characteristic manner involving segments of a distinctive pattern of sizes, and membrane integrity is lost after DNA fragmentation. The relative fraction of necrosis and apoptosis depends on the cell type and irradiation conditions. The mechanism of radiation induced apoptosis and its potential role in radiation therapy is the subject of much investigation[11,12,13].

4.3.2 Cell Survival Curves

Damage may be divided into that due to a single track or event and that due to two or more lesions which require interaction or accumulation to become effective. The latter lesions are often referred to as "sub-lethal damage" (SLD). Since the cell has many mechanisms to repair damage, the term "potentially lethal damage" (PLD) has been used to denote that type of damage which is not expressed if conditions after exposure are imposed which promote repair, such as maintaining cells in a resting phase or by impairment of cellular metabolism. Both single track and sub-lethal damage may be effected by the repair of potentially lethal damage. Since SSBs and isolated DSBs are fairly easily repaired by the cell, it is thought that two or more DSBs produced close together are required for cell lethality. However, multiple DSBs may be caused either by a single ionizing event (i.e. a single track lethal event resulting in the simultaneous production of two or more DSBs) or the random accumulation of isolated DSBs with sufficient spatial correlation to constitute local damage. Because single track damage results from a single event, the frequency of damage is linear with dose. Damage resulting from two DSBs produced by the random accumulation of isolated DSBs (sub-lethal damage) depends quadratically upon dose. Thus the overall frequency of lethal damage is described by

$$F(d) = a_1 d + a_2 d^2 \tag{4.1}$$

in which a_1 and a_2 are constants that depend on the cell type, radiation type and pre- and post irradiation conditions of exposure to dose d. The parameter a_1 can be though of as the frequency of single-track damage, while $\sqrt{a_2}$ represents the frequency of sub-lethal damage. Potentially lethal damage is a part of both terms. The surviving fraction $SF(d) = S(d)/S(0)$ of a large population of cells may then be described as

$$SF(d) = \mathrm{e}^{-(a_1 d + a_2 d^2)}. \tag{4.2}$$

The linear-quadratic cell survival curves for C3H 10T1/2 mouse embryo cells exposed to low, medium and high LET radiation are illustrated in Figure 4.4. For high-LET radiation, single track damage predominates, and the cell survival curve is described as a pure exponential. For low-LET radiation, single track events are comparatively rare, and the cell survival curve exhibits a considerable "shoulder" region. The shape of the cell survival curve depends on the cell type, LET, dose rate, presence of oxygen and radical scavenging agents, the relative position of cells within the cell cycle and post-irradiation conditions of cells that may promote or inhibit repair processes. If the dose is delivered in more than one fraction, cell survival also depends on dose per fraction and time between fractions. For cells in-vivo, survival also depends upon tissue level parameters such as vascular supply (effecting the oxygenation and nutritional status of cells) as well as the relative ability of cells to repopulate after depletion.

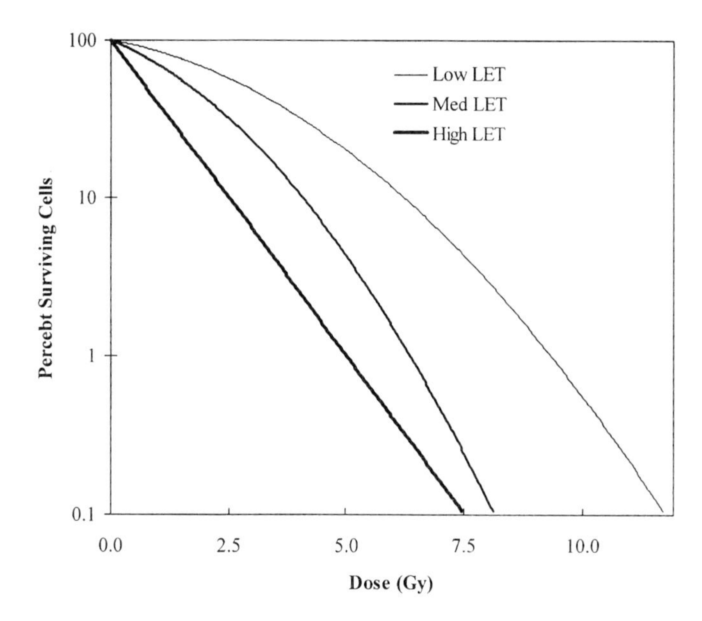

Figure 4.4. Survival curves for C3H 10T1/2 cells illustrating linear-quadratic cell survival for low, medium and high LET radiation. Cell survival curve parameters [a_1, a_2] are [0.12, 0.04] for low LET (2.6 keV/μ), [0.28, 0.07] for medium LET (50 keV/μ), and [0.92, 0] for high LET (128 keV/μ). Parameters taken from Barendsen[10].

The clinical goal of radiation therapy is to kill every cell in a tumor mass, using multiple fractions. A very simple model of cell survival of a population of cells is based on Poisson statistics. Eq. (4.2) may be generalized to describe the probability of survival of a single cell after irradiation to a fractionated total dose $D = nd$, where n is the number of fractions and d is the dose per fraction as

$$SF(D,n) = e^{-(a_1 D + a_2 D^2/n)}. \tag{4.3}$$

The number of surviving cells in a tumor having k cells is then $k \cdot SF(D)$, and the probability of zero surviving cells (probability of tumor cure) is

$$P(D,k,n) = e^{-k \cdot SF(D,n)} \tag{4.4}$$

Eq. (4.4) is plotted in Figure 4.5 for 10 000 000 cells exposed in 20 fractions. The sigmoid shaped dose response curve, having an apparent threshold and steeply sloped linear region followed by a shoulder, is typical of most bioeffects based upon cell killing. The simple model shown here does not include effects such as variability of a_1 and a_2 within tissues, dose dependent changes in a_1 and a_2, re-population and re-oxygenation of tumor cells. In more sophisticated models, cell survival data for various tumor and normal tissue types has been used in combination with cell population kinetic models to attempt to design optimal irradiation schedules for radiation therapy[14]. The goal of such models is to define the dose per fraction, total dose and fractionation schedule that results in maximal tumor cell killing and minimal normal cell killing.

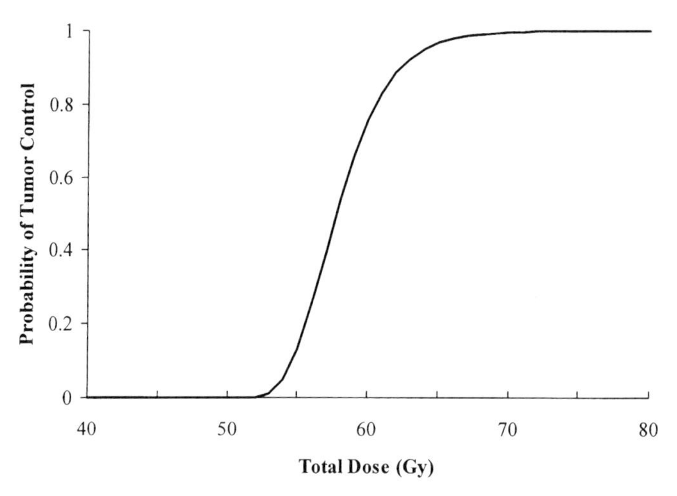

Figure 4.5. Probability of tumor control as a function of total dose. The simple Poisson model was used in combination with a linear quadratic cell survival curve (a_1 = 0.17 and a_2 = 0.04) to calculate the probability of control of 10 000 000 cells exposed in 20 fractions.

4.3.3 Effects on Tissues and Organs

The acute effects of ionizing radiation on organs depend upon the intrinsic radiation sensitivity of the cell type and upon the dose fractionation and total dose. In general, cell types that divide at higher rates and have less differentiation are most sensitive to cell killing, while cells that do not divide regularly and are highly differentiated are least sensitive to cell killing. For example, erythroblasts and intestinal crypts cells are very radiosensitive, while nerve and muscle cells are not[15]. When ionizing radiation is used for cancer treatment, the goal is kill all or most of diseased cells within a well defined tumor volume. The dose required to meet this goal depends not only upon the radiosensitivity of the tumor, total dose and dose fractionation, but also upon tumor volume (i.e. the number of tumor cells to be killed). The approximate range of curative doses[16] (assuming a dose per fraction of 1.8 to 2 Gy) for various tumor types and sizes is summarized in Table 4.1.

Table 4.1. Approximate curative doses for different tumor types and sizes. Reproduced from *Clinical Oncology*[16] with permission.

20–30 Gy
Seminoma
Acute lymphocytic leukemia
30–40 Gy
Seminoma
Wilm's tumor
Neuroblastoma
40–45 Gy
Hodgkin's disease
Non-Hodgkin's lymphoma
Skin cancer (basal and squamous)
50–60 Gy
Lymph nodes, metastatic (< 1 cm)
Squamous cell carcinoma (sub-clinical)
Breast cancer (< 1 cm)
Medulloblastoma
Retinoblastoma
Ewing's tumor
60–65 Gy
Larynx (< 1 cm)
Breast cancer, lumpectomy
70–75 Gy
Head and neck cancers (2–4 cm)
Bladder cancers
Cervix cancer
Uterine fundal cancer
Lymph nodes, metastatic (1–3 cm)
Lung cancer (< 3 cm)
80 Gy or more
Head and neck cancers (> 4 cm)
Breast cancer (> 5cm)
Glioblastomas
Osteogenic sarcomas
Melanomas
Thyroid cancer
Lymph nodes, metastatic (> 6 cm)

For the radiation therapy patient, adverse outcomes include lack of tumor control (due to inability to deliver a curative dose to the tumor) and complications resulting from irradiation of normal tissue. Cell killing in normal tissue may result in variety of undesirable effects, depending on the radiosensitivity of the tissue, the dose fractionation and total dose, as well as the volume of tissue irradiated. Table 4.2 summarizes the complication endpoints and the minimum tolerance dose resulting in a 5% complication rate within 5 years ($TD_{5/5}$) and a 50% complication rate with 5 years ($TD_{50/5}$) for various tissues and organs. This table is valid for supervoltage (>1 MeV) irradiation with the total dose delivered with a fraction size of approximately 2 Gy.

Table 4.2. Tolerance doses for fractionated doses to whole or partial organs. Reproduced from *Clinical Oncology*[16] with permission.

Dose Range / Target Cells	Complication End Point	$TD_{5/5} - TD_{50/5}$ (Gy)
2–10 Gy		
Lymphocytes and lymphoid	Lymphopenia	2–10
Testes spermatogonia	Sterility	1–2
Ovarian oocytes	Sterility	6–10
Diseased Bone Marrow	Severe leukopenia and thrombocytopenia	3–5
10–20 Gy		
Lens	Cataract	6–12
Bone marrow stem cells	Acute aplasia	15–20
20–30 Gy		
Kidney: renal glomeruli	Arterionephrosclerosis	23–28
Lung: Type II cells, vascular connective tissue stroma	Pneumonitis or fibrosis	20–30
30–40 Gy		
Liver: central veins	Hepatopathy	35–40
Bone marrow	Hypoplasia	25–35
40–50 Gy		
Heart (whole organ)	Pericarditis or pancarditis	43–50
Bone marrow micro-environments	Permanent aplasia	45–50
50–60 Gy		
Gastrointestinal	Infarction necrosis	50–55
Heart (partial organ)	Cardiomyopathy	55–65
Spinal cord	Myelopathy	50–60
60–70 Gy		
Brain	Encephalopathy	60–70
Mucosa	Ulcer	65–75
Rectum	Ulcer	65–75
Bladder	Ulcer	65–75
Mature bones	Fracture	65–75
Pancreas	Pancreatitis	>70

Skin tolerance and skin complications were a crucial element of patient treatment before the development of supervoltage X-ray sources. For 250 kVp X rays, moist desquamation may occur at a dose of 15 Gy given in a single fraction and 50 Gy given in 25 fractions. However, for X-ray energies above a few MeV, charged particle equilibrium does not occur until a depth of a centimeter or more beneath the skin, effectively sparing the skin from the maximum radiation dose. Although acute skin effects such as erythema and temporary desquamation are still seen with some types of radiation therapy regimens, the combination of skin sparing and dose fractionation have largely eliminated serious permanent skin complications. With the development of high dose rate fluoroscopy equipment and high frame rate digital imaging equipment used for interventional procedures such as neuroembolization and percutaneous transluminal angioplasty, acute skin reactions have been reported[17]. Skin effects from acute exposure (all dose delivered in 24 hours) to low energy (< 250 kVp) X rays are summarized in Table 4.3. The typical dose required for a given effect greatly increases when the dose is fractionated over a period of weeks.

Table 4.3. Skin effects from acute exposure to low energy X rays.

Skin Effect	Typical Threshold Dose (Gy)	Time to Occurrence
Early Transient Erythema	2	hours
Temporary Epilation	3	3–4 weeks
Main Erythema	6	2–3 weeks
Permanent Epilation	7	1–3 weeks
Dry Desquamation	10–15	4 weeks
Invasive Fibrosis	10	
Dermal Atrophy	11	> 14 weeks
Telangiectasis	12	> 52 weeks
Moist Desquamation	15–20	2–4 weeks
Late Erythema	15	6–10 weeks
Secondary Ulceration	20	> 6 weeks
Dermal Necrosis	> 25	2–10 weeks

All deterministic ionizing radiation effects result from cell killing. Except for special circumstances, particularly fetal development, a large number of cells must be killed for any damage to become apparent. Since the probability of killing a large population of cells follows a sigmoid dose response relationship, deterministic effects exhibit dose thresholds, below which no effect is seen. The risks of deterministic effects may be eliminated by keeping the dose well below individual organ thresholds, or risks may kept to acceptable levels by not exceeding a dose associated with a low severity of response. However, the occurrence and severity of deterministic effects in any one person cannot be predicted with absolute certainty, since factors such as radiation sensitivity, cellular repair and re-population exhibit considerable variation among individuals.

4.3 Effects of Low Level Ionizing Radiation

The term "low level" cannot be precisely defined. One definition of low doses and low dose rates is irradiation in which only one track or energy deposition event traverses the cell nucleus within the time interval during which repair mechanisms in the cell can operate. Using this definition, a low dose for mammalian cells in culture less than about 20 mGy[18]. As discussed above, the minimum whole body dose associated with acutely observable somatic effects is about 100 to 500 mGy delivered in one fraction. Many recommendations for occupational exposure limit the lifetime cumulative equivalent dose to the product of 10 mSv and age. Thus, doses below a few hundred mSv may be considered to be low level. Low level effects are those that are not appreciably due to cell killing, but instead arise from non-lethal cell damage, most likely cell mutation. Exposure to low level radiation results in "stochastic" effects. This term is used to describe effects whose probability of occurrence, rather than the severity in an affected individual, is a function of dose. A large group of individuals exposed to a "small" amount of radiation do not all develop a "small" amount of cancer. Instead, a small number of individuals will develop cancer while the rest will remain cancer free. Furthermore, it is impossible to predict, on an individual basis, which individuals will develop cancer. Low level ionizing radiation effects may occur long after irradiation and are sometimes termed "late" effects. The latent period between exposure and expression of the effect is due both to the manner in which damage is expressed and the time required for an effect to become clinically recognizable.

Low level radiation effects may be distinguished on the basis of the tissue of origin. Various types of cancer may result from radiation induced mutations in somatic cells. Fetal defects and childhood cancer may result from radiation induced mutations in embryonal and fetal cells. Genetic defects in future generations may result from radiation induced mutations in genetic stem cells.

4.4.1 Mechanisms of Damage from Low Level Ionizing Radiation

The mechanisms of damage from low level ionizing radiation are common with those discussed above for high level effects, and probably include more subtle mechanisms as well. Since high level effects result from cell killing, mechanisms of high level effects necessarily focus on DNA damage severe enough to cause cell necrosis or apoptosis. Since low level effects arise from mutational events in somatic, genetic or fetal tissue, low level effect mechanisms must result in irreparable but non-fatal DNA damage. While the formation of single and double DNA strand breaks is still of primary importance, DNA base alterations, DNA-DNA and DNA-protein cross links also may play and important role[19]. In addition to gross chromosomal abnormalities visible with light microscopy, more sophisticated molecular studies have revealed a wide range radiation induced mutations, including large deletions, large rearrangements detectable by restriction length polymorphism analysis, small deletions and/or insertions, base substitutions and frameshift mutations[20].

Cell mutations may be observed immediately in subsequent generations of irradiated cells, or may not be observed until many cell divisions have occurred. The latter type of mutations, which have been described as "delayed" mutations[21]. The two types of mutations are illustrated in Figure 4.6. Delayed mutations may be caused by "genomic instability." This term denotes an increased rate of acquisition of alterations in the genome, including chromosomal destabilization, gene amplification and mutation. In the theory of radiation induced genomic instability, the initial radiation event does not induce a specific mutation, instead, one or more genes are damaged that are crucial for maintaining the cell's ability to accurately reproduce the genome. A cell with reduced capacity to sense and repair DNA damage would be much more sensitive to subsequent mutational events and would exhibit an increased frequency of mutations. Experiments have shown that

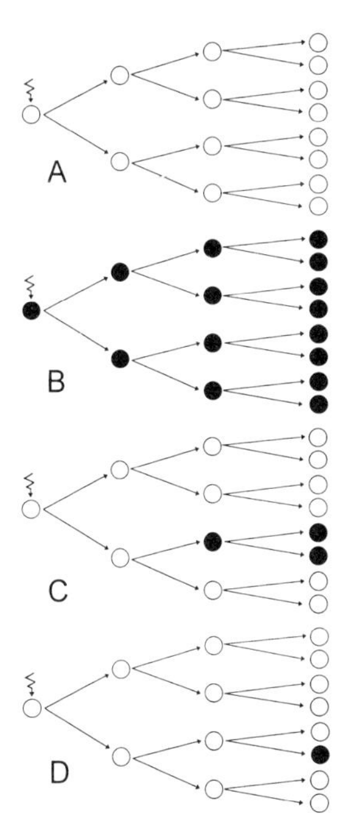

Figure 4.6. Illustration of direct and delayed mutations. Open circles represent normal cells and solid circles represent mutated cells. Panel A shows an irradiated cell with no mutations. Panel B is an example of a cell directly mutated by radiation exposure; all of the progeny cells will carry the mutation. Panels C and D are examples of delayed mutations. The original cell and its immediate progeny do not show the mutation, however certain of the later progeny are mutated. Genomic instability may explain delayed mutations. Reproduced from Little[21] with permission.

direct mutations exhibit 73% total or partial deletions and 27% point mutations, while delayed mutations exhibit 24% total and partial deletions and 76% point mutations[21]. Direct evidence for radiation induced genomic instability has been found[22]. Using an assay system based upon the ability of irradiated mouse mammary tissue to induce mammary tumors, it was found that a mutated p53 gene was present in 100% of cells after 27 cell division, but not present after 11 cell divisions. Thus, it was concluded that radiation did not directly induce a mutated gene, but induced instability in the genome that allowed the mutated gene to be expressed. While the target of direct mutations is a specific gene, target for genomic may be as much as 1% of the entire genome.

4.4.2 Carcinogenesis

Carcinogenesis is not a simple process. The conceptual framework for carcinogenesis includes initiation, promotion and progression. The initiation and promotion events constitute a multi-step process in which two or more events are required to transform a normal cell into a cancer cell[23]. The process of carcinogenesis may be subdivided as: (1) initial molecular damage to cellular DNA; (2) post irradiation modification of the damage leading to; (3) the generation of specific gene or chromosomal mutations in appropriate target somatic cells that initiate the oncogenic process; (4) early clonal expansion (promotion) of initiated cells generating pre-neoplastic lesions; and (5) the accumulation of additional genetic and epigenetic events leading to; (6) further clonal evolution-selection that drives progression and metastasis of the final malignancy[24]. Two classes of gene or chromosomal mutation may lead to initiation. One process involves activation of proto-oncogenes. This class of genes positively regulates cellular functions, either by reception and transduction of biological signals from the cell membrane to the cell nucleus or by directly controlling gene expression through transcription, such that activation leads to enhanced cell proliferation. The second process involves inactivation of tumor suppressor genes. This class of genes negatively regulates cellular functions such that loss of activity leads to enhanced cell proliferation or blocked differentiation. As discussed above, changes in gene function may occur as a result of direct mutation or as a result of radiation induced genomic instability. A promoter is an agent that, by itself, cannot induce cancer. In combination with an initiating agent, promoting agents synergistically enhance the process of malignant transformation. In principle, ionizing radiation can act as both an initiating and promoting agent, however its potential as an initiating agent is of greatest interest. Progression denotes the process by which transformed cells proliferate and take on an aggressive malignant character. Cell proliferation is an important mechanism involved both in the expression and fixation of damage. Agents that affect cell proliferation, such as hormones, can have a dramatic effect on carcinogenesis.

Since carcinogenesis involves multiple steps, each with many possible influencing agents, the relationship between the absorbed dose of ionizing radiation and the amount of induced malignant transformation can be quite complex. Damage at the microscopic scale, leading to DNA damage and hence initiating carcinogenesis, depends upon the amount of energy absorbed, the dose rate as which it is absorbed and upon the manner in which ionizing radiation deposits energy. Thus, similar to high level effects of radiation, the effect of a given exposure depends upon the dose, dose rate and linear energy transfer of the radiation. The capacity of the nucleus to repair DNA damage results in additional dependence on dose rate and dose fractionation.

The intrinsic (i.e. pre-irradiation) condition of the DNA also affects response. Several spontaneous chromosomal aberrations, such as Bloom's syndrome, ataxia telangiectasia and xeroderma pigmentosum, are known to be associated with increased sensitivity to radiation induced cancer[25]. Dependence on intrinsic DNA condition also provides a mechanism whereby ionizing radiation could interact synergistically with other carcinogens, such as alcohol and smoking. Evidence that genetic conditions increase sensitivity to radiation induced cancer may also be used to support the concept of relative risk, in which the risk of developing radiogenic cancer is proportional to the pre-existing spontaneous risk of cancer.

Agents that promote malignant transformation (such as TPA [tertradecanoyl phorbol acetate]) or suppress malignant transformation (such as protease inhibitors and vitamin A analogs) also can have dramatic effect on the magnitude of induced cancer[26]. The presence of TPA in the cell culture of irradiated cells can increase the frequency of malignant transformation by a factor of 100. The promoting effect of TPA is still observed even when administered long after irradiation. Tumor suppressing agents show similar dramatic effects in-vitro. Antipain, a protease inhibitor, has been shown to completely inhibit malignant transformation in cells irradiated with 600 cGy of X rays. Vitamin A analogs and, possibly, vitamin C, can suppress malignant transformation in-vitro. Beta carotene, vitamin E and vitamin C may affect malignant transformation via their action as oxygen free radical scavengers. Whether or not such tumor suppression agents have the same effect in humans is currently under study[27].

Hormones have long been known to be associated with cancer. Breast and endometrial cancer are caused by cumulative exposure to estrogen, while prostate cancer is caused by cumulative exposure to testosterone[25]. Thyroid hormones and growth factors have been shown to increase malignant transformation in-vitro[24]. Since hormonal status is gender and age dependent, gender and age can also affect the incidence of radiogenic cancer.

The existence of cellular repair mechanisms and tumor suppression agents raises the interesting question of whether low level ionizing radiation could give rise to a lower than spontaneous incidence of cancer. Many agents, such as heat, hypoxia, and toxic chemicals, that are toxic at high levels or concentrations apparently increase tolerance and even growth and longevity when administered at very low levels. This phenomenon, variously called adaptive response, hormesis or stress response, has been observed with ionizing radiation[28]. Adaptive response is studied by delivering an initial low dose of radiation (referred to as the conditioning, priming or adapting dose) followed by a high dose of radiation (referred to as the challenge dose). Such experiments have revealed a wide range of effects including reduced numbers of chromosome aberrations and sister chromatid exchanges in cultured human lymphocytes and fibroblasts, increased cell survival and proliferative capacity of many cell types, and decreased tumor growth in animals[18]. The mechanism of adaptive response phenomena involve homeostatic responses at the cellular level that permit cells to resist acute exposure to stress and subsequently to repair damage to the structure of the cell. Such results are obtained with broad strategies which include minimizing the amount of cell damage inflicted through the arrest of the cell cycle progression, the inhibition of expression of some classes of genes and the repair of protein and DNA damage by production of repair proteins[18,29]. Other possible mechanisms for the adaptive response effect include stimulation of proliferation, or stimulation of an immune response. Adaptive response in cells and animals has been demonstrated at very low priming doses. For instance, exposure of mouse embryo cells to 1 mGy has been shown to increase the ability of the cell to repair DSBs and to reduce the rate of malignant transformation from subsequent (24 hrs later) irradiation to doses of 10 and 100 mGy[30]. Adaptive response implies that, at least at some low exposure level, ionizing radiation is beneficial. If true for radiation carcinogenesis in human populations, adaptive response could have a profound effect on the choice of dose-response model. This possibility will be discussed further below.

From this brief summary, it can be seen that the number of factors that may influence the occurrence and magnitude of radiogenic cancer is quite large. Dose, dose rate, linear energy transfer, dose fractionation, DNA repair capacity, pre-existing genetic conditions and spontaneous cancer rate, tumor promoters, tumor suppressers, hormones and other proliferative agents, gender, age and presence of other carcinogens all can have an effect on response to ionizing radiation. Any of these factors also could represent a confounding agent in an epidemiological study. The task of estimating risk is to develop a model that either controls for, or includes as many factors as possible such that meaningful results can be obtained.

4.4.3 Dose Response Relationships

The number of excess tumors resulting from some amount of absorbed dose is expressed mathematically as the dose-response relationship. Since all of the factors discussed above can influence the shape of the dose-response relationship, no single, simple function is expected to be correct for all circumstances. The general types of functions that may be considered are illustrated in Figure 4.7, which plots the risk of cancer death versus absorbed dose. The data points are typical of A-bomb survivor data in the equivalent dose range below 2 Sv. Note that at zero dose, the risk is not zero, but is equal to the spontaneous risk of cancer death (0.2). Although it is more common to plot only excess cancer (i.e. the spontaneous incidence is subtracted out), the explicit inclusion of spontaneous incidence allows illustration of possible beneficial effects of low level exposure. The possible types of dose-response functions may be characterized as follows: The linear, no-threshold model assumes that response is linearly proportional to dose and that a detrimental response is present at any level of exposure. The linear-quadratic, no threshold model assumes that response has both a linear and a quadratic proportionality to dose and that a detrimental response is present at any level of exposure. Threshold models assume that there is no response up to some dose level, and that, thereafter, the response is proportional to the dose. Figure 4.7 shows a linear, threshold model in which, after the threshold dose is reached (0.2 Sv was assumed in this case), the detrimental response is linearly proportional to dose. Finally, dose-response functions can be developed that model possible beneficial effects (adaptive response) of low level exposure. An adaptive response dose response function results in no effect at zero dose, a less than spontaneous rate of cancer incidence at low dose levels (a beneficial response), and a detrimental response at higher dose levels.

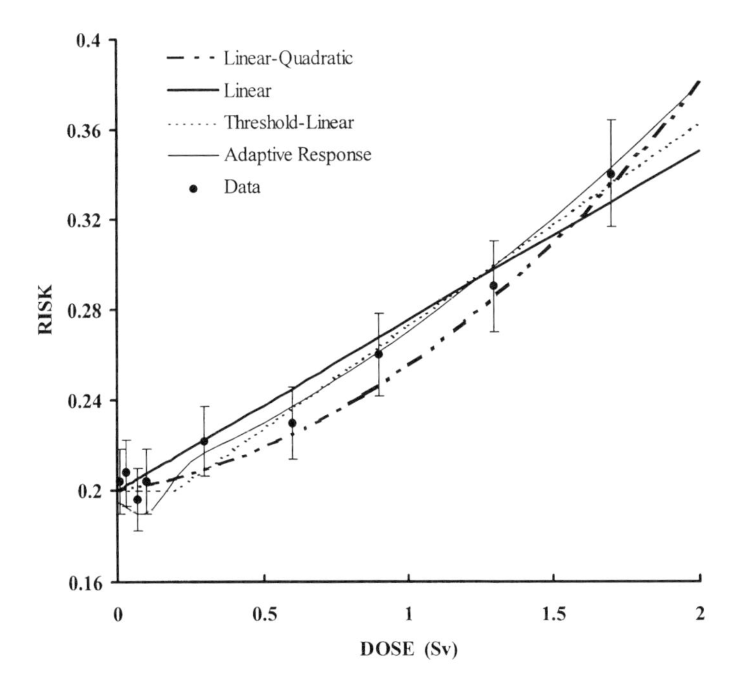

Figure 4.7. Illustration of the types of dose-response models that can be used to fit observed data. In this case, all models have approximately the same ability to fit the data, but yield greatly different response for doses less than 0.2 Sv.

The linear-quadratic, no threshold dose response model has been used to describe many radiation biology experimental results. In equation 4.1 above, the linear quadratic model was used to describe the frequency of single and two-track DSBs, as well as cell survival curves. In the context of dose-response models for malignant transformation or cancer incidence, the linear-quadratic dose response model may be written as,

$$S - S_0 = \alpha D + \beta D^2 \qquad (4.5)$$

where $(S - S_0)$ is the observed excess response above S_0, the spontaneous response, D is the absorbed dose and α and β are coefficients that depend on many factors such as cell type, dose rate and LET. Note that the linear, no threshold model is a subset of the linear-quadratic model. Assuming that cancer can arise from a single damaged cell (i.e. in the absence of repair processes), the microdosimetric properties of ionizing radiation yield strong evidence in favor or a no-threshold dose-response model. If the linear-quadratic model could be proved to correctly describe not only the formation of DNA damage such as SSBs and DSBs, but also expression of ultimate biological effect, a significant source of uncertainty could be removed from risk modeling. Without a fundamental scientific reason for its adoption, the linear-quadratic model becomes just another curve fitting function, with acceptance of the model based upon goodness of fit criteria.

Conventional linear-quadratic dose response models also can be modified to yield the adaptive response behavior illustrated in Figure 4.7[31]. Ionizing radiation still is assumed to cause physical damage at any dose level (i.e. physical damage is without threshold), however, this damage is assumed also to initiate repair processes that, at low doses, can lessen the effect of the radiation damage or even repair spontaneously occurring damage. As the radiation dose increases, the amount of damage overwhelms the effect of repair and the dose-response curve exhibits the conventional increase of detriment with increasing dose. Current evidence for adaptive response in human radiation carcinogenesis is equivocal. Studies of cancer incidence and mortality rates in regions of low and high (i.e. three times higher than the low background group) normal background radiation show no effect, either detrimental or beneficial, of low level radiation. However, such studies lack the statistical power to detect any proposed effect. Analysis of the A-bomb survivor data below 0.4 Sv, shown in Figure 4.8, show a possible adaptive response effect for leukemia, but not for solid tumors. Like natural background data, the A-bomb data does not allow determination of the dose response function with statistical precision. Proof that the adaptive response model is correct for human radiation risk modeling would have significant consequences, particularly in assessing risk from radiation sources that expose the populace to dose levels far below those typical of natural background radiation.

Examples of in-vitro dose response curves, as determined by the mouse fibroblast cell transformation assay technique[26], are shown in Figure 4.9. Three cases are illustrated: radiation given in a single fraction; a split dose

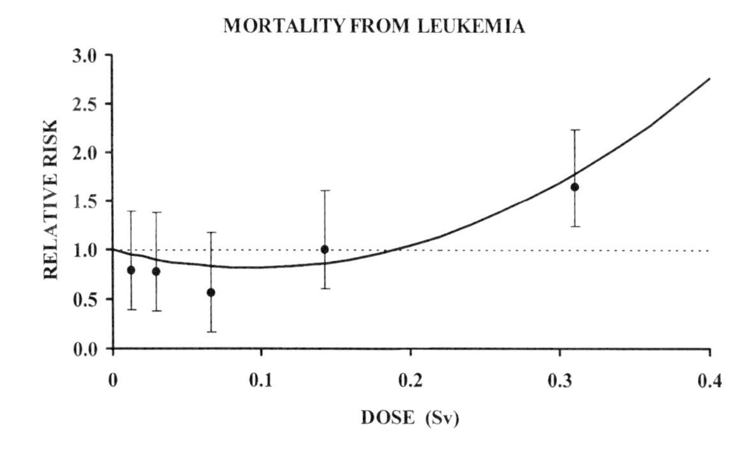

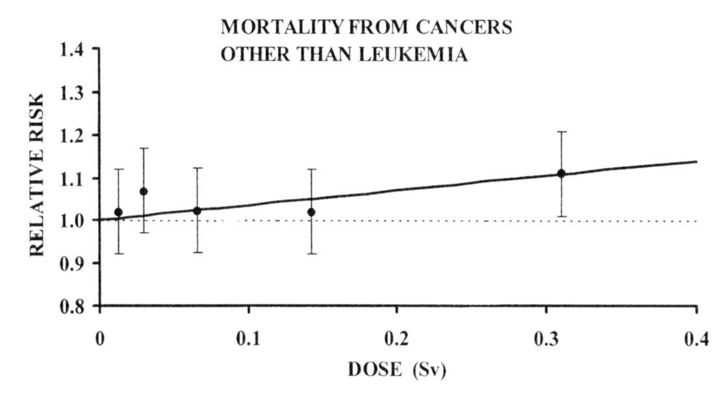

Figure 4.8. A-bomb survivor mortality from leukemia and all cancers other than leukemia for doses less than 0.4 Sv. The leukemia data show a linear-quadratic dose response function with a possible adaptive response effect, while the solid tumor data show a linear no-threshold dose response function with a possible supra-linear effect at low doses. Data reproduced from UNSCEAR[18] with permission.

regimen in which the dose is given in two equal fractions separated by 5 hours; and, a single dose of radiation followed by incubation of the cells in TPA, a tumor promoting agent. For comparison, a linear-quadratic fit to the single dose data also is shown. It is evident that, while the linear-quadratic model approximately describes the single dose data, the data show a complex behavior. Indeed, the dose-response data shows regions of three different slopes. In the low dose region the slope is approximately one and in the high dose region the slope is approximately two, both slopes being consistent with the linear-quadratic model. However, in the intermediate dose region (0.3–1 Gy) the slope is much less than unity, suggesting a plateau in which increasing dose has no effect. The split dose data show an even greater plateau effect, and hence even greater departure from the linear-quadratic model. Note that in the high dose region, the split dose decreases the transformation frequency, as would be predicted from cell survival experiments. However, in the low dose region, the split dose increases the transformation frequency. Such effects are not fully explained, but are thought to depend on repair mechanisms[26]. The presence of the tumor promoter TPA increases the magnitude of transformation and changes the shape of the dose-response curve from approximately linear-quadratic to pure linear.

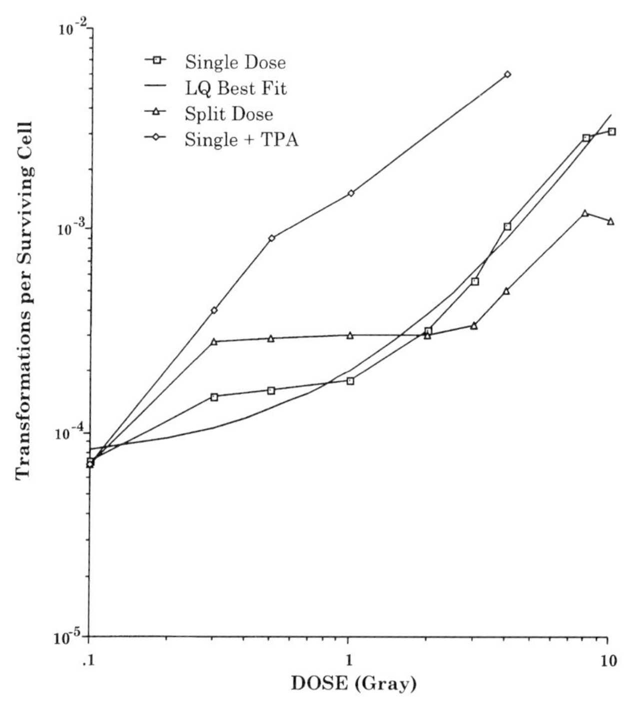

Figure 4.9. Measured dose response curves for in-vitro mouse cell fibroblast cell transformation showing the effects of fractionation and presence of the tumor promoting agent TPA. Data replotted from Chan and Little[26].

Radiobiological data such as that illustrated in Figure 4.9 lead to the conclusion that, although the linear-quadratic dose response model is a useful tool, the actual situation is more complex than such a simple model. Furthermore, factors such as cell type, fractionation, dose rate, LET, tumor promoters, tumor suppressers and proliferative agents can have large effects on values of the parameters of the model (i.e. α and β) and may result in behavior that cannot be accurately fit by the model. Human epidemiological data are not of much help in selecting a dose-response function. While much of the data comes from fairly high dose (greater than 0.1 Sv) irradiation, considerable data also are available in the low dose region. Unfortunately, the statistical power of the human data does not allow selection of the "correct" dose response model based solely on goodness of fit criteria. At the present time, the linear-quadratic model is the dose-response function used by scientific consensus organizations such as National Academy of Science Committee on the Biological Effects of Ionizing Radiation (BEIR) and the United Nations Scientific Committee on the Effects of Atomic Radiation (UNSCEAR). There is no reason to assume *a priori* that all types of radiogenic cancer follow the same dose-response model. As shown above in Figure 4.8, human leukemia data are better described by a linear-quadratic model, while solid cancer data are better described by a pure linear model. For the purpose of estimating the magnitude of stochastic effects of low doses in human populations, it is currently assumed that only non-threshold models are appropriate. This is the source of the often-quoted aphorism that "there is no safe level of radiation." A more precise statement is that the effects of low-level radiation cannot be precisely quantified because they are so small; hence it is prudent to extrapolate in a conservative manner with the assumption that even a small but finite amount of radiation has a small but finite effect. The linear no-threshold dose response model probably gives the upper limit of probable risk. The possibility that low doses have no effect or may even be beneficial cannot be excluded. This situation will change only when data from cellular and animal radiation biology experiments demonstrates sufficient evidence of a more accurate model.

4.4.4 Risk of Carcinogenic Effects

The risk of dying from cancer can be expressed as dimensionless number that is assumed to be equal to the observed number of cancer deaths per unit number of population. For instance, in a population of 100 000 individuals in the United States, 20 000 might eventually die from cancer. Hence the lifetime risk of cancer death for the U.S. population is 20 000/100 000 or 20 percent. One can also estimate the age specific risk of cancer death, which is assumed to be equal to the observed numbers of cancer deaths, at a given age, per unit number of population. Risk estimates or radiation carcinogenesis are, therefore, derived from observations of cancer death rates in exposed and unexposed populations. To quantify the relationship between dose and effect, it is necessary to distinguish radiogenic cancer from spontaneous cancer. Since this distinction cannot currently be achieved by any medical test, the only technique available is to detect an increase in the rate of cancer incidence above that normally expected in the population, and to presume that the excess is caused solely by radiation. The radiogenic increment will not be observable until some minimum latent period has passed and may not reach its peak value for some time after that. The total number of excess cancers is the difference in the areas underneath the incidence vs. time curves for irradiated and control populations. Hence, it is necessary to observe the irradiated and control populations until the incidence in the irradiated population returns to normal incidence. For cancers with long latent periods the incidence rate may never return to normal; and the total radiogenic excess will not be accurately known until all members of the irradiated population are followed for their entire lifetimes. As the dose decreases, it becomes increasingly difficult to distinguish the radiogenic increment with any statistical accuracy. Large control and irradiated populations are required to minimize the influence of normal random fluctuations in cancer incidence. To perform accurate experiments in the dose region that encompasses medical imaging (100 μGy to 100 mGy), millions of subjects would have to be irradiated and observed while an equally large population were maintained as a control.

The cancer risk in an irradiated population may relate to the spontaneous risk (i.e. natural incidence) in one of two possible ways. In the absolute risk model (also called the additive risk model), the additional risk associated with radiation exposure is independent of the spontaneous risk, while in the relative risk model the additional risk is proportional to the spontaneous risk. Absolute risk is expressed mathematically as

$$R(D, A, A_E, g) = R_0(A, g) + f(D) \cdot u(A_E, g) \tag{4.6}$$

where $R(D, A, A_E, g)$ is the age dependent total risk of cancer death in the irradiated population; R_0 is the spontaneous risk of cancer death that depends on A, the attained age, and, g, the gender; $f(D)$ is the dose-response function; and $u(A_E, g)$ is an excess risk function that depends on A_E, the age at exposure and gender. Relative risk is expressed mathematically as

$$R(D, A, A_E, g) = R_0(A, g) \cdot [1 + f(D) \cdot v(A_E, g)] \tag{4.7}$$

where R and R_0 are defined as above; $f(D)$ is the dose-response function, and $v(A_E, g)$ is an excess risk function that depends on A_E, the age at exposure and gender. The numerical value of the coefficients of the dose response model will be different, depending on the risk model. That is, the values of α and β, the coefficients of the linear-quadratic dose-response function will not be identical for both risk models. Note also that the values of u and v are numerically different, and in the simplest form are time constant; that is, they do not depend on time elapsed since exposure ($A - A_E$). However, the spontaneous risk of cancer is necessarily age and gender dependent.

Epidemiological studies use an unexposed control population to estimate $R_0(A, g)$, and an exposed population to estimate $R(D, A, A_E, g)$. The excess risk of cancer death, ER, is then given by

$$ER(D, A, A_E, g) = R(D, A, A_E, g) - R_0(A, g) \tag{4.8}$$

the relative risk of cancer death, RR, is given by

$$RR(D, A, A_E, g) = \frac{R(D, A, A_E, g)}{R_0(A, g)} \tag{4.9}$$

and the excess relative risk of cancer death, *ERR*, is given by

$$ERR(D, A, A_E, g) = \frac{R(D, A, A_E, g) - R_0(A, g)}{R_0(A, g)}. \tag{4.10}$$

For a study investigating only the total excess risk over the entire lifetime of a population, either risk projection model can be used to fit the data. However, as discussed below, the two models are distinguishable based upon how the excess risk and excess relative risk depend on attained age and how excess risk and excess relative risk differ among populations having different spontaneous cancer rates.

At present, there is no reason why one model should be better than the other based upon on fundamental grounds. The answer may become apparent as basic knowledge concerning carcinogenesis in general is obtained. For example, if radiation damage leading to activation of proto-oncogenes and inactivation of tumor suppresser genes is synergistic with spontaneous processes, then relative risk may be the most appropriate model. It could also turn out that, due to underlying cellular mechanisms, some tumors exhibit absolute risk, while others exhibit relative risk.

Selection of risk model can have profound consequences when incomplete data are used to project lifetime risk. Suppose an exposed population is observed to have an increased cancer rate, but the population is followed only for a short period of time following irradiation. Since cancer rates increase as a population ages, the relative risk model predicts that the number of excess cancers will continue to increase with time. The absolute risk model predicts that the number of excess cancers will remain constant, even as the spontaneous rate increases. Thus, when incomplete data are used to project lifetime risk, the relative risk model results in a larger number of cancers. Note that this problem occurs not because excess risk could be time dependent (see below), but because the spontaneous risk is time dependent. This problem does not occur when using data from a population that has been followed for its entire lifetime, since the model fitting parameters can be adjusted to be consistent with total cancer excess. The choice of risk projection model is particularly important in estimating risk for exposure at a young age. This is because the primary population upon which such a projection is based (atomic bomb survivors) is only now reaching the age at which spontaneous cancer is a significant cause of death. The model fitting procedures used by the BEIR III committee resulted in differences of up to a factor of four in projected total excess cancer deaths using relative risk as compared to absolute risk[32]. The BEIR V committee, using more complete data and a more sophisticated model, projected risks that differed only by a factor of two between the two models.

As illustrated in Figure 4.10, the absolute risk and relative risk models also yield divergent results when using data from one population to project risk for a different population. This difference occurs even when using data from a completely followed population that is not subject to the time projection

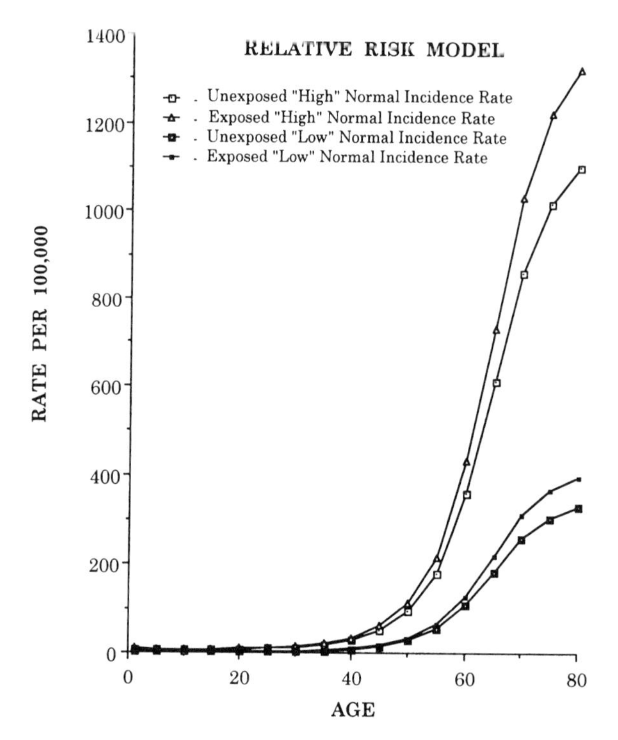

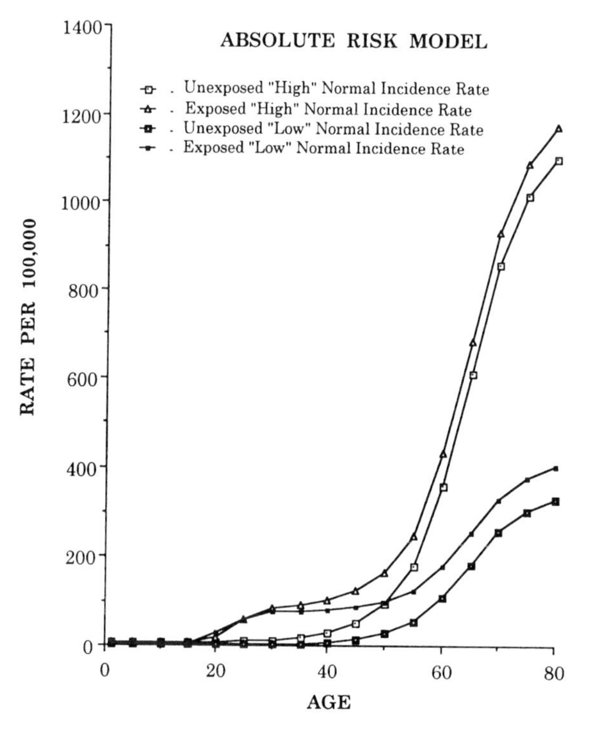

Figure 4.10. Age dependence of cancer incidence for the relative risk and absolute risk models. Two populations, one with a "high" normal cancer incidence rate and one with a "low" normal cancer incidence rate are exposed to the same dose of radiation. For the relative risk model (left), both exposed populations show the same fractional increase in rate of cancer deaths per 100 000 population. For a given population size (i.e. for 100 000 individuals), the exposed population with the greatest normal incidence rate will incur the greatest total number of excess cancer deaths. In the absolute risk model (right), excess cancer is independent of normal incidence. For a given population size (i.e. for 100 000 individuals), both exposed populations will suffer the same number of excess cancer deaths, regardless of the difference in normal incidence rates. Reproduced from *Health Effects of Exposure to Low Level Radiation*, with permission.

problem discussed above. The absolute risk model assumes that the number of excess cancers depends only on dose and not upon spontaneous cancer rate. Given two populations whose spontaneous cancer incidence rates differ by a factor of two, but receiving the same exposure, the absolute risk model predicts the same number of excess cancers in both. Given the same circumstances, but using the relative risk model, both populations would have the same fraction of excess cancers compared to the spontaneous rate. Hence, the same exposure causes twice the number of cancers in the population with twice the spontaneous cancer rate.

The attained age dependence of excess risk and the differences in excess risk among populations of different spontaneous risk provides a means of determining which model best describes reality. Since current knowledge of the fundamental processes of radiation carcinogenesis does not allow preference of one model over the other, the preferred model is the one that best fits observed data. The BEIR V committee[23] determined that the relative risk model best fits both the observed attained dependence of risk and the differences in risk observed in differing populations. The time independent absolute risk model is no longer used for risk estimation of human populations.

In either the relative risk or absolute risk models, further sophistication can be introduced by allowing the excess risk function to depend on time elapsed since exposure. The relative risk model can then be expressed as

$$R(D, A, A_E, g) = R_0(A, g) \cdot \left[1 + f(D) \cdot v(T, A_E, g)\right] \tag{4.11}$$

where all symbols are as previously defined and $v(T, A_E, g)$ is the relative risk function that depends on T, the time elapsed since exposure (i.e. $T = A - A_E$). By allowing the relative risk function to depend on time elapsed since exposure, it becomes possible for the model to take into account the possibility that radiation risk changes as time passes. For instance, no radiogenic leukemia is observed until at least two years has elapsed since exposure, while for solid tumors approximately ten years must elapse. The excess risk of leukemia declines between 25 to 30 years after exposure but the excess risk of solid tumors apparently never disappears. The risk of radiogenic breast cancer has a marked dependence on age at time of exposure, generally decreasing as age at time of exposure increases, while no such effect is seen for respiratory cancer.

Quantitative data for estimating the magnitude of radiation carcinogenesis in humans are based upon the study of comparatively few subjects. The National Academy of Sciences[23] and the United Nations Scientific Committee on the Effects of Atomic Radiation[18] have reviewed most relevant radiation epidemiology studies. In addition, several major epidemiological studies have been updated since the BEIR V and UNSCEAR 1994 reports. The largest population to receive exposure to all organs is the survivors of the atomic bombing of Hiroshima and Nagasaki. Of that population, 86 572 irradiated individuals have been carefully followed for over 40 years. As of 1990, in the 50 113 survivors receiving more than 0.005 Sv, a total of 421 excess cancer deaths (87 from leukemia and 334 from solid cancer) have been observed[33]. This is not an ideal group because precise dosimetry is difficult and because they were exposed to neutrons as well as γ and X rays. Since the A-bomb survivors were exposed to a single dose at a high dose rate, use of the data for risk estimation of fractionated and/or low dose rate exposure is problematic. Another important source of data is the approximately 15 000 patients who received medical irradiation for ankylosing spondylitis between 1935 and 1957. The dosimetry of this population is less uncertain; however, the exposure was localized and given over a period of weeks rather than instantaneously[34]. Additional data, particularly applicable to breast cancer[35] and lung cancer[36], are provided by 60 000 patients who received multiple fluoroscopies during pneumothorax treatment for tuberculosis or who received radiation therapy for benign breast disease. Tens of thousands of children treated by radiation therapy for tinea capitis between 1930 and 1955 have also provided important data, especially for thyroid cancer[37]. Effects of low doses and low dose rates in 96 000 nuclear industry workers have been analyzed by pooling data from the United States, the United Kingdom and Canada[38]. Several hundred thousand individuals were irradiated as a consequence of the Chernobyl accident, however, results of carefully controlled studies are just beginning to be published[39].

Overall, perhaps 300 000 individuals have been identified as having sufficient dosimetry, length of follow-up, and control populations to yield useful data. Many of these subjects received whole- or partial-body doses in excess of 0.5 Sv, although the A-bomb survivors and the nuclear industry workers provide over 120 000 individuals exposed to less than 0.1 Sv. The typical statistical quality of the data in the low dose region is illustrated in Figure 4.8. The relative risk for cancer incidence and mortality of the A-bomb survivors is illustrated in Figure 4.11. The relative risk of leukemia is high, and the uncertainty interval is large. Both effects are due in part to the fact that leukemia is a fairly rare disease. In the 1990 analysis of 86 572 A-bomb survivors, there were only 249 leukemia deaths, of which 87 were estimated to be radiation induced. One type of leukemia, chronic lymphocytic leukemia, is apparently not induced by radiation. Excess risk for leukemia is seen 2 to 3 years after exposure, appears to peak, and then declines with time. Populations, particularly those irradiated at a young age, have not been followed for a long enough period to determine if the excess risk ever returns to zero. The relative risk of solid tumors is much less than that of leukemia. However, since the normal incidence for all solid tumors is high, the number of radiation induced solid tumors exceeds that of leukemia. Risk for individual types of solid tumors is known with less certainty, due to the few number of cases for each disease type. There is good evidence for radiogenic breast, lung and stomach cancer; and suggestive evidence from many other sites. However, there is much less evidence for radiation induced cancer of the uterus, cervix, prostate and lymphomas. Excess risk of solid tumors is seen 5 to 10 years after exposure and remains elevated the remaining lifetime.

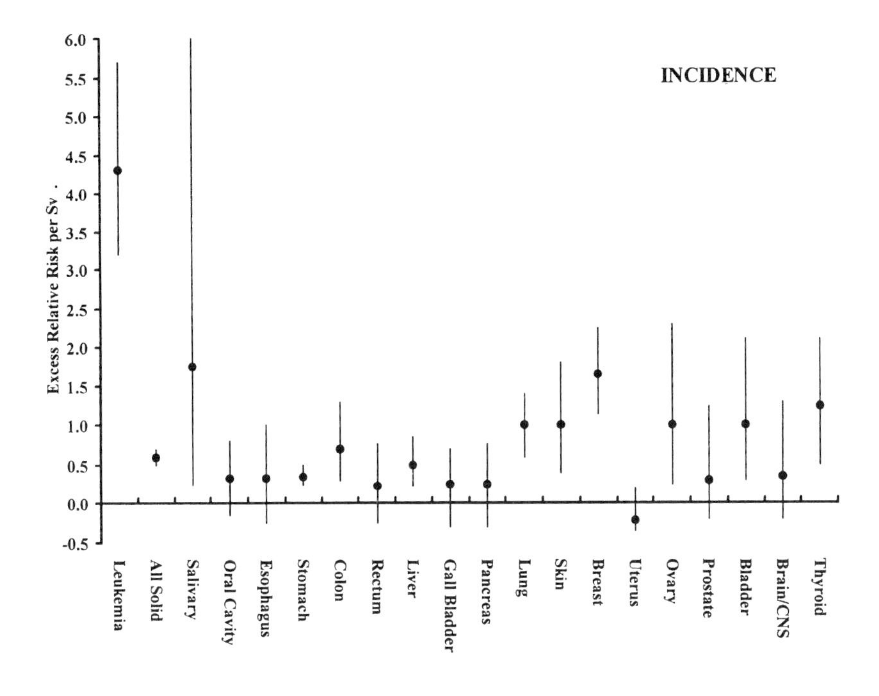

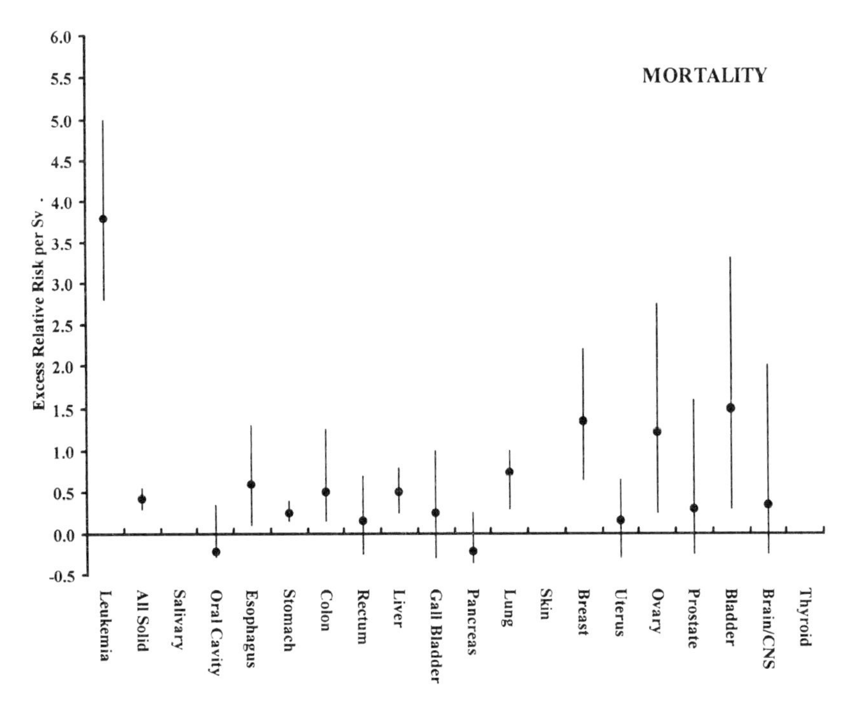

Figure 4.11. A-bomb survivor excess relative risk in incidence and mortality for various cancers. Data reproduced from UNSCEAR[18] with permission.

Risk models derived from epidemiological data are used to estimate radiation detriment over the lifetime of a population. One measure of lifetime risk, used by the BEIR committees, is the "excess lifetime risk" (ELR). Given two populations, one exposed to radiation and the other unexposed, excess lifetime risk is the difference in total cancer deaths, as counted after all members of the population have expired. For example, suppose two populations, each of 100 000 individuals distributed in age and gender similar to the total U.S. population, were followed for their entire lives. The unexposed population would suffer 100 000 deaths, of which around 20 000 would be due to cancer. An identical population exposed to a single acute equivalent dose of 0.1 Sv would also suffer 100 000 deaths, but about 20 800 deaths would be due to cancer. The projected lifetime risk is thus 800 excess cancer deaths or an increase of about 4% above normal.

Excess lifetime risk is calculated from the exposure-time-response models using standard life table (mortality table techniques). Life tables follow a population from one year to the next by tabulating mortality for each year. The number of members surviving from one attained age to one year older is equal to the number starting the year minus the mortality for that attained age. Eventually there are no surviving members to carry forth to the next attained age. In this manner, a population can be followed until all members have expired, and the number of deaths from all causes equals the initial number of the population. For radiation carcinogenesis risk estimates, the causes of mortality are divided into cancer and all other causes of mortality, with the assumption that radiation affects only cancer mortality. In an exposed population, the exposure-time-response model is used to modify the cancer mortality values for each attained age. The lifetime risk is the total excess cancer mortality (or percent excess cancer mortality) when all members of the population have expired and all sources of mortality have been tabulated.

Another method of estimating lifetime radiation detriment, used by the UNSCEAR committees[18], is the "risk of radiation-exposure induced death" (REID). REID also utilizes life table methodology, but measures a slightly different quantity. The risk of exposure-induced death can be interpreted as the risk that an individual will die from a cancer that has been caused by the exposure in question, while the excess lifetime risk is the difference between the proportion of people dying of cancer in an exposed population and the proportion dying of this cause in an otherwise identical, but unexposed, population. The difference between the REID and the ELR concerns the counting of cases who would have died from cancer in the absence of exposure, but who die of this cause at an earlier date following exposure. Such cases contribute to the risk of exposure-induced death, but are ignored in the calculation of excess lifetime risk. The ELR is always less than the REID. It can be shown that the ratio of ELR to REID is approximately equal to (1-B_r), where B_r is the lifetime risk for all exposure-induced causes among unexposed persons. Since the risk of dying from cancer is approximately 15% to 20% in most populations, the ELR is about 15% to 20% less than the REID. It has been shown that the quantitative similarity of risk estimates between the BEIR V report (using ELR) and the UNSCEAR 1988 report (using REID) happens as a result of approximate cancellation of a number of differences that could be quite large[40].

The BEIR committee reports, starting in 1972, and UNSCEAR committee reports, starting in 1977, have developed quantitative estimates of radiation carcinogenesis using the best epidemiological data, dosimetry, dose response and risk projection models available at the time of issuance of the report. Early estimates of excess lifetime risk were about 100 excess cancer deaths per 10 000 person-Sv, with a linear no-threshold dose response function for all cancer. As better dosimetry, and longer follow-up periods were available, the risk estimates have increased. The most recent reports give quantitative risk estimates only for populations exposed to an instantaneous dose (i.e. single fraction, high dose rate), since much of the data is based on this type of irradiation. Because the leukemia risk is best fit by a linear-quadratic dose response function, it is not possible to quote a single number of estimated excess cancer deaths per collective dose. Instead, risk estimates are developed for specific irradiation conditions. For instance, the BEIR V committee estimated the ELR for the U.S. population exposed to a single dose of 0.1 Sv, and a continuous lifetime exposure of 1 mSv/yr, while the UNSCEAR 1994 committee estimated the REID for a Japanese population exposed to a single dose of 0.2 and 1 Sv. Neither committee made risk estimates in low doses (< 0.1 Sv), in recognition of the uncertainty in the validity of dose response models a very low doses. Lifetime risk values calculated from the BEIR V and UNSCEAR 1994 reports are shown in Table 4.4 for whole body single exposures of 0.1, 0.2 and 1.0 Sv. The UNSCEAR values of REID exceed the BEIR V values of ELR by 20 to 30 percent, as may be expected. Note that below 0.2 Sv, both risk estimates do not deviate significantly from a linear dose dependence. This is because the linear term of the linear quadratic dose dependence for leukemia predominates in the low dose region. Solid tumors, best fit by a pure linear dose response function, outnumber leukemia by a wide margin at all dose levels.

Table 4.4. Cancer mortality in 100 000 individuals exposed to 0.1, 0.2 and 1.0 Sv as calculated from models given in the BEIR V and UNSCEAR 1994 Reports. Data are for a single exposure at high dose rate. The BEIR data are for excess lifetime cancer (ELR) in a U.S. population and the UNSCEAR data are for radiation-exposure induced death (REID) in a Japanese population, as discussed in the text.

	BEIR V			UNSCEAR 1994		
Dose (Sv)	Leukemia	Solid Tumors	Total	Leukemia	Solid Tumors	Total
0.1	95	695	790	65	950	1015
0.2	209	1390	1599	140	1900	2040
1.0	1804	6950	8754	1082	9500	10 582

The lifetime risk values given in Table 4.4 are for entire populations, distributed in age and gender according to the demographics of specific population. It is also useful to calculate risk for specific age and gender groups. The BEIR V committee calculated excess lifetime risk for males and females of various ages at time of exposure. The results, shown in Table 4.5, show that risk is highest at younger ages and continuously decreases with age at exposure. Females are at higher risk, due to greater sensitivity to breast and digestive cancer. Risk estimates for those aged 15 years or less at time of exposure are subject to greater uncertainty since there is inadequate follow-up to determine the best risk projection model for solid tumors. For solid tumors in those exposed at age 5 or less, the UNSCEAR 1994 report calculated that the REID varied by a factor of about 2, depending upon whether the risk was assumed to remain constant from 10 years after exposure or declined to zero risk at age 90.

In most cases in which ionizing radiation risk estimates are required, such as in evaluating risks of occupational exposure or risks of medical imaging, the exposure is delivered in many fractions at low to moderate dose rates. Risk estimates derived from single exposures at high dose rates may be modified for low dose rates and fractionated exposures by the application of a "Dose Rate Effectiveness Factor" (DREF). The dose rate effectiveness factor is the ratio of effects observed from an exposure delivered at high dose rates and single fractions to the effects observed from the same exposure delivered at low dose rates and multiple fractions. Thus, DREF is usually greater than one. To obtain risk estimates for low dose rates and/or fractionated exposure, the risk estimate from a high dose rate model, such as the BEIR V Report, must be divided by the value of DREF. Unfortunately, there is little accurate information concerning DREF for human carcinogenesis. For carcinogenesis in laboratory animals, DREF ranges from 2 to 10. The BEIR V committee suggested a value of 2 for leukemia, while analysis of nuclear industry workers[38] as compared to A-bomb survivors estimated the leukemia DREF is of the order of 1.7 with a lower limit of 0.6 and an upper limit of 28. Analysis of tuberculosis patients multiply irradiated during artificial pneumothorax treatments revealed little evidence for a fractionation effect for breast cancer[35]. However, analysis of lung cancer in the same patient group revealed essentially no excess lung cancer risk for fractionated exposure[36]. Using the 95% upper bound of risk for fractionated exposure, a DREF of 8.6 was estimated for lung cancer. The National Council on Radiation Protection and Measurements adopted a value of 2 for use in setting limitations on exposure[41].

For purposes of radiation protection of workers and the public, as well as for purposes of estimating the upper limit of risk to patients, the BEIR V and UNSCEAR risk estimates made at the lowest dose (i.e. 0.1 or 0.2 Sv) are adjusted by a conservative DREF and extrapolated using a linear no-threshold dose response model. As previously noted, there is no epidemiological evidence to exclude other dose response models, however, risk estimates derived in this manner are unlikely to underestimate the risk. At the 0.1 Sv level, risk estimates from BEIR V and UNSCEAR reports range from

Table 4.5. Cancer excess mortality by age at exposure and site for 100 000 individuals of each age exposed to 0.1 Sv in a single dose at high dose rate. "Average" refers to the risk to a heterogeneous population with an age distribution of the U.S. population. Data are reproduced from the BEIR V Report.

MALES							
Age at Exposure	Total	Leukemia	Solid Tumors	Breast	Respiratory	Digestive	Other
5	1276	111	1165	–	17	361	787
15	1144	109	1035	–	54	369	612
25	921	36	885	–	124	389	372
35	566	62	504	–	243	28	233
45	600	108	492	–	353	22	117
55	616	166	450	–	393	15	42
65	481	191	290	–	272	11	7
75	258	165	93	–	90	5	–
85	110	96	14	–	17	–	
Average	770	110	660		190	170	300

FEMALES							
Age at Exposure	Total	Leukemia	Solid Tumors	Breast	Respiratory	Digestive	Other
5	1532	75	1457	129	48	655	625
15	1566	72	1494	295	70	653	476
25	1178	29	1149	52	125	679	293
35	557	46	511	43	208	73	187
45	541	73	468	20	277	71	100
55	505	117	388	6	273	64	45
65	386	146	240	–	172	52	16
75	227	127	100	–	72	26	3
85	90	73	17	–	15	4	–
Average	810	80	730	70	150	290	220

7.9×10^{-2} to 12.7×10^{-2} Sv^{-1} (i.e. 790 to 1270 excess cancer deaths for a population of 100 000 exposed to 0.1 Sv). The average risk is not significantly different from 10×10^{-2} Sv^{-1}. Using a DREF of 2, the NCRP and ICRP have adopted a risk estimate for fatal cancer of 5×10^{-2} Sv^{-1} in a general population, and $4 \times 10^{-2}\ Sv^{-1}$ in an adult worker population[41]. The lesser risk in the worker population is because workers are assumed to be greater than 18 years of age.

4.4.5 Risk of Genetic Effects

Genetic effects of ionizing radiation encompass both serious and non-serious genetically carried conditions that are expressed in the progeny of irradiated individuals. It is important to distinguish genetic effects from fetal effects. Although both can cause birth defects, by definition genetic effects do not involve the irradiation of pregnant females. Somatic fetal damage cannot be transmitted to future generations, whereas genetic damage can. Furthermore, genetic effects cannot be caused by the irradiation of organs other than the gonads, since they are due solely to damage in the reproductive germ cell population. Early radiation protection standards were based, in part, on over-estimates of genetic risk. Early studies of radiation effects in human populations under-estimated cancer risk, due to insufficient follow-up, and over-estimated genetic risk, due to lack of human data and improper extrapolation from animals to humans. As reviewed by Sankaranarayanan[42], recent genetic risk estimates of the BEIR and UNSCEAR committees, as well as the most recent analysis of the progeny of A-bomb survivors, indicate that human genetic risks are much less significant than carcinogenic risks.

Genetic disorders can be classified into those that result from cytogenetically demonstrable chromosomal abnormalities and those that result from non-demonstrable chromosomal mutations. Chromosomal abnormalities, which are caused by meiotic nondisjunction and translocation, often result in severe birth defects. Trisomies and monosomies are the result of nondisjunction, with trisomy 21, or Down's syndrome, being a well-known example. There is very little evidence from insect or animal experiments to indicate that radiation induces trisomies. Reciprocal translocations, resulting from the exchange of broken-off pieces of chromosomes, are the most likely radiation-induced chromosomal abnormality. These have been detected in irradiated mice, but only at a very low rate of incidence compared to the total induced disorder rate. For these reasons, genetic disorders due to chromosomal aberrations are not a major contributor to the genetic effects of radiation. There are a great many genetic disorders that have not been associated with any chromosomal abnormality and hence are presumed to be caused by chromosomal mutation. Over 1800 naturally occurring genetic disorders have been identified. Some of these are of little consequence to the affected individual, whereas others are classified as serious (expected to cause serious handicap sometime during the individual's lifetime, or premature death). Genetic disease is not rare. As summarized in Table 4.6, approximately 1.6% of live births have Mendelian or chromosomal disease, 6% have congenital abnormalities and 65% have or will develop "multifactorial" conditions (such as heart disease or cancer) that are thought to have a genetic component.

Table 4.6. Prevalence and examples of genetic disease in human population. Data summarized from Sankaranarayanan[42].

Natural Prevalence of Genetic Diseases in Human Populations		
Disease Classification	Examples	Prevalence at Birth (%)
Mendelian		
Autosomal dominant	neurofibromatosis, Huntington disease, osteogenesis imperfecta	0.95
X linked	hemophilia, Duchenne muscular dystrophy, fragile-X mental retardation	0.05
Autosomal recessive	cystic fibrosis, phenylketonuria, Tay-Sachs disease	0.25
Chromosomal	Downs syndrome, Edwards syndrome, Patau syndrome	0.40
Congenital abnormalities	spina bifida, ventricular septal defect, cleft lip	6.00
Multifactorial Conditions	essential hypertension, diabetes mellitus, coronary heart disease, cancer	65.00

Extensive experiments with animals and insects have shown that radiation can cause genetic mutations; however, no radiation-induced mutations are seen that are not also seen normally. The same experiments have shown that the male is two or more times as sensitive as the female and that, in the low-dose region, the dose-response curve for mutations follows a linear non-threshold model. Also, fractionated or low-dose rate exposures have less effect than acute high-dose rate exposures. Because mature spermatozoa are more sensitive to the induction of genetic effects than are stem cells, the possible effects of male gonadal irradiation can be reduced by allowing a time interval between irradiation and conception. This period is approximately two months in mice and probably three to six times longer in humans.

Human genetic risks are estimated using two methods. In the "direct" method, the induction rate for a specific class of defects in mice, such as skeletal defects, is measured directly by using high dose rate radiation. The results are corrected for dose rate effects, extrapolated to all dominant mutations in mice and finally converted to a human population based upon the relative distribution of genetic disorders between the species. In the "doubling dose" method, risks are calculated from estimating the dose required double the naturally occurring prevalence of genetic disease. The doubling dose may be measured for animals and extrapolated to man, or estimated directly from human data, particularly that from the A-bomb survivors.

Although animal data are extensive and indisputable, no conclusive evidence has been found of increased genetic disorders in irradiated human populations. Studies of the progeny of the atomic bomb survivors have shown no statistically significant increase in serious genetic disorders; however, the data can be used to define the lower limit of the doubling dose. Based upon eight measures of genetic health and observations of the A-bomb population through 1990, the probable limits of the doubling dose for high dose rate single fraction human irradiation is 1.69 to 2.23 Sv. When corrected for chronic irradiation with low-level low-LET radiation the doubling dose for humans is more likely to be 3.4 to 4.5 Sv. However, the current genetic risk estimates of the BEIR and UNSCEAR committees are based upon an assumed doubling dose of 1 Sv.

Quantitative estimates of human genetic effects are, by necessity, derived from the extrapolation of animal data. As summarized Table 4.7, the BEIR V committee estimated the number of serious genetic disorders per 0.01 Sv per generation per million live born offspring, both for the first generation after exposure and at equilibrium, with the assumption that each successive generation also receives 0.01 Sv. Equilibrium means that the rate of genetic disorders is neither increasing nor decreasing; this condition is not achieved until several generations after an increase in exposure. The estimates of Table 4.7 assume a doubling dose of 1 Sv and a population distributed in age and gender similar to the U.S. population. At equilibrium, the risk of serious genetic disorders is about 100–200 additional cases per 10^6 live born offspring per 0.01 Sv, or about 1.5% per Sv. For comparison, the lifetime excess risk of cancer fatality at the 0.1 Sv level for high dose rate un-fractionated exposure is about 10% per Sv. UNSCEAR estimates of genetic risk are essentially the same as those of BEIR V. Neither the BEIR nor the UNSCEAR committees attempted to estimate the effect of ionizing radiation on multifactorial conditions. Since these conditions involve multiple genes, as well as environmental and life style factors, estimation of possible radiation effects is even more nebulous than for pure genetic disease. Assuming that 5% of multifactorial conditions are able to be effected by mutations, and that multifactorial conditions are 1/3 as severe as Mendelian or chromosomal conditions, the ICRP estimated that, at equilibrium, the risk of radiation induced multifactorial conditions is at most equal that of serious genetic conditions[42].

Table 4.7. Estimated risk from the genetic effects of radiation to the first generation and at equilibrium for a dose of 0.01 Sv per generation. Risk was calculated assuming a doubling dose of 1 Sv. Data reproduced from the BEIR V Report[23] with permission.

Estimated Genetic Risk of Ionizing Radiation			
Type of Disorder	Current Incidence per Million Liveborn Offspring	Additional Cases per 10^6 Liveborn Offspring per 0.01 Sv per Generation: First Generation	Equilibrium
Autosomal dominant			
Clinically severe	2500	5–20	25
Clinically mild	7500	1–15	75
X-Linked	400	< 1	< 5
Recessive	2500	< 1	Very slow increase
Chromosomal			
Unbalanced	600	< 5	Very little increase
Translocations			10–100
Trisomies	3,800	< 1	
Congenital abnormalities	20 000–30 000	10	10–100
Multi-factorial disorders of complex etiology			
Heart disease	600 000	Not estimated	Not estimated
Cancer	300 000		
Selected Others	300 000		

4.4.6 Risk of Fetal Effects

Fetal effects of ionizing radiation are defined as excess congenital malformations and cancer resulting from irradiation during the embryonic and/or fetal stages of development. One would expect that since the fetus is a rapidly proliferating cell system, its radiation sensitivity would be higher than that of the adult. This is indeed true. However, the magnitude of fetal risk is not so great as is often assumed by the public. As reviewed by Brent[43] and Stovall et. al.[44], the effects of embryonic and fetal irradiation have extensively studied in animals and man. The nature and magnitude of the teratogenic (producing physical defects in the developing embryo) effects of radiation are highly dependent upon the timing of the exposure with respect to gestational age. During the preimplantation and implantation stages (0 to 14 days after conception in humans), the embryo is relatively insensitive to the teratogenic and growth-retarding effects of radiation but is sensitive to lethal effects. Sufficiently high exposure during this period results in embryonic death and resorption, which, in the absence of controls, may appear simply to be failure to conceive. In mice, a dose of 0.1 Sv may result in a

1% to 2% increase in embryonic death. During the organogenesis stage (15 to 50 days after conception) the embryo is less sensitive to lethal effects but very sensitive to growth-retarding and teratogenic effects. Intrauterine growth retardation produces low weight at birth which may or may not be overcome by maturity. A wide range of congenital malformations has been associated with irradiation; however, central nervous system effects are the cardinal manifestations in humans. At a fetal dose of about 0.2 Sv and above, microcephaly, eye malformations, and mental retardation are the most commonly observed radiogenic CNS effects. During the fetal period (50 to 280 days after conception) the sensitivity to multiple organ malformations diminishes, although the CNS and growth-retardation sensitivity remains. Toward later fetal stages the sensitivity to gross malformations is very much reduced as the fetus begins to respond to radiation more like a fully developed individual.

Human data for estimating the magnitude of radiation-induced congenital malformations come primarily from atomic bomb survivors and therapeutically irradiated patients. These populations supply fewer than 1500 subjects exposed *in utero,* several hundred of which are suitable for analysis. The data suggest that congenital malformations may exhibit a threshold-type dose-response relationship, with the threshold dose in the range of 0.1 to 0.5 Sv, depending on the affected organ system. The possibility that fetal irradiation to comparatively low levels could increase the incidence of severe mental retardation has been reported. Otake and Schull initially reported that for the gestational period of 8–15 weeks, the risk of mental retardation increased with a slope of 40% per Sv with no threshold. Later analysis concluded that the data were more consistent with a threshold of 0.2 Sv. Nearly all animal and human data reveal no increased incidence of malformation or growth retardation from X-ray exposures below 0.05 Sv. Neither functional nor biochemical changes, such as to thyroid function, liver function, or fertility, have been observed at these low fetal doses. Doses at or below the 0.05 Sv range would thus not be expected to increase significantly the incidence of anatomic malformation, growth retardation, mental retardation, or spontaneous abortion.

Carcinogenesis is the fetal effect of greatest concern for exposure levels typically encountered in biomedical uses of ionizing radiation. The Oxford surveys have suggested that diagnostic exposure *in utero* at the 0.01–0.02 Sv level increases the relative risk of childhood cancer by as much as a factor of about 5 in the first trimester and 1.5 in the second and third trimesters. Based upon the Oxford data, quantitative estimates of excess risks in the first 10 years of life have been made. The risk of cancer death was estimated[43] to be 217 excess cancer deaths per 10^4 person-Sv and risk of cancer incidence was estimated to be 640 excess cancer cases per 10^4 person-Sv. A similar dose to a population age 0–10 years would increase the risk of cancer by a factor of only about 1.005. It should be noted that the carcinogenic effect of *in utero* radiation is not universally accepted. A-bomb survivor children exposed in utero showed no excess cancer in the first 10 years of life and over the first 35 years of follow-up showed an average annual excess cancer incidence rate of 6.6 cases per 10^4 person-Sv per year[43]. Animal experiments have also failed to confirm the effect. Finally, no firm dose-response relationship has been demonstrated, even from the studies that originally reported the effect. Until such time as conclusive evidence is obtained, it is prudent to assume that the upper limit of the increased relative risk of childhood cancer from 0.01–0.02 Sv *in utero* is approximately 1.5, while the lower limit is obtained by assuming that fetal sensitivity is identical to childhood sensitivity.

Occasionally it is necessary to expose a pregnant patient to ionizing radiation. In diagnostic imaging the circumstances often involve pregnancy discovered after a procedure has been performed, while for radiation therapy it may be necessary to provide treatment to treat cancer in the mother. Either circumstance requires careful evaluation. Consider the case of diagnostic imaging. It is first necessary to ascertain that the fetus was indeed irradiated, since diagnostic irradiation of areas other than the abdomen presents little hazard to the fetus. The stage of pregnancy at the time of irradiation should also be determined. It is also essential to calculate the dose to the fetus as accurately as possible. Methods for calculating fetal dose based upon radiographic technique factors and patient anatomy have been described[45]. The FDA has used Monte Carlo techniques to calculate ratios of fetal dose to entrance skin exposure for many radiographic[46] and fluoroscopic[47] examinations. With these data one can then make a quantitative estimate of the risk to the fetus for comparison with spontaneous risks. It is rare to encounter a diagnostic examination in which the fetal dose exceeds 0.01–0.02 Sv, since the overlying maternal tissues provide significant shielding. For the case of a fetal dose of about 0.01 Sv, the risk of serious malformations or mental retardation is minimal and the greatest concern is an increased risk of childhood cancer. If one assumes the upper limit of 1.5 for increased risk in the first trimester, the fetus may be subject to a 1 in 2000 chance of childhood cancer versus a normal incidence of about 1 in 3000. For comparison, the normal incidence of serious malformations is about 1 in 37 and the normal incidence of all serious genetic effects is about 1 in 10. The risks to the fetus associated with low level radiation are thus small in comparison to normal risks of childbirth.

Radiotherapy in pregnancy can expose the fetus to much higher levels of radiation. In the U.S., approximately 4000 pregnant women per year require treatment for a malignancy[44]. In some cases, such as conservation breast therapy, radiotherapy is the treatment of choice. Because the tumor volume is treated to a high (30–70 Gy) dose and because moderate levels of leakage and scatter radiation are present, the fetus can be irradiated to significant levels. Techniques for calculating and measuring fetal dose, as well as techniques of providing special shielding have developed[44]. Fetal dose depends upon the tumor dose, photon energy, field size, distance from the edge of the radiation field and the use of additional external shielding. In an example of treatment for Hodgkin's disease to a tumor dose of 38 Gy of 6 MV X rays, the fetal dose was measured to range from 0.42 Gy at the top of a fetus (15.5 cm from the nearest field edge) to 0.14 Gy at mid-fetus (28.5 cm from the nearest field edge). Use of five half-value layers external shielding reduced the dose to 0.17 Gy at the top of the fetus to 0.04 Gy at mid-fetus[44].

4.4.7 Summary of Risks from Low Level Ionizing Radiation

A summary of estimates of the carcinogenic, genetic, and fetal effects of low dose and low dose rate ionizing radiation on humans is given in Table 4.8. Also shown are estimates for the normal incidences of these effects. Note that the cancer risk is given in terms of excess cases and or deaths per unit equivalent dose. As such, it is directly applicable only to whole body exposure. Table 4.8 also assumes that the exposed population is distributed in age similar to the U.S. population. To estimate risk for circumstances of non-uniform exposure the dose to all organs must first be calculated and then converted to effective dose. Similarly, genetic risk is based upon uniform irradiation to the gonads of an entire population and fetal risk is based upon dose to the fetus.

Table 4.8. Summary of risks from low LET, low dose rate ionizing radiation.

Summary of Risks from Low LET, Low Dose Rate Ionizing Radiation		
Cancer: **Excess Lifetime Cancer Fatality** (Normal incidence: 200 000 deaths per million)		
Source of Estimate	Dose-Response Model	Quantitative Estimate (Excess cancer deaths per 10 000 person-Sv)
BEIR V (Assume DREF of 2.0 for solid tumors)	Linear-Quadratic (Leukemia) Linear (Solid Tumors)	460
Genetic: **Mendelian and Chromosomal Disease and Congenital Abnormalities** (Normal incidence: 47,300 serious genetic defects per million live born)		
Source of Estimate	Dose-Response Model	Quantitative Estimate (Excess Genetic Defects per 0.01 Sv per million live born)
BEIR V	Linear	15–35 first generation 110–200 at equilibrium
Genetic: **Multifactorial Disorders of Complex Etiology** (Normal incidence: 650 000 to 1 200 000 per million live born)		
Source of Estimate	Dose-Response Model	Quantitative Estimate (Excess Genetic Defects per 0.01 Sv per million live born)
BEIR V	Not Estimated	Not Estimated
ICRP	Linear	120 at equilibrium

Table 4.8 (continued).

Fetal: **Malformations** (Normal incidence: 27 serious malformations per thousand live born; 60–100 malformations of any type per thousand live born)		
Source of Estimate	Dose-Response Model	Quantitative Estimate (Excess malformations per 1.0 Sv per thousand live born)
BEIR V	Linear with probable thresholds	430 Severe mental retardation cases at 1.0 Sv, threshold may be 0.2 to 0.4 Sv
Brent	Linear with threshold	Negligible malformations for fetal dose 0.05 Sv Significantly increased risk above 0.2 to 0.5 Sv

Fetal: **Childhood Carcinogenesis (All Sites)** (Normal Incidence: 1400 per million incidence, 500 deaths per million in first ten years of life)		
Source of Estimate	Dose-Response Model	Quantitative Estimate (Excess Childhood Cancer per 10 000 fetus-Sv)
Stovall et. al. (Oxford)	None determined	640 (Incidence), 217 (Death)
Stovall et. al. (A-bomb)	Linear Quadratic	55 (Incidence)

Patient dose from diagnostic radiology procedures depends primarily upon X-ray energy, patient size, type of image receptor and number of images per procedure. Effective dose levels vary widely, ranging from less than 0.1 mSv for a chest and extremity procedures, to about 1.5 mSv for some body imaging procedures. An upper limit of risk of excess cancer fatality from imaging procedures thus ranges from about 5×10^{-6} to 70×10^{-6} (obtained by multiplying the effective dose by 460×10^{-4} Sv^{-1}). In the case of emission imaging, the dose to individual organs depends on the type of radiopharmaceutical, administered activity, and relative uptake as well as on attenuation. Doses from nuclear medicine procedures are dominated by "critical organs" that receive most of the dose from a given procedure. The effective dose is typically 10 per cent or less of the critical organ dose. With the exception of procedures using I-131 sodium iodide, the effective dose for nuclear medicine procedures ranges from about 0.5 mSv to 15 mSv. For cases in which definite medical indications exist for performing an imaging procedure, the radiation risk is usually insignificant compared to other risks faced by the patient. For the other cases, such as administratively required radiographs (e.g., pre-employment lumbar spine), admission chest radiographs, and examinations with questionable medical indications (e.g., a skull series for headache), the radiation risks become more significant because the benefit is questionable.

To estimate risk to groups of workers and the public for the purposes of radiation protection, it is desirable to have a single simplified figure that combines all stochastic risks. The ICRP and NCRP have adopted the concept of radiation "detriment," which includes risk from fatal cancer, non-fatal cancer and risk of genetic effects in future generations. Fetal effects are not included in radiation detriment, although special consideration is given to fetal effects when setting dose limits. The contribution of each source to total detriment is shown in Table 4.9. Adult workers and the general population have different radiation detriments due to the different age distribution of the two groups. Note also that a linear non-threshold dose response model has been assumed to apply to all detriments. Risk of solid tumors and genetic effects is best fit by a linear dose response, while the risk of leukemia in the low dose region is adequately fit by a linear dose response, even though at higher dose a linear-quadratic dose response is required. As previously noted, there is no epidemiological evidence that doses below 0.1 Sv cause a statistically significant increase in cancer deaths, cancer incidence, genetic effects, fetal malformations or childhood cancer. Hence the use of a linear non-threshold dose response model for estimating radiation detriment is conservative in the sense that it overestimates detriment.

Table 4.9. Summary of radiation detriment probability coefficients as used by the NCRP and ICRP. The difference in coefficients between populations is due to their different age distributions. Data taken from NCRP Report 116.

Exposed Population	Detriment			
	Fatal Cancer (10^{-2} Sv^{-1})	Nonfatal Cancer (10^{-2} Sv^{-1})	Severe genetic Effects (10^{-2} Sv^{-1})	Total Detriment (10^{-2} Sv^{-1})
Adult workers	4.0	0.8	0.8	5.6
Whole population	5.0	1.0	1.3	7.3

4.5 Conclusion

Ionizing radiation is an essential tool of modern biomedicine. Radiation therapy is used to treat nearly all types of cancer except leukemia, and plays a major role in the treatment of cancers of the breast, prostate and lung. Approximately half of all cancer patients will have radiation therapy during the course of overall treatment. Although variability among individuals results in some uncertainty, risks of high level irradiation are well characterized and predictable. Careful planning of radiation treatment, coupled with precise delivery, can avoid or minimize adverse high level radiation effects. Even for treatments in which there are unavoidable normal tissue effects, the benefits of treatment, such as elimination of the tumor or relief of pain, outweigh the risks.

Uses of ionizing radiation in diagnostic imaging also expose the patient to a presumed risk of detriment. Although there may be no excess risk associated with the radiation levels typical of most imaging procedures, the upper limit of risk to a population is summarized to Table 4.8. Radiation risk is one component that must be considered in discussion of risks and benefits of imaging procedures. Radiological risk is not the only risk to which the imaging patient is exposed. There may be equal or greater risk from other agents such as infectious disease or reaction to contrast agents. The risk of not performing an imaging procedure, i.e. the loss of opportunity to diagnose disease, also must be considered.

Elucidation of quantitative human risk of low level radiation carcinogenesis is a key element of modern radiation protection. Because current radiation protection standards are based on the concept of acceptable risk, it is impossible to set meaningful exposure limits or carry out risk-benefit analyses, without the best possible knowledge of radiation risk. However, it is unlikely that much new human data will become available in the near future. One large study that is still on-going is that of the atomic bomb survivors. The completion of this study is very important, since information must be obtained over the entire lifetime of an exposed population to assess effects of exposure in childhood and to distinguish between the absolute and relative risk models. Important new human effects information may become available from populations irradiated as a consequence of the Chernobyl disaster, but such studies will take years to complete. Human epidemiological studies capable of resolving questions about the precise form of the dose response relationship in the dose region below 0.01 Gy will probably be impossible to carry out.

Since there are significant impediments to new epidemiological data, future developments in risk modeling depend primarily upon developing a more comprehensive understanding of normal carcinogenesis and radiation carcinogenesis. Discovery of genetic markers of radiation carcinogenesis could allow detection of radiation induced cancers based on reasons other than statistical comparison of exposed and unexposed populations. In-vitro studies that have detected a unique spectrum of mutational events in irradiated mouse cells[48] may provide a basis for development of such markers. Understanding of the fundamental processes leading to cancer induction also could shed light on repair mechanisms and adaptive response mechanisms that may effect the shape of the dose response relationship as well as resolve questions about synergistic effects with other carcinogens.

Until a more sophisticated and complete understanding of human radiation carcinogenesis is obtained, reliance still must be placed in the risk modeling methods. These models must be used with full realization of both their strengths and weaknesses. In some cases, even the context of their application should be taken into account. For instance, the use of a linear non-threshold risk model with a DREF of two is reasonable in the context of risk estimation for radiation protection for workers, since a conservative approach is appropriate. Using the same model as a basis for proving causation in a legal suit is less appropriate, since the context requires greater degree of precision and certainty than the modeling process can assure. Use of the linear non-threshold model for risk assessment at dose levels below 0.1 mSv per year over a period of hundreds of years (as is sometimes done in assessing potential effects of wide spread release of very small quantities of radioactive materials) is not justified on the basis of any epidemiological or radiobiological data. It is the responsibility of all individuals performing radiation risk analysis to be aware of the limitations imposed by the modeling process, and to use the risk estimates derived from the models in a judicious fashion.

4.6 References

1. Committee for Review and Evaluation of the Medical Use Program of the Nuclear Regulatory Commission, *Radiation in Medicine: A Need for Regulatory Reform*, K.D. Gottfried and G. Penn, Eds., (National Academy Press, Washington, DC, 1996).
2. National Council on Radiation Protection and Measurements, *SI Units in Radiation Protection and Measurements NCRP Report 82*, (National Council on Radiation Protection and, MD, 1985).
3. International Commission on Radiological Protection, *1990 Recommendations of the International Commission on Radiological Protection, ICRP Publication 60*, Annals of the ICRP **21** (1–3) (Pergamon Press, Elmsford, NY,1991).
4. J.A. LaVerne and S.M. Pimblott, *Radiation Research* **141**, 208–215 (1995).
5. L. Shabason, in *Health Effects of Exposure to Ionizing Radiation*, W.R. Hendee and F.M. Edwards, Eds., (Institute of Physics Publishing, Philadelphia, PA, 1996), pp. 62–69.
6. J.F. Ward, *Progress In Nucleic Acid Research and Molecular Biology* **35**, 95–125 (1988).
7. R.E. Krisch, M.B. Flick and C.N. Trumbore, *Radiation Research* **126**, 251–259 (1991).
8. J.E. Moulder and J.D. Shadley, in *Health Effects of Exposure to Ionizing Radiation*, W.R. Hendee and F.M. Edwards, Eds., (Institute of Physics Publishing, Philadelphia, PA, 1996), pp. 74–76.
9. C. Le Sech, H. Frohlich, C. Saint-Marc and M. Charlier, *Radiation Research* **145**, 632–635 (1996).
10. G.W. Barendsen, *Radiation Research* **139**, 257–270 (1994).
11. H. Nakano and K. Shinohara, *Radiation Research* **140**, 1–9 (1994).
12. W.C. Dewey, C.C. Ling and R.E. Meyn, *Int. J. Radiation Oncology Biol. Phys.*, **33**, 781–796 (1995).
13. Q. Hu and R.P. Hill, *Radiation Research* **146**, 636–645 (1996).
14. M. Stuschke et. al., *Int. J. Radiation Oncology Biol. Phys.*, **32**, 395–408 (1995).
15. E.J. Hall, *Radiobiology for the Radiologist*, 4th ed., (Harper and Row, Philadelphia, PA, 1992).
16. P. Rubin and D.W. Siemann in *Clinical Oncology: A Multidisciplinary Approach for Physicians and Students*, P. Rubin, S. McDonald and R Qazi, eds. (W.B. Saunders Co. Philadelphia, PA, 1993), pp. 71–90.
17. W. Huda and K.R. Peters, *Radiology*, **193**, 642–644, (1994).

18. United Nations Scientific Committee on the Effects of Atomic Radiation, *Sources and Effects of Ionizing Radiation, UNSCEAR 1994 Report,* (United Nations Publications, New York, 1994).
19. W.F. Morgan, et. al., *Radiation Research* **146**, 247–258, (1994).
20 K. Sankaranarayanan, *Mutation Research* **258**, 75–97, (1991).
21. J.B. Little, *Radiation Research* **140**, 299–311, (1994).
22. C.S. Selvanayagam, C.M. Davis, M.N. Cornforth and R.L. Ullrich, *Cancer Research* **55**, 3310–3317, (1995).
23. National Academy of Sciences, National Research Council, *Health Effects of Exposure to Low Levels of Ionizing Radiation (BEIR V)*. (Washington, DC., National Academy of Sciences, 1990).
24. R. Cox, *Int. J. Radiation Biology*, **65**, 57–64, (1994).
25. S. Wolff and A.V. Carrano, in *Radiation Carcinogenesis* AC Upton, RE Albert, FJ Burns, RE Shore, eds. (New York: Elsevier Science Publishing Co.,1986).
26. G.L. Chan and J.B. Little in *Radiation Carcinogenesis* AC Upton, RE Albert, FJ Burns, RE Shore, eds. (New York: Elsevier Science Publishing Co.,1986).
27. B.E. Henderson, R.K. Ross, C.P. Malcolm, *Science* **254** 1131–1138, (1991).
28. T.D. Luckey, *Radiation Hormesis,* (CRC Press, Boca Raton, 1991).
29. P.K. Strudler, *Radiation Research* **145**, 107–117, (1996).
30. E.I. Azzam, S.M. de Toledo, G.P. Raaphorst, and R.E.J. Mitchel, *Radiation Research* **146**, 369–373, (1996).
31. T. Downs,. in: *Biological Effects of Low Level Exposures to Chemicals and Radiation,* E.S. Calabrese, ed.(Lewis Publishers, Boca Raton, 1992).
32. National Academy of Sciences, National Research Council, *The Effects on Populations Exposure to Low Levels of Ionizing Radiation 1980.* (Washington, DC., National Academy of Sciences, 1980).
33. D.A. Pierce, Y. Shimizu, D.L. Preston, et. al., *Radiation Research* **146**, 1–27, (1996).
34. H.A. Weiss, S.C. Darby, T. Fearn, and R. Doll, *Radiation Research* **142**, 1–11, (1995).
35. G.R. Howe and J. McLaughlin, *Radiation Research* **145**, 694–707, (1996).
36. G.R. Howe, *Radiation Research* **142**, 295–304, (1995).
37. E. Ron, J.H. Lubin, R.E. Shore, et. al., *Radiation Research* **141**, 259–277, (1995).
38. E. Cardis, E.S. Gilbert, L. Carpenter, et. al., *Radiation Research* **142**, 117–132, (1995).
39. P.D. Inskip, M. Tekkel, M. Rahu, et. al., in *Implications of New Data on Radiation Cancer Risk, Proceedings of the 32nd* Annual Meeting of the NCRP, (National Council on Radiation Protection and Measurements, Washington, DC., 1997.) pp. 123–141.
40. D. Thomas, S. Darby, F. Fagnani, et. al., *Health Physics*, **63**, 259–272, (1992).
41. National Council on Radiation Protection and Measurements *Limitations of Exposure to Ionizing Radiation NCRP Report 116*, (National Council on Radiation Protection and Measurements, Bethesda, MD, 1993).
42. K. Sankaranarayanan, in *Health Effects of Exposure to Ionizing Radiation*, W.R. Hendee and F.M. Edwards, Eds., (Institute of Physics Publishing, Philadelphia, PA, 1996), pp. 113–167.
43. R.L. Brent, in *Health Effects of Exposure to Ionizing Radiation*, W.R. Hendee and F.M. Edwards, Eds., (Institute of Physics Publishing, Philadelphia, PA, 1996), pp. 169–213.
44. M. Stovall, C.R. Blackwell, J. Cundiff, et. al., *Medical Physics*, **22**, 63–82, (1995).
45. L.K. Wagner, R.G. Lester and L.R. Saldana, Exposure of the Patient to Diagnostic Radiations, (J.B. Lippincott Co., Philadelphia, PA, 1985), pp. 131–175.
46. M. Rosenstein, *Handbook of Selected Tissue Doses for Projections Common in Diagnostic Radiology*, (Center for Devices and Radiological Health HEW Publication (FDA) 89–8031, Rockville, MA, 1989).
47. M. Rosenstein, O.H. Suleiman, R.L. Burkhart, and S.H. Stern, *Handbook of Selected Tissue Doses for the Upper Gastrointestinal Fluoroscopic Examination*, (Center for Devices and Radiological Health HEW Publication (FDA) 89–8031, Rockville, MA, 1989).
48. M.S. Turker, P. Pieretti and S. Kumar, *Mutation Research* **374**, 201–208, 1997.

5 Radiation Protection and Regulation

William R. Hendee

Medical College of Wisconsin

5.1 Introduction

Ionizing radiation includes, X rays, electrons, neutrons, protons and heavy ions from particle accelerators, alpha, beta and gamma rays from atoms that spontaneously undergo radioactive decay. Ionizing radiation can be biologically destructive in two ways: (1) the ionization produced by the radiation can induce changes in the composition and bonding of important molecules in cells (e.g., DNA and RNA), causing the cells to become dysfunctional; and (2) the ionization can produce intermediate chemical entities in the cell milieu, which in turn affect the composition and bonding of important molecules essential to cell integrity. These effects are known respectively as the *direct* effect and the *indirect* effect of radiation exposure. The two effects together constitute the biological basis of risk associated with exposure to radiation.

5.2 Quantitative Health Risk Evaluation

Health risk is characterized as any circumstance in which the health and wellbeing of a biological system (e.g., a human being) is threatened. Quantitative evaluation of the degree of health risk follows the multistep process that can be summarized in Figure 5.1.

Figure 5.1. Steps in the quantitative evalution of health risks

5.2.1 Hazard Identification

A hazard is any situation or agent that may not only adversely impact on the health and wellbeing of persons exposed to it. Hazard identification means that a hazard is thought to exist, but also that an individual or group of individuals may potentially be exposed to it. Hazard identification is easier if the underlying biological mechanisms by which the hazard affects exposed persons are understood. In the case of ionizing radiation, the direct and indirect effects of radiation are the underlying mechanisms.

5.2.2 Exposure Assessment

This step in quantitative risk evaluation involves identification of individuals or groups exposed to the hazard, and their magnitude and duration of exposure. It is also important to characterize the routes of exposure. For radioactive materials, for example, individuals may be exposed to sources of radiation that are outside the body (external exposure), or to sources taken accidentally (or purposefully in the case of medical uses) into the body internal exposure by inhalation, ingestion, injection, or absorption through the skin. In some situations, both routes of exposure may occur.

5.2.3 Exposure-Response Assessment

This step includes identification of the conditions under which exposure to a hazard places individuals at risk, and quantification of the degree of risk as a function of exposure. Exposure-response assessment usually requires reliance on a model of exposure/response that is based on experimental data acquired preferably on humans, but often with animals. Much of the present knowledge about cancer induction by exposure to chemicals (chemical carcinogens) has been acquired through experimentation with animals. In many situations, a response estimate is needed at exposure levels well beyond the range of experimental data. In these situations, the response must be estimated by extrapolating the model beyond the range of experimental data. The uncertainty of the estimate grows with increasing displacement of the extrapolation from the measured data. For example, exposure to ionizing radiation is thought to cause cancer in humans, but actual data supporting this hypothesis are available only at relatively high levels of exposure (0.1 sievert and above). To estimate the risk of cancer at low-level exposures, the exposure/response model must be extrapolated from data at high exposure levels. Various models for achieving this extrapolution are illustrated in Figure 5.2. For purposes of radiation protection, "low level" is usually defined as exposures below about 0.1 sievert (Sv).

5.2.4 Risk Characterization

This final step in quantitative risk evaluation involves use of the exposure/response model to estimate the degree of risk resulting from specific levels of exposure to a hazardous agent or situation. Often this process is accompanied

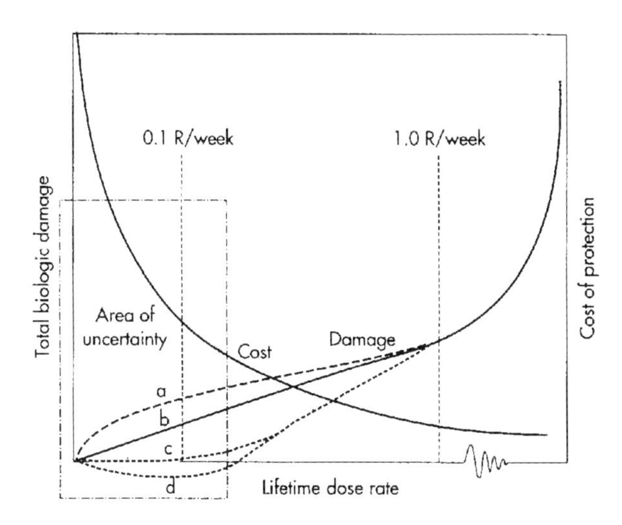

Figure 5.2. Model for total biological damage and cost of protection as a function of lifetime dose.

by considerable uncertainty, including that associated with estimates of the level of exposure, and with identification of the exposure/response appropriate for the level of exposure. In the characterization of risk, inclusion of the limits of uncertainty is important. In the case of exposure to ionizing radiation, every attempt to characterize risk at relatively low levels of exposure is subject to considerable uncertainty.

5.3 Radiation Injuries

In a biological system (e.g., the human body), two types of radiation effects can occur. These types are referred to as acute radiation injuries and late radiation injuries. Acute radiation injuries, also called prompt injuries, occur in response to relatively large exposures to radiation delivered over short periods of time. Acute injuries include erythema (skin reddening and burning), epilation (hair loss), gastrointestinal disorders (nausea, vomiting and diarrhea), blood changes (reductions in red and white blood cells), temporary or permanent sterility, organ atrophy, tissue fibrosis and, if the exposure is great enough, severe disability and death. Acute radiation injuries are said to be deterministic (nonstochastic), because both their likelihood and severity depend on the amount of exposure.

Late radiation injuries, also called long-term or delayed injuries, are limited to cancer and hereditary (genetic) effects. These effects can occur at high exposures, but unlike acute radiation injuries, also may result from much lower exposures. In addition, the exposures may occur over an extended period of time. Late radiation injuries are said to be probabilistic (stochastic), because their probability of occurrence, but not their severity, varies with the amount of exposure. For small exposures, only late radiation injuries are of concern. The most common model of late radiation injuries is the *linear, no-threshold* model. In this model the risk of injury increases from zero exposure in a straight-line fashion as a function of the amount of exposure. This model is widely used in efforts to quantify the risk of exposure to small amounts of radiation. The linear, no-threshold model is thought to be conservative because, if anything, it overestimates the risk at low levels of exposure. Hence its use is considered appropriate for establishing standards for protecting radiation workers and the public from the risk of radiation dose.

The risk of inducing a probabilistic effect is described by a *risk coefficient*, defined as the number of cases of a specific biological effect per unit exposure. For example, the risk coefficient for radiation-induced cancer of the thyroid can be stated as the number of thyroid cancers induced in a population of 10^6 persons, if each of the persons receives a dose to the thyroid of 1 Sv. The risk coefficient is a probabilistic statement that arises from the linear, no-threshold model of radiation injury. With this model, if a dose D produces X cancers in an exposed population of 10^6 persons, then a dose of (1/10) D would produce the same number of cancers in a population of 10^7 persons, and a dose of 10D would produce 10 X cancers in an exposed population of 10^6 persons.

In the most definitive analysis[1] to date of the risk of low-level exposure to radiation, the probability of induction of solid cancers is presumed to follow the linear, no-threshold model of radiation injury. For leukemia, however, the probability of induction by radiation is presumed to follow a no-threshold, linear-quadratic model in which the risk increases slowly at very low exposures, and more rapidly as the exposure increases beyond the initial low values. The linear and linear-quadratic models are compared in Figure 5.2. Also illustrated in this Figure is the concept of radiation hormesis in which an actual reduction of risk below baseline levels occurs at very low exposures. A hormetic effect has been detected by some investigators for certain radiation effects[2].

5.4 Human Study Limitations

The risk associated with exposure to most hazardous agents is estimated primarily from studies of animals intentionally exposed to the agents. Much animal work has been conducted over the years with ionizing radiation. With this particular agent, however, considerable data have also been acquired from epidemiological studies of humans exposed to radiation. Some of these studies have examined patients exposed to X rays and radiation from radioactive materials as a result of diagnostic and therapeutic medical procedures.

By far the greatest source of human data concerning the biological effects of radiation exposure has been the ongoing studies of survivors of the atomic explosions in Hiroshima and Nagasaki.

Shortly after the end of World War II, the Atomic Bomb Casualty Commission was established by the United States to identify long-term health effects in survivors of the Japanese atomic explosions. This epidemiological project eventually became a collaborative effort of the United States and Japan conducted under the aegis of the Radiation Effects Research Foundation (RERF).

The Japanese survivors of Hiroshima and Nagasaki are by far the largest and most closely followed of all human populations exposed to ionizing radiation. Still, many uncertainties exist with regard to the universal applicability of the Japanese results. The exposed and control populations were war-weary individuals with an underrepresentation of younger males. They were malnourished and without adequate social support and medical care following the explosions and the war itself. Hence, their comparison to normal populations of humans is difficult. The irradiated population was exposed to a single burst of neutrons and gamma rays released during nuclear fission. These features compromise the applicability of observed health effects to other groups exposed to other types of radiation under different conditions, including protracted exposures to much lower levels of radiation. Finally, in spite of extensive examination and simulation, considerable uncertainty remains about the actual radiation exposures received by the Japanese survivors.

Questions can be raised about any human epidemiological study. For example, when exposure to medical radiation is examined, study populations are frequently persons thought or known to be suffering from various diseases. Exposure estimates may have to be reconstructed or inferred, often with considerable uncertainty. Even in studies of occupationally-exposed individuals who wear personnel monitors to measure their radiation exposures, and who work in controlled areas where radiation exposures are well characterized, exposure estimates may still be uncertain because records are kept for protection rather than research purposes, and estimates of exposures may err on the high side as a conservative approach to radiation protection and risk reduction.

Confounding factors that might impact on risk estimates for ionizing radiation frequently are ignored in epidemiological studies because their presence cannot be verified or quantified. For example, radiation exposures received for medical diagnosis and therapy may not be considered in a study even though for some individuals such exposures may exceed those being examined epidemiologically. Factors such as tobacco and alcohol consumption, and exposure to chemical agents and other environmental toxins, may be impossible to identify and quantify and yet may be potentially more significant than radiation in influencing the health of exposed individuals. Although efforts are often made to compare injury in an irradiated population with that in a comparable control group, it usually is impossible to define a control population that is exactly identical with the study group in all aspects except radiation exposure. Finally, radiation effects such as cancer and hereditary defects are not different in expression from those due to "natural causes", so studies must focus on increases in incidence and mortality rather than on the mere presence of disease in the exposed population. This factor complicates the identification of health effects in populations exposed to low levels of radiation, and creates uncertainty about the extent of the effects. In many cases where adverse health effects are claimed, the occurrence of no effect, or even a beneficial health effect (*radiation hormesis*), is within the range of statistical uncertainty.

5.5 Early History

In late 1895, the German physicist Röntgen[3] recognized that a voltage applied across an evacuated Crookes tube caused the emission of a penetrating form of radiation. He named this new form of radiation *X rays*. A few months later, the French physicist Becquerel[4] observed that a photographic emulsion could be exposed by placing uranium-containing ore on an emulsion-containing glass plate. He concluded that the exposure was caused by a form of radiation (*beta radiation*) emitted by the ore. In 1898, *alpha particles* were detected by Curie[5] in her studies of radium, and *gamma rays* were discovered by Villard[6] in 1900. Within months of the announcement of the discovery of X rays, biological effects began to be reported by persons experimenting with X-ray sources[7–9]. By the turn of the century, only five years after the discovery of X rays, several acute effects from high-level exposure to radiation had been recognized, including dermatitis, erythema, epilation and the production of skin ulcers.

Intense research into the biological effects of ionizing radiation occurred over the two decades following the discovery of X rays. The research focused on acute effects that were produced by doses greater than 0.1 Sv, well above those considered today to be low-level. The ability of radiation to induce cancer in exposed humans was first observed in 1902 by Frieben[10] who reported cancer in the hand of a factory worker involved in producing X-ray tubes. Two years later, Thomas Edison's assistant Clarence Dally became the first documented cancer fatality attributable to radiation exposure[11]. Over the next few years, several X-ray induced malignancies were reported in humans, and a variety of experimental studies were conducted with animals in which skin cancers were induced by x- ray exposure. By 1911 at least 94 cases of X-ray induced skin cancer in humans had been reported in the literature[12]. In his 1911 monograph, Hesse offered some important observations about latency (the time between radiation exposure and the appearance of cancer). He noted that the latent period averaged 9 years and ranged from 4 to 14 years in the 94 cases he had uncovered. He also observed that the interval between radiation-induced dermatitis and the subsequent appearance of cancer ranged from 1 to 11 years, with an average of 4–5 years. His observations suggested that the period of latency decreases with increasing exposure, because the exposure (or exposure rate) to persons with dermatitis was likely to have been greater than that to individuals without dermatitis.

The cancer-inducing potential of smaller exposures to ionizing radiation was first suggested by Colwell and Russ[13] in the first edition of their book on radiobiology:

"The significant fact, therefore, is that repeated small doses of soft X rays, when applied to human tissues, produce gradual changes therein, which may cause such tissues to develop malignant features."

In 1921, Uhlig suggested that the relatively high frequency of lung cancer noted in miners of precious metals might be attributable to exposure to radium and radium emanation (radon), both naturally-occurring, alpha-emitting radioactive materials[14,15]. In the early years it was believed that actual tissue damage had to occur, as would be the case with relatively high radiation exposures, in order for cancer ultimately to develop[16,17]. Many observers believed that small doses of radiation might in fact be beneficial, with some citing the Arndt-Schulz Law as justification for their belief. The Arndt-Schulz Law, which serves as an underlying premise of homeopathy, was formulated in the late nineteenth century prior to the discovery of X rays and radioactivity. It proposes that small doses of a potentially toxic substance may produce a stimulatory effect, whereas larger doses will produce increasingly deleterious effects. This law is analogous to the concept of *radiation hormesis,* in which small exposures to ionizing radiation are considered beneficial. The belief that small exposures to radium or radium emanation are salutary was widely held through at least the first quarter of the twentieth century, and led to the popular use of radium spas (naturally-occurring hot springs where the water contains radium) and the ingestion of radium-containing nostrums for treatment of various physical ailments.

During the first few decades following the discovery of X rays and radioactivity, several other potential long-term effects of radiation exposure were identified. Significant among these were effects on the blood of exposed persons. As early as 1911, a cause-and-effect relationship was proposed in four cases of lymphatic leukemia observed in radiologists[13]. Other blood changes such as leukopenia in radiologists were also described. Although such effects were not noted by all investigators, awareness grew that continued exposure to radiation at levels considered to be within the range of safety might produce adverse long-term effects. This awareness was heightened by the deaths of some prominent radiation workers from aplastic anemia and other causes attributable to radiation exposure[18]. Blood effects served as the basis for early standards for protection from radiation exposure.

5.6 Tolerance Dose

Initial interpretations of radiation effects and the need to protect radiation workers evolved from the concept of radiation tolerance which in turn led to the concept of a tolerance dose. The underlying principle of radiation tolerance is that the human body can withstand exposure to ionizing radiation without ill effects, so long as the exposure is kept below a certain level defined as the tolerance dose. The tolerance dose was suggested by Arthur Mutscheller in 1924 before the American Roentgen Ray Society and published the following year[19]. It served as the basis for radiation protection for three decades. The tolerance dose was considered to be the level of radiation exposure that could be received by a person on a continuous basis without demonstrable ill effect. A demonstrable ill effect (i.e., harm) was defined in terms of what are now known as acute or deterministic effects. The risks of radiation exposure (or, more appropriately, the consequences of the hazard of exposure) were characterized in terms of the clinical manifestation of an effect such as a diminution in the concentration of circulating leukocytes or red cells in the blood. Even before 1920, film badges were used to monitor personnel exposures, and radiation workers were subjected to periodic blood tests to ensure that they had not been "over-exposed." The recommended frequency of blood tests varied from three to six months, and a significant drop in white or red cell count resulted in removal of a worker temporarily from an area where radiation exposure might occur.

At first, the tolerance dose was expressed as a fraction of the erythema dose, i.e., the exposure required to produce a perceptible reddening of the skin. The erythema dose is not an exact quantity of radiation exposure, because it depends on the energy of the radiation, the rate at which the exposure occurs, and the specific response of the individual to radiation exposure. Mutscheller's original tolerance dose corresponded to a whole body radiation dose of about 0.7 Sv per year, and was based on calculations of anticipated exposures to workers in well-run X-ray installations.

Over the next several years, the tolerance dose was reduced to a level equivalent to about 0.25 Sv per year, an annual level still significantly greater than that now considered a low exposure. This reduction was not stimulated by the discovery of adverse consequences at doses above 0.25 Sv. It reflected instead a recognition that exposures could be lowered without interfering with the exploration and application of ionizing radiation for useful purposes. Over this period, radiation protection evolved as a science in its own right[20]. Taylor[21], in discussing efforts by three prominent early investigators working independently to establish a quantitative level for the tolerance dose, stated

"... no one of these people, or anybody else, claimed that they had ever detected any injury due to radiation at levels above one-hundredth of an erythema dose per month."

In fact, there was evidence in the literature to the contrary. For example, Barclay and Cox[22] examined radiation risks to radiologists by making measurements under actual operating conditions. They failed to detect any ill effects in either of the two workers they followed, one of whom they estimated received 0.007 units of skin dose (i.e., erythema dose) daily for six years. Similar findings reported by other investigators reinforced the idea that adverse effects of radiation exposure simply did not occur, provided the exposure is kept to a relatively low level (i.e., below the tolerance dose). The tolerance dose expressed the concept of a threshold dose below which ill effects of radiation exposure do not occur. This was the prevailing philosophy of radiation risk for six decades after the discovery of X rays and radioactivity.

In the 1920s two voluntary organizations dedicated to radiation protection were formed. These organizations were the International Xray and Radium Protection Committee (1928), predecessor of the International Commission on Radiological Protection (ICRP), and the U.S. Advisory Committee on X-ray and Radium Protection (1929), forerunner of the National Council on Radiation Protection and Measurements (NCRP). Over the next decade or so, these bodies, as well as the League of Nations, various national governments and individual investigators, promulgated voluntary standards of radiation protection that relied on the concept of tolerance dose[23,24]. In 1931, the NCRP suggested a tolerance dose of 0.2 R/day, and the ICRP followed suit in 1934. In 1936, the NCRP reduced its recommended tolerance dose to 0.1 R/day. (In the 1930s, the erythema dose was replaced by radiation exposures expressed in the more quantitative unit roentgen (R), which, for photon and low-LET particulate radiations is roughly numerically equivalent to 100 times the absorbed dose in Gy or the effective dose-equivalent in Sv). In 1931, the League of Nations published a comprehensive report on radiation protection in which a tolerance dose of 10^{-5} R/s was assumed for radiation limitation based on an exposure of 8 hours per day for 300 days per year[25]. The report's authors noted that it was impossible to determine the exact greatest dose that is incapable of damaging cells, or "... exercising any stimulating action". They wrote in terms of a "harmless dose" that would produce no effects detectable by clinical examination.

Radiobiological studies and the establishment of standards for radiation protection were a significant feature of the Manhattan Project (development of the atomic bomb) of World War II. In large measure, radiobiological studies in animals were devoted to detection of deterministic effects (i.e., obvious clinical manifestations), with the objective of verifying basic protection criteria based on the concept of tolerance dose[26,27]. Results of these studies were in some cases quite unexpected. In a classic series of experiments, Lorenz[28]; Lorenz and coworkers[29] observed that mice exposed to 0.11R per day, approximately the accepted tolerance dose in the 1940s, outlived the equivalent control animals. This observation has never been satisfactorily explained. Clinical laboratory studies conducted on Manhattan Project workers failed to reveal any indications of long-term effects at low levels of exposure[30,31]. However, the population was not followed for a sufficient period to observe possible long-term effects that have a latency of several years.

5.7 Experience with Radium

The 1920s were notable for two disparate discoveries that in different ways have had a profound impact on the course of radiation protection standards and on the concept of risk associated with long-term effects of low-level exposure to ionizing radiation. The first of these discoveries occurred in 1924, when a New York dentist described a condition he termed 'radium jaw' observed among young women employed to paint luminous dials with radium-bearing paint. The saga of the dial painters, as well as of the iatrogenic use of radium to treat various debilitating conditions, is well documented in the scientific literature[15,32,33].

Radium dial painters and other persons with internal depositions of radium were studied extensively by Evans and coworkers at the Massachusetts Institute of Technology (MIT). They followed a group of more than 600 individuals, including radium dial painters, for over four decades[34]. No radiation-induced effects or significant skeletal abnormalities were found in the more than 500 individuals receiving less than an average skeletal dose of 200 Sv, well above the upper limit of 0.1 Sv defined for low level exposures. They interpreted their findings as unequivocal evidence of a threshold dose below which adverse effects of radiation exposure do not occur. At greater doses, the incidence of radiation-induced skeletal tumors rose steeply to about 30 percent of the group at 400 Sv. The incidence did not increase beyond this level with exposures up to an order of magnitude greater. Evans and his team examined more subtle skeletal effects, including spontaneous bone fractures, osteoporosis, and bone necrosis, and again identified a threshold at about 200 Sv. They also noted that the latency period for tumor formation was an inverse function of exposure, and concluded "... there would be a domain of small dosage for which the required tumor appearance time exceeds the maximum life span." In this manner they defined a *practical threshold dose.*

Extensive studies of radium dial painters and others with internal depositions of radium were also carried out by Finkel and coworkers at Argonne National Laboratory. The Argonne Radium Studies[35] confirmed the apparent threshold of the MIT study, by observing no skeletal tumors or blood dyscrasias in persons with total-body radium depositions below an activity of about 1.2 μCi. Above this threshold level, the incidence of malignancies and blood effects increased linearly with exposure to a level of 100 percent at a deposition of about 200 μCi. They did not observe a plateau at 30 percent incidence reported for the MIT cohort. The MIT and Argonne studies strongly endorse the presence of a threshold exposure (or at least a practical threshold exposure), below which malignancies are not seen in persons with internal depositions of radium.

5.8 Genetic Effects

A significant discovery of the 1920s, which became arguably the most important influence on radiation protection philosophy after World War II, was demonstration by the geneticist Müller that exposure to X rays produced mutations in the fruit fly *Drosophila melanogaster*[36]. Müller's demonstration of the mutagenicity of X rays was confirmed by Weinstein[37] and, in plants, by Stradler[38].

Also significant was the observation that exposure to X rays could induce mutations in cells other than those of the reproductive system[39,40]. These reports of somatic mutations, also made in fruit flies, were announced within months of Müller's initial report. The existence of X-ray induced somatic mutations not only offered a plausible explanation for the cancer-causing property of ionizing radiation, but also was consistent with the latency period associated with cancer induction.

Studies over the next decade confirmed the mutagenic properties of X rays and established that these properties are not limited to X rays, but also result from exposure to radium, alpha and beta radiation, and neutrons. While certain mutations were observed to occur more often than others, the yield of mutations was quite low even in genes most likely to mutate[41]. The induced mutationrate was found to be independent of exposure rate. The mutagenic effect of radiation seemed to exhibit no threshold dose, and the effect was postulated to be cumulative over a lifetime[42,43,44]. The work of the geneticists raised questions about the existence of a threshold exposure and the validity of the tolerance dose concept. This new idea would shape the direction of radiation research for many years.

Over the past several decades, genetic mutations caused by radiation exposure have been widely studied in various animal species. These studies are comprehensively reviewed in a number of recent publications by scientific bodies such as the NCRP[45,46,47] the National Academy of Sciences (NAS 1990)[1] and the United Nations Scientific Committee on the Effects of Atomic Radiation[48,49]. For chromosome aberrations, the data are suggestive of, or at least not inconsistent with a straight line relationship between exposure and mutation frequency.

Data verifying radiation-induced mutations in humans are limited. As a practical matter, radiation-induced mutations are probably not identifiable in a human population receiving a low-level exposure, because the frequency of spontaneous mutations not related to radiation exposure is simply too great. Nevertheless, there have been several genetic studies in human populations exposed to ionizing radiation. These include studies of the progeny of early radiologists and other radiation workers, which by and large have yielded equivocal results. The largest and most extensive study is the followup of survivors of the atomic explosions in Japan. This examination of the survivors and their progeny has been underway for a half-century. No statistically significant effects have been identified in this population for eight different indicators of genetic abnormalities[50], including mutational changes in proteins that would be expressed as recessive effects generations later. Evidence from the Japanese Study suggests that humans may be less sensitive to genetic effects of radiation than are some other species, including mice.

In conclusion, the genetic effects of exposure to low-level radiation are ambiguous. The concept of radiation-induced mutagenesis in humans has been largely inferred from studies of lower life-forms, including animals, fruit flies and plants.

5.9 Carcinogenesis

While animal investigations can provide insight into possible long-term effects of exposure of humans to ionizing radiation, it is human epidemiological studies that provide the most relevant data about such effects. Human epidemiological studies, however, are subject to substantial difficulties. These difficulties include the nature and small size of the populations available for analysis, the similarity of the effects observed to those that occur naturally (i.e., in the absence of radiation exposure), and the vagaries of the dosimetric measurements on which the studies are based. Consequently, it is challenging to assess the shape of the dose response curve and to estimate the level of human risk resulting from radiation exposure, particularly in view of the many confounding circumstances that may obscure the interpretation of data.

A major problem in assessing the degree of cancer risk associated with radiation exposure is related to population size. In small populations, there are likely to be few effects, and the incidence of radiation-induced changes may be overshadowed by the natural frequency of the effects. The difficulty of distinguishing between the natural incidence and a radiation-induced effect is a function of both population size and magnitude of exposure, and increases exponentially as the exposure is reduced. For effects that have a small likelihood of occurrence, the population size required for their demonstration may be prohibitively large. Consequently, there have been relatively few attempts to assess long-term effects of low-level exposures in humans, and most of those that have been conducted have been heavily criticized and disputed. Studies of radiation workers have been particularly controversial and contentiously criticized, frequently on methodological grounds.

The largest and most extensively studied population of exposed individuals, the Japanese atomic bomb survivors, have demonstrated a clear elevation of leukemia incidence at relatively high doses (>200 mSv) that peaks at 5–6 years after exposure. This group has also revealed an elevated incidence of various solid tumors that is correlated with the magnitude of exposure[51]. However, individuals exposed in the range of 10 to 90 mSv show a lower death rate from leukemia compared with controls[52].

5.10 The No-Threshold Model

In the 1950s a new model of radiation-induced injury replaced the concept of threshold dose for the purpose of establishing guidelines for radiation protection. This model assumed that the long-term effects of radiation exposure increase linearly with increasing exposure to radiation, and that a threshold exposure does not exist below which long-term effects do not occur. The model was justified in part by the work of Müller[36] and others that suggested that genetic effects are best described by a *no-threshold model* of radiation injury. Its acceptance was in part a response to concerns about the growing numbers of persons exposed to radiation in peacetime nuclear industries and from world-wide fallout released during atmospheric tests of nuclear weapons.

The no-threshold model of radiation risk is seductive in its mathematical simplicity and universal applicability. However, its adoption did not reflect any new data related to the occurrence of cancer or other somatic effects at low levels of radiation. These effects were demonstrable primarily in the Japanese populations only after relatively large exposures to radiation. The no-threshold model implied, without experimental verification, that the risk of such effects at low exposures could be estimated by extrapolation from high-exposure regions.

Implicit in the no-threshold model is the idea that the true risk of adverse health effects in the low-exposure region lies somewhere between zero and an upper limit defined by extrapolation, usually by extending a straight line (the linear, no-threshold model) or a curved line (the linear-quadratic no-threshold model) from effects at a relatively high-exposure to no effects at zero exposure. In the no-threshold model, the possibility of a threshold exposure was not eliminated. Instead it was included in the estimate of low-exposure health risks that ranged from zero to a maximum value obtained by extrapolation. Over the years, however, the concept of a range of health risks from zero to an upper-limit extrapolated value has been largely forgotten. Instead, the extrapolated upper limit has assumed an originally unintended prominence as a quantified measure of health risk related to low-level exposure. Often ignored in this assumption is its lack of support by evidence of somatic or hereditary effects in human populations exposed to ionizing radiation at low levels. Numerous studies of large populations of humans have failed to demonstrate with statistical significance any adverse health effects at low radiation exposures. In spite of these limitations, the no-threshold model has become widely adopted as the "preferred model" not only for estimating radiation risk, but also for quantifying the number of potentially injured persons in a population exposed to ionizing radiation. These interpretations of the model go far beyond the intentions of those originally responsible for its adoption.

Transition to the no-threshold model of radiation risk, and its utilization as a predictive model of radiation injury in exposed populations, occurred relatively rapidly after its introduction in the late 1940s. In 1954, the NCRP issued new guidance on radiation protection in which the tolerance dose was replaced by a new concept, the maximum permissible dose (MPD)[53]. Implicit in the MPD was rejection of the concept of tolerance dose and establishment of the idea of "acceptable risk" at low levels of exposure. In 1958 the United Nations Scientific Committee on the Effects of Atomic Radiation issued its first report on the effects of radiation exposure in humans[54]. This report estimated the risk of adverse effects of low-level radiation exposure using both a no-threshold and a threshold (i.e., tolerance dose) model of radiation risk. It also issued the following statement, which is as relevant today as it was in 1958:

"Present knowledge concerning long-term effects and their correlation with the amount of radiation received does not permit us to evaluate with any precision the possible consequence to man of exposure to low radiation levels. Many effects of radiation are delayed; often they cannot be distinguished from other agents; many will develop once a threshold dose has been exceeded; some may be cumulative and others not; and individuals in large populations or particular groups such as children and foetuses may have special sensitivity. These facts render it very difficult to accumulate reliable information about the correlation between small doses and their effects either in individuals or in large populations." [54]

With respect to radiation-induced leukemia identified in the Japanese populations exposed to atomic radiation well above the low-exposure limit, the UNSCEAR concluded that the threshold and no-threshold models of radiation injury have equal validity. This conclusion was contested by the Committee on Pathologic Effects of Atomic Radiation of the National Academy of Sciences/National Research Council (NAS/NRC), who stated unequivocally that "... a considerable body of experimental evidence ..." favors nonlinearity and hence presumably a threshold. The committee urged that nonlinear relationships between dose and effect should be given greater attention[55]. The following year, the short-lived U.S. Federal Radiation Council (FRC) observed that the linear, no-threshold model merely presents an extrapolated upper limit of radiation risk for low exposure levels[56]. In UNSCEAR reports in the 1960s, the committee emphasized that extrapolation of the no-threshold curve provides an upper limit to the risk of low-level exposures[57,58]. This position was endorsed by the ICRP[59].

In the late 1950s, the congressional Joint Committee on Atomic Energy conducted hearings on the potential long-term effects of exposure to low levels of ionizing radiation. These hearings had a major influence on the thinking of both the scientific community and the public with regard to radiation hazards. The hearings began in 1957 with an inquiry into the nature of radioactive fallout from weapons testing and its possible effects on humans[60]. Testimony from scientific experts led the committee to conclude that atmospheric testing of nuclear weapons constituted a hazard, but left unresolved the issue of the most appropriate model to estimate the degree of risk at low exposure levels. The JCAE revisited this issue in its 1959 hearings[61], and again left it unresolved. However, the committee's report included claims that certain bioeffects, including genetic mutations, leukemia induction, and life shortening occur without a threshold dose. Also influential was the testimony of the biologist Lewis who strongly supported the linear, no-threshold hypothesis as a model for radiation protection standards. Lewis conceptually proposed the protection philosophy of As Low As Reasonably Achievable (ALARA)[62]. In subsequent hearings over the 1960s on the hazards of uranium mining and fallout exposure, and in considerations of standards for the protection and compensation of radiation workers, the JCAE moved slowly to the no-threshold model of radiation risk.

Concern over long-term effects of low-level radiation exposure continued during the 1960s in spite of the cessation of atmospheric testing of nuclear weapons. In 1964 the NAS/NRC responded to an inquiry from the FRC by establishing an advisory committee on the Biological Effects of Atomic Radiation (BEAR) to examine issues related to radiation protection, including the shape of the dose-response curve at low exposures. In an effort to limit effects of radiation on future generations, the BEAR committee introduced the concept of restricting doses to the population as well as to radiation workers. The BEAR committee was renamed the NAS/NRC committee on the Biological Effects of Ionizing Radiation (BEIR), which issued its first report in 1972[63]. This report reviewed the literature on the effects of radiation at low exposure levels. The report sidestepped the issue of the shape of the dose-response curve. However, it did provide estimates of cancer risk at low exposures based on a linear extrapolation from cancer-mortality data at high exposures in Japanese survivors and other exposed groups. These estimates implied that radiation-induced cancer does not exhibit a threshold exposure, in spite of the absence of confirmatory experimental data. Also in 1972 the US Atomic Energy Commission (AEC, the forerunner of the U.S. Nuclear Regulatory Commission) introduced the concept of ALARA (then known as As Low As Practicable (ALAP)) in Appendix I to Title 10, Part 50 of the Code of Federal Regulations. The implication of ALARA is that no threshold exists for radiation bioeffects, and that any dose, no matter how small, is potentially injurious to exposed individuals. These actions of the NAS/NRC and AEC completed a major transition in the conceptualization of radiation risk at low exposures, and provided a foundation for the evolution of the discipline of health physics as a major industry devoted to the protection of workers and the public against small exposures to ionizing radiation.

In 1977 the ICRP announced its risk-based approach to the establishment of standards for radiation protection[64]. This approach was a radical departure from traditional exposure-based standards, and defined the concept of acceptable risk from radiation exposure of workers in terms of the fatal accident rate in so-called safe industries. In taking this approach, the ICRP compared real and measurable fatalities in other industries to hypothetical and unidentifiable deaths from cancers induced by low-level radiation that are predicted by the linear, no-threshold model. The ICRP also introduced a number of tissue-weighting factors to compute a new unit of radiation quantity – the effective dose equivalent – that expresses the risk of partial-body irradiation in terms of the equivalent risk of whole-body exposure. The risk-based approach to setting standards for radiation exposure has been refined and expanded not only by the ICRP[65], but also by the NCRP[46,66] and by several US regulatory agencies, including the Environmental Protection Agency[67], Department of Energy[68] and Nuclear Regulatory Commission[69].

In 1979 the NAS/NRC BEIR Committee released a new report (the "BEIR III" report) on the risks of exposure to ionizing radiation. Release of this report was characterized as premature by several committee members, and it was subsequently withdrawn. When finally published in 1980, the report was accompanied by two "minority opinions" of members who objected to use of the linear-quadratic, no-threshold model of radiation-induced cancer endorsed by the majority of the committee[70]. One minority opinion supported a linear, no threshold model, and the other endorsed a purely quadratic model of cancer induction. This dispute among committee members reflected more general disagreement within the scientific community about the most appropriate way to characterize radiation risk at low exposures. It also depicted concern over the growing practice of using dose-response models to estimate hypothetical cancer risks at exposures substantially below levels where epidemiological studies have confirmed injury.

The BEIR III report offered several important observations. The report noted that it was unknown and probably not determinable whether exposure rates of 1 mSv per year, on the order of those from background radiation, are detrimental to people. The report concluded that data presented by Sternglass[71] and others that purported to show an increased incidence of cancer in populations exposed to low levels of radiation were the result of flawed studies. The BEIR III committee recognized that different groups of people may exhibit different degrees of cancer risk for exposure to a specific amount of radiation, and that developmental effects from radiation exposure *in utero* may exhibit a threshold exposure. Finally, the report suggested that the linear, no-threshold model of radiation risk provides the best estimate of genetic risk. This suggestion had to be based on observations of genetic effects in exposed animals, since no hereditary effects have been documented in human populations exposed to ionizing radiation.

Two additional BEIR reports have been issued since the 1980 report of the BEIR III committee. The BEIR IV report, published in 1988, addresses the health risks of radon and other internally-deposited radionuclides[72]. The report offered several suggestions for further research that, collectively, called for intensified experimental efforts to characterize the shape of the dose-response curve for long-term health effects at low levels of exposure. The most recent report, BEIR V, once again considers the broad topic of adverse health effects from exposure to low levels of ionizing radiation[1]. As in previous reports, the committee noted the failure of epidemiological studies to demonstrate hereditary effects in humans exposed to low radiation levels. Nevertheless, the committee agreed with previous estimates of radiation-induced genetic risk in humans, and computed a mutation doubling dose of 1 Sv in agreement with the range of 0.2–2 Sv of BEIR I and 0.5–2.5 Sv of BEIR III. (The mutation doubling dose is the radiation dose predicted to cause a two-fold increase in the rate of genetic mutations in humans.) There was, however, a significant change in the BEIR V estimates of cancer risk from radiation compared with earlier BEIR reports. The new estimates were determined with the linear, no-threshold model and yielded a threefold increase in the risk of solid tumors and a fourfold increase for leukemia. Although the committee did not consider the rate of delivery of radiation in its estimates of cancer risk, it proposed a Dose Rate Effectiveness Factor (DREF) which, if applied, would reduce the lifetime cancer risk by a factor of two or more if the radiation were delivered over a protracted period.

5.11 Self-Regulation

Today in most countries, regulations to ensure personal safety and public health are considered primarily a governmental responsibility. For several decades after the discovery of ionizing radiation, however, individuals sought protection from the adverse effects of radiation primarily through guidelines developed by professional organizations. Citing the deaths of radiologists, the German Roentgen Society first provided guidelines in 1913 to reduce the hazards of radiation exposure to medical workers[73]. In 1915 and 1921, recommendations in Great Britain to protect X-ray operators included maximum work schedules, required amounts of leisure time, and special accommodations for the workers[74]. In the United States, the American Roentgen Ray Society and the American Radium Society were encouraged by George Pfahler to create specific guidelines for radiation protection. Dr. Pfahler suggested that:

1. The principles of radiology be thoroughly mastered, so that they can be adapted to the individual establishment, for the protection of both operator and patient;

2. A committee of the society be appointed to cooperate with other bodies of national organizations, to study and formulate definite directions and rules of protection;
3. The committee cooperate insofar as possible with the National Bureau of Standards in Washington in order to secure definite and permanent products, and possibly definite calibration of units;
4. Radium be insured against loss, so that men will not be suddenly hampered financially and prevented from carrying on the good work which has been started;
5. Every radiologist be provided with legal protection, with protection from insurance companies and the protection and cooperation of his county medical society; [and]
6. Every radiologist in the country associate himself with the American Radium Society, both for his good and the good of the Society[75].

The third of Pfahler's recommendations delineates a governmental role in the regulation and control of radiation sources. In 1927, the National Bureau of Standards began a voluntary program to inspect and calibrate radiation equipment and to send government representatives into laboratories to evaluate the safety of radiation sources, including X-ray machines. This program followed the model of a national inspection program initiated in 1921 by the National Physical Laboratory in Great Britain.

The International Commission on Radiological Protection (ICRP) held its formative meeting in 1928 and requested that each represented country develop a coordinated program of radiation control. The U.S. representative from the National Bureau of Standards formed the U.S. Advisory Committee on X-ray and Radium Protection, later named the National Committee (now Council) on Radiation Protection and Measurements (NCRP). The NCRP received a congressional charter in 1964.

By 1931, early philosophical constructs of radiation protection had been developed by the ICRP and NCRP, and the concept of tolerance dose was adopted as an upper limit for exposure of workers[76]. About the same time, the International Commission on Radiation Units and Measurements (ICRU) defined the unit "roentgen (R)" as the amount of radiation that would produce one electro static unit of ionization in a cubic centimeter of air at standard temperature and pressure (one esu/cm^3 at STP equals 2.58×10^{-4} coulombs/kilogram of air). In 1934 the ICRP established a tolerance dose of 0.2 R per day for exposure of workers to radiation, and in 1936 the NCRP reduced this limit to 0.1 R per day. This advisory limit was maintained through World War II and was applied to workers in the Manhattan Project[77]. The ICRP and, in the United States, the NCRP have remained to this day as the principal voluntary advisory agencies concerning radiation protection and limits for radiation exposure.

5.12 History of Government Regulation

5.12.1 Nuclear Regulatory Commission

The Hiroshima and Nagasaki explosions in 1945 brought the nuclear age abruptly into worldwide public consciousness. The end of World War II gave rise to a bitter dispute in the United States between military leaders and civilian officials concerning the best way to control nuclear energy and inhibit development of nuclear weapons by other countries. This dispute culminated in congressional passage of the Atomic Energy Act (AEA) of 1946. This act affirmed civilian control over nuclear energy while leaving weapons development with the military. The AEA created the Atomic Energy Commission (AEC) to oversee development of nuclear technologies. The AEC's principal function was to foster the continued development of nuclear weapons in the United States, including assurance of a sufficient supply of weapons-grade fissionable material. A secondary purpose of the AEC was to encourage peaceful uses of nuclear energy[78]. By 1954, the Isotopes Division of Oak Ridge National Laboratory had delivered 47,000 shipments of reactor-produced radioactive nuclides around the country, many of which were intended for medical uses.

The AEA also created a congressional Joint Committee on Atomic Energy (JCAE) to provide legislative oversight of the activities of the AEC. In 1952 the JCAE issued a document entitled Atomic Power and Free Enterprise that encouraged private enterprise to develop nuclear power for commercial purposes. Two years later President Eisenhower urged Congress to change the AEA to facilitate private development of nuclear technologies for peaceful purposes. In response, Congress passed the AEA of 1954 that endorsed Eisenhower's Project Plowshare to turn "atomic swords into plowshares." The 1954 AEA qualified the endorsement of peaceful applications of nuclear energy by limiting them "to the maximum extent consistent with the common defense and security and with the health and safety of the public." The act gave the AEC a difficult mandate: it was to encourage a private nuclear enterprise while regulating its activities to ensure compliance with national security and with personal and public health[79].

The use of radioactive material in medicine was included within the purview of the AEC's regulatory powers. The AEC was directed in the 1954 AEA to "exercise its powers in such manner as to insure the continued … research … in … utilization of … radioactive material for medical, biological [and] health … purposes." This encouragement was balanced by provisions to: "(1) protect health, (2) minimize danger to life or property, and (3) require the reporting and permit the inspection of work performed thereunder, as the Commission may determine." The AEC was authorized *to issue licenses to persons applying therefore for utilization facilities for use in medical therapy. In issuing such licenses the Commission is directed to permit the widest amount of effective medical therapy possible with the amount of special nuclear material available for such purposes, and to impose the minimum amount of regulation consistent with its obligations under this Act to promote the common defense and security and to protect the health and safety of the public.*

The dual mandate of the AEC to foster development of nuclear technologies and to protect national security and public health was viewed by many as contradictory objectives. After years of public debate, Congress passed the Energy Reorganization Act (ERA) in 1974 to separate the two objectives. A new Energy Research and Development Administration (ERDA, subsequently changed to the cabinet-level Department of Energy (DOE)), assumed "activities relating to research and development on the various sources of energy (and) other functions, including but not limited to the Atomic Energy Commission's military and production activities and its general basic research activities." The act also created the Nuclear Regulatory Commission (NRC) to continue the licensing and related regulatory functions of the Atomic Energy Commission. The NRC was to be directed by five commissioners appointed by the President with approval of the Senate.

5.12.2 Federal Radiation Council

In 1959, the Bureau of the Budget analyzed the radiation protection activities of various federal agencies. Perceiving a conflict of interest within the AEC between its promotion of weapons and energy development and its obligations to protect public health and safety, the Bureau recommended several changes to restore public confidence in federal control of radioactive materials. Among the recommendations was creation of the Federal Radiation Council (FRC), comprised of secretaries of the Departments of Commerce, Defense, and Health, Education and Welfare (DHEW), as well as the AEC chairperson.

The FRC was directed to rely on the expertise of the NCRP in proposing recommendations to the President on protection standards. These recommendations were to be limited to general standards and guidance; federal agencies retained their responsibilities for setting legally binding rules and regulations within their jurisdictions, and the NCRP retained its unofficial status as an independent voluntary agency. In proposing protection guidelines for workers and the general public, the FRC abandoned the concept of maximum permissible dose and substituted the term "radiation protection guide (RPG)." The FRC defined the RPG as "the radiation dose which should not be exceeded without careful consideration of the reasons for doing so; every effort should be made to encourage the maintenance of radiation doses as far below this guide as practicable." The FRC stated that "These guides are not intended to apply to radiation exposure resulting from natural background or the purposeful exposure of patients by practitioners of the healing arts."

In 1970, President Nixon issued Executive Order No. 3 to disband the FRC and place its functions within the newly created Environmental Protection Agency (EPA). The plan also transferred to the EPA certain functions of the DHEW Bureau of Radiological Health and of the AEC.

5.12.3 Other Executive Agencies

In 1980 the Radiation Policy Council (RPC) was established by President Carter. One year later, President Reagan disbanded it, creating in its place the Committee on Interagency Radiation Research and Policy Coordination (CIRRPC) to coordinate policy and resolve conflicts among agencies. No disputes were brought before the CIRRPC before its demise in 1995. Neither the RPC nor the CIRRPC had regulatory or enforcement authority, and their power as executive agencies was severely limited.

The U.S. Public Health Service (PHS) has been involved in radiation protection matters since the 1920s. Before World War II, its responsibilities included investigations of radium poisonings of luminous dial painters, radium and X-ray hazards in hospitals, and radiation safety programs for photofluorographic technicians. In 1948 the Radiological Health Branch of the PHS was established to help states limit hazards from radioisotopes and other industrial radiation sources and to assist the AEC in studying effects of nuclear waste in rivers and streams. It also surveyed the use of X rays in PHS hospitals, provided training in radiation protection for state personnel, and worked with the Department of Defense to monitor fallout from nuclear weapons testing in Nevada and the Pacific.

In 1958 the Surgeon General established the National Advisory Committee on Radiation, which recommended unification of all PHS activities related to radiation control. In response, the PHS established the Division of Radiological Health to conduct research on the effects of radiation, provide technical assistance and training to state radiological health programs, and coordinate its activities with other federal radiation programs. A year later, the Division assumed primary responsibility for the collection, analysis, and interpretation of data on environmental radiation levels, assessment of all forms of radiation exposure in the United States, and development of recommendations for acceptable levels of radiation exposure from air, water, milk, medical procedures, and the environment.

In 1967 the PHS Division of Radiological Health was renamed the National Center for Radiological Health. In 1968 the Center was renamed again as the Bureau of Radiological Health (BRH), a component of the PHS Environmental Control Administration. The Center was granted regulatory authority to implement the Radiation Control for Health and Safety Act (Public Law 90-602). This act called for a control program over electronic product radiation to include "the development and administration of performance standards to control the emissions of electronic product radiation from electronic products, and the undertaking by public and private organizations of research and investigation into the effects and control of such emissions." In support of its assistance program to states, the BRH endorsed establishment of the Conference of Radiation Control Program Directors (CRCPD), which continues today as an organization of state personnel involved in radiation control.

In 1971, the Office of Management and Budget transferred 318 persons and $7 million from the BRH to the newly created EPA. The remaining 389 individuals in the BRH were reassigned to the Food and Drug Administration (FDA) within the PHS. A decade later, the BRH and the Bureau of Medical Devices (BMD) were combined into the Center for Devices and Radiological Health (CDRH) within the FDA.

The CDRH assumed the BMD's responsibility for administering the Medical Device Amendments of the Food, Drug and Cosmetic Act of 1938, as well as continuing obligations of the BRH for control of radiation from electronic products and medical devices. In 1992 the CDRH was identified as the implementing agency for the Mammography Quality Standards Act (MQSA), to include publication of national standards for mammography and establishment of quality control criteria and a certification program for the more than 10,000 medical facilities providing mammography services in the United States. To accommodate the increasing demand for services in the medical device area, some CDRH programs directed to medical uses of radiation have been curtailed in recent years. Nevertheless, the CDRH maintains a continued interest in public health aspects of the medical uses of radiation.

5.12.4 Environmental Protection Agency

The Environmental Protection Agency was created in 1970 to accept certain functions and responsibilities from other federal agencies and departments. These responsibilities included establishment and enforcement of environmental protection standards consistent with national environmental goals, encompassing specific radiation criteria and standards originally placed in the AEC and BRH.

Since its inception, the EPA has maintained environmental radiation programs such as off-site monitoring around nuclear power plants and radioactive waste disposal sites. It also has been concerned with natural sources of radioactivity such as radon. It did not, however, exercise the authority transferred from the FRC until 1978, when it published a document on medical X-ray guidance for federal agencies. This document contained 12 recommendations that federal agencies were expected to implement. They covered several aspects of medical and dental radiology, including the need for quality assurance programs and procedures to ensure minimal exposure of patients to radiation. Other recommendations were directed at qualifications of prescribers and specifications on radiographic techniques, fetal exposures, proper collimation, and gonadal shielding. Under its derivative FRC authority, the EPA has published guidance for federal agencies on topics such as occupational radiation exposures and limits on radiation exposure for members of the general public.

5.12.5 Food and Drug Administration

Biological drugs first became subject to federal premarket approval for safety and effectiveness under the Biologics Act of 1902, which remains basically unchanged to this day. The 1902 act was administered by the Public Health Service until it was transferred to the Food and Drug Administration (FDA) in 1972.

The Pure Food and Drugs Act of 1906 imposed federal policing of all drugs (including biological drugs) to prevent adulteration or misbranding, but did not require premarket approval. The 1906 act was replaced by the Federal Food, Drug, and Cosmetics Act of 1938, which required premarket notification (but not approval) for the safety (but not effectiveness) of new drugs. The 1938 act was supplemented by the Drug Amendments of 1962 following the thalidomide tragedy. This action requires FDA premarket approval of all new drugs for safety and effectiveness.

Medical devices first became subject to general policing authority by the FDA under the 1938 act. This act was supplemented by the Medical Device Amendments of 1976 which imposed premarket notification regarding the safety and effectiveness of most medical devices, and premarket approval for safety and effectiveness of important new devices. All radiation products were included within these statutory requirements for drugs and devices administered by the FDA.

Congress enacted the Radiation Control for Health and Safety Act of 1968 to provide additional regulatory authority for all "electronic products" that emit ionizing or nonionizing radiation, including medical products. The responsibility for administering the 1968 act was transferred to the FDA in 1971 and combined with the FDA medical device program in 1982. Under this act, the FDA works with other federal and state agencies and private organizations to minimize exposure to electronic radiation, and has issued guidelines and recommendations regarding public exposure to ionizing and nonionizing radiation.

5.12.6 State Regulation

The responsibility for public health and safety has traditionally been assumed principally by the states. With the exception of the use of radioactive by-product material, this holds true for all applications of ionizing radiation in medicine. As early as 1949, California consulted the NCRP in an effort to regulate protection of workers exposed to ionizing radiation. In its response, the NCRP seemed more concerned that state regulations not disturb the national uniformity of radiation protection and less concerned that states move toward self-regulation[73].

When the 1954 AEA assigned to the AEC the responsibility for encouraging civilian applications of nuclear energy, together with the exclusive jurisdiction for ensuring health and safety associated with these applications, state governments raised objections. After extended debate, Congress in 1959 revised the 1954 AEA to establish the Agreement State Program in which AEC responsibilities for health and safety may be delegated to states. In 1962, Kentucky became the first Agreement State.

Several amendments to the Agreement State Program have been added over the years. In 1978, Congress instructed the NRC to review state programs periodically for compliance, and in 1980 gave the NRC power to suspend state programs that do not meet minimum standards. In 1995, the NRC applied its quality management rule to all Agreement States.

The CRCPD, established in 1968, is a voluntary network of state and local government officials responsible for radiation regulation and enforcement. The CRCPD periodically updates its *Suggested State Regulations for Control of Radiation*, a publication first issued in 1962 by the Council of State Governments. Most state and local radiation protection programs are based on these suggested guidelines. Although the CRCPD is composed of radiation regulators, it has no regulatory authority of its own.

As mentioned above, Congress acted in 1959 to provide a statutory framework for the federal government to relinquish to the states some of its regulatory authority concerning radioactive nuclear and byproduct materials. NRC authority is transferred to a state through a formal agreement between the governor and the NRC. The NRC must conclude that the state's radiation control program "... is in accordance with the requirements of [applicable parts of the AEA] and in all other respects compatible with the Commission's program for regulation of such materials, and that the State program is adequate to protect the public health and safety ..." States must pass enabling legislation compatible with NRC requirements to establish their authority to enter into these agreements. Once they have done so, and the NRC finds them capable of enforcing the requirements, state assumption of authority may become effective on the date the agreement is signed.

Although Agreement States regulate their own licensing and enforcement decisions, the NRC maintains significant authority over them. Biennially, the NRC's Management Review Board reviews each state's performance to ensure that its program is adequate and that it does not deviate significantly from the NRC's. The Management Review Board is composed solely of NRC staff, although Agreement States have requested representation on several occasions. As described in the NRC's "Final Statement of Principles and Policy for the Agreement State Program," the actions of the Management Review Board include (1) periodic assessments of Agreement State radiation control programs; (2) provision of assistance to help address weaknesses or areas within an Agreement State radiation program that needs improvement; (3) placement of a State on probationary status for serious program deficiencies that require heightened oversight; and (4) temporary suspension of an agreement and reassertion of NRC regulatory authority in an emergency if an Agreement State program experiences program difficulties that prevent it from providing adequate protection of the public health and safety. The NRC's actions must be based on a well defined and predictable process, and include a performance evaluation that is consistently and fairly applied.

In 1993, the General Accounting Office (GAO) published a report entitled *Better Criteria and Data Would Help Ensure Safety of Nuclear Materials*. This report reviewed the comparability of NRC's program for Agreement States to those of NRC-regulated states. The report concluded that the NRC lacks adequate criteria and data to evaluate the comparative effectiveness of the two programs. It stated further that the NRC does not have common performance indicators on inspection backlogs, radiation overexposures, or frequency of violations. The GAO concluded that the NRC cannot determine if its goals are being met or if the public is receiving at least a minimum level of protection in different states. The GAO found that the NRC's criteria for program evaluation are vague, and that evaluation depends on the professional judgment of the NRC staff. The GAO concluded that without specific criteria and procedures it is questionable whether the NRC can legitimately initiate the process to revoke its agreement with a state.

5.13 The Current Regulatory Framework

Regulatory authority over ionizing radiation in medicine is widely distributed over several government agencies at the federal, state, and local levels. At the federal level, the NRC and the FDA exercise primary regulatory authority over radiation use in medicine. In addition, the Environmental Protection Agency (EPA), the Occupational Safety and Health Administration (OSHA), and the Department of Energy (DOE) oversee exposure standards for the public and for workers. The transportation of radionuclides is regulated by the Department of Transportation (DOT). In some cases, regulatory standards are established at the federal level but are administered by the states. Where federal oversight is absent, some states regulate independently in their roles as protectors of the public health and safety. State laws and regulations often differ, as does the rigor with which state laws and regulations are enforced.

5.13.1 Nuclear Regulatory Commission

The Atomic Energy Act (AEA) of 1954 authorized the Atomic Energy Commission (AEC) (now the NRC) to regulate three types of radiation sources: byproduct materials, source materials, and special nuclear materials. Byproduct materials used in medicine are subject to regulation under Title 10 of the Code of Federal Regulations (CFR) Part 35. Other forms of ionizing radiation used in medicine, such as radium, accelerator-produced materials, and machine-produced radiation were not included under the 1954 act and are not subject to NRC regulation. The NRC regulates byproduct materials by directly licensing manufacturers and approving the marketing of products containing byproduct materials following evaluation of their safety and effectiveness. Under its Medical Use Program, the NRC exercises licensing authority over physicians who use products and sets criteria for determining their proper use. For radiation sources that are not subject to NRC jurisdiction, there is no parallel to the NRC's Medical Use Program. Hence, the use of some products in radiation medicine is rigorously controlled at the federal level, while the medical use of others is regulated to varying degrees at the state level.

5.13.2 The NRC's Medical Use Program

The NRC licenses facilities, authorizes physician users, develops radiation safety regulations, sets criteria for defining misadministration of byproduct materials in medical use, orders prompt reporting of misadministrations, conducts compliance inspections, applies sanctions for infractions of its regulations, and assesses and collects fees and fines. The program is administered through two different approaches. In 29 states known as Agreement States, the NRC formally delegates authority to regulate byproduct material to the state government. In the remaining 21 states (Non-Agreement States), the NRC directly licenses, monitors, inspects, and enforces its regulations for approximately 2,000 licensed users and institutions.

Part 35 of Title 10 of the CFR, called "Medical Use of Byproduct Material," contains provisions to protect patients and workers from devices and radiation beams and sources. To protect patients scheduled for radiation procedures, Part 35 requires implementation of quality management (QM) procedures (section 35.32), measurement of each dose prior to administration (35.53), survey of the patient after removal of temporary implants (35.406), and safety checks of teletherapy machines and rooms (35.615). Other sections of Part 35 pertain to protection of the public and patients not scheduled for radiation procedures. These provisions include surveys before returning radiation areas to unrestricted use (35.315, 35.415), criteria for releasing patients who have received doses of radioactivity (35.75), and QM redundancy procedures for verifying patient identity (35.32).

The NRC oversees medical use licensees through its inspection, investigation, and enforcement programs. Inspections involve (1) periodic unannounced visits by NRC personnel to licensed facilities, and (2) special inspections to follow up an incident or major noncompliance finding. Inspections are intended to provide assurance that licensed programs are conducted in accordance with NRC requirements, specific provisions of the license, and requirements for the health and safety of workers and the public.

Enforcement actions may be taken against licensees when violations of NRC regulations are discovered. Violations may range from failure to follow procedures detailed in a QM program to actual threats to public health and safety. Sanctions include more frequent inspections, release of negative publicity to the media, civil fines and penalties, and license revocation.

5.13.3 Misadministration Rule

In 1968, six years before the Energy Reorganization Act split the AEC into two separate agencies, a patient died when exposed to 1,000 times the intended therapeutic dose of radioactive gold (Au-198). This error led the U.S. General Accounting Office (GAO)[80] in 1972 to issue the report *Problems of the Atomic Energy Commission Associated with the Regulation of Users of Radioactive Materials for Industrial, Commercial, Medical and Related Purposes*. Noting the lack of information on overexposures, and tallying approximately 20 incorrect doses brought to the AEC's attention over the prior 10 years, the GAO criticized the AEC's "lax oversight" of byproduct material licensees. In response, the AEC proposed several revisions in its procedures. However, the AEC failed to act on the proposed revisions, and they gradually faded into obscurity.

For several months in 1975 and 1976, a cobalt (Co-60) teletherapy unit was miscalibrated at the Riverside Methodist Hospital in Columbus, Ohio. During this period, almost 400 radiation therapy patients were overexposed by as much as 40 percent. By the time the error was reported, two patients had died as a direct result of the miscalibration. Over the next several months, eight additional deaths occurred that were probably attributable to the mistake. The NRC modified Riverside's byproduct materials license to require full annual calibrations, monthly spot checks, and detailed recordkeeping.

In 1979 the NRC extended the Riverside requirements to all Co-60 teletherapy licensees. It also implemented the GAO's 1972 recommendation that misadministrations of radiation be reported to the NRC and brought to the attention of the patient or family. After public hearings on these actions, the NRC issued a "Medical Policy Statement," which proposed three actions:

1. The NRC will continue to regulate the medical uses of radioisotopes as necessary to provide for the radiation safety of workers and the general public.
2. The NRC will regulate the radiation safety of patients where justified by the risk to patients and where voluntary standards, or compliance with these standards, are inadequate.
3. The NRC will minimize intrusion into medical judgments affecting patients and into other areas traditionally considered to be a part of the practice of medicine[81].

In 1980 the NRC issued its rule on "Misadministration Reporting Requirements," that broadened reporting requirements to encompass both diagnostic and therapeutic procedures. The misadministration rule was incorporated into 10 CFR 35.41 and 35.45 and required licensees to (1) keep records of all misadministrations; (2) promptly (within 24 hours) report all therapy misadministrations to the NRC, referring physicians, and the patient or responsible relative or guardian; and (3) report diagnostic misadministrations quarterly to the NRC. The NRC estimated that the cost of the misadministration rule would be about $1.2 million.

A misadministration is defined by the NRC as the administration of some radioactive substance in an amount that exceeds the prescribed dosage by a certain percentage. The percentage calculation depends upon the substance in question. A misadministration may also be the administration of a correct dosage, but of the wrong substance or to the wrong patient.

Over the following four years, NRC licensees reported 27 therapy misadministrations, or about 7 per year. Sixteen of these incidents involved teletherapy equipment; five, brachytherapy treatment; and six, radiopharmaceutical therapy. In analyzing these incidents, the NRC identified three basic causes: inadequate training, inattention to detail, and lack of procedural redundancy. The NRC noted that

although professional medical groups involved with radiotherapy and related government agencies encourage quality assurance programs in radiotherapy facilities, no government agency or non-governmental accrediting body requires that radiotherapy facilities have quality assurance programs that conform to the programs recommended by professional medical groups. Thus, many facilities may not have quality assurance programs that are consistent with recommendations of medical professional groups involved with radiation therapy.[82]

The NRC instructed its Office of Nuclear Material Safety and Safeguards to: *dispense the information contained in its report to affected licensees; contact appropriate professional organizations to encourage and support the initiation of a voluntary industry-directed quality assurance program for radiotherapy facilities; determine the effectiveness of the voluntary program within two years; and consider the possibility of imposing a quality management rule if substantial progress toward completion of the voluntary program, including a final completion date, had not been demonstrated at the end of two years.*

In January 1986, Washington Hospital Center in Washington, D.C., reported that a patient was administered 150 rads of radiation with no request or desire for treatment from the referring physician. This event caused the NRC commissioners to direct staff to develop rulemaking that would initiate quality assurance (OA) programs to reduce the chance for therapy misadministrations. A year later the NRC published a proposed rule of "Basic Quality Assurance in Radiation Therapy" and an advance notice of proposed rulemaking, which called for a comprehensive QA program for any medical use of radioactive byproduct material. The NRC claimed that voluntary QA programs may not adequately assure public health and safety, but it limited the scope of the proposed prescriptive rule to radiation therapy and diagnostic procedures involving radioactive iodine.

5.13.4 Quality Management Rule

In January 1992, the NRC implemented the final version of its QM rule[83]. The rule calls upon NRC licensees to establish a QM program in compliance with 10 CFR 35.32 and 35.33 if they administer: (1) radiation from sealed sources containing byproduct material for therapy (brachytherapy); (2) cobalt teletherapy, or (3) therapeutic unsealed radionuclides. The rule also applies to any diagnostic administration of greater than 30 microcuries of sodium iodide I-125 or 1-131. Moreover, it requires NRC licensees to submit written certification that they have implemented a QM program. Whereas NRC licensees have been following this rule since January 1992, agreement states were not required to follow suit until January 1995.

The QM rule is a performance-based approach to quality management. This approach includes five specific objectives:

1. Prior to an administration, a written directive must be prepared.
2. Prior to each administration, the patient's identity must be verified by more than one method as the individual named on the written directive.
3. Final plans for treatment and related calculations for brachytherapy, teletherapy, and gamma stereotactic radiosurgery must be in accordance with the respective written directives.
4. Each administration must be in accordance with the written directive.
5. Any unintended deviation from the written directive must be identified and evaluated, and appropriate action must be taken.

The FDA exercises direct authority to determine the safety and effectiveness, and to approve the marketing, of all radiation products used in medicine. No radiation product may be investigated in humans without an investigational new drug (IND) application submitted to the FDA. Also, no new radiation product may be marketed without FDA approval of a new drug application (NDA), a product license application (PLA) if it is a biological product, or one of two types of medical device approvals if it is a device (PMA or 510K). The FDA must approve the labeling of all products prior to marketing, and the manufacturing procedures of all products except medical devices that are substantially equivalent to those that have been previously marketed.

After a radiation product is approved by the FDA for marketing, it remains subject to the same stringent postmarketing requirements as all other drugs and devices. The product must be manufactured in compliance with good manufacturing practice, and is subject to drug and device reporting requirements in cases of product defects and adverse reactions. Once a drug or device reaches the market, the FDA imposes surveillance and reporting requirements but does not control the medical use of the product. The FDA's policy is that physicians have legal authority to use approved drugs and devices (including radiation products) for unapproved uses.

The 1991 Safe Medical Device Act (SMDA) requires that institutions must report all adverse events that occur because of equipment malfunction defects or user error. Adverse events resulting in serious bodily injury are to be reported within 10 days to the device manufacturer, and on a semiannual basis to the FDA. Deaths must be reported immediately to the FDA. The definition of a "device" does not cover pharmaceuticals or radioactive sources, but does include equipment involved in the administration of radioactive sources and radiopharmaceuticals.

The FDA is responsible for implementing the Mammography Quality Standards Act (MQSA) of 1992, and has promulgated regulations establishing quality control standards and a certification program for mammography services.

For radiation products subject to both FDA and NRC jurisdiction, there are overlapping regulatory requirements. A memorandum of understanding delineates the differences between the NRC's Medical Use Program and the FDA's procedures for approving pharmaceuticals (including radiopharmaceuticals) and devices (including radiation-emitting and radiation-producing devices). The FDA requires the registration of manufacturing establishments, and the NRC licenses these establishments. Both the FDA and the NRC separately approve the method of manufacture, require continuing compliance with good manufacturing practice requirements, and inspect the establishments to assure compliance for the same radiation products. As already noted, the FDA and the NRC both impose reporting requirements regarding adverse reactions for the same radiation products.

For medical uses of radiation products that are subject to FDA requirements but are not regulated by the NRC, many states have imposed their own controls. Although these requirements are not uniform from state to state, there is a trend toward consistency emanating from the Conference of Radiation Control Program Directors.

5.13.5 Environmental Protection Agency

The EPA establishes and enforces environmental protection standards consistent with national environmental goals. It also has responsibility, once vested in the AEC and Federal Radiation Council, for establishing radiation exposure criteria and standards. The EPA maintains active programs related to environmental radiation, nuclear power plants, radioactive waste dumps, and natural sources of radioactivity such as radon. An example of EPA's standard setting is its 1978 guidance on medical X rays. This guidance contained 12 recommendations for federal agencies that covered several aspects of medical and dental radiology, including QA programs, qualifications of prescribers, technique specifications, fetal exposures, proper collimation and gonadal shielding. The EPA has published guidance on occupational radiation exposure, and proposed guidance on radiation exposure for members of the general population.

5.13.6 States

States have broad authority and responsibility under the Constitution to protect citizens in matters of public health and safety. States have authority to regulate the use of X-ray equipment in medicine and industry, as well as particle accelerators and naturally occurring and accelerator-produced radioactive material (NARM). States also regulate the practice of medicine, and license physicians and supporting health professionals such as radiologic technologists and radiation therapists.

States have the greatest share of regulatory responsibility for overseeing medical uses of radiation. States achieve a relative level of uniformity in many areas of their regulatory responsibility through cooperative, voluntary, and informal arrangements. An example of such arrangements is the publication *Suggested State Regulations for Control of Radiation* (SSRCR). This publication forms the basis for most state and local radiation protection programs. The SSRCR was first issued in 1962 by the Council of State Governments, with advice and assistance from the AEC and the U.S. Public Health Service. It has been regularly revised and updated to reflect changes in standards and technology and to incorporate alterations in mandatory and voluntary federal regulations. Today, these model regulations are prepared and published by the CRCPD, a not-for-profit network of state and local government radiation regulators.

Another nongovernmental organization influential in the regulation of ionizing radiation in medicine is the United States Pharmacopoeia (USP). Standards for production and acceptable uses of marketed radiopharmaceuticals are described in USP's National Formulary. The USP standards are used by state pharmacy boards, professional medical societies, and the FDA.

5.14 Current Status of Radiation Protection Guidelines

In 1977 the ICRP developed the philosophy that radiation protection guidelines should yield a risk of radiation-induced cancer in occupationally-exposed individuals that is no greater than the risk of a fatal accident of workers in so-called "safe" industries[64]. The NCRP[46] adopted this philosophy a few years later. Intrinsic in this adoption are several issues:

1. Uncertainty of at least a factor of two in the risk per unit radiation dose for exposures at high dose and dose rate.
2. Uncertainty of two or more in the extrapolation of risks from high exposures to exposures at low dose and dose rate. At low doses and dose rates it is conceivable that the risk is zero, or even that a hormetic effect may be present.
3. Although many nominally safe industries have annual fatal accident rates of 10^{-4} or less, substantial morbidity from nonfatal injuries and work-related diseases may occur[84].
4. Comparisons of mortality from cancer with mortality from accidents are not fixed, since both mortality rates have continued to decrease over the years.

Shown in Table 5.1 are the fatal accident rates for two specific years for selected industries in the United States. These rates have been decreasing at the rate of nearly 3 percent per year. The NCRP recommends that radiation protection guidelines should limit the annual risk of radiation-induced fatal cancer to 10^{-4} or less. Since these guidelines define the maximum doses to be received by workers, it is reasonable to state that the risk associated with these doses can be greater than 10^{-4}, perhaps between 10^{-4} and 10^{-3} per year. This approach yields a risk to the *average* worker of 1/4 to 1/6 of this value (i.e., between 2×10^{-5} and 2×10^{-4}).

Table 5.1. Fatal accident rates in various U.S. industries, 1976 and 1991.

	Mean rate 1976[a] (10^{-4} y^{-1})	Mean rate 1991[b] (10^{-4} y^{-1})
All groups	1.42	0.90
Trade	0.64	0.40
Manufacture	0.89	0.40
Service	0.86	0.40
Government	1.11	0.90
Transport and Public utilities	3.13	2.20
Construction	5.68	3.10
Mines and quarries	6.25	4.30
Agriculture (1973–80)	5.41	4.40

[a] NSC (1977 [85])
[b] NSC (1992 [86])

5.14.1 Units for Radiation Protection Guidelines

Different types of ionizing radiation vary in their ability to produce biological damage. This variability is expressed as the relative biological effectiveness (RBE) of each type of radiation. In radiation protection, the RBE is accounted for by use of the concept of equivalent dose ($H_{T,R}$) which is the average absorbed dose ($D_{T,R}$) in a tissue or organ (T) multiplied by a radiation weighting factor (w_R) for each type of radiation (R) delivering the dose.

$$H_{T,R} = w_R \, D_{T,R}$$

The radiation weighting factor w_R is a dimensionless quantity that accounts for differences in the overall biological effectiveness of different types of radiation. Its value is influenced by the type and energy of the radiation. The equivalent dose for a particular type of radiation is conceptually different from the dose equivalent. The dose equivalent H represents the absorbed dose at a "point" in tissue weighted by a distribution of quality factors Q that are related to the LET of the radiation at the point, whereas the equivalent dose H_T is based on an average absorbed dose in the tissue or organ adjusted by the radiation weighting factor w_R for the particular radiation. When the exposure is due to radiations of different types and energies, the equivalent dose H_T is the sum of the average doses from each of the component radiations multiplied by their respective w_R values.

$$H_T = \sum_R w_R \, D_{T,R}$$

Recommended radiation weighting factors w_R for different types and energies of radiation are provided in Table 5.2.

Table 5.2. Radiation weighting factors w_R [from NCRP 116 (NCRP 1993[66]) as adapted from the ICRP (ICRP 1991[65])].

Type and energy range		w_R
X and γ rays, electrons, positrons and muons[b]		1
Neutrons, energy	<10 keV	5
	10 keV to 100 keV	10
	>100 keV to 2 MeV	20
	>2 MeV to 20 MeV	10
	>20 MeV	
Protons, other than recoil protons and energy > 2MeV		2[d]
Alpha particles, fission fragments, nonrelativist heavy nuclei		20

a All values relate to the radiation incident on the body or, for internal sources, emitted from the source.
b Excluding Auger electrons emitted from nuclei bound to DNA since averaging the dose in this case is unrealistic. The techniques of microdosimetry are more appropriate in this case.
c In circumstances where the human body is irradiated directly by > 100 MeV protons, the RBE is likely to be similar to that of low-LET radiations and, therefore, a w_R of about unity would be appropriate for that case.
d The w_R value for high energy protons recommended here is lower than the recommended value in ICRP (1991a [65]).

The effective dose E is a concept that yields the same probability of cancer and genetic effects irrespective of whether the radiation exposes the whole body uniformly, part of the body or only a specific organ. The effective dose provides a means for accommodating situations of nonuniform and partial-body irradiation.

The effective dose E is the sum of weighted equivalent doses for all exposed tissues and organs. Appropriate weighting is accomplished by use of a tissue weighting factor w_T that accounts for radiation effects on various organs and tissues, including different mortality and morbidity risks from cancer, the risk of severe hereditary effects for all generations, and the length of life lost because of these effects. The risks for all effects will be the same whether the whole body is irradiated uniformly or nonuniformly if

$$E = \sum_{T} w_T H_T \tag{5.1}$$

Values of w_T are given in Table 5.3. These values have been developed for a reference population of equal numbers of both genders and a wide range of ages, and should not be used to obtain specific estimates of potential health effects for a given individual.

The following simplification can be made:

$$E = \sum_T w_T H_T = \sum_T w_T \sum_R w_R D_{T,R} \tag{5.2}$$

where w_R is independent of the tissue or organ, and w_T is independent of the radiation type or energy.

Table 5.3. Tissue weighting factors w_T for different tissues and organs [from NCRP 116 (NCRP 1993[66]) as adopted from ICRP 1991[65]).

0.01	0.05	0.12	0.20
Bone Surface	Bladder	Bone marrow	Gonads
Skin	Breast	Colon	
	Liver	Lung	
	Esophagus	Stomach	
	Thyroid		
	Remainder[bc]		

a The values have been developed for a reference population of equal numbers of both genders and a wide range of ages. In the definition of effective dose, they apply to workers, to the whole population and to either sex. These w_T values are based on rounded values of the organ's contribution to the total detriment.

b For purposes of calculation, the remainder is composed of the following additional tissues and organs; adrenals, brain, small intestine, large intestine, kidney, muscle, pancreas, spleen, thymus and uterus. The list includes organs which are likely to be selectively irradiated. Some organs in the list are known to be susceptible to cancer induction. If other tissues and organs subsequently become identified as having a significant risk of induced cancer, they will then be included either with a specific w_T or in this additional list constituting the remainder. The remainder may also include other tissues or organs selectively irradiated.

c In those exceptional cases in which one of the remainder tissues or organs receives an equivalent dose in excess of the highest dose in any of the 12 organs for which a weighting factor is specified, a weighting factor of 0.025 should be applied to that tissue or organ and a weighting factor of 0.025 to the average dose in the other remainder tissues or organs [see ICRP (1991a)[65]].

Radiation doses from radionuclides deposited in an organ or tissue will be delivered over time depending upon the effective half-life of the radionuclide. The effective half-life considers both radioactive decay and biological elimination of the radionuclide from the organ or tissue. The concept of committed dose considers the ongoing delivery of radiation dose from deposited radionuclides. The committed equivalent dose H_T(t) is the integral over time of the equivalent dose-rate in a specific tissue (T) following intake of a radionuclide into the body. For a single intake of radionuclide at time t_0, $H_T(\tau)$ is

$$H_T(\tau) = \int H_T \, dt \tag{5.3}$$

where the limits of integration are from t_0 to $t_0+\tau$. The committed effective dose E(t) for each internally-deposited radionuclide is the sum of products of the committed equivalent doses and the appropriate w_T values for all irradiated tissues:

$$E(\tau) = \sum_T w_T \, H_T(\tau) \tag{5.4}$$

Unless otherwise specified, the recommended integration time after intake is 50 years for occupationally-exposed persons and 70 years for members of the public. Methods for handling radionuclides with specific ranges of half-lives are provided in NCRP Report 116[66].

Annual limits on intake (ALI) are quantities (expressed in activity units of becquerels (Bq)) of radionuclides present continuously in the body that will yield a committed effective dose of 20 mSv per year. The NCRP refers to these values as annual reference levels of intake (ARLI). The derived air concentration (DAC) [expressed as the derived reference air concentration (DRAC) by the NCRP] is the concentration of a radionuclide in air which, if breathed by a standard man inspiring 0.02 m^3 per minute for a working year, would yield an intake of one ALI (or ARLI by the NCRP). The DRAC is computed as:

$$\text{DRAC} = (\text{ARLI}) \div [50 \text{ wk/yr} \times 40 \text{ hr/wk} \times 60 \text{ min/hr} \times 0.02 \text{ m}^3/\text{min}] \tag{5.5}$$

The DRAC provides an approach to controlling airborne exposures in the workplace. They are intended for use with occupationally-exposed individuals and not for members of the public.

5.14.2 Occupational Dose Limits

Certain assumptions are necessary in establishing an effective dose limit for exposure of occupationally-exposed individuals to ionizing radiation. These assumptions are that, for a single uniform whole-body equivalent dose of 0.1 Sv, the nominal lifetime excess risk is:

4.0×10^{-3} for fatal cancer;
0.8×10^{-3} for severe genetic effects
0.8×10^{-3} for nonfatal cancer

for a total radiation-induced detriment of 5.6×10^{-3}. It also is assumed that cancer deaths on the average result in 15 years of life lost. Data[45] of monitored radiation workers reveal that the average annual dose equivalent was about 2.1 mSv in 1980. The total radiation-induced detriment in these workers is

$$[(5.6.x\ 10^{-3} / 0.1Sv) \times (2.1 \times 10^{-3}Sv / y)] \sim 1 \times 10^{-4}\ y^{-1} \qquad (5.6)$$

This total detriment is consistent with the average risk of accidental death in "safe" industries, as mentioned earlier.

In its 1993 report of occupational radiation limits[66], the NCRP recommends that

"... the numerical value of the individual worker's lifetime effective dose in tens of mSv be limited to the value of his or her age in years (not including medical and natural background exposure)"

and

"the annual occupational effective dose be limited to 50 mSv (not including medical and natural background exposure)."

With these two criteria applied to worst-case circumstances, the lifetime fatal cancer risk would be approximately 3×10^{-2}, comparable to the worst-case situation for accidental death in "safe" industries

$$[(5 \times 10^{-4}\ y^{-1}) \times (50\ y)] = 2.5 \times 10^{-2} \qquad (Eq.\ 5.7).$$

Recommendations of the ICRP differ somewhat from those of the NCRP. The ICRP[65] suggests a limit of 100 mSv over 5 years and no more than 50 mSv in any single year. Limits of the ICRP result in a lifetime dose limit of about 1 Sv, in contrast with a limit of approximately 0.7 Sv for the NCRP. Details of the NCRP limits are somewhat more flexible than those of the ICRP, but require maintenance of cumulative exposure records over lifetimes rather than the 5-year periods needed to meet ICRP recommendations.

A third recommendation of the NCRP has generated considerable controversy within radiation protection circles. This recommendation, offered in the spirit of ALARA, is

"... all new facilities and the introduction of all new practices should be designed to limit annual exposures to individuals to a fraction of the 10 mSv per year limit implied by the cumulative dose limit."

This recommendation is controversial because it undermines the flexibility provided by the preceding dose limits of the NCRP.

Finally, all effective dose limits apply to the sum of the effective dose from external irradiation and the committed effective dose from internal exposures.

Since ALARA is intrinsic to a radiation protection program, annual dose limits by themselves are not considered by the NCRP to be wholly sufficient for radiation design and control purposes. Instead, reference dose levels should be established that are achievable within institutional practice and which, if followed, will yield annual exposures well below annual dose limits. If exposures exceed reference levels, remedial action should be taken. Reference levels are normally site-specific depending on the types and usage patterns of sources at the site.

It is conceivable that in some circumstances relative high doses (e.g., several Sv/y) could be delivered to specific regions without exceeding the effective dose limits described above. To protect against deterministic effects occurring as a result of such doses, the NCRP has recommended dose limits for certain regions. The limits are (1) 150 mSv for the crystalline lens of the eye, and (2) 500 mSv for localized areas of the skin, hands and feet.

Among atomic bomb survivors exposed *in utero*, an increase in severe mental retardation occurred in the group exposed 8–15 weeks after conception, and to a reduced extent in the group exposed 16–25 weeks after conception. No increase in mental retardation was noted in groups younger or older in gestational age. The data suggest a linear relationship (slope ~0.4 Gy^{-1}) between dose and effect above a threshold of 0.1–0.2 Gy. The threshold is raised to ~4 Gy if two children with Down's syndrome (a genetic condition that is clearly not a developmentally-induced injury) are excluded. Also, some epidemiological evidence suggests an association between medical X-ray exposure and an increase in childhood cancer, although other evidence refutes this contention.

The sensitivity of the embryo-fetus should always be considered in exposures of pregnant or possibly- pregnant women. The NCRP recommends that

"... a monthly equivalent dose limit of 0.5 mSv to the embryo-fetus (excluding medical and natural background radiation) once the pregnancy is known."

The NCRP does not recommend special limits or restrictions for occupationally-exposed women who are not known to be pregnant. Internally-deposited radionuclides pose special problems with regard to protection of the embryo-fetus. The NCRP (and the ICRP) recommend that, once pregnancy is known, the intake of radionuclides be limited to 1/20 of the values of the ARLI for radiation workers.

Annual dose limits are intended to control the maximum lifetime radiation-induced risk, and average exposures of individuals should remain well below the limit. However, exposures slightly in excess of annual dose limits have little biological risk, especially since the lifetime risk is influenced by past lower exposures or reduced exposures in future years. The principal importance of excessive exposures is to identify and eliminate their causes within the workplace. For persons whose cumulative effective dose exceeds the age-related limit, the NCRP recommends that they

"... should be restricted in their exposures to no more than 10 mSv per year until the age-related lifetime limit is met."

If more flexibility is warranted, particularly for older workers whose effective dose may approach or exceed 10 mSv × age, then, on a case by case basis, with dialog between the employee and employer, an exposure limit of up to an average of 100 mSv in 5 years and 50 mSv in any year may be considered.

Unusual occupational situations could conceivably exist in which the worker population cannot fulfill required functions under recommended annual limits and in which risks other than radiation may be much greater than normal. Long-duration space missions and some mining activities might be examples of such situations. For these special circumstances, the NCRP recommends that

"... consideration be given to establishing special dose limits for those selected occupational groups requiring higher exposures to accomplish needed activities."

Except in life-saving situations, individuals should never receive doses significantly above the annual effective dose limit. When life-saving emergencies occur, older workers with low lifetime accumulated doses should be chosen from individuals volunteering for action. If emergency conditions arise that do not involve saving lives, and where exposures cannot be kept within occupational dose limits, a limit of 0.5 Sv effective dose and an equivalent dose of 5 Sv to the skin should be adopted[65].

5.14.3 Dose Limits for Members of the Public

As guidelines for controlling exposure of members of the public to radiation, dose limits are defined as 1/10 of those applied to radiation workers. These limits are

"For continuous (or frequent) exposure, it is recommended that the annual effective dose not exceed 1 mSv. Furthermore, a maximum annual effective dose limit of 5 mSv is recommended to provide for infrequent annual exposures[66].

These recommendations apply to all radiation sources except natural background and medical radiation exposure.

For education and training purposes, it may infrequently be necessary (and desirable) to accept occasional radiation exposure of persons under 18 years of age. These exposures should be permitted only when assurance is high that the resulting annual effective dose can be restricted to no more than 1 mSv, and that the dose equivalent can be maintained below 15 mSv to the lens and 50 mSv to the hands, feet and skin. Intentional exposure of trainees (e.g., practicing radiographic exposures on each other) should be avoided.

5.14.4 Negligible Individual Risk Level

The Negligible Individual Risk Level (NIRL) was introduced by the NCRP[46] in 1987 as the level of average annual excess risk of fatal health effects attributable to radiation below which efforts to reduce radiation exposure to the individual is unwarranted. Several factors influence the identification of negligible risk, including:

- magnitude of dose
- difficulty in detecting and measuring dose and resultant health effects
- natural risk for the same health effects
- estimated risk of natural background radiation levels
- risks to which people are accustomed
- perception of, and behavioral response to, risk levels

Based on these criteria, a NIRL of 10^{-7}/y was adopted, corresponding to an annual effective dose equivalent of 0.01 mSv. In 1993 the NCRP reiterated that[66]

"... an annual effective dose of 0.01 mSv be considered a Negligible Individual Dose (NID) per source or practice."

This recommendation has not been accepted by regulatory agencies such as the Nuclear Regulatory Commission.

5.15 No-Threshold Model: An Analysis

The no-threshold model of radiation injury has had major consequences for human health and public policy in the United States and around the world. If the model is correct in portraying radiation risk at low levels of exposure, then its use in setting limits for radiation exposure of occupationally-exposed persons and members of the public would certainly be justified. Furthermore, the model would support implementation of the ALARA concept of radiation exposure limitation, wherein exposures are reduced to the lowest possible levels consistent with economic and societal considerations. On the other hand, if the no-threshold model is overly conservative, and if risks are less (or non-existent) at low levels of exposure, then the costs of radiation protection may be higher than necessary, and the intrusiveness of regulations into the beneficial applications of radiation may be excessive. Furthermore, public policy decisions that cost large sums of taxpayer money, or that deprive the public of beneficial applications of radiation, would be unjustified. For example, the U.S. Department of Energy (DOE) has embarked on a program of radiation cleanup at DOE facilities that is estimated to cost as much as 200 billion dollars. Whether this immense expenditure will reduce the risk of the public to any measurable degree is highly debatable. As another example, no new repository for low-level radioactive waste has been sited anywhere in the country, and the one remaining site that has been recently reopened (to all states except North Carolina) is subject to closure at any time. This problem jeopardizes the continued availability of medical procedures using radioactive sources at reasonable cost. Both of these issues exist because of the assumption that there is no threshold below which adverse effects of radiation exposure do not occur.

The dilemma for public policy is that existing data are simply inadequate to establish the validity (or lack thereof) of the no-threshold model of radiation-induced injury [87,88]. Furthermore, no reasonable experiment or epidemiological study of human exposures can be conceived that would answer the question. At low exposure levels, the number of persons required for both the exposed and control populations are simply too great to be attainable under any practical circumstances. What can be said is that the no-threshold model defines an upper limit for the response of humans to radiation exposure. What cannot be said is whether the actual response might be significantly lower than this upper limit – even zero, or perhaps less than zero if beneficial effects occur at low levels of exposure (radiation hormesis).

Replacement of the concept of tolerance dose by the no-threshold model of radiation response, as described earlier in this chapter, reflects several influences that worked together after World War II to enhance the perceived legitimacy of the no-threshold model. First, certain approaches to biological modeling of radiation injury at the cellular level support the no-threshold theory. These approaches evolved from early studies of radiation-induced mutations in fruit flies in which Müller and others were unable to demonstrate an exposure threshold below which no effects were observed. The cellular model of radiation injury that grew from this work assumed that damage was caused by deposition of small, discrete amounts of energy ("hits") in sensitive "targets" within the cell. This hypothetical approach, known as the target theory of radiation injury[41], assumed that the hits occur randomly in the cell so that even the smallest dose would have a probability greater than zero of hitting a target and producing harm. The target theory suggests that even very low exposures might be harmful, and supports the no-threshold model and the restriction of radiation exposures to the lowest practical level.

In the 1950s, it became apparent that mutations in somatic (body) cells may be responsible as the first step in the development of cancer in exposed individuals. As studies began to show an increased incidence of leukemia and, later, other forms of cancer in Japanese survivors, the target theory gained prominence as a model of cellular radiation injury that leads ultimately to cancer. Over the years, studies of Japanese survivors and other exposed populations have failed to reveal an increased incidence of genetic abnormalities in the offspring of members of the exposed population. At the same time, an increased incidence of a variety of cancers have been detected with varying levels of statistical significance. Today, the induction of cancer, rather than genetic effects, is widely recognized as the major risk of radiation exposure. However, it is important to emphasize that the populations that yield this increased incidence of cancer have all been exposed to relatively large amounts of radiation. For example, individuals in the Japanese populations studied by the RERF all received doses of 0.5 Sv or more.

Sagan[89] has suggested two additional reasons why the no-threshold model of radiation response became the dominant explanation of radiation injury after WWII. One of these reasons comes from engineering. When engineers design a structure (e.g., a bridge or building) for safety, they construct a plan that accommodates maximum loads and stresses, and then add a relatively large safety factor (often a factor of two or more) for good measure. This approach results in essentially a zero probability of failure of the structure if the design is followed and materials of high physical integrity are used. But in designing a nuclear facility (e.g., a nuclear reactor), nuclear engineers reject the notions of absolute safety and risk thresholds. Instead, they assume that the risk of failure is always finite. Hence, backup safety systems are required to reduce the likelihood and consequences of failure to an acceptable minimum which may approach but never reach zero. This no-threshold model of failure has served the public well in several instances, the most notable of which was the Three Mile Island incident. However, it also sends the message that failure can always occur, and that radiation and radiation-producing facilities are never without some level of risk, no matter how many safety features are built into their design.

Sagan also believes that the "new environmentalism" movement introduced in the late 1950s has had a major influence on the emergence of the no-threshold model of injury for exposure to a wide spectrum of toxic agents. A turning point in this movement was the book Silent Spring by Rachel Carson[90]. Carson not only depicted the ecological consequences of environmental exploitation, but also suggested that trace quantities of chemicals in the environment can cause cancer in humans. At about the same time, there was intense public debate about the possible health effects of radioactive fallout from atmospheric tests of nuclear weapons. Opponents of weapons testing emphasized the possible health effects of fallout at very low concentrations, even though no data existed to substantiate these effects. Although more political than scientific, these statements largely went unchallenged by the scientific community, and added to the public's concern that exposure to even the smallest amounts of toxins such as ionizing radiation could lead to significant health effects.

5.15.1 No-Threshold Model as an Operational Paradigm

Several reasons exist to explain why the no-threshold model has become the operational paradigm for estimating the risk of exposure to low levels of radiation (as well as to other potentially toxic agents). These reasons, all rather hypothetical, have foundations in science (target theory and susceptible biological molecules), engineering (finite risks of disastrous consequences), and social policy (elimination of environmental contaminants by governmental action). There may be other reasons as well, some of which may reflect innate biases, preferred judgments, and social values of the principals who endorse the use of the no-threshold model.

It is widely accepted in public policy circles that protection standards to control risks to individuals and populations should always err on the conservative side. That is, if the magnitude of risk is uncertain, then the risk should be estimated so that if the estimate is wrong, it overestimates the risk. In this manner, protection standards and procedures can be established that reflect a cautious, perhaps even over-cautious, approach to protecting people against toxic agents.

However, it is one thing to use a conservative model to estimate the magnitude of risk at low levels of exposure. It is something else if the risk estimate is then used to compute the number of persons affected in a population exposed to low levels of a toxic agent. Finally, it is indefensible to assign significance to this number as if the effects actually exist. For example, suppose that an upper limit of lifetime risk of cancer of 1/10,000 per unit exposure is estimated by linear extrapolation for a toxic agent (e.g., ionizing radiation) thought to induce cancer at high levels of exposure. Next, suppose that this upper limit is established as a standard for maximum exposure of the public to the agent. Then assume that each person in a population of 250 million (e.g, the US population) receives on the average one unit of exposure over a lifetime. By multiplying the risk times the population size, the number 25,000 is computed. Not infrequently, this approach is used, wholly inappropriately, to determine the number of hypothetical "cancers" induced in a population as a result of exposure to a toxic agent. Almost as frequently, this number is reported in the public media without inclusion of "hypothetical" as a qualifying adjective. This attempt to estimate numbers of persons affected is disingenuous and a perversion of the efforts of scientists to establish an upper limit of exposure for purposes of risk estimation and standards setting.

Why, until recently, have so few scientific challenges been directed against acceptance of the no-threshold model of injury following low-level exposure to toxic agents, including radiation? One reason may be that the no-threshold model suits the purposes of influential people. For example, the model appeals to the paranoia of many who fear the "dark side" of technologies that they do not fully understand. The professional community of protection experts benefits from this fear because it enhances job opportunities and security. Researchers benefit from the continued flow of funds to identify and quantify risks associated with exposure to toxic agents. Attorneys benefit from increased litigation resulting from the public's conviction that low levels of exposure cause cancers and other forms of illness and abnormality. Regulators have their budgets and authorities enhanced through legislative actions taken in response to the public's fear of exposure.

Whether the public truly benefits from these responses to the fear of low-level exposures is questionable, especially since whether an actual risk exists cannot be demonstrated. The responses are developed at significant public expense and at the further cost of reinforcing the public's fear of the unknown. Money spent to address these suspected but unproven risks cannot be used to prevent or correct problems such as industrial and domestic accidents, personal violence, tobacco use, alcohol abuse, and other known threats to human health. Although there is no good way to quantify the consequences of these decisions, their social implications may be enormous.

5.15.2 Values in Science

There is considerable uncertainty about the relevance of the no-threshold model to estimate the risk of low-level exposure to toxic agents such as ionizing radiation. Yet this model has become widely accepted as the preferred paradigm for risk estimation at low exposures. Some of the reasons for acceptance of this model have been described above. But there is yet another reason which may override all the others in explaining why this model has been adopted with such universality. The reason is that the model is consistent with the expectations of society that scientists, and the designers of public policies and governmental regulations, should provide a "safe" margin of error in situations where risk cannot be quantified. These expectations coincide with the values of the scientific community that is looked to for leadership in cases where risk is uncertain.

The concept of values in the formulation of scientific guidance may seem paradoxical. For many (perhaps most) persons, science is assumed to function as a value-free enterprise, and scientists are perceived as searching for "truth" in a value-free universe. As Sagan[89] suggests, scientists are perceived as being trained to

"... observe the world dispassionately and collect data in a scrupulously objective fashion, which they then dutifully report in peer-reviewed journals. These reports then become the substance of an ever-expanding knowledge."

But in fact, scientists are subject, just as everyone else, to value judgments in everything that they do, including constructing hypotheses for scientific experiments, selecting approaches and materials for testing hypotheses, acquiring data from experiments, and interpreting the data according to one or more prevailing paradigms that are intrinsic to the hypotheses. These paradigms influence the way scientists think about their work and the results that it produces. It is probably true that science is more objective than some other courses of study that are not so verifiable through experimental protocols and rigid methods of statistical analysis. But it is certainly not value-free.

In his book **The Structure of Scientific Revolutions**, Kuhn[91] suggests that scientific thinking is guided by the use of accepted models of truth that he describes as scientific paradigms. These models are developed to explain natural phenomena, and they gradually become so ingrained into scientific thinking that they achieve a state of unquestioned acceptance. In this condition, the underlying assumptions and universal applicability of the paradigms are seldom challenged. When persons raise questions about the paradigms, they are likely to be labeled as "quacks" or "pseudoscientists," and their opinions are usually discarded as outlandish and foolish by other scientists, journal editors and committees of peers empowered to recommend research funding. Almost always, experimental evidence that is not consistent with a scientific paradigm is ignored because it is considered outside the bounds of conventional knowledge. Such evidence may be censored as heretical, at least until it becomes so overwhelming that a new model that encompasses it arises as a replacement for prior "acceptable thinking." The acceptance of this new model represents, in Kuhn's terms, a "paradigm shift" in scientific thinking.

Kuhn emphasizes that scientific thinking often progresses not so much through gradual unidirectional evolution, but instead by infrequent but dramatic paradigm shifts that most often are driven by persons outside of the "scientific establishment." Major paradigm shifts have occurred throughout the history of science. Examples include the transition from the earth to the sun as the acknowledged center of the solar system, the replacement of Newtonian mechanics by quantum physics as the preferred model to explain atomic phenomena, and natural selection as a paradigm of biological evolution. But not all paradigm shifts withstand scientific scrutiny over the course of time. An example of a paradigm shift that ultimately proved fraudulent is Lysenkoism. This theory, widely accepted in the Soviet Union, stated that inherited characteristics of individuals are not mediated by physical properties of chromosomes, and that acquired behaviors of individuals can be passed on to future generations. In the field of radiation biology and protection, the transition after World War II from the concept of tolerance dose to the no-threshold model of radiation injury represents a paradigm shift in scientific thinking. Whether the new model of radiation injury will hold up over time remains to be seen. It currently is receiving intense scrutiny.

The scientific understanding of the causes of cancer has advanced well beyond the simplistic model of cellular changes resulting exclusively from single exposure to a mutagenic agent[92]. Ames, once an outspoken critic of environmental pollution, has challenged the prevailing notion that industrial contaminants in the environment are a major source of mutations leading to increased cancers in exposed populations[93]. The challenge is not whether some of the contaminants are theoretically capable of producing somatic mutations leading to cancer. Instead, the issue is that the concentrations of the contaminants are very low compared with mutagenic agents that occur naturally in the environment. For example, many plants contain naturally-occurring pesticides that are at least as mutagenic as industrial contaminants. Ames estimates that the average person consumes about 1500 milligrams of natural pesticides daily. This amount overwhelms the small amount (<0.1 milligram) of synthetic pesticides that finds its way into the human diet. Many seasonings and spices (e.g., pepper, curry and cinnamon) are mutagenic, and cooking, including frying and baking, adds to the concentration of dietary mutagens. Charred meats and toasted bread yield plentiful concentrations of mutagenic agents, as do coffee, tea and many soft drinks.

Today it is accepted that cells in the body are exposed each day to thousands of potentially damaging events. These events do not yield disastrous consequences because the body has elegant processes to repair the damage they may cause. It also is understood that the effectiveness of repair mechanisms declines with age, with the decrease occurring more rapidly in some individuals than in others. In all persons it is probably true that the integrity of repair processes, rather than the exposure to mutagenic agents, is the more significant influence on the susceptibility of individuals to cancer and many other diseases.

For the no-threshold paradigm of radiation injury, several observations justify its reexamination as the preferred model for estimating health risks of exposure to low levels of radiation. Among these observations are the following:

1. There are few data available concerning actual effects in human or animal populations following exposure to low levels of radiation, and the data that do exist are contradictory, with some suggesting an adverse consequence and some yielding a beneficial effect (radiation hormesis).
2. The concept of hormesis at low exposures is not unique to radiation. Many agents that are toxic at high concentrations are known, or at least thought, to be beneficial at low concentrations. Examples are aspirin, wine, and even water, as well as a variety of other chemicals[94].
3. Each person is exposed daily to amounts of naturally-occurring mutagenic agents that overwhelm the small amounts introduced by synthetic chemical agents and by radiation in exposed individuals.
4. The current understanding of the action of mutagenic agents suggests that past models of cancer induction are overly simplistic, and that repair mechanisms at the cellular level play a dominant role in determining the susceptibility of individuals to cancer, including that possibly induced by radiation.
5. The no-threshold model of radiation injury has been widely adopted by the press, and politicians, and has led to a number of social policies and governmental regulations to protect radiation workers and the public from even minute amounts of radiation exposure.
6. These policies and regulations have increased the cost and decreased the availability of beneficial uses of radiation in a wide spectrum of applications, including medicine and electricity generation.

These observations suggest that the disadvantages of the no-threshold model of radiation injury may outweigh the advantages by a substantial margin. They also suggest that reexamination of the merits of this model may be appropriate. However, reexamination will not be easy. As Barker has emphasized,

"New paradigms put everyone practicing the old paradigm at great risk. And, the higher one's position, the greater the risk. The better you are at your paradigm, the more you have invested in it. To change your mind is to lose that investment."[95]

5.16 Acknowledgements

The work of two individuals have contributed substantially to the ideas presented in this article. Those individuals are Ronald Kathren of Washington State University at Tri-Cities in Richland, Washington, and Leonard Sagan, deceased.

5.17 References

1. National Academy of Sciences/National Research Council (NAS/NRC), *Health Effects of Exposure to Low Levels of Ionizing Radiation: BEIR V.* (National Academy Press, Washington, DC, 1990).
2. T.D. Luckey, *Radiation Hormesis.* (CRC Press, Boca Raton, Florida, 1991).
3. W. Röntgen,*Uber eine neue Art von Strahlen* (vorlaufige Mitteilung). (Sitzungs Berichte der Physikalisch-Medicinischen Gesellschaft zu Würzburg 9, 1895).
4. H. Becquerel,*Compt. Rend.* **122**, 240 (1896).
5. M. Curie,*Traité de Radioactivité.* (Gauthier-Villars, Paris, 1910).
6. P. Villard,*Compt Rend* **130**, 1010 (1900).
7. T.A. Edison, *Nature*, London **53**: 421 (1896).
8. W.J. Morton, *Nature*, London **53**: 421 (1896).
9. W. Marcuse, *Deutsche Med.Wochschr.* **21**, 681 (1896).
10. A. Frieben, *Fortschritte Gebiete Röntgenstrahlen* **6**, 106–111 (1902).
11. P. Brown, *American Martyrs to Science through the Roentgen Ray.* (Charles C. Thomas, Springfield, 1936) pp. 32–42.
12. O. Hesse, Symptomologie, Pathogenese and Therapie des Röntgenkarzinoms. (J.A. Barth, Leipzig, 1911).
13. H.A. Colwell, S. Russ, *X-ray and Radium Injuries.* (B. Gell and Sons, London, 1915).
14. M. Uhlig,Virchosw. Arch. Pathol. Anat. Physiol. **230**, 76–98 (1921).
15. J.N. Stannard, *Radioactivity and Health: A History.* (Pacific Northwest Laboratory, US Department of Energy, 1988).
16. J. Belot, Bull. Mem. Soc. Radiol. Med. Paris. **2**, 34–41 (1909).
17. S.R. Wolbach, *J. Med. Res.* **21**, 415–449 (1909).
18. P.S. Henshaw, *J. Nat. Cancer Inst.* **1**, 789–805 (1941).
19. A. Mutscheller, *Am. J. Roentgenol.* **13**, 65–69 (1925).
20. R.L. Kathren, P.L. Ziemer, in *Health Physics: A Backward Glance*, R.L. Kathren and P.L Ziemer, Eds. (Pergamon Press, New York, 1980) pp. 1–9.
21. L.S. Taylor, in *Health Physics: A Backward Glance*, R.L. Kathren and P.L. Ziemer, Eds. (Pergamon Press, New York, 1980) pp. 109–122.
22. A.E. Barclay, S. Cox, *Amer. J. Roengenol. Radium Ther.* **19**, 551–558 (1928).
23. L.S. Taylor, *Organization for Radiation Protection: The Operations of the ICRP and NCRP 1928–1974.* (National Technical Information Service, Springfield, 1979), U.S. Department of Energy Report DOE/TIC-10124.
24. S.C. Bushong, The Development of Radiation Protection in Diagnostic Radiology. (CRC Press, Cleveland, 1973).
25. H. Wintz, W. Rump, Protective Measures Against Dangerous Results from the Use of Radium, Roentgen and Ultra-Violet Rays. Official Publication C.H 1054. (League of Nations, Geneva, 1931).
26. S.T. Cantril, in *Industrial Medicine on the Plutonium Project.* (McGraw-Hill, New York, 1951) pp. 36–74.

27. J.J. Nickson, in *Industrial Medicine on the Plutonium Project*. (McGraw-Hill, New York, 1951) pp. 75–112.
28. E. Lorenz, in *Biological Effects of External X and Gamma Radiation*. (McGraw Hill, New York, 1954), pp. 24–148.
29. E. Lorenz, J.W. Hollcroft, E. Miller, C.C. Congdon, R. Schweisthal, *J. Nat. Canc. Inst.* **15**: 1049–1058 (1955).
30. L.O. Jacobson, E.K. Marks, in *Industrial Medicine on the Plutonium Project*. (McGraw-Hill, New York, 1951) pp. 113–139.
31. L.O. Jacobson, E.K. Marks, E. Lorenz, in *Industrial Medicine on the Plutonium Project*. (McGraw-Hill, New York, 1951), pp. 140-196.
32. J.C. Aub, R.D. Evans, L.H. Hemplemann, H.S. Martland, *Medicine* **31**, 221–329, 1952.
33. H.S. Martland,*Collection of Reprints on Radium Poisoning*. (U.S. Atomic Energy Commission Technical Information Service, Oak Ridge, 1951).
34. R.D. Evans, A.T. Keene, M.M. Shanahan, in *Radiobiology of Plutonium*, B.J. Stover and W.S.S. Jee, Eds., (J.W. Press, Salt Lake City, 1972), pp. 431–468.
35. A.J. Finkel, C.E. Miller, R.J. Hasterlik, in *Delayed Effects of Bone-seeking Radionuclides*, C.W. Mays, W.S.S. Jee, R.D. Lloyd, B.J. Stover, J.H Dougherty and G.N. Taylor, Eds. (University of Utah Press, Salt Lake City, 1969) pp. 195–225.
36. H.J. Müller, *Science* **66**, 84–87 (1928).
37. A. Weinstein, *Science* **67**, 376–377 (1928).
38. L.J. Stadler, *Proc. Nat. Acad. Sci.* **14**, 69–75 (1928).
39. J.T. Patterson, *Science* **68**: 41–42 (1928).
40. N.W. Timofeev-Ressovsky, *Amer. Naturalist* **63**, 118–122 (1929).
41. D.A. Lea, *Actions of Radiation on Living Cells*. (Macmillan, New York, 1947) p. 132.
42. C.P. Oliver, *Zeitschr. Indukt. Abstammungsl.* **61**, 447 (1932).
43. N.W. Timofeev-Ressovsky, R.G. Zimmer, M. Delbruck, *Nach. Gesellschaft Wissenschaften* (Gottingen) (1935), pp. 189–245.
44. D.E. Uphoff, C. Stern, *Science* **100**, 609–611 (1949).
45. National Council on Radiation Protection and Measurements (NCRP), *Influence on Dose and its Distribution in Time on Dose-Response Relationship for Low-Let Radiations*. (National Council on Radiation Protection and Measurements, Bethesda, MD, 1980), NCRP Report No. 64.
46. National Council on Radiation Protection and Measurements (NCRP), *Recommendations on Limits for Exposure to Ionizing Radiation*. (National Council on Radiation Protection and Measurements, Bethesda, MD, 1987), NCRP Report No. 93.
47. National Council on Radiation Protection and Measurements (NCRP),*The Relative Biological Effects of Radiations of Different Quality*. (National Council on Radiation Protection and Measurements, Bethesda, MD, 1990), NCRP Report No. 104.
48. United Nations Scientific Committee on the Effects of Atomic Radiation (UNSCEAR). *Sources, Effects and Risks of Ionizing Radiation*. (United Nations, New York, 1988), Sales No. E.88 IX.7, ISBN 92-1-142143–8.
49. United Nations Scientific Committee on the Effects of Atomic Radiation (UNSCEAR). *Sources and Effects of Ionizing Radiation*. 1994 Report to the General Assembly with Scientific Annexes. (United Nations, New York, 1994).
50. J.V. Neel, W.J. Shull, A.A. Awa, C. Satoh, H. Kato, M. Otake, Y. Yoshimoto, *Am. J. Hum. Genetics* **46**, 1053–1072 (1990).
51. H. Kato, Y. Shimizu, in *Health Effects of Atomic Radiation*. (Proceedings of Japan-USSR Symposium on Radiation Effects Research, Tokyo, June 25–29, 1990) pp. 225–236.
52. S. Kondo, *J. Radiat, Res.* **31**, 174–188 (1990).
53. National Council on Radiation Protection and Measurements (NCRP), *Permissible Dose from External Sources of Ionizing Radiation*. (US Department of Commerce, National Bureau of Standards Handbook 50, Washington, D.C., September 24, 1954), NCRP Report No. 17.
54. United Nations Scientific Committee on the Effects of Atomic Radiation (UNSCEAR), *Report of the United Nations Scientific Committee on the Effects of*

Atomic Radiation. (United Nations, New York, 1958), Thirteenth Session Supplement No. 17 (A/3838).
55. National Academy of Sciences/National Research Council (NAS/NRC) Committee on Pathologic Effects of Atomic Radiation, *A Commentary on the Report of the United Nations Scientific Committee on the Effects of Atomic Radiation.* (National Academy of Sciences/National Research Council, Washington, D.C., 1959), NAS/NRC Publication 647.
56. Federal Radiation Council (FRC), *Background Material for the Development of Radiation Protection Standards.* (Federal Radiation Council, Washington, DC, 1960), Report No. 1.
57. United Nations Scientific Committee on the Effects of Atomic Radiation (UNSCEAR), *Report of the United Nations Scientific Committee on the Effects of Atomic Radiation.* (United Nations, New York, 1962), Seventeenth Session Supplement No. 16 (A/5216).
58. United Nations Scientific Committee on the Effects of Atomic Radiation (UNSCEAR), *Report of the United Nations Scientific Committee on the Effects of Atomic Radiation.* (United Nations, New York, 1964), Nineteenth Session Supplement No. 14 (A/5814).
59. International Commission on Radiological Protection (ICRP),*The Evaluation of Risks from Radiation* . (Pergamon Press, Oxford, 1966), ICRP Publication 8.
60. Joint Committee on Atomic Energy (JCAE) of the Congress of the United States, *Hearings on the Nature of Radioactive Fallout and Its Effects on Man.* (US Government Printing Office, Washington, DC, 1957), May 27–29 and June 3–7 (2 Vols.).
61. Joint Committee on Atomic Energy (JCAE) of the Congress of the United States, *Hearings on Fallout from Nuclear Weapons Tests.* (US Government Printing Office, Washington, DC, 1959), May 5–8, (3 volumes and summary).
62. Joint Committee on Atomic Energy (JCAE) of the Congress of the United States, *Selected Materials on Radiation Protection Criteria and Standards: Their Basis and Use.* (US Government Printing Office, Washington, D.C., May 1960).
63. National Academy of Sciences/National Research Council (NAS/NRC),*The Effects on Populations of Exposure to Low Levels of Ionizing Radiation.* (National Academy Press, Washington, D.C., 1972).
64. International Commission on Radiological Protection (ICRP), *Recommendations of the International Commission on Radiological Protection.* Ann. ICRP **1** (1977); also (Pergamon Press, Oxford, 1977), ICRP Publication 26.
65. International Commission on Radiological Protection (ICRP), *Recommendations of the International Commission on Radiological Protection.* Ann. ICRP **21** (1991); also (Pergamon Press, Oxford, 1991), ICRP Publication 60.
66. National Council on Radiation Protection and Measurement (NCRP), Limitations for Exposure to Ionizing Radiation. (National Council on Radiation Protection and Measurements, Bethesda, MD, 1993), NCRP Report No. 116.
67. United States Environmental Protection Agency (USEPA), *Federal Register* **52**, 2822–2834 (1987).
68. United States Department of Energy (USDOE), *Radiation Protection of Occupational Workers.* (US Department of Energy, Washington, D.C., December 21, 1988), Order 5480.11.
69. United States Nuclear Regulatory Commission, *Federal Register* **56**, 23360-23474 (May 21, 1991).
70. National Academy of Sciences/National Research Council (NAS/NRC),*The Effects on Populations of Exposure to Low Levels of Ionizing Radiation: BEIR III.* (National Academy Press, Washington, D.C., 1980).
71. E.J. Sternglass, *Health Phys.* **15**, 202 (1968).

72. National Academy of Sciences/National Research Council (NAS/NRC), Committee on the Biological Effects of Ionizing Radiation (BEIR-IV), *Health Effects of Radon and Other Internally Deposited Alpha-Emitters.* (National Academy Press, Washington, D.C., 1988).
73. L.S. Taylor, *Health Physics* **41**, 227–232 (1981).
74. British X-ray and Radium Protection Committee, *Journal of the Roentgen Society* **17**, 100 (1921).
75. L.S. Taylor, Organization for Radiation Protection: The Operations of the ICRP and NCRP 1928–1974. Assistant Secretary for Environment, Office of Health and Environmental Research and Office of Technical Information, U.S. Department of Energy, 1979.
76. W.R. Hendee, in *Health Effects of Exposure to Low Level Ionizing Radiation* (2nd edition) W.R. Hendee and F.M. Edwards, Eds. (Institute of Physics, London, 1996) pp. 495–525.
77. W.R. Hendee, *Medical Physics* **20**, 1303–1314 (1993).
78. G.T. Mazuzan, J.S. Walker, *Controlling the Atom: The Beginnings of Nuclear Regulation 1946–1962.* (University of Califomia Press, Berkeley, CA, 1994).
79. Institute of Medicine (IOM), *Radiation in Medicine: A need for Regulatory Reform.* (Washington, D.C.: National Academy Press, 1966).
80. General Accounting Office (GAO), Problems of the Atomic Energy Commission Associated with the Regulation of Users of Radioactive Materials for Industrial, Commercial, Medical and Related Purposes. (General Accounting Office, Washington, D.C., August 18, 1972), B-164105.
81. United States Nuclear Regulatory Commision (USNRC), *Federal Register* **44**, 8242 (1979).
82. United States Nuclear Regulatory Commission (USNRC), Case Study Report on the Therapy Misadministrations Reported to the NRC Pursuant to 10 CFR 35.42. (USNRC, Rockville, MD, 1985).
83. United States Nuclear Regulatory Commission (USNRC), in *Federal Register*, Vol 56. (National Archives and Records Administration, Washington, D.C., 1992), p 34104.
84. International Commission on Radiological Protection (ICRP), *Annals of the ICRP* **15**, 3, (Pergamon Press, Elmsford, New York, 1985), ICRP Publication 45.
85. National Safety Council (NSC), *Accident Facts, 1976 edition.* (National Safety Council, Chicago, 1977).
86. National Safety Council (NSC), *Accident Facts, 1992 ed*ition (National Safety Council, Chicago, 1992).
87. C.E. Land, in *Advances in Radiation Biology. Effects of Low Dose and Dose Rate Reduction,* O.F. Nygard and J.F. Lett, Eds., Vol 16. (Academic Press, San Diego, 1993), pp. 259–272.
88. E.S. Gilbert, Occupational Medicine: State of the Art Reviews **6**, 665–680 (1991).
89. L. Sagan, *BELLE Newsletter*, **2**, 1–7, (1993).
90. R.L. Carson, *Silent Spring*. (Houghton Mifflin, Boston, MA, 1962).
91. T. Kuhn, *The Structure of Scientific Revolutions.*, 2nd edition. (University of Chicago Press, Chicago, 1970).
92. W.R. Hendee, in *Health Effects of Exposure to Low-Level Ionizing Radiation*, W.R. Hendee and F.M. Edwards, Eds, 2nd edition. (Institute of Physics, London, England, 1996), pp. 103–111.
93. B.N. Ames, L.S. Gold, *Proc. National Academy of Sciences* **87**, 7772 (1990).
94. E.J. Calabrese, M.E. McCarthy, E. Kenyon, *Health Phys.* **52**, 531, (1987).
95. J.A. Barker, *Discovering the Future: The Business of Paradigms*. (Illinois Press, St. Paul, IL, 1988).

6 Radiation Protection Design and Shielding of Diagnostic X-ray Facilities

Douglas J. Simpkin
St. Luke's Medical Center

Benjamin R. Archer
Baylor College of Medicine

Robert L. Dixon
Bowman Gray School of Medicine of Wake Forest University

Disclaimer: The data, parameter values and methodology given in this paper represent only the views and opinions of the authors and of the authors of the references cited. This work should not be construed as having either the implied or actual endorsement of the National Council of Radiation Protection and Measurements or the American Association of Physicists in Medicine.

6.1 Introduction

Report No. 49 of the National Council of Radiation Protection and Measurements[1] (hereafter referred to as NCRP49) has effectively remained the primary guide for diagnostic X-ray structural shielding design in the U.S. for more than a quarter of a century. The data and shielding methodology presented therein have served the radiology community extremely well. However, many aspects of the practice of diagnostic radiology have changed since the publication of the report. In 1989, Task Group No. 9 of the American Association of Physicists in Medicine (AAPM) Diagnostic Imaging Committee was formed to investigate the applicability of the methods and data in NCRP49 to the modern clinical setting. Several of the reports of this Task Group critically examine the conservatism built into the NCRP49 methodology. These provide new and more accurate estimates of the shielding parameters as well as innovative approaches for computing barrier requirements. At the time of this writing an NCRP committee is in the process of updating NCRP49.

This chapter will present new data for shielding diagnostic facilities, investigate the impact of recent regulatory changes, and introduce new, more accurate shielding methodologies. A list of the symbols used in this chapter is contained in Table 6.1.

Table 6.1. Symbols used in this chapter.

a_1	scatter fraction per cm^2 scaled by 10^6
B	broad-beam transmission
B_P	broad-beam transmission of primary beam
$B_{housing}$	transmission of leakage radiation through tube housing
B_{pre}	transmission of primary beam through image receptor and its supporting structures
B_s	broad-beam transmission of scattered X-ray beam
B_{sec}	broad-beam transmission of secondary radiation
D	radiation dose
D_L	radiation dose in an occupied area due to leakage radiation
$\dot{D}_L$	leakage dose rate at 1 m from source
$\dot{D}_0$	dose in air at 1 m per unit workload due to raw primary beam
D_P	radiation dose in an occupied area due to primary radiation
$D_P{}^1$	primary radiation dose at 1 m calculated for a workload distribution of total workload W_{norm}
D_S	radiation dose in an occupied area due to scatter radiation

D_{sec} radiation dose in an occupied area due to secondary radiations

$D_{sec}{}^{1}$ unshielded secondary radiation dose at 1 m calculated for a workload distribution of total workload W_{norm}

D_{tot} total radiation dose in an occupied area

d_F primary beam distance at which primary beam field size is F

d_L leakage distance (m) from X-ray tube to occupied area

d_P distance (m) travelled by primary beam from X-ray tube to occupied area

d_S scatter distance (m) from center of patient to occupied area

d_{sec} secondary distance (m) (assumed = $d_L = d_S$)

F primary beam field size at primary beam distance d_F

H_E effective dose equivalent

HVL half value layer for broad X-ray beam after having been hardened by significant amounts of overlying material

i subscript indicating a particular X-ray tube

I_{max} highest X-ray tube current that can be sustained at kVp_{max}

L maximum permitted leakage dose rate at 1 m when X-ray tube is operated at its leakage technique factors kVp_{max} and I_{max}

m type of barrier material

N_{pat} number of patient procedures performed in an X-ray facility per week

P design dose limit

T occupancy factor

θ scattering angle (measured from original primary beam direction)

U use factor

kVp X-ray tube operating potential

kVp_{max} maximum X-ray tube operating potential at which continuous operation is possible

W X-ray tube workload

W_{norm} total workload to which unshielded dose is calculated in Tables 6.10 and 6.11. For workload distributions following the AAPM TG-9 survey, W_{norm} is the total average workload per patient

X radiation exposure

x thickness of shielding barrier material

x_{acc} thickness of barrier material that decreases dose in occupied area to an acceptable level

x_{pre} thickness of barrier material that yields transmission B_{pre}

6.2 Basic Concepts

Operation of an X-ray tube in a diagnostic radiologic facility gives rise to three distinct forms of photon radiation, each of which requires analysis for shielding requirements. These are primary, scatter, and leakage radiation. Scatter and leakage are said to be secondary radiations.

Primary radiation is that generated by bremsstrahlung or fluorescent emission in the anode of the X-ray tube, emanating from the X-ray tube portal and directed toward the patient and the image receptor. The energies of the primary photons range from ~15 keV to an energy in keV equal to the maximal X-ray tube operating potential in kV. This peak potential or kVp is determined by the demands of radiographic contrast and image receptor speed. The dose rate from primary radiation is dependent on the operating potential, is proportional to the X-ray tube current, or mA, and inversely proportional to the square of the distance from the X-ray tube focal spot to the occupied area.

Scatter radiation is created by Compton and coherent interactions in objects struck by the primary photons, such as the patient and the patient-supporting radiographic or fluoroscopic table. Scatter radiation is thus an unavoidable consequence of the primary X-ray beam. The dose due to scatter is proportional to the number of primaryX rays striking the scattering medium, and inversely proportional to the square of the distance from the scattering medium to the occupied area. The direction of the scattered X rays is fairly random, with some preference to forward and backscattering. The distribution of photon energies due to scatter is assumed to match that of the primary beam.

Leakage is radiation generated at the X-ray tube anode, emitted not through the X-ray tube portal but in a random direction through the tube housing. This hardens the transmitted radiation to the point that only the most energetic photons are assumed to constitute leakage. The exposure rate due to leakage radiation is limited by regulation to not exceed 0.1 R h^{-1} at 1 m from the source when the X-ray tube is operated at its leakage technique factors. These are the maximum potential, kVp_{max} at which continuous tube operation is permitted, and the highest tube current, I_{max}, that can be sustained at this potential. The emission of leakage radiation is assumed to be isotropic.

The primary, scatter, and leakage radiation therefore differ in their intensity, point of origin, and photon energy distribution and thus transmission. The total dose to an occupied area, $D_{tot\ i}\ (x,m)$, due to radiation from the ith X-ray tube in a radiologic room, shielded by a barrier of thickness x of material m, is the sum of the primary, scatter, and leakage dose contributions from this tube. Let $D_{P\ i}\ (x,m)$ be the dose from primary radiation transmitted to the occupied area from the ith X-ray tube. The dose delivered to this shielded area from scatter radiation generated by this X-ray tube is $D_{S\ i}\ (x,m)$, while that from leakage is $D_{L\ i}\ (x,m)$. Then

$$D_{tot\,i}(x,m) = D_{P\,i}(x,m) + D_{S\,i}(x,m) + D_{L\,i}(x,m)\,. \tag{6.1}$$

From all X-ray tubes in a room, the total dose to the occupied area is

$$D_{tot}(x,m) = \sum_i D_{tot\,i}(x,m). \tag{6.2}$$

A radiation barrier of thickness x_{acc} is deemed acceptable when the total dose to the occupied area does not exceed the appropriate dose limit, P. New values for dose limits are discussed in Section 6.3. For cases where an individual only occupies the shielded area a fraction T of the time that the X-ray tube is activated, the transmitted dose may be allowed to exceed P by the factor T^{-1}. Therefore, an acceptable radiation shielding barrier thickness, x_{acc}, is defined by

$$D_{tot}(x_{acc}, m) = \frac{P}{T}\,. \tag{6.3}$$

In order to estimate *a priori* the dose in an occupied area due to primary, scatter, and leakage radiation from a radiologic installation, assumptions need to be made of the room's layout, the magnitude of the X-ray tube's use, and the efficiency with which a barrier attenuates the photons from that tube. Estimates of the distances from the X-ray tube and the patient to the occupied room environs should be accurate and conservatively short. Fig. 6.1 shows a hypothetical diagnostic X-ray imaging room, with the applicable distances defined. The magnitude of the X-ray tube's use is stated as the tube's workload. The transmission is the fraction by which the dose is decreased due to the presence of the shielding barrier. The workload and transmission are typically strong functions of the tube's operating potential.

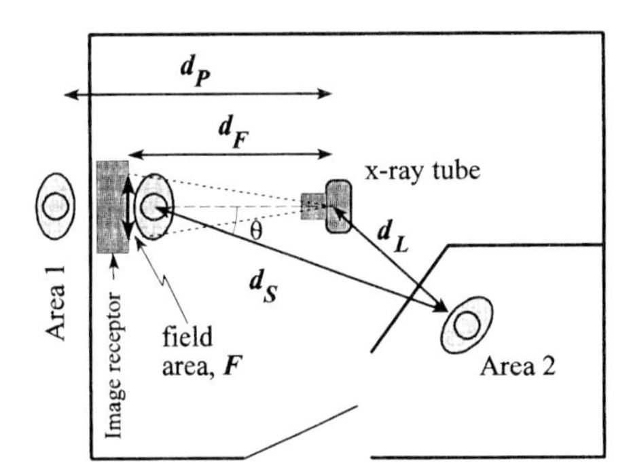

Figure 6.1. A typical diagnostic X-ray room layout. For the indicated tube orientation, the individual in Area 1 would need to be shielded from the primary beam, with the distance from the X-ray source to the shielded area equal to d_P. The person in Area 2 would need to be shielded from scatter and leakage radiations, with the indicated scatter distance d_S and leakage distance d_L. The primary X-ray beam has area F at distance d_F.

The workload of a diagnostic X-ray tube is the time integral of the X-ray tube current over a one week period, typically quoted in units of mA·min per week. The workload is a direct measure of the number of electrons incident on the X-ray tube anode, and at a given operating potential, is directly proportional to the radiation dose in the primary beam at a specified distance. While it is convenient for shielding calculations to assume that the X-ray tube only operates at a single potential, it is often far from realistic. A recent survey of diagnostic X-ray facilities conducted by Task Group 9 of the X-ray Imaging Committee of the AAPM[2] (hereafter referred to as AAPM-TG9) shows that the workload in a diagnostic X-ray room is spread over a range of operating potentials, with most tube operation occurring in the 60–100 kVp range. With the exception of chest radiography, only a small fraction of the workload exceeds 110 kVp. This distribution of workloads over operating potential kVp constitutes the workload distribution[3]. Let the workload distribution from the ith X-ray tube be $W_i(kVp)$. Table 6.2 summarizes the findings of the AAPM-TG9 survey, listing the average workload per patient and the average number of procedures performed in typical modern imaging facilities. These workloads may be assumed when more detailed information on a particular installation is unavailable. The concept of workload distribution is discussed in more detail in Section 6.4.1.

The broad-beam transmission, $B(x,m)$, of X rays through a shielding barrier of thickness x of material m is defined as the ratio of the dose from a wide X-ray beam to an occupied area when shielded, $D(x,m)$, to that in an unshielded condition, $D(0)$.

$$B(x,m) = \frac{D(x,m)}{D(0)}. \tag{6.4}$$

Transmission will therefore depend on the energies of the X radiation and the thickness and material of the shielding barrier. The scatter transmission is assumed to be equal to that of the primary beam, since to first approximation the energy spectrum of scattered photons is the same as that for primary photons generated at less than 150 kVp. The transmission of leakage radiation is assumed exponential, since penetration through the tube housing will have removed all but the highest energy X rays generated in the tube. For tube operation at a given potential, the leakage transmission will therefore exceed the primary and scatter transmission. The half value layer of the leakage radiation is obtained from the primary beam transmission at high attenuation.

Before examining the current theory of diagnostic shielding it is essential to review the new requirements and latest data that have shaped the development of this theory.

Table 6.2. Distribution of workload (mA·min) normalized per patient, as from survey of Task Group 9 of the AAPM.[2] The kVp refers to the highest potential in the 5 kVp-wide bin. The three columns under *Radiography Room* tabulate the workload spectrum for all barriers in the room, for just the wall holding the chest board, and for all other barriers exclusive of the wall with the chest board. The number of patients wk^{-1} is the mean value from the survey.

	Radiography Room								
kVp	Rad Rm (all barriers)	Rad Rm (chest bucky)	Rad Rm (floor/others)	Fluoroscopy Tube (R&F)	Radiog Tube (in R&F)	Chest	Mammo-graphy	Cardiac Angio	Peripheral Angio
25	0	0	0	0	0	0	9.25×10^{-1}	0	0
30	0	0	0	0	0	0	4.67	0	0
35	0	0	0	0	0	0	1.10	0	0
40	1.38×10^{-4}	0	1.38×10^{-4}	0	0	0	0	0	0
45	7.10×10^{-4}	0	7.10×10^{-4}	0	5.78×10^{-4}	0	0	0	0
50	8.48×10^{-3}	6.78×10^{-3}	1.70×10^{-3}	0	7.65×10^{-4}	0	0	0.34	8.94×10^{-2}
55	1.09×10^{-2}	4.56×10^{-4}	1.04×10^{-2}	7.02×10^{-2}	7.26×10^{-4}	0	0	0.42	3.98×10^{-2}
60	9.81×10^{-2}	8.96×10^{-3}	8.91×10^{-2}	1.13×10^{-1}	1.52×10^{-2}	0	0	1.96	6.99×10^{-1}
65	1.04×10^{-1}	3.42×10^{-2}	7.00×10^{-2}	1.87×10^{-1}	2.52×10^{-2}	0	0	4.55	15.0
70	4.58×10^{-1}	7.25×10^{-2}	3.85×10^{-1}	1.45×10^{-1}	8.89×10^{-2}	2.02×10^{-2}	0	6.03	12.2
75	5.01×10^{-1}	9.53×10^{-2}	4.05×10^{-1}	1.94×10^{-1}	2.24×10^{-1}	2.36×10^{-3}	0	8.02	15.3
80	5.60×10^{-1}	1.40×10^{-1}	4.20×10^{-1}	1.72	4.28×10^{-1}	0	0	25.4	11.0
85	3.15×10^{-1}	6.62×10^{-2}	2.49×10^{-1}	2.19	2.18×10^{-1}	7.83×10^{-4}	0	40.3	4.09
90	1.76×10^{-1}	1.41×10^{-2}	1.62×10^{-1}	1.46	5.33×10^{-2}	0	0	21.0	3.43
95	2.18×10^{-2}	3.51×10^{-3}	1.82×10^{-2}	1.15	4.89×10^{-2}	0	0	10.6	6.73×10^{-1}
100	1.55×10^{-2}	8.84×10^{-4}	1.46×10^{-2}	1.12	5.87×10^{-2}	3.01×10^{-2}	0	7.40	1.53
105	3.48×10^{-3}	1.97×10^{-3}	1.51×10^{-3}	9.64×10^{-1}	1.05×10^{-2}	0	0	7.02	9.27×10^{-2}
110	1.05×10^{-2}	9.91×10^{-3}	5.51×10^{-4}	7.47×10^{-1}	6.46×10^{-2}	2.14×10^{-2}	0	6.59	3.05×10^{-2}
115	4.10×10^{-2}	3.74×10^{-2}	3.69×10^{-3}	1.44	2.90×10^{-2}	9.36×10^{-2}	0	13.8	0
120	6.99×10^{-2}	5.12×10^{-2}	1.87×10^{-2}	9.37×10^{-1}	1.04×10^{-1}	4.74×10^{-2}	0	3.35	0

kVp	Radiography Room Rad Rm (all barriers)	Rad Rm (chest bucky)	Rad Rm (floor/others)	Fluoroscopy Tube (R&F)	Radiog Tube (in R&F)	Chest	Mammo-graphy	Cardiac Angio	Peripheral Angio
125	4.84×10^{-2}	4.81×10^{-2}	3.47×10^{-4}	1.38×10^{-1}	8.13×10^{-2}	0	0	2.75	0
130	1.84×10^{-3}	1.71×10^{-3}	1.25×10^{-4}	1.53×10^{-1}	4.46×10^{-2}	0	0	3.1×10^{-2}	0
135	7.73×10^{-3}	7.73×10^{-3}	0	1.46×10^{-1}	9.47×10^{-3}	0	0	0	0
140	0	0	0	1.92×10^{-2}	4.26×10^{-3}	0	0	0	0
Total Workload:	2.45	0.601	1.85	12.9	1.51	0.216	6.69	160	64.1
Patients wk^{-1}:	112	112	112	17.6	23.3	206	47.4	19.1	21

6.3 New Design Dose Limits

A barrier interposed between the source and individual to be protected must decrease the radiation level in the occupied area to a dose not exceeding the design dose limit. This limit represents the maximum radiation dose that a particular barrier is designed to transmit. In NCRP49, the exposure limits selected for shielding design are 100 mR wk $^{-1}$ for occupationally exposed persons (controlled areas) and 10 mR wk^{-1} for non-controlled areas. Designing a facility to these limits means that in theory, a person could receive the maximum dose limit of 5 rem (controlled areas) or 0.5 rem (non-controlled areas) over a year.

Report Number 116 of the NCRP[4] has lowered these design values significantly. Regarding occupational exposures, the report states that *"all new facilities and the introduction of new practices should be designed to limit annual exposures to individuals to a **fraction** of the 10 mSv per y limit implied by the cumulative dose limit"*. The fraction suggested in this chapter is one-half. This brings the dose limit to areas occupied by radiation workers in line with that for a pregnant worker. A worker may therefore occupy any controlled area so shielded without regard to her pregnancy status.

NCRP 116 also states that *"for the design of new facilities or the introduction of new practices, that the radiation protection goal in such cases should be that no member of the public would exceed the 1 mSv annual effective dose limit from all manmade sources ..."*. Table 6.3 summarizes the proposed changes in the design dose limits. These reduced values have generated controversy in the literature.[5,6]

Table 6.3. Comparison of annual and weekly design dose limits. The proposed design dose limits are based on NCRP Report 116.[4]

	NCRP49		Proposed	
	mSv y^{-1}	mSv wk^{-1}	mSv y^{-1}	mSv wk^{-1}
Controlled Area	50	1	5	0.1
Uncontrolled Area	5	0.05	1	0.02

The proposed design limits are a factor of ten for controlled areas and a factor of five for non-controlled areas below those promulgated by NCRP49. Shielding to these conservative dose limits using the conservative assumptions and methodology presented in NCRP49 will obviously generate barriers thicker than those currently in use in diagnostic facilities. Based on evidence from years of film badge records, these barriers have proven to be sufficient to reduce doses to the lower levels. To avoid costly and wasteful overshielding, it is prudent to use more realistic and accurate estimates of the shielding parameters and a common sense approach to the shielding process. The following sections will examine some of these.

It is of some import to understand the radiation intensity parameters and units used for shielding calculations. The dose limit P is stated as an effective dose equivalent, while x-radiation intensity is routinely measured by ionization chambers. The vast body of the literature on shielding is presented in terms of the exposure, X, determined from these measurements. This includes the ratios derived from ionization chamber measurements, including transmission and scatter fraction. While the dose in air (in mGy) is directly proportional to the exposure in R, given by

$$D_{air}(mGy) = 8.76\ X(R), \tag{6.5}$$

the effective dose equivalent, H_E, derived from exposure is a strong function of photon energy and incident beam direction on the exposed individual. This is due to shielding of individual organs by overlying tissues. Figure 6.2 shows the data of ICRP Report 51[7] for the conversion of exposure to H_E. Similar response is expected for the conversion of exposure (or dose in air) to effec-

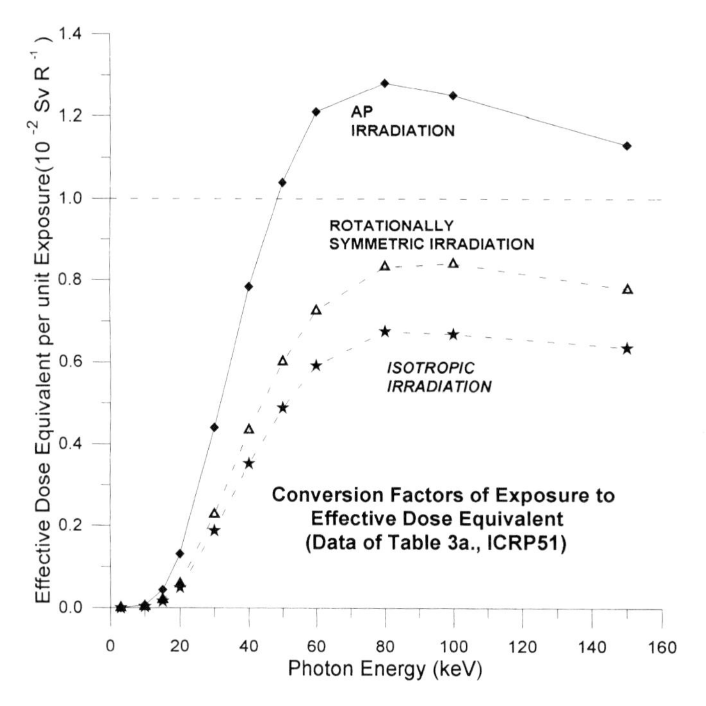

Figure 6.2. The data of ICRP Report 51[7] for the conversion of photon exposure to effective dose equivalent, H_E. Note that the effective dose equivalent is a strong function of both photon energy and incident direction. For low energyX rays, such as those used for mammography, the theory in this chapter will be conservative by factors of ~2.

tive dose. It has been shown[8] that while unshielded ambient H_E (in cSv) near radiologic facilities are smaller than the corresponding exposures (in R), this is offset by beam hardening in the shielding barrier, causing the transmission of H_E to exceed that of exposure. For tungsten (W) anode tubes operated above ~50 kVp, in typical installations these effects approximately cancel. It is therefore assumed that H_E is equal to D_{air} for tube operating potentials at and above 50 kVp, and that Eq. (6.3) holds for P written in Sv and D in Gy. For mammography at ≤ 30 kVp, from Fig. 6.2 it is conservative to assume that H_E (in cSv) is 45% of the exposure (in R), or 0.45/0.876 ≈ 50% of the dose in air (in cGy). Then for potentials ≤ 30 kVp, Eq. (6.3) is conservative by a factor of ~2. Finally, in the absence of more complete information, it is assumed that the data in the literature for transmission and scatter fraction, measured originally as ratios of exposures or doses to air, are equivalent to the ratios of H_E.

6.4 New Data for Shielding

6.4.1 Workload and Workload Distribution

In 1996, Simpkin[2] published the results of the first national survey to attempt to measure workload and use factor data. The data was primarily gathered by members of AAPM-TG9. Workloads at 14 medical institutions involving approximately 2,500 patients and seven types of radiology installations were determined. The workload survey results for the workload per patient, the average number of patients imaged per week, and the total workload per week for each type of installation are shown in Table 6.2.

NCRP reports have traditionally assumed that the entire workload in an installation is performed at a single kVp, for example 1000 mA·min wk^{-1} at 100 kVp. This is a crude model which ignores the fact that the diagnostic workload is typically spread over a wide range of X-ray tube potentials. For example, in a typical general purpose radiographic room, extremity exams (about one third of the total exams done in the room) are normally performed at about 50–60 kVp, while abdominal exams are done at 70–80 kVp and chest exams at more than 100 kVp. Because the dose in air as well as the barrier transmission exhibit a strong kVp dependence, for shielding design the distribution of kVp is more important than the magnitude of the workload. The radiation level transmitted through a lead barrier varies exponentially with kVp (three orders of magnitude over the range of 60–100 kVp), but only linearly with the workload. Leakage radiation shows an even more dramatic decrease with falling kVp. From the regulatory limit of 100 mR hr^{-1} at 150 kVp, leakage rates fall by more than eight orders of magnitude over the range from 150 to 50 kVp. This significant reduction in leakage radiation with kVp is not considered in the shielding model used in NCRP49.

The surveyed workload distributions in Table 6.2 provide a new approach to the shielding design of diagnostic X-ray examination rooms. Although the actual workload distribution for a given X-ray room may differ from that in Table 6.2 and will vary from facility to facility, and even from week to week in the same facility, the average distribution obtained from the workload survey represents a more realistic model than the single kVp approximation. A kVp distribution of workloads provides a more accurate estimate of the quality and quantity of radiation produced in a diagnostic X-ray room. Furthermore, a multiple kVp distribution can be used to calculate an attenuation curve analogous to the attenuation curve measured at a single kVp. As will be shown, this is no more difficult to use than the curve from a single kVp distribution. Figure 6.3 compares the workload distribution for the walls and floor of a general radiographic room with the single 100 kVp 'spike' that results with the assumption that all exposures are made at the same kVp.

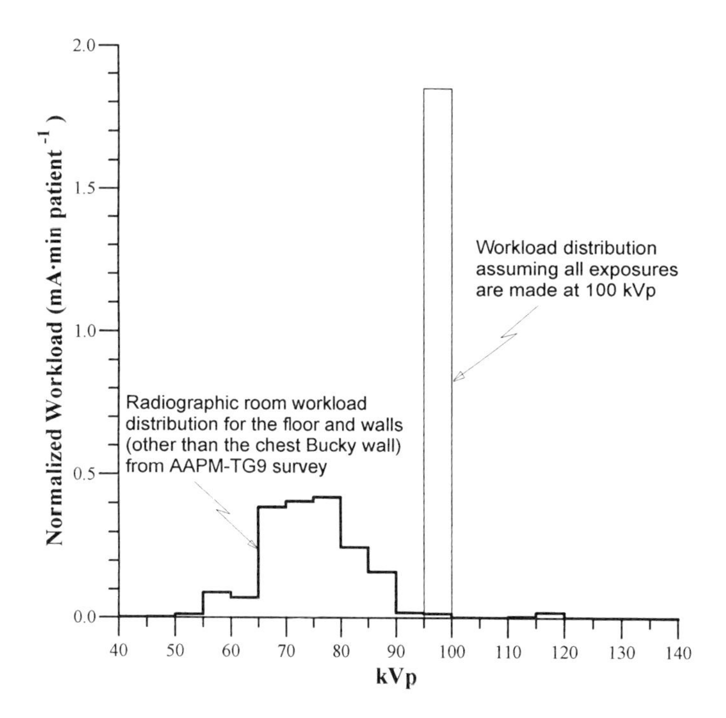

Figure 6.3. The *Radiographic Room (floor/other walls)* workload distribution observed by the AAPM-TG9 survey[2] compared to the single 100 kVp 'spike' that results with the assumption that all exposures are made at 100 kVp. The total workload of both distributions is 1.85 mA·min per patient.

A radiographic X-ray room typically contains a chest image receptor on one wall and a table on which overhead and cross table procedures are performed. The kVp workload distribution for the chest wall, which includes chest exams and other upright procedures, is also substantially different from the workload distribution of the walls and floor. That workload directed toward the vertical cassette assembly, termed *Radiographic Room (chest bucky wall)* in Table 6.2, occurs at far greater potentials than that directed at the other barriers in the room, *Radiographic Room (floor/ other barriers)*. Likewise, the radiographic and fluoroscopic tubes in a radiographic/ fluoroscopic room are considered separately. Figure 6.4 shows the lead transmission curve computed for the *Radiographic Room (floor/ other barriers)* workload distribution and compares it to the transmission curve measured[9] at 100 kVp.

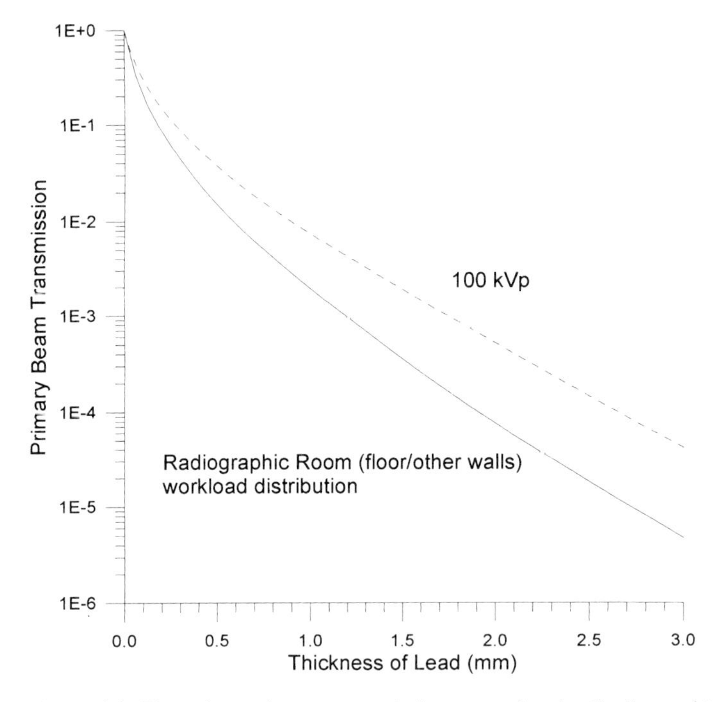

Figure 6.4. The primary beam transmission curve for the *Radiographic Room (floor/other barriers)* workload distribution observed by the AAPM-TG9 survey[2] compared to that measured[9] for 100 kVp X rays.

Since the workload per patient (mA·min per patient) is assumed to be known in the AAPM-TG9 survey model, the total workload can now be expressed in terms of the number of patients per week (N_{pat}) instead of the traditional, but often more confusing units of mA·min per week. Table 6.4 lists patient workload values of diagnostic facilities including hospitals and clinics with different patient volume levels that the authors feel to be typical. These values may be employed if more accurate information on patient workload is not available. The shielding designer must keep in mind, however, that current patient levels could change in the future and the design should incorporate the potential for these change.

Table 6.4. Suggested total workloads in various radiologic installations in clinics and hospitals.

Room Type	Number of Patients per week		Total Workload (mA·min wk^{-1})	
	Average	Busy	Average	Busy
Radiographic Rm (chest bucky)	120	160	75	100
Radiographic Rm (floor/other barriers)	120	160	240	320
Dedicated Chest Room	200	400	60	120
Fluoroscopic Tube in R&F room	20	30	250	400
Radiographic Tube in R&F room	20	40	30	60
Mammographic Suite	50	100	350	700
Cardiac Angiographic Suite	20	30	3200	2000
Peripheral Angiographic Suite	20	30	1300	2000

6.4.2 Use Factors

The use factor, U, is the fraction of the primary beam workload that is directed to a particular barrier. The recommended use factors for primary barriers given in NCRP49 are $U = 1$ for floors and $U = ¼$ for each wall. The AAPM-TG9 survey[2] shows that the primary beam is directed to the non-chest walls much less often than the fraction suggested by NCRP49. Table 6.5 displays the primary use factors found from the survey. In a radiographic room, note that there are two use factors approximately equal to unity for the workload distribution model because this model invokes two different workload distributions. For the *Radiographic Room (chest bucky)* workload distribution, $U =$ 1 for the wall containing the upright image receptor and 0 for all other barriers. Additionally, it is conservatively simple to round the very large value of the use factor for the *Radiographic Room (floor/ other barriers)* workload distribution for the floor to unity. As the X-ray beam is only very rarely directed at the control booth, U for this wall in a radiographic room is safely assumed to be zero.

Since the image receptor for mammography and image-intensified fluoroscopy is required by regulation to act as primary beam stop, $U = 0$ in these rooms.

Table 6.5. Primary beam use factors, *U*, defined as the fraction of total workload for which this barrier serves as primary beam stop, from the survey of Task Group 9 of the X-ray Imaging Committee of the AAPM.[2] In the radiographic room, *Wall #3* is a wall other than the cross-table wall or that holding the upright image receptor.

Room Type	Use Factor, *U*			
	Floor	Wall with upright image receptor	Cross-table Wall	Wall #3
Radiographic Room As fraction of *Rad Rm* *(all barriers)*	0.69	0.22	0.07	0.02
As fraction of *Rad Rm* *(chest bucky)*	0	1.0	0	0
As fraction of *Rad Rm* *(floor/other)*	0.89	0	0.09	0.02
Dedicated Chest Room	0	1.0	0	0
Mammography		0 for all barriers		
Fluoroscopy Tube		0 for all barriers		
Cardiac/Peripheral Angiography *(image intensified imaging)*		0 for all barriers		

6.4.3 Occupancy Factors

The occupancy factor, *T*, is defined as the fraction of time that a maximally present individual is in the shielded area while the X-ray beam is on. The occupancy factor for an area can be found by estimating the maximum amount of time (in hours) a given person would be likely to occupy that space in an eight hour working day (averaged over a year), divided by 8 hr. Assuming that an X-ray unit is randomly used during the day, the occupancy factor is the fraction of an eight-hour working day that a given person would occupy the area, averaged over the year. For example, an outdoor area adjacent to an X-ray room having an assigned occupancy factor of 1/40 would imply that a given member of the general public would spend an average of one hour per week in that area every week for a year. A factor of 1/40 would certainly be conservative for most outdoor areas used only for pedestrian or vehicular traffic (e.g., sidewalks, streets, vehicular drop-off areas, or lawn areas with no benches or seating).

The shielding designer must recognize that use of $T < 1$ allows the average radiation dose in a partially occupied area to be higher than that for a fully occupied area by a factor of T^{-1}.

Note that the occupancy factor for an area is not the fraction of the time that it is occupied by *any* persons, but rather is the fraction of the time it is occupied by the *single* individual who spends the most time there. Thus a waiting room might be occupied at all times during the day, but have a very small occupancy factor since no single person is likely to spend more than 10 hours per year in a given waiting room. Occupancy factors in uncontrolled areas for non-occupationally exposed persons would rarely be determined by visitors to the facility or its environs who might be there only for a small fraction of a year, but rather by non-radiological employees of the facility itself or employees or residents of an adjacent facility.

The shielding designer should make reasonable and realistic assumptions concerning occupancy factors, since each facility will have its own particular circumstances. The occupancy factors for uncontrolled areas given in Table 6.6 are values which may be utilized if more detailed information on occupancy is not available. The designer of a new facility should, however, keep in mind that the function of adjacent areas may change over time. For example, a storage room may be converted into an office without anyone reconsidering the adequacy of the existing shielding (particularly if the conversion is made in an adjacent uncontrolled area).

Table 6.6. Suggested occupancy factors, T. These values may be assumed in the absence of site-specific information.[a]

Location	T
Uncontrolled Areas	
Offices, shops, living quarters, children's indoor play areas, occupied space in nearby buildings, laundry, attended waiting room[c]	1
Patient exam and treatment rooms, nurses stations, kitchens, cafeterias	1/2
Patient rooms[b], corridors, employee lounge	1/8
Rest rooms or bathrooms, unattended vending areas, storage rooms, outdoor areas with seating	1/20
Outdoor areas with only transient pedestrian or vehicular traffic, unattended parking lots, vehicular drop off areas (unattended), attics, unattended waiting rooms, stairways, unattended elevators, patient dressing room, janitors closets	1/40
Controlled Areas	
X-ray control booth, film reading area, ultrasound exam room, nuclear medicine scan, offices, workroom, employee lounge, adjacent X-ray room	1
Barium kitchen	1/2
Rest rooms, corridor, patient holding areas	1/4
Patient dressing rooms	1/8

[a] Care should be taken when assuming a low occupancy factor for an area immediately adjacent to an X-ray room. More distant areas may have significantly higher occupancy factors and may therefore represent the limitation for shield design.
[b] Limited by attending nursing staff – not by patients and families.
[c] Limited by attendant

Care must also be taken when assigning a low occupancy factor to an uncontrolled area such as a corridor, immediately adjacent to an X-ray room. The actual limitation for shielding design may be a fully occupied area further removed from the X-ray room (such as an office across the corridor). The designer must therefore take a larger view of the facility in arriving at the appropriate limitations for shielding design.

The minimum occupancy factor recommended for an uncontrolled area is 1/40 in order to ensure that an individual member of the general public would receive no more than 20 μSv in a one-hour period. Shielding an uncontrolled area with an occupancy factor of 1/40 to the design dose limit, *P*, of 20 μSv per week would then deliver not more than 20 μSv in any one hour of a 40-hour work week.

Occupancy factors less than unity for occupationally exposed persons in controlled areas may be utilized. However, the shielding designer should validate the assumption of partial occupancy. For example, it is reasonable to assume that a worker would not spend (on average) more than 2 hours per day in a rest room. However, it should be realized that shielding an area with an assumed partial occupancy may result in a worker receiving a dose greater than the design dose limit of 0.1 mSv wk^{-1}. However, given the factor of 10 difference in the controlled design dose limit (5 mSv yr^{-1}) and the worker dose limit (50 mSv yr^{-1}), the worker's dose most probably would be well below the maximum permissible dose for *T* not less than 1/10. A suitably conservative approach would be to assign the occupancy factor for a controlled area based on the fraction of time it is occupied by *any* radiation worker. Some suggested values of *T* in controlled areas are shown in Table 6.6.

6.4.4 Pre-shielding of the Primary Beam by the Image Receptor and Supports

For primary barrier shielding calculations, it is traditionally assumed that the raw or unattenuated primary beam is incident on the floor or walls which constitute primary barriers. In fact, the primary beam is normally attenuated by the patient and image receptor hardware, commonly including the cassette, grid, and radiographic table or wall-mounted cassette holder. Since the attenuation through the patient is spatially variable, with raw primary radiation striking parts of a properly exposed radiograph, it is prudent to ignore attenuation in the patient. Dixon[10] measured the primary beam attenuation provided by the image receptor and its supporting structures. Figure 6.5 summarizes the measured primary beam transmission as a function of operating potential for film cassette and grid assemblies, such as used in cross-table radiography, and various grids, cassettes, radiographic tables and wall-mounted cassette holders. The image receptor and its support are seen to provide 1 to 3 tenth value layers to the primary radiographic beam.

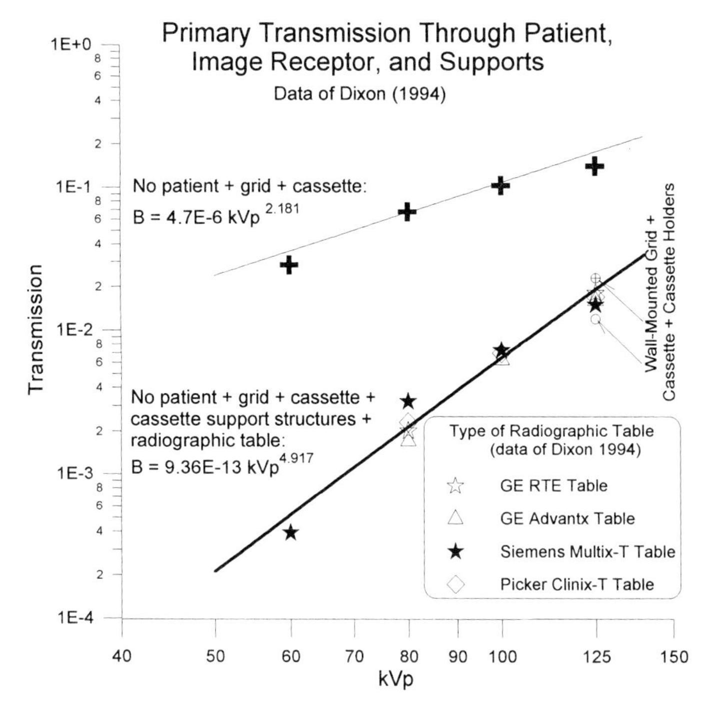

Figure 6.5. Transmission through primary image receptors. Data of Dixon[10]. The top curve shows the primary beam transmission through a typical radiographic grid and cassette. The bottom curve shows that through the grid, cassette, cassette support tray, and radiographic table or wall-mounted "bucky" image receptor.

It is reasonable[11] to ascribe this image receptor attenuation to its thickness in barrier material, x_{pre}. For the workload distributions in the AAPM-TG9 survey[2] a typical radiographic table or wall-mounted image receptor with a cassette in the cassette tray has a Pb equivalence in excess of 0.8 mm and a concrete equivalence in excess of 7 cm. Table 6.7 shows the values of x_{pre} in a number of shielding materials calculated from these workload distributions. For X-ray tube operation at 60–120 kVp, the values of x_{pre} are essentially independent of the details of the kVp workload distribution.

Table 6.7. Shielding equivalencies of image receptors, x_{pre}, (mm). Note that x_{pre} ignores the significant attenuation in the patient. These values may be conservatively used with any of the workload distributions from the Task Group 9 survey of the AAPM.[2]

Image Receptor	Lead	Concrete	Steel	Plate Glass	Gypsum Wallboard
Grid & cassette	0.3	30	2	35	90
Grid, cassette, cassette tray, and radiographic table or wall-mounted chest bucky	0.85	72	7	83	230

In practice, the thickness of the primary structural barrier required is determined by subtracting x_{pre} from the primary barrier thickness calculated assuming that the raw primary beam impinges directly on the barrier. This approach gives a considerable margin of conservatism even if the shield designer does not have detailed information about the X-ray table construction, since patient attenuation has been ignored. The degree of conservatism in this approach has been measured by Dixon, who measured the primary beam dose transmitted through the radiographic table of 0.015 mSv wk^{-1} for a workload of 120 patients wk^{-1}. In this case the patient and image receptor itself provided an adequate primary barrier.

6.4.5 Transmission Data

The X-ray transmission curves published in NCRP49 were generated at least a quarter century ago using single-phase X-ray equipment and various measurement techniques. More current transmission data for modern three-phase and constant potential X-ray generation are now available. The set of measurements of Archer et al.[9] will be assumed to represent primary broad beam transmission of diagnostic X rays from modern equipment operated between 50 and 150 kVp in lead, steel, plate glass, gypsum wallboard, lead acrylic and wood barriers. For concrete, the primary transmission data of Legare et al[12] will be employed. In the mammographic energy range (25–35 kVp), transmission values of Simpkin[13] will be used.

The transmission, B, of broad X-ray beams through a variety of shielding materials in diagnostic X-ray applications has been found to be well described by a mathematical model published by Archer et al.[14]. This model has the form

$$B = \left[\left(1 + \frac{\beta}{\alpha} \right) e^{\alpha \gamma x} - \frac{\beta}{\alpha} \right]^{-\frac{1}{\gamma}}, \tag{6.6}$$

where x is the thickness of shielding material and α, β, and γ are the fitting parameters. This may be inverted to find the barrier thickness x as a function of transmission B.

$$x = \frac{1}{\alpha \gamma} \ln \left[\frac{B^{-\gamma} + \frac{\beta}{\alpha}}{1 + \frac{\beta}{\alpha}} \right]. \tag{6.7}$$

This form proves useful in describing both primary and secondary transmission curves for X rays generated at both single potentials and those generated following clinical workload distributions.

Fitting parameter values due to Simpkin[15] for the primary beam three-phase transmission for Al-filtered W-anode and Mo-anode and filtered mammography X-ray beams for a variety of shielding materials are shown in Table 6.8. These were obtained by interpolation of published transmission data[9,12,13]. Figures 6.6 through 6.11 show the transmission curves for Pb, concrete, gypsum wallboard, steel, plate glass, and wood.

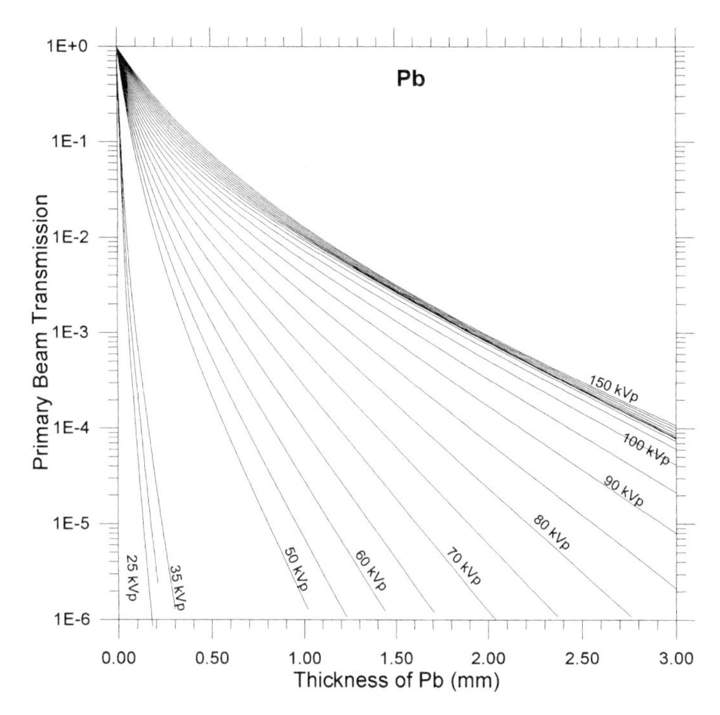

Figure 6.6. The broad primary beam transmission through Pb, plotted for increments of 5 kVp. W-anode, Al-filtered beam data of Archer et al.[9] and Mo-anode, Mo-filtered mammography beam calculations of Simpkin[13] as interpolated in Simpkin[15].

For large values of x, known as the high attenuation condition, the transmission curves tend toward an exponential which decreases with constant half value layer. The half value layers at high attenuation, $HVL(m,kVp)$, are shown in Figure 6.12 for a variety of shielding materials. This value may be extracted from the transmission fitting parameters in Table 6.8. From the asymptotic form of Eq. (6.6), $HVL(m,kVp) = (ln\ 2)/\alpha$ for $x \longrightarrow \infty$. For a kVp distribution of workloads, the half value layer at high attenuation may be conservatively assumed to be that occurring at the highest potential in the distribution.

The primary transmission curves for the workload distributions which utilize primary beams were calculated by summing the incremental dose in each kVp interval transmitted through a given barrier thickness and dividing that

Table 6.8. Fits of transmission curves for broad primary X-ray beams to Eq. (6.6), due originally to Archer et al.[9], Légaré et al.[12], and Simpkin[13], as interpolated by Simpkin.[15] Thicknesses input in mm. The 25–35 kVp data are for Mo-anode X-ray tubes. All other data are for W-anode tubes.

kVp	Pb			Concrete			Gypsum		
	α (mm^{-1})	β (mm^{-1})	γ	α (mm^{-1})	β (mm^{-1})	γ	α (mm^{-1})	β (mm^{-1})	γ
25	4.952×10^{1}	1.940×10^{2}	3.037×10^{-1}	3.904×10^{-1}	1.645	2.757×10^{-1}	1.576×10^{-1}	7.175×10^{-1}	3.048×10^{-1}
30	3.880×10^{1}	1.780×10^{2}	3.473×10^{-1}	3.173×10^{-1}	1.698	3.593×10^{-1}	1.208×10^{-1}	7.043×10^{-1}	3.613×10^{-1}
35	2.955×10^{1}	1.647×10^{2}	3.948×10^{-1}	2.528×10^{-1}	1.807	4.648×10^{-1}	8.878×10^{-2}	6.988×10^{-1}	4.245×10^{-1}
40				1.297×10^{-1}	1.780×10^{-1}	2.189×10^{-1}			
45				1.095×10^{-1}	1.741×10^{-1}	2.269×10^{-1}			
50	8.801	2.728×10^{1}	2.957×10^{-1}	9.032×10^{-2}	1.712×10^{-1}	2.324×10^{-1}	3.883×10^{-2}	8.730×10^{-2}	$5.105 \times 10^{-}$
55	7.839	2.592×10^{1}	3.499×10^{-1}	7.422×10^{-2}	1.697×10^{-1}	2.454×10^{-1}	3.419×10^{-2}	8.315×10^{-2}	5.606×10^{-1}
60	6.951	2.489×10	4.198×10^{-1}	6.251×10^{-2}	1.692×10^{-1}	2.733×10^{-1}	2.985×10^{-2}	7.961×10^{-2}	6.169×10^{-1}
65	6.130	2.409×10^{1}	5.019×10^{-1}	5.528×10^{-2}	1.696×10^{-1}	3.217×10^{-1}	2.609×10^{-2}	7.597×10^{-2}	6.756×10^{-1}
70	5.369	2.349×10^{1}	5.881×10^{-1}	5.087×10^{-2}	1.696×10^{-1}	3.847×10^{-1}	2.302×10^{-2}	7.163×10^{-2}	7.299×10^{-1}
75	4.666	2.269×10^{1}	6.618×10^{-1}	4.797×10^{-2}	1.663×10^{-1}	4.492×10^{-1}	2.066×10^{-2}	6.649×10^{-2}	7.750×10^{-1}
80	4.040	2.169×10^{1}	7.187×10^{-1}	4.583×10^{-2}	1.549×10^{-1}	4.926×10^{-1}	1.886×10^{-2}	6.093×10^{-2}	8.103×10^{-1}
85	3.504	2.037×10^{1}	7.550×10^{-1}	4.398×10^{-2}	1.348×10^{-1}	4.943×10^{-1}	1.746×10^{-2}	5.558×10^{-2}	8.392×10^{-1}
90	3.067	1.883×10^{1}	7.726×10^{-1}	4.228×10^{-2}	1.137×10^{-1}	4.690×10^{-1}	1.633×10^{-2}	5.039×10^{-2}	8.585×10^{-1}
95	2.731	1.707×10^{1}	7.714×10^{-1}	4.068×10^{-2}	9.705×10^{-2}	4.406×10^{-1}	1.543×10^{-2}	4.571×10^{-2}	8.763×10^{-1}
100	2.500	1.528×10^{1}	7.557×10^{-1}	3.925×10^{-2}	8.567×10^{-2}	4.273×10^{-1}	1.466×10^{-2}	4.171×10^{-2}	8.939×10^{-1}
105	2.364	1.341×10^{1}	7.239×10^{-1}	3.808×10^{-2}	7.862×10^{-2}	4.394×10^{-1}	1.397×10^{-2}	3.815×10^{-2}	9.080×10^{-1}
110	2.296	1.170×10^{1}	6.827×10^{-1}	3.715×10^{-2}	7.436×10^{-2}	4.752×10^{-1}	1.336×10^{-2}	3.521×10^{-2}	9.244×10^{-1}
115	2.265	1.021×10^{1}	6.363×10^{-1}	3.636×10^{-2}	7.201×10^{-2}	5.319×10^{-1}	1.283×10^{-2}	3.271×10^{-2}	9.423×10^{-1}
120	2.246	8.950	5.873×10^{-1}	3.566×10^{-2}	7.109×10^{-2}	6.073×10^{-1}	1.235×10^{-2}	3.047×10^{-2}	9.566×10^{-1}
125	2.219	7.923	5.386×10^{-1}	3.502×10^{-2}	7.113×10^{-2}	6.974×10^{-1}	1.192×10^{-2}	2.863×10^{-2}	9.684×10^{-1}

	Pb			Concrete			Gypsum		
130	2.170	7.094	4.909×10^{-1}	3.445×10^{-2}	7.160×10^{-2}	7.969×10^{-1}	1.155×10^{-2}	2.702×10^{-2}	9.802×10^{-1}
135	2.102	6.450	4.469×10^{-1}	3.394×10^{-2}	7.263×10^{-2}	9.099×10^{-1}	1.122×10^{-2}	2.561×10^{-2}	9.901×10^{-1}
140	2.009	5.916	4.018×10^{-1}	3.345×10^{-2}	7.476×10^{-2}	1.047	1.088×10^{-2}	2.436×10^{-2}	9.964×10^{-1}
145	1.895	5.498	3.580×10^{-1}	3.296×10^{-2}	7.875×10^{-2}	1.224	1.056×10^{-2}	2.313×10^{-2}	9.987×10^{-1}
150	1.757	5.177	3.156×10^{-1}	3.243×10^{-2}	8.599×10^{-2}	1.467	1.030×10^{-2}	2.198×10^{-2}	1.013

	Steel			Plate Glass			Wood		
kVp	α (mm^{-1})	β (mm^{-1})	γ	α (mm^{-1})	β (mm^{-1})	γ	α (mm^{-1})	β (mm^{-1})	γ
25	9.364	4.125×10^{1}	3.202×10^{-1}	3.804×10^{-1}	1.543	2.869×10^{-1}	2.230×10^{-2}	4.340×10^{-2}	1.937×10^{-1}
30	7.406	4.193×10^{1}	3.959×10^{-1}	3.061×10^{-1}	1.599	3.693×10^{-1}	2.166×10^{-2}	3.966×10^{-2}	2.843×10^{-1}
35	5.716	4.341×10^{1}	4.857×10^{-1}	2.396×10^{-1}	1.694	4.683×10^{-1}	1.901×10^{-2}	3.873×10^{-2}	3.732×10^{-1}
50	1.817	4.840	4.021×10^{-1}	9.721×10^{-2}	1.799×10^{-1}	4.912×10^{-1}	1.076×10^{-2}	1.862×10^{-3}	1.170
55	1.493	4.515	4.293×10^{-1}	8.552×10^{-2}	1.661×10^{-1}	5.112×10^{-1}	1.012×10^{-2}	1.404×10^{-3}	1.269
60	1.183	4.219	4.571×10^{-1}	7.452×10^{-2}	1.539×10^{-1}	5.304×10^{-1}	9.512×10^{-3}	9.672×10^{-4}	1.333
65	9.172×10^{-1}	3.982	4.922×10^{-1}	6.514×10^{-2}	1.443×10^{-1}	5.582×10^{-1}	8.990×10^{-3}	6.470×10^{-4}	1.353
70	7.149×10^{-1}	3.798	5.378×10^{-1}	5.791×10^{-2}	1.357×10^{-1}	5.967×10^{-1}	8.550×10^{-3}	5.390×10^{-4}	1.194
75	5.793×10^{-1}	3.629	5.908×10^{-1}	5.291×10^{-2}	1.280×10^{-1}	6.478×10^{-1}	8.203×10^{-3}	6.421×10^{-4}	1.062
80	4.921×10^{-1}	3.428	6.427×10^{-1}	4.955×10^{-2}	1.208×10^{-1}	7.097×10^{-1}	7.903×10^{-3}	8.640×10^{-4}	9.703×10^{-1}
85	4.355×10^{-1}	3.178	6.861×10^{-1}	4.721×10^{-2}	1.140×10^{-1}	7.786×10^{-1}	7.686×10^{-3}	1.056×10^{-3}	1.015
90	3.971×10^{-1}	2.913	7.204×10^{-1}	4.550×10^{-2}	1.077×10^{-1}	8.522×10^{-1}	7.511×10^{-3}	1.159×10^{-3}	1.081
95	3.681×10^{-1}	2.654	7.461×10^{-1}	4.410×10^{-2}	1.013×10^{-1}	9.222×10^{-1}	7.345×10^{-3}	1.133×10^{-3}	1.116
100	3.415×10^{-1}	2.420	7.645×10^{-1}	4.278×10^{-2}	9.466×10^{-2}	9.791×10^{-1}	7.230×10^{-3}	9.343×10^{-4}	1.309
105	3.135×10^{-1}	2.227	7.788×10^{-1}	4.143×10^{-2}	8.751×10^{-2}	1.014	7.050×10^{-3}	6.199×10^{-4}	1.365
110	2.849×10^{-1}	2.061	7.897×10^{-1}	4.008×10^{-2}	8.047×10^{-2}	1.030	6.921×10^{-3}	1.976×10^{-4}	3.309
115	2.579×10^{-1}	1.922	8.008×10^{-1}	3.878×10^{-2}	7.394×10^{-2}	1.033	6.864×10^{-3}	-3.908×10^{-4}	6.469×10^{-1}

Table 6.8 continued

	Steel			Plate Glass			Wood		
120	2.336×10^{-1}	1.797	8.116×10^{-1}	3.758×10^{-2}	6.808×10^{-2}	1.031	6.726×10^{-3}	-8.308×10^{-4}	1.006
125	2.130×10^{-1}	1.677	8.217×10^{-1}	3.652×10^{-2}	6.304×10^{-2}	1.031	6.584×10^{-3}	-1.214×10^{-3}	1.192
130	1.969×10^{-1}	1.557	8.309×10^{-1}	3.561×10^{-2}	5.874×10^{-2}	1.037	6.472×10^{-3}	-1.539×10^{-3}	1.285
135	1.838×10^{-1}	1.440	8.391×10^{-1}	3.481×10^{-2}	5.519×10^{-2}	1.049	6.306×10^{-3}	-1.731×10^{-3}	1.465
140	1.724×10^{-1}	1.328	8.458×10^{-1}	3.407×10^{-2}	5.145×10^{-2}	1.057	6.191×10^{-3}	-1.849×10^{-3}	1.530
145	1.616×10^{-1}	1.225	8.519×10^{-1}	3.336×10^{-2}	4.795×10^{-2}	1.063	6.115×10^{-3}	-1.869×10^{-3}	1.498
150	1.501×10^{-1}	1.132	8.566×10^{-1}	3.266×10^{-2}	4.491×10^{-2}	1.073	6.020×10^{-3}	-1.752×10^{-3}	1.483

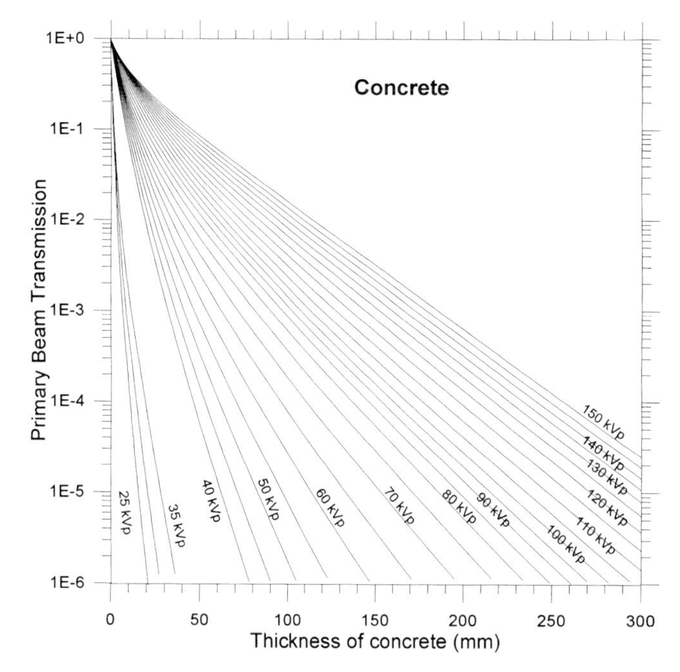

Figure 6.7. The broad primary beam transmission through concrete, plotted for increments of 5 kVp. W-anode, Al-filtered beam data of Légaré et al.[12] and Mo-anode, Mo-filtered mammography beam calculations of Simpkin[13] as interpolated in Simpkin.[15]

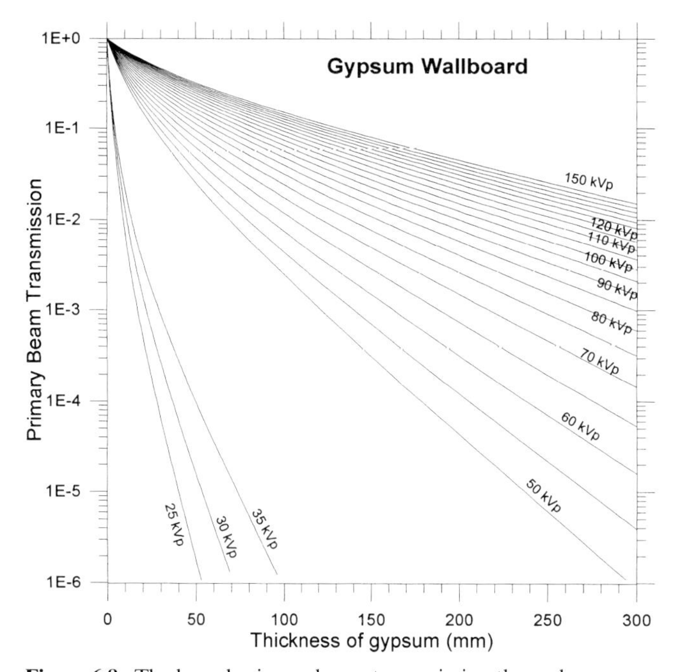

Figure 6.8. The broad primary beam transmission through gypsum wallboard. Data as in Fig. 6.6.

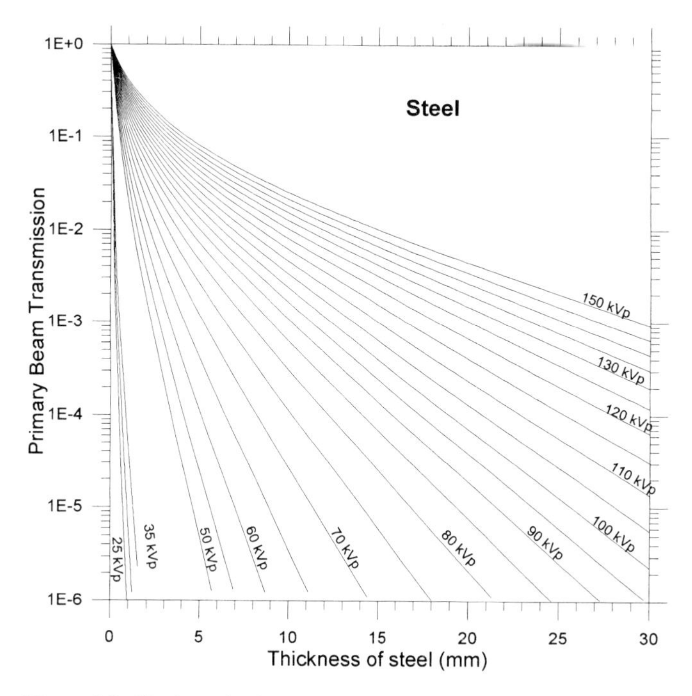

Figure 6.9. The broad primary beam transmission through steel. Data as in Fig. 6.6.

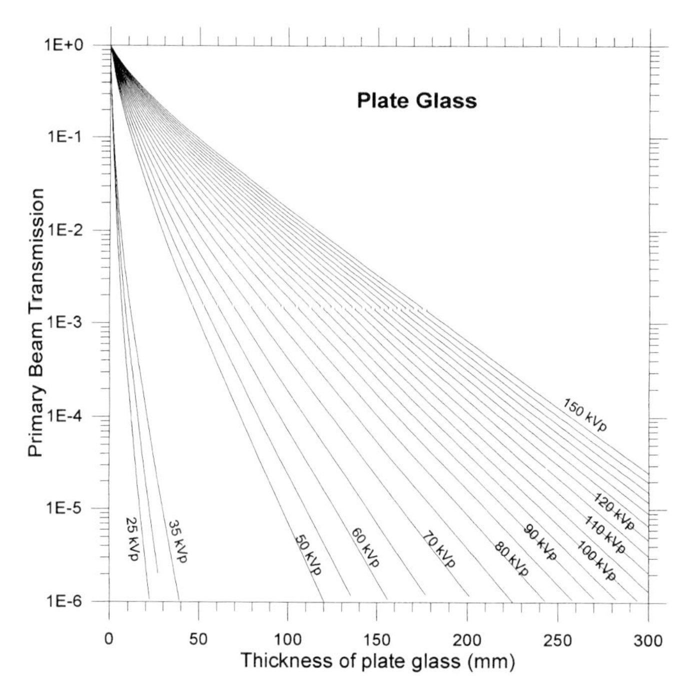

Figure 6.10. The broad primary beam transmission through plate glass. Data as in Fig. 6.6.

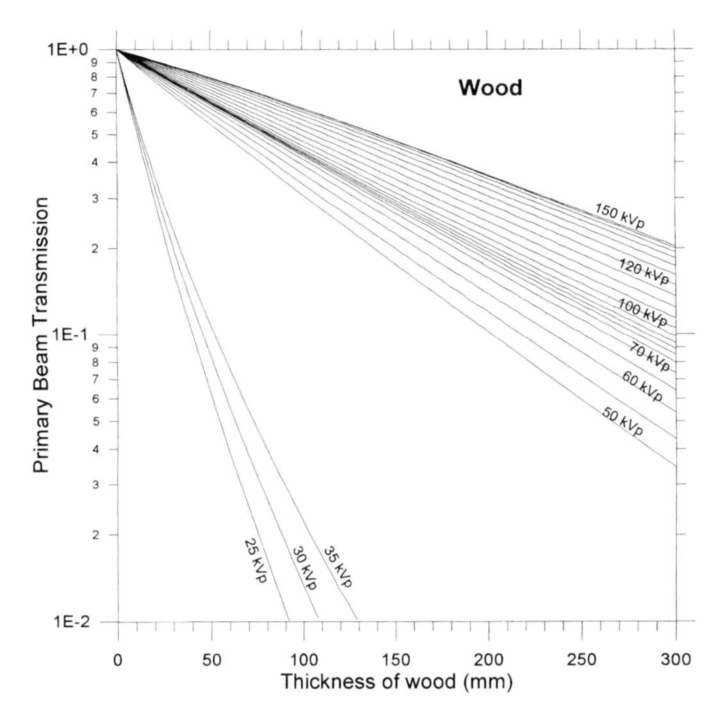

Figure 6.11. The broad primary beam transmission through wood. Data as in Fig. 6.6.

by the dose expected with no barrier. These workload distribution-specific primary beam transmission curves are shown in Figures. 6.13 through 6.18 for the six materials listed above. Table 6.9 lists the fitting parameters for these curves to Eq. (6.6). Note that these transmission curves are no more difficult to utilize than the transmission curves for X rays generated at a single kVp.

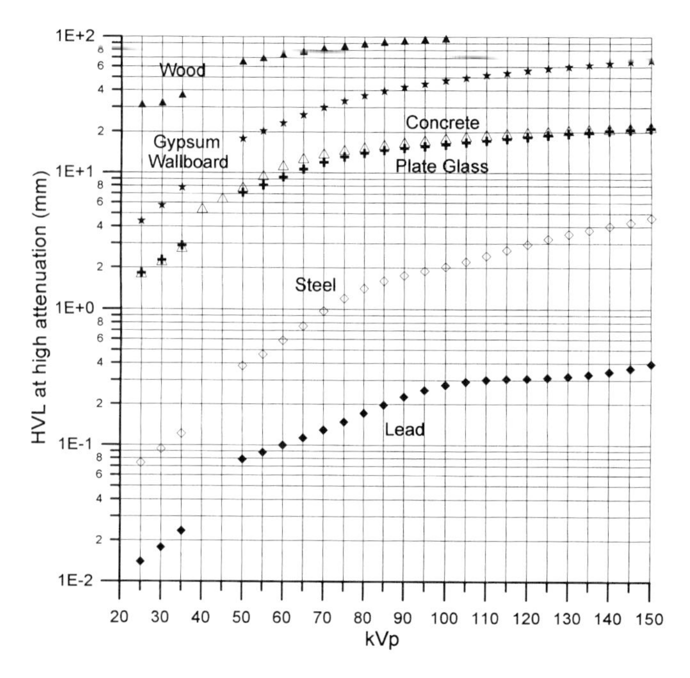

Figure 6.12 The half value layers at high attenuation for diagnostic X rays. Calculated from data in Figs. 6.6–6.11.

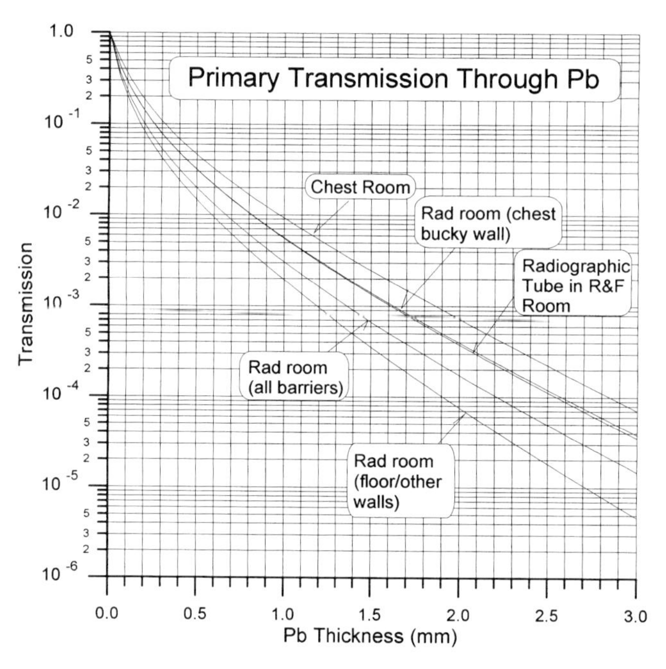

Figure 6.13. The broad primary beam transmission calculated in Pb for the clinical installations surveyed by Task Group 9 of the X-ray Imaging Committee of the AAPM[2].

Table 6.9. Fitting parameters for primary X-ray beams to Eq. (6.6). Thickness, x, is input in mm. The workload distributions are those surveyed by Task Group 9 of the X-ray Imaging Committee of the AAPM[2], described in Table 6.2.

Workload	Pb			Concrete			Gypsum Wallboard		
Distribution	α (mm^{-1})	β (mm^{-1})	γ	α (mm^{-1})	β (mm^{-1})	γ	α (mm^{-1})	β (mm^{-1})	γ
Rad Rm (all)	2.346435	15.90001	0.498151	0.036259	0.142917	0.493217	0.014202	0.057814	0.744523
Rad Rm (chest)	2.26393	13.08274	0.560009	0.035524	0.11774	0.600743	0.012776	0.048481	0.860891
Rad Rm (floor)	2.650888	16.55986	0.458469	0.039935	0.144765	0.423108	0.01679	0.061243	0.735586
Fluoro (R&F)	2.346529	12.66798	0.614868	0.036157	0.097205	0.518564	0.013402	0.042834	0.879589
Rad tube in R&F	2.29544	13.00436	0.55734	0.035485	0.116368	0.57736	0.013002	0.047779	0.848509
Chest	2.282586	10.73989	0.63703	0.036216	0.077657	0.540419	0.012859	0.035048	0.935623
Mammo	30.59843	177.6198	0.330824	0.257743	1.765204	0.364371	0.09148	0.708976	0.345912
Cardiac Angio	2.389411	14.26411	0.594799	0.037167	0.108673	0.487921	0.01409	0.048142	0.841865
Periph Angio	2.727855	18.51677	0.461402	0.042919	0.153833	0.423637	0.017742	0.064492	0.715766

Workload	Steel			Plate Glass			Wood		
Distribution	α (mm^{-1})	β (mm^{-1})	γ	α (mm^{-1})	β (mm^{-1})	γ	α (mm^{-1})	β (mm^{-1})	γ
Rad Rm (all)	0.216344	3.101258	0.574503	0.039068	0.106912	0.594029	0.007616	0.000767	1.027461
Rad Rm (chest)	0.217863	2.677046	0.720856	0.037616	0.097508	0.786729	0.007142	0.000308	1.616676
Rad Rm (floor)	0.25346	2.74	0.429725	0.043607	0.108217	0.546344	0.007915	0.00088	0.979
Fluoro (R&F)	0.232284	2.190426	0.650927	0.039012	0.085884	0.808147	0.007089	0.000474	1.580188
Rad tube in R&F	0.212586	2.567713	0.678777	0.037776	0.093653	0.748277	0.007162	0.000411	1.54149
Chest	0.250049	1.988594	0.77207	0.038662	0.077214	0.984324	0.00765	–0.00098	0.080834
Mammo	5.998262	42.91407	0.392749	0.246708	1.654902	0.369382	0.019138	0.041657	0.285818
Cardiac Angio	0.253345	2.461373	0.624316	0.040246	0.094822	0.752281	0.007303	0.000722	1.204809
Periph Angio	0.367025	3.260039	0.503562	0.04642	0.120297	0.57627	0.008103	0.000844	0.975373

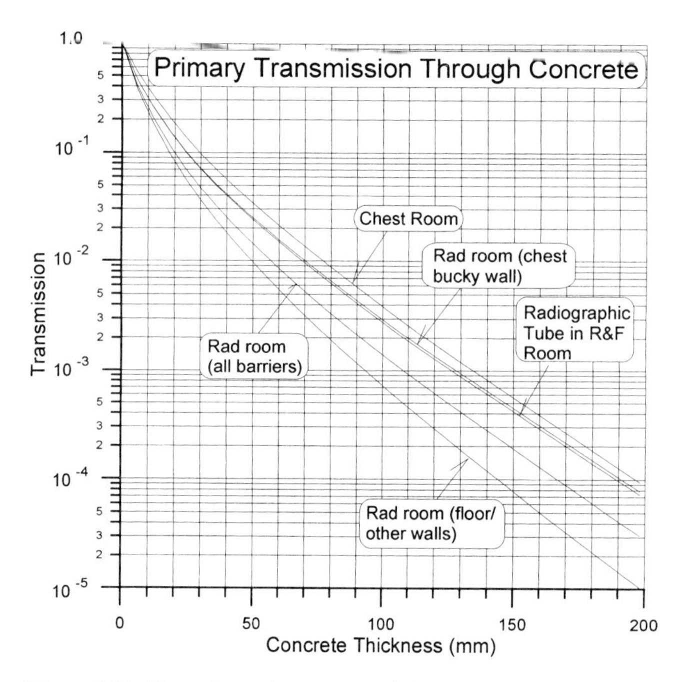

Figure 6.14. The primary beam transmission calculated in concrete. Data as in Fig. 6.13.

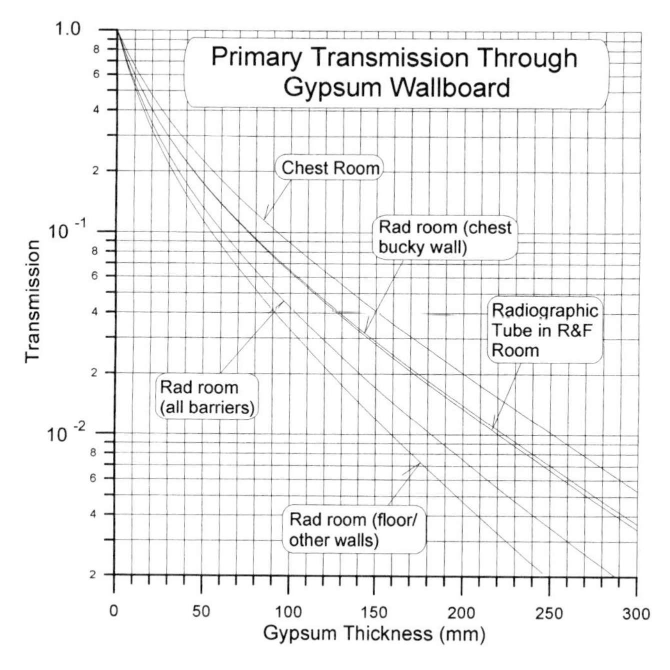

Figure 6.15. The primary beam transmission calculated in gypsum wallboard. Data as in Fig. 6.13.

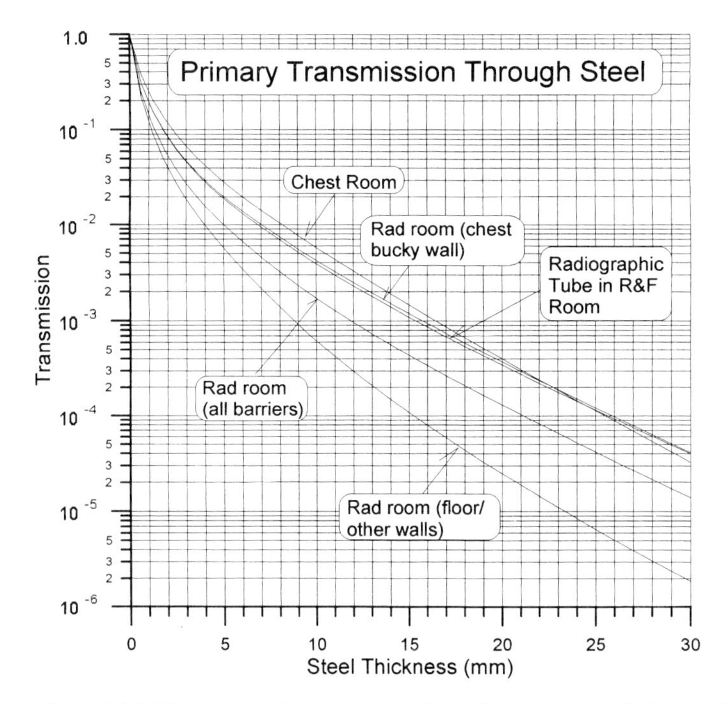

Figure 6.16. The primary beam transmission calculated in steel. Data as in Fig. 6.13.

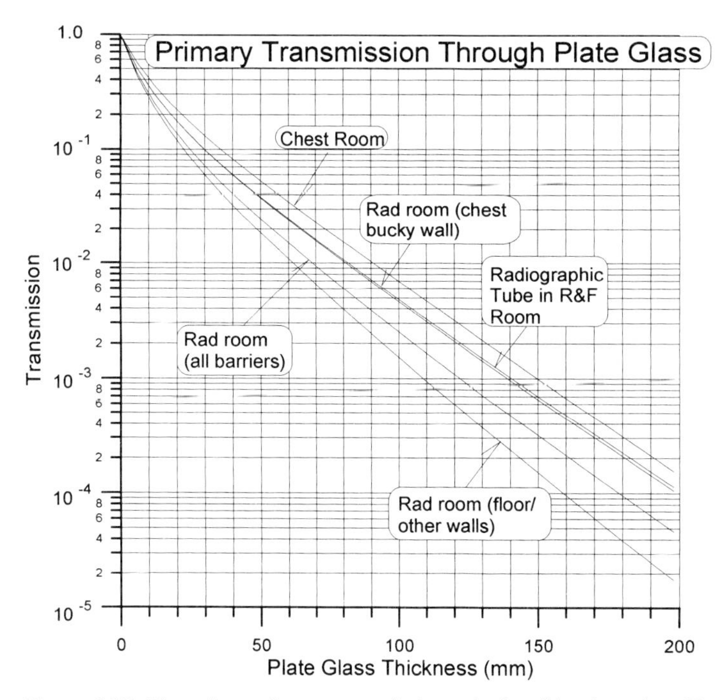

Figure 6.17. The primary beam transmission calculated in plate glass. Data as in Fig. 6.13.

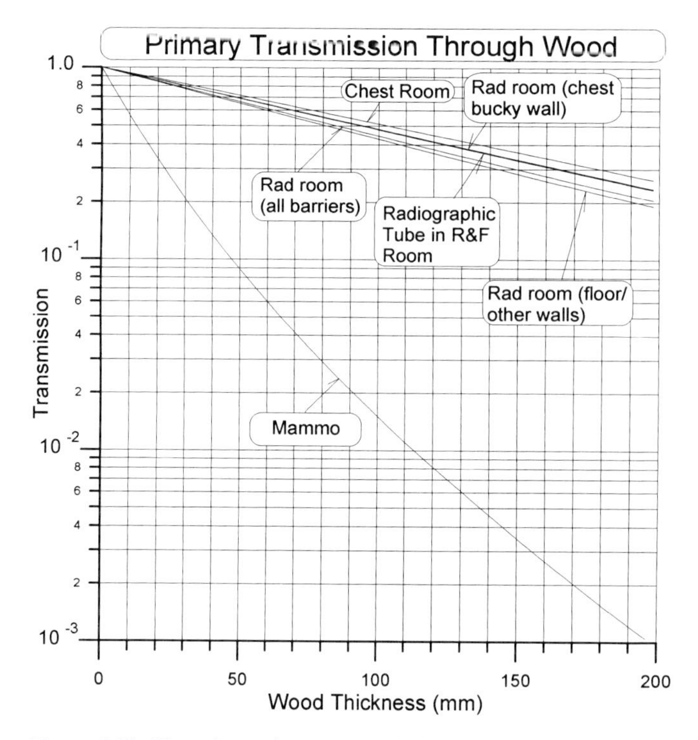

Figure 6.18. The primary beam transmission calculated in wood. Data as in Fig. 6.13.

6.5 Primary Barriers

6.5.1 Computation of Primary Barrier Thickness

The primary beam is the intense, spatially restricted radiation field which emanates from the X-ray tube portal and is incident upon the patient and image receptor. Primary protective barriers are found in radiographic rooms, dedicated chest installations and radiographic/ fluoroscopic rooms. Primary barriers include the wall on which the vertical cassette holder assembly is mounted and the floor and those walls toward which the primary beam may be occasionally directed. Since the image intensifier in general fluoroscopy, cardiac and peripheral angiography and the breast support tray in mammography are required by regulation to act as primary beam stops, these rooms do not normally contain primary barriers.

Let $\dot{D}_0(kVp)$ be the primary beam dose rate per workload (mGy per mA·min) at 1m from the X-ray source operated at potential kVp. Values of $\dot{D}_0(kVp)$ for individual X-ray tubes will depend on the generator voltage waveform, anode material, filtration, and anode angle. Figure 6.19 shows $\dot{D}_0(kVp)$ for typical Mo-anode, Mo-filtered mammography beams at and below 35 kVp, and a typical three-phase twelve pulse generated W-anode, Al-filtered radiographic beam at above 40 kVp.[9] In what follows, these beams will be taken as representative of modern clinical practice.

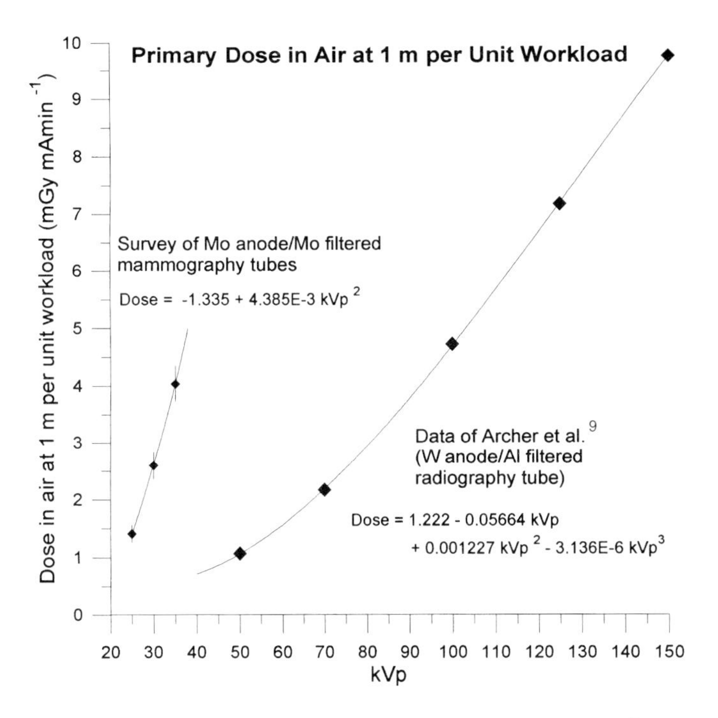

Figure 6.19. The primary beam dose in air per unit workload, $\dot{D}_0(kVp)$, measured at 1 m. Below 40 kVp a Mo anode, Mo filtered mammography beam is assumed. The graphed data come from a survey of modern mammographic units. Above 40 kVp the W anode, Al filtered radiographic data of Archer et al.[9] are shown.

Assume that the workload for this X-ray tube, $W(kVp)$, is known at operating potentials kVp. The unattenuated primary dose 1 m from the source due to the workload of this X-ray tube at this potential is just

$$D_P^1(0,\ kVp) = \dot{D}_0\,(kVp)\,W(kVp). \tag{6.8}$$

At distance d_p from the focal spot of an X-ray tube, the total primary dose due to use factor-corrected workload $U\ W(kVp)$ is

$$D_P(0,\ kVp) = \frac{\dot{D}_0\ (kVp)\ U\ W(kVp)}{d_P^2}. \tag{6.9}$$

Behind a barrier of thickness x whose transmission to primary X rays of this potential is $B_P(x,\ kVp)$, the shielded dose is

$$D_P(x,\ kVp)=\frac{\dot{D}_0\ (kVp)\ U\ W(kVp)}{d_P^2}\ B_P(x,\ kVp), \tag{6.10}$$

and the total shielded dose, $D_p(x)$, in the occupied area due to all potentials in a workload distribution is

$$D_P(x)=\sum_{kVp} D_P(x,\ kVp)=\sum_{kVp}\frac{\dot{D}_0\ (kVp)\ U\ W(kVp)}{d_P^2}\ B_P(x,\ kVp). \tag{6.11}$$

Now consider an X-ray tube whose workload follows one of the surveyed distributions[2], with total workload W_{norm} equal to the average per patient. The unshielded primary dose per patient at 1 m for the workload distributions, D_P^1, may then be calculated.

$$D_P^1=\sum_{kVp} D_P^1(kVp)=\sum_{kVp} \dot{D}_0(kVp)W(kVp). \tag{6.12}$$

Table 6.10 gives the unshielded primary beam dose per patient at 1 m calculated for the surveyed workload distributions, as well as the workload per patient, W_{norm}.

For an X-ray tube whose total workload W_{tot} is due to N_{pat} patient procedures,

$$W_{tot}=N_{pat}\ W_{norm}\,. \tag{6.13}$$

If the primary beam is directed at the occupied area only a fraction U of the time, the total workload W_{tot} (or, equivalently, the number of patients, N_{pat}) must be scaled by the use factor. At distance d_p from the focal spot of the X-ray tube, the unshielded primary dose, $D_p(0)$, is then

$$D_P(0)=\frac{D_P^1\ N_{pat}\ U}{d_P^2}=\frac{D_P^1\ W_{tot}\ U}{W_{norm}\ d_P^2}\,. \tag{6.14}$$

If the occupied area is shielded by a barrier of a given material and thickness x having primary transmission $B_p(x)$, then the dose to the occupied area is

$$D_P(x)=\frac{D_P^1\ N_{pat}\ U}{d_P^2}\ B_P(x)=\frac{D_P^1\ W_{tot}\ U}{W_{norm}\ d_P^2}\ B_P(x). \tag{6.15}$$

An acceptable barrier of thickness x_{acc} will reduce the primary dose $D_p(x_{acc})$ at d_p to the design dose limit corrected for the occupancy factor, P/T. Thus,

$$B_P(x_{acc})=\frac{P\ d_P^2}{D_P^1\ U\ T\ N_{pat}}=\frac{P\ W_{norm}\ d_P^2}{D_P^1\ W_{tot}\ U\ T}\,. \tag{6.16}$$

The shielding task is then to find the barrier whose thickness, x_{acc}, satisfies Eq. (6.16). This may be achieved graphically using the transmission curves in Figs. 6.6–6.11 and 6.13–6.18. Equivalently, the acceptable primary barrier thickness can be calculated in closed form by substituting the transmission from Eq. (6.16) into Eq. (6.7).

$$x_{acc} = \frac{1}{\alpha\gamma} \ln \left[\frac{\left(\frac{D_P^1 N_{pat} U T}{P d_P^2} \right)^{\gamma} + \frac{\beta}{\alpha}}{1 + \frac{\beta}{\alpha}} \right] = \frac{1}{\alpha\gamma} \ln \left[\frac{\left(\frac{D_P^1 W_{tot} U T}{P d_P^2 W_{norm}} \right)^{\gamma} + \frac{\beta}{\alpha}}{1 + \frac{\beta}{\alpha}} \right] \quad (6.17)$$

If, as discussed in Section 6.4.4, the image receptor is available to provide attenuation of the primary beam before it strikes the structural barrier, the thickness of the required structural barrier is only

$$x_{req} = x_{acc} - x_{pre} \,. \quad (6.18)$$

6.5.2 Example Calculation for Primary Barriers; Floor of a Radiographic Room

A typical poured concrete floor slab has a periodic structure on a steel deck. The minimum thickness of the concrete, which is less than the nominal thickness often quoted by architects and construction personnel, should normally be used in calculating the barrier equivalence. The designer should also be aware of the possibility that "light-weight" (e.g. low-density) concrete (1.8 g cm^{-3}) may be used in modern building construction rather than normal density (2.35 g cm^{-3}) concrete. Hence the required thickness of "light-weight" concrete is 31% greater than that of normal density concrete to achieve the same shielding ability. The tables and figures in this chapter are for normal density concrete. The thin steel decking provides negligible additional shielding, since beam hardening in the concrete removes all but the highest energy X rays in the primary beam.

Assume that the fully occupied area beneath the proposed normal weight concrete floor of a radiographic room is an uncontrolled area, as shown in Fig. 6.20. The workload of the room is 125 patients per week. The distance to a 2 m tall person standing in the area below is 3.8 m. Determine the required concrete thickness.

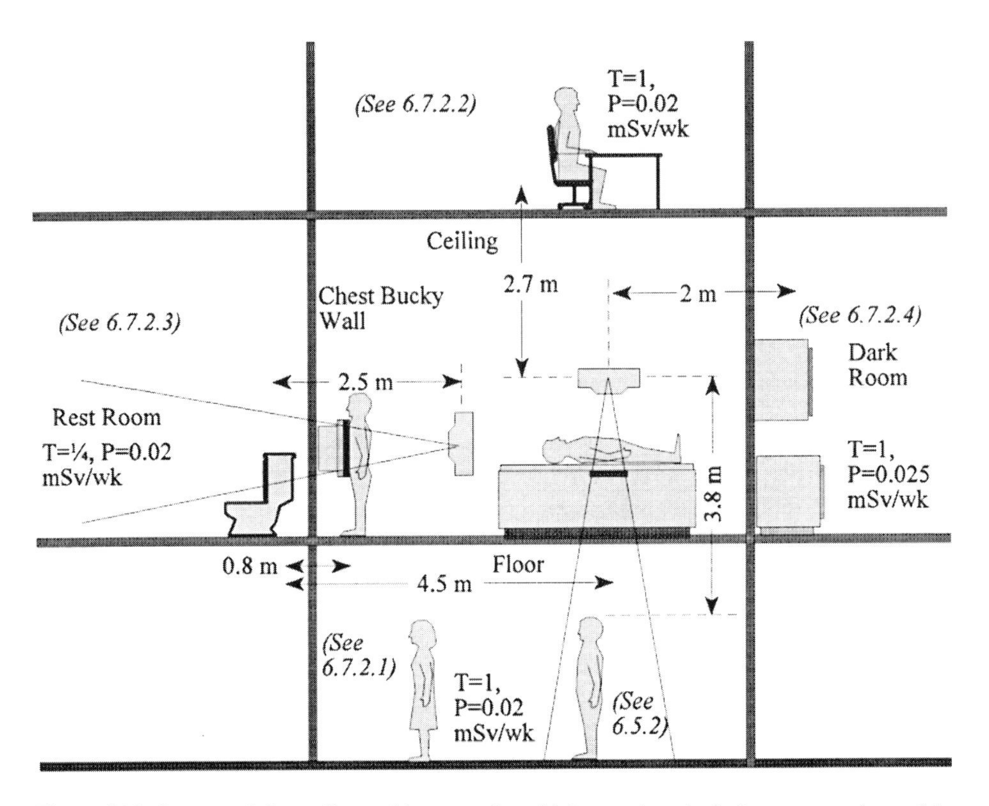

Figure 6.20. Layout of the radiographic room for which sample calculations are performed in the text. Assume 125 patient exams are performed in this room. See Sections 6.5.2 and 6.7.2 for details of the shielding calculations in each of the indicated areas.

The protocol for determining the required concrete floor thickness in a radiographic room is to compute the required primary barrier thickness, assuming that the primary beam impinges directly on the floor, and then subtract 72 mm of concrete or 0.8 mm of lead to account for the shielding capability of the image receptor and radiographic table. Table 6.10 gives the unshielded primary beam X-ray dose per patient at 1 m, D_p^1, in mSv. The shielding parameters are $N_{pat} = 125$, $P = 0.02$ mSv wk^{-1}, $D_p^1 = 5.15$ mSv, $T = 1$, $d_p = 3.8$ m, with U conservatively rounded up from 0.9 to 1. Either a graphical or algebraic method of determining the required shielding may be used.

Table 6.10. Unshielded primary dose, D_p^1, (mSv) for the indicated workloads, W_{norm}, and workload distributions, normalized to primary beam distance d_p = 1 m. These primary doses ignore the attenuation available in the radiographic table and image receptor. For the indicated clinical installations, W_{norm} is the average workload per patient, and the workload distributions are those surveyed by Task Group 9 of the X-ray Imaging Committee of the AAPM[2], described in Table 6.2.

Workload Distribution	W_{norm} Total Workload per patient (mA·min patient^{-1})	D_p^1 Unshielded Primary Dose per patient at d_p = 1 m (mSv patient^{-1})
Radiographic Rm (all barriers)	2.45	7.41
Radiographic Rm (chest bucky wall)	0.601	2.25
Radiographic Rm (floor/other walls)	1.85	5.15
Overhead radiographic tube in Rad/Fluoro Suite	1.51	5.85
Chest Room	0.216	1.21

To find the solution graphically, first calculate the unshielded dose to the occupied area. From Eq. (6.14),

$$D_P(0) = \frac{D_P^1 N_{pat} U}{d_P^2} = \frac{5.15 \text{ mSv pat}^{-1} \times 125 \text{ pat wk}^{-1} \times 1}{(3.8 \text{ m})^2} = 44.6 \text{ mSv wk}^{-1}.$$

The barrier transmission required to reduce this to the design dose limit is

$$B_p = \frac{0.02 \text{ mSv wk}^{-1}}{44.6 \text{ mSv wk}^{-1}} = 4.5 \times 10^{-4}.$$

Using the primary transmission curve for concrete (Fig. 6.14) for the *Radiographic Room (floor / other barrier)* workload distribution, a 110 mm concrete barrier is required if attenuation by the patient, X-ray table, and cassette assembly were ignored. From Table 6.7, the attenuation provided by a typical radiographic table/image receptor (ignoring patient attenuation) is equivalent to 72 mm concrete for this workload distribution. Thus the net concrete thickness required in the floor under the X-ray table to attenuate the primary beam to 0.02 mSv wk^{-1} is 110 mm–72 mm = 38 mm or 1.5 inches.

The solution may equivalently be found algebraically by using Eq. (6.17). The fitting parameters for the primary beam for the *Radiographic Room (floor / other barrier)* workload distribution in concrete are given in Table 6.9. Here α = 0.039935 mm^{-1}, β = 0.144765 mm^{-1}, and γ = 0.423108. Substituting into Eq. (6.17) yields the required total concrete thickness.

$$X_{acc} = \frac{1}{0.0399 \times 0.4231} \ln \left[\frac{\left(\frac{5.15 \times 125 \times 1 \times 1}{0.02 \times 3.8^2} \right)^{0.4231} + \frac{0.14477}{0.0399}}{1 + \frac{0.14477}{0.0399}} \right] = 110 \text{ mm}.$$

Subtracting the preshielding thickness of 72 mm gives the same result obtained above, 38 mm.

It is instructive to compare these results with those obtained using the methods of NCRP49. Using a workload of 1 000 mA·min wk^{-1} at 100 kVp, as recommended in NCRP49 for a radiographic room with a patient load of 125 patients wk^{-1}, and assuming that the primary beam impinges directly on the floor (ignoring table attenuation), the required concrete thickness is 180 mm (7.1 inches). Even with a more realistic total workload of 125 patients wk $\times$ 1.85 mA·min $patient^{-1}$ = 231 mA·min wk^{-1}, as in our example, the NCRP49 calculation at 100 kVp requires 144 mm (5.7 inches) concrete. From the example, if the realistic kVp workload distribution for the floor is used but X-ray table attenuation is ignored, the required concrete thickness is reduced to 110 mm (4.3 inches). Finally, with X-ray table attenuation included, the required concrete thickness is only 38 mm (1.5 inches). Use of these more realistic (yet still conservative) assumptions allows a significant savings in concrete thickness in the floor. By the time the shielding designer is asked to participate, the floor slab design has usually already been established by the structural engineers, with the minimum concrete thickness often not exceeding 3 inches. The shielding designer and architect are then faced with the difficult construction problem of adding lead to the floor, which is both costly, and as shown here, unnecessary.

6.6 Secondary Barriers

6.6.1 Secondary Radiation

Secondary radiation is an unavoidable consequence of the primary X-ray beam. Barriers which are otherwise never struck by the primary beam must therefore serve as adequate shields against scatter and leakage. In some radiologic imaging situations regulations require the primary beam to be completely intercepted by an absorbing barrier behind or incorporated into the image receptor. This is the case for operation of an image intensifier and dedicated mammography systems. The dose to an occupied area from primary radiation is thus assumed nil, and that from scatter and leakage will predominate.

6.6.2 Scatter Radiation

The intensity of x-radiation scattered off the patient is dependent on the scattering angle, θ (defined from the direction of the center of the primary beam to a ray pointing to the occupied area), the number of primary photons incident on the phantom, the primary beam photon energy, and the location of the X-ray beam on the phantom. It is presumed that, all else being equal, the number of primary photons incident on the phantom varies linearly with the X-ray beam field size. Thus for fixed kVp, mAs, and collimator jaw opening, the scatter intensity is independent of the distance from the primary X-ray source to the phantom.

Trout and Kelley[16] made a series of widely accepted radiographic scatter measurements 100 cm from the center of a phantom, which were then related to the primary dose at 1 m. This ratio of scatter to primary dose, when divided by the primary beam field size at 1 m primary distance, defines the scatter fraction. Unfortunately, the filtration of the X-ray beams used by Trout and Kelley at 50 and 70 kVp are not typical of X-ray systems used today, invalidating their results at these lower potentials. Dixon[10] repeated their measurement for 90° scatter over a range of potentials. His results indicate a linear increase in scatter fraction with kVp. The Trout and Kelley data at 100, 125, and 150 kVp have been reanalyzed[17] for scatter fraction, a_1, measured per cm^2 of primary beam area, scaled by 10^6. Scatter fractions at lower potentials were obtained using linear extrapolation in kVp. The scatter fraction a_1 is broadly distributed over a range of beam sizes, with coefficients of variation in a_1 with field size on the order of 30%. Figure 6.21 shows a_1 determined from the Trout and Kelley publication at the mean plus one standard deviation level as a function of scattering angle and operating potential. Figure 6.21 also shows a_1 for mammographic beams measured by Simpkin[18].

Consider the primary beam from an X-ray tube incident on a patient. At 1 m primary distance, with an area of 1 cm^2, this tube delivers primary dose D_p^1 at potential kVp. By the definition of the scaled scatter fraction, a_1, at scattering angle θ, the unshielded dose 1 m from the center of the patient due to scatter radiation is

$$D_S(\theta, kVp) = D_P^1(kVp) \times a_1(\theta, kVp) \times 10^{-6}. \tag{6.19}$$

Note that the scaled scatter fraction a_1 read from Figure 6.21 has simple values between 0.1 and 8. At secondary distance d_S (meters) from the center of the patient, the scatter dose is modified by d_S^{-2}. As noted above, it is assumed that the scatter dose scales linearly with primary X-ray beam area. If the primary beam area is F (cm^2) at primary distance d_F (in meters), then the field size at 1 m primary distance is $F \times d_F^{-2}$. In diagnostic radiology it is convenient to take F as the image receptor area and d_F as the source-to-image receptor distance (SID). Thus the unshielded scatter dose D_s at secondary distance d_S from the patient is given by

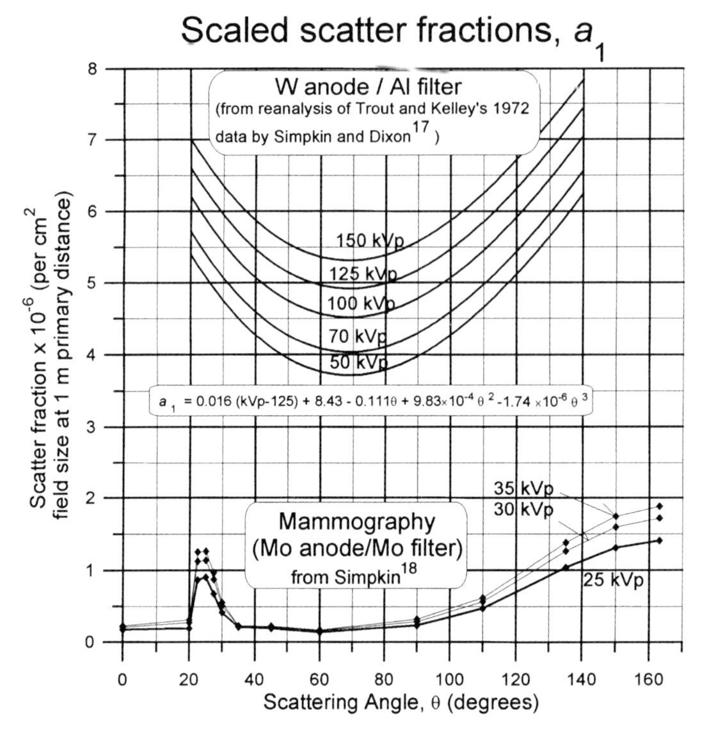

Figure 6.21. The scatter fraction per cm^2 of primary beam area scaled by 10^{-6}, a_1. This is defined as the ratio of the scatter dose at 1 m distance to the primary dose at 1 m, per primary field area. The scaled scatter fractions for the W anode / Al anode are derived from the data of Trout and Kelley [16]. The data at the mean plus one standard deviation level are shown for 100, 125, and 150 kVp. The 125 kVp data have been fit to a cubic equation. This fit is extrapolated to other potentials assuming the linear kVp variation of the 90° scatter data of Dixon[10] and Trout and Kelley. The Mo anode/Mo filter mammography scatter fractions were measured by Simpkin.[18]

$$D_S(\theta, kVp) = \frac{D_P^1(kVp) \times a_1(\theta, kVp) \times 10^{-6}}{d_S^2} \frac{F}{d_F^2}. \tag{6.20}$$

Note that, as in Eq. (6.8), $D_P{}^1(kVp)$ is simply $\dot{D}_0(kVp)\ W(kVp)$. Behind a shielding barrier of thickness x having transmission $B(x,kVp)$, assumed identical to that of the primary beam, the scatter dose is

$$D_S(x, \theta, kVp) = \frac{\dot{D}_0(kVp) W(kVp) a_1(\theta, kVp) \times 10^{-6}}{d_S^2} \frac{F}{d_F^2} B(x,kVp). \tag{6.21}$$

If a fraction U of the X-ray tube's workload is expended as a primary beam directed at this barrier, the workload available to generate scatter radiation on this barrier should be reduced[19] from $W(kVp)$ to $(1-U)W(kVp)$. Then

$$D_S(x, \theta, kVp) = \frac{\dot{D}_0(kVp)(1-U)\ \ W(kVp) a_1(\theta, kVp) \times 10^{-6}}{d_S^2} \frac{F}{d_F^2} B(x,kVp). \tag{6.22}$$

For tube operation over a range of potentials, the total scatter dose is simply the sum over the operating potentials.

$$D_S(x,\theta) = \sum_{kVp} D_S(x,\theta,kVp). \tag{6.23}$$

6.6.3 Leakage Radiation

Leakage radiation is limited by regulation to 0.1 R h^{-1} at 1 m at the maximum operating potential $kVp_{\max}$ and tube current $mA_{\max}$ at which the tube can be operated continuously. For radiographic tubes this is typically 150 kVp at 3 to 5 mA, and 50 kVp at 5 mA for mammographic tubes. The amount of shielding required in the housing to limit transmission to the regulatory limit is based on these techniques even though the tube is rarely operated at these techniques.

A model predicting the leakage dose for tube operation at potential kVp can be developed by assuming that the leakage dose rate for operation with *no* tube housing matches the primary beam dose rate. The thickness of the Pb-lined housing can then be specified by knowing the leakage technique factors, the primary transmission curves, the leakage dose limit L, and assuming that the primary dose rate varies as kVp^2. The leakage dose rate 1 m from the X-ray tube operated at potential kVp and tube current I is then

$$\dot{D}_L(kVp) \propto kVp^2\, I\, B_{housing}(kVp), \tag{6.24}$$

where $B_{housing}$(kVp) is the transmission through the tube housing. For leakage technique factors of 150 kVp at 3.3 mA, a Pb-lined housing 2.32 mm thick is required to reduce the leakage intensity rate at 1 m to 0.1 R hr^{-1} = 0.876 mGy hr^{-1} = 0.0146 mGy min^{-1}. The ratio of the leakage dose rate at 1 m at clinical parameters kVp and I to that at the leakage technique factors $kVp_{\max}$ and $I_{\max}$ yields

$$\dot{D}_L(kVp) = \frac{L\, kVp^2\, B_{housing}(kVp)\, I}{kVp_{\max}^2\, B_{housing}(kVp_{\max})\, I_{\max}}. \tag{6.25}$$

Note that this assumes the highest allowed dose rate at the leakage technique factors. This is usually assumed to be conservative by factors of at least 2 to 10. Integrating Eq. (6.25) over time yields the leakage dose, D_L, accumulated due to operation at potential kVp and workload $W(kVp)$. The workload is simply the time integral of the tube current. Consider an area located at leakage distance d_L from the X-ray tube. The transmission of leakage radiation through a shielding barrier of thickness x will be $exp(-(ln\ 2)\ x\ /\ HVL(kVp))$, where HVL is the half value layer through the barrier material at high attenuation. The values of HVL are shown in Figure 6.12. If a fraction U of the X-ray tube's workload is expended as a primary beam directed at this barrier, the workload available to generate leakage radiation on this barrier should be reduced[19] from $W(kVp)$ to $(1-U)W(kVp)$. The leakage dose to this shielded area is then

$$D_L(x,kVp)=\frac{L\,kVp^2\,B_{housing}(kVp)(1-U)\;W(kVp)}{kVp_{max}^2\,B_{housing}(kVp_{max})\,I_{max}}\,\frac{e^{\left(\frac{-\ln 2\,x}{HVL(kVp)}\right)}}{d_L^2}, \tag{6.26}$$

with the total leakage dose equal to the sum over the operating potentials in the workload.

Note that this devolves to Eq. 5m in NCRP49 should the clinical operating parameters match the leakage technique factors. For a tube with typically-assumed leakage technique factors of 150 kVp at 3.3 mA, the NCRP49 model overpredicts the unshielded leakage dose by factors of 10^6, 500, and 4.5 for operation at 60, 80, and 100 kVp, respectively. The NCRP49 model predicts leakage doses about 20 times greater than those calculated for the workload distribution for radiographic tubes. Even larger disparity is found for leakage doses behind typical shielding barriers. These differences are due to the far greater transmission of radiation at the leakage technique factor as compared to that at typical clinical operating potentials.

6.6.4 The Total Secondary Barrier and Secondary Transmission

Consider a radiation barrier that is not struck by primary radiation. The total radiation dose behind this secondary radiation barrier will then be due only to secondary radiations. The total secondary dose, $D_{sec}(x)$, behind a secondary barrier of thickness x is simply the sum of the scatter and leakage doses.

$$D_{sec}(x)=D_S(x)+D_L(x). \tag{6.27}$$

The unshielded secondary radiation dose is predicted by Eq. (6.27) with $x = 0$. Table 6.11 shows the scatter, leakage, and total secondary radiation doses at 1 m calculated for a variety of workload distributions at typical X-ray beam sizes and leakage technique factors. The total unshielded secondary doses in this table, D_{sec}^1, were calculated for scatter at 90° ("side-scatter") and 135° ("forward and back-scatter") assuming distances $d_S = d_L = d_{sec}$. The workload distributions for the clinical sites are taken from the AAPM-TG9 survey, with total workload, W_{norm}, equal to the average workload per patient reported by the survey. The unshielded secondary dose is seen to be due almost exclusively to scatter. It should, however, be anticipated that beam hardening in the barrier will substantially increase the contribution of leakage to the dose in the shielded area, so it is prudent to not ignore leakage radiation in a shielding calculation.

Table 6.11. Unshielded secondary doses, $D_{sec}{}^1$, (mSv) for the indicated workload distributions at $d_S = d_L = 1$ m. The workload distribution and total workload W_{norm} for the indicated clinical sites are the average per patient surveyed by Task Group 9 of the X-ray Imaging Committee of the AAPM[2], listed in Table 6.2. The primary field size F (cm^2) is known at primary distance d_F. Side scatter is calculated for 90° scatter. Forward and backscatter are conservatively calculated for 135° scatter. Leakage technique factors are: 150 kVp at 3.3 mA to achieve 0.1 R h^{-1} for all tubes save mammography, which assumes leakage technique factors of 50 kVp at 5 mA.

	W_{norm}			$D_{sec}{}^1$, Unshielded Secondary Dose (mSv) per workload W_{norm} at 1 m				
Workload Distribution	Total Workload (mA·min)	F (cm^2) at	d_F (m)	Leakage	Side-scatter	Side-scatter Total	Forward/ Back Scatter	Forward/ Back Total
50 kVp (W anode)	1.0	1000	1.00	1.23×10^{-11}	4.24×10^{-3}	4.24×10^{-3}	6.34×10^{-3}	6.34×10^{-3}
70 kVp (W anode)	1.0	1000	1.00	4.70×10^{-7}	9.44×10^{-3}	9.44×10^{-3}	1.38×10^{-2}	1.38×10^{-2}
100 kVp (W anode)	1.0	1000	1.00	9.90×10^{-4}	2.24×10^{-2}	2.34×10^{-2}	3.17×10^{-2}	3.26×10^{-2}
125 kVp (W anode)	1.0	1000	1.00	2.56×10^{-3}	3.73×10^{-2}	3.98×10^{-2}	5.14×10^{-2}	5.39×10^{-2}
150 kVp (W anode)	1.0	1000	1.00	4.42×10^{-3}	5.44×10^{-2}	5.88×10^{-2}	7.36×10^{-2}	7.80×10^{-2}
Radiographic Rm (all barriers)	2.45	1000	1.00	5.32×10^{-4}	3.37×10^{-2}	3.42×10^{-2}	4.83×10^{-2}	4.88×10^{-2}
Radiographic Rm (chest bucky wall)	0.60	1535[a]	1.83	3.88×10^{-4}	4.91×10^{-3}	5.30×10^{-3}	6.94×10^{-3}	7.33×10^{-3}
Radiographic Rm (floor/other barriers)	1.85	1000	1.00	1.44×10^{-4}	2.30×10^{-2}	2.31×10^{-2}	3.31×10^{-2}	3.32×10^{-2}
Fluoroscopy Tube in R & F Rm	12.9	730[b]	0.80	1.16×10^{-2}	0.314	0.326	0.443	0.455
Radiographic Tube in R & F Rm	1.51	1000	1.00	9.42×10^{-4}	2.78×10^{-2}	2.87×10^{-2}	3.92×10^{-2}	4.02×10^{-2}
Chest Room	0.216	1535[a]	2.00	3.81×10^{-4}	2.31×10^{-3}	2.69×10^{-3}	3.22×10^{-3}	3.60×10^{-3}
Mammography Suite (Mo anode)	6.69	720[c]	0.58	1.14×10^{-5}	1.13×10^{-2}	1.13×10^{-2}	4.89×10^{-2}	4.89×10^{-2}
Cardiac Angiography	160	730[b]	0.90	8.83×10^{-2}	2.61	2.70	3.70	3.79
Peripheral Angiography	64.1	730[b]	0.90	3.38×10^{-3}	0.655	0.658	0.946	0.950

Notes: [a] The area of a 14″ × 17″ field.
[b] The area of a 12″ diameter image intensifier.
[c] The area of a 24 cm × 30 cm cassette.

The ratio of the secondary dose behind a barrier of thickness x to the unshielded dose defines the secondary transmission $B_{sec}(x)$

$$B_{sec}(x) = \frac{D_{sec}(x)}{D_{sec}(0)}. \tag{6.28}$$

Figures 6.22 through 6.27 show the transmission of secondary radiation through Pb, concrete, gypsum, steel, plate glass, and wood for 25–150 kVp operation for leakage technique factors of 50 kVp at 5 mA for 25–35 kVp Mo anode X rays, and 150 kVp at 3.3 mA for the W anode X rays. Figures 6.28 through 6.33 show the secondary transmission for the AAPM-TG9 surveyed clinical workload distributions. The fitting parameters of the transmission curves in Figures 6.22 through 6.33 to Eq. (6.6) are given in Table 6.12. These curves assume 90° scatter, that the transmission of scatter radiation matches that of the primary beam, and that distances $d_S = d_L$. Note that the choice of 90° scatter is conservative, in that the scatter fraction at 90° is relatively small, leading to a smaller scatter contribution to the secondary dose. The secondary transmission will therefore be greater with the small value of the scatter fraction, since the more penetrating leakage radiation will thereby have a greater

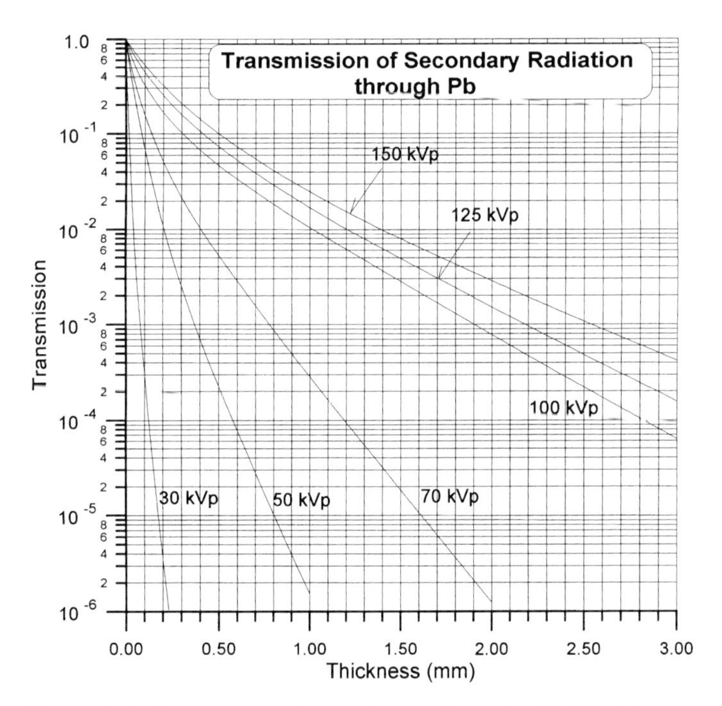

Figure 6.22. Transmission of secondary radiation for 30 kVp Mo anode mammography X rays, and 50, 75, 100, 125, and 150 kVp W anode X rays in Pb. This assumes 90° scatter, a primary beam size of 1000 cm^2 at 1 m, $d_s = d_L$, and leakage technique factors of 50 kVp, 5 mA for 30 kVp operation and 150 kVp, 3.3 mA for all others.

Table 6.12. Fitting parameters of the secondary transmission curves in Figs. 6.22 through 6.33 to Eq. (6.6). Thicknesses input in mm. The 30 kVp and *Mammography room* data are for molybdenum-anode X-ray tubes. All other data are for tungsten-anode tubes. The appropriateness of the fits should not be assumed for barrier thicknesses beyond those plotted in Figs. 6.22–6.33. For completeness, it should be added that the 25 and 35 kVp molybdenum-anode X-ray tube secondary transmission curves in wood are fit by the (α, β, γ) parameters of (0.02229, 0.04341, 0.1937) for 25 kVp and (0.01882, 0.03878, and 0.3825) for 30 kVp.

	Lead			Concrete			Gypsum wallboard		
Workload Distribution	α (mm^{-1})	β (mm^{-1})	γ	α (mm^{-1})	β (mm^{-1})	γ	α (mm^{-1})	β (mm^{-1})	γ
30 kVp	38.79	180.0	0.3560	0.3174	1.725	0.3705	0.1198	0.7137	0.3703
50 kVp	8.801	27.28	0.2957	0.0903	0.1712	0.2324	0.0388	0.0873	0.5105
70 kVp	5.369	23.49	0.5883	0.0509	0.1697	0.3849	0.0230	0.0716	0.7300
100 kVp	2.507	15.33	0.9124	0.0395	0.0844	0.5191	0.0147	0.0400	0.9752
125 kVp	2.233	7.888	0.7295	0.0351	0.0660	0.7832	0.0120	0.0267	1.079
150 kVp	1.791	5.478	0.5678	0.0324	0.0775	1.566	0.0104	0.0202	1.135
Rad Rm (all barriers)	2.298	17.38	0.6193	0.0361	0.1433	0.5600	0.0138	0.0570	0.7937
Rad Rm (chest bucky)	2.256	13.80	0.8837	0.0356	0.1079	0.7705	0.0127	0.0445	1.049
Rad Rm(floor/other barriers)	2.513	17.34	0.4994	0.0392	0.1464	0.4486	0.0164	0.0608	0.7472
Fluoro tube R&F	2.322	12.91	0.7575	0.0363	0.0936	0.5955	0.0133	0.0410	0.9566
Rad tube R&F	2.272	13.60	0.7184	0.0356	0.1114	0.6620	0.0129	0.0457	0.9355
Chest Room	2.288	9.848	1.054	0.0364	0.0659	0.7543	0.0130	0.0297	1.195
Mammography Room	29.91	184.4	0.3550	0.2539	1.8411	0.3924	0.0883	0.7526	0.3786
Cardiac Angiography	2.354	14.94	0.7481	0.0371	0.1067	0.5733	0.0139	0.0464	0.9185
Peripheral Angiography	2.661	19.54	0.5094	0.04219	0.1559	0.4472	0.01747	0.06422	0.7299

Table 6.12. (continued).

Workload Distribution	Steel			Plate Glass			Wood		
	α (mm^{-1})	β (mm^{-1})	γ	α (mm^{-1})	β (mm^{-1})	γ	α (mm^{-1})	β (mm^{-1})	γ
30 kVp	7.408	42.49	0.4061	0.3060	1.620	0.3793	0.02159	0.03971	0.2852
50 kVp	1.817	4.840	0.4021	0.09721	0.1799	0.4912	0.010764	0.001862	1.170
70 kVp	0.7149	3.798	0.5381	0.05791	0.1357	0.5968	0.00855	0.000539	1.194
100 kVp	0.3424	2.456	0.9388	0.04279	0.08948	1.029	0.00723	0.000894	1.316
125 kVp	0.2138	1.690	1.086	0.03654	0.05790	1.093	0.006587	–0.00114	1.172
150 kVp	0.1511	1.124	1.151	0.03267	0.04074	1.134	0.006027	–0.00163	1.440
Rad Rm (All barriers)	0.2191	3.490	0.7358	0.03873	0.1054	0.6397	0.007552	0.000737	1.044
Rad Rm (chest bucky)	0.2211	2.836	1.123	0.03749	0.0871	0.9086	0.007058	0.000229	1.875
Rad Rm(floor/other barriers)	0.2440	3.012	0.5019	0.04299	0.1070	0.5538	0.007887	0.000877	0.980
Fluoroscopic Tube R&F	0.2331	2.213	0.8051	0.03886	0.08091	0.8520	0.007057	0.000422	1.664
Radiographic Tube R&F	0.2149	2.695	0.8768	0.03762	0.08857	0.8087	0.007102	0.000345	1.698
Chest Room	0.2518	1.829	1.273	0.03866	0.06270	1.128	0.007485	–0.00081	0.09459
Mammography	5.798	44.12	0.4124	0.2404	1.709	0.3918	0.018882	0.04172	0.2903
Cardiac Angiography	0.2530	2.592	0.7999	0.04001	0.09030	0.8019	0.007266	0.000674	1.235
Peripheral Angiography	0.3579	3.466	0.5600	0.04612	0.1198	0.5907	0.008079	0.000847	0.9742

contribution to the secondary dose. At low (< 100 kVp) potentials, the leakage contribution through the tube housing is nil, so that the secondary transmission is little different from the primary. At higher potentials the increased penetration of leakage radiation makes the secondary transmission exceed the primary. This is most pronounced for Pb and steel. For 100 kVp X rays, the secondary transmission through a typical 1.58 mm (1/16 inch) Pb barrier exceeds the primary transmission by 50%.

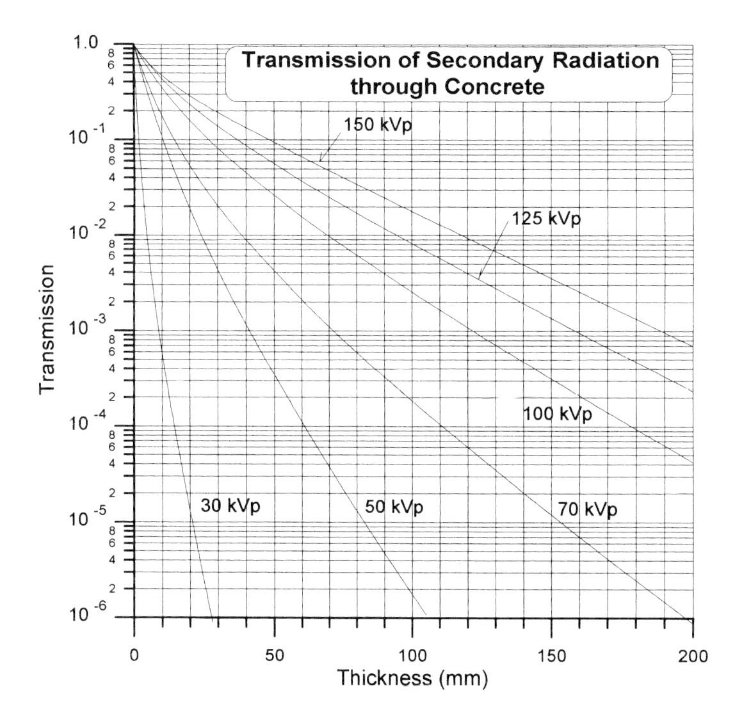

Figure 6.23. Transmission of secondary radiation in concrete. Data as in Fig. 6.22.

Given knowledge of the secondary dose per patient (or known workload W_{norm}) and the secondary transmission, a simple shielding protocol for secondary radiations may be developed analogous to Eqs. (6.14) through (6.17). The unshielded secondary dose at 1 m for N_{pat} patients (or equivalently, total workload W_{tot}) is simply $D^1_{\sec} N_{pat} = D^1_{\sec} \dfrac{W_{tot}}{W_{norm}}$. An inverse square correction is needed to extrapolate the unshielded dose to distance $d_{\sec}$.

$$D_{\sec}(0) = \frac{D^1_{\sec} N_{pat}}{d^2_{\sec}} = \frac{D^1_{\sec} W_{tot}}{W_{norm}\, d^2_{\sec}} . \tag{6.29}$$

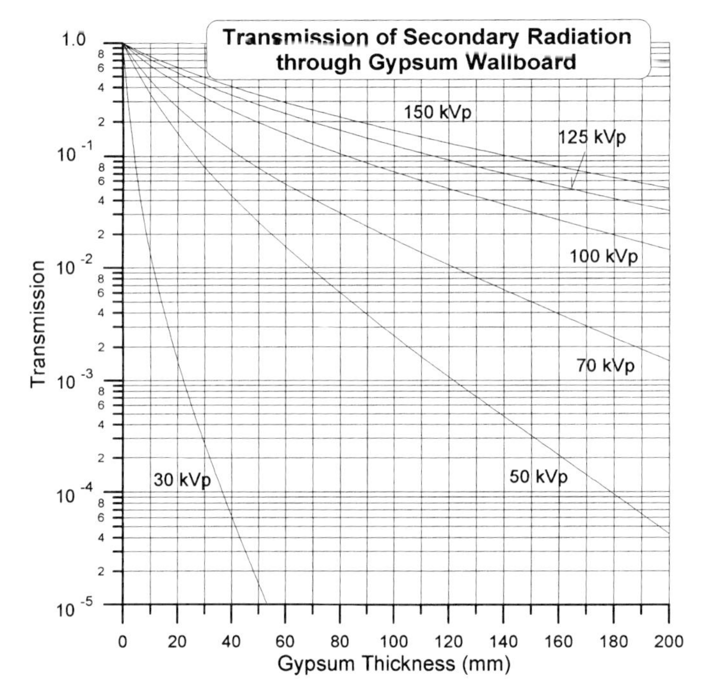

Figure 6.24. Transmission of secondary radiation in gypsum wallboard. Data as in Fig. 6.22.

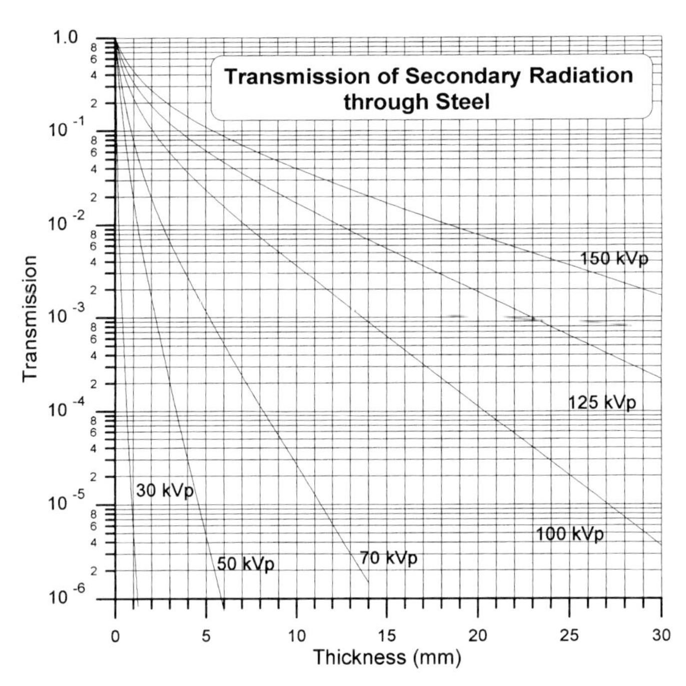

Figure 6.25. Transmission of secondary radiation in steel. Data as in Fig. 6.22.

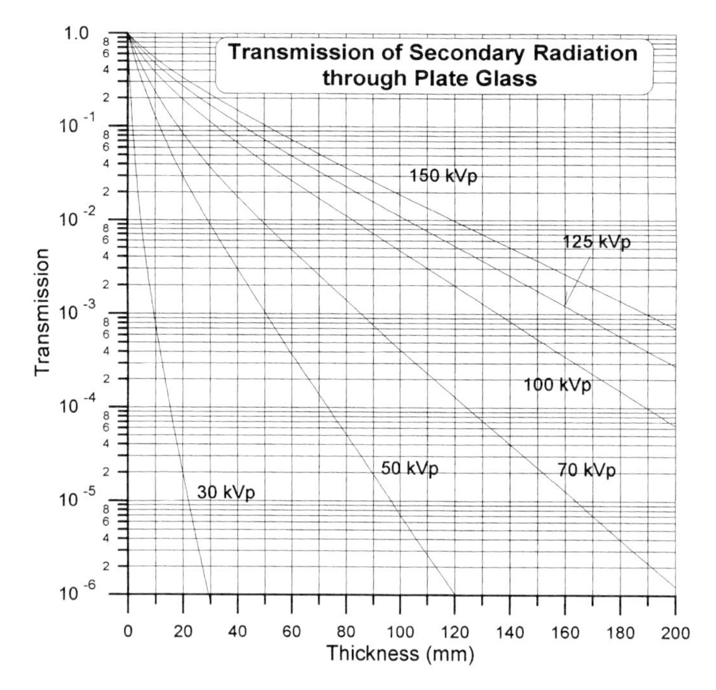

Figure 6.26. Transmission of secondary radiation in plate glass. Data as in Fig. 6.22.

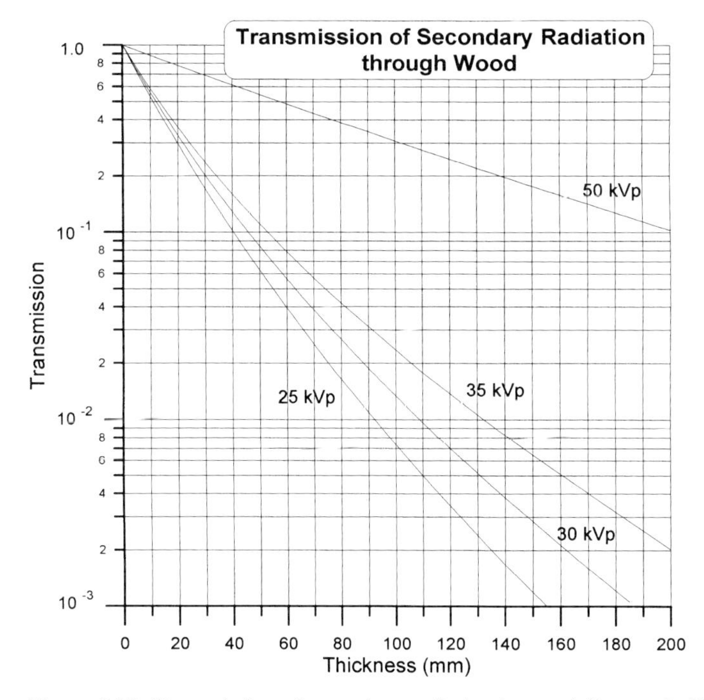

Figure 6.27. Transmission of secondary radiation in wood. Data as in Fig. 6.22.

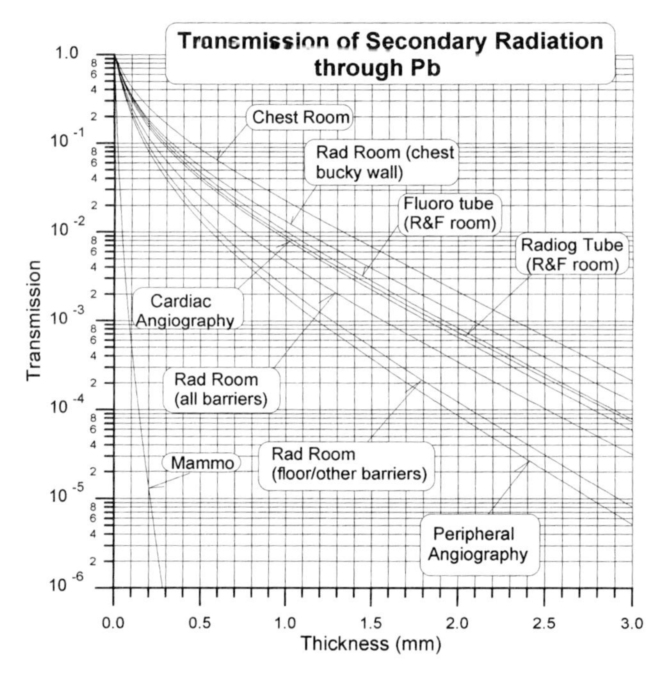

Figure 6.28. Transmission of secondary radiation in Pb for the clinical installations surveyed by Task Group 9 of the X-ray Imaging Committee of the AAPM.[2] This assumes 90° scatter, primary beam sizes listed in Table 6.11, $d_s = d_L$, and leakage technique factors of 150 kVp at 3.3 mA.

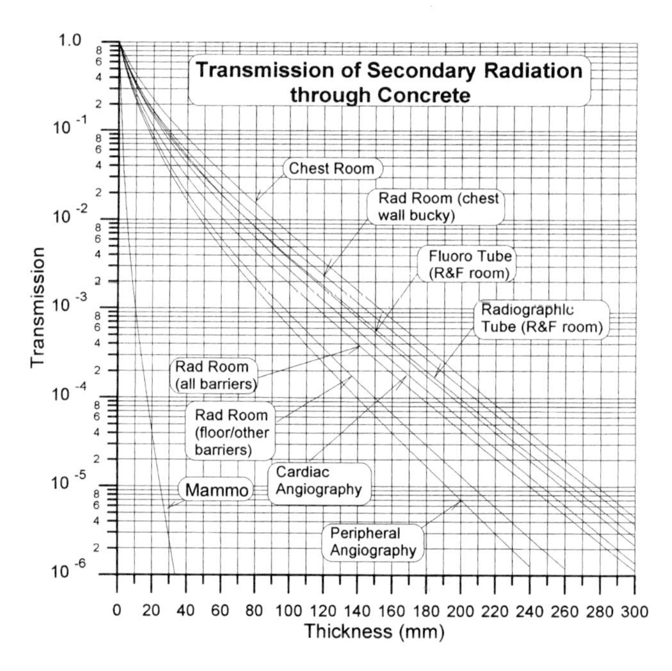

Figure 6.29. Transmission of secondary radiation in concrete for clinical installations. Data as in Fig. 6.28.

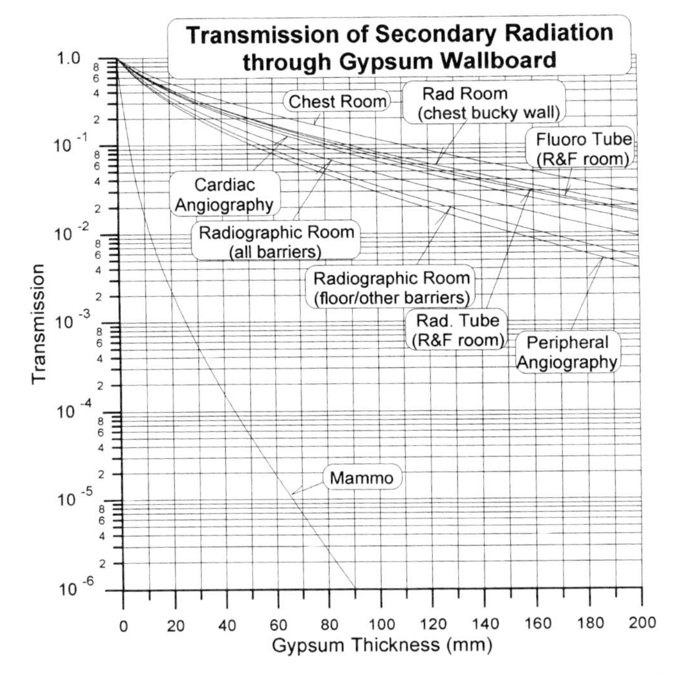

Figure 6.30. Transmission of secondary radiation in gypsum wallboard for clinical installations. Data as in Fig. 6.28.

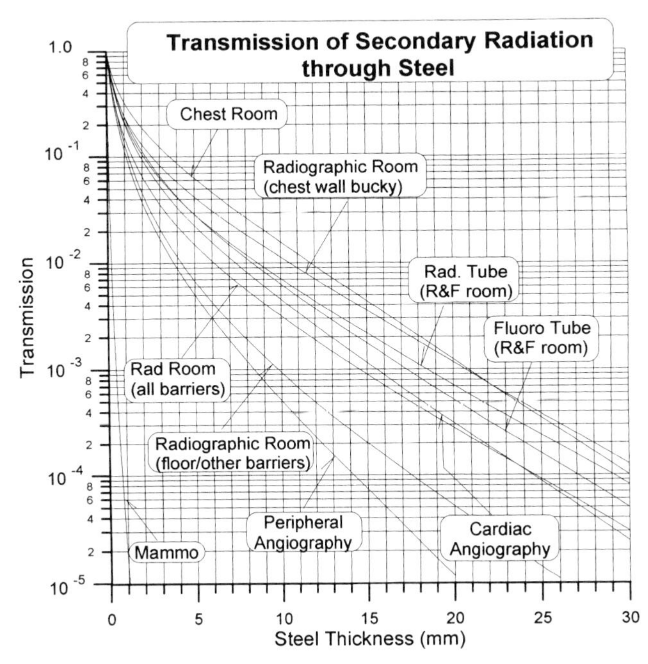

Figure 6.31. Transmission of secondary radiation in steel for clinical installations. Data as in Fig. 6.28.

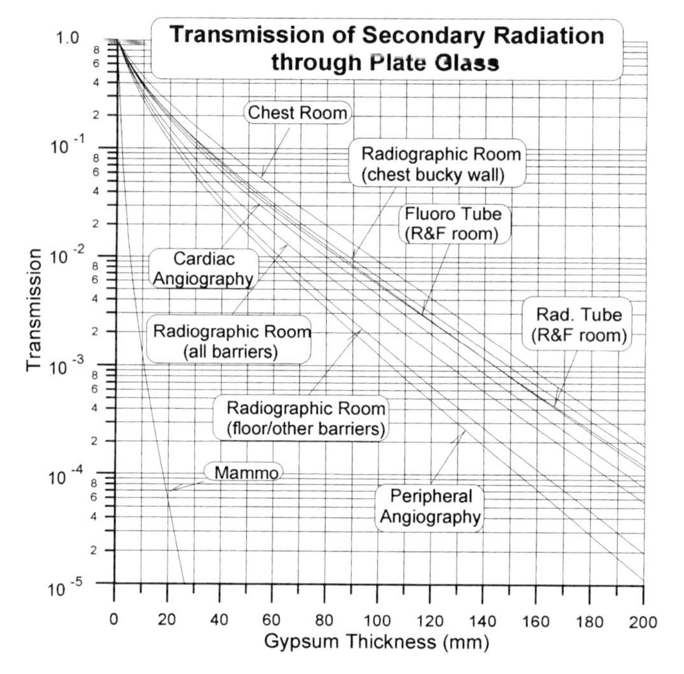

Figure 6.32. Transmission of secondary radiation in plate glass for clinical installations. Data as in Fig. 6.28.

The secondary radiation will be attenuated in a barrier of secondary transmission $B_{sec}(x)$, so that the transmitted secondary dose through the shielding barrier is

$$D_{\sec}(x)=\frac{D_{\sec}^{1}N_{pat}}{d_{\sec}^{2}}B_{\sec}(x)=\frac{D_{\sec}^{1}W_{tot}}{W_{norm}\,d_{\sec}^{2}}B_{\sec}(x). \tag{6.30}$$

An acceptable barrier thickness, x_{acc}, is one which limits the transmitted secondary dose to P/T, so that

$$B_{\sec}(x_{acc})=\frac{P}{T}\frac{d_{\sec}^{2}}{D_{\sec}^{1}N_{pat}}=\frac{P}{T}\frac{W_{norm}\,d_{\sec}^{2}}{D_{\sec}^{1}W_{tot}}. \tag{6.31}$$

Using the fitting parameters to Eq. (6.6) in Table 6.12 to describe the transmission curves, the required barrier thickness may be determined algebraically.

$$x_{acc}=\frac{1}{\alpha\gamma}\ln\left[\frac{\left(\frac{D_{sec}^{1}\,N_{pat}\,T}{P\,d_{sec}^{2}}\right)^{\gamma}+\frac{\beta}{\alpha}}{1+\frac{\beta}{\alpha}}\right]=\frac{1}{\alpha\gamma}\ln\left[\frac{\left(\frac{D_{sec}^{1}\,W_{tot}\,T}{P\,d_{sec}^{2}\,W_{norm}}\right)^{\gamma}+\frac{\beta}{\alpha}}{1+\frac{\beta}{\alpha}}\right]. \tag{6.32}$$

6.6.5 Example Calculations for Secondary Barriers; Cardiac Angiography Laboratory

Consider first a cardiac angiography laboratory in which 25 patient procedures are performed per week. Assume a 12″ diameter image intensifier and a 0.9 m SID. An uncontrolled ($P = 0.02$ mSv wk^{-1}) fully occupied ($T = 1$) area is $d_{sec} = 4$ m from the X-ray unit. Conservatively assume 135° scatter. From Table 6.11 the unshielded dose per patient at 1 m for this case is 3.79 mSv.

The unshielded dose in the occupied area is calculated from Eq. (6.29) to be

$$D_{sec}(0) = \frac{3.79\ \text{mSv pat}^{-1} \times 25\ \text{patients wk}^{-1}}{(4\,\text{m})^2} = 5.92\ \text{mSv wk}^{-1},$$

of which just 2% is due to leakage radiation. The transmission of the shielding barrier must not exceed

$$B_{sec}(x_{acc}) = \frac{0.02\ \text{mSv wk}^{-1}}{5.92\ \text{mSv wk}^{-1}} = 0.0034,$$

which, from Fig. 6.27, is provided by $x_{acc} = 1.3$ mm Pb. The contribution of leakage radiation to the transmitted dose has increased to 32% since the leakage radiation is harder than the scatter and therefore more penetrating.

This barrier thickness could equivalently be determined algebraically. From Table 6.12 for secondary radiation generated by an cardiac angiography unit, $\alpha = 2.354$ mm^{-1}, $\beta = 14.94$ mm^{-1}, and $\gamma = 0.7481$ for transmission through lead. Substituting into Eq. (6.32),

$$x_{acc} = \frac{1}{2.354 \times 0.7481} \ln\left[\frac{\left(\dfrac{3.79 \times 25 \times 1}{0.02 \times 4^2}\right)^{0.7481} + \dfrac{14.94}{2.354}}{1 + \dfrac{14.94}{2.354}}\right] = 1.3\ \text{mm}.$$

6.7 The Mixed Radiation Barrier

6.7.1 Overview

The techniques in the preceding sections provide simple methods for calculating shielded and unshielded dose contributions to an occupied area from an X-ray tube, given assumptions about the tube's use, room geometry, and type of occupancy in the shielded area. Only in the simplest situations will it be possible to precisely prescribe the thickness of the shielding barrier that will decrease the transmitted dose to the design dose limit. These cases include single X-ray tubes in fixed geometry, or multiple X-ray tubes each having the same transmission through the barrier.

In the more general case, multiple X-ray tubes, or single tubes used in different locations generating X rays of varying transmission, will each contribute dose through a barrier. The thickness of that barrier required to decrease the transmitted dose to an acceptable level can be found by a number of approximation techniques, all of which tend to achieve accuracy only at the price of complexity.

A simple technique is to use the methods of sections 6.5.1 and 6.6.4 to 1) calculate the unshielded doses from each X-ray tube, 2) sum the doses to determine the total unshielded dose, 3) take the ratio of (P/T) to the total unshielded dose as the required barrier transmission, and 4) from the most penetrating of the transmission curves for the X-ray sources in the room, graphically estimate the barrier thickness required to give this transmission. This procedure will generally be conservative and prove accurate if the transmission curves of the various X-ray sources are similar.

A useful, although computationally intensive, technique is to iteratively find the barrier thickness which decreases the sum of the transmitted dose contributions to (P/T). Consider two test barrier thicknesses, x_1 and x_2, for which the total transmitted dose has been calculated to be D_1 and D_2, respectively. Assume x_1 and x_2 bound the solution, such that $x_1 < x_2$ and $D_1 > P/T > D_2$. From the shape of the transmission curves, it is reasonable to use exponential interpolation to estimate the thickness x_{est} at which the dose is P/T. That is,

$$x_{est} \approx x_1 + (x_2 - x_1) \frac{\ln\left(\frac{P/T}{D_1}\right)}{\ln\left(\frac{D_2}{D_1}\right)}. \tag{6.33}$$

The dose $D(x_{est})$ is then compared to P/T, and x_{est} used as a new upper (or lower) bounding thickness. This procedure may be used iteratively to find the value x_{est} yielding $D(x_{est})$ that approaches P/T to any desired precision.

6.7.2 Example Calculations for a Radiographic Room

Accurately determining the minimum barrier thicknesses needed for a simple radiographic room with a single X-ray tube is surprisingly complicated. The room will have typically three different image receptors, namely the radiographic table, the wall-mounted upright Bucky film holder, and the cross-table grid and cassette holder. As evidenced by the AAPM-TG9 survey results[2], exposures made on the wall-mounted image receptor will occur at significantly higher potentials than those used elsewhere in the room. The shielding afforded in the cross-table image receptor is significantly different than that available in the table or wall-mounted image receptor assembly.

Consider as an example a radiographic room in which $N_{\mathrm{pat}} = 125$ patient exams are performed every week. The workload distribution is assumed to follow that of the radiographic room from the AAPM-TG9 survey. An elevation plan for this room and its surroundings is shown in Fig. 6.20.

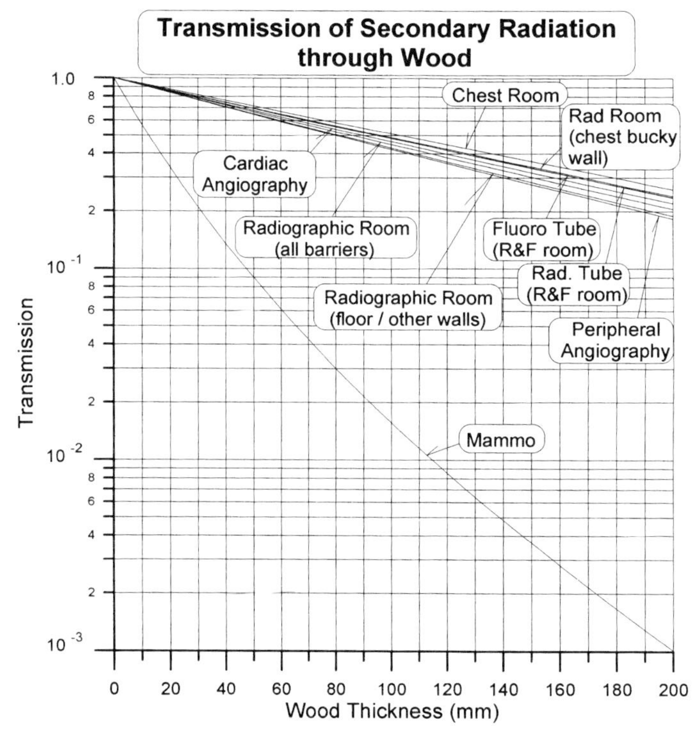

Figure 6.33. Transmission of secondary radiation in wood for clinical installations. Data as in Fig. 6.28.

6.7.2.1 Secondary Barrier Calculation for the Floor of a Radiographic Room

Since the floor of the radiographic room must act not only as a primary barrier under the radiographic table (see Section 6.5.2 above), but also as a secondary barrier for exposures directed at barriers other than the table, the floor should be analyzed for its suitability as a secondary barrier for the floor areas beyond the X-ray table. For example, we must shield the location of the woman in Fig. 6.20 Note that this secondary radiation will impact the floor directly and not be attenuated by the table-mounted image receptor hardware. For simplicity, we will assume that all scatter and leakage is produced following the workload distribution labeled *Radiographic Room (all barriers)*. Conservatively assume that the X-ray tube is located so that the scatter and leakage distances are $d_s \approx d_L \approx d_{sec} = 3$ m. From Table 6.11, the total unshielded secondary dose per patient at 1 m, assuming 90° scatter, is 3.42×10^{-2} mSv per week. Thus, the weekly unshielded secondary dose for 125 patients is

$$D_{sec}(0) = \frac{3.42 \times 10^{-2}\,\text{mSv patient}^{-1} \times 125\ \text{patients wk}^{-1}}{(3\ \text{m})^2} = 0.48\ \text{mSv wk}^{-1}.$$

To reduce this to 0.02 mSv wk^{-1} a barrier of transmission

$$B = \frac{0.02\ \text{mSv wk}^{-1}}{0.48\ \text{mSv wk}^{-1}} = 4.2 \times 10^{-2}$$

is required. From Figure 6.29 for the transmission of secondary radiation through concrete and utilizing the curve for the *Radiographic Room (all barriers)* workload distribution, the required concrete thickness is found to be 33 mm. Hence the 36 mm concrete thickness required in Section 6.5.2 under the radiographic table will suffice for the entire floor

6.7.2.2 Secondary Barrier Calculation for the Ceiling of a Radiographic Room

This area is uncontrolled ($P = 0.02$ mSv wk^{-1}) with an occupancy factor $T = 1$. This barrier is purely a secondary barrier. Assume, as above, that only one X-ray tube location is needed, with $d_L = 2.7$ m and $d_s = 3.5$ m, and conservatively set $d_s = d_L = d_{sec} = 2.7$ m. Assuming the more conservative forward/backscatter value for the *Radiographic Room (all barriers)* workload distribution, the unshielded dose per patient from Table 6.11 is 4.88×10^{-2} mSv. The unshielded total dose is then

$$D_{sec}(0) = \frac{4.88 \times 10^{-2}\,\text{mSv patient}^{-1} \times 125\ \text{patients wk}^{-1}}{(2.7\ \text{m})^2} = 0.84\ \text{mSv wk}^{-1}.$$

To reduce this to 0.02 mSv wk^{-1} requires a secondary barrier of transmission

$$B = \frac{0.02\ \text{mSv wk}^{-1}}{0.84\ \text{mSv wk}^{-1}} = 2.4 \times 10^{-2}.$$

Figure 6.29 indicates that a concrete ceiling at least 44 mm thick will provide this transmission.

6.7.2.3 Primary Barrier Calculation for the Wall Behind the Chest Bucky

Assume that the area behind the wall-mounted bucky is a bathroom in the (controlled) X-ray department, and hence $P = 0.1$ mSv wk^{-1}. From Table 6.6 the suggested occupancy factor for radiation workers in a toilet is $T = 1/4$. Hence P/T is 0.4 mSv wk^{-1}. Regions of this bathroom will receive primary radiation penetrating through the image receptor, while other areas of the bathroom not immediately behind the upright Bucky will receive only secondary radiations. It is prudent to analyze the shielding requirements of these areas separately.

From Eq. (6.14) and Table 6.10 with the *Radiographic Room (chest bucky wall)* workload distribution, the weekly unshielded primary dose at 2.5 m from the chest tube is

$$D_P(0) = \frac{2.25\ \text{mSv patient}^{-1} \times 125\ \text{patients wk}^{-1} \times 1}{(2.5\,\text{m})^2} = 45\ \text{mSv wk}^{-1}.$$

Note that $U = 1$ is used here, as all of the workload in the *Radiographic Room (chest bucky wall)* distribution is directed at the upright bucky. The primary barrier transmission required of this wall is

$$B = \frac{0.4\ \text{mSv wk}^{-1}}{45\ \text{mSv wk}^{-1}} = 8.9 \times 10^{-3}.$$

Using the *Radiographic Room (chest bucky wall)* workload distribution transmission curve in Figure 6.13, the total required lead thickness if patient and image receptor attenuation were ignored would be 0.84 mm. From Table 6.7 the attenuation provided by a typical wall-mounted cassette holder is equivalent to 0.85 mm of lead. Therefore, no primary barrier shielding is required for this barrier. This is conservative since we have ignored attenuation of the primary beam by the patient.

6.7.2.4 Secondary Barrier Calculation for the Wall Behind the Chest Bucky

There are two sources of scatter and leakage radiation contributing secondary radiation dose to the bathroom. One is the secondary radiation generated by the over-table exposures. The other is the secondary radiation from exposures made against the upright bucky itself.

The unshielded secondary radiation dose from the over-table X-ray tube location can be calculated from Eq. (6.29) and Table 6.11. Assume 90° scatter, and the *Radiographic Room (floor/other barriers)* workload distribution with $d_{sec} = 4.5$ m. Then

$$D_{sec}(0) = \frac{2.31 \times 10^{-2}\ \text{mSv patient}^{-1} \times 125\ \text{patients} \times \text{wk}^{-1}}{(4.5\ \text{m})^2} = 0.14\ \text{mSv wk}^{-1}.$$

The scatter and leakage due to exposures made against the upright bucky must also be considered. Let the scatter distance from the patient against the upright bucky to the occupied area be $d_S = 0.8$ m. The leakage distance from the X-ray tube to this area is $d_L = 2.5$ m. As the assumption that d_S equals d_L is not met in this case, consider the scatter and leakage dose contributions independent of one another. From Table 6.11, the unshielded scatter and leakage doses are

$$D_{\text{sec}}(0) = \left(\frac{4.91 \times 10^{-3} \dfrac{\text{mSv}}{\text{patient}}}{(0.8 \text{ m})^2} + \frac{3.88 \times 10^{-4} \dfrac{\text{mSv}}{\text{patient}}}{(2.5 \text{ m})^2} \right) \times 125 \, \frac{\text{patients}}{\text{wk}}$$

or 0.96+0.008 = 0.97 mSv wk^{-1}. To this must be added the secondary radiation from the over-table tube location. The total unshielded secondary dose is then

$$D_{\text{sec}}(0) = 0.97 + 0.14 = 1.1 \text{ mSv wk}^{-1}.$$

Conservatively assuming that the more penetrating *Radiographic Room (chest bucky wall)* workload distribution is appropriate for both sources, then from Figs. 6.28 and 6.30 a barrier of 0.12 mm Pb or 27 mm of gypsum wallboard is required to limit the dose to 0.4 mSv wk^{-1}. A more exact calculation using the correct location for each scatter/leakage source and a 30° scattering angle for the chest source with the correct workload distribution for each tube location yields 0.092 mm Pb and 27.4 mm gypsum. Since the normal construction material of two layers of 5/8″ gypsum board is barely adequate in this case, the designer should probably opt for adding 1/32″ lead to this wall if it is new construction. Note that 1/32″ is the minimum commercially available lead thickness.

6.7.2.5 Dark Room wall, Protection of X-ray Film

NCRP49 recommends that shielding for radiographic film limit the exposure to the film to 0.02 mGy per storage period. Suleiman et al.[20] have shown that for typical modern X-ray films radiation levels of about 50 to 300 times greater than the NCRP49 limit are required to produce fog levels of 0.05 optical density. Based on these studies, a reasonable new limit for the exposure of medical X-ray film is 0.1 mGy per storage period. If a storage period of one month is assumed, then the limit is 0.025 mGy per week, nearly the same dose limit as for a fully-occupied uncontrolled area. Note that since film storage shelving is often located higher than the usual shielding height of 2.1 m (7 ft) above the floor, it is desirable to extend the shielding in the darkroom wall to at least 2.4 m (8 ft) above the floor.

For the radiographic room in Fig. 6.20, assume that no exposures are made against the darkroom wall, so that only secondary radiation need be shielded against. For simplicity, assume that all secondary doses are generated with the X-ray tube over the table for a secondary distance of 2 m. The *Radiographic Room (all barriers)* workload distribution is appropriate in this case. Then from Table 6.11 with 90° scatter, the unshielded secondary dose in the darkroom is

$$D_{sec}(0) = \frac{3.42 \times 10^{-2} \text{ mSv patient}^{-1} \times 125 \text{ patients wk}^{-1}}{(2 \text{ m})^2} = 1.07 \text{ mSv wk}^{-1}.$$

The required barrier will have transmission

$$B = \frac{0.025 \text{ mSv wk}^{-1}}{1.07 \text{ mSv wk}^{-1}} = 2.3 \times 10^{-2}.$$

The required lead shielding from Figure 6.28 is therefore 0.53 mm lead.

The film passbox between the dark room and the radiographic room will typically contain unexposed film loaded in cassettes, thereby greatly increasing the sensitivity of the film to radiation-induced fogging. Assuming that all the cassettes in the passbox will be recycled in one day (eg., once every 25 patients), the unshielded dose to a cassette in the passbox would then be

$$D_{sec}(0) = \frac{3.42 \times 10^{-2} \text{ mSv patient}^{-1} \times 25 \text{ patients wk}^{-1}}{(2 \text{ m})^2} = 0.213 \text{ mSv}.$$

Assuming 0.5 μSv will fog a film in a cassette[20], the shielding for this cassette must provide a transmission of not more than 0.0005 / 0.213 = 0.0023. From Figure 6.28, this requires a lead thickness of 1.3 mm in the front door of the film passbox. The Suleiman study[20] points out that unexposed X-ray film loaded into a cassette is about 100 times more sensitive to fogging from radiation than film not in a cassette. Thus, in-room areas where loaded cassettes are stored, such as inside the passbox or inside the operator's control booth, may require special consideration.

The practice of using lead discs or tabs to cover the drywall screws or nails used to attach lead-bonded gypsum wallboard has been evaluated by D. Shearer and J. Gray (personal communications). They found that only very small perforations in the sheet lead are made when the screw is inserted normally and that the attenuation of the steel screw compensates for the lead displaced from the hole. However, if lead disks are to be used, the screw is often pounded into the wall with a hammer and an indention made in the drywall to allow the disc to be glued flush with the wall surface. This process typically causes a major rip in the lead, compromising the integrity of the barrier. It is therefore concluded that the only time that lead disks are required is when the surface is prepared to receive the lead disks.

6.8 Evaluation of Shielding Adequacy

The final steps in the shielding process involve the acceptance test of the shielded X-ray facility. A survey should be conducted to assure the adequacy of the installed shielding in terms of the barrier thickness and location. Once the shielding has been applied but before the walls are closed, a visual inspection will assure barrier adequacy and placement. Radiation surveys of the completed installation may be performed. The presence of voids in the shielding may be determined by placing an unshielded gamma-emitting radionuclide source in the room, and surveying the room environs with a sensitive detector such as a scintillation survey meter. Local nuclear licensing and regulatory concerns would have to be addressed before this is attempted. A completed installation may also be surveyed using radiographic exposures inside the room. An X-ray source, either the installed unit or a portable radiographic device, and patient-equivalent phantom may be used to simulate room use. Radiation doses at points of concern outside the room may be measured with an exposure meter, such as an ionization survey meter operating in the 10^{-2} Gy (μR) range. The weekly dose may then be estimated by scaling the measured dose by the ratio of the assumed weekly workload to the workload used for the measurement. At the time of this writing, the details of these procedures are being addressed by the NCRP Committee updating NCRP49.

6.9 References

1. National Council on Radiation Protection and Measurements. Bethesda, MD: NCRP: NCRP Report No. 49; (1976).
2. D.J. Simpkin, *Med. Phys.* **23**, 577 (1996).
3. D.J. Simpkin, *Health Phys.* **61**, 259, (1991).
4. National Council on Radiation Protection and Measurements. Bethesda, MD: NCRP; NCRP Report No. 116; (1993).
5. R. L. Tanner, American Association Of Physicists in Medicine Newsletter, **21,** 12 (May/June 1996).
6. J. R. Cameron, American Association Of Physicists in Medicine Newsletter, **21,** 8 (September/October 1996).
7. ICRP Publication 51, International Commission on Radiological Protection. (International Commission on Radiation Protection, Pergamon Press, Oxford, 1987).
8. D.J. Simpkin, *Med. Phys.* **21**, 893 (1994).
9. B.R. Archer, et al., *Med. Phys.* **21**, 1499 (1994).
10. R.L. Dixon, *Med. Phys.* **21**, 1785 (1994).
11. R.L. Dixon, D.J. Simpkin, *Health Phys.* **74**, 181 (1998).
12. J.M Légaré, et al., *Radioprotection* **13,** 79 (1977).
13. D.J. Simpkin, *Health Phys.* **53**, 267 (1987).
14. B.R. Archer, J.I. Thornby, S.C. Bushong, *Health Phys.* **44**, 507 (1983).
15. D.J. Simpkin, *Health Phys.* **68**, 704 (1995).

16. E. D. Trout and J.P. Kelley, *Radiology* **104**, 161 (1972).
17. D. J. Simpkin, R.L. Dixon, *Health Phys.* **74**, 350 (1998).
18. D.J. Simpkin, *Health Phys.* **70**, 238 (1996).
19. D.J. Simpkin, *Health Phys.* **52**, 431 (1987).
20. O.H. Suleiman, et al., *Med. Phys.* **22** (10), 1691 (1995).

7 Ultrasound (Including Doppler)

Carolyn Kimme-Smith

UCLA School of Medicine, Radiology

7.1 Introduction

Medical ultrasound, with its immediate visual feedback and lack of biological effect (at diagnostic power levels) is an enticing area of specialization for the medical physicist. Because the physics is based on mechanical energy propagation and wave theory, it is a pleasant diversion from photon interactions and the production of ionizing radiation. Furthermore, the application of radar principles to ultrasound has given us unique capabilities for measuring in-vivo blood flow using Doppler methods. For the hospital based medical physicist, a knowledge of medical ultrasound quality control procedures will ensure that equipment is suitable for the intended medical task, and that quantitative measurements are accurate.

Medical ultrasound is used throughout most hospitals and clinics, not just in the radiology department. It is usual to find imaging ultrasound in cardiology, obstetrics, ophthalmology, and vascular laboratories. It has become common for upper gastrointestinal, prostate, breast, and sinus examinations. It can be used to direct a core needle biopsy; intraoperative use during neurological surgery is less experimental than it was five years ago. With all these applications, most hospital based physicists will find between 15 and 40 ultrasound units at their facility. Hospital administrators are often not aware that the medical physicist employed for radiation safety or X-ray equipment calibration is also trained in ultrasound and can be a valuable consultant when purchasing or repairing ultrasound equipment. For this reason, medical physicists should understand the physical principles behind ultrasonic imaging.

7.2 Physical Principles of Medical Ultrasound

In order to understand resolution and scattering properties of medical imaging ultrasound (US), it is initially necessary to model US as a continuous sonic wave with a single frequency traveling through a liquid medium. Because the wave travels through a non-viscous medium, it is unattenuated. A liquid medium also ensures that only longitudinal waves will propagate; no shear waves perpendicular to the longitudinal waves need be considered in our model as they would need to be considered if the tissue were bone, for instance. Once modeled, we can study the changes that must occur in our model when we pulse the US wave, producing a spectrum of frequencies, and change the liquid medium to one of soft tissues, containing cells and boundaries between tissues with acoustic impedance differences.

7.2.1 Wave Representation of Mechanical Energy

Acoustic theory usually begins with derivation of the wave equation for longitudinal waves (in the direction of wave propagation) from Hooke's law, with

$$\frac{\delta^2 u}{\delta z^2} = \frac{1}{c^2}\frac{\delta^2 u}{\delta t^2} \tag{7.1}$$

where c = propagation velocity of sound in the nonviscous medium, u = the amount of displacement of a particle in the medium, and z is the distance traveled in the medium in time t[1]. The solution of this equation is $u = u_o \sin k(ct - z)$, where k is the wave number, equal to $2\pi/\lambda$ (λ is the wavelength at frequency f). Note that because the medium is lossless, u_o is constant. If energy is attenuated in the medium, then the solution becomes

$$u = u_o e^{-\alpha z} \sin 2\pi\,(ft - z/\lambda) \tag{7.2}$$

where α is the attenuation coefficient in the medium.

With this solution in place, we can model the surface of a source of continuous wave ultrasonic energy as a number of point sources, vibrating at frequency f. Huygen's principle allows us to calculate the diffraction pattern that results from this model when the transducer face is a circular disk of radius r. Along z, perpendicular to the surface of the disk and at the center of the disk, the intensity of the wave changes proportional to $\sin^2(\pi/\lambda - z)$.

When normalized by the initial intensity, and in a nonattenuating medium, the maxima occur at

$$z = \frac{4r^2 - \lambda^2(2m+1)^2}{4\lambda(2m+1)} \quad \text{where } m = 0, 1, 2 \ldots \tag{7.3}$$

The last maximum occurs at $m = 0$, when $z_{\text{last}} = (4r^2 - \lambda^2)/4\lambda$.

7.2.1.1 Transducer Beam Profile Model

When the radius is much larger than the wavelength (which occurs when medical ultrasound frequencies prevail), then the position of the last maximum can be approximated by $T = r^2/\lambda$. This is a transition point between what we call the near field or Fresnel zone and the far field, or Fraunhofer zone. In the near field, intensity oscillates and in a lossless medium does not decrease in peak value. However, when it reaches a distance T from the source, it begins to spread and decreases according to $1/z^2$. In the far field, we describe the wave as spherical, while in the near field, the model is a plane piston source. Unfortunately, very few transducers which produce medical ultrasound waves are circular. Most transducers are fabricated in small rectangular or annuli sections which, when combined, approximate larger surfaces. When a rectangular transducer vibrates, the local intensity maxima increase until the last maximum value, which is less than that of a circular transducer and which occurs farther from the face of the transducer than the maximum intensity of a circular transducer.

This model assumes that the wave has a single frequency and that it propagates in a nonattenuating, liquid medium. Medical imaging ultrasound is pulsed, rather than continuous, so it consists of a mixture of frequencies, and soft tissue consists of tissue cell interfaces and an attenuating medium which produce scattering and impedance mismatches that form images when reflected ultrasound is received by the transmitting transducer.

7.2.1.2 Scattering Models

Backscatter from cellular interfaces has been modeled in several ways, depending on whether discrete or continuous models of tissue are used[2]. For each type of tissue model, the scattering model is considerably simplified if the Born approximation holds. This assumption states that the amplitude from scattering is so small when compared to the amplitude of the incident wave that the effects of multiple scattering and interference with the incidence wave can be neglected.

One of the earliest scattering models described Rayleigh scattering, which occurs when the scattering objects are discrete points arranged with equal spacing. The spacing must also be far enough apart so that diffraction effects do not occur. This tissue model assumes that each point scatters the incidence wave to produce an isotropic spherical wave with pressure

$$p_s(z) = \frac{Ae^{i(kz-2\pi f)}}{z} \tag{7.4}$$

The sum of all scattered waves forms the scatter model equation. It assumes that the diameter of the scatterers, a, is very much smaller than a wave length, λ; that is, a $\ll \lambda$. It also assumes that the distance between scatterers, d, is very large, so that $d \gg \lambda$. As this spacing decreases, but remains constant throughout the tissue, then interference phenomena must be included in the total scattering equation, and the Born approximation no longer holds. If $d \gg \lambda$ and varies over the tissue model, while the scatterers also vary in diameter, but still remain much less than λ, then ray tracing, using a classical Boltzmann integro-differential equation, can be used to sum the scattering effects from many scatterers. The intensity obtained from these models is always proportioned to $1/\lambda^4$.

When the distance between scatterers is close to a wavelength, then there is correlation between the scatter, and the Born assumption does not hold. Because of this, Foldy[3] developed a method of describing the interactions of the incident and scattered waves in terms of coherent and incoherent parts. He was able to show that single scattering events were coherent with respect to the incident wave and proportional to N^2, where N is the number of scatterers per unit volume. Multiple scattering events, proportional to N, contribute the incoherent or "out of phase" components to the scattering model.

When we try to include other physical properties of the tissue in the scattering models, we derive a different set of scattering equations. Continuous tissue models usually incorporate terms for the mean density and mean velocity (or mean compressibility or mean bulk modulus). They also assume that deviations from these means will be small. Some tissue models use a correlation function rather than the mean to characterize properties in the tissue. For example

$$R(x) = \mathrm{e}^{-x^2/d^2}$$

describes the changes in R as x changes with respect to a correlation distance. The constant d represents the distance at which the tissue property is uncorrelated.

This equation serves as a basis to derive a scattering cross section dependent on the frequency f, compressibility $1/\beta$, attenuation α and cell size a:

$$s \sim \frac{f^2 a}{\beta^2}\left(1 - \mathrm{e}^{-(2\pi\frac{f}{a})^2}\right) - \alpha^2 f^3 \left(\mathrm{e}^{-(2\pi\frac{f}{a})^2}\right) \tag{7.5}$$

Note that cell size is based on an average cell size over the correlation distance R, rather than a discrete cell size as was used in the previous models. When average cell size is very small with respect to frequency, the amount of compressibility in the tissue controls scattering, while for larger average cell sizes, attenuation also contributes. The higher the frequency, the more scatter can be expected. These models illustrate the complexity of ultrasound interactions with tissue and the need to understand these complexities in order to predict the behavior of medical ultrasound in tissue.

7.2.2 Concepts Needed for Ultrasound Imaging

7.2.2.1 Impedance and Reflections

Scattering, as modeled above, consists of reflections off surfaces that are the size of a wavelength or smaller. Specular reflection, on the other hand, is the reflection of US from surfaces which are 10λ or greater in size[1]. Investigation of these effects are based on more realistic assumptions than the models we have previously examined. First we must note that reflection only occurs when differences between the two tissues create a boundary. The significant factors of this boundary consist of the speed of sound (propagation velocity), c, in the two tissues and their densities, ρ. The product of $c\rho$ is called acoustic impedance, Z.

Propagation velocity, c, differs for different tissues because it depends on tissue density and bulk modulus, or elasticity. Thus $c = \sqrt{\beta/\rho}$. So for very stiff tissue, such as bone, we can expect velocity to be high (despite its increased density), while for fat, which at body temperature is soft, it is low. These factors also affect the wavelength of US, since $\lambda = c/f$. In bone, with a velocity of 3500 m/sec, a 5 MHz frequency has a wavelength of 0.8 mm, while for the same frequency in fat, with a velocity of 1450 m/sec, the wavelength is 0.3 mm. Most soft tissue has very similar density, but varying elasticity, so acoustic impedance changes mainly because of this factor. Since reflection and transmission of US between different tissues depends on acoustic impedance, an understanding of velocity differences in tissues is critical to understanding the behavior of US. Refraction does not affect US imaging as much as reflection because it is a reflected rather than transmitted image but, because it accounts for some interesting artifacts, it must also be defined. When the angle of incidence of the US wave is not 90°, but is θ_i, then the reflected wave has the same angle, $\theta_i = \theta_r$, and the separation between them is $2\,\theta_i$ (see Figure 7.1).

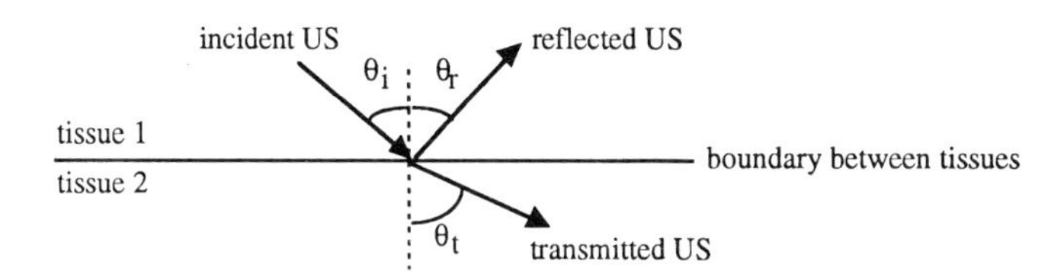

Figure 7.1. Relationship of incidence, reflected and transmitted wave from a specular reflector.

Note that in Figure 7.1, $\theta_\tau \neq \theta_r$, because refraction has occurred, as it will if the propagation velocities in the two tissues differ. As in optics, the amount of deflection follows Snell's law:

$$\frac{\sin\theta_i}{\sin\theta_t} = \frac{c_1}{c_2} \tag{7.6}$$

where c_1 is the propagation velocity in tissue 1.

The derivation of the reflection and transmission coefficients, which gives the fraction of the incident US pressure or intensity that is reflected at the boundary, depends on the continuity of the ultrasound at the boundary. Particle velocity, V, unlike propagation velocity, which is a constant, is a function of where in the wave oscillation it is measured. Thus twice in each wavelength, when the particles in the medium move from rarefied to compressed, particle velocity is zero. When a boundary between two tissues is encountered, some of the particles in the tissue will be reflected and some will continue on into the second medium. Similarly, the pressure created by the US wave will also be partially transmitted through the boundary, while the remainder is reflected. In fact, if P_r is the reflected pressure,

$P_i + P_r = P_t$ and $V_i \cos\theta_i - V_r \cos\theta_r = V_t \cos\theta_t$

because of continuity across the boundary. Since pressure can be calculated from particle velocity, that is $P = c\rho V$, then we can derive equations for the fraction of pressure in the reflected US:

$$\frac{P_r}{P_i} = \frac{Z_2\cos\theta_i - Z_1\cos\theta_t}{Z_2\cos\theta_i + Z_1\cos\theta_t} \tag{7.7}$$

where $Z_1 = \rho_1 c_1$ and $Z_2 = \rho_2 c_2$.

Similarly, the fraction of transmitted pressure is:

$$\frac{P_t}{P_i} = \frac{2Z_2\cos\theta_i}{Z_2\cos\theta_i + Z_1\cos\theta_t} \tag{7.8}$$

When θ_i is 90°, the cos θ terms drop out.

Let us look at what this implies clinically. Table 7.1 gives the acoustic impedance values for some common tissues[4].

Table 7.1 Acoustic Impedancies for Different Tissues

Tissue	Acoustic impedance (in Rayls⁺) × 10^6
Liver	1.64
Muscle	1.70
Fat	1.38
Kidney	1.62
Bone	7.80
Lung	0.26

⁺ a Rayl is a kg/m^2s

Example: Is it easier to scan the liver of a fat man or the liver of a very muscular man? Assume that muscle and fat layers in the two men are perpendicular to the propagation direction of the US wave. Then, for the fat man, the reflected fraction is:

$$(Z_2 - Z_1) / (Z_2 + Z_1) = (1.64 - 1.38) / (1.64 + 1.38) = 0.26 / 3.02 = 0.086$$

For the muscular man, the reflected fraction is:

$$(1.64 - 1.70) / (1.64 + 1.70) = -0.06 / 3.34 = -0.018$$

So while 8.6% of the US beam is reflected back (and 91.4% is transmitted into the liver to make an image) when the fat man is scanned, only 1.8% is reflected back for the muscular man, leaving much more US in the liver to make an image. When the US beam returns to the scanning transducer (which must occur if an image is to be formed), the same interfaces are encountered and the same proportions of the US wave are reflected back into the liver.

Note that the reflected US pressure changes sign at the boundary in the muscular man. This indicates a phase reversal in pressure at the boundary. If the wave was in its compression phase, it became a wave in the rarefaction phase when reflected.

Note also that if $\theta_i \neq 90°$, the difference in reflection and transmission will depend on the amount of refraction present between the two media. That is, if $\theta_i = \theta_t$, then the cos θ terms cancel and reflection is the same as it would be at $\theta_i = 90°$. Thus, only if the propagation velocity in the two tissues does not match does the angle of incidence affect the reflection coefficient. Note also that refraction may not be present even if two tissues have different acoustic impedances if their propagation velocities are the same.

In the above discussion and in the example, we have dealt exclusively with changes in pressure caused by the ultrasound wave. In fact, medical ultrasound equipment is more likely to measure voltage changes, analogous to pressure, or intensity differences, proportional to P^2. So the equations for reflection and transmission coefficients can also be written without any change in terms of V_r / V_i for voltage, or as squared terms for I_r / I_i when intensity is measured.

In order to understand why the lung, stomach, and intestines cannot be imaged with ultrasound, let us calculate from Table 7.1 the transmission coefficient for a muscle/lung interface in terms of intensity, assuming $\theta_i = 90°$.

$$I_t / I_i = 4z^2 / (z_2 + z_1)2 = 1.04 / (1.96)^2 = 0.27$$

Thus only 27% of the US wave intensity is transmitted into the lung, the other 73% is reflected back to the transducer. Since this loss occurs to the 27% left in the lung on its return trip, very little of the US wave is left to make an image of lung structure.

7.2.2.2 Attenuation

Attenuation does not include the amount of US energy reflected from interfaces, but is a measure of the US energy absorbed (by conversion to other modes such as heat) and scattered (by sub wavelength structures) in a single type of tissue. It is measured in dB or nepers, both expressed as exponentials: In dB, when pressure or voltage attenuation is measured, attenuation, $\alpha_{v,} = 20 \log_{10} (V_2/V_1)$, while if intensity attenuation is measured, $\alpha_I = 10 \log_{10} (I_2/I_1)$. On the other hand, nepers measures power or intensity attenuation as $\log_e (I_2/I_1)$ so that 1 neper = 8.686 dB. Because ultrasound attenuation varies with frequency, soft tissue attenuation coefficients are usually expressed in dB/cm/

MHz, although most frequency dependence in the imaging range of 2.5 to 10 MHz is $f^{1.3}$ rather than $f^{1.0}$ as implied by the units. Note that the attenuation coefficients of liquids, including fat, tend to have an f^2 frequency dependence.

The proportion of absorption to scattering component varies between tissues. Because cell size and tissue elasticity affects scattering, as we saw when scattering models were discussed earlier, these factors control the proportion of absorption and scattering. An additional factor is resonance at frequencies where the wavelength is resonant with molecular or cellular sizes, which increases the attenuation for that ultrasound frequency and contributes to nonlinear attenuation.

Since absorption is primarily the conversion of ultrasonic energy to heat, due to viscous drag or friction as the wave particles move against each other, absorption is also frequency dependent. It is of primary concern for biological damage in US, although very high power levels may generate free radicals during gas bubble cavitation, which also can produce bioeffects.

Different tissues have different attenuation coefficients, but all assume a linear dependency on frequency. Table 7.2 gives examples which are derived from in vitro experiments with human tissue[4].

Table 7.2 Attenuation Coefficients for Different Tissues

Tissue	α (dB/cmMHz)
Fat	0.6
Muscle (transverse)	3.5
Muscle (longitudinal)	1.2
Liver (normal)	0.9
Liver (cirrhotic)	1.2
Amniotic fluid	0.005

Example: As an example, 3 cm of fat, when imaged with a 3 MHz US transducer, would attenuate 5.4 dB$(0.6 \times 3 \times 3)$.

These attenuation values may vary when measured experimentally, depending on the ages of the donors, whether in vitro or in vivo measurements were performed and, if in vitro, whether the tissue was fixed or was fresh and at room temperature, or had been degassed before being measured. Because US attenuation is dependent on tissue compressibility, the amount of collagen in tissue will affect its attenuation. As people age, the amount of collagen in their tissues increases, which increases that tissue's acoustic attenuation. For the same reason, fixed tissue is stiff, affecting its acoustic properties (impedance, propagation velocity, attenuation) also. The temperature of the tissue sample will affect the pliability of the tissue, which in turn affects velocity and, to a lesser extent, attenuation. The extreme example of this is the change in acoustic parameters of frozen tissue. Perfusion of tissue is difficult to achieve in vitro; the lack of blood in excised tissue will also affect acoustic parameters. Finally, once tissue has been excised, it begins to decay, producing gas bubbles. This gas causes excessive scattering, which increases attenuation. Therefore, to measure tissue attenuation, the tissue sample must be placed in a vacuum for 10 to 20 minutes (depending on the amount of vacuum) before measurements are made.

Because the US image consists of the image of a slice of tissue oriented from the surface of the scanning probe to the limits of the transducer's imaging depth, at each additional cm of greater depth the tissue is imaged with f less acoustic intensity than it had the previous cm of depth and so will have less amplitude on the monitor. To correct for this loss of intensity, the returning signal is amplified by a varying amount, depending on the depth (or time delay) of the returning signal. This variable amplification is called time-gain compensation (TGC) or depth-gain compensation (DGC).

7.2.2.3 Piezoelectric Transducers

Before describing the electronic signal processing applied to the returning wave, it is necessary to describe the principle of piezoelectric transducers[5]. Most US transducers that operate in the frequency range of 2–5 MHz are made of lead zirconate titanate (PZT). This is a magnetostrictive ceramic (a magnetic field can change the physical dimensions of it) which can be made piezoelectric (an electric field will change the physical dimensions of it) by the following process: after electrodes are placed at each end of a wafer of PZT, it is heated to above its Curie temperature (about 330°C), attached to a 2kV electric source and allowed to cool slowly over a 2–3 day period in a thermostatically controlled cooling oven. The heating "loosens" the molecules, which are dipoles in a disorganized orientation. The voltage applied to the wafer aligns these dipoles and the slow cooling while the voltage is applied freezes the dipoles in their desired orientation. At the conclusion of this process, the PZT is piezoelectric, so that when an oscillating voltage is applied to the wafer, it vibrates, and when an acoustic wave hits the wafer it generates a voltage. The amount of vibration that occurs for a given voltage is governed by several factors: d_{ii} is the transmitting constant, c_{ii} is the elastic stiffness constant, g_{ii}, the voltage output or receiving coefficient, and e_r is the relative dielectric constant. The last two of these factors determine the efficiency of the material's conversion of mechanical energy to electrical energy and therefore the sensitivity of the material, while the efficiency of converting voltage to acoustic power depends on the stress coefficient, e_{ii}, where $e_{ii} = d_{ii}c_{ii}$. Another factor to be considered is the acoustic impedance of the material. Because PZT has a propagation velocity of about 4000 m/sec, it does not match the Z of soft tissue, which is about 1.64×10^6 Rayls, so that much of the US energy generated in the wafer is reflected back into the wafer rather than entering the body.

Another problem is that PZT is a rather fragile, stiff ceramic and to achieve the high propagation wave frequencies needed for medical imaging, the wafers must be cut to $\lambda/2$ thickness. The half wave requirement takes advantage of the reinforcement of the particular frequency which has a wavelength of λ in a $\lambda/2$ thick wafer (see Figure 7.2).

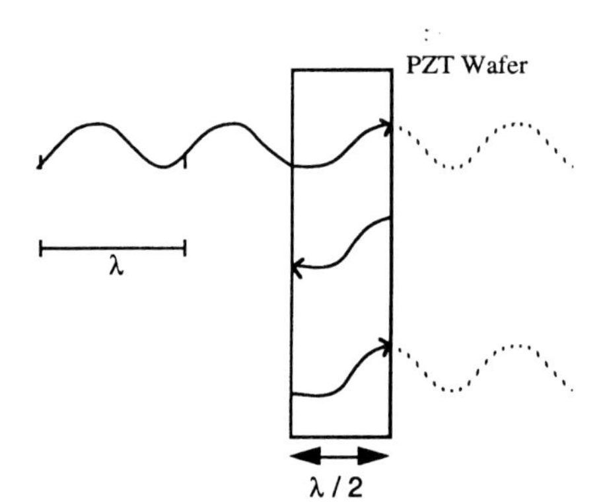

Figure 7.2. Wave amplification due to half wavelength thickness of a transducer wafer.

However, for a 5 MHz transducer, the wavelength in PZT is 0.8 mm, so the wafer must be 0.4 mm thick, a difficult fabrication with fragile ceramic material. Therefore, for transducers 5 MHz and higher, a different material, polyvinyledene fluoride (PVDF) is more frequently used. This material is made up of layers of flexible plastic sheets which are then made piezoelectric by a more complex process than the one employed for PZT. The thinness of the "wafer" required is less than that of PZT because the propagation velocity of PVDF is less than that of PZT, but because of the flexibility of the material this is not the problem that it is for PZT. The further advantage of PVDF is that it can be shaped into curved, lens-like wafers, and it has an acoustic impedance of 2.5×10^6 Rayls compared to 30×10^6 Rayls for PZT. Unfortunately, it is not as efficient at converting voltage into acoustic power as PZT, which has an e_{ii} of 9.2 compared to PVDF's e_{ii} of 0.069.

7.2.2.4 Transducer Natural Focus

The transducer beam profile for these single wafer transducers depends on the frequency, wafer diameter, and wafer shape, as we saw when we modeled a transducer with a circular wafer. By "beam profile," we mean the inherent focusing characteristics of the wafer. Where does the constructive and destructive wave interference come together in a focused beam and how fast does it diverge?

From the model of the plane piston source we found that, for a circular transducer, the last maximum intensity value occurred at $T = r^2/\lambda$, where r was the radius of the transducer aperture. The last peak intensity point marks the center of the "unfocused" transducer's focal zone. Thereafter the intensity diverges as $(z - T)^2$, where z is the distance from the transducer. How abrupt this divergence is depends on r and λ. The divergence angle, θ, is given by the formula $\sin \theta = 0.61\ \lambda/r$, but since θ is small for large aperture transducers, $\theta \sim 0.61\ \lambda/r$.

Figure 7.3 illustrates how these formulas predict the behavior of naturally focused transducers. However, medical transducers are rarely designed unfocused, since resolution would be as large as the aperture, even at the focal zone, and low frequency transducers need large apertures in order to place their maximum intensity points deep in tissue. The design of focused transducers and fabrication of medical transducers will be covered when we have described the signal processing needed to form an image from the reflected ultrasound image.

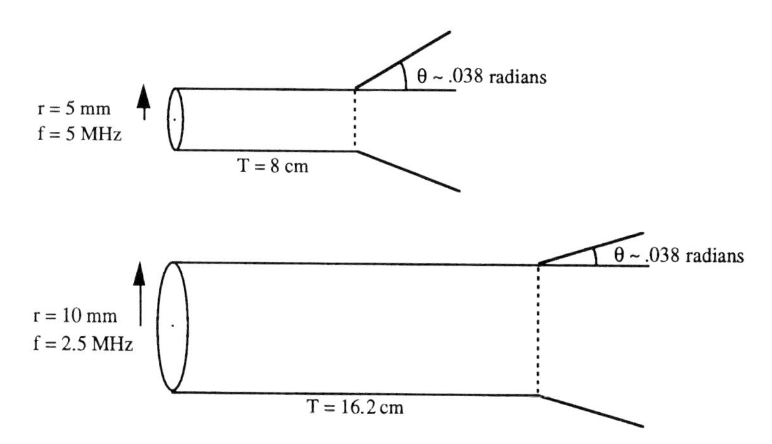

Figure 7.3. Beam profiles for two transducers.

7.3 Medical Ultrasound Imaging Equipment Design

There are four methods to display the reflected ultrasound signal, only two of which are currently widely used. A-mode imaging was the first use of ultrasound, but is currently not used. It is simply the depth dependent amplification of the RFsignal returning from tissue to the transducer. Scatter appears as low amplitude signal, while specular reflectors are peaks. When used for ophthalmology, with the transducer pressed against the closed eye, the lens and retina were seen as peaks. Any intervening peaks represented masses. Similarly, when diagnosing brain injury, the transducer was placed alternatively on each temple and the distance to the midline was measured. Any difference indicated unilateral brain swelling. These uses have been replaced by conventional real-time ultrasound imaging for ophthalmology applications, and by CT or MR for brain imaging.

C-mode imaging consists of reconstructing a plane made up of pixels from a number of ultrasound slices. It has never been more than a research tool until recently, when three- dimensional ultrasound has been moved from the laboratory to clinical testing at a few universities. M-mode, standing for movement mode, is used extensively in cardiology. The transducer is placed in a stationary position, and a series of A-mode lines are mapped together over time. The amplitude of each A-line is translated into a gray scale. Thus when the heart beats, the movement of valves or expansion of the left ventricle can be measured quantitatively so long as the transducer is positioned correctly. Echocardiology technologists use real-time imaging to position the transducer before the M-mode scan is initiated.

7.3.1 Brightness Mode Display

Real-time images consist of a series of B-mode frames, where B-mode stands for brightness mode. Each frame is a series of adjacent A-lines, with RF amplitude transformed into a gray scale. The number of A-lines in a frame is the *line density* of the image and is dependent on the depth that the ultrasound must penetrate, the frame rate, focusing algorithms and, for electronic focusing, the construction of the transducer (see Figure 7.4).

Initially, the transducer must be made to vibrate in order to produce an ultrasound wave. This is accomplished by providing a short burst of voltage (50–200 V) to the transducer wafer. The burst is short to ensure that the packet of ultrasound waves entering the body will be no longer than two to three wavelengths so that good axial resolution will result. Since we use the returning wave front to measure where various tissue interfaces are located, we must have a short interval of ultrasound in order to locate interfaces accurately. By restricting the pulse width to 3λ, we can ensure resolution in the direction of the ultrasound beam of 3λ. For a 5 MHz transducer, this is 0.9 mm. In general, axial resolution is superior to lateral resolution, perpendicular to the axial direction, which depends on the transducer aperture, distance from the transducer, frequency and transducer focusing.

Because this short burst of voltage can only be generated by combining a number of frequencies in the voltage burst, the ultrasound generated also contains a mix of frequencies. However, because it is cut $\lambda/2$ thick, the center frequency, f_o, will predominate. The spread of frequencies about f_o is the frequency bandwidth, and it will determine the axial resolution and dynamic range of the transducer[3] (see Figure 7.5). Electrical engineers relate the bandwidth and center frequency in a "quality" factor, Q, defined as

$$Q = f_o / (f_2 - f_1) \tag{7.9}$$

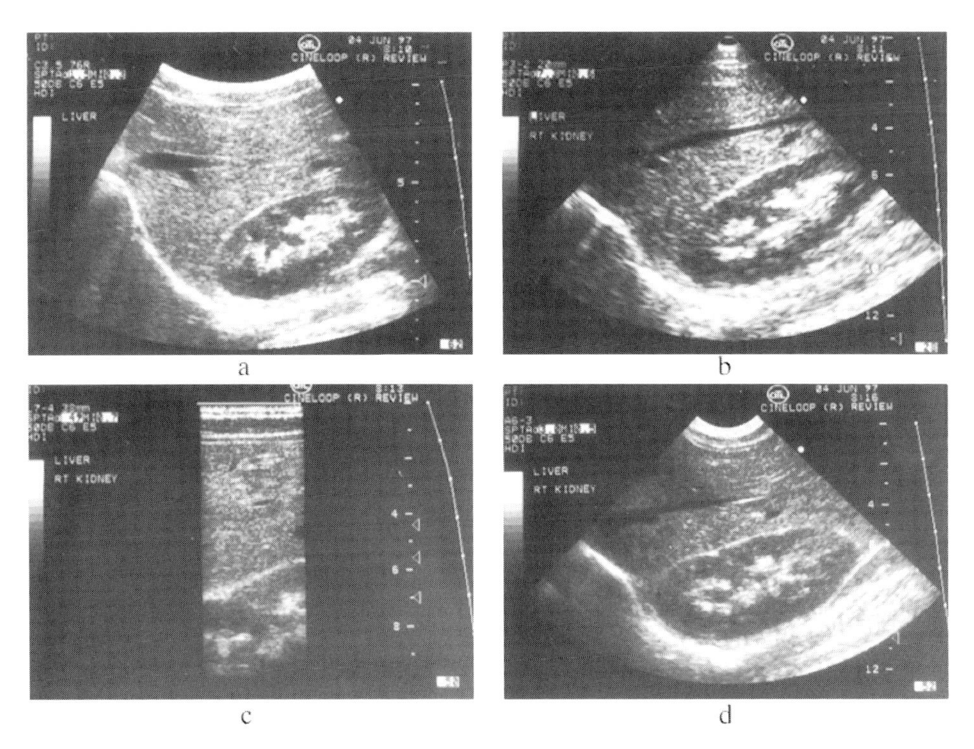

Figure 7.4. Each transducer forms an image with a different shape: a) 5.0 MHz curved linear, b) 2.25 MHz sector, c) 7.5 MHz linear, d) 5.0 MHz annular sector.

where f_2 and f_1 are the full width, half maximum (FWHM) frequencies of the bandwidth. Typical Q factors for B-mode imaging are 1 to 2. Just as we "pay" for the superior resolution of higher frequency transducers by their higher attenuation, so also must we pay for their broad bandwidth by a lack of sensitivity when we wish to amplify the returning signal. Amplifiers work more efficiently if the Q of the signal is high; that is, there is a narrow band of frequencies to be amplified. Just as high fidelity sound systems are expensive because of the range of frequencies they must amplify, so also are ultrasound amplifiers. Because the high frequency component (f_2) will be more attenuated by tissue than the low frequency component (f_1), the bandwidth will be

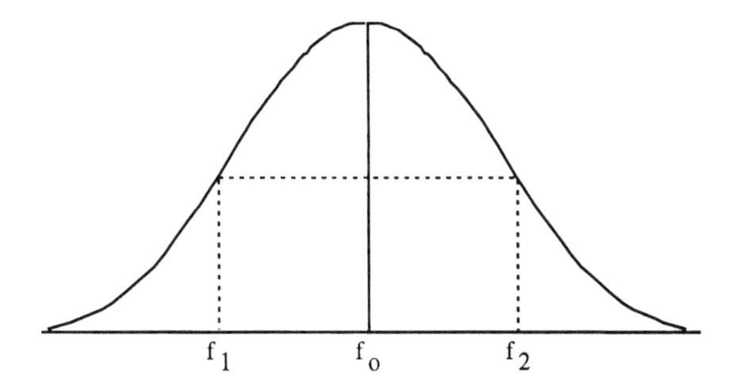

Figure 7.5. Bandwidth of the transducer frequency spectrum.

reduced and f_0 will also tend to move toward lower frequencies. This helps the amplifier problem, but reduces resolution in the axial direction at deep depths. For lateral direction resolution, beam profile effects are much greater (since they depend on r^2) than the small reduction in f_o, so these changes are less apparent.

7.3.1.1 Pre-Processing

After being pulsed for about 0.6 μsec (2–3λ), the transducer cannot receive returning echoes for 1–2 μsec (3–8 mm). This is "dead time," represented by a band at the top of the image. The depth selected by the operator sets the transducer receive time, which is at least as long as a round trip to the depth selected. All American-manufactured equipment is set for a propagation velocity of 1540 m/sec, so if 5 cm is selected as the depth, the transducer waits at least 32 μsec for returning signals. Modern ultrasound equipment needs very little processing time, so the transducer can easily pulse again at 50 μsec after moving to a new position. Each received A-line is logarithmically amplified, sometimes in the transducer scan head to avoid loss over the cable attaching the transducer to the ultrasound unit (Figure 7.6). Amplification is logarithmic so that lower amplitude signals, often from deep in the body, are not lost before the TGC can be applied. Some expensive ultrasound units amplify in stages to increase high frequency preservation. Others digitize the RF at this stage and use digital amplifiers, but most apply the TGC amplification before digitization. If digitization is applied before demodulation and envelope detection, the digitization rate of the analog to digital conversion

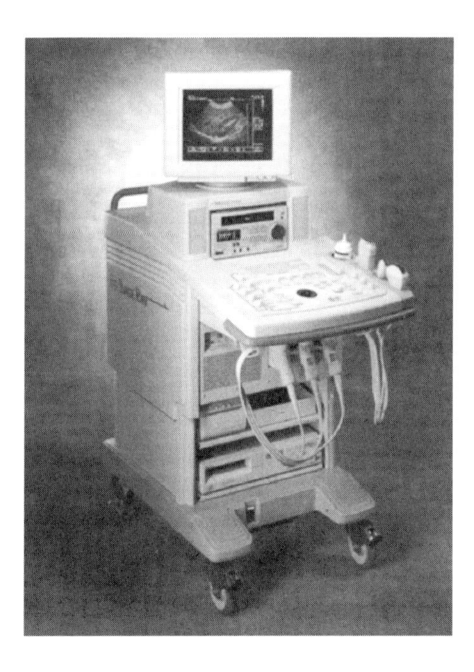

Figure 7.6. Typical ultrasound unit. Note cabling between transducers and the unit.

(A/D) will need to be twice the highest frequency remaining in the bandwidth (the Nyquist frequency). If digitization occurs after envelope detection, the rate can be set by the minimum display pixel size. The range of voltages covered by digitization is the dynamic range of the system. Usually, the TGC requires about 60 dB to make up for tissue attenuation. This leaves up to a 60 dB range to be mapped into the 20 dB of the display device. If the full 60 dB is displayed, tissue texture and small intensity interfaces will have little contrast because they will be assigned to very few gray values (see Figure 7.7). If, however, we assume that the top 20 dB echoes are all from specular interfaces, we can map them into one gray value and spread the small intensity echoes over a large gray level range, thus giving them more contrast. This is accomplished by reducing the dynamic range of the signals digitized from 60 to 40. Most ultrasound units now have this option. It can only be implemented while acquiring an image, not after the image has been frozen on the monitor, so it is called a pre-processing option.

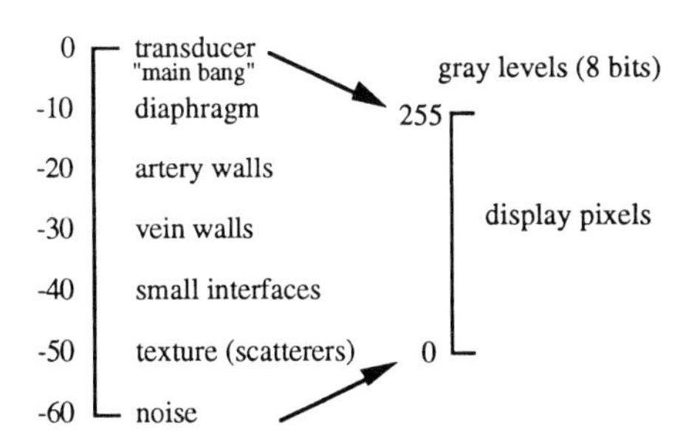

Figure 7.7. Mapping 60 dB to 8 bits during digitization.

7.3.1.2 Post-Processing

When the gray scale is changed after digitization, the image may be frozen and several different look-up tables (LUTs) can be applied to the same frame. This is called a post-processing option.

The memory where the image is displayed is called the digital scan converter (DSC). Locations on this memory are often organized in a lookup table format to increase display speed by reducing computation. Thus a particular A-line will have a series of pixels in the DSC assigned to it. As the time intervals are received, amplified, and digitized, they are put in sequential positions in their assigned locations. When the transducer moves to a new position, ready to send the next A-line, the next memory locations are ready to be filled. Because transducers have different shapes and different firing patterns, the DSC may have a series of memory location templates, depending on the transducer selected.

7.3.2 Transducer Design

7.3.2.1 Focusing Single Element Probes

We can begin to understand the difficulties of designing an ultrasound transducer if we examine specific problems caused by the focusing properties of this equipment[5]. To avoid divergence in the far field, we wish to use a large aperture transducer, particularly for frequencies below 5 MHz, since the divergence angle is approximately 0.61 λ/r. Thus for a 3 MHz probe, we would use a 2 cm aperture (which will still fit between the ribs), giving a natural focus of $r^2/\lambda = 100/0.5 = 20$ cm. At 20 cm or 40 cm round trip, assuming an attenuation coefficient of 0.7 dB/cm/MHz, the ultrasound beam will be –84 dB, which must be compensated for by time gain compensation. This level of attenuation will leave very little dB difference for display, and 20 cm penetration is thicker than most abdomens. If instead, an aperture of 1 cm is used, the natural focus is now at 5 cm, too shallow a depth for effective imaging with a 3 MHz transducer. For a 5 MHz transducer, these apertures (1 and 2 cm) give focal depths of 8 and 33 cm, respectively; both are beyond the expected penetration depth of 5–6 cm for a 5 MHz transducer.

A solution for this problem is to place a focusing lens on the piezoelectric wafer. Because the lens material has a greater propagation velocity than tissue, it curves toward the tissue in order to refract the ultrasound beam to a focal point. The geometry of focused transducers is shown in Figure 7.8.

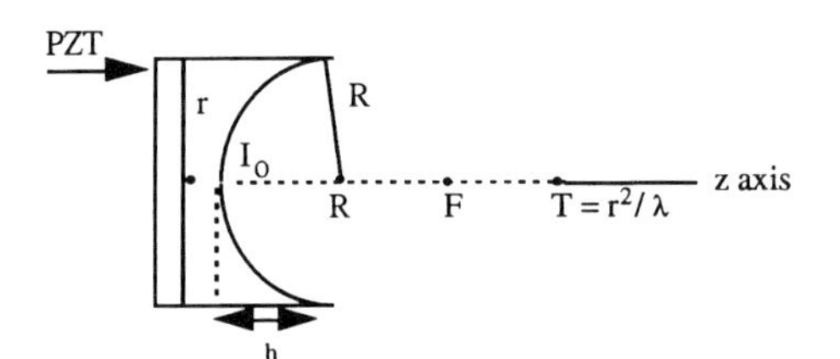

Figure 7.8. An external lens of radius R focuses the ultrasound beam, naturally focused at T, at F, where $R \leq F \leq T$.

The intensity of the beam after T, the last peak intensity, decreases as $1/(z - T)^2$. At R, for a focused transducer, the intensity has dropped to $I_R/I_o \sim (2\pi h/\lambda)^2$, so higher frequency transducers are more efficiently focused than lower frequency transducers and, as expected, the greater the curvature, the higher the intensity amplification. Using Snell's law and ray tracing, the new focal point, F, is now at approximately $R/(1 - c_2/c_1)$, so long as $h < 0.1R$, i.e., the lens is weakly focused. Here c_2 is the propagation velocity of tissue and c_1 is the propagation velocity of ultrasound in the lens. Since most acoustic lenses are made of polystyrene, with a $c_1 = 2350$ m/s, c_2/c_1 is 0.655. The multiplying factor is therefore about 2.9, so the focusing lens radius should be about one-third the desired focal distance.

Lenses are usually manufactured at $\lambda/2$ thickness to improve transmission of the fundamental frequency. A polystyrene lens for a 5 MHz transducer requires a lens thickness of 0.235 mm. Designers of focusing lenses use several approximations to predict the dimensions of the focal zone (see Figure 7.9).

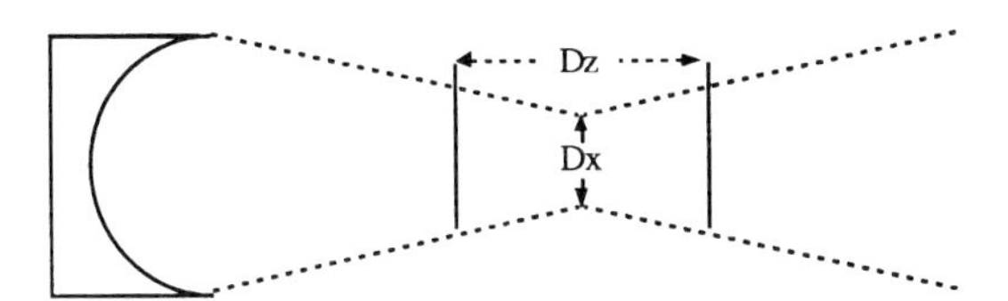

Figure 7.9. Dimensions of the focused focal zone.

The height, Dx, or lateral resolution of the ultrasound beam at the –3 dB level, can be approximated by $\lambda F/2r$, where λ is measured in tissue and the focusing is not too acute. Thus for a 5 MHz transducer with a focal lens that provides a focus, F, at 3 cm, and with a 1 cm aperture, lateral resolution should be about 0.9 mm. A more weakly focused 5 MHz transducer, say one focused at 5 cm, would have lateral resolution of 1.5 mm. The extent of the focal zone is no greater than 15 Dx, measured at the –3 dB point where peak intensity has dropped 50%. So, for the medium focus at 3 cm, Dz = 1.4 cm, while for the more weakly focused case, Dz = 2.3 cm. These examples illustrate the trade-offs necessary for a versatile transducer. Imaging a child with a medium focus transducer would give a band of good resolution of about 1.5 cm, between 2.3 and 3.8 cm in depth. Resolution deeper than 3.8 cm would degrade quickly.

7.3.2.2 Matching Layer

In addition to an external lens placed on the transducer, a matching layer is added between the lens and patient to prevent the high reflection coefficient that would occur between these two materials. Matching layers are usually made $\lambda/4$ thick to facilitate transmission of the fundamental frequency. The material of the matching layer is usually an epoxy with a Z equal to the geometric mean of the acoustic impedances of the two adjacent materials.

7.3.2.3 Phased Arrays

Instead of an external lens, a shaped piezoelectric wafer may act as a lens. For 2 to 3 MHz transducers, such internal lenses can be made with PZT but, for high frequency transducers, shaping PZT is too difficult and PVDF is used.

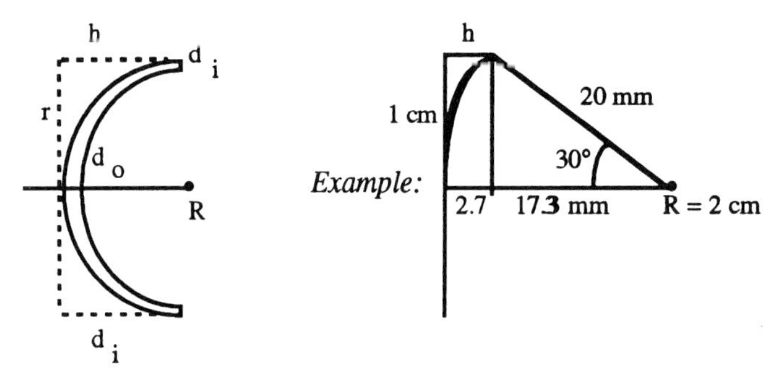

Figure 7.10. Internally focused lens with parameters labeled.

The concept of an internally focused lens allows us to model focused ultrasound as a series of oscillating points along the edge of the lens[4]. Because each point on the lens is positioned at a distance d_i from the central point d_o (see Figure 7.10), it also begins to oscillate at a time different from t_o equal to d_i/c, where $c = 1540$ m/sec. Note that $d_i = h$ for an externally focused lens.

This observation leads to the design of multiple element transducers, which can be fired in a pre- arranged order with delays representing the model of an internally focused transducer. Thus if we wish to simulate a lens with $R = 2$ cm and $r = 10$ mm, then the h of such a lens is 2.7 mm (see example, Figure 7.10). At 1540 m/sec propagation velocity, d_o will begin oscillating 17.5 μsec after d_i. Note that the lens described is sharply focused, and that most differences are less than a wavelength or are a fraction of the phase of the ultrasound oscillation. When the transducer is made up of many separate elements, instead of a single wafer, each element can be fired at a different time, thus making an artificial lens. The firing time differences are usually fractions of the phase, and so such transducers are called phased arrays. In addition to focusing by delaying the firing of centrally located elements, a linear array of elements, similar to Figure 7.4c, can be fired sequentially, simulating the movement of the array across the body of the patient. If the array is 5 cm long, and each element is separated from its neighbor by 0.25 mm, then the 200 elements can be fired 40 at a time to simulate a 1 cm aperture transducer. For a 3 MHz transducer, this would ensure a natural focus at 5 cm. To focus closer than this, some delays at the center elements in each aperture will be necessary. If a deeper focus is desired, a larger aperture can be simulated. If 80 elements are fired at the same time, then the natural focus is 20 cm and greater delays will be needed to focus the beam at 5–10 cm depth. Without this variable aperture feature, shallow depths require greater delays than deeper depths to simulate different focusing lenses.

Electronic delays can also be used to steer the wave front, thus simulating a single element transducer which is being rocked from side to side. Delays on the right side of the array simulate movement of the A-line to the right. The extent of the sweep is controlled by the amount of delay. The sector or pie-shaped image (with the transducer at the apex) that results (Figure 7.4b) is particularly useful when scanning between ribs to image the heart, kidneys, spleen, or gallbladder (for patients with small livers).

7.3.2.4 Intercavitary Transducers

Some organs are simply not located conveniently for ultrasound imaging. The prostate, the ovaries, the early term fetus, the ventricles of the heart, and the esophagus and stomach are examples of "problem" organs. Fortunately, we are constructed with openings near these organs which will accommodate ultrasonic transducers. Endorectal and endovaginal transducers are mounted on wand-like supports which, when inserted into the rectum or vagina, can clearly image the prostate, ovaries or first trimester fetus. These transducers may be single element wafers, rocked mechanically by a motor, or they may be swept and focused electronically. Because they can be positioned close to the organ of interest, they are usually 5 MHz transducers. Esophageal probes are mounted on a flexible fiber optic bundle so that the transducer can be positioned visually before scanning begins. These single element transducers rotate in a circular pattern and only penetrate about a centimeter when fired at 12 MHz. Cardiac imaging through the esophagus produces much better resolution than external imaging, since 5 MHz transducers can replace 2 MHz ones.

7.3.3 Real-Time Imaging

Real-time imaging means that slices of tissue are produced automatically and quickly enough to show physiological motion, if present. Either the transducer is moved mechanically by a motor, or it is swept electronically. In either case, the frame-rate, depth of penetration, d, and number of A-lines in a frame are interdependent. If we assume that processing can be performed in parallel with pulsing and receiving the next A-line, then the frame rate for an L line image is $fr = 154\,000 / 2\,Ld$. For example, penetrating 10 cm in a 100 line/frame system will allow 77 frames/sec. The number of lines per frame can be used to increase the field of view for a sector format transducer, or increase the line density for linear and curved linear arrays.

Mechanical oscillation is found mainly on annular arrays, such as in Figure 7.4d, or intercavitary probes. An annular array consists of an electronically focused array with annular elements which can be fired independently. Their advantage is that the lateral electronic focus will be the same as the slice thickness focus. From 5 to 8 elements are common. The probe is enclosed in a liquid medium acoustically matched to soft tissue and is either oscillated in an arc or swept in a circle.

7.4 Ultrasound Imaging Artifacts

7.4.1 Variations in Tissue Velocity

Ultrasound equipment assumes that propagation velocity in soft tissue is 1540 m/sec when, in fact, velocities ranging from 1460 (brain) to 1580 (blood) are common. The ultrasound beam is refracted between tissues of different velocities for non-perpendicular interfaces but, because it is similarly refracted on the return trip, accurate imaging still occurs. However, when localizing a tumor for biopsy, refraction can sometimes cause small, deep masses to be displaced 2 to 3 mm. Similarly, when placing electronic calipers in the axial direction, 1 to 2 mm errors may result if the structure does not have a propagation velocity close to 1540 m/sec. For instance, a carcinoma, with its stiff structure, may have a velocity as high as 1590 m/sec. For a 2 cm mass, an error of 0.6 mm will result. A more serious consequence to variations in velocity is the defocusing that occurs in electronically focused transducers. Since the delays are calculated based on 1540 m/sec, travel through tissues with other velocities will disrupt the focusing of the beam.

Finally, velocity variations can cause artifacts behind a higher or lower velocity structure. A common artifact is the misalignment of the diaphragm behind the liver when a large liver tumor intervenes. Another artifact of diagnostic significance is the "twin" artifact. When scanning the gravid uterus through the abdomen, a combination of muscle and fat layers can so refract the beam that twin concepti are imaged when only one fetus is present.

7.4.2 Attenuation Effects

When a more attenuating structure larger than 10λ intervenes in a homogeneous medium, it "uses up" the ultrasound intensity, which causes an ultrasound shadow to be imaged behind it. Renal stones larger than 5 mm, large breast calcifications, and some malignant tumors can be more easily identified because of this posterior shadowing. Since attenuation includes scattering, small anechoic regions will often have low level echoes. This is particularly troublesome when distinguishing cystic from solid masses, and is exacerbated by the volume effect if the cyst is not in the slice thickness focus of the transducer (see Figure 7.11). In addition to posterior shadowing, a cyst or other low attenuation tissue may not "use up" the amplification provided by the TGC. The posterior echoes will then be brighter than the surrounding tissue, causing *posterior enhancement*.

7.4.3 Transducer Caused Artifacts

All electronically focused transducers, except annular arrays, are focused in only one direction: the inplane lateral direction.

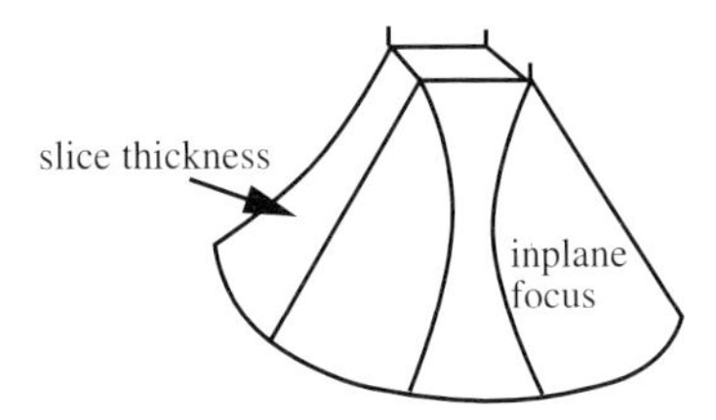

Figure 7.11. A sector format phased array showing the inplane and slice thickness focus directions.

7.4.3.1 Volume Effects

In order to reduce the slice thickness, an external focusing lens is placed on all the elements. This lens is designed for the clinical tasks assigned to the transducer. A 5 MHz breast transducer will have a slice thickness lens focused between 1.5 to 3 cm depth, while a pediatric 5 MHz transducer will not be as sharply focused; the slice thickness focus will range from 2 to 4.5 cm depth. When a mass or other fine structure is smaller than the slice thickness at the depth where it is located, scatter from the surrounding tissue is likely to obscure the structure so that it cannot be imaged, creating a volume effect.

Other problems may occur if output power (the voltage initially applied to the transducer) or gain is set too high. This will cause the interference patterns (side lobes and grating lobes, Figure 7.12) at the side of the aperture to be amplified.

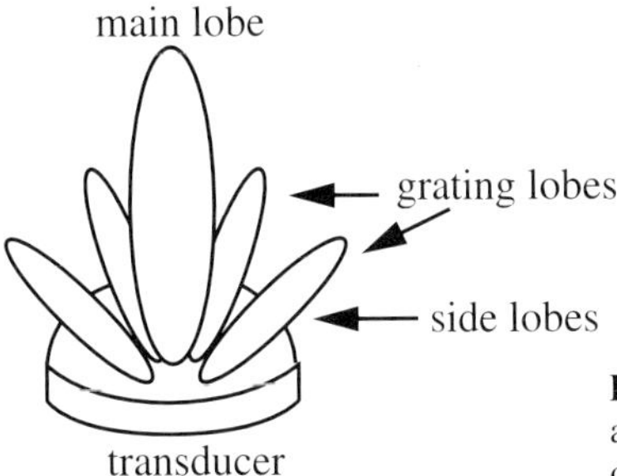

Figure 7.12. Side lobes are present on all transducers, but are of low amplitude (–60 dB). Grating lobes only occur on electronically focused transducers and are of higher amplitude (–20 dB).

7.4.3.2 Apodization

Side lobes are usually below the noise level of the transducer and so do not cause problems. Grating lobes occur because of the finite size of the elements of an electronically focused transducer. When the spacing between elements is greater than λ, grating lobes will occur. The angle at which the primary grating lobe occurs is θ = arc sin (λ/s), where s is the spacing between elements. Thus if $s = 2\lambda$, a grating lobe will occur at 30°. Because of grating lobes, elements are generally made less than λ, sometimes by subdicing the elements into electronically isolated components. Grating lobes, if present, are inherent in an electronically focused transducer and cannot be removed, although they can be made less obvious by keeping gain and output power low. Manufacturers may reduce grating lobes by a process called apodization. The elements on the outside of any given aperture will be fired with a lower voltage than that applied to the elements in the center of the aperture.

Sometimes electronically focused transducers will "lose an element", either because air gets under the pad between the patient and transducer or because electrical connections to that element fail. In this case, a narrow shadow will extend in the image from the transducer to the bottom of the image. It will always be in the same position on the image, even when the transducer moves to a new area.

7.5 Doppler and Blood Flow Imaging

The Doppler effect was first described by Christian Doppler in 1842 as it applied to the color of a moving luminous body. While widely used in astronomy, our interest is in its medical application when an acoustic frequency changes because an object is moving with respect to an incident ultrasound wave. If a person is moved rapidly toward the source of a continuous wave of sound, he will encounter compression and rarefaction waves more closely spaced than if he were stationary. Effectively, the sound would have a higher pitch. The increase in the incident frequency, f_0, is called the Doppler frequency, f_D. It is dependent on the velocity of movement into the sound field, v, the angle at which the movement occurs, θ, the velocity of sound (in air for this example), c, and the incident frequency of the sound being broadcast[6]. That is:

$$f_D = \frac{2Vf_0\cos\theta}{c} \tag{7.10}$$

If instead of a person moving into an audible sound field, we consider blood flowing toward the transducer, then we can calculate the velocity of the blood from the change in frequency of the ultrasound wave returning to the transducer. How this is implemented in medical ultrasound equipment is the subject of the next section.

7.5.1 Continuous Wave Doppler (CW)

The simplest implementation of Doppler ultrasound uses a continuous wave. Here a single frequency AC voltage is produced by an oscillator, applied to a transducer which then emits a continuous ultrasound wave. If the ultrasound wave encounters a moving interface or moving scatterers, the reflected ultrasound wave will change in frequency. A receiving transducer, usually located near the transmitting transducer, picks up and amplifies the resulting RF signal, which is then demodulated to remove the incident frequency. If a beat frequency results, it is audibly amplified or displayed as a Doppler frequency. The voltage applied to the transmitting transducer is usually low: 2–5V. The two transducers may be fabricated by splitting a circular transducer in half and offsetting one of the semicircles (called a "D") at an angle to define a region of sensitivity.

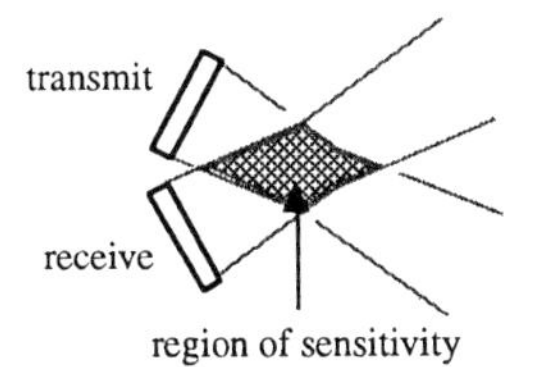

Figure 7.13. The angle between the transmit and receive transducers of a continuous wave Doppler unit determines the depth and extent of the area most sensitive to motion detection.

Doppler probes, like real-time probes, require a matching layer between the piezoelectric elements and the patient to improve sensitivity. Since the received Doppler signals range from 10 mV to 100 μV, amplification up to 80 dB may be needed. Because the continuous ultrasound wave interrogates all the tissue contained in the overlapping area of the transmit and receive transducers, it will identify a mix of Doppler frequencies from the physiological movement of soft tissue (due to bowel, cardiac or breathing motion), the wall motion of the vessel, and the mix of frequencies represented by flowing blood. The detection of the Doppler frequency is based on discovering the "beat frequency" that results when waveforms from two different frequencies are combined. Because the reflected CW wave also contains Doppler frequencies from slow moving vessel walls and physiological movement, the demodulated RF is filtered before being displayed as the Doppler spectrum. When the received signal is multiplied by the transmitted waveform, the demodulation is coherent. After this operation, the waveform is low-pass filtered to remove the beat (or Doppler) frequency. Further low-pass filtering will remove non-blood movement artifacts. This filtering is variously referred to as a "wall" filter, "clutter" filter, or "thump" filter.

Two types of flow have been extensively modeled: laminar and turbulent. Laminar flow assumes that the flow profile is parabolic, with the velocity in the center of the vessel higher than that at the sides of the vessel. This type of flow assumes that low velocity blood is represented as a Newtonian fluid with linear viscosity. Normal human (and porcine) blood has a viscosity near 10^{-2} Pascal. As velocity increases, turbulent flow prevails. The threshold between laminar and turbulent flow varies depending on the density of blood (ρ ~1.06/ m^3), vessel diameter (d = 0.5 to 2.5 cm), velocity (v = 5 – 150 cm/sec) and viscosity, μ. The Reynolds number, R, is used to describe this threshold, where

$$R = \rho dv/\mu \tag{7.11}$$

When $R < 2000$, laminar flow is usual. Thus if only one vessel is interrogated by the CW probes, the peak Doppler frequency will represent the peak velocity in the returning mix of Doppler frequencies. However, without knowledge of θ, the angle of the ultrasound beam to the vessel, no accurate estimate of blood velocity can be made. Note that the Doppler frequency for the f_o ultrasound range of 2–5 MHz and blood velocity from 5–150 cm/sec is in the audible range. For this reason, most Doppler systems supply an audio signal representing the blood flow detected by the ultrasound. Most technologists and physicians become very adept at recognizing anomalies in this blend of acoustic frequencies.

7.5.2 Pulsed Doppler Systems

7.5.2.1 Sample Volume and Doppler Angle

Unfortunately, veins and arteries are often located adjacent to one another, and their Doppler signals are mixed together in the signal from a CW Doppler system. In order to identify the velocity of a particular vessel, a more accurate Doppler method is needed. If a B-mode image is formed, and the vessel of interest identified visually, then a pulse of ultrasound can be time-gated (range-gated) to interrogate just the depth of interest where the vessel lies. This technique is called pulse wave (PW) or duplex Doppler because both Doppler and real-time ultrasound are used. In addition to restricting the Doppler signal to the vessel, the angle between the ultrasound beam and flow in the vessel can be measured, so cos θ can be incorporated into the Doppler equation.

The length of time over which the Doppler signal is sampled is the sample volume; it usually ranges from 1 mm to 8 mm. Clearly, if the transducer frequency is less than 3 MHz, axial resolution is greater than 1 mm and such a small sample volume is unrealistic. Similarly, the angle between the beam and blood flow (Figure 7.14) is restricted to about a 20° variation between 45° and 65° because of the following factors: angles less than 45° will be specularly reflected away from the transducer and will not be transmitted into the center of the vessel. Angles greater than 65° will lack sensitivity, since cos θ, where $90° > \theta > 65°$, ranges from 0 to 0.42, reducing the Doppler signal to noise levels[7].

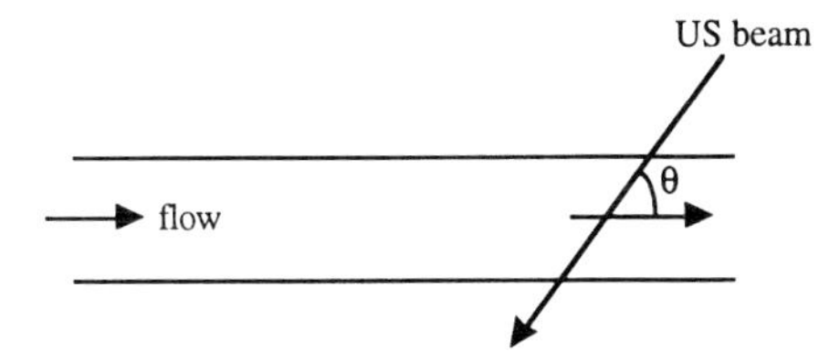

Figure 7.14. θ is the angle between the flow vector and the ultrasound beam. It is measured within the sample volume.

7.5.2.2 Wall Filter

As was noted for CW, PW also includes clutter from vessel walls, respiration, etc., which should be removed with a high pass filter. The cut-off for this wall filter should change with transducer frequency and the location of the interrogated vessel. Thus vessels in a fetus must have a high wall filter to remove the effects of the mother's heartbeat, while breathing artifacts can be filtered at 50 Hz. These filters cause the frequencies near the $f = o$ axis to be free of Doppler frequency. If blood flow is very slow and a high wall filter is selected, the Doppler signal may disappear, filtered out of existence.

7.5.2.3 Blood Flow Approximations

The display of the Doppler signal for both CW and PW can be represented by a frequency spectrum sweep, with the Doppler frequency mapped on the ordinate axis and the time displayed on the abscissa axis (Figure 7.15). The frequency axis can be converted to the peak velocity of blood when an angle correction is made by the technologist. By measuring peak velocity, Vp, an approximation of average velocity, $\overline{V}$, can be made, if laminar flow is assumed, from the formula

$$\overline{V} = 1/2\ Vp \tag{7.12}$$

The volume of blood, F, flowing through a vessel can then be approximated from the diameter, d, of the vessel and the average velocity:

$$F = \frac{\pi d^2 \overline{V}}{4} \tag{7.13}$$

These approximations will be accurate only if 1) flow is laminar; i.e., the vessel is relatively straight and unoccluded; 2) the sample volume is centered in the vessel so it includes peak flow; 3) the diameter is accurately measured (this assumes that the transducer is aligned with the center of the vessel and is perpendicular to it if a transverse cut is used for measurement purposes); 4) the angle θ has been accurately measured; and 5) the Doppler gain is high enough so peak velocity can be accurately measured.

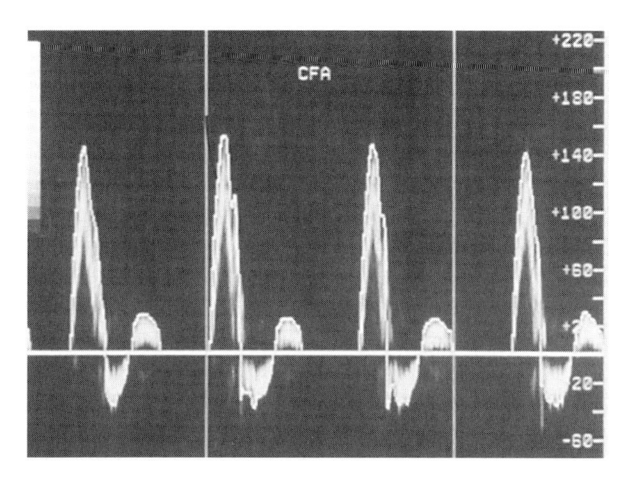

Figure 7.15. Example of a carotid artery frequency spectrum with automatic tracing of the waveform.

7.5.2.4 Heterodyne and Quadrature Phase Demodulation

In order to display both positive (toward the transducer) and negative (away from the transducer) flow, heterodyne demodulation or quadrature phase demodulation is applied to the returning RF signal[8]. The coherent demodulation described in the CW section destroys information about blood flow direction, since only a difference from the original frequency is detected. To correct this problem using heterodyne demodulation, the returning echo is shifted up in frequency by f_h, which is high enough to encompass any negative velocity flow which would cause a negative flow in the $f_o + f_h$ signal. The second method, quadrature demodulation, phase shifts the reference signal by 90° (or a quarter wavelength) and processes the returning wave with both a shifted and non-shifted frequency. If the flow was toward the transducer, the signal from the non-phase shifted demodulation will lead the second, phase-shifted demodulation signal. If the flow is away from the transducer, the opposite situation will prevail. Each condition can be detected electronically by a comparator so that the spectral trace can be placed on the time axis with the correct directional velocity.

7.5.2.5 Aliasing

When demands on the PW system exceed its capabilities, aliasing may result. PW Doppler requires that the sample volume must be sampled twice for each cycle of Doppler frequency in order to accurately predict the Doppler frequency. If the ultrasound beam cannot reach the depth of the sample volume for two round trips in the time required, then the Doppler frequency will be calculated as less than the true Doppler frequency. This artifact is easy to correct by either changing the Doppler angle to a larger angle (and thus reducing the Doppler frequency), reducing the interrogating transducer frequency, f_o, or increasing the range of velocities expected by the equipment, which will increase the pulse repetition frequency (PRF). The aliasing artifact appears as a "wraparound" spectrum at the bottom of a forward direction flow.

Example: An example of the limitations of aliasing can provide an understanding of the causes of this artifact. Consider a Doppler sample volume placed at 10 cm depth, offset 60° from the direction of blood flow. A 3 MHz transducer is used to find the peak velocity of this flow. How high a velocity can be measured under these circumstances without aliasing?

It takes 0.26 msec to sample the blood flow twice at 10 cm depth, so the Doppler frequency wavelength must be at least 400.4 mm. At 3 MHz, this means the peak blood velocity must be less than 197 cm/sec. A compact equation calculates maximum allowable velocity $V_{max} = c^2/8f_o d$, where d is the depth of the sample volume[9].

7.5.2.6 Doppler Sensitivity

Doppler sensitivity is affected by the attenuation of the beam before it is reflected from the small red blood cells that represent most of the scattered signal. To avoid the loss of sensitivity resulting from the higher attenuation of high frequency transducers, PW often transmits at a higher frequency for imaging than for Doppler. This can be accomplished by reserving specific transducer elements for the Doppler A-line. Sensitivity is also affected by the Doppler angle, with angles outside of the preferred range (45 to 65°) decreasing sensitivity. Therefore, most linear arrays sweep the Doppler pulse to avoid angles near 90°. Similarly, sample volume size and placement can affect sensitivity. A small sample volume may not sample the center of the vessel even though the sample volume is positioned there because the expected propagation velocity does not match that of the patient. If the transducer is not centered over the vessel, the sample volume may not be in the peak flow area, and the resulting low velocity will also be harder to detect. Electronically focused transducers should be focused at the depth of the sample volume to increase sensitivity.

7.5.3 Color Flow

Visually, the most exciting medical ultrasound is found in real-time images with blood vessels color-coded to represent the velocity of the blood flowing in them. Physicians use these images to position the PW sample volume accurately and to identify small vessels not seen on the real- time image.

Instead of a single sample volume, each A-line in the color flow image is partitioned into sample volumes, time-gated to a particular resolution of posi tion accuracy. If the Doppler frequency amplitude calculated for each time gate falls below a pre-set threshold, then the real-time image does not color the pixels within the time gate, but leaves them in their original shade of gray. Some color flow systems also test the gray level value of the real-time display at that position. If the gray value is above a certain level (i.e., if it represents a bright echo), no color will be written there on the theory that blood contains small scatterers and so is usually relatively echo-free[10].

7.5.3.1 Autocorrelation to Obtain Average Velocity

The frequency spectrum of PW systems is produced by a Fast Fourier Transform (FFT), but this method is too slow for color flow systems[11]. Two methods are available in current equipment: an autocorrelation method and a time domain process. The autocorrelation method consists of comparing each time gate position with the same time gate position of the preceding A-line, pulsed in the same position. Only time gate positions which have changed produce a changed signal. Since the autocorrelation is in the frequency domain, change will be a Doppler frequency with an amplitude. However, the average change will be a mean frequency, so a mean velocity will be estimated. In order to assign a direction to the velocity, the quadrature method is used. Thus at least three A-lines must be compared to establish directionality. The accuracy of this measurement can be increased by sampling more A-lines, with seven considered the minimum to give adequate sensitivity to the color image. The number of A-line samples is called "color sensitivity", "ensemble length", or "packet size", depending on the system.

7.5.3.2 Time Domain Processing to Obtain Peak Velocity

A second method uses cross-correlation in each time interval to recognize the pattern of scatterers there and predict their movement[12]. If the cross-correlation is highest at the same position for each A-line sample, then no movement has occurred. If the value of the cross- correlation is higher for a fixed distance, R, then for the correlation distance at zero the scatterers are presumed to have moved $R \cos \theta$, and the velocity can be computed from the pulse repetition frequency. About 10 A-lines are needed for an accurate estimate. This method is considered more accurate since it can be corrected for the blood flow angle, estimates peak velocity rather than mean velocity, and is faster because it is less computationally intense. Originally, it was thought to avoid aliasing, which is a problem that occurs for the autocorrelation method when the PRF is too low. However, if the blood velocity is too high during the time domain processing, the scatterers have moved out of the sample volume and a random velocity will be calculated.

7.5.3.3 Color Flow Scanning Parameters

Color Doppler has a number of options which make it difficult for the medical physicist to be versatile on unfamiliar equipment. To avoid inappropriate parameter selection, most manufacturers offer a variety of menu options which pre-set these color flow parameters. A few manufacturers do not allow users to change these pre-set color parameters. We have described the frequency amplitude, gray scale threshold combination which must be satisfied before the color pixel overwrites a gray scale pixel. This is often referred to as a "read/write priority." Also, the number of A-lines for each color line can be adjusted on some equipment. When too many A-lines are selected in order to increase color sensitivity, the frame rate decreases dramatically and some color "bleeding" outside the vessels will occur. High wall filters can dramatically reduce sensitivity, as can selecting too high a peak velocity for the clinical application being studied. Just as for PW Doppler, an inappropriately large angle θ can decrease color flow sensitivity while correct focusing to the depth of interest will improve sensitivity. Frame rates of color flow examinations are usually much lower than real-time examinations. This problem can be minimized if the area containing the color Doppler image is reduced and if a large pixel size is selected. For some systems, color pixels the same size, two times the size, and four times the size of real-time pixels are available.

Both color flow methods offer blood flow calculations which are 15 to 30% accurate, depending on the skill of the operator and the vessel being measured. Time domain processing requires a "collection time" and automatically recognizes the walls of the vessel during this time so that pulsatility of the vessel can be included in the estimate. Autocorrelation flow methods differ between manufacturers and are usually less accurate.

Although time domain processing is able to estimate quantitative information in the image, color flow is regarded as qualitative, able to display very low flow rates, identify the location of vessels, and distinguish partial occlusions and turbulent flow. However, even experienced operators are sometimes frustrated by the low sensitivity of color flow systems. For this reason, two innovations have dramatically changed sensitivity in recent years: power Doppler and ultrasound contrast agents.

7.5.4 Power Doppler

This alternate display method does not color code different velocities, but codes any velocity into a single color based on the power of that velocity. Thus the peak velocity selected does not affect the power Doppler display[13]. Fine vascularity is uniformly colored while regions with more power are rendered in brighter hues. The operator need not select parameters carefully when using this modality. Some radiologists will progress from power Doppler to color Doppler to PW Doppler, thus using each modality to obtain ever more specific information about the clinical diagnosis.

7.6 Contrast Media for Ultrasound

Ultrasound contrast media which is not harmful to the patient and which remains visible throughout an ultrasound scan has been sought for many years[14]. Even with color flow imaging, vessels are not always visible, and fine vascularity is usually obscured because it contains slow flow and is likely to be eliminated by wall filters and other clutter removal methods. In recent years, four contrast agents have been developed and tested on animals and humans covered by Human Subject Protection protocols. All are either through FDA testing or are almost through, so we may expect these aids to be clinically available very soon.

7.6.1 Impedance Mismatch Materials

These materials are based on the belief that an ideal ultrasound contrast agent will have both an increase in the acoustic impedance of the contrast material over that of soft tissue, and a small size, so the media can recirculate. If Rayleigh scattering describes the contrast media, then the radius of the scatterers, r, must be much smaller than the interrogating ultrasound wavelength. In fact, to pass through the pulmonary capillaries, $2r < 5$ μm is necessary.

7.6.1.1 Current Specifications of Contrast Materials

One of the contrast agents under investigation is a perfluorochemical agent (Imagent™ Ultrasound, Alliance Pharmaceuticals, San Diego CA) which can be seen in the liver for several days after injection. The acoustic impedance of this material is 1140 compared to that of soft tissue (1540), primarily because of its low velocity[14].

All other ultrasound contrast agents are various materials surrounding microbubbles. The materials are designed to provide uniform bubble size and to increase the effective half-life of the bubbles so they can be injected intravenously. One manufacturer (ImaRx Pharmaceuticals, Tucson AZ) coats nitrogen bubbles with a lipid to produce 2 μm size Aerosomes™. The only FDA-approved ultrasound contrast agent (at this time) is Albunex™ (Molecular Biosystems Inc., San Diego CA). This contrast medium consists of air bubbles coated with albumin to produce short lived 4 μm (average size) particles which are cleared from circulation by the liver within 3 minutes of injection. A longer lived agent has been developed; Levovist™ (Schering AG, Berlin, Germany) is made from saccharide covered gas bubbles with a range of sizes from 2 to 8 μ. It can be imaged in the blood vessels for more than three minutes.

All of these contrast agents are injected intravenously in very small quantities. Once injected, vessels can be seen using color Doppler imaging because the increased number of scatterers increases the color Doppler sensitivity (Figure 7.16) making them brighter than their surrounding tissue, rather than being anechogenic as they are without contrast. Using a high resolution option for the color flow imaging, vasculature within the kidneys can be appreciated.

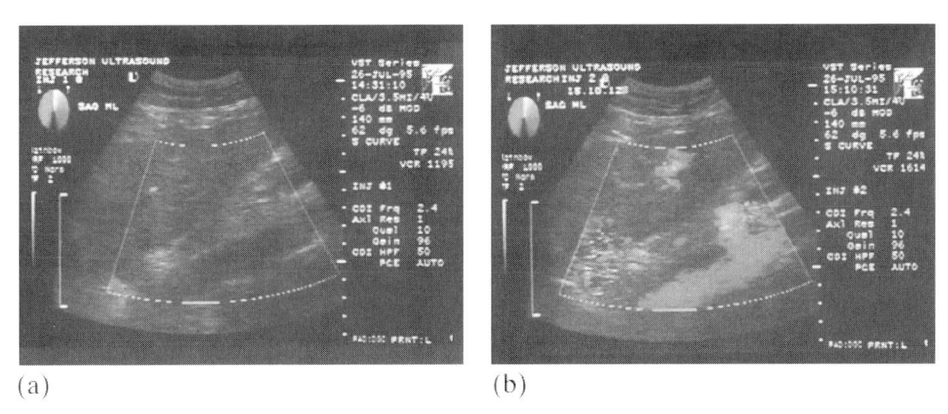

Figure 7.16. Before (a) and after (b) ultrasound contrast injection. This is a black and white print of a color Doppler image.

7.6.1.2 Harmonic Ultrasound

For those imaging tasks that still cannot be enhanced using contrast agents conventionally, harmonic ultrasound is being investigated using Levovist. The size of bubbles in this agent causes resonance for certain frequencies of ultrasound. By sending the ultrasound at one frequency (usually 2–3 MHz) and receiving at twice that frequency (the first harmonic), the locations containing contrast are enhanced by over 1000 times. If the adjacent non-resonate and resonate frames are subtracted, regions containing contrast are left, similar to digital subtraction angiography (although with much less resolution).

7.6.2 Current Uses of Contrast Materials

Because the liver phagocytizes some of the particles in ultrasound contrast materials, it retains the material longer than other tissues. For this reason, it is clinically useful to improve the search for liver metastases using contrast agents. Because Levovist remains active so long, it has been used to image the portal circulation as well as small vessels in the kidneys. Tumors can be recognized by their increased vascularization, while cysts and benign tumors tend to be anechoic. Contrast agents have been used to study breast masses, so that benign masses can be differentiated from malignant ones.

Contrast can cause scanning difficulties, particularly when using Doppler or color flow imaging. For a Doppler examination, contrast increases the amplitude of the frequency spectrum. This increases the volume of sound emitted from the loudspeakers during Doppler examinations. Furthermore, either because the Reynolds number has changed with the addition of contrast, so that turbulent rather than laminar flow results, or because we can better see the low amplitude, high frequency component of the Doppler spectrum, peak velocity measurements of contrast enhanced blood are generally higher than unenhanced blood. For color flow imaging, all the factors which increase sensitivity: packet size, velocity setting, color/gray scale write priority, must be adjusted to prevent blooming or bleeding near vessel walls. We are all inexperienced when using ultrasound contrast. Much experimentation remains to be done with these interesting agents.

7.7 Ultrasound Quality Control Procedures

Now that the American College of Radiology has initiated certification programs for Medical Ultrasound, there is increased interest in establishing a minimum set of equipment test procedures for quality control. In addition to this minimum set, most hospital based medical physicists should be acquainted with a set of acceptance tests for newly acquired medical ultrasound equipment. As for most medical imaging equipment, specialized test objects are needed and can be purchased from Gammex (RMI), Nuclear Associates, ATS, or CIRS. A tissue equivalent medium is required which is stable over at least two years, and which has scattering properties similar to liver. Embedded in these test objects is a series of filaments, cylinders, cones, and spheres which simulate idealized structures in the body. One such (generic) test object is illustrated in Figure 7.17.

A test object of this type can be used for both acceptance testing and for routine monthly or quarterly testing. The regions marked A, B, C, D, and E in Figure 7.17 refer to various quality control tests described below. Before scanning the top surface of the test object, a film of scanning gel should be applied to the surface. When acceptance testing a unit, all the transducers which have been purchased to operate on this unit should be tested for the following factors[15].

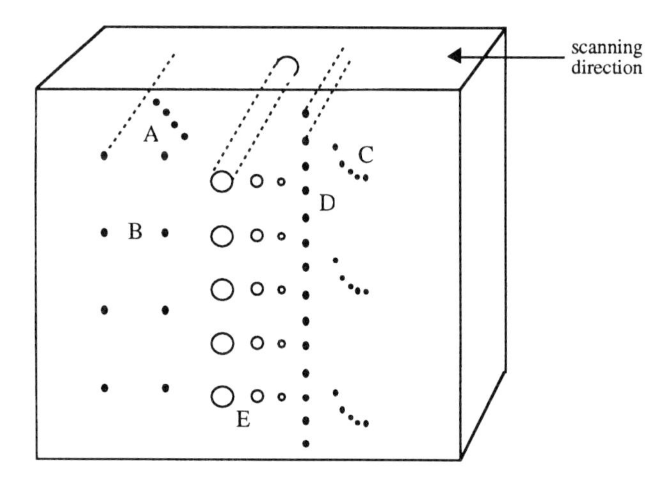

Figure 7.17. A gray scale test object (GSTO) with embedded filaments and cylinders which, when scanned perpendicular to their long axis, are imaged as points and circles.

7.7.1 Depth of Penetration

Area E of the test object (or lesser depths near the anechogenic rods for higher frequencies) should be centered in the ultrasound image. The focal zone and depth should be as deep as possible (6–8 cm for a 5 MHz transducer, 8–12 cm for lower frequencies), Gain, TGC, and output power should be adjusted to maximize penetration without saturating the image at more shallow depths. The settings should be recorded. If the TGC is controlled by a sliding scale, the position of these slides can be recorded by cutting paper or cardboard to fit against each slide. When testing penetration subsequently, this template can be used to adjust the TGC to the same position as was achieved on the previous tests. For knobs without labels, the physicists may wish to mark them with enamel paint for the various transducers tested. Permanent felt pen marks may not be permanent when germicide cleansers are used daily by the technologists, so enamel is recommended. Once the maximum depth has been established by observing the deepest cyst-like structure visible in the image of the test object, a photograph of the image should be made on the multiformat camera or digital camera attached to the ultrasound unit. For at least one transducer, this image should be compared to the image on the monitor to ensure that the same cyst visibility exists on both images and that the same number of steps are visible on any step wedge displayed on the monitor and the film.

7.7.2 Field Uniformity

Scan the GSTO after placing the focal zone(s) at about half the maximum penetration depth. Adjust the TGC, gain and output power so there are no horizontal streaks. If this is not possible, the unit may have hardware problems which must be repaired. Check that there are no vertical shadows extending from the surface near the transducer down to the depth of the focal zone. If a streak is found, move the transducer on the test object to establish if the streak moves also. If it doesn't, the test object has air in it and may need to be replaced. If the streak is associated with a particular position on the face of the transducer, the transducer needs to be repaired or replaced.

7.7.3 Ring Down

All transducers have a period of time that they are recovering from the initial pulse of voltage applied to them. This time period represents the "ring down" depth where they do not receive echoes returning to the transducer. The points in region A are spaced from the top surface of the test object. The point first imaged indicates the depth where ring down has stopped. Thus if the points are spaced at 1, 6, 11, and 16 mm depth, and the 6 mm point is the first that can be seen, then the ring down depth will be less than 6 mm. These points will be more visible if gain, TGC, and output power are reduced so the background is very dark. Most transducers can image the 6 mm point.

7.7.4 Axial Resolution

Groups of filaments similar to those in area C are imaged to determine resolution in the axial (or ultrasound beam) direction. The spacing of the five filaments is traditionally 3, 2, 1, and 0.5 mm. If all 5 points can be seen, then axial resolution is 0.5 mm, while if only 4 points are resolved, 1 mm axial resolution is established. Most 5 MHz transducers can resolve all 5 points. Using the zoom features available on most units will help evaluate this test, as will reducing gain and output power. Moving the focal zone to the level of the axial resolution filaments will improve the accuracy of this measurement. Any one of the three groups of filaments in the phantom can be imaged. Higher frequency transducers will perform better on groups near the transducer.

7.7.5 Lateral Resolution

The points near region D are used for lateral resolution measurements. Image these filaments with the focal zone positioned at half the penetration depth. Freeze the image and zoom in on that depth. Place electronic calipers at each end of the filament image underneath the narrowest point (see Figure 7.18).

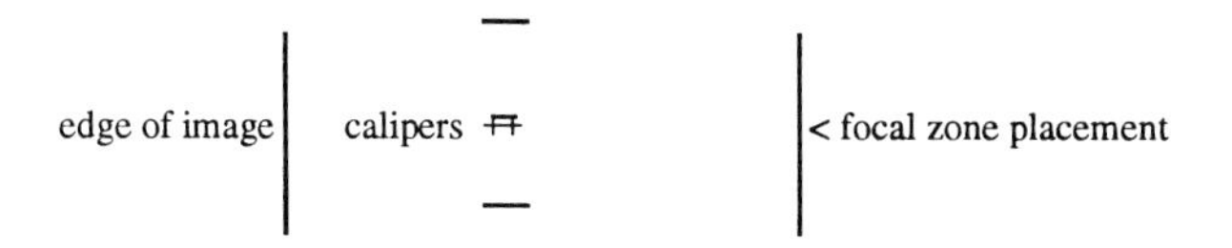

Figure 7.18. Caliper placement for lateral resolution measurement.

The measurement gives the resolution perpendicular to the beam axis. Some medical physicists also measure the range of good focus available in transducers when only one focal zone is used. This is established by finding those pins whose dimensions are no more than 50% larger than the length of the narrowest point. The axial distance between the last two points on each side of the central point which meet this criterion is the focal zone range.

7.7.6 Vertical and Horizontal Caliper Accuracy

Because the mechanisms which ensure vertical accuracy (the calibrated propagation velocity of the unit in soft tissue) are different from those that ensure horizontal caliper accuracy (spacing of the transducer elements and, for phased or curved arrays, the depth of the measurement), both directions must be tested. The points near region B can be used for horizontal caliper accuracy testing. As for the lateral resolution test, calipers should be placed underneath the filament image, but centered on the image. For vertical caliper accuracy, the filaments near region D should be measured, and here the calipers should be placed at the same depth, but beside the filament image (see Figure 7.19).

At least 3 cm should be measured for each test. Vertical accuracy should be within 1.5 mm or 5% if less than 3 cm is measured, while horizontal accuracy may be less because of the poorer lateral resolution, and errors of 2 mm or 6% are acceptable.

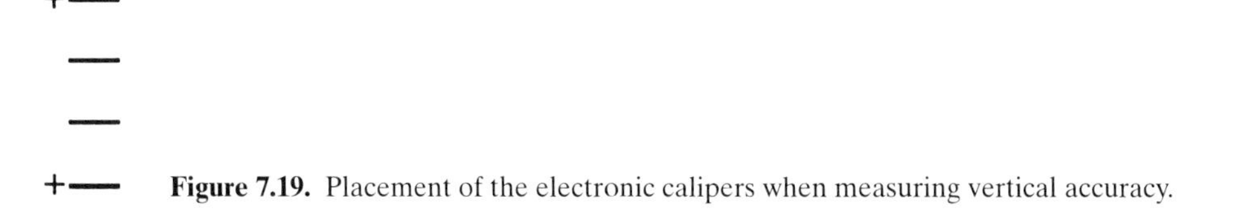

Figure 7.19. Placement of the electronic calipers when measuring vertical accuracy.

7.7.7 Slice Thickness Resolution and Range

Slice thickness of the "out of image" resolution depends on the mechanical focusing lens attached to each element of the transducer. To assess where this lens is focused and the range of good focus requires a specialized phantom. If such a phantom is not available, a less rigorous assessment can be made by scanning each of the smallest anechogenic cylinders in the GSTO at different depths with the transducer perpendicular to the usual orientation. A band, representing the thickness of each cylinder, will then be imaged. If the band is anechogenic (i.e., black), then the transducer is in slice thickness focus at the depth of that cylinder. Where the edges of the cylinders have scattered echoes within areas that should be anechogenic, the transducer is not focused at the depths of these cylinders. A more rigorous and quantitative test can be performed using a "beam profile" test object (Figure 7.20).

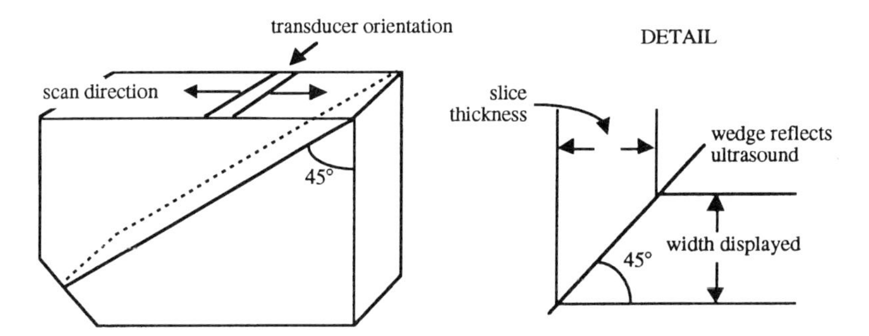

Figure 7.20. Transducer scans across the phantom producing an image of the wedge.

By placing the transducer perpendicular to the usual GSTO orientation, the slice thickness of the beam at different depths will reflect from the 45° wedge. When the transducer is moved across the top of the test object from right to left, the beam encounters the wedge at different depths. Because the wedge is positioned at an angle of 45°, the image of the resulting stripe is the same thickness as the slice thickness of the transducer at that depth. Find the depth where the stripe is narrowest. Put the focal zone at this depth. Measure the thickness of the stripe and then find the two depths where the stripes are 50% larger than this thickness. These depths represent the range of the slice thickness focus. This is an important measurement to report to the radiologist who will use the equipment, since small pathologies outside this range are likely to be missed.

7.7.8 Film Processing

Those ultrasound units which depend on transparent film (rather than Polaroid or paper prints) should establish processor quality control procedures for their equipment. Mammography processor quality control manuals[16] give complete directions for these procedures.

7.7.9 Duplex Doppler Quality Control

These procedures require a calibrated pump which can produce peak velocities in 5 to 7 cm inner diameter tubing of from 10 to 60 cm/sec. Commercial Doppler phantoms are available, but are expensive to purchase and maintain. Blood substitute material is also expensive, but temporary fluid with scatterers can be made with measured, small amounts (less than 1t) of corn starch dissolved in a quart of water. The peak velocity at a calibrated setting of the pump should match two times the calculated average flow rate within 20% with the pump at one midrange velocity. The precision of the duplex Doppler unit can then be measured at a variety of pump settings for several transducers. Because inter- and intra-observer accuracy is no more than 10% (6 cm/sec at 60 cm/sec peak flow)[17], greater accuracy cannot be expected.

If a commercial flow phantom is not used, then the tubing carrying the simulated blood will have to traverse a salt water bath (so the propagation velocity is adjusted to that of soft tissue), or the tubing will need to be embedded in a gelatin surround. The sample volume can then be placed at physiologically realistic depths. Doppler quality control is not rigorous, since vibrations from the pump, clutter removal and other non-clinical phenomena tend to distort quality control measurements. When possible, ensure that all the transducers in the facility measure similar peak velocities so that serial examinations will not show significant changes unless those changes have occurred in the patient.

7.7.10 Routine Quality Control

The tests described above are for acceptance testing and would be performed once in the lifetime of an ultrasound unit or transducer. When transducers are replaced, the new transducer should be acceptance tested. Routine testing, usually done by the technologist, should include penetration, uniformity, vertical calipers, photography and axial resolution. Of course, processor control should be an ongoing concern. The frequency of these tests varies from monthly to semi- annually at different facilities.

7.8 Conclusions

Ultrasound offers a novel blend of electronic innovation, digital image processing, and sophisticated mathematical algorithms. It is a challenge for the medical physicist who must explain ultrasound physics to residents and medical physics students, who must negotiate with vendors over innovative designs, and who must ensure that equipment is clinically useful and reliable. The information in this chapter should aid the medical physicist in acquiring this knowledge.

7.9 References

1. PNT Wells, *Biomedical Ultrasonics* (Academic Press, New York, 1977) pp. 13–18, 27–40.
2. CR Hill, *Physical Principles of Medical Ultrasonics* (Ellis Horwood Ltd. from John Wiley and Sons, New York, 1986) pp. 225–257.
3. LL Foldy, *Phys Rev* **67**: 107–119 (1945).
4. WR Hedrick, DL Hykes, DE Starchman, *Ultrasound Physics and Instrumentation* (Mosby, New York, 1995) pp. 7, 19, 98–109.
5. DA Christensen, *Ultrasonics Bioinstrumentation* (John Wiley and Sons, New York, 1988) pp. 69–118.
6. P Atkinson, JP Woodcock, *Doppler Ultrasound and Its Use in Clinical Measurement* (Academic Press, New York, 1982) pp. 45–124.
7. DH Evans, WN McDicken, R Skidmore, JP Woodcock, *Doppler Ultrasound, Physics, Instrumentation, and Clinical Applications* (John Wiley and Sons, New York, 1989) pp. 87–106.
8. WN McDicken, *Diagnostic Ultrasonics* (Churchill Livingston, New York, 1991) pp. 67–75.
9. L Hatle, B Angelsen, *Doppler Ultrasound in Cardiology* (Lea and Febiger, Philadelphia, 1985) p. 38.
10. JW Taylor, PN Burns, PNT Wells, *Clinical Applications of Doppler Ultrasound* (Raven Press, New York, 1988) pp. 42–44.
11. HF Routh, *IEEE Engineering in Medicine and Biology*, 31–40 (Nov/Dec 1996).
12. O Bonnefous, P Pesqúe, *Ultrasonic Imaging* **8**:73–85 (1986).
13. JM Rubin, RO Bude, PL Carson, RL Bree, RS Adler, *Radiology* 190:853–856 (1994).
14. BB Goldberg, J-B Liu, F Forsberg, *Ultrasound in Medicine and Biology* **20**: 319–332 (1994).
15. MM Goodsitt, P. Carson, S. Witt, D. Hykes, J. Kofler, *Real-time B-mode ultrasound quality control test procedures. Medical Physics* 25: no. 8 (1998).
16. *Mammography Quality Control Manual.* American College of Radiology, 1994.
17. FN Tessler, C Kimme-Smith, ML Sutherland, VL Schiller, RR Perrella, EG Grant, *US Med Biol* **16**:653–657 (1990).

8 Magnetic Resonance: Principles and Spectroscopy

William R. Riddle and Haakil Lee

Vanderbilt University

8.1 Introduction

Magnetic resonance involves absorption of very low-energy radio frequency photons by atomic nuclei in an applied magnetic field. The phenomenon of nuclear induction was first reported in 1946 by physicists Bloch, Hansen, and Packard[1]. In the 1950s, organic chemists started using nuclear magnetic resonance (NMR) spectroscopy to identify and analyze organic compounds. After it was realized in 1973 that magnetic resonance could be used to create images[2], the field of magnetic resonance imaging (MRI) has developed into a widely used modality for imaging soft tissue.

Ever since studies of tissue metabolism with phosphorous NMR were reported in 1974[3], investigators have been exploring *in vivo* magnetic resonance spectroscopy (MRS). There are numerous techniques currently available for *in vitro* studies with NMR on analytical spectrometers, but many of these methods cannot be used for *in vivo* MRS studies on clinical imager/spectrometers. This chapter discusses some of the basic principles of magnetic resonance, then presents some of the current techniques used in clinical MRS.

8.2 Basic Principles

8.2.1 Macroscopic Magnetization

The nucleus of an atom is comprised of protons and neutrons. The positively charged proton creates a magnetic field called a magnetic dipole. Even though a neutron has no net charge, it does have charge inhomogeneities which create a magnetic dipole. The total magnetic moment (μ) of an atom is a function of the number of protons and neutrons in the nucleus. The nuclear spin angular momentum quantum number (I) may be zero, a multiple of ½, or a multiple of 1, depending on the number of unpaired nucleons[4]. When the sum of protons and neutrons (atomic mass, A) and the sum of protons (atomic number, Z) are both even, I is equal to zero. When A is even and Z is odd, I is a whole number. When A is odd, I is a multiple of ½. A nuclide with a spin number of I can assume $2I+1$ discrete (also called Zeeman) energy states. When I is equal to zero, the nuclide has one energy state, and the magnetic moment is equal to zero. When I is not equal to zero, the nuclide has multiple energy states with a non-zero magnetic moment. When a nuclide with a non-zero spin number is placed in an external magnetic field, the nuclide's magnetic field interacts with the external field. This interaction causes the nuclide to precess about its axis at a frequency (f) proportional to the magnetic field strength (B_0), and is described by the Larmor equation:

$$f = \gamma B_0 \tag{8.1}$$

where γ is the gyromagnetic ratio.

From quantum mechanics, a photon associated with the emission or absorption of a particular frequency (v) can have but one energy (E_v) and is given by the Einstein-Planck equation[5]:

$$E_v = hv \tag{8.2}$$

where h is Planck's constant (6.625×10^{-34} joule-sec). In magnetic resonance, this v is equal to the Larmor frequency (γB_0) and E_v is the change in energy between two energy states. When nuclei with a magnetic moment are placed in an external magnetic field, they become magnetized and align in the magnetic field. For nuclei with quantum spin of ½, the alignment is either with the magnetic field (parallel) or against the magnetic field (anti-parallel). The energy of a group of atoms in thermal equilibrium (E_T) is equal to

$$E_T = kT \tag{8.3}$$

where k is the Boltzmann constant (1.38×10^{-23} joule/°K) and T is temperature in °K.

Protons have a quantum spin number of ½ with two Zeeman energy levels. The number of protons with the lower energy state (N_{m-1}) is defined by the Boltzmann distribution[6] as:

$$N_{m-1} = N_m \exp\left(\frac{E_v}{E_T}\right) \tag{8.4}$$

where N_m is the number of protons with the higher energy state. Since $N_m \approx N_{m-1}$ and $E_v \ll E_T$, the excess number of protons in the lower energy state (ΔN) can be simplified to

$$\Delta N = N_m \left(\frac{E_v}{E_T}\right) \tag{8.5}$$

The excess number of nuclei in the lower energy state, per one million protons, is equal to $E_v/E_T \times 500\,000$. In the absence of an external magnetic field (E_v=0), the dipoles are distributed randomly and have a net magnetization of zero. In the case of protons at 310 °K in the earth's magnetic field (0.5 gauss, or 0.00005 tesla), ΔN is 0.00016 protons per million. For protons at 310 °K in a 1.5 T external field, ΔN increases to 4.9 protons per million.

Magnetic properties of elements that are biologically relevant are listed in Table 8.1. The nuclide with the highest concentration, ^1H (proton), is used for magnetic resonance imaging. There are about 6.0×10^{19} protons per cubic millimeter in tissue, and with ΔN equal to 4.9 ppm at 1.5 T, there will be 3.0×10^{14} more protons in the parallel direction per cubic millimeter of tissue. For a ^{1}H metabolite with a concentration of 20 mM in a 1.5 T field, there will be 6.0×10^{10} more protons in the parallel direction per cubic millimeter. For a ^{31}P metabolite with a concentration of 20 mM in a 1.5 T magnet, there will be 2.4×10^{10} more phosphorous atoms in the parallel direction per cubic millimeter.

Table 8.1. Characteristics of Biologically Relevant Elements

Nucleus	Quantum Spin (I)	Physiological Concentration (M)	Magnetic Moment (μ)	Gyromagnetic Ratio (γ) (MHz/T)
^{1}H	1/2	100	2.79	42.58
^{13}C	1/2	–	0.69	10.71
^{23}Na	3/2	0.080	2.22	11.26
^{31}P	1/2	0.075	1.13	17.24

In the presence of a magnetic field (B), the average magnetic moment per unit volume (M) is the number of spins per unit volume times the average magnetic moment[7]. The change in M with respect to time in a magnetic field is given by

$$\frac{\mathrm{d}M}{\mathrm{d}t} = \gamma(M \times B) \tag{8.6}$$

When M is parallel to B, the volume is at equilibrium and $\mathrm{d}M/\mathrm{d}t$ is equal to zero because there is no torque. With the application of energy at the Larmor frequency, M will precess about B with a constant angular velocity and at a fixed angle. When B is along the z axis, equations in cartesian coordinates are:

$$\begin{aligned} \frac{\mathrm{d}M_z}{\mathrm{d}t} &= 0 \\ \frac{\mathrm{d}M_x}{\mathrm{d}t} &= \gamma M_y B_z \\ \frac{\mathrm{d}M_y}{\mathrm{d}t} &= -\gamma M_x B_z \end{aligned} \tag{8.7}$$

One possible solution to these equations is

$$\begin{aligned} M_z &= M_{\parallel} = constant \\ M_x &= M_{\perp} \cos(-2\pi f_0 t) \\ M_y &= M_{\perp} \sin(-2\pi f_0 t) \end{aligned} \tag{8.8}$$

where f_0 is the cyclic Larmor frequency in Hz. If it is assumed that thermal agitation causes M_z to return to equilibrium (M_0) by first-order kinetics, then the rate that M_z changes is equal to

$$\frac{dM_z}{dt} = \frac{1}{T_1}(M_0 - M_z) \tag{8.9}$$

where T_1 is a time constant called the longitudinal relaxation time.

If it is also assumed that the x and y components of M disappear exponentially, then the rates at which M_x and M_y change are equal to

$$\begin{aligned} \frac{dM_x}{dt} &= -\frac{M_x}{T_2} \\ \frac{dM_y}{dt} &= -\frac{M_y}{T_2} \end{aligned} \tag{8.10}$$

where T_2 is a time constant called the transverse relaxation time. Combining these approximation equations for relaxation in the absence of an applied magnetic field with the change in M in a magnetic field (Eq. 8.6) gives the Bloch equations:

$$\begin{aligned} \frac{dM_z}{dt} &= \left(\frac{1}{T_1}\right)(M_0 - M_z) + \gamma(M \times B)_z \\ \frac{dM_x}{dt} &= -\frac{M_x}{T_2} + \gamma(M \times B)_x \\ \frac{dM_y}{dt} &= -\frac{M_y}{T_2} + \gamma(M \times B)_y \end{aligned} \tag{8.11}$$

These equations are the result of a simple model of the magnetic resonance phenomenon. They are not exact, especially concerning the T_2 time constant, but they are useful in explaining nuclear spin in magnetic resonance. The solution to Eq. (8.11) for a static field B along the z axis after a 90° pulse is

$$M_z = M_0\left[1 - \exp(-t/T_1)\right]$$

$$M_x = M_0 \exp(-t/T_2)\cos(-2\pi f_0 t) \tag{8.12}$$

$$M_y = M_0 \exp(-t/T_2)\sin(-2\pi f_0 t)$$

where f_0 is the cyclic Larmor frequency.

8.2.2 Laboratory and Rotating Frames

Two frames of reference are used in magnetic resonance: the laboratory frame and the rotating frame. The laboratory frame, from the observer's point of view, is a stationary reference frame where the protons precess about the z axis with circular projections in the xy plane. Figure 8.1 illustrates tipping M towards the xy plane in the laboratory frame with a circular polarized B_1 field. The displacement (α), in degrees, of the M_0 vector from its original position after a radio frequency (RF) pulse is called the flip angle. The rotating frame spins at the Larmor frequency (f_0). In this frame, spins at the Larmor frequency are stationary, and spins above and below f_0 rotate slowly about the z axis. Figure 8.2 illustrates the rotating frame with M tipping towards the X′ axis with a B_1 field applied at the Larmor frequency in the Y′ direction.

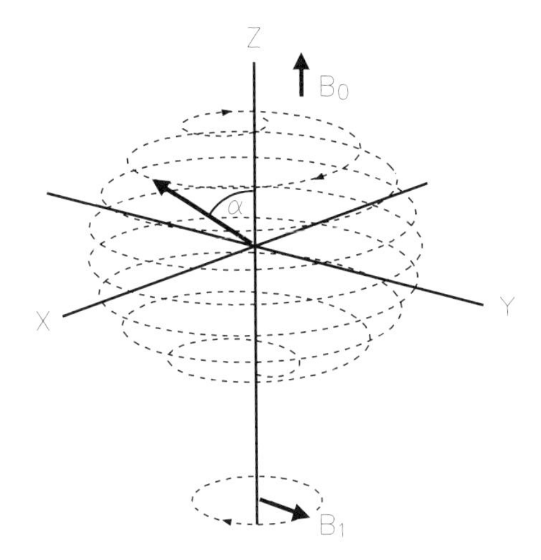

Figure 8.1. This diagram shows the tipping of M towards the xy plane in the laboratory frame with a circular polarized B_1 field. The vector M is spinning at the Larmor frequency.

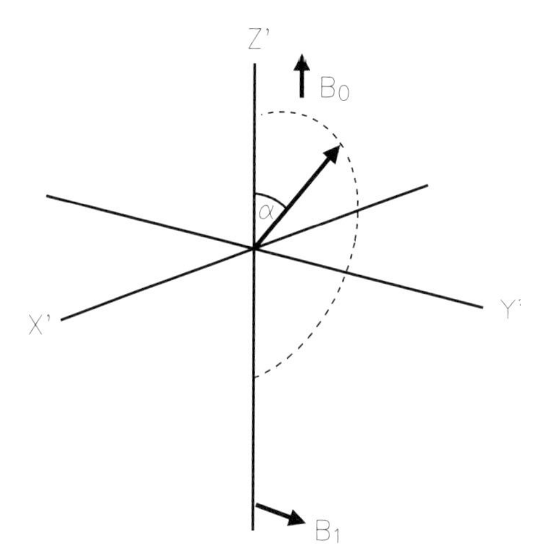

Figure 8.2. This diagram shows the tipping of M in the rotating frame, which is spinning at the Larmor frequency.

8.2.3 Relaxation Times

T_1 is the longitudinal relaxation time constant in the Bloch equations. It is also known as the thermal or spin-lattice relaxation time. Regrowth of longitudinal magnetism requires transfer of energy from the nuclear spin system to its environment. T_1 relaxation occurs when a proton encounters another magnetic field near its frequency, which is usually another proton or electron. T_1 values increase as magnetic field strength increases. For most biological tissues, T_1 increases approximately as $B_0^{0.333}$.[8]

T_2 is the transverse relaxation time constant in the Bloch equations. It is also known as spin-spin relaxation. T_2 relaxation can take place with or without transfer of energy. T_2 relaxation results from any process that causes the spins to lose their phase coherence, such as local field inhomogeneities or a varying magnetic field. Molecules found in membranes and most macromolecules have very short T_2 values. T_2 values remain approximately unchanged over the field strengths used in MRI and MRS.

T_2^* is the effective T_2 in the presence of inhomogeneities in the main magnetic field. These inhomogeneities may result from defects in the magnet, susceptibility-induced field distortions, or field modulations due to eddy currents. The relationship between T_2^*, T_2, and field distortions (ΔB) is approximately

$$\frac{1}{T_2^*} = \frac{1}{T_2} + \gamma \Delta B \tag{8.13}$$

8.2.4 Magnetic Field Gradients

A magnetic field gradient is defined as the rate at which the magnetic field strength changes with position. Gradient coils create a spatially varying magnetic field which is superimposed on the main magnetic field. The spatial encoding necessary for MRI and localized MRS is provided by three orthogonal linear gradients. Depending on how the gradients are used in the sequence, they are called the frequency encoding gradient, the phase encoding gradient, and the slice selection gradient. The Z gradient (in direction of B_0) is obtained with a pair of loops called a Maxwell pair. The X and Y gradients are obtained with four saddle coils for each gradient.

8.2.5 Chemical Shift

If all the atoms of a specific nuclide had the same Larmor frequency, then magnetic resonance spectra would be limited to a single peak. However, the magnetic field of a nucleus can be affected by shielding from the electrons surrounding the nucleus. That is, the same nuclide with different chemical neighbors will have different resonance frequencies. The Larmor frequency with chemical shift is

$$f_{cs} = \gamma B_0(1-\sigma) \tag{8.14}$$

where σ is a screening constant and $\sigma \ll 1$. This small change in the resonance frequency is the basis for magnetic resonance spectroscopy.

8.3 Magnet Technology

The important parameters for clinical MRI and MRS systems are 1) field strength, 2) field homogeneity, 3) field stability, 4) access to the region of homogeneity, and 5) stray magnetic field[9]. The minimum field strength needed for MRS is roughly 0.5 tesla, but since the separation of the peaks is directly proportional to field strength, more is better. For whole body applications, the magnet needs to have horizontal access with enough clearance to move subjects in and out of the magnet. The types of magnets used in magnetic resonance include permanent, resistive, superconductive, and hybrid (permanent plus resistive). With the requirement of a one meter bore, the maximum field strength is about 0.5 T for a permanent magnet, 0.2 T for a resistive magnet, and 4.0 T for a superconductive magnet. Therefore, superconductive magnets are the magnet of choice for whole body magnetic resonance spectroscopy.

8.4 Implementation Considerations

Several factors need to be considered before acquiring MRS data. These include choice of RF coil, type of RF pulse, and adjustment of system parameters.

8.4.1 Radio Frequency Coils

Radio frequency coils are necessary and essential components of every MR system. They are used for exciting nuclear spins in a sample and for detection of the resulting nuclear precession[10]. During excitation, the coil converts RF power into a transverse rotating RF magnetic field, called B_1. During reception, the coil measures the precessing nuclear magnetization that was induced in the sample. While it is desirable for a RF coil to be both efficient and have spatially uniformity, efficiency and uniformity cannot be maximized simultaneously. Increasing a coil's spatial uniformity decreases both the coil's efficiency and the received signal. The majority of the energy deposited in the sample with magnetic resonance is in the form of heat. Only a very small fraction of the energy is transferred to the nuclear spins. RF coils used in MRI and MRS fall into two groups, volume coils and surface coils. Volume coils have high spatial uniformity; surface coils have high sensitivity that is strongly dependant on spatial coordinates.

Currently, the most widely used design for a volume coil is usually referred to as a "birdcage coil"[11]. The birdcage design consists of a series of identical loops on the surface of a cylinder that are connected together with capacitors. Birdcage coils with 8, 16, 32, or more elements are used for head and body imaging, and have essentially equal sensitivity over their entire volume.

The sensitivity of a surface coil is a function of distance from the coil. The expected signal for a single loop receive-only coil along the center of the coil is given by the following equation[12]:

$$B(y) \propto \frac{1}{(r^2 + y^2)^{1.5}} \tag{8.15}$$

where r is the coil radius and y is the distance from the center of the coil. Following a rectangular pulse, the signal at distance y from a transmit/receive coil is given by the following equation[12]:

$$S(y) \propto B(y) \sin \left\{ \frac{\alpha_{\text{center}}}{[1 + (y/r)^2]^{1.5}} \right\} \tag{8.16}$$

where α_{center} is the pulse flip angle at the coil center. Figure 8.3 illustrates the signal from a surface coil as the flip angle and distance are changed.

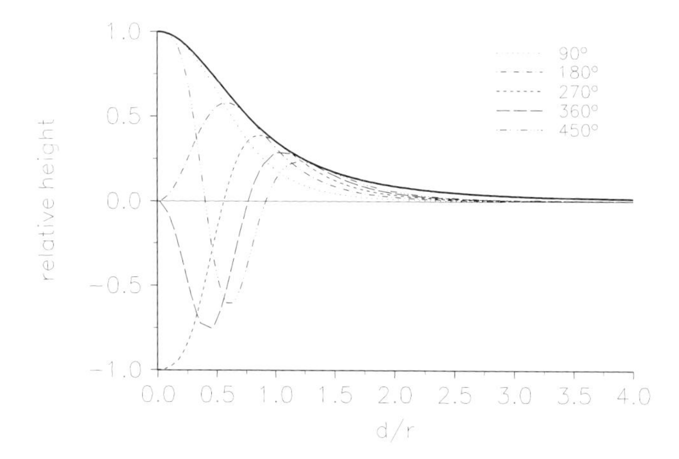

Figure 8.3. This illustrates the signal from a surface coil as the flip angle at the center of the coil and distance from the coil are changed. The solid line is the coil sensitivity, and the broken lines show the response as the flip angle is changed.

RF coils are designed to operate at a Larmor frequency. In a 1.5 T magnet, a ^{1}H RF coil operates at 63.9 MHz and a ^{31}P RF coil operates at 25.8 MHz. A double-tuned coil has two resonate frequencies, for example, 63.9 MHz for ^{1}H and 25.8 MHz for ^{31}P. Such a coil can be used to obtain both proton and phosphorous spectra, or to obtain proton images and phosphorous spectra.

8.4.2 Radio Frequency Pulses

MR systems use RF pulses to excite nuclear spins. These pulses are classified as non-selective and selective. A non-selective RF pulse excites the whole volume being studied and is applied without a field gradient. A selective pulse has a specific frequency and is applied with a gradient to excite a specific volume (slice) of tissue. The displacement, in degrees, of the *M* vector from its original position after an RF pulse is called the flip angle.

8.4.2.1 Rectangular Pulses

The flip angle (α) after a rectangular pulse of duration t_p is given by the following equation:

$$\alpha = \gamma B_1 t_p \tag{8.17}$$

where γ is the gyromagnetic ratio and B_1 is the strength of the RF field. As a first approximation, the bandwidth of this pulse is $1/t_p$ Hz[13]. A rectangular pulse is non-selective.

8.4.2.2 Adiabatic Pulses

If the rate of change of the flip angle of a non-selective pulse is sufficiently slow to allow the magnetization induced by the pulse to remain parallel to B_{eff}, the pulse is termed adiabatic[14]. An adiabatic half-passage (AHP) pulse will cause a flip angle of 90°, and an adiabatic full-passage (AFP) pulse will cause a flip angle of 180°. When a simple rectangular pulse is used for excitation with a surface coil, the inhomogeneity of the RF field produces a wide range of flip angles with nonuniform excitation of spins. When an adiabatic pulse is used for excitation with a surface coil, there is more uniform excitation of the spins with more uniform flip angles. The first two reported adiabatic pulses were the sin/cos amplitude- and frequency-modulated AHP pulse[15] and the sech/tanh modulation scheme for AFP pulse[16]. When phase is modulated instead of frequency, the sin/cos AHP pulse can be generated with the following equation:

$$B_1(t) = \sin\left(\frac{\pi t}{2pl}\right)\exp\left[4 \times OF \times pl \times \sin\left(\frac{\pi t}{2pl}\right)\right] \tag{8.18}$$

where pl = pulse length (sec) and OF = initial offset frequency (Hz). The first sine term is the amplitude modulation, and the exponential term is the phase modulation of the AHP pulse. Figure 8.4 illustrates a 2.56 msec AHP pulse with OF = 1000 Hz. Panel A shows the real, imaginary, and magnitude components of the AHP pulse; Panel B shows the phase of the AHP pulse; and Panel C shows the slope of the phase expressed in Hz, which is $d\alpha/dt$. The value of $d\alpha/dt$ starts at the specified initial offset frequency (OF) and ends at zero. Excitation with this AHP pulse will excite the spins uniformly over several coil radii[12].

8.4.2.3 Gaussian Pulses

Sometimes it is desirable to excite only a narrow range of frequencies, as in water or fat presaturation. This selective saturation can be provided by a Gaussian RF pulse. An n point Gaussian pulse, with $-n/2 \leq k < n/2$, is defined by the following equation:

$$\text{gauss}(k) = \exp[-3.5602(k \times BW \times \Delta t)^2] \tag{8.19}$$

where BW is the bandwidth of the selective excitation and Δt is the time between each point. To minimize the edge effects, the Gaussian pulse is multiplied by a modified Hamming window. The modified Hamming window (HW) for point k is defined by the following equation:

$$\begin{aligned} &\text{for } |k| \leq n/4: \text{HW}(k) = 1.0 \\ &\text{for } |k| > n/4: \text{HW}(k) = 1 + \cos(4\pi k/n + \pi) \end{aligned} \tag{8.20}$$

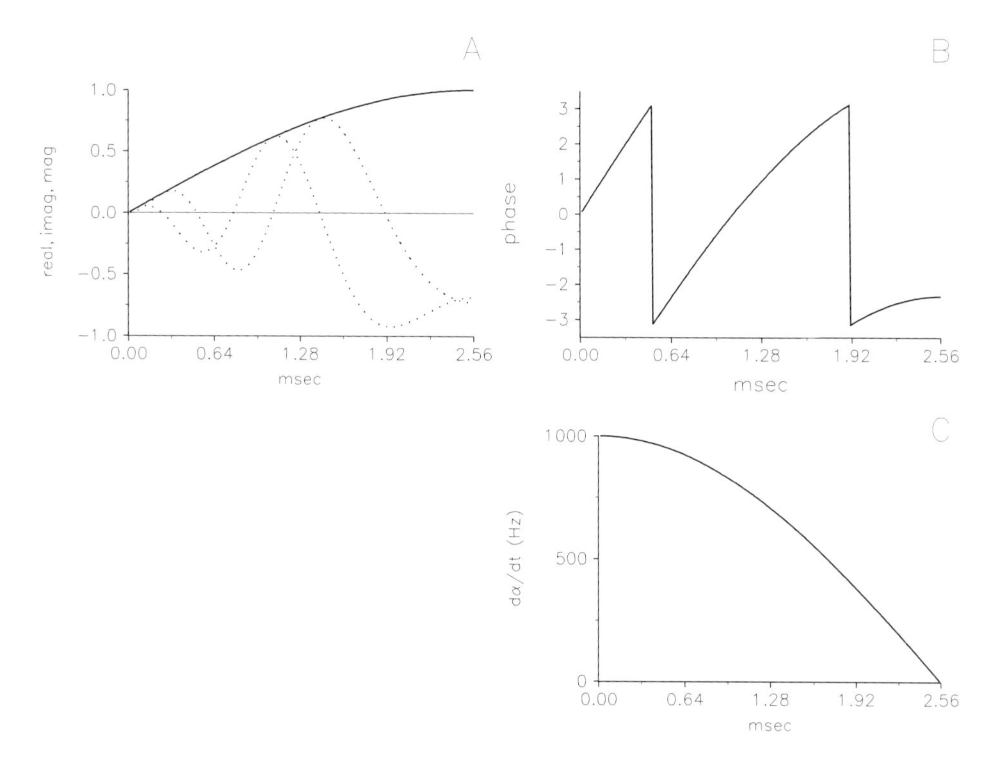

Figure 8.4. 256-point sin/cos amplitude- and phase-modulated adiabatic half-passage (AHP) pulse with $\Delta t = 0.00001$ sec and $OF = 1000$ Hz. Panel A shows the real, imaginary, and magnitude components of the AHP pulse. Panel B shows the phase of the AHP pulse, and Panel C shows the slope of the phase expressed in Hz, which is $d\alpha/dt$.

To represent the Gaussian pulse as a complex number, the value of gauss(k) $\times$ HW(k) is placed in the real component, and the imaginary component is set to zero. This will center the Gaussian pulse at the center frequency of the transmitter. The Gaussian pulse can be shifted with the following equation:

$$\text{offsetgauss}(k) = \exp[-2\pi \times OF \times \Delta t \times (k+n/2)] \times \text{gauss}(k) \tag{8.21}$$

where OF is the shift in Hz. An example of a 256-point Gaussian pulse with Δt – 0.0001 sec, $BW = 60$ Hz, $OF = -225$ Hz is shown in Figure 8.5.

8.4.2.4 SINC Slice Selective Pulses

A selective RF pulse has a specific frequency and is applied under a gradient. A selective pulse excites a specific volume (slice) of tissue. Since a SINC function in the time domain produces a rectangular pulse in the frequency

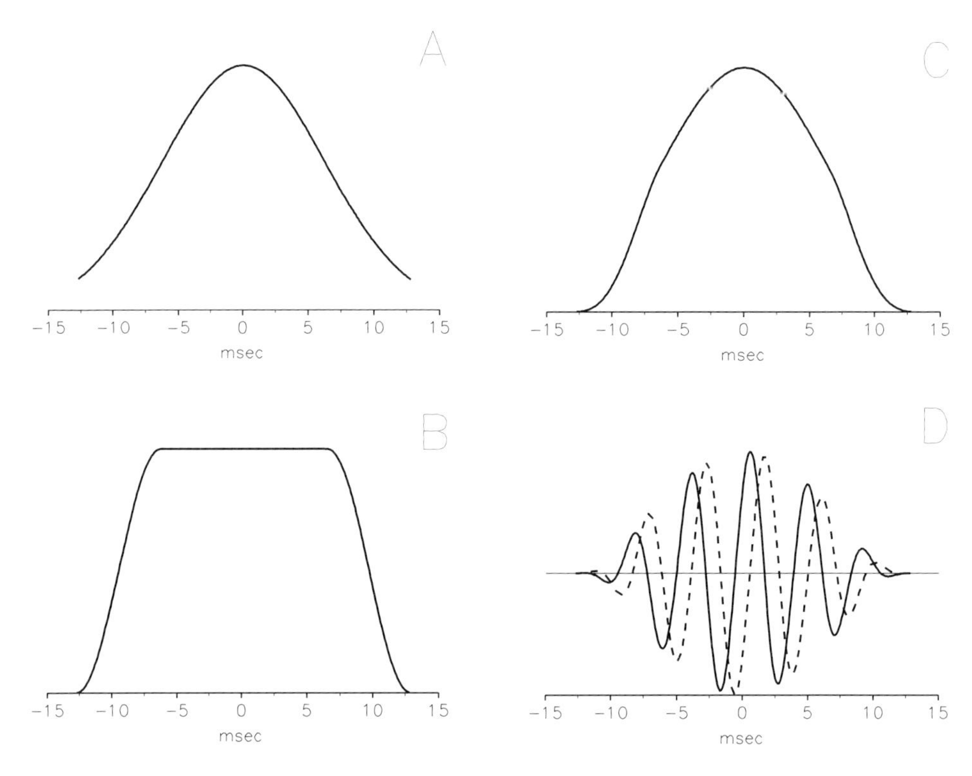

Figure 8.5. 256-point Gaussian pulse with $\Delta t = 0.0001$ sec, $BW = 60$ Hz, $OF = -225$ Hz. Panel A is the Gaussian pulse generated with Eq. (8.19). Panel B is the modified Hamming window in Eq. (8.20). Panel C is the Gaussian pulse after the Hamming window. Panel D is the real (solid line) and imaginary (dashed line) components of the Gaussian pulse offset −225 Hz with Eq. (8.21).

domain, a common method for generating a slice selective RF pulse is with a SINC function. An n point SINC pulse, with $-n/2 \leq k < n/2$, is defined by the following equation:

$$\begin{aligned} &\text{for } k \neq 0: \quad \text{sinc}(k) = \frac{\sin(\pi D G \gamma \Delta t k)}{\pi D G \gamma \Delta t k} \\ &\text{for } k = 0: \quad \text{sinc}(k) = 1.0 \end{aligned} \tag{8.22}$$

where γ is the gyromagnetic ratio in MHz/T, D is the slice thickness (in mm), G is the gradient strength (mT/m), and Δt is the time between each point. To minimize the edge effects, the SINC pulse is multiplied by the modified Hamming window in Eq. (8.20). To represent the SINC pulse as a complex number, the value of $\text{sinc}(k) \times \text{HW}(k)$ is placed in the real component, and the imaginary component is set to zero. This will specify a slice at the isocenter. The selected slice is moved P mm from the isocenter by the following equation:

$$\text{offsetsinc}(k) = \exp[-2\pi \times OF \times \Delta t \times (k+n/2)] \times \text{sinc}(k) \tag{8.23}$$

where $OF = PG\gamma$. An example of a 256-point SINC pulse with $\Delta t = 0.00001$ sec, $D = 40$ mm, $G = 2.0$ mT/m, $\gamma = 42.5759$ MHz/T, and $P = 20$ mm is shown in Figure 8.6.

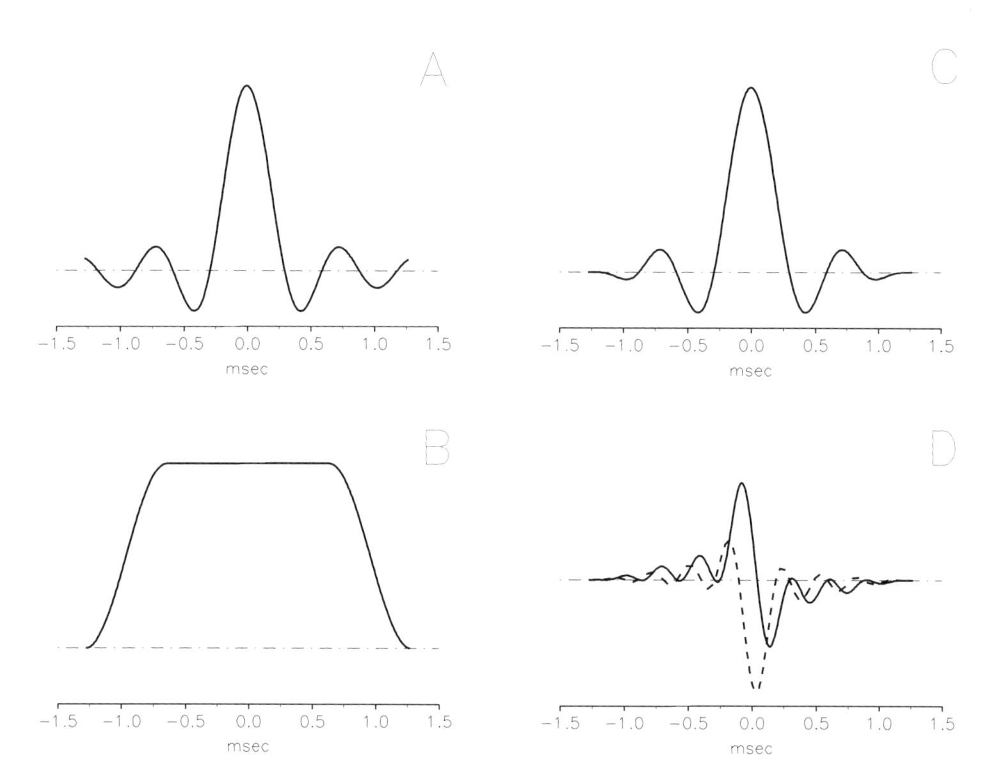

Figure 8.6. 256-point SINC pulse with $\Delta t = 0.00001$ sec, $D = 40$ mm, $G = 2.0$ mT/m, $\gamma = 42.5759$ MHz/T, and $P = 20$ mm. Panel A is the SINC pulse generated with Eq. (8.22). Panel B is the modified Hamming window in Eq. (8.20). Panel C is the SINC pulse after the Hamming window. Panel D is the real (solid line) and imaginary (dashed line) components of the SINC pulse offset 20 mm with Eq. (8.23).

8.4.2.5 Shinnar-Le Roux Slice Selective Pulses

Another method for designing a slice selective pulse is with the Shinnar-Le Roux algorithm[17]. This algorithm is based on a discrete approximation of the spin domain version of the Bloch equation. The RF pulse design problem is mapped into a digital filter design problem that reduces to two complex polynomials. After these polynomials are defined, the solution is then mapped back into the RF pulse. RF pulses designed with the Shinnar-Le Roux algorithm have better defined slice profiles in the frequency domain and do not have the sidelobes found with a Hamming windowed SINC pulse.

8.4.3 System Adjustments Before Acquiring Spectra

After placing the sample in the magnet, it is necessary to perform several adjustments before spectra can be acquired. The operator must tune the coil; adjust transmitter frequency, transmitter gain, and receiver gain; homogenize the magnetic field by shimming; and, in the case of proton spectroscopy, optimize water suppression. The operator must also decide whether or not to use CYCLOPS or phase cycling.

8.4.3.1 Coil Tuning

After placing the RF coil and the sample in the magnet, the coil's tuning capacitors are adjusted to make the impedance of the coil match the source impedance of the transmitter. When impedances are matched, the reflected power is minimized. The tuning process for some coils has been automated, but other coils still require manual tuning.

8.4.3.2 Adjusting Frequency

In proton spectroscopy, the center frequency of the transmitter is usually placed at the frequency of water. This is accomplished by acquiring an FID, transforming it to the frequency domain, measuring the displacement (δ) of the water peak from zero, then adjusting the transmitter frequency by δ. This will make the offset frequency of water equal to zero. For phosphorous spectroscopy, the center frequency of the transmitter is usually placed at the phosphocreatine (PCr) peak.

8.4.3.3 Transmitter Gain

The transmitter gain is adjusted by determining the voltage that produces a 90° non-selective pulse. When using slice selective RF pulses, the transmitter voltage is equal to the voltage of the non-selective 90° pulse divided by the area under the selective pulse.

8.4.3.4 Receiver Gain

Adjustment of the receiver gain is necessary because a low gain setting can cause the signals to become corrupted by the digitization of the analog-to-digital converter (ADC), and an excessive gain setting can cause signals to exceed the range of the ADC. Since the gain of the receiver is linear, the correct gain can be determined by acquiring an FID with the receiver gain at its lowest setting (G_0) and measuring the maximum of the magnitude of the time domain signal (S_0). The optimum gain (G_{rec}) is then determined by

$$G_{rec} = G_0 \frac{S_1}{S_0} \tag{8.24}$$

where S_1 was chosen as 30 000 ADC units (for a 16-bit, 2's complement ADC, the range of the ADC is +32,767 to –32,768). In the event the receiver gain cannot be adjusted to reach G_{rec} from Eq. (8.24), the gain is simply set to the maximum gain.

8.4.3.5 CYCLOPS And Phase Cycling

Physical differences between the two channels of the quadrature receiver can introduce artifacts into a magnetic resonance image or magnetic resonance spectrum. There will be ghosting of the peaks mirrored about zero frequency if the phase difference between the two channels is not 90° or the amplifier gains of the two channels are not equal. This is often called a quadrature ghost. If the DC offsets of the two channels are not equal to zero, there will be a spike at zero frequency. CYCLically Ordered Phase Sequences (CYCLOPS) can be used to correct for these artifacts[18]. CYCLOPS involves rotating the transmitter phase by 0°, 90°, 180°, and 270°, then adding and subtracting the two receiver channels. Figure 8.7 shows the first 50 msec of the real (X) and imaginary (Y) channels of a simulated FID with imbalan-

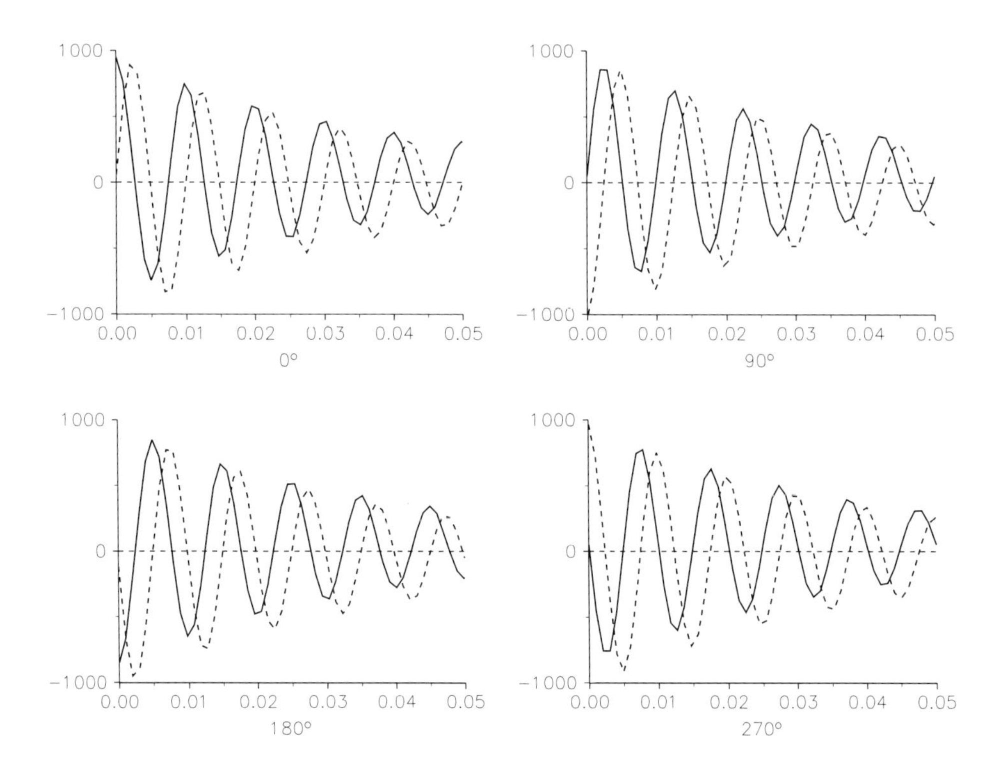

Figure 8.7. First 50 msec of simulated FIDs with misadjustments in the receiver channels with the phase of the transmitter shifted 0°, 90°, 180°, and 270°. The quadrature phase was 85°, the channel gains were unequal, and the DC offsets were not zero.

ced gains, DC offsets, and a phase difference of 85°. For CYCLOPS, the corrected X and Y channels for each point are calculated with the following equations:

$$X_{\mathrm{corr}} = X(0°) - Y(90°) - X(180°) + Y(270°) \tag{8.25}$$

$$Y_{\mathrm{corr}} = Y(0°) + X(90°) - Y(180°) - X(270°) \tag{8.26}$$

A reduced implementation of CYCLOPS that toggles the transmitter phase between 0° and 180° is called phase cycling. For phase cycling, the corrected X and Y channels for each point are calculated with the following equations:

$$X_{\mathrm{corr}} = X(0°) - X(180°) \tag{8.27}$$

$$Y_{\mathrm{corr}} = Y(0°) - Y(180°) \tag{8.28}$$

Figure 8.8 shows spectra without phase cycling or CYCLOPS (Panel *A*), with phase cycling (Panel B), and with CYCLOPS (Panel C). Phase cycling corrects for the DC artifact but does not correct for the quadrature ghost. CYCLOPS corrects for both the DC artifact and the quadrature ghost. Phase cycling requires a multiple of two acquisitions, while CYCLOPS requires a multiple of four acquisitions.

8.4.3.6 Shimming

An essential component of any MRS study is shimming the magnet to minimize the static field inhomogeneities. When large electromagnets were used for nuclear magnetic resonance measurements, the field homogeneity was adjusted by placing thin brass shim stock between the magnet and the pole faces to make them parallel. This mechanical method of shimming has since been replaced by adjusting electrical currents in special "shim" coils, but the term "shimming" has remained.

Increasing the static field homogeneity with shimming will improve spectral sensitivity and resolution. A better shim will produce a larger integral in the time domain and a taller, narrower peak in the frequency domain. There have been reports in the literature describing automated shimming methods in both the time domain[19–21] and the frequency domain[22–26]. Time domain methods maximize the integral of the received signal, and frequency domain methods minimize the spread of frequencies on a properly prepared image.

Figure 8.9 shows how the integral (sum of magnitude points) changes in a clinical imager/spectrometer as the X, Y, Z, and Z^2 shim currents were changed. Another way to present these responses is to plot 1/integral versus shim current. The value of 1/integral is directly proportional to the FWHM,

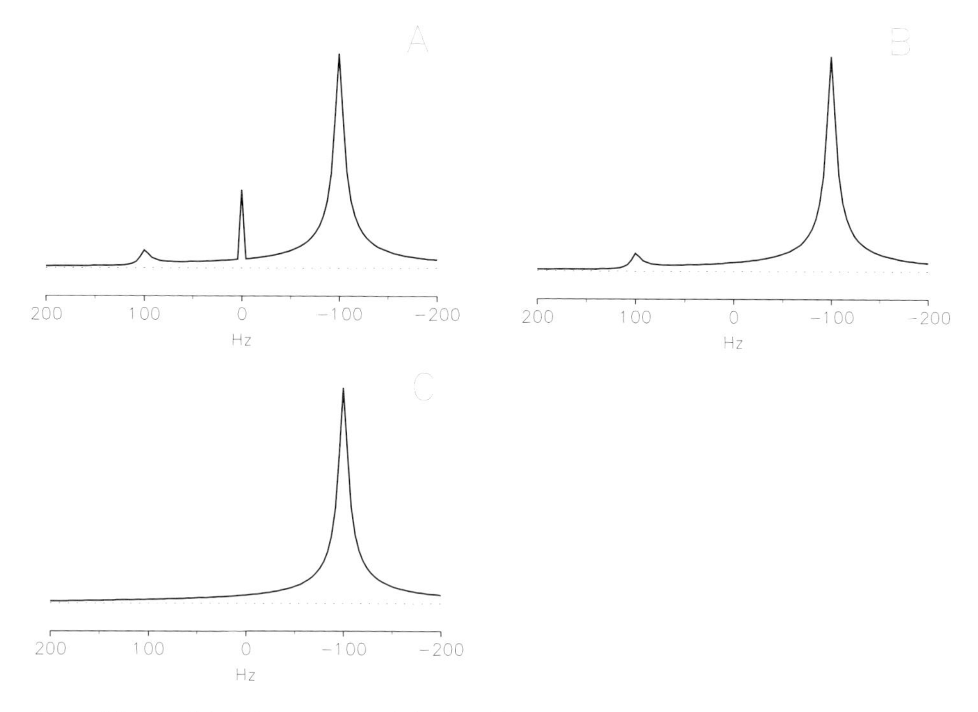

Figure 8.8. Panel A is the spectrum of simulated FID with the receiver misadjustments. There is a spike at 0 Hz and a quadrature ghost at 100 Hz. Panel B is the spectrum with phase cycling. The spike at 0 Hz was removed, but the quadrature ghost is still present. Panel C is the spectrum with CYCLOPS. The spike at 0 Hz and the quadrature ghost have been removed.

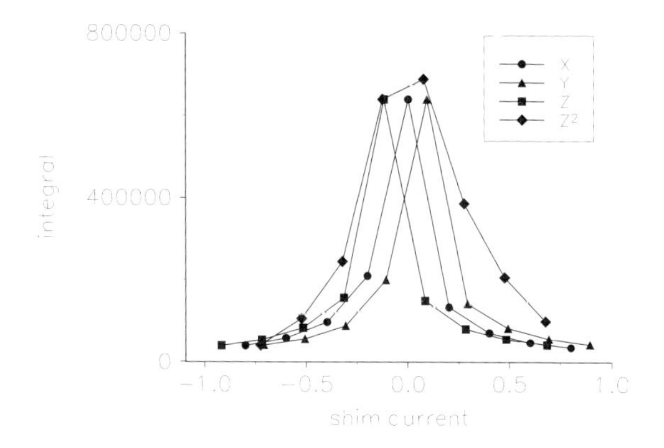

Figure 8.9. Sum of magnitude points (integral) from the free induction decay as the X, Y, Z, and Z^2 shim currents are changed in a clinical imager/spectrometer.

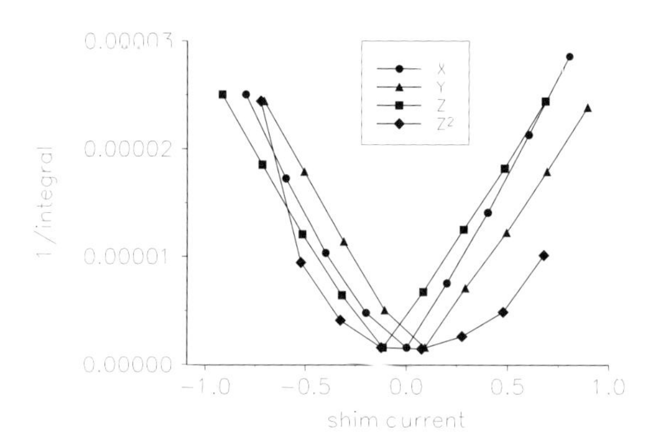

Figure 8.10. 1/integral as the X, Y, Z, and Z^2 shim currents are changed.

so maximizing the integral or minimizing 1/integral will minimize the FWHM. Figure 8.10 plots 1/integral versus the shim current in the X, Y, Z, and Z^2 coils. Since these curves are approximately parabolic, a method for shimming is to determine 1/integral at three different shim currents to define a parabola (ax^2+bx+c), then set the shim current to the minimum of the parabola ($-b/2a$). Figure 8.11 illustrates the 1/integral points for the *Z* shim currents from Figure 8.10 and the parabola generated from three points. This approach can be used in an automated shimming procedure.

To show the effect of localized shimming, a spin echo imaging sequence was modified to provide a two-dimensional "topographical" map of the magnetic field. A PRESS sequence (TE=135 msec, TR=3 sec) was used to acquired the signal with a 2 × 2 × 2 cm volume of interest (VOI) within a cylindri-

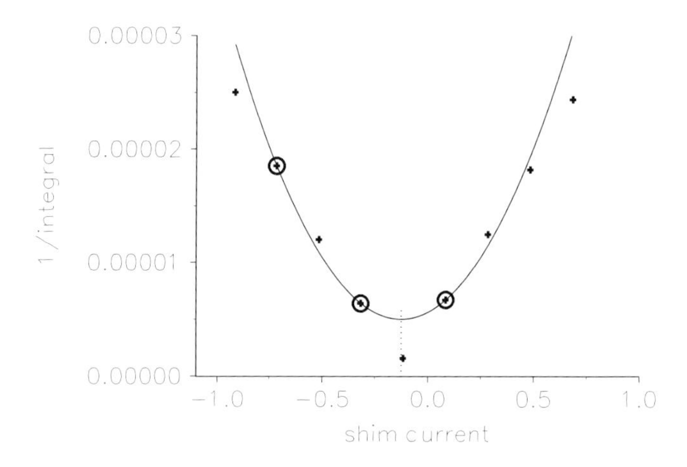

Figure 8.11. 1/integral points (+) for the Z shim currents from Figure 8.10 and the parabola generated from three points (⊕). The dotted line indicates the shim current defined from the minimum of the parabola.

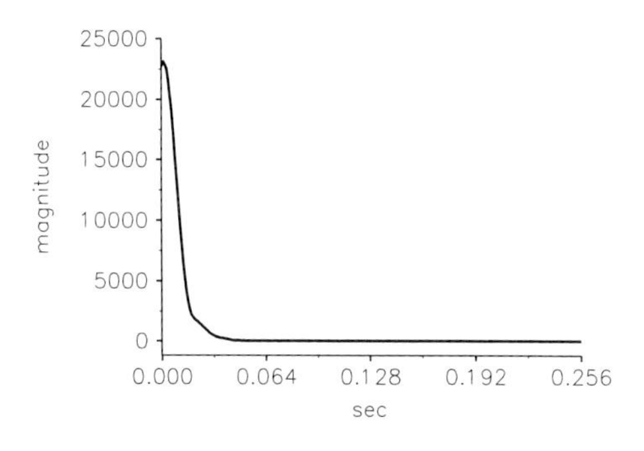

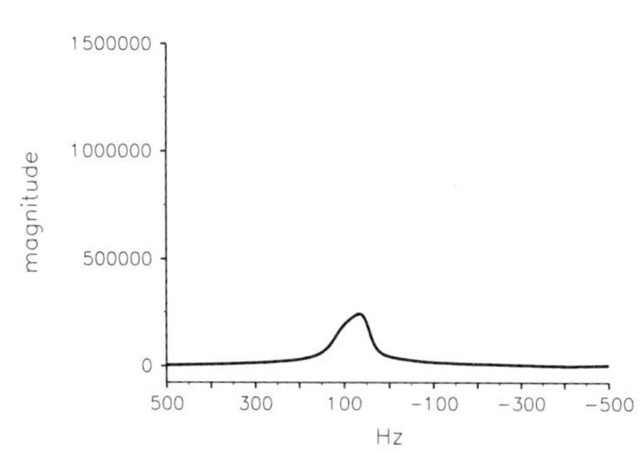

Figure 8.12. Time domain signal, frequency domain signal, and a field map with 10 Hz contours from a cylindrical water phantom acquired with the imaging shim currents. The box in the image indicates the 2 × 2 × 2 cm VOI. The FWHM of the water peak is 31.2 Hz.

cal water phantom. Figure 8.12 shows the time domain signal, the frequency domain signal, and the field map acquired with the imaging shim currents. The magnet was then shimmed to the VOI using the X, Y, Z, and Z^2 shim currents. Figure 8.13 shows the improvement in the time and frequency domain signals and the shift of the "sweet spot" in the field map. The FWHM of the water peak in the VOI improved from 31.2 Hz in Figure 8.12 to 3.3 Hz in Figure 8.13.

8.4.3.7 Water Suppression

The concentration of protons in water is greater than 100 M, while the concentration of *in vivo* metabolites is 20 mM or less[27]. When performing proton spectroscopy *in vivo*, water suppression is required because of the dynamic range limitation of the ADC. One method of water suppression is the use of chemical-shift-selective (CHESS) excitation pulses to saturate the water signal[28]. These CHESS pulses are narrow bandwidth Gaussian pulses. For lo-

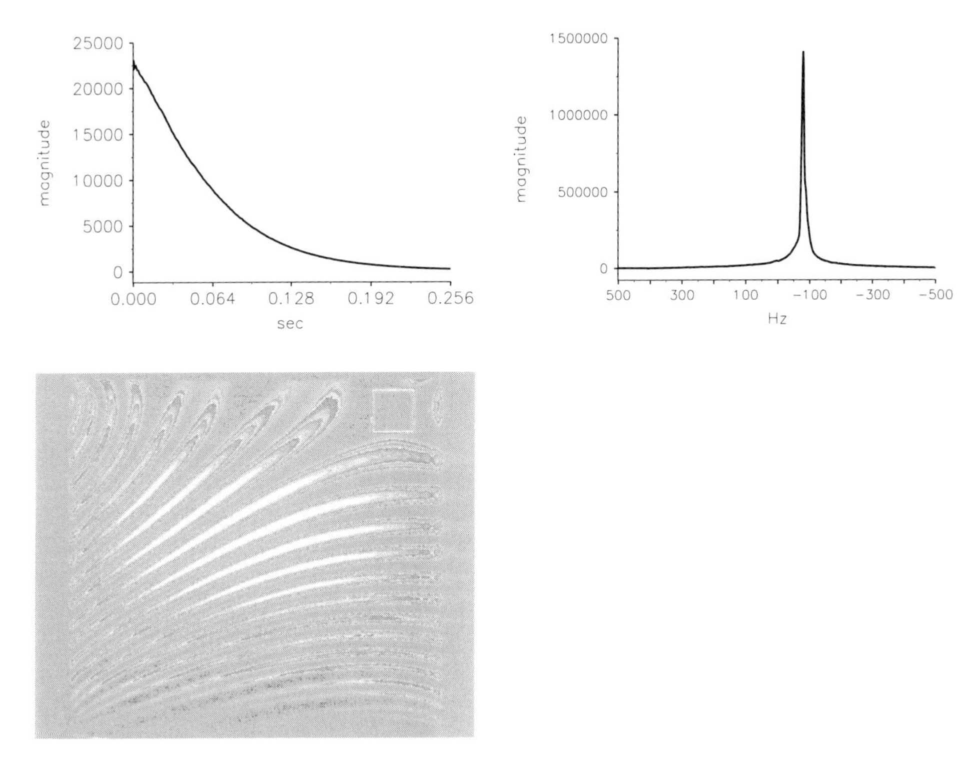

Figure 8.13. Time domain signal, frequency domain signal, and a field map with 10 Hz contours from a cylindrical water phantom acquired after optimizing the X, Y, Z, and Z^2 shim currents to the VOI. The box in the image indicates the 2 × 2 × 2 cm VOI. The FWHM of the water peak is 3.3 Hz.

calized single voxel spectroscopy, STEAM or PRESS sequences with one or three CHESS pulses are used. To achieve maximal water suppression, it is necessary to adjust the transmitter voltage (flip angle) of the CHESS pulse(s).

Figure 8.14A shows the effect of varying the transmitter voltage of the CHESS pulse in a PRESS sequence (TE=135 msec) with a 2 × 2 × 2 cm VOI at the center of a 2-liter spherical phantom. The transmitter voltage for a 90° CHESS pulse is equal to the voltage of a non-selective 90° pulse divided by the area under the CHESS pulse. However, due to T_1 relaxation between the effective center of the CHESS pulse and the subsequent excitation, a true 90° CHESS will not minimize the water signal. The minimums at 3.4, 6.5, 9.8, 12.7, 15.7, and 18.7 volts correspond to effective flip angles of 90°, 270°, 450°, 630°, 810°, and 990° in the VOI. As the CHESS amplitude passes through the minimums, there is a reversal in the polarity of the real and imaginary components of the FID. Another method of plotting this data is to evaluate the phase at the beginning of each FID. When the phase of the FID with zero volts is selected as zero, a plot of the initial phase versus transmitter voltage has a sharp π shift at each minimum point (Figure 8.14B). The integral can be

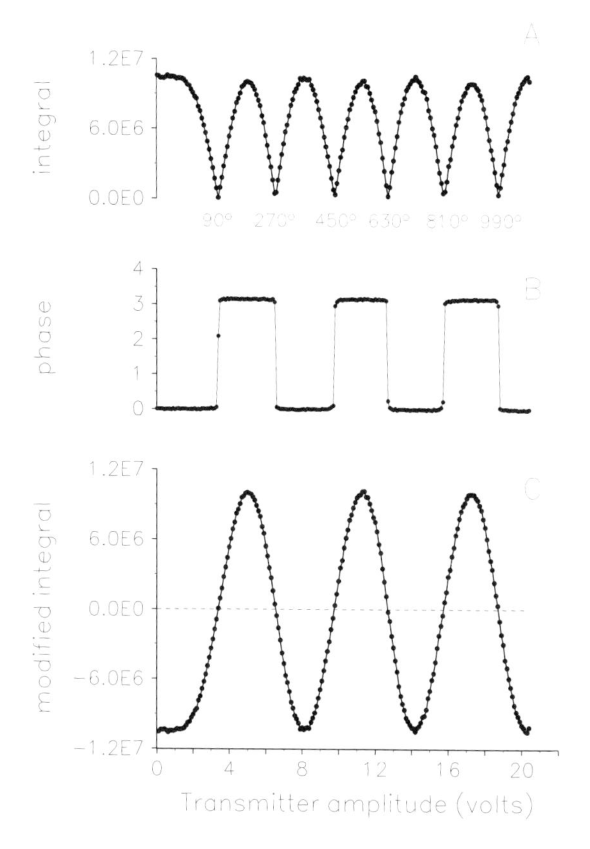

Figure 8.14. Effect of varying the transmitter amplitude of a CHESS pulse in a PRESS 135 sequence. Panel A is the integral (sum of magnitude points). Panel B is the phase of each FID at time=0. Panel C shows the modified integral that includes the phase in Panel B.

modified to incorporate phase by setting the integral negative whenever phase is less than $\pi/2$ and positive when phase is greater than $\pi/2$. This modified integral becomes a sine wave, as shown in Figure 8.14C. This modified integral can be used in an automated water suppression procedure.

8.5 Spectroscopy Sequences

Producing a signal on an MR system requires running a program, often called a sequence, that controls the RF transmitter, the magnetic field gradients, and the ADC. A sequence timing diagram shows the RF pulses, the X, Y, and *Z* field gradients (slice selection, phase encoding, and frequency encoding), and the ADC. The following sequences are used in MRS.

8.5.1 FID

An FID sequence transmits a single RF pulse, then turns on the ADC to "listen" to the free induction decay. When the amplitude of the RF pulse is adjusted to maximize the amplitude of the FID, this pulse is called a 90° or $\pi/2$ pulse. The pulse is non-selective and can be a simple rectangular pulse or an adiabatic half-pass pulse. There is no localization with an FID sequence. Figure 8.15 is a timing diagram for an FID sequence with a rectangular RF pulse.

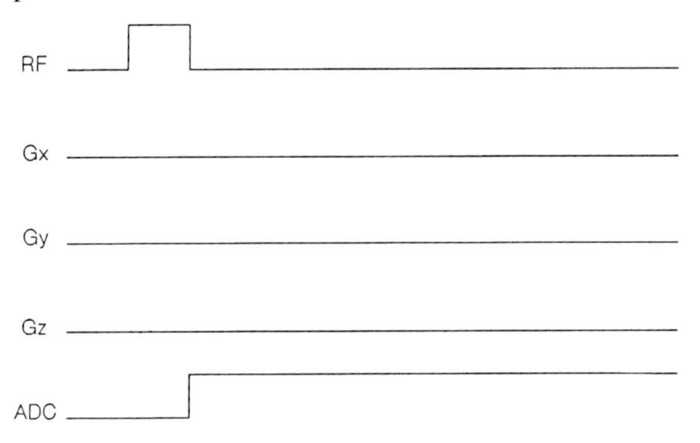

Figure 8.15. Timing diagram for an FID sequence. There is a single rectangular pulse followed by turning on the ADC.

8.5.2 ISIS

An ISIS (Image Selective *In vivo* Spectroscopy) sequence provides a means for acquiring spatially selective signals[29]. ISIS uses selective inversion of the spin population before a non-selective 90° pulse and data acquisition. In a one-dimensional ISIS sequence, a slab is selected with two acquisitions. The first acquisition uses only the non-selective 90° pulse. The second acquisition inverts the spin population in a selected slice with a selective pulse followed by the non-selective 90° pulse. Subtraction of the two FIDs results in signal from the spins perturbed by the inversion pulse.

In a two-dimensional ISIS sequence, four acquisitions are used to specify a bar. The first acquisition uses only the non-selective pulse. The second and third acquisitions invert the spin population in one of two different orthogonal axes followed by the non-selective pulse. The fourth acquisition inverts the spins in two axes followed by the non-selective pulse. The spins in the bar are obtained by adding the first and fourth FIDs and subtracting the second and third FIDs.

In a three-dimensional ISIS sequence, eight acquisitions are required to select a rectangular parallelepiped (box). Figure 8.16 shows the eight acquisitions and the spin states in the box after each acquisition. The spins in the

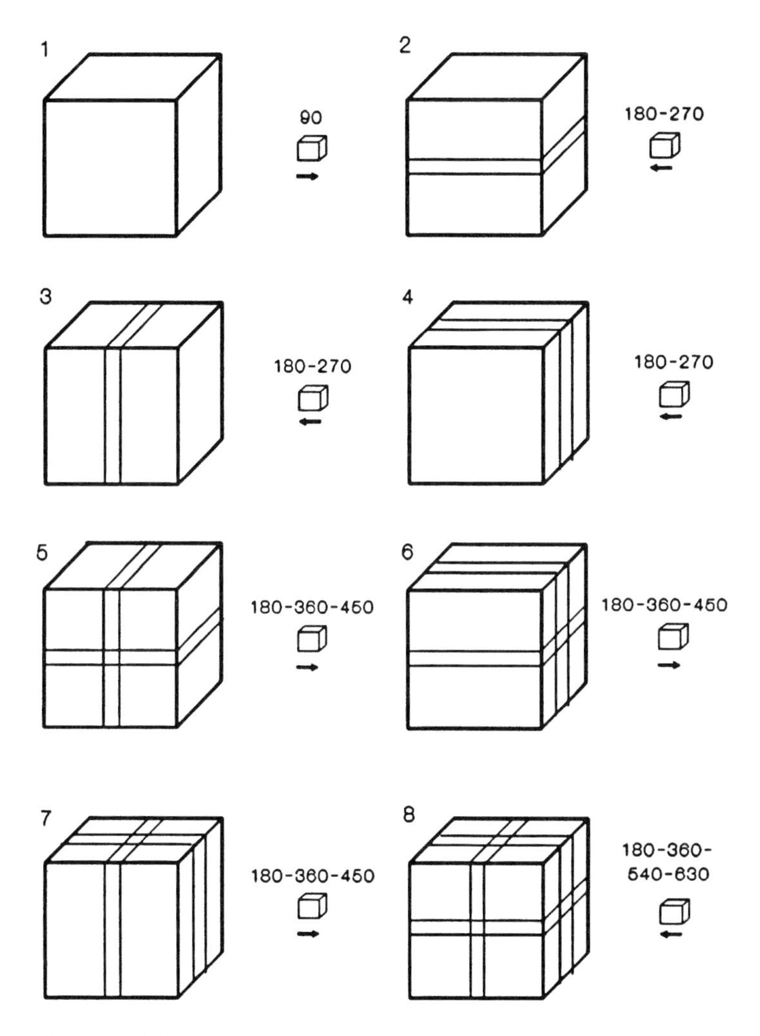

Figure 8.16. This illustrates the regions affected during the eight acquisitions in a 3-D ISIS sequence. The thin slices indicate the regions that receive the 180° spin inversions. The spins in the VOI are shown adjacent to each large cube.

box are obtained by adding the first, fifth, six, and seventh FIDs and subtracting the second, third, fourth, and eighth FIDs. Figure 8.17 is a timing diagram for the eighth acquisition in a 3-D ISIS sequence. The other seven acquisitions can be obtained by omitting one or more of the selective inversion pulses.

The flip angle of the inversion pulses in ISIS is not critical, but 180° pulses will maximize the amplitude of the final FID. When using an ISIS sequence, the receiver gain must be set using the acquisition without spin inversion pulses, then held constant for all acquisitions. ISIS is based on signal subtraction, and the signal in the selected slab, bar, or box is a small fraction of the total

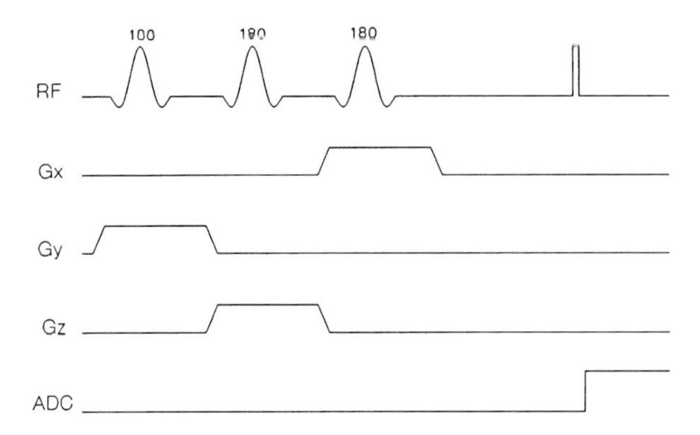

Figure 8.17. Timing diagram for the eighth acquisition in a 3-D ISIS sequence. The other seven acquisitions can be obtained by omitting one or more of the 180° selective inversion pulses.

signal. Figure 8.18 shows images acquired in a head coil of an oil and water phantom and the selected 2 × 2 × 2 cm cube. The volume of the water bottle was 500 mL and the volume of the oil bottle was 946 mL. Figure 8.19 shows the spectrum without any inversion pulses (FID) and the 3-D ISIS spectrum from the cube.

A potential problem with ISIS is that motion occurring during the sequence cycle could introduce volume localization artifacts[30]. For the spins to respond correctly to the inversion pulses, the TR of an ISIS sequence should be at least three times the T_1 of the substance being studied.

Figure 8.18. Image of an oil and water phantom and the selected 2 × 2 × 2 cm cube in the water bottle. The volume of the water bottle was 500 mL, and the volume of the oil bottle was 946 mL.

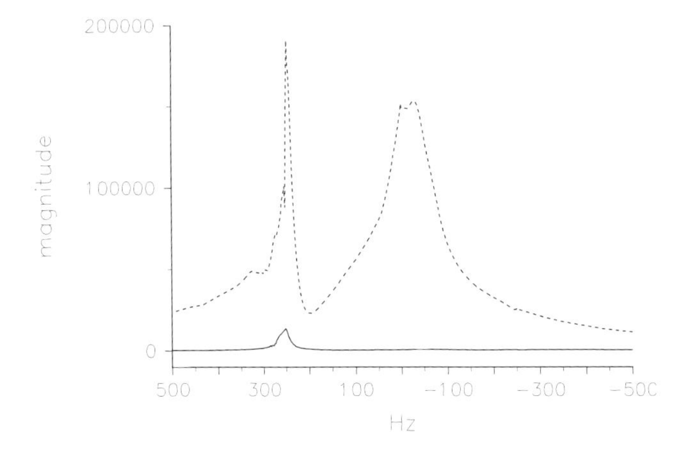

Figure 8.19. Spectra acquired with the 3-D ISIS sequence. The dashed line is the spectrum from the first acquisition (without any inversion pulses) which is a non-selective FID. The solid line is the spectrum after combining the eight ISIS acquisitions.

8.5.3 FID with Presaturation

Since the sensitivity of a surface coil decreases to 9% by one coil diameter, surface coils can be used to obtain localized MRS signals. However, there are circumstances when only signal from some of the region seen by the surface coil is desired. One method for doing this is with a 1-D ISIS sequence. Another method uses slab presaturation followed by excitation with a non-selective pulse to acquire signal from the region that was not presaturated[31]. Figure 8.20 shows a timing diagram for an FID with presaturation sequence. This is similar to an ISIS sequence with only one inversion pulse, except it has a 90° saturation pulse instead of a 180° inversion pulse.

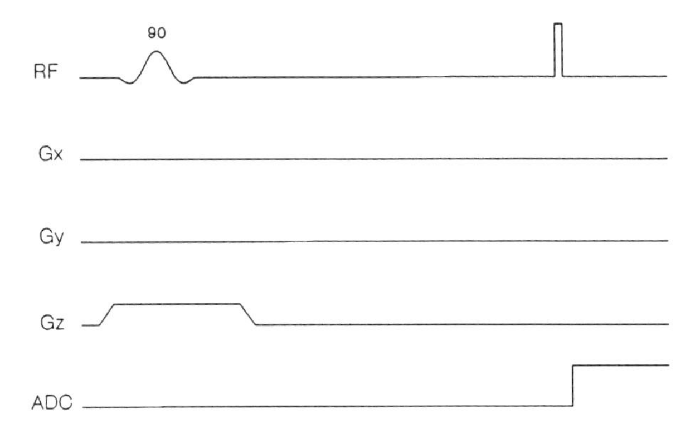

Figure 8.20. Timing diagram for an FID with presaturation. This is similar to an ISIS sequence with only one inversion pulse, except it has a 90° saturation pulse instead of a 180° inversion pulse.

Figure 8.21 shows an image of a phantom with two containers. The bottom vial contains 34 mL of 100 mM methylenediphosphonic acid, and the top vessel contains 225 mL of 37.4 mM inorganic phosphate. The box around the bottom vial indicates the thickness of the slab that is to be presaturated. To determine the transmitter voltage for the presaturation pulse, the reference voltage of the presaturation and non-selective pulses was varied from 4 to 15 volts. A plot of the "integral" of the FID versus transmitter voltage (Figure 8.22) shows a dip between 8 and 9 volts, which corresponds to the voltage that maximizes saturation. After the voltage for the presaturation pulse is set, the voltage of the non-selective pulse was varied from 8 to 48 volts to determine the voltage that gives the maximum "integral" (Figure 8.23). Figure 8.24 shows spectra with and without the presaturation pulse. While a 1-D ISIS sequence requires two acquisitions, an FID with presaturation sequence requires only one acquisition.

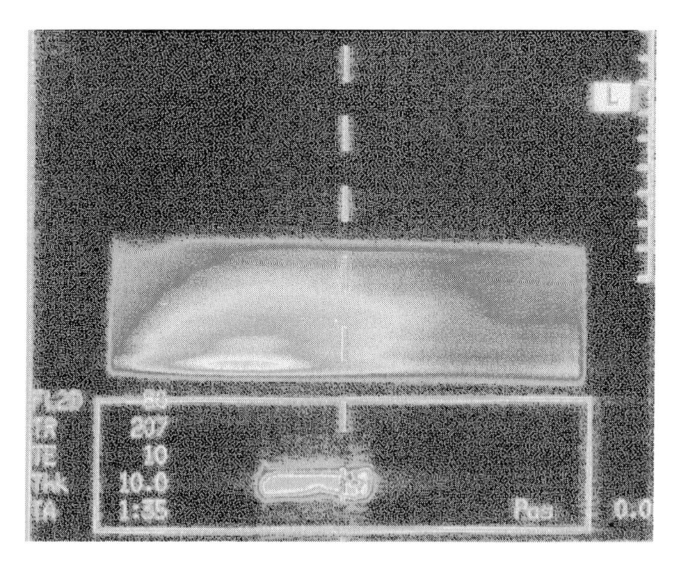

Figure 8.21. Image of a phantom with two containers. The bottom vial contains 34 mL of 100 mM methylenediphosphonic acid and the top vessel contains 225 mL of 37.4 mM phosphate. The box around the bottom vial indicates the thickness of the slab that is to be presaturated.

8.5.4 STEAM

STEAM (STimulated Echo Acquisition Mode)[32,33] sequences provide localized *in vivo* spectroscopy from a user-defined volume. Hahn showed that a stimulated echo (STE) occurs after excitation by three consecutive RF pulses[34]. STEAM achieves localization in a single acquisition with three selective RF pulses in the presence of orthogonal magnetic field gradients and has the form 90°–TE/2–90°–TM–90°–TE/2–STE. Figure 8.25 shows a volume of inter-

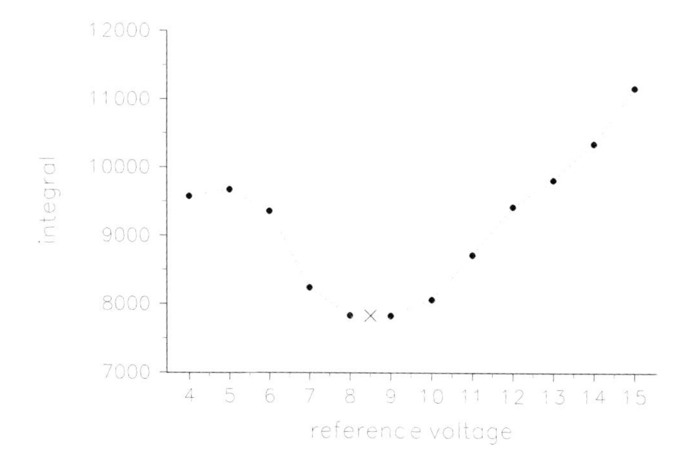

Figure 8.22. Sum of magnitude points of FID versus transmitter reference voltage for FID with presaturation sequence. The dip between 8 and 9 volts corresponds to the voltage that maximizes saturation. The × at 8.5 volts indicates the reference voltage used.

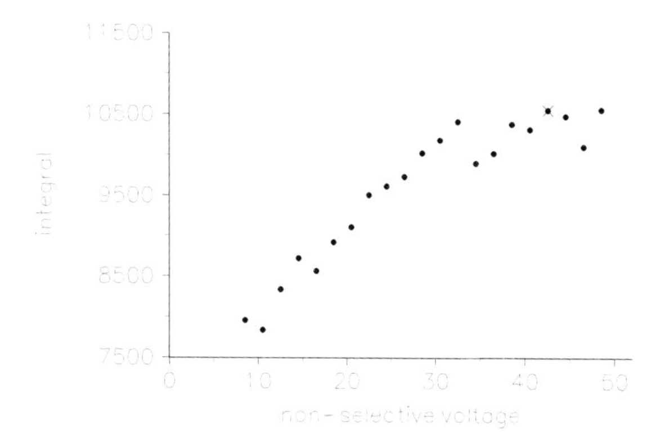

Figure 8.23. Sum of magnitude points of FID versus voltage of the non-selective pulse for FID with presaturation sequence. The × at 42 volts indicates the transmitter voltage used.

est (VOI) which is the intersection of the slabs selected by the X, Y, and Z gradients. This example shows a cube, but a general description of the VOI is a rectangular parallelepiped. An important feature of this sequence is that only half of the transverse magnetization prepared by the first 90° pulse is transformed into longitudinal magnetization by the second 90°pulse. During the TM period, longitudinal magnetization decays with T_1 rather than T_2. The third RF pulse transforms the longitudinal magnetization stored by the second RF pulse back into the transverse plane in the form of a stimulated echo. Besides the stimulated echo, the three RF pulses also generate multiple spin echoes[35]. The signal induced by the first RF pulse is refocused by the second RF pulse at 2×TE/2 and by the third RF pulse at 2×(TE/2+TM). The

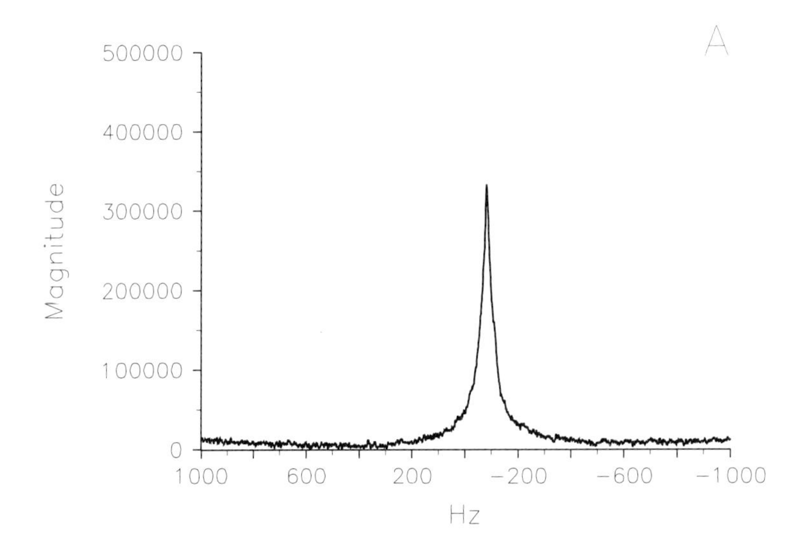

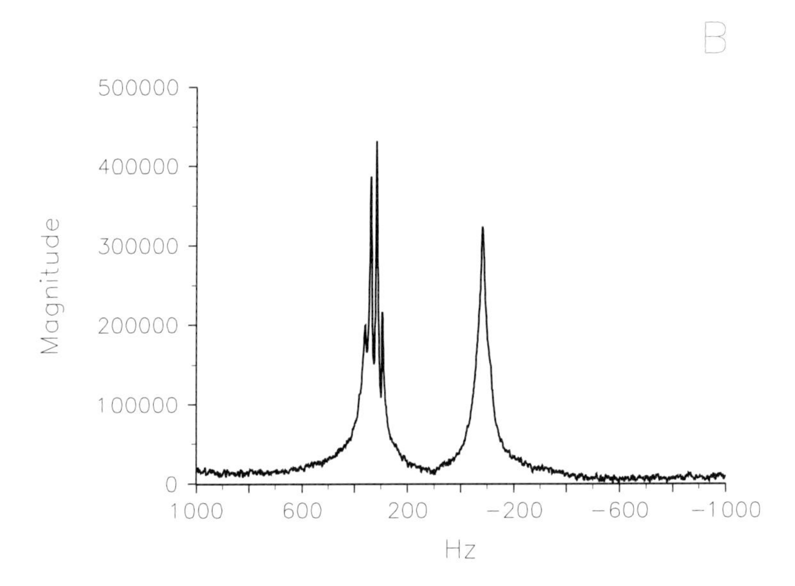

Figure 8.24. Spectra with presaturation (Panel A) and without presaturation (Panel B). All other settings remained unchanged.

signal induced by the second RF pulse is refocused by the third RF pulse at TE/2+2×TM. Extra gradient pulses are added to dephase these spin echoes, to reduce the signal attenuation due to motion and diffusion, and to optimize the stimulated echo[28]. Figure 8.26 shows the timing diagram of a STEAM sequence with one CHESS pulse for water suppression, and Figure 8.27 shows a spectrum from a STEAM acquisition.

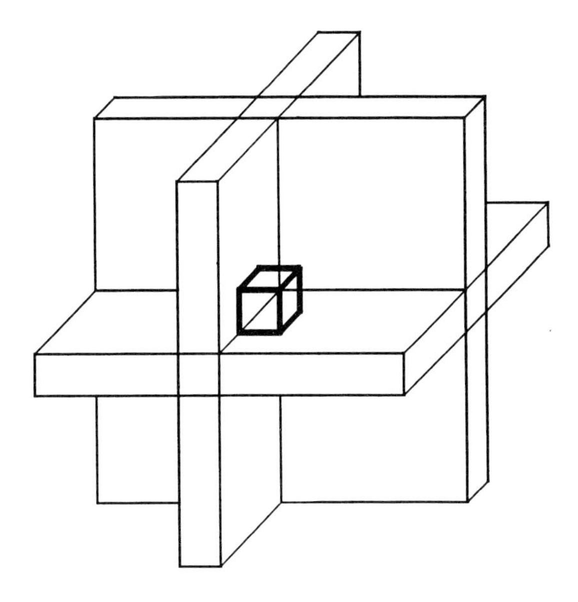

Figure 8.25. Volume of interest (VOI) selection with three orthogonal slabs. This method for defining a VOI is used by STEAM and PRESS sequences.

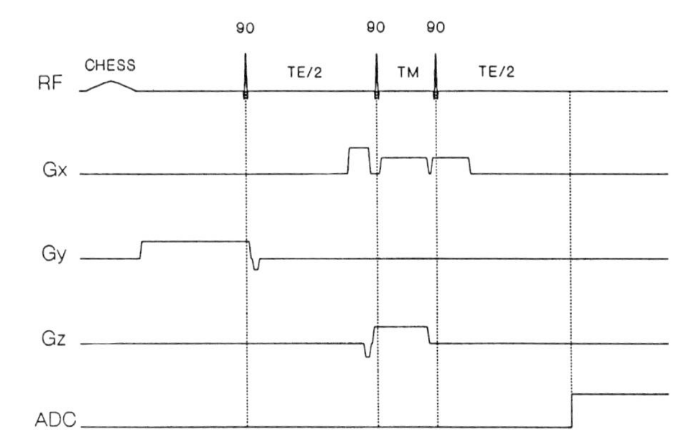

Figure 8.26. Timing diagram for a STEAM sequence with one CHESS pulse for water suppression (TE=135 msec, TM=30 msec).

8.5.5 PRESS

PRESS (Point RESolved Spectroscopy)[30] sequences provide localized *in vivo* spectroscopy from a user-defined volume. PRESS achieves localization with three selective RF pulses in the presence of orthogonal magnetic field gradients, and has the form $90°-\tau_1-\ 180°-\tau_1+\tau_2-180°-\tau_2$–SE, where TE = $2(\tau_1+\tau_2)$. The signal in a PRESS sequence decays with T_2 during the entire TE period and recovers all the available magnetization. Figure 8.28 shows the timing diagram of a PRESS sequence with one CHESS pulse for water suppression, and Figure 8.29 shows a spectrum from a PRESS acquisition. This spectrum was obtained with the same phantom, VOI, TE, and TR as the STEAM spectra in Figure 8.27. The peak areas with the PRESS sequence will be at least twice those observed with a STEAM sequence.

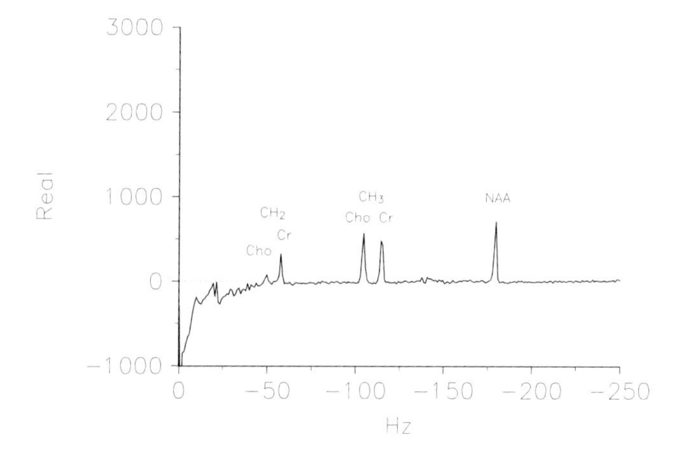

Figure 8.27. Spectrum from a STEAM acquisition (TE=135 msec, TM=30 msec, TR=3 sec) with a 2 × 2 × 2 cm VOI, 32 averages, and eddy current correction. Phantom was a 2 liter round bottom flask containing 45 mM in methyl protons of choline (Cho), creatine (Cr), and N-acetyl-aspartate (NAA). The FWHM of the water signal was 0.9 Hz.

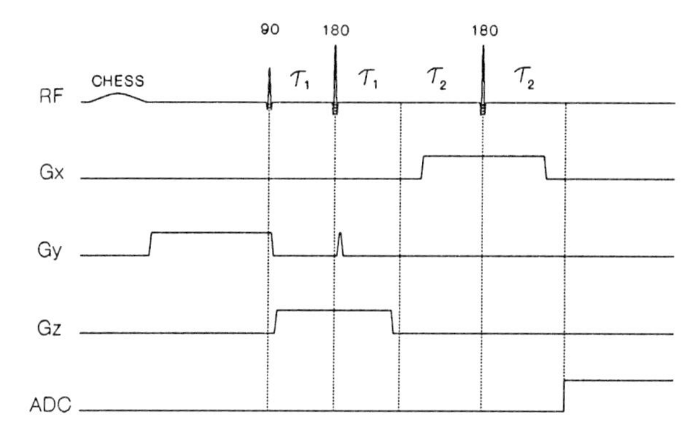

Figure 8.28. Timing diagram for a PRESS sequence with one CHESS pulse for water suppression. The TE is equal to $2(\tau_1+\tau_2)$.

8.5.6 Chemical Shift Imaging

Chemical shift imaging (CSI)[36] sequences provide localized spectra in one, two, or three dimensions. CSI uses phase encoding gradients prior to turning on the ADC to encode spatial information into the signal along with the frequency information of the chemical shifts. One-dimensional CSI obtains spectra from slabs, two-dimensional CSI obtains spectra from a grid in a selected slab, and three-dimensional CSI obtains spectra from cubes in the volume being studied. Figure 8.30 shows a timing diagram of a 1D-CSI sequence with phase encoding gradients in the Gz gradient. The amplitude of the phase encoding gradient varies from $-G_{max}$ to $+G_{max}$, which is determined by the following equation:

$$G_{max} = \frac{N_P/2}{\gamma t_P \mathrm{FOV}} \tag{8.29}$$

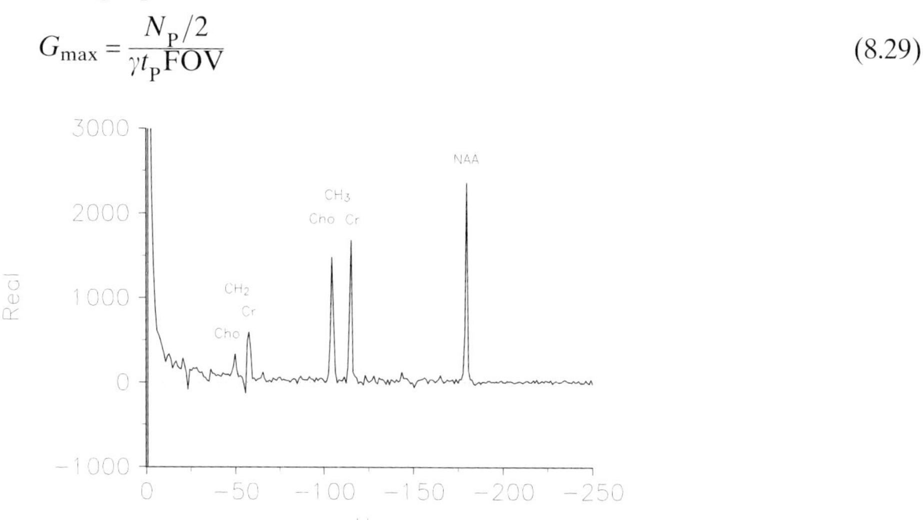

Figure 8.29. Spectrum from a PRESS acquisition (TE=135 msec, TR=3 sec) with a 2 × 2 × 2 cm VOI, 32 averages, and eddy current correction. Phantom was a 2 liter round bottom flask containing 45 mM in methyl protons of choline (Cho), creatine (Cr), and N-acetylaspartate (NAA). The FWHM of the water signal was 0.8 Hz.

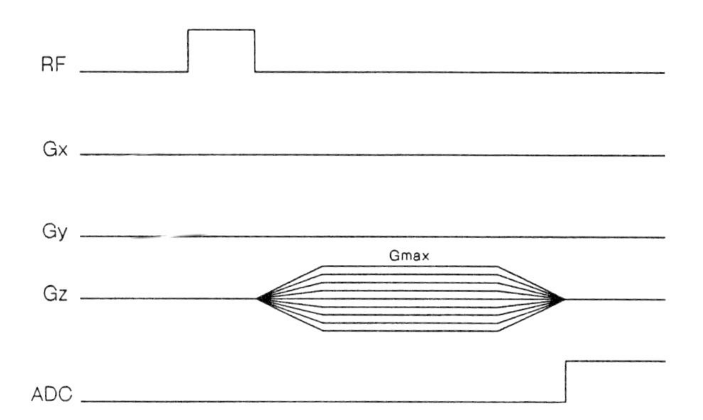

Figure 8.30. Timing diagram of a 1D-CSI sequence with phase encoding gradients in the Gz gradient. The amplitude of the phase encoding gradient varies from $-G_{max}$ to $+G_{max}$.

where N_P is the number of phase encoding steps, γ is the gyromagnetic ratio, t_P is the duration of the gradient, and *FOV* is the total slab thickness. When a surface coil is used with a CSI sequence, replacing the 90° square pulse with an adiabatic half-passage pulse will provide better results[12]. A 2D-CSI sequence has phase encoding in two directions, and a 3D-CSI sequence has phase encoding in all three directions.

Because the phase encoding is between the RF pulse and the ADC sampling window, spectra acquired with CSI require significant first order phasing. An alternative to using a 90° hard pulse is to acquire CSI data with a special

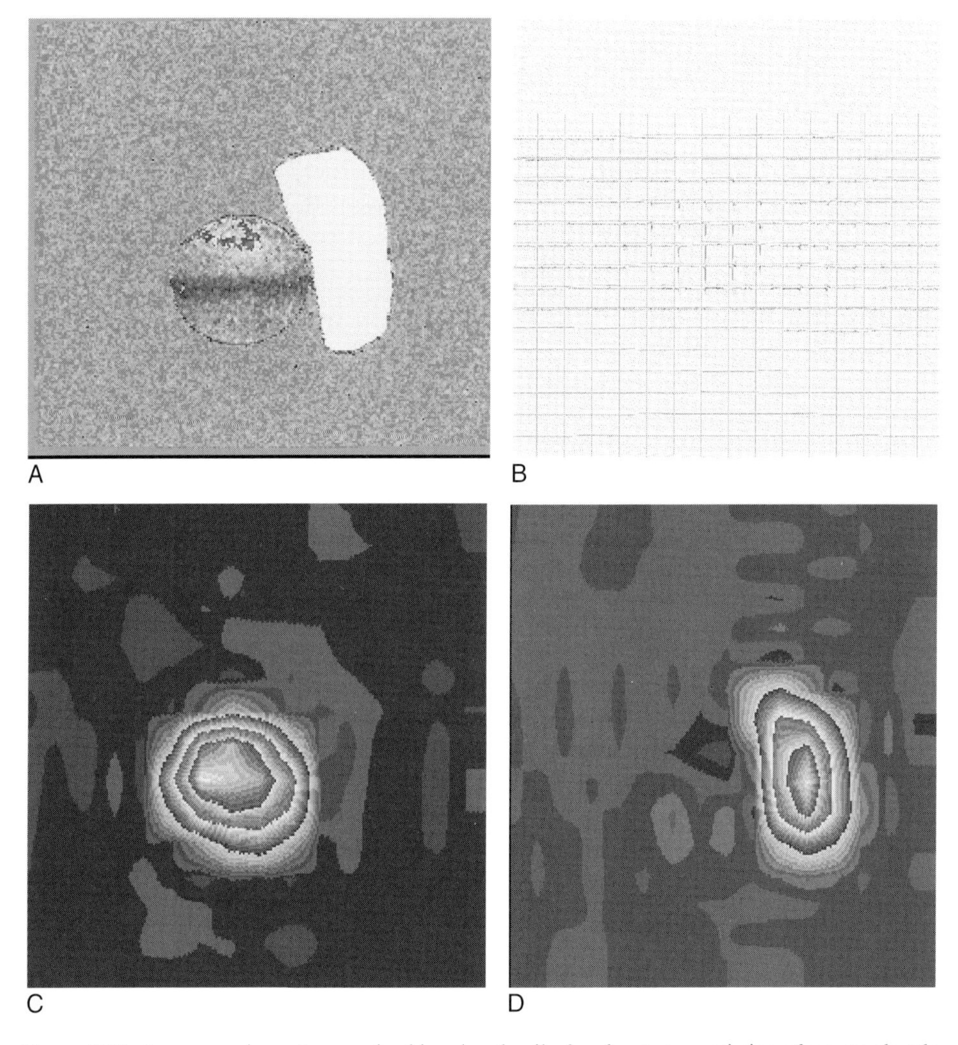

Figure 8.31. Images and spectra acquired in a head coil of a phantom consisting of a water bottle and a bag filled with oil. Panel A is a high resolution MR image. Panel B shows the 2D-CSI spectra acquired with two 16 step phase encoding gradients. Panel C is a 16 × 16 metabolite image of the water peak. Panel D is a 16 × 16 metabolite image of the fat peak.

spin echo pulse that contains both the excitation pulse and the 180° refocusing pulse[37]. Spectra acquired with this pulse require only zero order phasing. Figure 8.31 shows images acquired in a head coil of a phantom consisting of a water bottle and a bag filled with oil. Figure 8.31A is a high resolution MR image. Figure 8.31B shows the 2D-CSI spectra acquired with two 16-step phase encoding gradients. Another method for displaying the spectra is as metabolite images. Figure 8.31C is a 16 × 16 metabolite image of the water peak, and Figure 8.31D is a 16 × 16 metabolite image of the fat peak.

8.6 Mathematics Used in Spectroscopy

Angular frequency (radians/sec) is equal to linear, or cyclic, frequency (Hz) times 2π. Since RF transmitters specify cyclic frequency, the equations in this section are in terms of cyclic frequency.

8.6.1 Free Induction Decay in Time Domain

The free induction decay [$S(t)$, detected as a complex data set] is a three-dimensional signal. The three components are the two channels of the receiver [$X(t)$ and $Y(t)$] and time(t).

$$S(t) = [X(t), Y(t)] \tag{8.30}$$

The FID can be converted to magnitude [$M(t)$] (complex absolute value) and phase [$\Phi(t)$] by the following equations:

$$M(t) = \sqrt{X(t)^2 + Y(t)^2} \tag{8.31}$$

$$\Phi(t) = \arctan\left[\frac{Y(t)}{X(t)}\right] \tag{8.32}$$

The two channels of the FID can be reconstructed from $M(t)$ and $\Phi(t)$ with the following equations:

$$X(t) = M(t)\cos[\Phi(t)] \tag{8.33}$$

$$Y(t) = M(t)\sin[\Phi(t)] \tag{8.34}$$

The FID may also be represented in complex notation in the following form:

$$S(t) = M(t)\, e^{i\Phi(t)} \tag{8.35}$$

The signal received from a single spin compound, whether it is a free induction decay or the last half of an echo (i.e., STEAM, PRESS), can be described mathematically as a damped complex exponential

$$S(t) = \beta e^{i(-2\pi f_0 t+\Phi)-t/T_2^*} \tag{8.36}$$

where
β is the magnitude of the FID at t = 0
f_0 is the frequency of the rotating frame (Hz)
T_2^* is the decay constant in an inhomogeneous field (sec)
ϕ is phase offset at $t = 0$.

The two channels of this complex signal are

$$X(t) = \beta \times \cos(-2\pi f_0 t+\Phi) \times \exp(-t/T_2^*) \tag{8.37}$$

$$Y(t) = \beta \times \sin(-2\pi f_0 t+\Phi) \times \exp(-t/T_2^*) \tag{8.38}$$

The magnitude of this signal at time t is equal to

$$M(t) = \beta \exp(-t/T_2^*) \tag{8.39}$$

The integral of $M(t)$ from zero time to infinity is

$$\int_0^{\infty} M(t)\, dt = \beta\, T_2^* \tag{8.40}$$

In the digitized MR signal, the sum of the time domain magnitude points (often called the integral) for N points is equal to

$$\sum_{i=1}^{N} M_i = \frac{\int_0^{\infty} M(t) dt}{\Delta t} = \frac{\beta T_2^*}{\Delta t} \tag{8.41}$$

where M_i is the complex magnitude of the signal at time $= i \times \Delta t$ and Δt is the dwell time.

When substances with n spectral lines are measured, the two receiver channels are the sum of the individual lines

$$X(t) = \sum_{i=1}^{n} \beta_i \times \cos(-2\pi f_{0_i} t+\Phi_i) \times \exp(-t/T_{2_i}^*) \tag{8.42}$$

$$Y(t) = \sum_{i=1}^{n} \beta_i \times \sin(-2\pi f_{0_i} t + \Phi_i) \times \exp(-t/T^*_{2_i}) \tag{8.43}$$

Since phase [$\Phi(t)$] of the signal received from a single spin compound is equal to $-\omega t$, the slope of phase [$d\Phi(t)/dt$] is equal to $-\omega$. This ω is angular frequency in radians/sec and can be converted to frequency (f) in Hz by dividing the slope by 2π. Because of the 2π discontinuities of phase, an offset needs to be utilized when calculating the slope. This offset is calculated by the following rules:

for $|\Phi_n - \Phi_{n+1}| \leq \pi$: offset = 0
for $(\Phi_n - \Phi_{n+1}) > \pi$: offset = -2π
for $(\Phi_n - \Phi_{n+1}) < -\pi$: offset = $+2\pi$.

The slope of the phase, or frequency, at point n may be determined by the following equation:

$$f_n = \frac{\Phi_n - \Phi_{n+1} + \text{offset}}{2\pi\,\Delta t} \tag{8.44}$$

where Δt is the time interval between point n and point $n+1$, or the dwell time. This frequency is the difference between the rotating frame (transmitter frequency) and the precession frequency (γB_0). During the acquisition of the FID, the transmitter frequency and the gyromagnetic ratio (γ) are constant, therefore changes in the slope of $\Phi(t)$ are indicative of changes in $B_0(t)$. This slope [$d\Phi(t)/dt$], in the absence of gradients, is a direct measure of the B_0 inhomogeneity. Changes in $d\Phi(t)/dt$ after a gradient are indicative of changes in the detected resonance frequency due to eddy currents.

8.6.2 Free Induction Decay in Frequency Domain

The time domain free induction decay (damped complex exponential) can be transformed to the frequency domain with a Fourier transform. Spectroscopists call this peak a Lorentzian. When the phase offset (ϕ) is zero, the real and imaginary components are absorption mode [$A(f)$] and dispersion mode [$D(f)$] spectra

$$A(f) = \frac{1}{\Delta t}\,\frac{\beta T^*_2}{1 + \left[2\pi T^*_2 (f - f_0)\right]^2} \tag{8.45}$$

$$D(f) = \frac{1}{\Delta t}\,\frac{\beta 2\pi (T^*_2)^2 (f - f_0)}{1 + \left[2\pi T^*_2 (f - f_0)\right]^2} \tag{8.46}$$

where f_0 is the center frequency of the peak. Several relationships between the time domain and the frequency domain with a single spin compound are summarized in Table 8.2.

Table 8.2. Mathematical Relationships of a Single Spin Compound

	Time Domain	Frequency Domain
Initial Amplitude of FID	β	$\frac{A(f_0)\Delta t}{T_2^*}$
"Integral" of FID $\left(\sum_{i=1}^{n} M_i\right)$	$\frac{\beta T_2^*}{\Delta t}$	$A(f_0)$
Amplitude of Peak	$\sum_{i=1}^{n} M_i$	$A(f_0)$
Peak Area $\left(\int_{-\infty}^{\infty} A(f)\mathrm{d}f\right)$	$\frac{\beta}{2\Delta t}$	$\frac{A(f_0)}{2T_2^*}$
FWHM	$\frac{\beta}{\pi\Delta t \sum_{i=1}^{n} M_i}$	$\frac{1}{\pi T_2^*}$

With substances containing n spectral lines, the real and imaginary channels are the sum of the individual lines

$$A(f) = \frac{1}{\Delta t}\sum_{i=1}^{n} \frac{\beta_i T_{2_i}^*}{1+\left[2\pi T_{2_i}^* (f-f_{0_i})\right]^2} \tag{8.47}$$

$$D(f) = \frac{1}{\Delta t}\sum_{i=1}^{n} \frac{\beta_i 2\pi (\mathrm{T}_{2_i}^*)^2 (f-f_{0_i})}{1+\left[2\pi T_{2_i}^* (f-f_{0_i})\right]^2} \tag{8.48}$$

General equations for the real and imaginary channels which include ϕ are given by the following equations:

$$Re(f) = \cos(\Phi)A(f) + \sin(\Phi)D(f) \tag{8.49}$$

$$Im(f) = \cos(\Phi)D(f) - \sin(\Phi)A(f) \tag{8.50}$$

Magnitude mode spectra can be generated using Eq. (8.31). Magnitude mode spectra do not require phasing, but they include dispersion mode data, which makes the peaks broader. Absorption mode spectra are usually used in magnetic resonance spectroscopy, even though they require phasing.

8.7 Graphical Presentation of Spectroscopic Data

For the plots in Figures 8.32, 8.33, and 8.34, a 512-point FID was generated using Eq. (8.36) with a dwell time of 0.001 seconds, β = 100, f_0 = 5.86 Hz, ϕ = 0, and T_2^* = 0.2 sec. The FID was transformed to the frequency domain without line broadening or zero filling.

8.7.1 Time Domain

The conventional display of a quadrature-received FID is as two-dimensional plots of real versus time (Figure 8.32A) or imaginary versus time (Figure 8.32B). Six alternative presentations of an FID are shown in Figure 8.33. The three-dimensional representation of the FID is shown as the heavy line in Panel *A* and emphasizes that the MRS signal is a complex value. The X and Y projections are the two orthogonal channels of the FID shown in Figure 8.32. The second representation (Panel B) plots magnitude (from Eq. 8.31) against time. This plot represents the distance of the FID from the [0,0] line in Panel *A* and contains no frequency information. A companion to the magnitude representation is phase (from Eq. 8.32) versus time (Panel C); this plot is independent of magnitude and contains frequency information only. Since ω in this FID is constant, the plot of phase versus time is linear. The slope of the phase, expressed in frequency, is shown in Panel D. Since phase is constant, this is simply a flat line. A fifth presentation (Panel E) plots

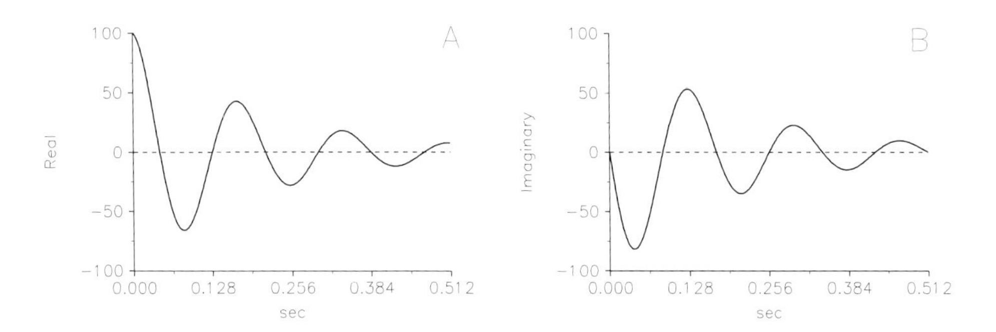

Figure 8.32. Real and imaginary channels of a 512–point FID with Δt=0.001 sec, β=100, f_0=5.86 Hz, ϕ=0, and T_2^*=0.2 sec.

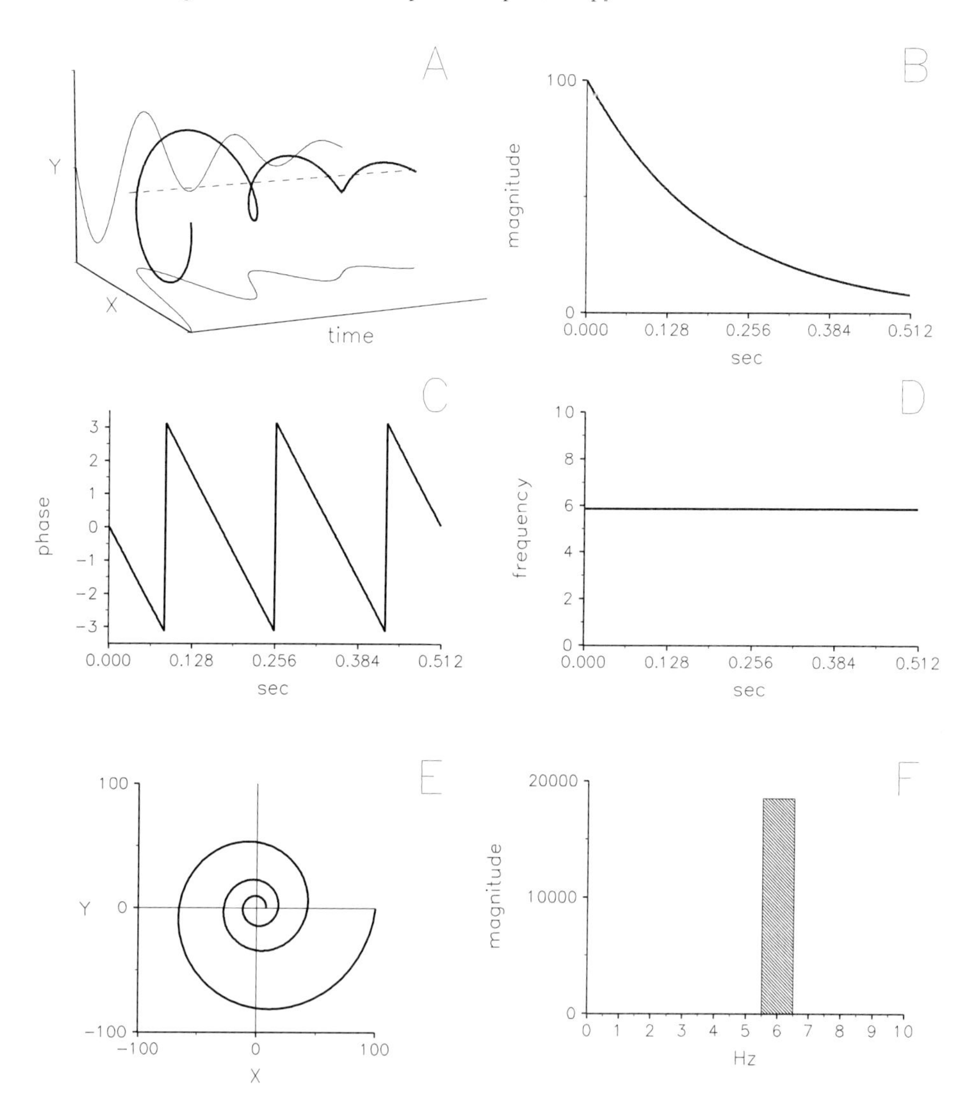

Figure 8.33. Alternative presentations in the time domain of the FID in Figure 8.32. (Panel A) The heavy line is the three-dimensional representation of the FID. The thin lines are the X and Y projections of the FID and correspond to the real and imaginary channels shown in Figure 8.32. (Panel B) Magnitude of FID versus time. (Panel C) Phase of FID versus time. (Panel D) Slope of the phase expressed in Hz versus time. (Panel E) Real channel versus imaginary channel. This is identical to a polar plot of magnitude versus phase. (Panel F) Histogram generated with frequency from the slope of the phase and sum of magnitudes at the specific frequency.

the X channel against the Y channel. This is the projection of the 3-D FID in the X-Y plane and is identical to a polar plot of $M(t)$ versus $\Phi(t)$. Panel F shows a histogram where the abscissa is frequency (from the slope of the phase) and the height of the bar is the sum of the magnitudes for the frequencies contained in the bar. Since the phase is constant, the histogram has only one bar. This histogram is similar to a magnitude spectrum after transforming the FID to the frequency domain.

8.7.2 Frequency Domain

Before the introduction of the fast Fourier transform[38], almost all spectrometers employed continuous-wave irradiation which swept either the applied field or the transmitter frequency. The abscissa for spectra went from low field to high field. Now almost all spectrometers use the fast Fourier transform to produce spectra. To display these spectra from low field to high field requires varying the abscissa from positive frequency to negative frequency. Since ϕ is equal to zero, the real data in Figure 8.34A is an absorption mode spectrum, and the imaginary data in Figure 8.34B is a dispersion mode spectrum. Taking the complex absolute value of the real and imaginary data produces a magnitude mode spectrum (Panel C). When ϕ is not equal to zero,

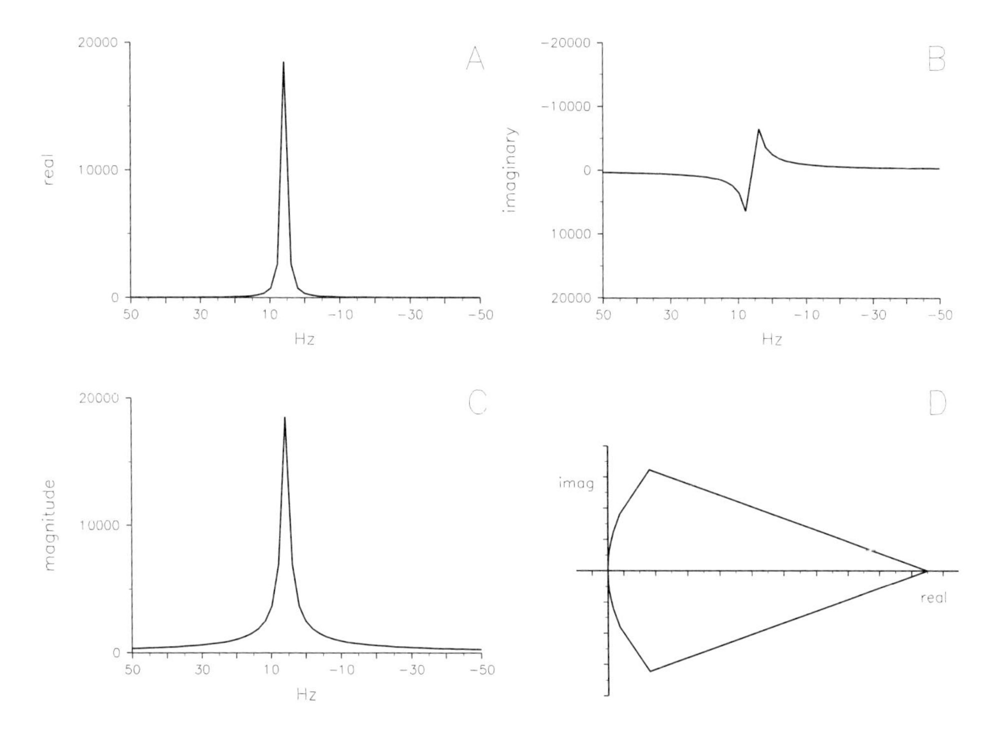

Figure 8.34. Four presentations in the frequency domain of the FID in Figure 8.33. (Panel A) Real data which, in this case, is an absorption mode spectrum. (Panel B) Imaginary data which, in this case, is a dispersion mode spectrum. (Panel C) Magnitude mode spectrum. The peak amplitude of the absorption mode and magnitude mode spectra is equal to the "integral" (sum of magnitude points) in the time domain. Since the FID has not decayed to zero in the 512 msec sample window, these peaks are not equal to $\beta T_2^*/\Delta t$.

the real and imaginary channels contain mixtures of the absorption mode and dispersion mode data, but the magnitude mode is unchanged. A DISPA graph[39] (Panel D) plots the real data versus the imaginary data.
The frequency between peaks in a spectrum is proportional to the field strength (B_0). To compare spectra obtained at different field strengths, the abscissa of spectra is expressed in parts-per-million (ppm). Frequency (in Hz) is converted to ppm by the following equation:

$$\mathrm{ppm} = \frac{f}{\gamma B_0} \tag{8.51}$$

where γ is the gyromagnetic ratio in MHz/T and B_0 is the field strength in tesla.

8.8 Spectroscopy Post-processing

Quantifying the acquired MR data requires several post-processing techniques. These techniques include eddy current correction, line broadening, solvent suppression, zero filling, Fourier transformation, phasing, baseline correction, evaluating areas under the peaks, and calculating concentration.

8.8.1 Eddy Current Correction

Major factors that degrade the MRS signal are static inhomogeneities and temporal B_0 shifts from eddy currents. Before an FID is acquired, static homogeneity is improved by shimming, and eddy currents are minimized with gradient waveform shaping and/or self-shielded gradient coils. After the signal is acquired, there may still be some residual eddy current distortions which are manifested as non-linear phase in the time domain signal. The detected phase $\Phi(t)$ may be defined as

$$\Phi(t) = \psi(t) + \epsilon(t) \tag{8.52}$$

where $\psi(t)$ is the correct phase [if $B_0(t)$ is constant] and $\epsilon(t)$ is the phase distortion caused by a changing $B_0(t)$[40,41]. In a water FID observed at the Larmor frequency, $\psi(t)$ would be equal to zero at every t. When $\Phi(t)$ is not equal to zero, the observed phase is equal to $\epsilon(t)$. In an FID from a substance containing multiple spectral lines, $\psi(t)$ will not be equal to zero, but $\epsilon(t)$ is equal to $\Phi(t)$ from the water FID. Subtracting the phase of a water FID from the phase of the water-suppressed FID and then reconstructing the water-suppressed FID with the corrected phase will remove the distortions caused by the changing $B_0(t)$[42].

To illustrate the results with post-processing eddy current correction, localized FIDs were obtained from a 2-liter spherical phantom containing 45 mM (in methyl protons) of choline, creatine, and N-acetylaspartate. A PRESS sequence (TE = 135 ms, TR = 3000 ms) with one CHESS pulse for water suppression was used with a 2 × 2 × 2 cm voxel in a magnet without shielded gra-

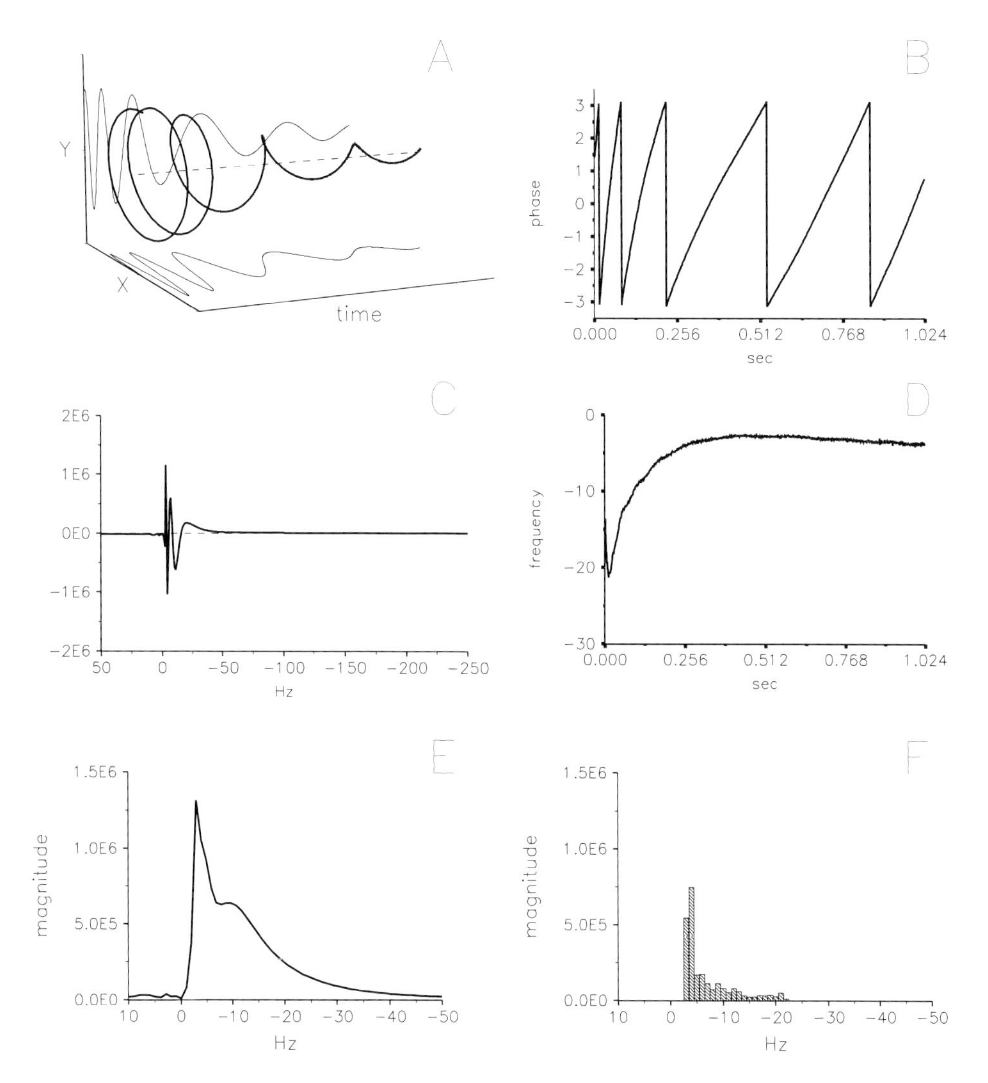

Figure 8.35. Water FID with a PRESS 135 sequence and a 2 × 2 × 2 cm VOI from a 2 liter round bottom flask containing 45 mM in methyl protons of choline, creatine, and N-acetylaspartate. (Panel A) Three-dimensional plot of the FID. (Panel B) Phase plot of the FID. (Panel C) Real spectra after Fourier transformation. (Panel D) Slope of the phase. The frequency varies for –21 Hz to –3 Hz due to the eddy currents. (Panel E) Magnitude mode spectrum. (Panel F) Histogram generated with frequency from the slope of the phase and sum of magnitudes at the specific frequency.

dients. The water FID (Figure 8.35) was acquired with the CHESS pulse turned off, and the sample FID (Figure 8.36) was acquired with the CHESS pulse turned on. The slope of the phase plot (Figure 8.35D) shows the shifting frequency of the water signal due to the eddy currents. The histogram in Figure 8.35F is similar to the magnitude mode spectrum in Figure 8.35E. Notice the similar nonlinear phase in Figures 8.35B and 8.36B. The distortions of the peaks in Figure 8.36C are similar to the distortions of the water peak in Figure 8.35C. Figure 8.37A shows the FID after eddy current correction, and Figure 8.37B shows the more linear phase. After eddy current correction, the distortions in Figure 8.37C are greatly diminished.

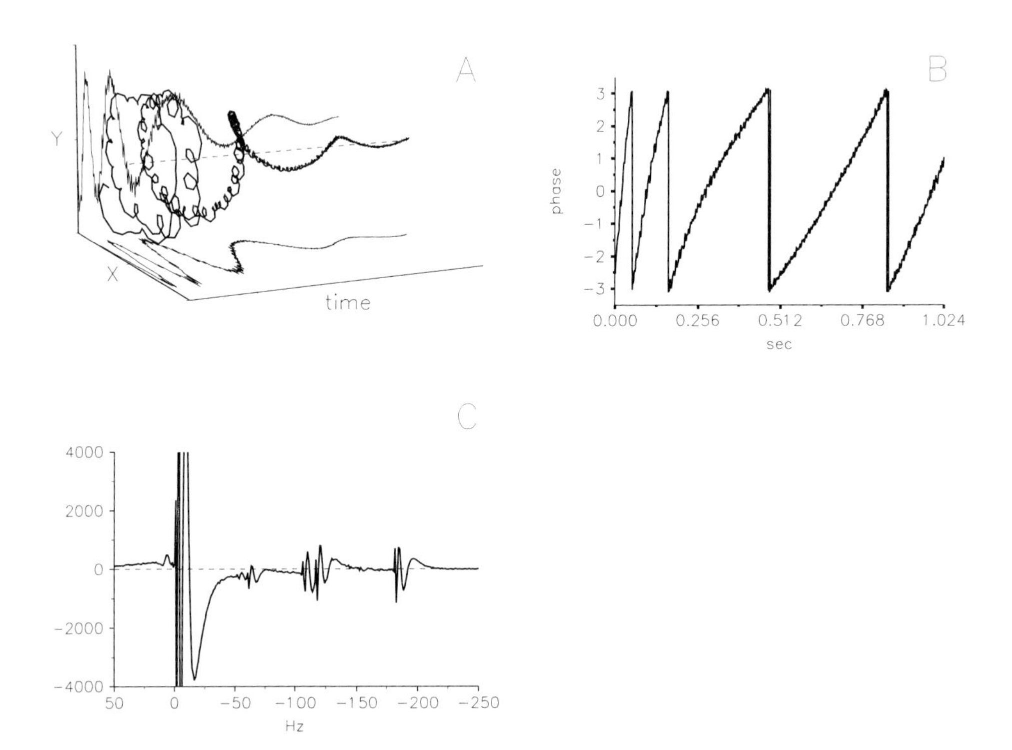

Figure 8.36. Sample FID acquired with same sequence and VOI as Figure 8.35 but with water suppression. The small spirals in Panel A are from the methyl and methylene protons in the phantom. The phase in Panel B is similar to the phase in Figure 8.35B, and the real spectrum in Panel C has the same distortions as the distortions in Figure 8.35C.

8.8.2 Convolution Filters

Convolution is a special transform that can be used to describe the result of combining processes, such as using a filter to modify a waveform. Convolution in the frequency domain corresponds to multiplication in the time

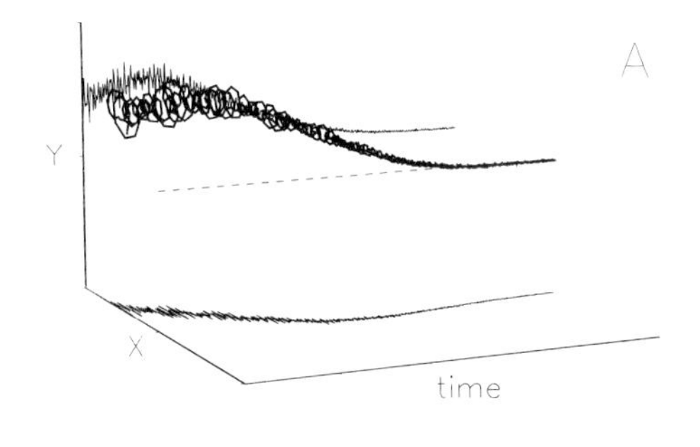

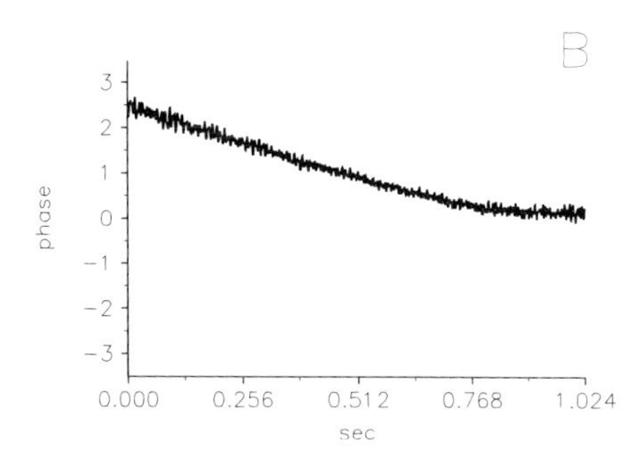

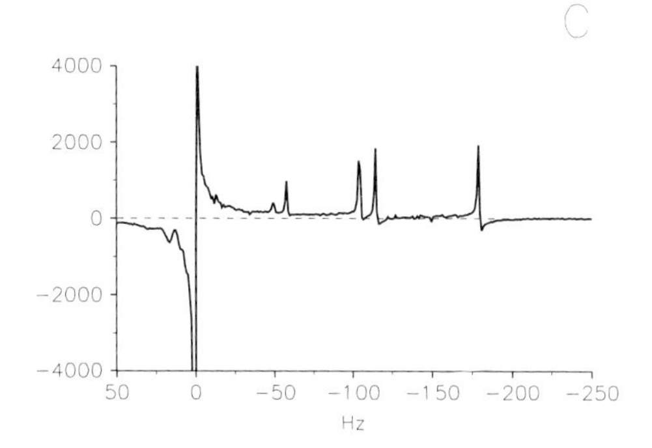

Figure 8.37. Sample FID from Figure 8.36 after eddy current correction. The large spiral in Figure 8.36A has been removed (Panel A), and the phase (Panel B) is more linear. Panel C shows the residual water at 0 Hz and more Lorentzian methyl and methylene peaks.

domain; convolution in the time domain corresponds to multiplication in the frequency domain. In MRS, convolution filters are used for suppressing the noise component (line broadening filters, or apodization filters) and for solvent suppression (convolution difference filters).

8.8.2.1 Apodization Filters

The signal in a free induction decay contains the signal from the compound being studied and the noise in the detector channels. A line broadening filter decreases the received signal at the end of the sampling window, which increases the signal to noise ratio in a spectrum but increases the linewidth of the peak in the frequency domain. The filters typically used are exponential and Gaussian. These filters multiply the time domain FID by the filter before transforming to the frequency domain. This weighing of the time domain data is known as apodization in the frequency domain.

An exponential filter has the following form:

$$E(t) = \exp(-t/TC) \tag{8.53}$$

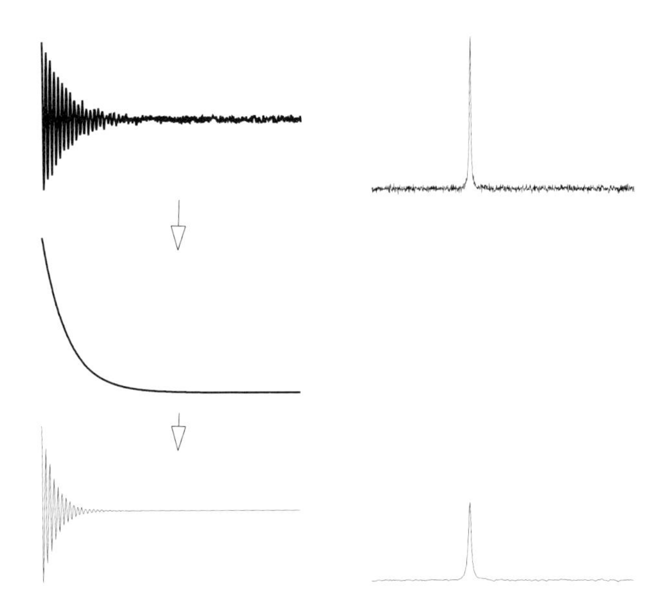

Figure 8.38. Matched exponential filter. The top left and right plots show the time domain and frequency domain representations of a single resonance with noise. After apodization with an exponential filter with the same T_2, the baseline noise is reduced and the height of the peak after filtering is 50% of the original height, the FWHM of the peak is doubled, and the area under the peak is unchanged.

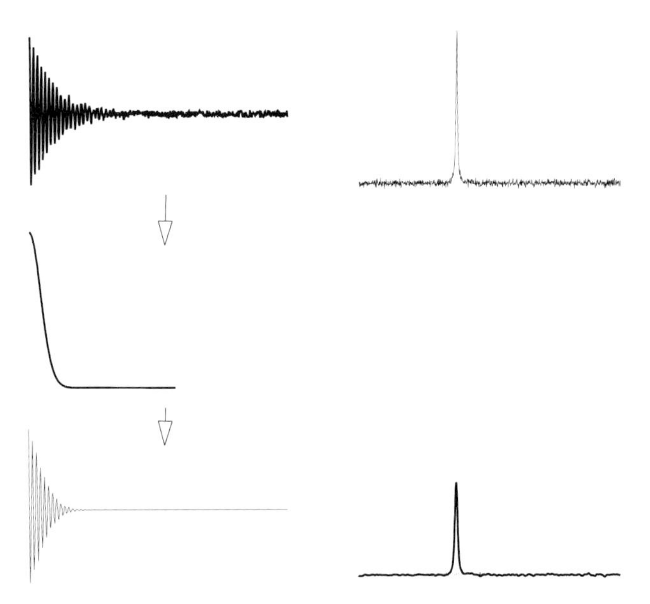

Figure 8.39. Matched Gaussian filter. The top left and right plots show the time domain and frequency domain representations of a single resonance with noise. After apodization with a Gaussian filter with the same T_2, the baseline noise is reduced and the height of the peak after filtering is 60% of the original height, the FWHM of the peak is doubled, and the area under the peak is unchanged.

and a Gaussian filter has the following form:

$$G(t) = \exp\left[-\frac{(t/TC)^2}{\ln 4}\right] \tag{8.54}$$

where TC is a time constant representing the filter bandwidth. The amount of line broadening, in terms of full width at half maximum (FWHM), is TC = 1/(π FWHM). A matched filter has a time constant equal to the time constant of the FID and will increase the FWHM by a factor of two. Figure 8.38 shows a matched exponential filter, and Figure 8.39 shows a matched Gaussian filter. Multiplying a free induction decay by a Gaussian filter will result in peaks that are Gaussian rather than Lorentzian[13].

8.8.2.2 Solvent Suppression Filters

Often the water suppression used with proton spectroscopy does not completely eliminate the water signal. Post-processing with a convolution difference filter[43] can be used to eliminate any residual water signal. This filter applies a low-pass filter to the FID, then subtracts the filtered signal from the original data in the time domain. The low frequency component is obtained by convolution with either a sine-bell window function

$$f(k) = \cos\left(\frac{k\pi}{2K+2}\right) \tag{8.55}$$

or a Gaussian window function

$$f(k) = \exp\left(\frac{-4k^2}{K^2}\right) \tag{8.56}$$

where K is related to the cutoff frequency (in Hz) of the filter. For the sine-bell filter, K is defined by

$$K = \mathrm{int}\left(\frac{0.5}{\Delta t \times \mathrm{cutoff}}\right) \tag{8.57}$$

Regular convolution of finite signals assumes periodicity of the signal[44]. Since an FID is not periodic, the first K points and last K points are calculated by linear extrapolation. For an FID with N points, the low-pass signal $[L(i)]$ is calculated by the following equation:

$$\text{for } K+1 \le i \le N-K: L(i) = \frac{\sum_{k=-K}^{K} f(k)S(i+k)}{\sum_{k=-K}^{K} f(k)}$$

$$\text{for } 1 \le i \le K: \quad L(i) = L(K+1) + (K-i+1)\frac{L(K+1)-L(2K+1)}{K} \tag{8.58}$$

$$\text{for } N\text{–}K\text{+}1 \le i \le N : L(i) = L(N\text{–}K) + (i\text{–}N\text{+}K)\,\frac{L(N-K)-L(N-2K)}{K}$$

The frequency response of a sine-bell convolution difference filter with a cutoff frequency of 40 Hz is shown in the Bode diagram in Figure 8.40. The ordinate of the Bode diagram is the ratio of output amplitude divided by the input amplitude, expressed in decibels (db). This filter is a second-order high pass filter (–40 db/decade roll-off) with slight under-damping[45]. Figure 8.41 shows processing with a convolution difference filter. Panel A shows the two channels of the FID in Figure 8.37; Panel B shows the low-pass components calculated with a 50 Hz sine-bell filter and Eq. (8.58); and Panel C shows the two channels of the FID after the convolution difference filter. The low frequency component and baseline distortion evident in Figure 8.37 has been removed in Figure 8.42.

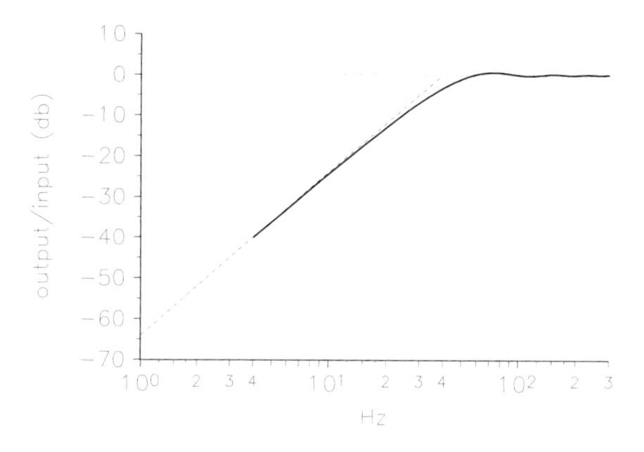

Figure 8.40. Bode diagram of the response of the convolution difference filter with a cutoff frequency of 40 Hz. Dashed line shows ideal second-order high pass filter with –40 db/decade roll-off.

8.8.3 Zero Filling

An N point FID has N real and N imaginary, or $2N$ points. After Fourier transformation, there are N absorption points and N dispersion points with frequency spacing equal to $1/(N\ \Delta t)$. Zeros can be added to the end of the FID to decrease the frequency spacing but only if the FID is, in fact, zero-valued over the interval where the zeros are added[46]. Figure 8.43 illustrates an absorption peak drawn with Eq. (8.45) (solid line) and data points after fast Fourier transformation of an FID with 256 points and zero filling to 512, 1024, and 2048 points.

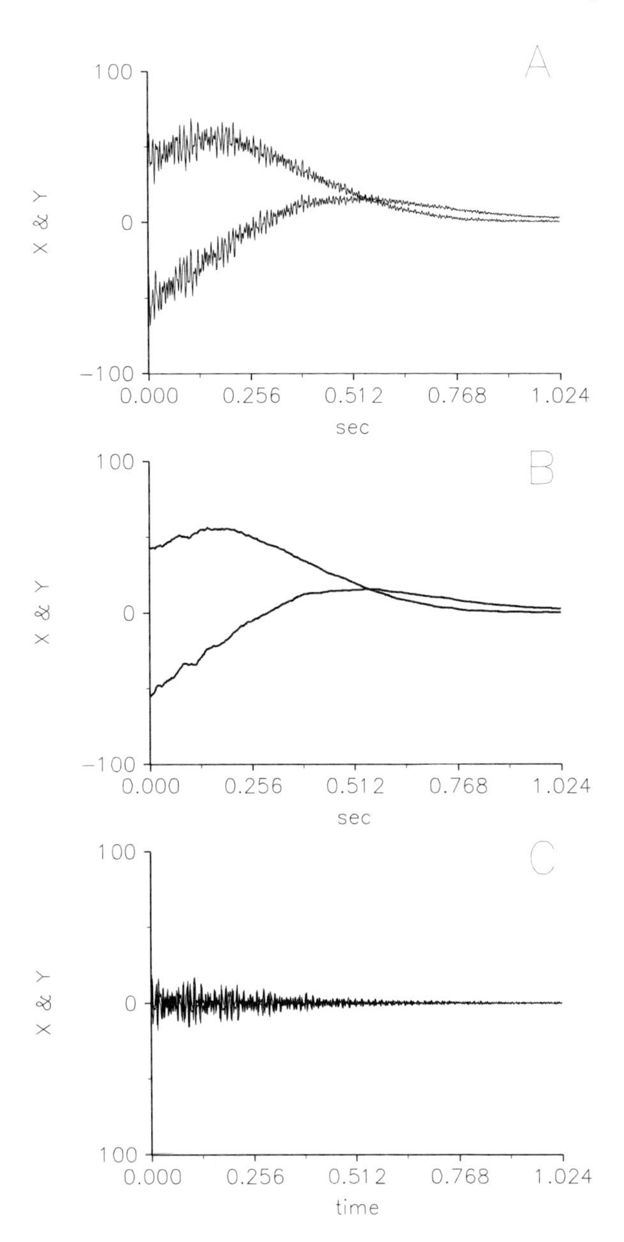

Figure 8.41. This demonstrates results of 50 Hz convolution difference filter. Panel A shows the two channels of the sample FID after eddy current correction in Figure 8.37. Panel B shows the results of the low pass filter which is the signal from the residual water. Panel C shows the results of the convolution difference filter.

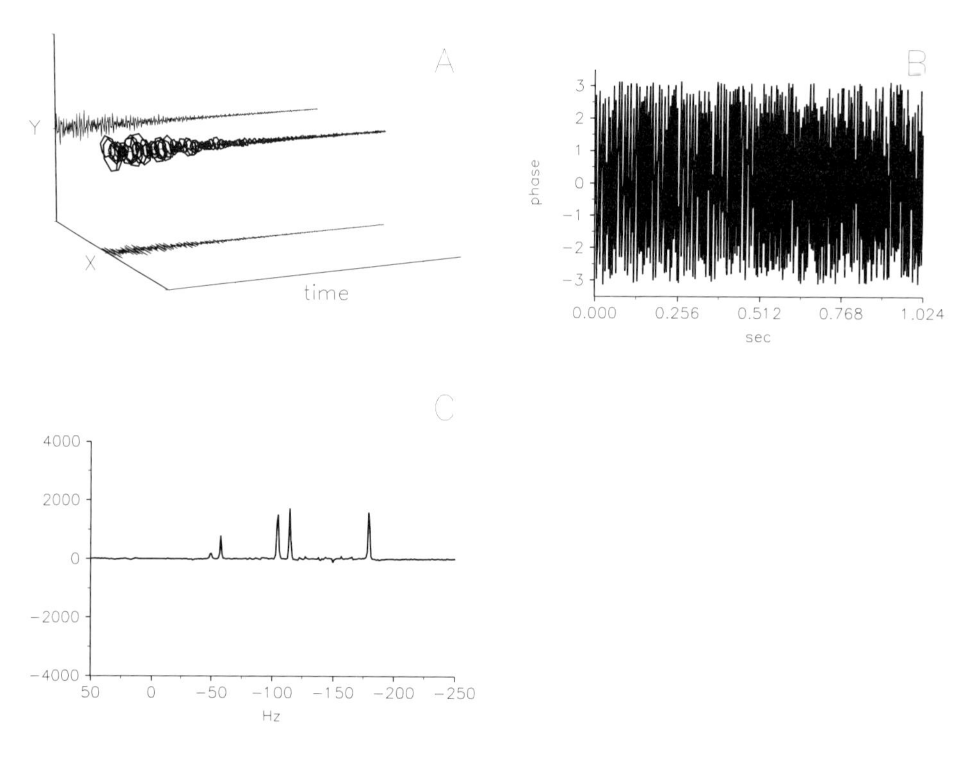

Figure 8.42. Three dimensional and time domain phase of sample FID after eddy current correction and convolution difference filtering. The spectrum in Panel C shows Lorentzian shaped peaks for the methyl and methylene protons and no residual water peak.

8.8.4 Fast Fourier Transform

Any periodic function can be expressed as the sum of sinusoids of different frequencies and amplitudes. The Fourier transform is a method for evaluating the frequencies and amplitudes of these sinusoids. An algorithm for calculating a Fourier transform with a computer was published in 1965 by Cooley and Tukey and is now known as a fast Fourier transform (FFT)[38]. The FFT treats the signal as a periodically repeated function, whether it is or not[44]. When the signal is a free induction decay, there is a sudden jump between the last point (which is usually equal to zero) and the first point (which has the greatest amplitude). This sudden jump in the time domain will produce a DC offset in the frequency domain equal to half the first point of the FID. If the first point of the FID is divided by two before transformation, this DC offset will be eliminated[47].

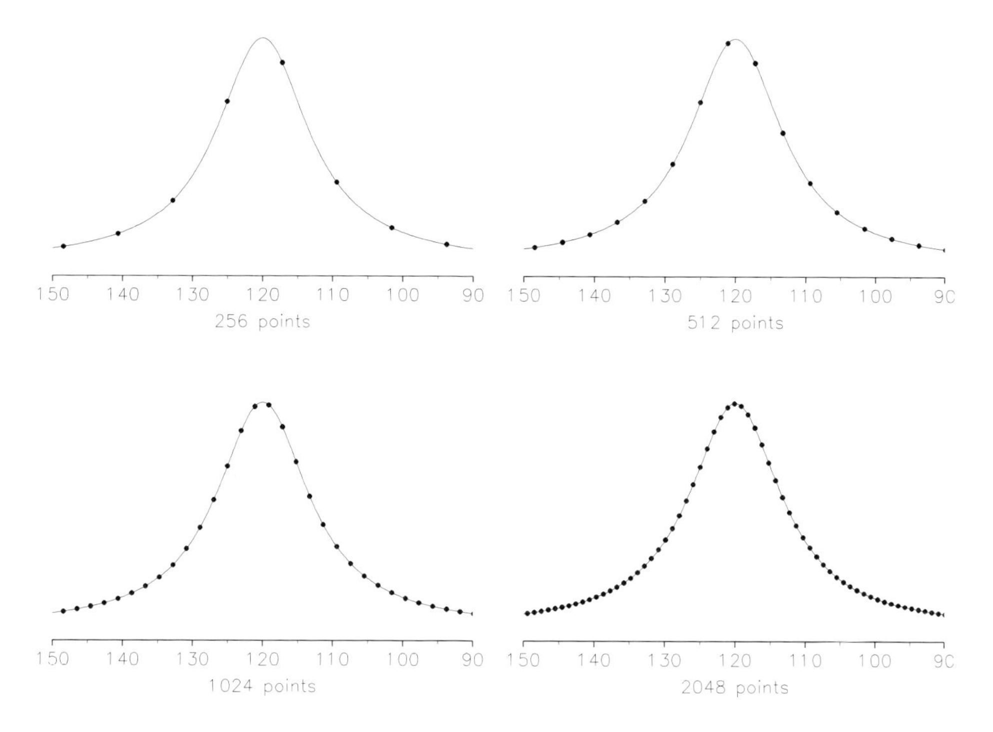

Figure 8.43. Effects of zero filling. The solid line in each plot was generated with Eq. (8.45), while the points in each plot were generated with an FFT of a time domain FID with 256 points. The 256-point plot was without zero filling, and the other plots were zero filled to 512, 1024, and 2048 points.

8.8.5 Phasing

Delay in the time domain corresponds to a phase shift in the frequency domain. Whenever the initial phase of an FID (ϕ in Eq. (8.36) is not zero, the real and imaginary channels after Fourier transform will contain mixtures of absorption mode and dispersion mode spectra. Phasing a spectrum sorts the real and imaginary channels into absorption mode and dispersion mode spectra.

$$\text{Absorption }(f) = Re(f)\cos\left(\frac{\theta\pi}{180^\circ}\right) + Im(f)\sin\left(\frac{\theta\pi}{180^\circ}\right) \tag{8.59}$$

$$\text{Dispersion }(f) = Im(f)\cos\left(\frac{\theta\pi}{180^\circ}\right) - Re(f)\sin\left(\frac{\theta\pi}{180^\circ}\right) \tag{8.60}$$

Phase (θ) is specified as the zero order phase (Z) (constant for all frequencies) and the first order phase (F) (linear with frequency) and is defined by the following equation:

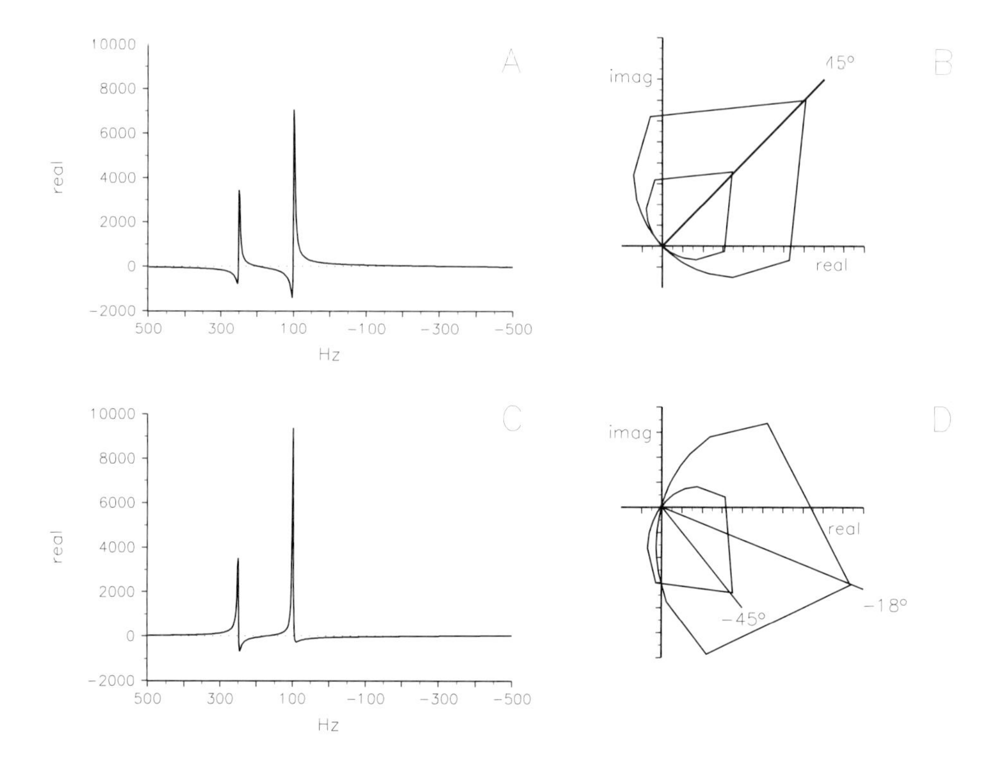

Figure 8.44. Example of phase errors in spectra. Panel A is the real spectrum from transforming an FID with two peaks and the initial phase equal to 45°. With the zero order phase shift, both of the loops in the DISPA plot (Panel B) are rotated 45°. Panel C is the real spectrum from transforming an FID with two peaks with a 0.0005 sec delay. With the first order phase shift, the loops in the DISPA plot (Panel D) are rotated different amounts. The peak centered at 100 Hz peak is rotated –18° and the peak centered at 250 Hz is rotated –45°.

$$\theta = Z + F\left(\frac{f - f_{ref}}{SW}\right) \tag{8.61}$$

The zero order phase term corrects for the initial phase of the FID. For a single spin compound, θ in Eq. (8.59) is equal to ϕ in Eq. (8.36) and can be determined by taking the arctangent of the first data point of the time domain FID. The first order phase term is necessary whenever the start of the A/D sampling window does not start at the peak of the damped exponential. The "best" phase is that which produces the most symmetrical absorption mode peaks and has a zero rotation angle in a DISPA plot[39].

Zero and first order phasing are illustrated in Figure 8.44. The spectrum in Panel *A* was created from an FID with $\phi = 45°$. With the constant phase shift, both loops in the DISPA plot (Panel B) are rotated 45°. This spectrum can be phased by Eq. (8.59) with $Z = 45°$ and $F = 0°$. The spectrum in Figure 8.41C was created from an FID after a 0.0005 sec delay. The phase of the 100 Hz

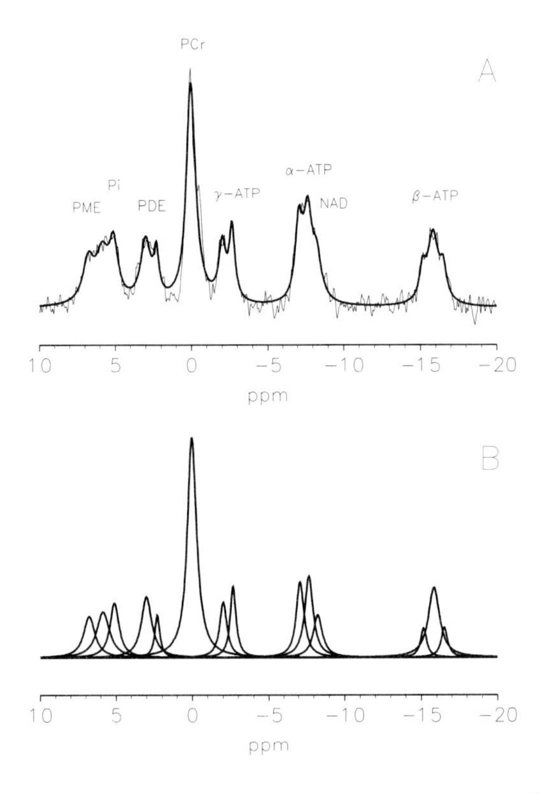

Figure 8.45. (Panel A) The thin line is the observed ^{31}P spectrum, and the thick line is the fitted spectrum with 14 Lorentzian peaks. The lines in Panel B are the 14 individual Lorentzian peaks used for the fit.

peak is –18° (100 cycles/sec × 0.0005 sec × 360°/cycle), and the phase of the 250 Hz peak is –45° (250 cycles/sec × 0.0005 sec × 360°/cycle). In the DISPA plot (Panel D), the loop corresponding to the 100 Hz peak is rotated –18°, and the loop corresponding to the 250 Hz peak is rotated –45°. This spectrum can be phased by Eq. (8.59) with $Z = 0°$, $F = -180°$, $SW = 1000$ Hz, and $f_{ref} = 0$ Hz. When phasing a sampled spectrum, set the first order term to zero, apply a zero order phase to get a specific peak phased, set the reference frequency to the frequency of the specific peak, then phase the rest of the spectrum with the first order term.

When correcting water-suppressed spectra for eddy currents with post-processing (see section 8.8.1), the phase of a water FID is subtracted from the phase of a water-suppressed FID. Besides linearizing the phase, this subtraction also applies a zero order phase to the spectra because the phase offset of the water FID is equal to the phase offset of the water-suppressed FID. If the A/D sample window is placed correctly in a STEAM or PRESS sequence, first order phasing is not required, and the eddy-current corrected spectrum is properly phased.

8.8.6 Baseline Correction

The quantification of MR spectroscopy requires evaluating the area under the peaks in the absorption mode spectrum. Distortions of the baseline around these peaks may greatly affect the accuracy of these areas. Methods for defining the baseline of a spectrum include DC offset correction, correction with linear tilts, and cubic-splines. This baseline is subtracted from the spectrum before calculating the areas.

8.8.7 Peak Areas

Evaluating the area under a single peak can be done with the traditional running integral. However, evaluating the areas with multiple, overlapping peaks can not be evaluated this way[48]. With overlapping peaks, it is necessary to deconvolve the individual peaks to find their areas. Such deconvolution involves fitting a spectrum with Lorentzian, Gaussian, or a Lorentzian-Gaussian mixture of line shapes so that the difference between the observed and fitted spectra is approximately equal to the noise. The fitting process usually utilizes the Levenberg-Marquardt method for non-linear least-squares to determine the f_0, $A(f_0)$, and T_2^* parameters for each peak. Figure 8.45 shows an observed spectrum, the fitted spectrum using Lorentzian line shapes, and the individual peaks of the fitted spectrum. Deconvolution is necessary to evaluate the area under the Pi peak with the overlapping PME peaks. Fitting can also be done on the unprocessed time domain data[49–51].

8.8.8 Calculating Concentration

The measured area under a spectral peak is directly proportional to the mass of the metabolite. Since spectroscopy sequences do not have $TR = \infty$ and $TE = 0$, the measured area must be corrected for saturation and relaxation effects by the following equation:

$$\text{area}_{corrected} = \frac{\text{area}_{measured}}{\exp\left(\frac{-TE}{T_2}\right)\left[1-\exp\left(\frac{-TR}{T_1}\right)\right]} \tag{8.62}$$

where
- TE is the sequence echo time
- TR is the sequence repetition time
- T_1 is the T_1 of the metabolite
- T_2 is the T_2 of the metabolite.

When studying metabolites at different temperatures and with different numbers of magnetically equivalent atoms, the corrected area should also include terms for temperature and number of equivalent atoms per metabolite. Concentration (C) is equal to mass ÷ volume, and can be calculated by the following equation:

$$C = \frac{K \times \text{area}_{corrected}}{G_{rec} \times V} \tag{8.63}$$

where
K is a calibration constant relating the MRS signal to concentration
G_{rec} is the receiver gain
V is the water volume in the VOI.

Methods have been reported that calculate concentration using an internal water signal[52–54], an internal creatine reference[55], an external reference[56,57], the amplitude of a nonselective 90° reference pulse[58], and the amplitude of a modified water-suppression pulse[59,60]. Barantin has written a well-referenced review of absolute quantification[61].

8.9 Acknowledgements

The authors want to thank Ronald R. Price, Ph.D., Cynthia B. Paschal, Ph.D., S. Julian Gibbs, D.D.S., Ph.D., and David R. Pickens, Ph.D. for assistance in preparing this chapter.

8.10 List of Symbols

A(*f*)	Absorption mode spectrum
ADC	Analog to digital converter
AFP	Adiabatic full-passage
AHP	Adiabatic half-passage
ATP	Adenosine triphosphate
B	Magnetic field
B_{eff}	Effective magnetic field from the RF coil
B_0	Main magnetic field (tesla)
B_1	Radio frequency field
C	Concentration
CHESS	CHEmical Shift Selective
Cho	Choline

CH_2	Methylene protons
CH_3	Methyl protons
Cr	Creatine/phosphocreatine
CSI	Chemical Shift Imaging
CYCLOPS	CYCLically Ordered Phase Sequences
$D(f)$	Dispersion mode spectrum
D	Slice thickness of a SINC RF pulse (mm)
E_T	Energy of a group of atoms in thermal equilibrium
E_ν	Energy of frequency ν from Einstein-Planck equation
$E(t)$	Exponential convolution filter
F	First order phase
f_{cs}	Chemical shift frequency (Hz)
f_0	Frequency of the rotating frame and center frequency of a peak (Hz)
FFT	Fast Fourier Transform
FID	Free Induction Decay
FOV	Total slab thickness is CSI
FWHM	Full Width at Half Maximum
G	Gradient strength (mT/m)
G_{max}	Maximum amplitude of phase gradient in CSI
G_{rec}	Receiver gain
$G(t)$	Gaussian convolution filter
h	Planck's constant (6.625×10^{-34} joule-sec)
I	Nuclear spin angular momentum quantum number
ISIS	Image Selective *In vivo* Spectroscopy
k	Boltzmann constant (1.38×10^{-23} joule/°K)
K	Calibration constant relating the MRS signal to concentration
$L(i)$	Low-pass filter data, used with convolution difference filter
M	Average magnetic moment per unit volume
$M_{\|\|}$	Magnetic moment parallel to B_0
$M_{\perp}$	Magnetic moment perpendicular to B_0
$M(t)$	Complex magnitude of the time domain signal
N_m	Number of protons in the higher energy state
N_{m-1}	Number of protons in the lower energy state
N_p	Number of phase encoding steps in CSI
NAA	N-acetylaspartate
NAD	Nicotinamide adenine dinucleotide
OF	Initial offset frequency of an AHP pulse or offset of a Gaussian pulse
P	Slice shift of a SINC RF pulse (mm)
pl	Pulse length of adiabatic half-passage pulse
PCr	Phosphocreatine
PDE	Phosphodiesters
Pi	Inorganic phosphate
PME	Phosphomonoesters
PPM	Parts-per-million

PRESS	Point RESolved Spectroscopy
$S(t)$	Complex received quadrature signal
STEAM	STimulated Echo Acquisition Mode
SW	Spectral Width (Hz) = $1/\Delta t$
t_p	duration of a RF or gradient pulse
TE	Echo time of a sequence
TM	Mixing time between second and third RF pulses of a STEAM sequence
TR	Repetition time of a sequence
T_1	Longitudinal or spin-lattice relaxation time
T_2	Transverse or spin-spin relaxation time
T_2^*	T_2 decay constant in an inhomogeneous field
V	Water volume in a VOI
VOI	Volume Of Interest
$X(t)$	One of two channels of MR receiver
$Y(t)$	One of two channels of MR receiver
Z	Zero order phase
α	Flip angle of an RF pulse
β	Complex magnitude of an FID at zero time
γ	Gyromagnetic ratio (MHz/tesla)
δ	Displacement or chemical shift
ΔN	Excess number of protons in lower energy state
Δt	Dwell time
$\varepsilon(t)$	Phase distortion caused by a changing $B_0(t)$
θ	Combined zero and first order phase correction
μ	Total magnetic moment
ν	Frequency associated with a photon
σ	Screening constant
ϕ	Initial phase of FID
$\Phi(t)$	Measured phase of complex value
$\psi(t)$	Correct phase if $B_0(t)$ is constant
ω	Angular frequency (radians/sec)

8.11 References

1. F. Bloch, W.W. Hansen, M. Packard, *Physical Review* **70**, 474 (1946).
2. P.C. Lauterbur, *Nature* **242**, 190 (1973).
3. D.I. Hoult, S.J.W. Busby, D.G. Gadian et al., *Nature* **252**, 285 (1974).
4. R.R. Price, W.H. Stephens, C.L. Partain, in *Magnetic Resonance Imaging*, C.L.Partain, R.R. Price, J.A. Patton, M.V. Kulkarni, A.E. James, Eds. (W.B. Saunders, Philadelphia, 1988), p. 976.
5. W.C. Reynolds and H.C. Perkins, *Engineering Thermodynamics*. (McGraw-Hill, New York, 1970), p. 140.
6. J. Morgan, *Introduction to University Physics*. (Allyn and Bacon, Boston, 1969), p. 862.
7. R.K. Hobbie, *Intermediate Physics for Medicine and Biology*. (Wiley, New York, 1988), p. 519.
8. A.D. Elster, *Questions and Answers in Magnetic Resonance Imaging*. (Mosby, St. Louis, 1994), p. 40.
9. K.G. Dobson, in *NMR in Medicine: The Instrumentation and Clinical Applications*, S.R. Thomas and R.L. Dixon, Eds. (American Institute of Physics, New York, 1986), p. 85.
10. C.E. Hayes and W.A. Edelstein, in *NMR in Medicine: The Instrumentation and Clinical Applications*, S.R. Thomas and R.L. Dixon, Eds. (American Institute of Physics, New York, 1986), p. 142.
11. J.F. Schenck, in *The Physics of MRI. 1992 AAPM Summer School Proceedings*, M.J. Bronskill and P. Sprawls, Eds. (American Institute of Physics, Woodbury, NY, 1993), pp. 98–134.
12. T.R. Brown, S.D. Buchthal, J. Murphy-Boesch J et al., *J. Magn. Reson.* **82**, 629 (1989).
13. S.W. Homans. *A Dictionary of Concepts in NMR*. (Oxford University Press, New York, 1992), pp. 58, 141.
14. C.P. Slichter. *Principles of Magnetic Resonance, 3rd ed.* (Springer-Verlag, New York, 1990), p. 24.
15. M.R. Bendall, D.T. Pegg, *J. Magn. Reson.* **67,** 376 (1986).
16. M.S. Silver, I. Joseph, D.I. Hoult, *Phys. Rev.* **31,** 2753 (1984).
17. J. Pauly, P. Le Roux, D. Nishimura, et al., *IEEE Trans. Med. Imaging* **10,** 53 (1991).
18. D.I. Hoult, R.E. Richards, *Proc. R. Soc. Lond.* A. **344,** 311 (1975).
19. S.N. Deming, S.L. Morgan, *Analytical Chemistry* **45,** 278A (1973).
20. D. Holz, D. Jensen, R. Proksa et al., *Med. Phys.* **15,** 898 (1988).
21. P.G. Webb, N. Sailasuta, S.J. Kohler et al., *Magn. Reson. Med.* **31,** 365 (1994).
22. I.S. Mackenzie, E.M. Robinson, A.N. Wells et al., *Magn. Reson. Med.* **5,** 262 (1987).
23. D.M. Doddrell, G.L. Galloway, I.M. Brereton et al., *Magn. Reson. Med.* **7**, 352 (1988).
24. M.G. Prammer, J.C. Haselgrove, M. Shinnar et al., *J. Magn. Reson.* **77**, 40 (1988).
25. J. Tropp, K.A. Derby, C. Hawryszko et al., *J. Magn. Reson.* **85**, 244 (1989).
26. P. Webb, A. Macovski, *Magn. Reson. Med.* **20**, 113 (1991).
27. P.A. Bottomley, *Radiology* **170**, 1 (1989).
28. J. Frahm, H. Bruhn, M.L. Gyngell et al., *Magn. Reson. Med.* **9**, 79 (1989).
29. R.J. Ordidge, A. Connelly, J.A.B. Lohman, *J. Magn. Reson.* **66,** 283 (1986).
30. P.A. Bottomley, *Ann. N. Y. Acad. Sci.* **508,** 333 (1987).
31. R. Sauter, S. Mueller, H. Weber, *J. Magn. Reson.* **75,** 167 (1987).
32. J. Frahm, K.D. Merboldt, W. Hänicke, *J. Magn. Reson.* **72,** 502 (1987).
33. J. Granot, *J. Magn. Reson.* **70,** 488 (1986).
34. E.L. Hahn, *Physical Review* **80,** 580 (1950).
35. J. Frahm, K.D. Merboldt, W. Hänicke et al., *J. Magn. Reson.* **64,** 81 (1985).
36. T.R. Brown, B.M. Kincaid, K. Ugurbil, *Proc. Natl. Acad. Sci.* **79,** 3523 (1982).
37. K.O. Lim, J. Pauly, P. Webb et al., *Magn. Reson. Med.* **32,** 98 (1994).
38. J.W. Cooley, J.W. Tukey, *Math. of Comp.* **19,** 297 (1965).

39. A.G. Marshall, F.R. Verdun. *Fourier Transforms in NMR, Optical, and Mass Spectroscopy.* (Elsevier Science Publishing, Amsterdam, 1990), pp. 11–12, 54–55, 86.
40. R.J. Ordidge, I.D. Cresshull, *J. Magn. Reson.* **69,** 151 (1986).
41. U. Klose, *Magn. Reson. Med.* **14,** 26 (1990).
42. W.R. Riddle, S.J. Gibbs, M.R. Willcott, *Medical Physics* **19,** 501 (1992).
43. D. Marion, M. Ikura, A. Bax, *J. Magn. Reson.* **64,** 425 (1989).
44. R.N. Bracewell. *The Fourier Transform and Its Applications.* (McGraw-Hill, New York, 1978), p 362.
45. P. Dransfield, *Engineering Systems and Automatic Control.* (Prentice Hall, Englewood Cliffs, NJ, 1968), p. 179.
46. E.O. Brigham. *The Fast Fourier Transform and Its Applications.* (Prentice Hall, Englewood Cliffs, NJ, 1988), p. 172.
47. G. Otting, H. Widmer, G. Wagner et al., *J. Magn. Reson.* **66,** 187 (1986).
48. P.A. Bottomley, *Radiology* **181,** 344 (1991).
49. H. Barkhuijsen, R. de Beer, W.M.M.J. Bovée et al., *J. Magn. Reson.* **61,** 465 (1985).
50. D. Spielman, P. Webb, A. Macovski, *J. Magn. Reson.* **79,** 66 (1988).
51. J.W.C. van der Veen, R. de Beer, P.R. Luyten et al., *Magn. Reson. Med.* **6,** 92 (1988).
52. P. Christiansen, O. Henriksen, M. Stubgaard et al., *Magn. Reson. Imaging* **11,** 107 (1993).
53. P.B. Barker, B.J. Soher, S.J. Blackband et al., *NMR Biomed.* **6,** 89 (1993).
54. B.J. Soher, R.E. Hurd, N. Sailasuta et al., *Magn. Reson. Med.* **36,** 335 (1996).
55. J. Frahm, H. Bruhn, M.L. Gyngell et al., *Magn. Reson. Med.* **11,** 47 (1989).
56. T. Ernst, R. Kreis, B.D. Ross, *J. Magn. Reson. B* **102,** 1 (1993).
57. R. Kreis, T. Ernst, B.D. Ross, *J. Magn. Reson. B* **102,** 9 (1993).
58. T. Michaelis, K.D. Merboldt, H. Bruhm et al., *Radiology* **187,** 219 (1993).
59. E.R. Danielsen, O. Henriksen, *NMR in Biomedicine* **7,** 311 (1994).
60. E.R. Danielsen, T. Michaelis, B.D. Ross, *J. Magn. Reson. B* **106,** 287 (1995).
61. L. Barantin, S. Akoka, *J. Magn. Reson. Anal.* **3,** 21 (1997).

9 Magnetic Resonance Imaging: Principles, Pulse Sequences, and Functional Imaging

Rasmus M. Birn, Kathleen M. Donahue, and Peter A. Bandettini

Biophysics Research Institute, Medical College of Wisconsin

9.1 Introduction

Over the past decade magnetic resonance imaging (MRI) has developed into a very powerful and versatile medical diagnostic technique. It has advanced rapidly from the creation of the first images in 1973[1] to the current state of providing detailed information about both anatomy and function. This explosive growth is not an accident. Unlike X-ray techniques, a multitude of tissue parameters can affect the MR signal the most significant of which are the tissue relaxation times. Signal acquisition can be manipulated in a variety of ways enabling the user to control image contrast. More recently, advances in scanner hardware have enabled the collection of an entire image in 50 msec or less. Consequently, MRI has evolved from being able to provide images with superb soft tissue contrast to one which is also capable of imaging fast physiologic processes.

The goal of this chapter is to provide the conceptual background for understanding the acquisition of MR images with a particular emphasis on functional imaging methods and applications. An understanding of the basic principles of the magnetic resonance phenomenon as described in Chapter 8, is assumed. The first section introduces the basic principles of MR imaging, i.e., the creation of spatial information using magnetic field gradients. Next, a brief overview of conventional and fast and echo planar imaging sequences will be presented. Finally, a discussion of the use of these sequences, or variations of them, to evaluate function will be presented. The entire last section will be devoted to explain the use of MRI to observe human brain function.

9.2 Basic Principles of Magnetic Resonance Imaging

9.2.1 Magnetic Resonance Phenomenon

9.2.1.1 Nuclei in a Magnetic Field

The first step in creating a magnetic resonance image is placing the subject in a strong magnetic field. The magnet setup is shown in Figure 9.1. This field is typically in the range of 0.5 to 3 Tesla, which is ten to sixty thousand times the strength of the earth's magnetic field. Recall from Chapter 8, that the presence of such a strong magnetic field causes the nuclear spins of certain atoms within the body, namely those atoms that have a nuclear spin dipole moment, to orient themselves with orientations either parallel or antiparallel to the main magnetic field (B_o). The nuclei precess about B_o with a frequency, called the resonance or Larmor frequency (ν_o), which is directly proportional to B_o:

$$\nu_o = \gamma B_o \qquad \text{(Eq. 9.1)}$$

where γ is the gyromagnetic ratio, a fundamental physical constant for each nuclear species. Since the proton nucleus (^1H) has a high sensitivity for its MR signal (a result of its high gyromagnetic ratio, 42.58 MHz/Tesla) and a high natural abundance, it is currently the nucleus of choice for magnetic resonance imaging (MRI). Because the parallel state is the state of lower energy, slightly more spins reside in the parallel configuration, creating a net magnetization represented by a vector, M_o.

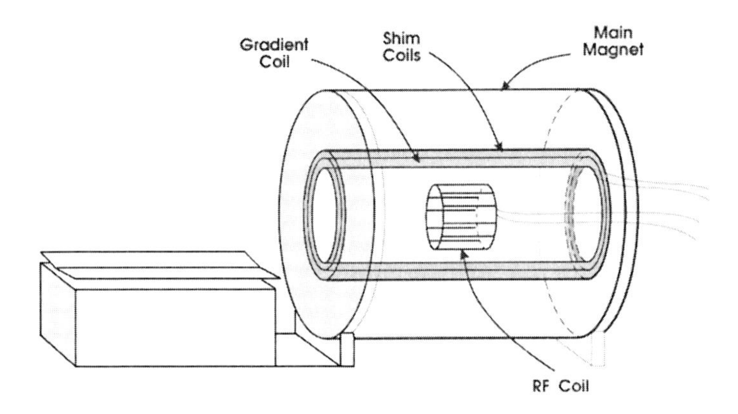

Figure 9.1. A schematic of a typical MR imaging system. The essential components include the magnet producing the main magnetic field, shim coils, a set of gradient coils, an RF coil, and amplifiers and computer systems (not shown) for control of the scanner and data acquisition.

9.2.1.2 Radiofrequency Field

Magnetic *resonance* occurs when a radiofrequency (RF) pulse, applied at the Larmor frequency, excites the nuclear spins raising them from their lower to higher energy states. Classically this can be represented by a rotation of the net magnetization, M_o, away from its rest or equilibrium state. The amount of this rotation is given in terms of the flip angle which depends on the strength and duration of the RF pulse. Common flip angles are 90°, where the magnetization is rotated into a plane perpendicular to B_o, thereby creating transverse magnetization (M_T), and 180° where the magnetization is inverted or aligned antiparallel to B_o. A vector diagram of a 90° pulse is schematically shown in Figure 9.2. Once the magnetization is deflected, the RF field is switched off and the magnetization once again freely precesses about the direction of B_o. According to Faraday's Law of Induction, this time dependent precession will induce a current in a receiver coil, the RF coil. The resultant exponentially decaying voltage, referred to as the free induction decay (FID), constitutes the MR signal. The FID is shown in Fig. 9.3. Since precession occurs at the Larmor frequency, the resulting MR signal also oscillates at a frequency equal to the Larmor frequency.

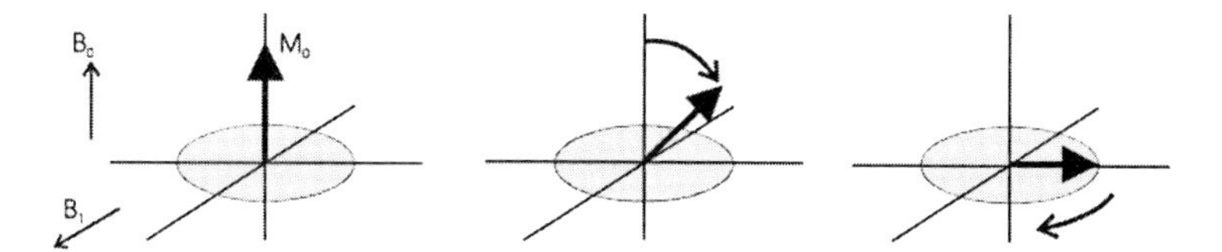

Figure 9.2. A series of vector diagrams illustrating the excitation of a collection of spins by applying an alternating magnetic field, in this case a 90° radio-frequency (RF) pulse (represented here as B_1). B_0 indicates the direction of the main magnetic field. The first 2 vector diagrams are in a frame of reference rotating with the radio-frequency pulse. As a result, the alternating magnetic field can be represented by a vector in a fixed direction. Application of the RF pulse flips the magnetization into the transverse plane, after which the magnetization continues to precess about the main magnetic field.

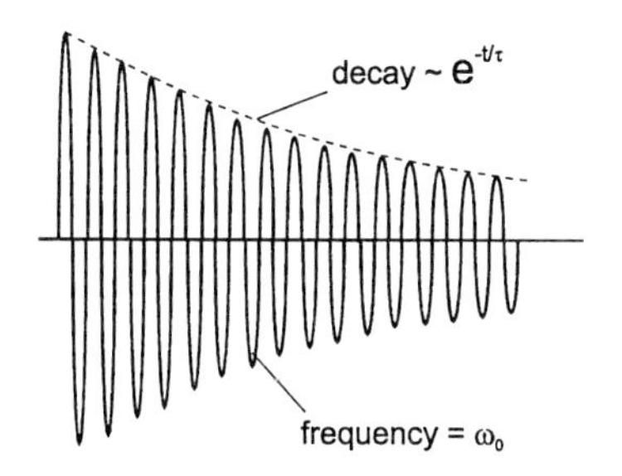

Figure 9.3. The signal acquired after excitation in the absence of applied magnetic field gradients is a decaying sinusoid, called the free induction decay (FID). This signal is characterized by two parameters – the amplitude and the frequency, which depend on the number and type of spins being studied and the magnetic environment that the spins are in.

During the period of free precession the magnetization returns to its original equilibrium state by a process called relaxation which is characterized by two time constants, $T1$ and $T2$. $T1$ and $T2$ depend on certain physical and chemical characteristics unique to tissue type, therefore contributing substantially to the capability of MRI to produce detailed images of the human body with unprecedented soft tissue contrast. A more thorough review of these time constants follows.

9.2.1.3 Relaxation Phenomenon

Spin-Lattice Relaxation ($T1$):
Radiofrequency stimulation causes nuclei to absorb energy, lifting them to an excited state. The nuclei in their excited state can return to the ground state by dissipating their excess energy to the lattice. This return to equilibrium is termed spin-lattice relaxation and is characterized by the time constant, $T1$, the spin-lattice relaxation time. The term lattice describes the magnetic environment of the nuclei. In order for the lattice field to be effective in transferring energy from the excited spins to the lattice, the lattice must fluctuate and the fluctuations must occur at a rate that matches the precessional frequency of the excited spins – the Larmor frequency. As energy is released to the lattice the longitudinal magnetization (magnetization along the z-axis, M_Z) returns to its equilibrium value. This return to equilibrium is characterized by

the time constant $T1$. To better understand $T1$ relaxation consider the following example. Suppose in the equilibrium state M_o is oriented along the z-axis. A 90° rf pulse rotates M_o completely into the transverse plane so that M_Z (the z-component of M_o) is now equal to zero. After one $T1$ interval M_Z = 0.63 M_o. After 2 $T1$ intervals M_Z = 0.86 M_o and so on. Thus the $T1$ relaxation time characterizes the exponential return of the M_Z magnetization to M_o from its value following excitation.

The inversion recovery sequence is the most common pulse sequence used to measure $T1$. It consists of a 180° rf pulse followed by a delay, TI, the inversion time, which in turn is followed by a 90° rf pulse and signal acquisition (AQ). It is denoted by:

$$180° - TI - 90° - AQ \quad (9.2)$$

The experiment to measure $T1$ is described as follows. At time t = 0 M_o is inverted by a 180° pulse after which M_Z (= M_o) lies along the negative z-axis. Because of spin-lattice relaxation M_Z will increase in value from $-M_o$ through zero and back to its full equilibrium value of $+M_o$. A 90° detection pulse is applied at a time TI after the initial 180° pulse. The 90° pulse rotates the partially recovered magnetization, M_Z, into the transverse plane resulting in a detectable MR signal or FID. The FID reflects the magnitude of M_Z after a time TI. The process is then repeated with a different inversion time. By varying the TI, the rate of return of M_Z to its equilibrium position can be monitored, as shown in Figure 9.4. If it is assumed that M_Z is initially equal to $-M_o$ after the 180° pulse and recovers with an exponential decay rate $1/T1$, the equation describing the recovery of M_Z is given by:

$$M_Z(t) = M_o[1-2e^{-t/T1}] \quad (9.3)$$

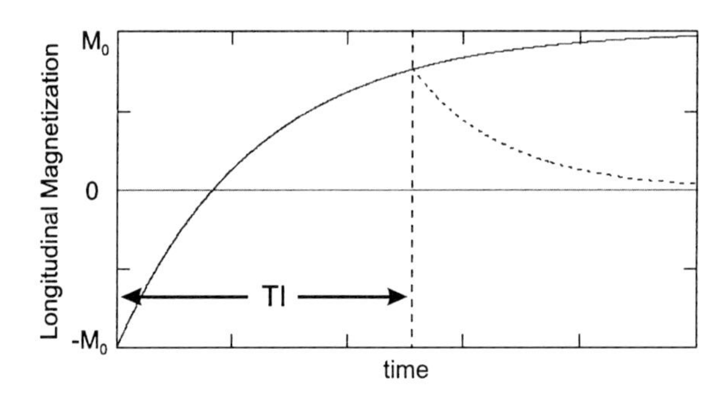

Figure 9.4. In an inversion recovery sequence, an initial 180° RF pulse flips the magnetization along the -z axis. The magnetization then relaxes back to its equilibrium state with a time constant $T1$. At a time TI after the 180° pulse, a 90° pulse is applied, flipping the partially recovered magnetization into the transverse plane. This acquired signal (dotted line) is modulated by the $T1$ relaxation of the tissue.

Spin-Spin Relaxation ($T2$, $T2^*$):
Immediately after an rf pulse, the magnetic moments (or spins) are in phase. Because of natural processes that cause nuclei to exchange energy with each other, the moments begin to spread out in the transverse plane and lose their phase coherence. As a result, the net transverse magnetization (M_T) decays to zero exponentially with time, hence spin-spin relaxation. This decay is characterized by the time constant $T2$. However processes other than inherent spin-spin interactions also cause the spins to dephase. The main magnetic field is not perfectly homogeneous. So, nuclei in different portions of the sample experience different values of B_o and precess at slightly different frequencies. This is described in more detail later. When both natural processes and magnetic imperfections contribute to M_T decay, the decay is characterized by the time constant $T2^*$ which is less than $T2$. Typically, Both $T2^*$ and $T2$ are much less than $T1$.

The spin-echo pulse sequence was designed to correct for the transverse decay due to field inhomogeneities. It consists of a 90° rf pulse followed by a 180° rf pulse and signal acquisition (AQ):

$$90^\circ - \tau - 180^\circ - \tau - AQ \tag{9.4}$$

As illustrated in Figure 9.5, following the 90° rf pulse, spins experiencing the slightly higher fields precess faster than those experiencing the lower fields. Consequently, the spins fan out or lose coherence. Then at some time τ

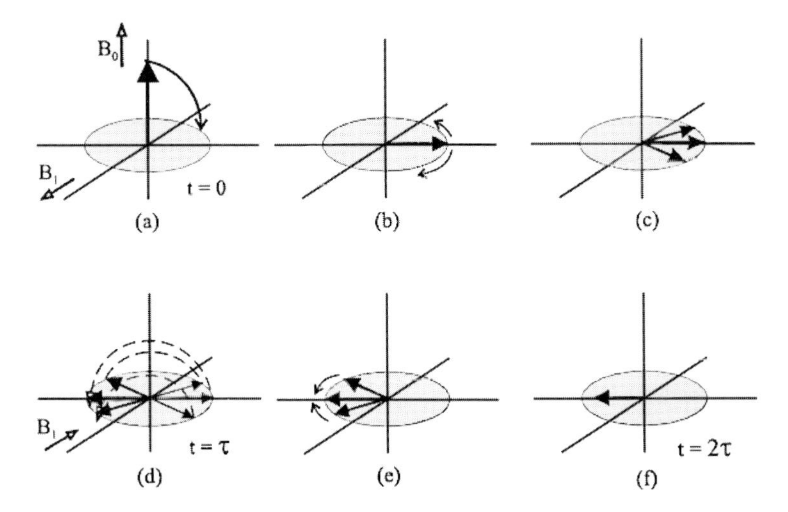

Figure 9.5. A series of vector diagrams illustrating the formation of a spin echo. The diagrams are shown in a frame of reference rotating with the resonance frequency of water. The magnetization is excited by an RF pulse, flipping it into the transverse plane (a). Due to magnetic field inhomogeneities, the spins dephase, shown here as a "fanning out" of the vector (b). At a time t (c), a 180 degree RF pulse is applied that flips the spins to the other side of the transverse plane (d). The spins then continue to precess as before, but now the slower precessing spins are ahead of the faster ones (e). The spins refocus, forming an echo of the original transverse magnetization at time 2t (f).

after application of the 90° pulse an 180° pulse is applied and the spins will be flipped into mirror image positions, i.e. the fast spins will now trail the slow spins. So at a time τ later the fast spins will have caught up with the slow spins so that all are back in phase and a *spin echo* is created. The total period between the initial 90° pulse and the echo is denoted the echo time ($TE = 2\tau$). Thus, the spin-echo reflects the magnitude M_T after time TE.

Spins lose phase coherence not only because of field inhomogeneities but also because of the natural processes responsible for spin-spin relaxation. These natural processes are irreversible and cannot be refocussed. Therefore, the spin-echo signal amplitude at time TE reflects $T2$ decay. Consequently, as the value of TE is increased the echo amplitudes will decrease. This is shown in Figure 9.6 and can be simply described by:

$$M_T(t) = M_o e^{-t/T2} \tag{9.5}$$

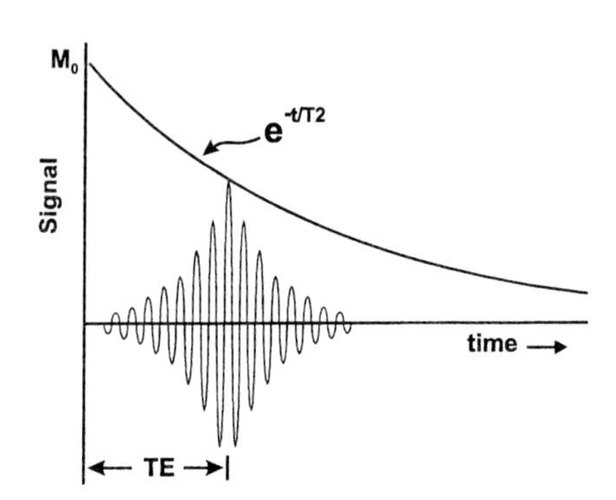

Figure 9.6. In a spin echo sequence, the amplitude of the acquired signal (shown here as a spin echo) is modulated by the $T2$ relaxation of the spins. Signals acquired at a longer TE will be smaller.

9.2.2 Imaging Concepts

The basic goal of MR imaging is to measure the distribution of magnetization within the body, which depends on both the variation in the concentration of water and the magnetic environment between different tissues. In a completely uniform field, all of the hydrogen protons of water resonate at the same frequency. The RF coil used to detect the signal, as shown in Figure 9.1, is only sensitive to the frequency, amplitude and phase of the precessing magnetization, not to the spatial location. It cannot distinguish two spins at different locations that are precessing at the same frequency. It can only distinguish spins precessing at different frequencies. To make an image it is necessary to make the spin's precessional frequency depend on the location of the spin. This is accomplished by superimposing linear magnetic field gradients on the main magnetic field. The term "gradient" designates that the magnetic field is altered along a selected direction. Referring to the Larmor Eq. (Eq. 9.1), it can be seen that if the field is varied linearly along a certain direction, then the resonance frequency also varies with location, thus providing the information necessary for spatial localization.

The conventional method by which gradients are applied to acquire a two-dimensional image are reviewed first. Understanding these principles will aid in the understanding of more advanced techniques, such as fast gradient echo and echo planar imaging, described in the following sections.

Obtaining a two dimensional image requires three steps. The first step is to excite only the spins in the slice of interest, called "slice selection." The next steps are to localize the spins within that slice using techniques called "frequency encoding" and "phase encoding." For convenience let "z" denote the direction for slice selection, "x" denote the direction for frequency encoding, and "y" denote the direction for phase encoding. This designation is arbitrary and is unrelated to the physical orientation of the x, y, and z gradient coils. These concepts are introduced by building up a conventional spin-echo imaging sequence, which consists of a combination of RF and gradient pulses.

9.2.2.1 Slice Selection

The first step is the selection of a slice, which is achieved by applying a magnetic field gradient along the z-axis (G_z) during a 90° RF pulse of a specific frequency bandwidth (period 1 of Figure 9.7). When the slice select gradient, G_z, is applied along the z-axis, the resonance frequencies of the protons become linearly related to position along the z-axis. Individual resonance frequencies correspond to individual planes of nuclei. In this example, these planes are oriented perpendicular to the z-axis. When the frequency-selective 90° pulse is applied while G_z is on, only nuclei in the plane with corresponding frequencies will be excited; thus a slice will be selected. This is indicated as the dark gray area in Figure 9.8. The frequency bandwidth of the excitation pulse, together with the gradient, confines the excitation to the nuclei in the slice. No signals are excited or detected from areas outside the defined slice.

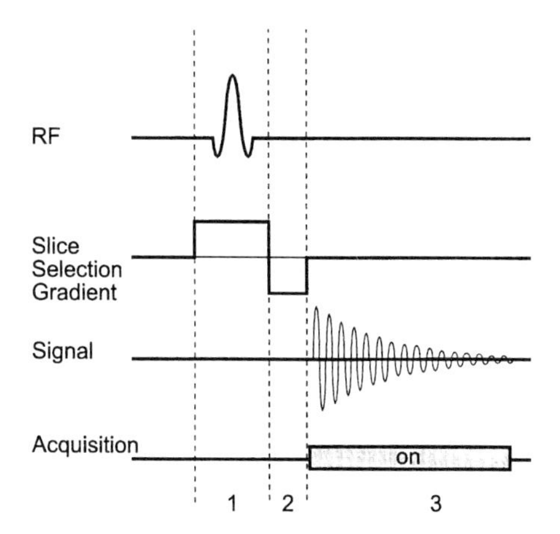

Figure 9.7. The sequence of RF power and gradient strength used for slice selection. To excite only one slice, a magnetic field gradient is applied during the excitation RF pulse.

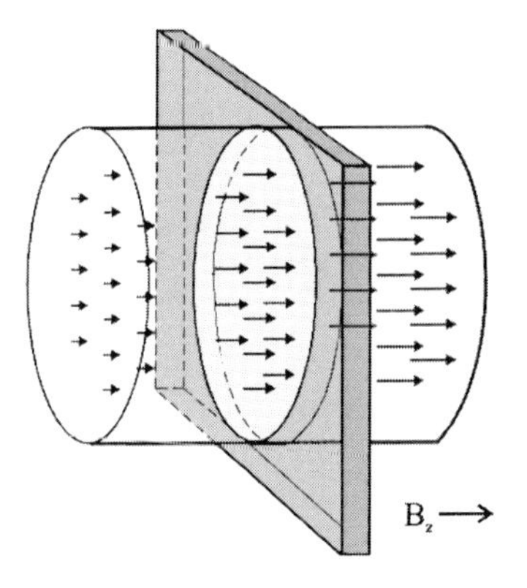

Figure 9.8. The application of a magnetic field gradient in slice selection creates a stronger magnetic field at one end of the sample than at the other end, shown here as arrows of varying length. When the RF pulse is transmitted into the sample, only those spins whose precessional frequency matches the frequencies in the RF pulse are excited, shown here as a dark gray slab.

The RF pulse that is transmitted to the patient contains not just one frequency, but a narrow range, or bandwidth, of frequencies. Quantitatively, the thickness of the excited slice (Δz) in cm is related to the gradient amplitude, G_Z, and RF bandwidth Δf as follows:

$$\Delta z = \Delta f / \gamma G_Z \tag{9.6}$$

If Δf is increased such that more frequencies are present in the RF pulse, then a larger slice will be excited. Alternatively, when the strength of the gradient is decreased, then more spins are resonating in a given range of frequencies, and again a larger slice is excited. Therefore the thickness of the slice excited can be varied in two ways – either by varying the bandwidth of the transmitted RF pulse, or by changing the strength of the gradient, as indicated in Figure 9.10. The location of the excited slice can be varied by transmitting an RF pulse of a different frequency, as shown in Fig. 9.9.

The slice-selection gradient, G_Z, has two effects on the MR signal, the desired one of aiding in spatial localization and the unwanted one of dephasing the signal (since the phase of the spins is also proportional to field strength). Therefore, after the slice-selection gradient (period 1), a negative

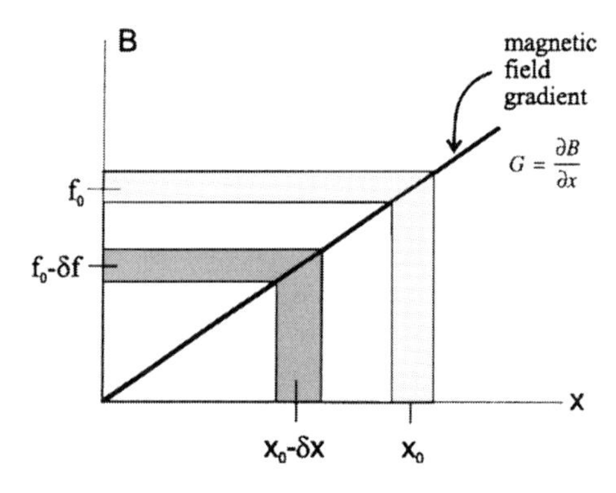

Figure 9.9. The position of the excited slice can be varied by changing the frequency of the transmitted RF pulse.

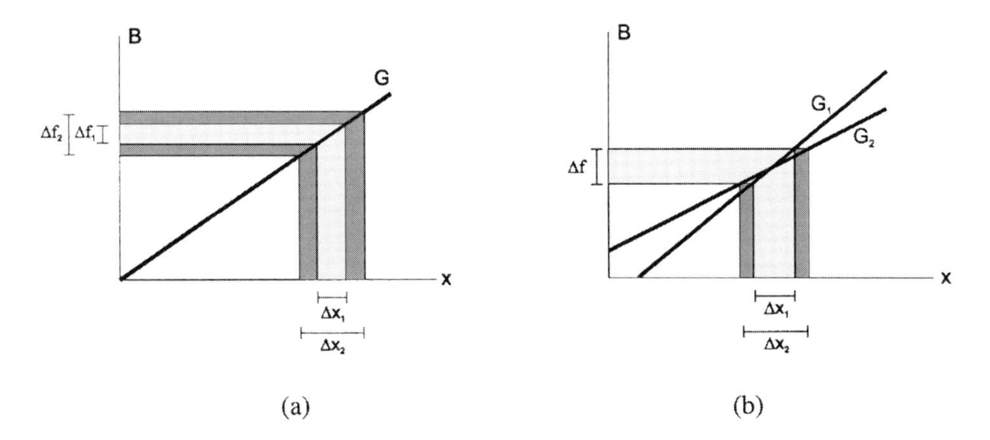

Figure 9.10. The thickness of the excited slice can be varied either by (a) changing the bandwidth of the transmitted RF pulse or by (b) changing the amplitude of the slice selection gradient.

z-gradient follows (period 2 of Figure 9.7) to compensate for the dephasing effects of the slice-selection gradient. Ideally, this gradient will result in an accumulated phase that is equal and opposite to the phase accumulated from the initial slice-select gradient thereby canceling its dephasing effects. This type of gradient is often referred to as a time-reversal or rephasing gradient.

9.2.2.2 Frequency Encoding

After slice selection, the next task is to distinguish signal from different spatial locations within this slice. This is accomplished in the x-direction by applying a gradient (G_x), the frequency-encoding gradient, during the acquisition of the signal. (time period 3 in Figure 9.11). Since the MR signal is sampled during the time that G_x is on, this period is also commonly referred to as the read-out period and G_x as the read gradient. This signal can come from either the FID or a spin-echo, the latter of which is formed by applying a 180° pulse at a time $TE/2$ after the 90° excitation pulse, as shown in Figure 9.11b. Sequences that collect the signal from the FID are known as gradient-echo (GRE) sequences, whereas sequences that collect the signal from the spin echo are known as a spin-echo (SE) sequences. The two differ in the contrast that they provide. Because of the refocussing pulse, spin echo sequences, for example, are less susceptible to magnetic field inhomogeneities and thus reflect differences in $T2$ relaxation times between the tissues, rather than $T2^*$. These differences will be discussed in detail later. The next few sections will deal mainly with the spin-echo sequence.

The way in which the linear gradient encodes the spatial information can be more easily seen by considering a sample consisting of 2 vials of water, aligned with the y-axis and placed some distance apart in the x-direction. Please refer to Figure 9.12. All of the signal comes from these 2 sources of

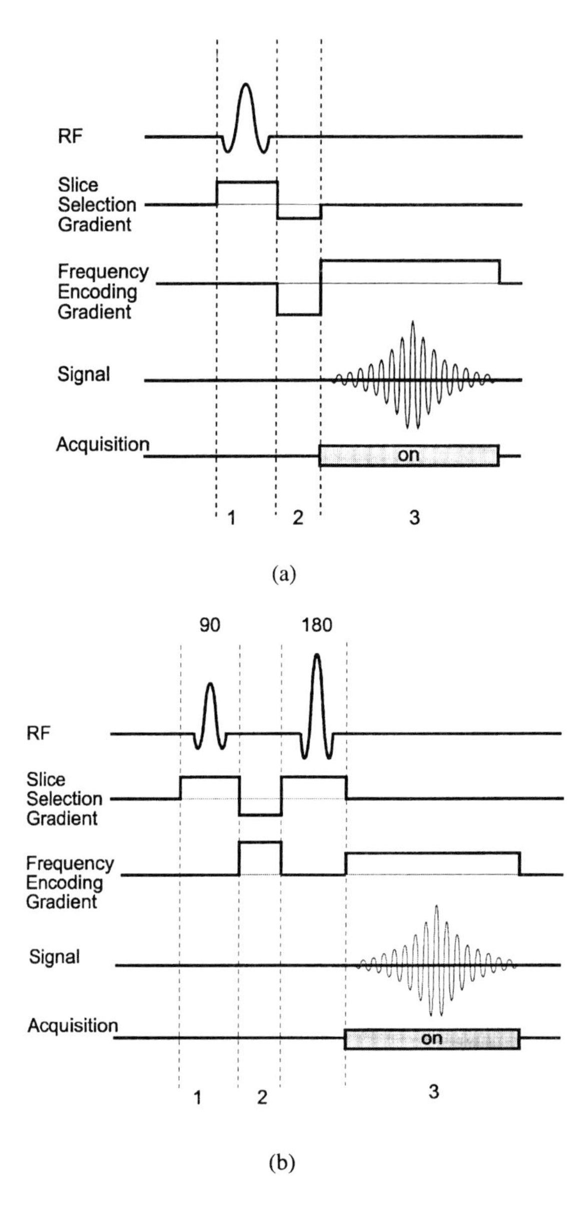

Figure 9.11. The sequence of RF power and gradient amplitudes used to excite one slice and encode the positions of the spins within that slice into the signal. In this "frequency encoding," the positions of the spins encoded by applying a magnetic field gradient in one of the directions in the excited slice during the acquisition. Note that the signal can come either from the FID (a) or from a spin echo (an echo of the FID) (b). Sequences using signals from the FID are called 'gradient-echo' sequences, while sequences using signals from the spin echo are called 'spin-echo' sequences.

water. If signal from this sample is collected without the application of any gradients, both areas are precessing at the same frequency. Consequently the signal will appear as a pure sinusoid, and applying a mathematical process called the Fourier transform will show that it contains only one frequency. (Whereas the FID represents the time evolution of M_{xy}, the Fourier Transform of the signal represents its frequency distribution.) The amplitude of this frequency peak corresponds to the total amount of water from both vials. If a gradient is applied during the acquisition of the signal, however, the spins in one vial are in a slightly higher magnetic field than those in the other vial.

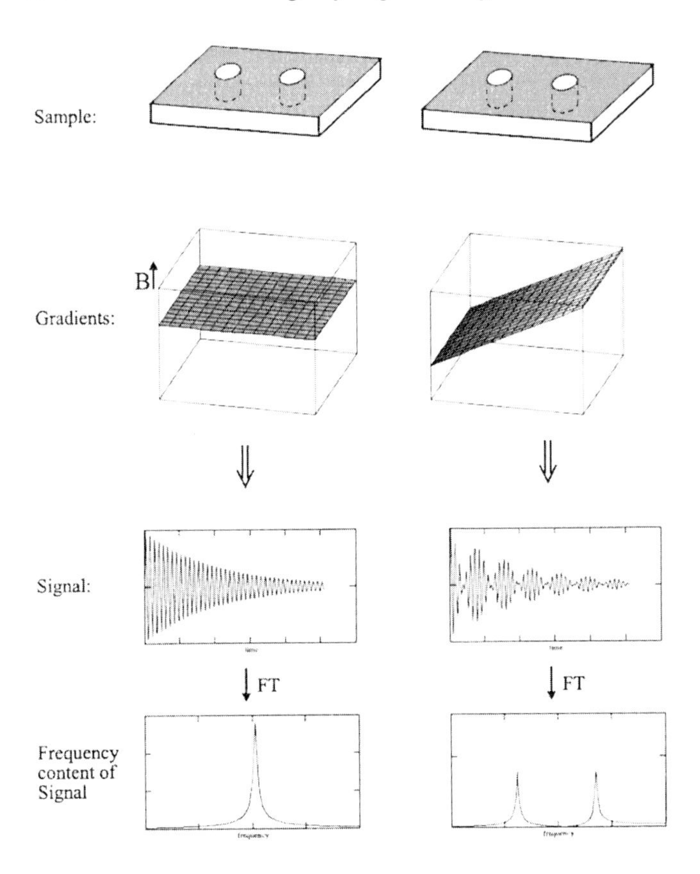

Figure 9.12. A series of steps illustrating the concept of frequency encoding to distinguish the signal coming from two point sources of magnetization, e.g. small vials of water, in an object.
(left) When no gradient is applied, both sources of magnetization resonate at the same frequency and the signal is a simple decaying sinusoid. When this signal is Fourier transformed, the signal is shown to contain only one frequency.
(right) When a gradient is applied, one of the sources of magnetization precesses at a higher frequency than the other. The resulting signal is an interference pattern of the two frequencies and is shown to contain by Fourier transformation to contain two distinct frequencies. Notice that the Fourier transformed signal is the projection of the amount of magnetization along the axis along which the gradient was applied. That is, in this one dimensional case, the frequency content of the signal *is* the image.

According to the Larmor relation, one group of spins will precess faster than the other group, and the signal will be an interference pattern, or combination, of both of these frequencies. If a Fourier transform is applied to this signal, the signal is found to contain two distinct frequencies. Since a spatially linear gradient was applied, the frequencies of these peaks exactly correspond to the position of the vials. Also, the amount of signal at a given frequency is determined by the number of spins precessing at that frequency, and is thus directly related to the amount of magnetization at a given location. In other words, the Fourier transform of the signal is simply a projection of the distribution of magnetization onto the frequency encoding axis.

Figure 9.13 shows the phases of the magnetization vectors in one slice at three time points during frequency encoding. The presence of the gradient causes spins at one end to precess faster than those at the other end, causing an increasing amount of phase shift along this direction. As time progresses (when the gradient has been applied for a longer duration) the amount of "phase twisting" is increased. One effect of this is that the peak of the signal (when it is least dephased) will be at the beginning of the acquisition. To move the peak signal to the center of the acquisition window, a negative gradient lobe (sometimes referred to as a time reversal gradient) with exactly half the area of the frequency encoding gradient is applied just before the frequency encoding gradient. This initially dephases the spins, which are then brought back in phase by the applied frequency encoding gradient (see Figure 9.14). In the spin echo sequence the gradient lobe is positive and occurs before the 180° inversion pulse.

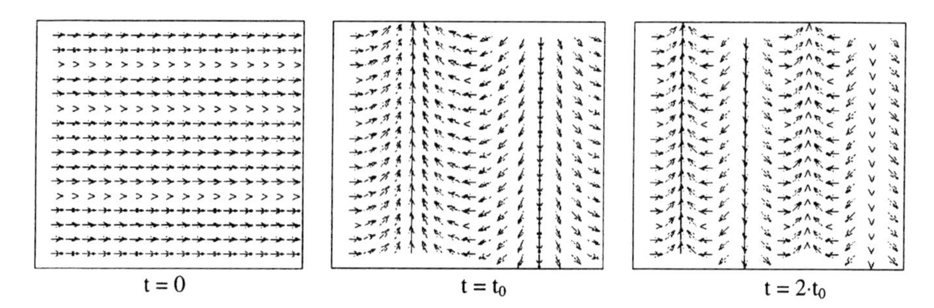

Figure 9.13. At each time increment when the signal is acquired in frequency encoding, the gradient has been applied for a longer period of time. This causes an increasingly greater variation in the phase in the direction in which the gradient was applied.

The details of the frequency encoding procedure dictate the field of view (image size in cm) along the x-axis (FOV_x):

$$FOV_x = \frac{BW}{\gamma G_x} \tag{9.7}$$

where BW is the receiver bandwidth. Note that the receiver bandwidth should not be confused with the excitation RF bandwidth, which dictates the slice thickness (Eq. 9.6). Here the BW is the effective range of frequencies that can be properly detected (as determined by the Nyquist criterion[2]). The BW is controlled by the digital sampling rate, which in turn is determined by the number of points on the signal to digitized (N_X) and the length of time the receiver is on, the acquisition time (AQ):

$$BW = \frac{N_X}{AQ} \tag{9.8}$$

Accordingly, from these two equations, the pixel size along the frequency encoding axis can be derived:

$$\text{pixel size} = \frac{\text{FOV}_X}{N_X} = \frac{1}{\gamma G_X AQ} \tag{9.9}$$

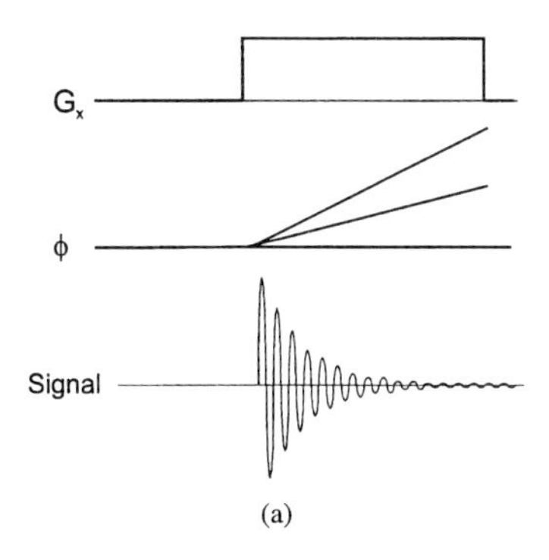

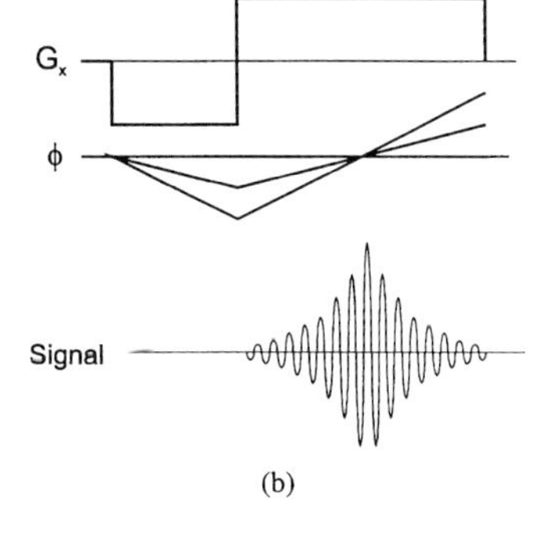

Figure 9.14. Diagrams showing the gradient amplitude, phase of two spins subjected to these gradients, and the profile of the resulting signal. When an initial negative gradient is applied as in (b), the spins are in phase in the center of the acquisition window. This leads to a greater net signal.

9.2.2.3 Phase Encoding

The final spatial dimension can be encoded into the signal by applying a programmable phase-encoding gradient, G_y, simultaneous with the rephasing gradient in period 2, in the time between the excitation and the acquisition, as shown in Figure 9.15. During the phase encoding period, nuclei in each column of voxels along the y direction experience different magnetic fields. Nuclei subjected to the highest magnetic field precess fastest. This is no different from the effect of the frequency-encoding pulse. However, the state of magnetization during the phase-encoding pulse is less important the phase shift accumulated after the phase-encoding gradient has been turned off. When the gradient is on, the nuclei that experience the highest field advance the farthest and therefore acquire a phase angle, ϕ_y, that is larger than that in voxels experiencing smaller magnetic fields. After G_y is turned off, the nuclei revert to the resonance frequency determined by the main magnetic field. However, they "remember" the previous event by retaining their characteristic y-coordinate-dependent phase angles. Similar to Eq. 9.9, the field of view in the y-direction (FOV_y) is quantitatively defined as:

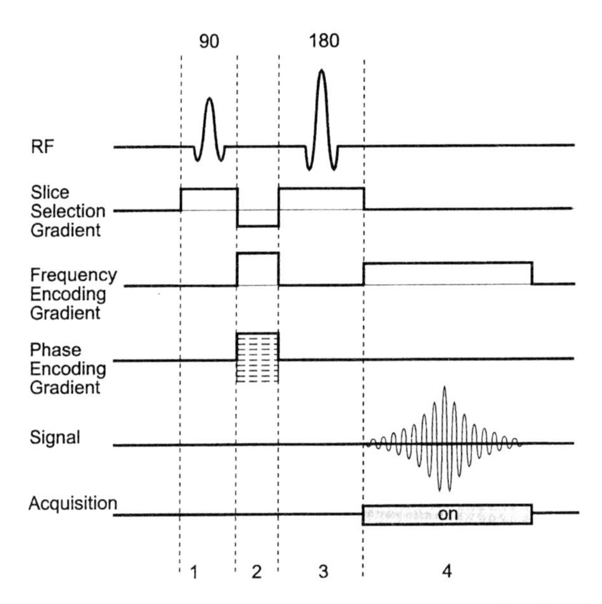

Figure 9.15. A complete pulse sequence diagram for the spin echo sequence. Spatial locations of the spins are encoded into the signal by applying three orthogonal gradients, techniques that are called slice selection, frequency encoding, and phase encoding. In period 1, a 90 degree pulse and a slice selection gradient excite one slice. In period 2, the initial frequency encoding gradient and the phase encoding gradient are applied. In period 3, a 180 degree pulse is applied, along with a slice selection pulse (such that only the spins in the same excited slice are "flipped,") and in period 4 the frequency encoding gradient is applied and the signal is acquired. The sequence shown here is repeated numerous times (128, 256, 512, etc. depending on the desired resolution) each time with a different strength of the phase encoding gradient.

$$\mathrm{FOV}_y = \frac{1}{\gamma T_y G_{ymax}} \tag{9.10}$$

where T_y is the duration and G_{ymax} is the maximum amplitude of the phase-encoding gradient.

Although the signal obtained from one acquisition (slice selection, phase encoding and frequency encoding) contains information from all voxels in the imaging slice, the information gathered from one iteration of this sequence is not sufficient to reconstruct an image. Consequently, the sequence has to be repeated with different settings of the phase-encoding gradient G_y.

When a phase encoding gradient of a particular value has been applied, the effect of that gradient is to shift the phases of the spins by an amount depending on their position in this case in the y-direction and the amplitude of the phase encoding gradient. Spins near the isocenter, for example, experience no phase shift, whereas spins at positions off center are shifted by a certain amount depending on their distance from the center. The net result of this spin dephasing is simply a decrease in the signal. It is only by varying the amount of this dephasing (thus varying the amount of signal decrease) by stepping through phase encoding gradient's range of amplitudes that the location of structures along the phase encoding gradient can be identified.

If the data at each cycle of the G_y setting were plotted, it would show sinusoidal curves with a frequency dictated by the rate of phase change (between each iteration of the pulse sequence), which, in turn, depends on location. A curve much like that is derived during frequency encoding, but with a difference: each sample along this curve originates from a different MR signal. Each of these MR signals follows a phase-encoding gradient pulse of different amplitude. However, similar to frequency encoding, the frequency components of the curve are identified by the Fourier transform, and the magnetization ascribed to a given location.

In summary, for a matrix of size $N_y \cdot N_x$ the required number of iterations is N_y. The N_y signals, each corresponding to a different value of G_y, are sampled N_x times during the read period. Subsequent two-dimensional Fourier Transformation yields the intensity values of each of the $N_y \cdot N_x$ pixels.

9.2.2.4 Image Formation Mathematics: K-Space

The key to image formation is encoding the location of the magnetization in the phase of the MR signal. It is worthwhile to look at this encoding process in more detail. Consider the encoding of spatial information along one dimension within the plane after the slice has been excited. A collection of spins along one dimension can be thought of as a column of vectors, as shown in Figure 9.16. After the slice has been excited, all of the spins within the slice are in phase. Once a magnetic field gradient is applied, the spins at will precess at different frequencies, depending on their location. At any given point in time, certain spins will have accumulated more phase than others. These gradients can thus be thought of as "twisting" the initially aligned column of spins. This "twisting" of the magnetization vectors by the gradients can be expressed as a rotation of the magnetization by an angle ϕ which depends on the strength of the magnetic field that the magnetization at that particular location experiences.

$$M = M_T(x, y)e^{-i\phi} = M_T(x, y)\, e^{-i\int_0^t \gamma B \, dt'} \tag{9.11}$$

At each point in time, the RF coil integrates this magnetization over the entire volume, and thus the signal at a given point in time can be expressed as,

$$S(t) = \int M_T(x, y)\, e^{-i\int_0^t \gamma B \, dt'} \, dx \, dy \tag{9.12}$$

For imaging linear gradient fields are applied. Therefore the magnetic field, B, experienced by the spins can be rewritten as,

$$\begin{aligned} B &= \int G_x \, dx + \int G_y \, dy \\ &= G_x x + G_y y \end{aligned} \tag{9.13}$$

If it is assumed that the position of the magnetization with respect to the coils does not change with time (patient does not move), then the signal can be written as,

$$S(t) = \int M_T(x, y)\, e^{-i\left(x\gamma\int_0^t G_x \, dt' + y\gamma\int_0^t G_y \, dt'\right)} \tag{9.14}$$

If the following substitution is made,

$$\begin{aligned} k_x &= \gamma \int_0^t G_x \, dt' \\ k_y &= \gamma \int_0^t G_y \, dt' \end{aligned} \tag{9.15}$$

then Eq. (9.14) becomes,

$$S(k_x, k_y) = \int M_T(x, y)\, e^{-i(k_x x + k_y y)} \, dx \, dy \tag{9.16}$$

This signal is a Fourier transform of the magnetization, by virtue of the gradients applied. A measure of the magnetization, M(x,y), can be obtained by taking a 2-dimensional Fourier transform of the signal.

Because of the way the gradients are applied during the imaging scan, it is natural to think of the MR signal as being collected in spatial frequency space, or "k-space"[3–5], as implied by the terms in Eq. (9.15). This representa-

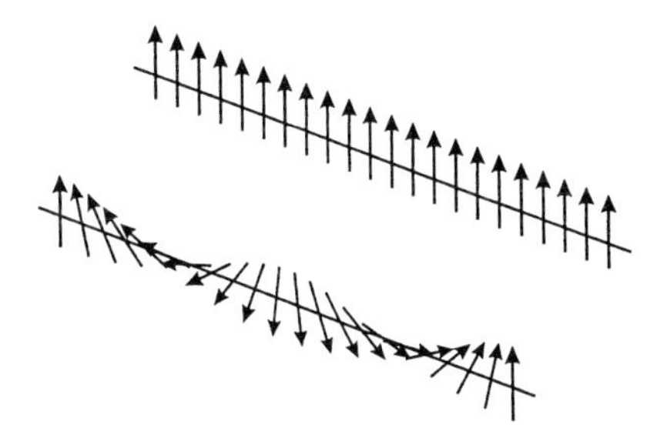

Figure 9.16. The application of a magnetic field gradient can be thought of as twisting the initially aligned column of spins. These spins are then summed at every point in time using the RF coil.

tion is often much more convenient in discussing the details of pulse sequences. In this space, usually plotted in 2 dimensions, each point describes the amount of a particular spatial frequency present in the imaged object.

The strongest signal of imaged objects is typically in the center of k-space, where all gradient values are equal to zero. The regions farther out in k-space correspond to higher spatial frequencies, which are especially important in discerning sharp differences in signal, such as at edges. Therefore, the highest spatial frequency sampled (or the furthest sample from the center of k-space) determines the resolution of the final image. The further out, the higher the resolution. In contrast, the interval between the samples in k-space, or the resolution in k-space, determines the field of view of the image – in other words, the largest spatial extent that can be acquired. The smallest sample interval corresponds to a large field of view. Care must therefore be taken to acquire samples both finely enough such that the entire region of interest is imaged, and far enough out in k-space to obtain the desired resolution.

Sampling different points in k-space is accomplished by applying magnetic field gradients, as demonstrated in Eq. (9.15). The center of k-space corresponds to the time immediately after excitation and immediately prior to the application of magnetic field gradients. Application of a magnetic field gradient causes the phases of the spins to twist by an increasing amount corresponding to the amplitude and duration of the gradients, as implied by Eq. (9.15). Collecting the signal at this time will cause increasingly higher spatial frequencies to be sampled. In other words, the gradients allow movement in k-space, as shown in Figures 9.17 and 9.18.

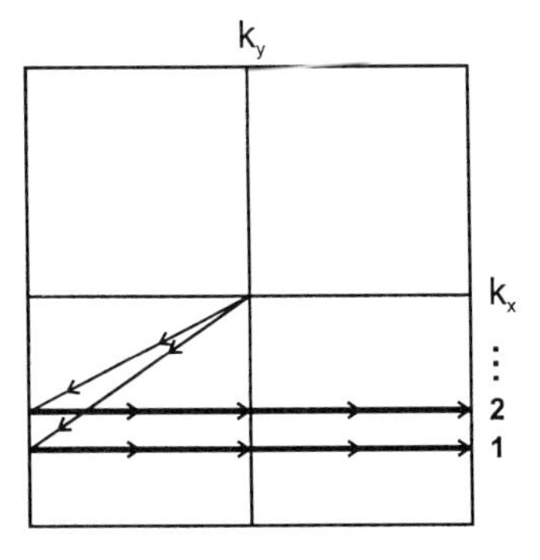

Figure 9.17. A k-space diagram showing the path through k-space taken to acquire the signal for the gradient echo (GRE) sequence described above. For each excitation, the phase encoding gradient moves us a fixed distance in the negative ky direction and the initial negative frequency encoding gradient moves us the in the negative kx-direction. The signal is then sampled moving in the positive kx-direction as the frequency encoding gradient is applied. The signal is then allowed to relax, and the sequence is repeated with a different value for the phase encoding gradient. In this manner, a sufficient range of k-space can be scanned.

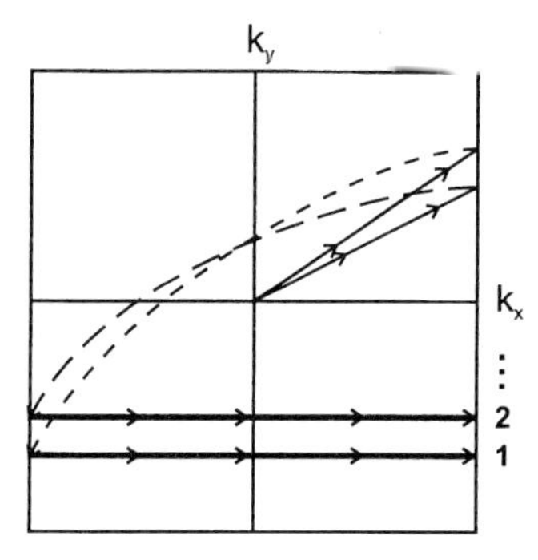

Figure 9.18. A k-space diagram for the spin-echo (SE) sequence. After the spins are excited, gradients in the positive x and y directions are applied, moving us in the positive-kx and positive-ky direction. The 180 degree pulse flips us through the center of k-space to the negative-kx, negative-ky direction, after which the positive frequency encoding gradient moves us in the positive-kx direction, allowing us to sample the frequencies as before. This sequence of steps is repeated with different values for the phase encoding gradient.

Following is a brief example of how pulse sequences are commonly described in the context of k-space. In the gradient-echo sequence described earlier, we started at the origin of k-space after the excitation. The initial negative x-gradient lobe moves us to the left (negative x frequency) and the phase encoding gradient moves us a specific amount in the y direction of k-space. The final application of an x-gradient moves us in the positive x direction during which time we acquire the signal. The signal is then allowed to relax, and with the next excitation, the value of the phase encoding gradient is changed, allowing us to scan a different line in k-space (see Figure 9.17). For the spin-echo sequence, the initial positive x-gradient lobe moves us in the positive x-direction, and the initial phase encoding gradient moves us a specified amount in the positive y-direction. The application of a 180° RF pulse flips us through the origin of k-space to the negative-x, negative-y direction, after which point the signal acquisition occurs just as in the gradient echo sequence (see Figure 9.18). In this manner a large range of spatial frequency space is sampled.

9.2.2.5 Image Contrast

Though $T1$, $T2$ and proton density are intrinsic tissue parameters over which the user has no control, the operator can alter tissue contrast and signal to noise (S/N) by the choice of the pulse sequence parameters. Specifically, images can be obtained in which tissue contrast is primarily determined by (i.e., weighted toward) $T1$, $T2$ or proton density characteristics. For example, with the spin-echo imaging sequence the type of image weighting is determined by the repetition time (TR) and the echo time (TE). The effects of TR and TE on image weighting is depicted schematically in Figure 9.19 for the case of two tissues with different $T1$ and $T2$ relaxation times. TR determines the extent of $T1$ relaxation. The initial 90° rf pulse completely tips the existing longitudinal magnetization into the transverse plane leaving zero longitudinal magnetization. If the spins were again excited at this time, no signal would be produced. Therefore, a time interval (TR) is allowed to elapse between excitations, so that the spins can undergo $T1$ relaxation and recover at least part of their longitudinal magnetization. It is apparent from Figure 9.19 that the maximum $T1$ contrast between tissues occurs when TR is greater than 0 and less than some time when both tissues have completely recovered their longitudinal magnetization. A long TR ($>> 5T1$) allows enough time to elapse so that almost complete $T1$ relaxation occurs and therefore signal intensity is not a function of $T1$. The maximum magnetization to which the signal returns is determined by proton density. Likewise, the amount of $T2$ contrast is dictated by the choice of TE. The longer the time interval TE the greater the extent of $T2$ relaxation. Therefore, spin-echo images acquired with short TR ($TR \sim T1$) and short TE ($TE < T2$) are $T1$-weighted. With shorter TR values tissues such as fat which have short $T1$ values appear bright, whereas tissues that have longer $T1$ values, such as tumors and edema, take more time to relax towards equilibrium and therefore appear dark. The short TE value diminishes the importance of tissue $T2$ differences. Similarly, images acquired with long TR (to diminish $T1$ differences) and long TE ($TE \sim T2$) are $T2$-weighted. Therefore, tissues with long $T2$, such as tumors, edema, and cysts, appear bright, whereas tissues that have short $T2$, such as muscle and liver, appear dark. Images acquired with long TR ($TR >> 5T1$) and short TE ($TE < T2$) are called proton-density weighted images. Tissues with increased proton density appear moderately bright. It should be noted that both $T1$ and $T2$-weighted images are always partly weighted toward proton density as well.

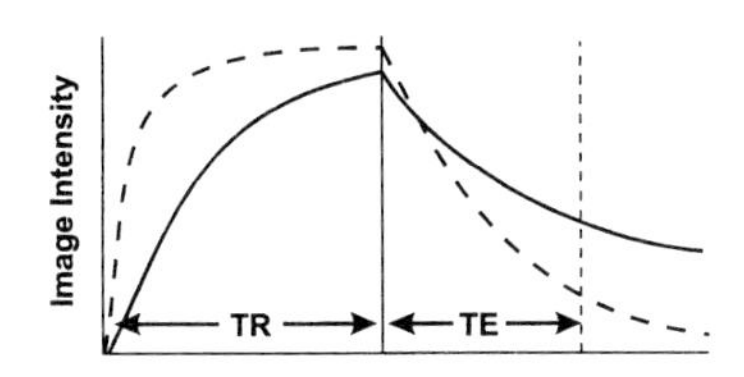

Figure 9.19. Schematic depicting the effects of TR and TE on the weighting of image intensity. The solid and dashed curves represent two tissues with different $T1$ and $T2$ values. The choice of TR (position of solid vertical line) dictates the degree of $T1$-weighting while the choice of TE (position of dashed vertical line) determines the amount of $T2$ weighting.

9.2.2.6 Sequence Timing

Conventional spin-echo and gradient-echo sequences are repeated at time intervals equal to TR, the repetition time. The number of times the sequence is repeated (for one average) is determined by the desired spatial resolution (proportional to the number of voxels) along the phase encoding direction and is equal to the number of phase encoding steps (N_y). For NEX (number of excitations) averages the total time required to obtain an image slice is:

$$TR \times N_y \times \text{NEX} \tag{9.17}$$

Typical parameters for conventional spin-echo sequence are a TR of 2.0 sec, 128 phase encoding steps and 2 averages giving a total acquisition time of 8.5 minutes.

To decrease imaging time one or more of these parameters can be decreased. Decreasing the number of averages by two halves imaging time, but has the additional effect of decreasing the signal to noise ratio by $\sqrt{2}$ or

41% thereby increasing the graininess of the image. Motion artifacts which are also decreased by averaging could become significant if imaging time were decreased by decreasing the number of averages. Decreasing the image matrix size or the number of phase encoding steps decreases imaging time at the expense of spatial resolution. However, the larger pixels result in an increased S/N. The simplest way to speed up an ordinary SE scan would be to drastically reduce *TR*. However, the signal produced depends on the amount of *T*1 relaxation that occurs during the interval *TR*, and therefore the available signal for the next excitation. A short *TR* relative to *T*1 would result in significant signal losses. Consequently, the *T*1 relaxation times of tissue protons limit the degree to which the pulse repetition times (*TR*) can be shortened. Two techniques which overcome *TR* limitations include gradient echo (GRE) and echo planar imaging (EPI) techniques. These fast imaging sequences are discussed in the following sections.

Alternatively, to speed up the acquisition of spin-echo images a procedure was developed in which several 180-degree pulses follow each 90° RF excitation pulse creating several spin-echoes, with each echo differently phase-encoded. Consequently, if four spin-echoes follow each 90° excitation pulse the total acquisition time would be 1/4 of what it is using the conventional approach of acquiring one phase encoding step per excitation pulse. This principle underlies the RARE (rapid acquisition with relaxation enhancement) imaging technique[6]. Obviously, acquisition of signals at different effective echo times lends strong *T*2 weighting to RARE images.

One final note, in conventional spin-echo imaging, when a profile is measured after the excitation of a slice, one has to wait until the spins are in equilibrium again, for that slice. This waiting time is about 3 × *T*1, which is on the order of seconds. Yet, the technical minimum time between excitations is the time needed to collect one profile; the echo time plus the second half of the acquisition time and the first half of the selection pulse. It is therefore possible to collect many profiles in one *TR* period. This procedure, in which more than one slice is measured per *TR*, is called interleaved multi-slice imaging. If the time to collect the desired number of slice profiles does not exceed the difference between *TR* and the minimum time between excitations, then multi-slice imaging does not add any additional time to the total acquisition time.

9.2.3 Pulse Sequence and Contrast Topics

9.2.3.1 Fast Gradient-Echo Imaging

In its most basic form, the GRE pulse sequence, as shown in Figure 9.11a, consists of one RF pulse with a flip angle α, followed at some time later by the acquisition of the gradient echo. The time between the excitation and the acquisition of the gradient echo is defined as the echo time, *TE*:

$$\alpha \text{ degrees} - TE - \text{(gradient-echo)} \tag{9.18}$$

Because GRE sequences lack a 180° refocussing pulse, images generated with these sequences are sensitive to artifacts from magnetic field inhomogeneities i.e., $T2^*$ effects.

Gradient echo sequences are typically used as fast sequences because data are acquired before the dephasing of spins from previous application of the pulse sequences is complete, i.e. $T2^*$ decay is not complete. In most cases, the TR is less than the time for more than 90 percent of the spins to dephase (3 × the $T2$ time). Consequently, GRE sequences may be further divided into two categories according to how they handle the residual magnetization after the data acquisition: those that attempt to maintain it in a steady-state condition and those that eliminate it. Those that maintain it (e.g., refocussed FLASH (fast low angle shot) FISP (fast imaging with steady-state precession); or GRASS (gradient-recalled acquisition steady state)) rephase spins along one or more axes prior to reapplication of the next RF pulse.

GRE sequences that eliminate the residual transverse magnetization (e.g., spoiled FLASH or spoiled GRASS sequences) typically use a "spoiler" pulse to accelerate the dephasing (Fig. 9.20). Specifically, a high-amplitude, long-duration gradient ruins, or spoils, the residual transverse magnetization by disturbing the local magnetic field homogeneity. The best results occur when

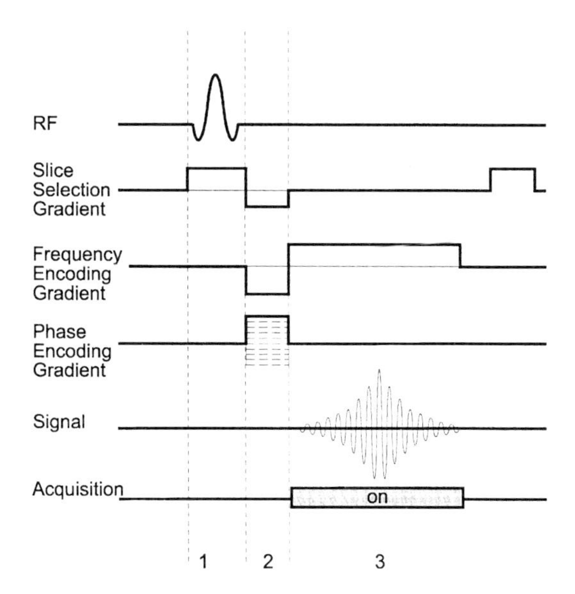

Figure 9.20. A complete pulse sequence diagram for a gradient echo sequence. Spatial locations of the spins are encoded into the signal by applying three orthogonal gradients. The sequence shown here is repeated numerous times (128, 256, 512, etc. depending on the desired resolution) each time with a different strength of the phase encoding gradient. The gradient echo sequence has an advantage over the spin echo sequence for fast imaging in that it does not use a 180° pulse, and since it does not rely on a 90° excitation pulse.

the spoiler gradient is applied across the slice-selection direction. Other spoiling schemes include the use of random RF pulse phases (RF spoiling), and variable *TR*. See reference 7 for a more thorough review of these and other fast gradient echo sequences.

9.2.3.2 Gradient-Echo Image Contrast

In SE imaging, tissue contrast may be manipulated by changes in the *TR* and *TE*, as described above. With GRE sequences the image contrast is varied by changing *TR*, *TE* and the flip angle (α), depending on the pulse sequence[8]. The amount of *T*2-weighting is dictated by the *TE*, *TR* and type of sequence. A short *TE*, long *TR* and transverse-spoiled sequence all serve to decrease the degree of *T*2-weighting. Low flip angles increase proton density weighting, while high flip angles increase *T*1 weighting all at a given *TE*. At very short *TR*'s however, the images become weighted toward *T*2/*T*1., i.e., structures with larger *T*2/*T*1 ratios (e.g.. liquids) appear bright. Yet, with very fast GRE sequences ($TR \sim 3$ ms) and $\alpha < 5°$, soft tissue contrast almost disappears[9]. The signal becomes dominated by spin density. However, if conventional MR experiments are placed before the whole GRE imaging sequence, images of any desired contrast can be achieved without changing the measuring time. The turbo-FLASH imaging technique is one such technique that implements this idea.

The turbo-FLASH method employs an initial 180° RF pulse to invert the spins. Next, an inversion delay (*TI*) is allowed to elapse, during which differences in longitudinal magnetization (*T*1 contrast) evolve depending on the *T*1 relaxation times of various tissues. Finally a very rapid gradient-echo acquisition using an ultrashort *TR* (e.g., 4 msec) and an ultrashort *TE* (e.g., 2 msec) is performed. The total time for data acquisition (32 phase-encoding steps) is on the order of 100 msec. When using a contrast agent, an appropriate *TI* value can be selected so that signal from the tissue that does not receive contrast agent is eliminated so wash-in of a contrast agent can be easily visualized[10].

9.2.3.3 Gradient-Echo Timing

Like the spin-echo sequence, the GRE sequence is repeated at time intervals equal to *TR*, with the total time required to obtain an image slice being $TR \times N_y \times$ NEX. However, because the *TR*s used in GRE imaging are typically very much shorter ($\approx$ 10 ms) than those used in SE imaging ($\approx$ 1 sec), GRE images can be acquired in seconds rather than minutes. For example, for an image matrix of 128 × 128, a $TR = 10$ ms, and 2 averages the total image acquisition time is 2.56 seconds. However, because the often-used very short *TR*s preclude an interleaved multislice acquisition, as discussed for spin-echo imaging, there simply is not enough time within *TR* for excitation/detection in other slices. Therefore the total acquisition time for multi-slice imaging is defined by the following product:

$$\text{Number of slices} \times \text{Number of views per slice} \times \text{NEX} \times TR \tag{9.19}$$

9.2.3.4 Echo Planar Imaging Sequences

Echo planar imaging (EPI) is significantly different from standard two-dimensional Fourier transform (2DFT) imaging methods. With 2DFT methods, only one projection (or line in k-space) is acquired with each *TR* interval, so that the image acquisition time is relatively lengthy. In contrast, the EPI method acquires k-space lines needed to create an image after a single RF excitation (hence, one "plane" is acquired with one RF excitation and subsequent "echo"). First, as in a 2DFT SE sequence, a spin echo is produced by application of a 90° and 180° RF pulse, with the echo peaking at the echo time (*TE*). However, rather than apply a single phase-encoding gradient and a constant frequency-encoding gradient, the frequency-encoding gradient is rapidly oscillated during the build-up and decay of the spin-echo. A series of gradient echoes are thereby produced, each of which is separately phase-encoded by application of a very brief phase-encoding gradient pulse. Because all of the data are acquired after a single rf pulse, the images are free from *T*1 weighting and can be strongly *T*2 weighted, with the degree of *T*2 weighting dependent on the value of *TE*.

In addition to spin-echo EPI images, it is possible to obtain gradient-echo EPI images. The acquisition method is similar to that for spin-echo EPI, except that the series of separately phase-encoded gradient echoes are acquired under the envelope of a gradient-echo signal produced by a single RF pulse.

The measuring time of EPI methods lies between 32 and 128 ms. EPI requires special hardware to allow for rapid gradient switching whereas gradient-echo techniques can be readily implemented on standard imaging systems. EPI sequences will be described in considerably more detail in Section 9.3.

9.2.3.5 Other Factors Affecting The MR Image

There are several additional intrinsic and extrinsic factors that influence the MR image. A few of the more common factors follow.

Diffusion, macroscopic flow

Water molecules, which make up approximately 70% of the body, are in constant random motion called diffusion. Diffusion sensitivity on MR images can be obtained by applying a pair of matched gradients[11]. Consequently, the amount of signal loss caused by diffusion-sensitizing gradients increases with the amount of molecular motion. Protons with slow diffusion will show little change in signal intensity, those with high diffusion will show more significant signal decreases, and bulk flow shows the most signal attenuation. The strength and duration of the diffusion-sensitizing gradients tends to be greater than those used for standard imaging sequences. Unfortunately, the diffusion gradients also make the sequence sensitive to any motion, resulting in artifacts as well as an overestimation of the apparent diffusion coefficient. However, with ultrafast and EPI diffusion sequences excellent results in quantification of diffusion despite motion and pulsation artifacts can be obtained.

Flowing blood can result in both increased and decreased signal intensities. Decreased signal intensity, often called a flow void, will result with high velocity, turbulence or dephasing while increased signal will result with an approach termed even-echo rephasing, flow-related enhancement and diastolic pseudogating As an example of decreases in signal intensity consider imaging high-velocity flow with a spin-echo sequence. For maximum signal, protons must experience both the 90° slice-selective pulse and the 180° refocussing pulse. Protons that acquire the 90° pulse and leave the section before acquiring the 180° pulse emit no signal, resulting in a flow void. Similarly, protons flowing into the section following selective 90° pulse also emit no signal. The magnitude of the signal loss therefore depends on the flow velocity, slice thickness and echo time. Conversely, as an example of flow-related signal increases, when using a multi-echo spin-echo sequence with steady laminar flow, dephasing due to flow seen at the first echo can be reconstituted on the second echo. These effects are covered in more detail elsewhere[12–15].

Susceptibility
Magnetic susceptibility, the source of contrast in much of functional MRI, represents the tendency of a substance to become magnetized. The susceptibility is primarily determined by the magnetic properties of the electrons, which have magnetic moments 1000 times greater than protons. There are several types of magnetic susceptibility: diamagnetic, paramagnetic, superparamagnetic and ferromagnetic. *Diamagnetic* substances, which contain paired electrons only, weakly repel the main magnetic field. Although most tissues are diamagnetic, changes in their signal intensity due to this factor are overwhelmed by much larger effects from other sources, such as relaxation parameters. *Paramagnetic* substances, which contain unpaired electrons, align with the magnetic field. Paramagnetic agents have received the most attention as useful contrast agents. (Contrast agents are exogenous agents which alter the natural tissue contrast.) In addition, oxygenated blood, which is diamagnetic, becomes paramagnetic upon deoxygenation. This endogenous contrast mechanism underlies the signal changes used to detect functional brain activation, as described in more detail below. *Superparamagnetic* substances more strongly align with the magnetic field. They, therefore, have more potent magnetic effects than do paramagnetic substances. An example of a naturally occurring superparamagnetic substance is hemosiderin. In addition, exogenously administered iron oxide contrast agents, which are also superparamagnetic, are currently coming into greater use in MR imaging[16]. Finally *ferromagnetic* substances remain permanently magnetized after being removed form a magnetic field. These substances include a number of iron and cobalt-containing metal alloys. Like superparamagnetic agents, ferromagnetic agents align strongly with the magnetic field.

Variations in magnetic susceptibilities within a voxel produce local inhomogeneities in the magnetic field. These inhomogeneities produce dephasing, which in turn results in signal loss and image distortion[17]. Signal loss also occurs at the border between two regions with differing magnetic susceptibilities, such as between tissue and air-containing sinuses. Although susceptibility differences can be a source of artifacts, they are also useful in the imaging of brain activation changes, which rely on the susceptibility effects of deoxygenated blood, as described below.

Contrast Agents
Exogenous substances which alter natural tissue contrast are contrast agents. In MRI, contrast agents are used to enhance image contrast between normal and diseased tissue and/or indicate the status of organ function or blood flow. There are several types of contrast agents in clinical use or under development. These include *T*1-active agents (agents that primarily shorten *T*1), *T*2-agents, which predominantly shorten *T*2, and non proton agents that contain no hydrogen. Paramagnetic agents have received the most attention as useful contrast agents. These agents enhance both *T*1 and *T*2 decay, with a predominant effect on *T*1 at low doses, and *T*2 or *T*2* at high doses when it is also compartmentalized. The magnitude of the changes in relaxation times is influenced primarily by the magnetic field strength and concentration of the paramagnetic agent.

Gadolinium agents such as gadolinium diethylenetriamine pentaacetic acid [Gd-DTPA] (*Magnevist*, Berlex Laboratories, Wayne, New Jersey), which are paramagnetic, are the most widely used clinical MR contrast agent. More recently, superparamagnetic contrast agents, which primarily affect *T*2 decay, are coming into greater use in MR imaging[16].

Shortening of both *T*1 and *T*2 by paramagnetics creates very complex changes in the MRI signal, which is dependent on the chosen RF pulse sequence. In addition, unlike electron-absorbing contrast agents used in nuclear medicine, the contrast agents used in MRI are not directly imaged. Rather, it is their indirect effect on NMR relaxation rates that is detected in the images as MRI signal intensity changes. As a result the rate of motion of water within and between tissue compartments can have significant effects on the resulting image contrast and accordingly the accuracy in quantification of tissue parameters using contrast agent[18].

9.3 Functional MRI

The human brain is likely the most complex and least understood system known. The understanding of its workings is a naturally inspiring goal, and the development of new methods to further this understanding is fundamental to the pursuit of this goal.

New methods for understanding human brain function can be extensively applied. Clinically, these methods can allow for faster, cheaper, and more effective diagnoses and treatments of neurological, cognitive, or neurophysiologic pathologies. In neuroscience research these can complement and add to the vast current efforts towards understanding the human brain – ranging from molecular to systems levels. Imaging of the healthy human brain during learning, reasoning, visualization, language, and creative functions may give insights into the dynamic structures of these emergent processes, therefore helping to uncover principles of cognition.

9.3.1 Brain Activation

Brain activation fundamentally consists of an increase in the rate at which action potentials are generated. The action potential, the unit of information in the brain, is a transient and cascading change in neuronal membrane polarity. At neuronal junctions, neurotransmitter synthesis, release, and uptake takes place, causing modulation of the action potential propagation[19–21]. When a population of neurons experiences these membrane polarity changes during activation, measurable electrical and magnetic changes in the brain are created[19–25]. Because of the energy requirements of membrane repolarization and neurotransmitter synthesis, brain activation also causes a measurable increase in neuronal metabolism[19–21, 26–31]. Through incompletely understood mechanisms[19, 32–43] these changes are accompanied by changes in blood flow[19–21, 32–49], volume[50–53], and oxygenation[52–56]. All techniques for assessing human brain function are based on the detection and measurement of these electrical, magnetic, metabolic, and hemodynamic changes that are spatially and temporally associated with neuronal activation.

The most recently developed brain activation imaging methods to emerge have been those which use magnetic resonance imaging (MRI). These MRI-based techniques have been collectively termed functional MRI (fMRI).

Use of fMRI has grown explosively since its inception[51, 57–60]. Among the reasons for this explosive growth are the non-invasiveness of fMRI, the wide availability of MR scanners capable of fMRI, and the relative robustness and reproducibility of fMRI results. With these reasons for using fMRI came a proportional need for caution. The technology can be easily misused and results can be over-interpreted. A solid understanding of the basics of fMRI is necessary. In this section, basic concepts behind of fMRI are clarified, several practical issues related to its use are discussed, and potential innovations regarding fMRI use are suggested.

This section of the chapter is organized into five parts. First, an introduction to magnetic susceptibility contrast is given. Second, the types of hemodynamic contrasts observable with fMRI are described. Third, ongoing issues of fMRI implementation are discussed. Fourth, several of the most common platforms for performing fMRI are described. Lastly, current fMRI applications are mentioned.

9.3.2 Magnetic Susceptibility Contrast

MRI emerged in the 1970's and 80's as a method by which high–resolution anatomical images of the human brain and other organs could be obtained non–invasively[1, 61–64]. The first types of image contrast used in MRI were proton density, spin–lattice relaxation ($T1$), and spin-spin relaxation ($T2$) contrast[65–69]. The large number degrees of freedom in MR parameter space has allowed MR contrast types to expand from physical to physiological[70]. The types of intrinsic MRI physiological contrast that have since been discovered and developed have included blood flow[70–73], diffusion[11, 70, 74–77], perfusion[70, 75–84], and magnetization transfer[70, 85, 86]. Chemical shift imaging has been able to provide information about relative concentrations and distributions of several chemical species[70, 87, 88].

The effects of endogenous and exogenous paramagnetic materials and, more generally, of materials having different susceptibilities, have also been characterized. An understanding of susceptibility contrast is an essential prerequisite to the exploration of fMRI contrast mechanisms.

Magnetic susceptibility, χ, is the proportionality constant between the strength of the applied magnetic field and the resultant magnetization established within the material[89]. In most biologic materials, the paired electron spins interact weakly with the externally applied magnetic field, resulting in a small induced magnetization, oriented opposite to the applied magnetic field, that causes a reduction of field strength inside the material. These materials are diamagnetic, and have a negative magnetic susceptibility.

In materials with unpaired spins, the electron magnetic dipoles tend to align parallel to the applied field. If the unpaired spins are in sufficient concentration, this effect will dominate, causing the induced magnetization to be aligned parallel with the applied field, therefore causing an increase in magnetic field strength inside the material. These materials are paramagnetic. Figure 9.21 is an illustration of magnetic field flux through diamagnetic and paramagnetic materials.

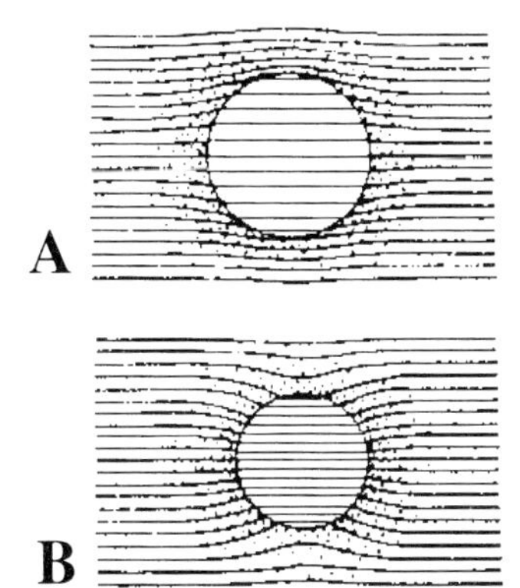

Figure 9.21.. Illustration of the magnetic field flux through **A** diamagnetic and **B** paramagnetic materials. Within diamagnetic materials, the net flux is less. Within paramagnetic materials, the net flux is greater. Magnetic field distortions created around the material are propotional to the object geometry and the difference in susceptibility between the object and its surroundings.

As mentioned earlier in this chapter, because of the Larmor relationship, spins will process at a faster frequency when experiencing a higher magnetic field. In the presence of a magnetic field perturber having a different susceptibility from surrounding tissue, spins will process at different frequencies, depending on their location, relative to the perturber. In such a situation, the spins will rapidly become out of phase and the MRI signal will resultingly be decreased. When the susceptibility differences between the perturber and its surroundings are large, the field distortions are large. Correspondingly, when the susceptibility of the perturber becomes more similar to its surroundings, the field distortions decrease, therefore causing more protons to have similar precession frequencies – allowing them to stay in phase longer. Increased phase coherence increases the MRI signal by decreasing the $T2^*$ and $T2$ decay rate. As an example, Figure 9.22 shows two plots of MRI signal intensity, using the simplified gradient-echo signal intensity relationship $S(TE) = S_o\, e^{-TE/T_{2^*}}$, where $S\,(TE)$ is the signal as a function of echo time (TE). $T2$ is the signal decay rate. Here, the two $T2^*$ values used are 48 and 50 ms. $R2^* = 1/T2^*$.

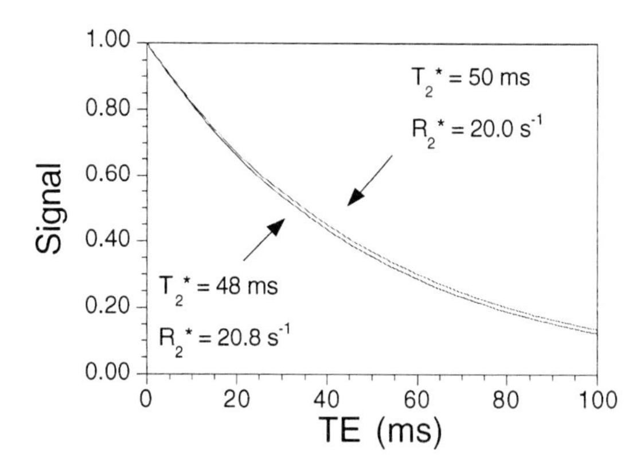

Figure 9.22. Plot of Signal vs. TE. The two curves represent typical values of $R2^*$ in the brain. The difference in relaxation rates represent typical differences between resting (20.8 s^{-1}) and activated (20.0 s^{-1}) $R2^*$ in the brain (–0.8 s^{-1}). These signals are referred to as *Sr* (resting signal) and *Sa* (active signal) in the discussion below. MR signal, in general, is *S*.

Considering that:

$$Ln(S) = TE/T2^* \tag{9.20}$$

$$Ln(S)/TE = 1/T2^* = R2^* \tag{9.21}$$

$R2^*$ may be obtained by the slope of $Ln(S)$ vs. TE, as shown in Figure 9.23.

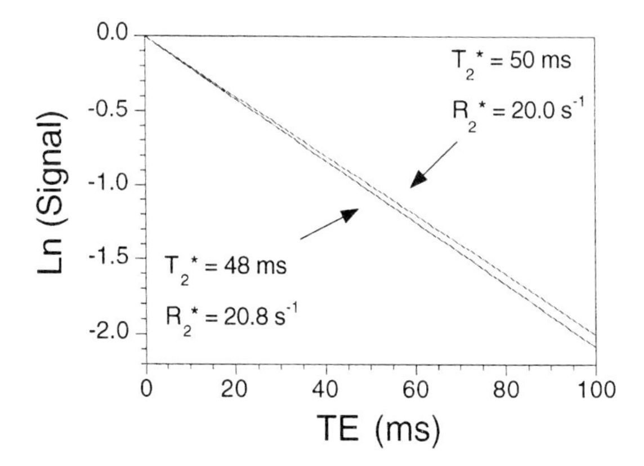

Figure 9.23. *Ln(S)* vs. *TE*. Transverse relaxation rates (*R*2 and *R*2*) are measured by applying a linear fit to curves such as these. Here, activation–induced changes in So are considered zero and single exponential decays are assumed.

Assuming that signal changes are affected by changes *only* in *R*2*, the change in relaxation rate, Δ*R*2*, may be estimated by measurement of *Sr* and *Sa* at single *TE* values and using the expression:

$$-Ln(Sa/Sr)/TE = \Delta R2^* \tag{9.22}$$

The expression relating percent change to Δ*R*2* is:

$$\text{percent signal change} = 100\,(e^{-\Delta R2^*\,TE} - 1) \tag{9.23}$$

Figure 9.24 is a plot of the percent signal change vs. *TE* between the synthesized resting and activated curves. An approximately linear fractional signal increase with *TE* is demonstrated.

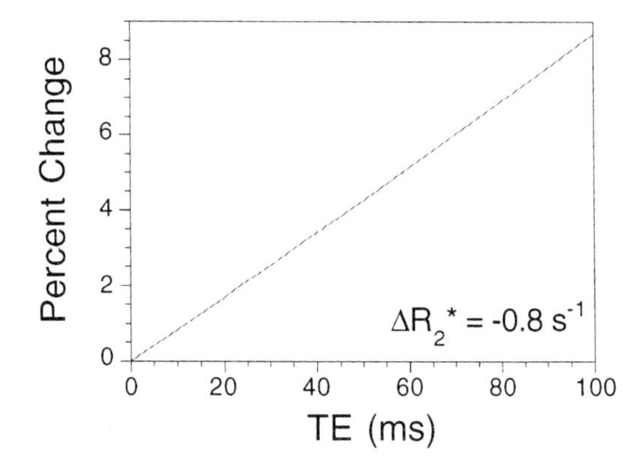

Figure 9.24. Percent change vs. *TE* from the same synthesized data set as shown above. Given a Δ*R*2* value typically obtained, a linear dependence of percent change on *TE* is observed in the *TE* range typically used.

If $\Delta R2^*$ is small relative to $R2^*$, the signal difference between the two curves will be maximized at $TE \sim T2^*$ (gradient–echo) or $T2$ (spin–echo), as demonstrated below. Contrast between two signal intensities (Sa and Sr), having a difference in relaxation rate equal to $\Delta R2^*$, can be approximated by:

$$Sa - Sr = e^{-TE\,(\,\Delta R2^* \,+\, R2^*r\,)} - e^{-TE\,(R2^*r)} \tag{9.24}$$

where $R2^*r$ is the relaxation rate associated with a measured Sr at a given TE value. The TE value at which Eq. (9.25) is maximized is given by:

$$TE = Ln(\,(\,\Delta R2^* + R2^*r\,)/R2^*r\,)/\Delta R2^* \tag{9.25}$$

In the limit that $\Delta R2^*$ approaches 0, the TE value at which contrast is maximized approaches $1/R2^*r$ or $T2^*r$. A graphical demonstration of this contrast maximization is shown in Figure 9.25 Even though the percent change increases, as shown in Figure 9.24, the contrast or signal *difference* does not increase monotonically with TE.

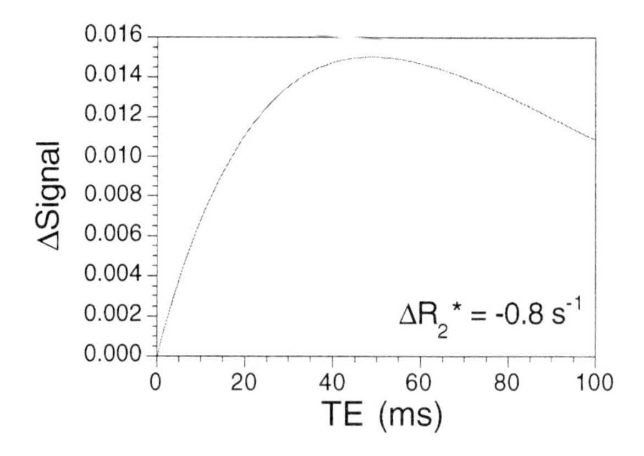

Figure 9.25. Plot of ΔS vs. TE from the same synthesized data sets as shown in the previous figures. A maximum is reached at $TE \approx T2^*r$ ($\approx$ 48 ms).

Bulk susceptibility changes (either endogenous or exogenous) lead to MRI signal changes primarily in the manner described above. A more detailed description of the precise effects of susceptibility perturbers will be provided later in the chapter.

In a typical 70-kg human body, paramagnetic materials include iron (3–5 g), copper (70–120 mg), manganese (12 mg), nickel (10 mg), chromium (2 mg), cobalt (0.3 mg), vanadium (2 mg), molybdenum (13 mg), and tungsten (trace). Iron is approximately 30 times more abundant than all the other transition elements in a typical human body. Much of the iron is contained in red blood cells. In biological tissues, additional sources of paramagnetism include materials, which because of peculiarities in their chemical bonding, have unpaired spins. These include oxygen, O_2, and nitric oxide, NO[89].

9.3.2.1 Endogenous Susceptibility Contrast

One of the three fMRI contrast mechanisms described in this section, (blood oxygenation level dependent contrast: BOLD), is based on the understanding that blood has oxygenation–sensitive paramagnetic characteristics[89–92]. Hemoglobin is the primary carrier of oxygen in the blood. Hemoglobin that is not bound to oxygen, called deoxyhemoglobin (deoxy-Hb), contains paramagnetic iron, while hemoglobin that is carrying oxygen, called oxyhemoglobin (oxy-Hb), contains diamagnetic oxygen-bound iron[89–92]. The modulation in the magnetic susceptibility of blood by oxygenation changes is the basis of BOLD contrast. Using MR susceptometry[93], the susceptibility of completely oxygenated red blood cells was measured to be $-0.26 \pm 0.07 \times 10^{-6}$ (cgs units). With this technique, blood susceptibility was also shown to be linearly proportional to blood oxygenation (it decreases linearly as oxygenation increases). The susceptibility of completely deoxygenated red blood cells is $0.157 \pm 0.07 \times 10^{-6}$. The susceptibility difference between completely oxygenated and completely deoxygenated red blood cells is therefore 0.18×10^{-6}. The profound effects of blood oxygenation changes on MR signal intensity have been demonstrated since 1981[91–102]. The precise mechanisms for this effect on MR signal will be described later in the chapter.

9.3.2.2 Exogenous Susceptibility Contrast

Exogenous paramagnetic substances, which include Gd(DTPA) and Dy(DTPA) can give useful information regarding several aspects of organ function[103]. In the brain, these intravascuar agents, when injected, can give information on blood volume and vascular patency[93, 102–109]. The effects of these agents on tissue $T1$, $T2^*$, and $T2$, are highly dependent on chemical environment and compartmentalization, as has been observed[93, 102–109] and modeled[103, 105–108, 110–123].

One mechanism of action for these compounds is dipolar interaction, having an effect on intrinsic $T1$ and $T2$ relaxation times[103, 105]. This effect relies on the direct interaction of water with unpaired spins. Homogeneous distributions of solutions containing paramagnetic ions display relaxivity changes that can be predicted by the classical Solomon–Bloembergen equations[105], but in the healthy brain, these agents, upon injection, remain compartmentalized within the intravascular space, which contains only about 5% of total brain water. The extent of agent–proton interaction is reduced by the limited rate at which diffusing or exchanging protons in the other 95% of brain water pass through the intravascular space, which is also less accessible due to the blood brain barrier. These combined effects greatly limit the agent–induced

$T1$ effects, which rely on direct interaction of protons with the paramagnetic agents. In this case, $T2^*$ and $T2$ shortening effects, caused by contrast agent induced bulk susceptibility differences between intravascular and extravascular space[105–108, 110–119] dominate over classical dipolar relaxation effects. A detailed description of $T2^*$ and $T2$ shortening effects is given below.

9.3.2.3 Exchange Regimes

The effect on transverse relaxation by magnetic field inhomogeneities can be characterized by[102, 106–108, 110, 111]:

$$1/T2^* = 1/T2 + 1/T2' \tag{9.26}$$

The relaxation rate, $1/T2^*$, also termed $R2^*$, is the rate of free-induction decay, or the rate at which the gradient-echo amplitude decays. The relaxation rate, $1/T2$, also termed $R2$, is the rate at which the spin-echo amplitude decays. The relaxation rate, $1/T2'$, also termed $R2'$, is the water resonance linewidth, which is a measure of frequency distribution within a voxel. The key concept to understand is that the relative magnitude of $R2'$ is not only proportional to the susceptibility of the magnetic field perturber, but to the dimensions of the perturber relative to the local proton dynamics.

More specifically, in the presence of a magnetic field perturber, the relative $R2$ and $R2^*$ relaxation rates depend on: the diffusion coefficient (D) of spins in the vicinity of induced field inhomogeneities, the radius (R) of the field perturber, and the variation in the Larmor frequency at the surface of the perturber[102, 106-108, 110–113]. These two physical characteristics (R and D) can be collapsed into one term, the proton correlation time, τ, which can be described as:

$$\tau = R^2/D \tag{9.27}$$

The variation in the Larmor frequency ($d\omega$), at the perturber surface is:

$$d\omega = \gamma(\Delta\chi)B_o \tag{9.28}$$

where γ is the gyromagnetic ratio, $\Delta\chi$ is the susceptibility difference, and B_o is the strength of the applied magnetic field. Depending on the relative values of these variables, intravoxel dephasing effects are commonly described by three regimes, termed the fast, intermediate, and slow exchange regimes[102, 106–108, 110–113]. The exchange regimes are summarized in Table 9.1 and shown graphically in Figure 9.26.

In the fast exchange regime[110, 111, 113–119], the high diffusion rate causes all spins to experience a similar range of field inhomogeneities within an echo time, therefore causing a similar net phase shift of all spins, and a minimal

Table 9.1. Summary of the exchange regimes commonly referred to when the effects of magnetic field perturbations on transverse relaxation rates are described.

Slow exchange	$\tau(\delta\omega) >> 1$	$\Delta R2^* >> \Delta R2$ $\Delta R2' \approx \Delta R2^*$
Intermediate exchange	$\tau(\delta\omega) \approx 1$	$\Delta R2^* > \Delta R2$ $\Delta R2' < \Delta R2^*$
Fast exchange	$\tau(\delta\omega) << 1$	$\Delta R2^* \approx \Delta R2$ $\Delta R2' \approx 0$

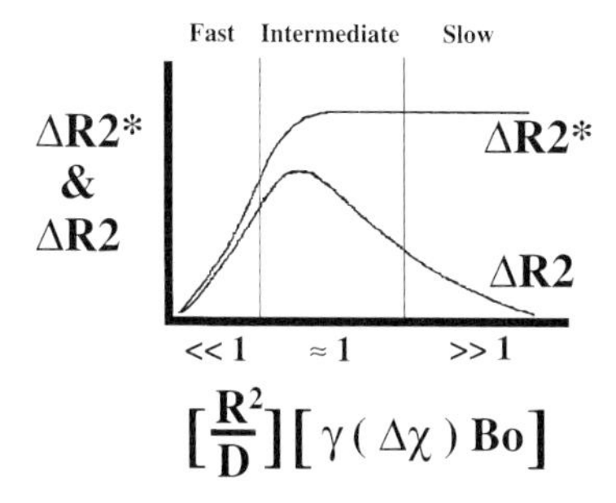

Figure 9.26. Plot of typical $\Delta R2^*$ and $\Delta R2$ due to susceptibility perturbations. The essential variables are the radius of the perturber (R), the proton diffusion coefficient (D), and the frequency shift caused by the perturber ($\gamma\chi B_o$). In the fast exchange regime, the ratio of $\Delta R2^*$ to $\Delta R2$ approaches 1. In the slow exchange regime, the ratio is greater than about 10.

loss of phase coherence as well as a similar loss of phase coherence between gradient–echo and spin–echo sequences. The fast exchange regime is relevant in two particular cases.

In the human brain, the dephasing experienced in the presence of susceptibility–induced gradients in the vicinity of capillaries and red blood cells has been described as being in the fast exchange regime[102, 107, 111].

In the slow exchange regime, the magnetic field experienced by any spin can be approximated as a linear gradient or, in the extreme case, an off-resonance static field. This exchange regime would apply to situations where magnetic field gradients are created at large interfaces of tissues having susceptibility differences (e.g. bone, air, tissue), or in the space surrounding large vessels or trabecular structure. Because of the large spatial scale of the frequency offset, spin diffusion distances in an echo time would be relatively insignificant .

The effects of off resonance effects near large interfaces of materials having different susceptibilities, have been characterized and imaged[17, 123–127]. Dephasing effects created by susceptibility–induced gradients in the vicinity of bone trabecular structure and generalized structures have been modeled[128–133], and experimentally studied[134, 135]. The $R2^*$ of bone was found to be proportional the trabecular density[134, 135]. A decrease in $R2^*$ with age and with osteoporosis was also demonstrated[134]. Given a change in trabecular density, the change in $R2$ was not perceptible, while the change in $R2^*$ was pronounced[135]. The gradients induced in the vicinity of the trabecular structure, which contribute to a large $R2'$ effect, are also likely to be the reason why bone is brighter in spin–echo images having similar parameters as gradient–echo images. All dephasing effects that occur are refocussed, when using a spin-echo sequence, by the 180° because no significant irreversible diffusion related dephasing occurs. These slow exchange regime effects have been also modeled in the context of susceptibility differences between lung tissue and air[136].

Because the assumption is made, in the slow exchange regime, that spins either diffuse through linear gradients or experience a shifted resonance frequency, analytic expressions for these effects on decay rate have been derived[112, 131]. The effects have also been simulated[110, 117, 118, 120–122, 132, 137]. The dependence of relaxation rate change upon induced frequency shift has been found to be linear[131].

In the intermediate exchange regime, diffusing spins neither completely experience nor minutely sample the complicated gradients induced by the field perturbers. Analytic solutions are difficult to derive due to the large spatial heterogeneity of the induced field gradients. Therefore numerical simulation methods are required. These have included Monte Carlo techniques[110, 117, 118, 120, 121], and deterministic methods[122, 138].

The dephasing effects of spins in the vicinity of much of the human microvasculature, which has vessels ranging in radius from 2.5 μm in capillaries to 100 μm in pial vessels[43, 139], have been described as being within the intermediate exchange regime[102, 106–108, 110, 117, 118, 120–122, 138].

9.3.3 Hemodynamic Contrast

Several types of cerebrovascular information can be mapped using MRI. The tomographic information that can be obtained include: a) maps of cerebral blood volume[51, 70, 106, 108, 140, 141] and cerebral perfusion[79, 83, 142–146], and b) maps of *changes* in blood volume[51], perfusion[58, 82, 83, 143–145, 147, 148], and oxygenation[57–60, 137, 149–155]. Below is a description of how these various hemodynamic properties are selectively detected using fMRI.

9.3.3.1 Blood Volume

A technique developed by Belliveau and Rosen et al.[106, 108, 140] utilizes the susceptibility contrast produced by intravascular paramagnetic contrast agents and the high speed imaging capabilities of echo planar imaging (EPI) to create maps of human cerebral blood volume (CBV). A bolus of paramagnetic contrast agent is injected (the technique is slightly invasive) and $T2$ or $T2^*$ – weighted images are obtained at the rate of about one image per second using echo-planar imaging (EPI) [83, 156–158]. As the contrast agent passes through the microvasculature, magnetic field distortions are produced. These gradients, which last the amount of time that it takes for the bolus to pass through the cerebral vasculature, cause intravoxel dephasing, resulting in a signal attenuation which is linearly proportional to the concentration of contrast agent[106, 108, 120], which, in turn is a function of blood volume.

Changes in blood volume that occur during hemodynamic stresses or during brain activation (≈ 30% change) can then be observed by subtraction of two maps: one created during a "resting" state and one created during a hemodynamic stress or neuronal activation[51]. The use of this method marked the first time that hemodynamic changes accompanying human brain activation were mapped with MRI.

9.3.3.2 Blood Perfusion

An array of new techniques now exist for mapping cerebral blood perfusion in humans. The MRI techniques are similar to those applied in other modalities such as positron emission tomography (PET) and single photon emission computed tomography (SPECT) in that they all involve arterial spin labeling. The MRI based techniques hold considerable promise of high spatial resolution without the requirement of contrast agent injections. They use the fundamental idea of magnetically tagging arterial blood outside the imaging plane, and then allowing flow of the tagged blood into the imaging plane. The RF tagging pulse is usually a 180° pulse that "inverts" the magnetization.

Generally, these techniques can be subdivided into those which use continuous arterial spin labeling, which involves continuously inverting blood flowing into the slice[142], and those which use pulsed arterial spin labeling, periodically inverting a block of arterial blood and measuring the arrival of that blood into the imaging slice. Examples these techniques are: 1) "echo planar imaging with signal targeting and alternating RF," (EPISTAR), schematically illustrated in Figure 9.27a which involves alternately inverting slabs of magnetization above and below the imaging slice[82, 83], and 2) "flow-sensitive alternating inversion recovery," (FAIR), schematically illustrated in Figure 9.27b, which involves the alternation between slice selective and non slice selective inversion. The latter was introduced by Kwong et al.[144, 148, 159] and referred to as FAIR by Kim et al.[143]. Recently, a pulsed arterial spin labeling technique known as "quantitative imaging of perfusion using a single subtraction," (QUIPSS), has been introduced[145, 146]. In the case of the pulsed techniques, pairwise subtraction of sequential images, illustrated in Figure 9.27c with and without the application of the RF tag outside the plane gives a perfusion related signal.

Variation of the delay time between the inversion or tag outside the imaging plane and the acquisition of the image gives perfusion maps highlighting blood at different stages of its delivery into the imaging slice. Because there is necessarily a gap between the proximal tagging region and the imaging slice, there is a delay in the time for tagged blood to reach the arterial tree, this delay time can be highly variable, ranging from about 200 ms to about 1 sec

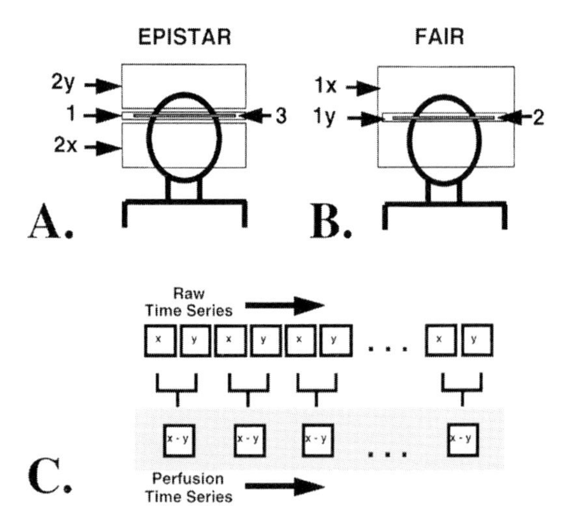

Figure 9.27. A. Schematic illustration of EPISTAR "echo planar imaging with signal targeting and alternating RF." First, the imaging slice is presaturated with a saturation pulse [1]. Second, protons above the imaging plane and below the imaging plane are alternately inverted or tagged [2x and 2y]. Third, the image is collected after a delay time, *TI*, to allow the tagged protons to perfuse into the imaging plane [3]. Alternate images collected in the sequential time series correspond to either the tag below [2x] or above [2y] the plane. **B.** Schematic illustration of FAIR "flow – sensitive alternating inversion recovery." First protons either within the plane or everywhere are alternately inverted or tagged [1x and 1y]. Second, the image is collected after a delay time, *TI*, to allow the tagged protons [1x] to perfuse in to the imaging plane. Alternate images collected in the sequential time series correspond to either the tag everywhere [1x] or only within [1y] the imaging plane. **C.** The method by which the time series of perfusion images is created from the pulse sequences shown in A. and B. The alternate images, x and y, are collected in time. These images, with different tags applied, are different only in the degree to which flowing spins contribute to the signal. Therefore, a perfusion-signal-only time series of images is created by pairwise subtraction of the images.

for a gap of 1 cm. At 400 ms, typically only blood in larger arteries has reached the slice and the pulsed arterial spin labeling signal is dominated by focal signals in these vessels, while at 1000 ms, tagged blood has typically begun to distribute into the capillary beds of the tissue in the slice. Images acquired at late inversion times can be considered qualitative maps of perfusion. Figure 9.28 shows perfusion maps created at different *TI* times using both the FAIR and the EPISTAR technique. As *TI* is lengthened, tagged blood distributes from large arteries into smaller vessels and capillary beds. In the capillaries, the tagged blood water exchanges almost completely with tissue water. To quantify perfusion using these techniques, it is necessary to more carefully model the phenomena and relevant variables[143, 144, 160, 161]. For quantification, a minimum of two subtractions at different *TI*'s are required in order to calculate the rate of entry of tagged blood into the slice (perfusion)[161].

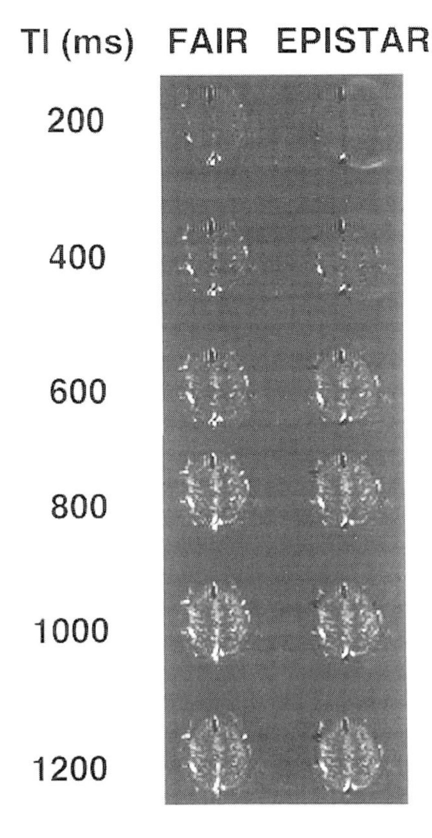

Figure 9.28. Comparison of EPISTAR and FAIR at corresponding *TI* values. As *TI* is lengthened, tagged blood distributes from large arteries into smaller vessels and capillary beds. In the capillaries, the tagged blood water exchanges almost completely with tissue water. Short *TI*'s highlite rapidly flowing blood, and the long *TI*'s highlight capillary bed perfusion.

For the application of mapping of human brain activation, (i.e. to only observe activation-induced *changes* in blood perfusion), a more commonly used flow sensitive method is performed by application of the inversion pulse always in the same plane. In this case, the intensity of all images obtained will be weighted by modulation of longitudinal magnetization by flowing blood and also by other MR parameters that normally contribute to image intensity and contrast (proton density, $T1$, $T2$). Therefore, this technique allows only for observation of *changes* in flow that occur over time with brain activation. A local perfusion change of 50% (typical with brain activation) would create a $T1$-weighted MRI signal change of approximately 2%. This technique was first implemented by Kwong et al.[58] to observe activation – induced flow changes in the human brain. In this seminal paper, activation – induced signal changes associated with local changes in blood oxygenation were also observed.

9.3.3.3 Blood Oxygenation

In 1990, pioneering work of Ogawa et al.[137, 151, 152] and Turner et al.[153] demonstrated that MR signal in the vicinity of vessels and in perfused brain tissue decreased with a decrease in blood oxygenation. This type of physiological contrast was coined “blood oxygenation level dependent” (BOLD) contrast by Ogawa et al.[152].

The use of BOLD contrast for the observation of brain activation was first demonstrated in August of 1991, at the 10′th Annual Society of Magnetic Resonance in Medicine meeting[162]. The first papers demonstrating the technique, published in July 1992, reported human brain activation in the primary visual cortex[58, 59] and motor cortex[57, 58]. Two[57, 58] of the first three reports of this technique involved the use of single shot EPI at 1.5 Tesla. The other[59] involved multishot “fast low angle shot” (FLASH) imaging at 4 Tesla. Generally, a small (2%) local signal increase in activated cortical regions was observed using gradient echo pulse sequences – which are maximally sensitive to changes in the homogeneity of the main magnetic field.

The working model constructed to explain these observations with susceptibility contrast imaging is that an increase in neuronal activity causes local vasodilatation which, in turn, causes an increase in blood flow (≈ 50%). This results in an excess of oxygenated hemoglobin beyond the metabolic need, thus *reducing* the proportion of paramagnetic deoxyhemoglobin in the vasculature. The oxygen saturation of venous hemoglobin is thought to change from 75% saturated to about 90% saturated. This hemodynamic phenomenon was previously suggested using non-MRI techniques[53–55]. A reduction in deoxyhemoglobin in the vasculature causes a reduction in magnetic susceptibility differences in the vicinity of veinuoles, veins and red blood cells within veins, thereby causing an increase in spin coherence (increase in $T2$ and $T2^*$), and therefore an increase in signal in $T2^*$ and/or $T2$ – weighted sequences.

Presently, the most widely used fMRI technique for the non-invasive mapping of human brain activity is gradient-echo imaging using BOLD contrast. The reasons for this are that a) gradient-echo $T2^*$ – sensitive techniques have demonstrated higher activation-induced signal change contrast, by about a factor of two to four, than $T2$ – weighed, flow-sensitive, or blood volume-sensitive techniques, and b) BOLD contrast can be obtained using more widely available high speed multi-shot non-EPI techniques. c) While $T2^*$ – weighted techniques are sensitive to blood oxygenation changes in vascular structures that include large vessels that may be spatially removed from the focus of activation, for most applications the sacrifice in functional contrast to noise ratio in techniques more sensitive to microvascular structures does not outweigh the necessity for the highest possible contrast to noise in functional images. This last issue will be discussed further below.

A summary of the cascade of hemodynamic events that occur on brain activation and of their effects on the appropriately weighted MRI signal are shown in Figure 9.29.

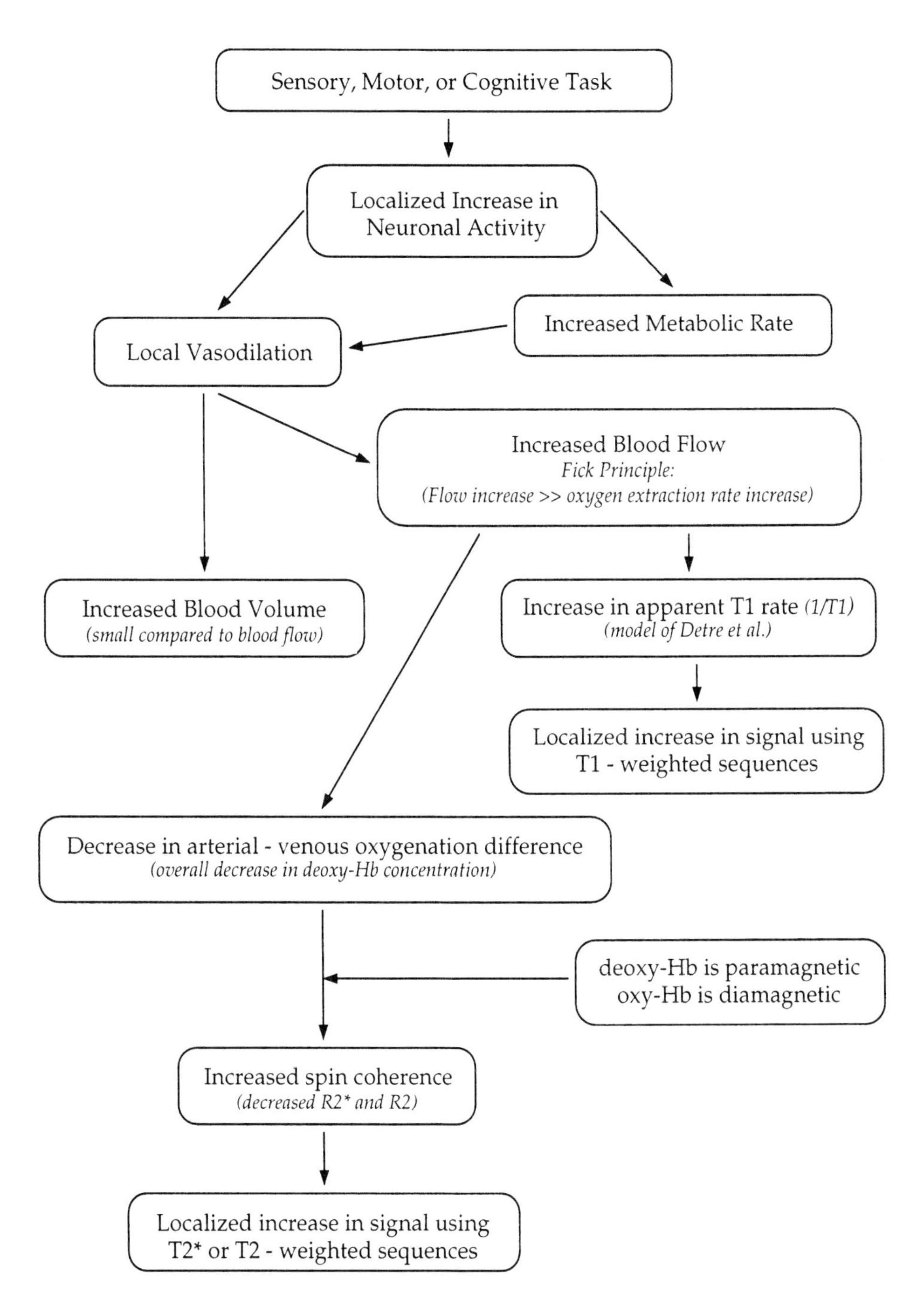

Figure 9.29. Flow chart summarizing the cascade of hemodynamic events that occur with brain activation and their corresponding effects on the appropriately sensitized MRI signal.

9.3.4 Issues in fMRI

Although progress is being rapidly made, many issues in fMRI remain incompletely understood. Below is a description of the current state of understanding regarding some general fMRI issues, categorized into: interpretability, temporal resolution, spatial resolution, dynamic range, sensitivity, and some unknowns.

9.3.4.1 Interpretability

The question of interpretability regards the concern of exactly what the relationship is between the fMRI signal and underlying neuronal activation. Two "filters" separate direct observation of neuronal processes using fMRI. The first is the relationship between neuronal activation and hemodynamic changes, and the second is the relationship between hemodynamic changes and MR signal changes.

In the past five years, considerable progress has been made in the characterization of the second relationship: that between activation – induced hemodynamic changes and the fMRI signal changes. Below the issue of MRI – achievable hemodynamic specificity is discussed. Also discussed are the upper limits of temporal and spatial resolution, and the dynamic range of fMRI.

A high priority in fMRI is to accurately correlate activation – induced MR signal changes with underlying neuronal processes. It is generally accepted that perfusion and oxygenation changes in capillaries are closer in both space and time to neuronal activation than those arising from arteries or veins As mentioned, different pulse sequences can be made sensitive to specific populations of vessel sizes, blood flow velocities, and contrast mechanisms.

The fMRI pulse sequence that gives the highest functional contrast to noise ratio is a $T2^*$-weighted gradient-echo sequence, which is likely to have contrast weighting which includes large draining vein effects and, in the case of short TR – high flip angle sequences (short TR values are required for non-EPI fMRI sequences), large vessel arterial inflow effects. Sequences that may be able to more selectively observe capillary oxygenation or perfusion effects are less robust. They have a lower functional contrast to noise ratio, are generally less time efficient, and may not allow extensive multislice imaging. The tremendous need for high fMRI contrast to noise ratio, high image acquisition speed, and high flexibility such as multi-slice imaging has to date outweighed the need, in most cases, for selective observation of capillary effects for most applications. Enhancements in fMRI sensitivity may allow these hemodynamically selective pulse sequences to be more commonly used. The strategies for achieving hemodynamic specificity not only include pulse sequence modifications but also simple vein and artery identification strategies or even activation strategies which remove draining vein effects. Below, several of the more common pulse sequences and paradigm strategies for obtaining higher hemodynamic specificity are listed in alphabetical order and described. These methods can be considered as relevant to the goals summarized in Table 9.2.

Table 9.2. Goals regarding the achievement of hemodynamic specificity in fMRI and corresponding methods that have been proposed in the literature. The method numbers correspond to the methods listed below.

Hemodynamic Specifity	
Goal	Method Number
Separation of flow and oxygenation effects	flow: 6,15, oxygenation: 11, 10, 14 both: 13, 18
Identification of large arteries and veins	veins: 4, 5, 7, 11, 14 arteries and veins: 1, 16, 19
Reducation of large artery or vein effects	veins: 2, 4, 8, 17 arteries and veins: 9, 10
Selective imaging of capillary effects	flow: 13, combination of 3 and 13. oxygenation: 12, combination of 3 and 17

1. Angiography:[73, 163] Use of standard high resolution angiographic techniques can identify rapidly flowing blood. *Advantages*: It can be performed relatively quickly and independently of the functional imaging series. *Disadvantages*: Blood in larger arteries are visualized but slowly flowing venous blood may be missed.

2. Asymmetric spin-echo:[164] This technique involves the use of a spin-echo, but with the readout window shifted from the spin-echo center (asymmetrically located) so that similar susceptibility ($T2'$) weighting as a gradient echo sequence is achieved. *Advantages*: Rapidly flowing blood does not experience the 180° pulse, applied about 40 ms after the 90° pulse, therefore does not contribute to the signal. This phenomenon also reduces some of the pulsatile fluctuations over time. *Disadvantages*: The use of a spin-echo increases imaging time by about 100 ms, which may limit the number of slices (in space) obtained in a *TR* using EPI. This time cost for non-EPI sequences (with the possible exception of fast spin echo[165]) is practically prohibitive. This sequence is also as equally sensitive as regular gradient-echo sequences to intravascular effects ($T2^*$ dephasing of the blood) from large vessels that have slowly flowing spins and to extravascular effects (spin dephasing that occurs outside of the veins as a result of magnetic field gradients extending from the vessels due to the difference in magnetic susceptibility between the vessels and tissue) of large vessels with intravascular signal that has been removed by the 180° pulse.

3. Diffusion weighting:[166, 167] This technique incorporates additional magnetic field gradients between RF excitation and data acquisition to selectively dephase signal from faster moving populations of spins. Blood having rapid incoherent motion (i.e. blood in larger vessels) within a voxel is dephased, and therefore removed from contributing to the fMRI signal change. *Advantages*: Intravascular large vessel effects that cannot be seen using other techniques (possibly because they may be subvoxel in size) are reduced with this technique. *Disadvantages*: The addition of diffusion weighting reduces the image signal to noise ratio and the functional contrast to noise ratio, and increases the motion sensitivity over time. This technique can only realistically be performed using EPI. Also, while large vessel intravascular effects are not eliminated. Lastly, large vessel extravascular dephasing effects ($T2^*$ contrast) are unaffected and therefore may still contribute to fMRI signal changes.

4. High field strength:[154, 163, 168, 169] In the context of fMRI, a field strength above 2 Tesla is considered high. *Advantages*: Signal to noise theoretically increases linearly with field strength. BOLD based functional contrast may increase from linearly or sublinearly[170] to almost quadradically[121, 154]. Because $T1$ relaxation rates become longer at high field strengths, flow imaging techniques [82, 83, 142–146, 148, 160] also benefit because of decreased decay of the tag signal. Higher field strength also allows detection of more subtle effects, higher spatial resolution, and/or less need for averaging over time. Also the $T2^*$ difference between deoxygenated blood and gray matter becomes greater allowing clear identification of veins as dark spots in high resolution $T2^*$ – weighted images[163, 171]. *Disadvantages*: High field magnets do not have a large market, therefore are not as tried and true as lower field clinical workhorses. (i.e. More troubleshooting is needed.) The primary practical problem at high fields is the increased field distortion due to magnetic susceptibility effects. This field distortion causes both image distortion and signal dropouts, but because, on a microscopic scale, it is also the mechanism of BOLD contrast, techniques that are sensitive to BOLD contrast are inherently sensitive to these other deleterious effects. These problems make magnetic field shimming more important at high fields. Because the field distortions can only be partially removed by shimming, they often preclude whole brain imaging, and imaging of structures at the base of the brain. Lastly, physiological fluctuations may increase with field strength, which, if not filtered, can increase the noise and nullify the inherent signal to noise advantages of high fields. Alternatively, an increase in physiological fluctuations may translate to an advantage if the fluctuations prove to contain useful physiologic or neuronal information.

5. Hypercapnia normalization:[172] Since the fractional signal change using BOLD contrast is highly weighted by the distribution of blood volume across voxels, a uniform oxygenation increase, concomitant with a hypercapnia – induced flow increase, would cause the BOLD signal increase in each voxel that is in proportion to underlying hemodynamic variables, and primarily venous blood volume. Maps of venous blood volume distribution can be made in this manner. Assuming that hypercapnia and activation cause similar hemodynamic events[58, 102]; one global and the other localized to neuronal activation, then division of a "percent change during brain activation" image by a "percent change during hypercapnia" image would give a ratio map of task-induced signal activation which is normalized to the signal change accompanying global vasodilatation. *Advantages*: This technique has the potential for normalizing for all hemodynamic variations over space that can modulate the signal given a constant oxygenation change – and not just remove large vessel effects. *Disadvantages*: Division of percent change images obtained in different imaging runs reduces the signal to noise significantly and is also highly sensitive to systematic variations over time. Also, giving a hypercapnic stress before or after every fMRI study is impractical from a time, convenience, and safety viewpoint.

6. Inversion recovery:[58] As described above, an inversion-recovery sequence allows maximum sensitivity to activation – induced perfusion related $T1$ changes. Used with minimally $T2$ or $T2^*$ sensitive imaging (i.e. short TE spin-echo acquisition), exclusive sensitivity to perfusion is achieved. *Advantages*: Used with minimally $T2$ or $T2^*$ sensitive imaging (i.e. short TE spin-echo acquisition), exclusive sensitivity to flow changes is achieved. *Disadvantages*: This technique can only be practically used with EPI because the waiting period (TI) is too long for standard multishot fMRI techniques. Also, it has lower sensitivity to functional changes than gradient-echo sequences.

7. Latency mapping:[173] It is thought that, on activation, larger vessels "downstream" from the activated region become oxygenated at a slightly later time than capillaries or veinuoles. This technique uses this vessel size-specific BOLD contrast latency to identify draining veins. *Advantages*: This technique can be applied in a post hoc manner, and is relatively easy to implement. *Disadvantages*: Because of functional contrast to noise limitations, latency differences on the order of one second require significant averaging to be differentiated. The latency differences between large veins and capillaries may vary, and, in many cases be less than 1 or 2 seconds, therefore making the technique somewhat unreliable. Also, while unlikely, it is possible that, some neuronal processes may have latency differences (or hemodynamically transmitted latency differences) on the order of a second[174], therefore confounding the technique.

8. Latency tagging:[175–178] This is useful for high resolution mapping of subtle spatial differences in the hemodynamic response as the cortical representation of the stimulus is continuously varied in time. This technique lends itself to high resolution mapping of contiguous cortical regions. *Advantages*: Large vessel effects may be reduced since the stimulus is continuously "on" but spatially modulated. Large vessels, receiving flow from a relatively large area, will be in a steadily more oxygen-saturated state. The "spillover" of oxygenated blood is constant, therefore allowing a higher functional spatial resolution by having all the "spillover" effects subtracted out. The highest fMRI "functional" resolution reported has been with the use of this technique[177]. The functional contrast per unit time is optimized because the entire time course has information embedded within it. *Disadvantages*: This technique does not lend itself to the mapping of regions in which a continuous variation in the stimuli does not cause a continuous variation in the cortical regions activated. (i.e. those cortical representations of a time varying stimuli that do not vary continuously over space)

9. Long TR (high flip angle) or Short TR (low flip angle):[179, 180] This is a method by which arterial inflow effects are minimized. Differences in steady – state magnetization between the imaging plane and outside of the imaging plane are minimized. Effects elicited by changes in activation – induced inflow (activation causes fresh un-RF-saturated spins to enter the imaging plane at a higher rate) are reduced. *Advantages*: These techniques are simple to implement and well understood. *Disadvantages*: The long *TR* technique ($TR > 1$ sec) can only be practically achieved using EPI. Multi-shot techniques generally need use a short *TR* to collect images in a practically feasible time. If a short *TR* is necessary, reduction of the flip angle below the Ernst angle is sub optimal from a signal to noise standpoint.

10. Outer volume saturation:[181–183] This technique is similar, in principle, as technique #9, but instead of the difference in plane – out of plane magnetization being decreased by an increase in the in plane magnetization, the out of plane magnetization is reduced. This technique reduces signal not only from inflowing arterial spins but also inflowing large venous vessel spins. Therefore only smaller (slower flowing) vein intravascular BOLD effects and large (rapidly flowing) vein extravascular BOLD effects are observed. *Advantages*: Implementation is straightforward. *Disadvantages*: The saturation slice profile may interfere with the signal from the slices of interest. Rapidly flowing blood arriving from outside of the saturation plane remains unaffected.

11. Phase shift mapping[184–186] If a single vein having a single orientation is located within a voxel, then, during a change in oxygenation, the resonant frequency within that vessel will change, causing a coherent phase shift within the voxel, depending on the *TE*. These phase shift effects are not present in voxels containing only randomly oriented capillaries. Visualization of resting state phase shifts or phase dispersions and activation – induced phase shifts can be used to identify large vessel effects. *Advantages*: This technique is easy to implement. NMR phase images simply need to be created. *Disadvantages*: The technique works best with very small voxels, but may miss large vessels due to its sensitivity to vessel orientation.

12. Pre-undershoot "dip":[168, 187] Several studies have shown an initial decrease in the fMRI signal 0.5[187] to 2 sec[168] after the stimulus onset but immediately prior to the increase in signal that is typically observed. These changes are hypothesized to be caused by an increase in oxidative metabolic rate[55, 188] and/or change in the ionic environment of the neurons[187] occurring at the regions of neuronal activity prior to subsequent flow and oxygenation increases. *Advantages*: Assuming that the hypothesized origins of this signal behavior are substantiated, observation of this signal would allow localization of neuronal activity with a high degree of spatial and temporal specificity. *Disadvantages*: This transient signal can only be observed with high speed imaging (EPI) or by functional spectroscopy. Secondly, this is an extremely subtle effect and has not been extensively reproduced. High contrast to noise ratio with extensive averaging and physiologic noise reduction may be essential to observe this. The pre-undershoot has not yet been demonstrated in any other cortical region but visual cortex.

13. Spin tagging techniques:[82, 83, 142–146, 148, 160] These include the array of techniques mentioned in section 9.3.3.2. Flowing spins are imaged by inverting or saturating spins outside the imaging plane, waiting a time period for the tagged spins to flow into the imaging plane, then imaging. Both resting state perfusion and activation – induced perfusion changes can be imaged. *Advantages*: This is a non-invasive and robust technique by which quantifiable maps of flow and flow changes can be created. The pulse sequence can be adjusted so that capillary perfusion is selectively imaged. Also, the flow images created are insensitive to oxygenation effects, which translates to a potentially more direct measure of the degree of neuronal activation. Also, because pair-wise subtraction is performed, the images are sensitive to motion occurring only in the brief interval (≈ 2 sec) between successive images, and much less sensitive to typically problematic motion occurring on longer time scales. Lastly, if each of the image pairs is oxygenation – sensitive (i.e. $T2^*$ or $T2$ weighted), oxygenation effects can be assessed by observation of every other image in the time series[143, 146, 147], therefore giving both flow and oxygenation information simultaneously. *Disadvantages*: Presently, only one or a very few imaging planes can be imaged at one time. This techniques also involves a relatively long waiting period (TR at least 2 sec) for each image, and requires that pairs of images are subtracted, therefore reducing the contrast to noise per unit time.

Figure 9.30 shows a comparison of a spin-tagging technique (FAIR) with BOLD contrast functional imaging. Low resolution (64 × 64) and high resolution (128 × 128) anatomical and functional (correlation maps) BOLD – contrast images (Gradient-echo, TE = 40 ms) were obtained of an axial slice through the motor cortex. Single shot EPI was performed using a local gradient coil[189] and a 3T /60 Bruker Biospec scanner. The images were 5 mm thick and the FOV was 20 cm. The task was bilateral finger tapping. Resting and active state perfusion maps, created using FAIR (TI = 1400, TR = 2 sec, spin-echo TE = 42 ms), are also shown. A functional correlation maps using BOLD contrast at the two different resolutions are compared with a functional correlation map using the FAIR perfusion time course series. The magnified images, shown in Figure 9.31, illustrate that the areas of activation obtained using FAIR and BOLD contrast generally overlap, but also have some significant differences. These spatial shifts in activation are likely to be due to the differences in hemodynamic sensitizations of the two sequences. FAIR imaging using a TI of 1400 ms is optimally sensitized to imaging capillary perfusion, as shown in the resting and active state flow maps. BOLD contrast functional images are strongly weighted by large draining vein effects.

14. Tailored RF gradient-echo sequence:[190] This technique uses a tailored RF pulse that dephases static and flowing tissue in homogeneous fields, but does not dephase tissue in the presence of field inhomogeneities[191] created around vessels containing deoxygenated blood. *Advantages*: When used in

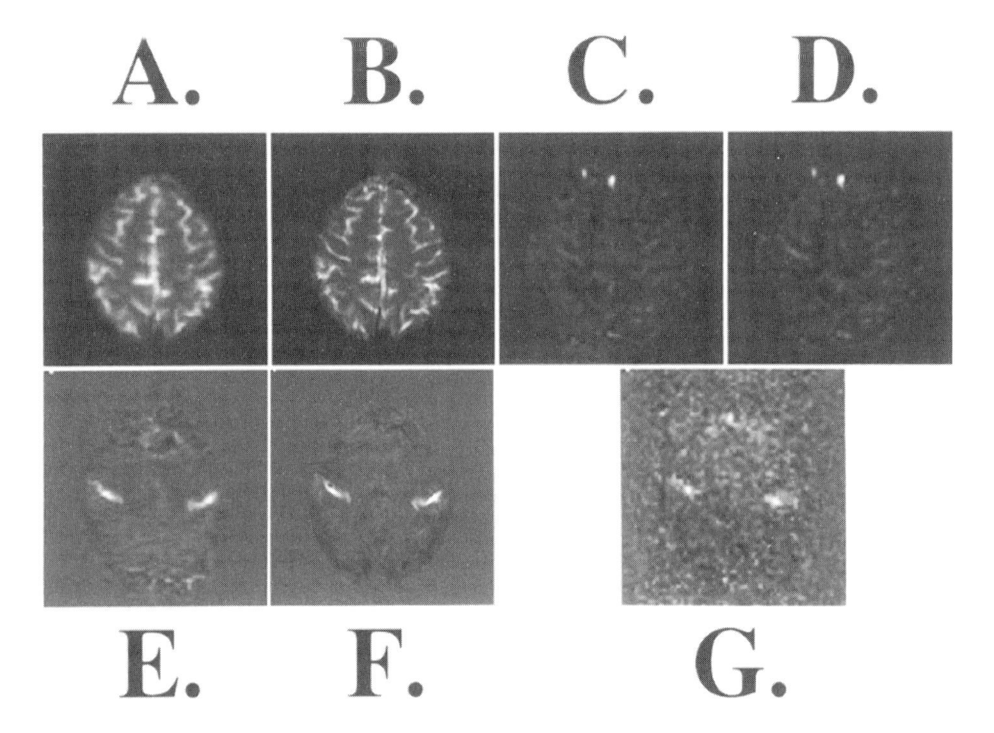

Figure 9.30. Comparison of perfusion-weighted and BOLD – weighted functional echo planar images at 3 Tesla. Echo planar imaging was performed using a Bruker 3T/60 scanner and a local head gradient coil. All images were created of the same plane in the same experimental session. The slice thickness was 5 mm and the FOV was 20 cm. An axial plane was chosen which contained the motor cortex.
A. 64 × 64 gradient – echo anatomical image ($TE = 50$ ms, $TR = \infty$),
B. 96 × 96 gradient-echo anatomical image ($TE = 50$ ms, $TR = \infty$),
C. Perfusion image created during the resting state using a FAIR time course series. ($TI = 1400$ ms, Spin-echo $TE = 60$ ms, $TR = 2$ sec.)
D. Perfusion image created from the same time course series as C. during bilateral finger tapping.
E. 64 × 64 BOLD contrast functional correlation image created from the time series of images in which image A. was the first of the series. Bilateral finger tapping was performed.
F. 96 × 96 BOLD contrast functional correlation image created from the time series of images in which image B. was the first of the series. Bilateral finger tapping was performed.
G. 64 × 64 perfusion-only functional correlation image created from the same time series of perfusion images from which the resting state and active state images (C. and D.) were created. Note the difference in spatial location of the area of activation between the flow-weighted and perfusion-weighted functional images. The "hot spot" in the BOLD contrast images is likely to be a draining vein which does not appear in the perfusion-weighted functional image created using FAIR.

conjunction with techniques requiring a short TR, inflow effects are suppressed. It may be less sensitive to motion because signal from static tissue is suppressed, therefore slight misregistration of images will not cause large signal changes. Also, use of this technique allows direct visualization of subvoxel inhomogeneities, giving the potential to directly visualize veins. *Disadvantages*: It is not clear how implementation of this for fMRI is an improvement over simple flip angle reduction for reducing inflow effects. Since no functional images using the technique have yet been published, the robustness of the technique has not been demonstrated.

15. Short TR, short TE spin-echo: This technique is a simple method for achieving $T1$ weighting and therefore flow-sensitive contrast. A short TE spin-echo is minimally sensitive to oxygenation changes and a short TR gives increased sensitivity to flow changes. *Advantages*: This is more time efficient than inversion recovery sequences. It is useful in multi-shot imaging and when using EPI to sample transient hemodynamic events. *Disadvantages*: This technique has half the flow sensitivity of inversion recovery imaging.

16. Small voxels with high SNR: Reduction of the voxel size make it more likely that a large vein will completely fill one or several voxels (100% blood volume), while the blood volume per voxel from capillaries remains the same (2% to 5% blood volume). With higher resolution and with high enough signal to noise at high resolution to visualize subtle capillary effects ($\approx$ 1% signal change at 1.5T), a greater stratification of vessel effects (increase with higher resolution[192]) from capillary effects (insensitive to resolution) is achieved. *Advantages*: It is relatively easy to interpret high resolution and high functional contrast to noise functional images. *Disadvantages*: Because of signal to noise demands, this technique is likely to be achievable only at higher field strengths or with significant data averaging.

17. Spin-echo with long TE:[165, 193] Instead of signal being collected immediately after the 90° pulse (during the free induction decay (FID)), data is collected during the echo that occurs after a refocussing (180° degree) pulse is applied. Activation – induced changes in $T2$ instead of $T2^*$ are imaged. Macroscopic susceptibility gradients are refocussed by the spin-echo but susceptibility gradients on the spatial scale of the distance that a water molecule diffuses in an echo time ($\approx 10\ \mu$m) are not refocussed. It is for this reason that spin-echo sequences are thought to be sensitive to susceptibility gradients (and activation-induced changes in susceptibility gradients) caused by small compartments such as red blood cells and capillaries. *Advantages*: Extravascular large vessel effects are not seen because the refocussing pulse eliminates the effect of gradients on a spatial scale significantly larger than large vessels. This technique also has the same advantages as technique #2. *Disadvantages*: With this technique, activation-induced intravascular signal from blood in large vessels flowing slow enough to still experience the 180° pulse remains present. Secondly, the functional contrast to noise of this technique is about 1/4 that of gradient-echo sequences[193, 194].

18. TE stepping:[163, 193, 195, 196] This technique involves the systematic incrementation of the echo time, allowing acquisition of two types of hemodynamic information simultaneously. TE stepping allows direct measurement of $R2^*$ (from the slope a monoexponential fit to the decay curve) and measurement of inflow effects (from the intercept of the monoexponential fit to the decay curve). *Advantages*: The simultaneously provided information is useful in that systematic errors that come from measures across trials are avoided. This is useful for studies which require direct registration of individual voxels and for studies in which successive course series can never be identical – such as those involving a hemodynamic stress such as hypercapnia. *Disadvantages*: The time cost for this technique is high. The sensitivity of the technique for measuring flow ($TE = 0$ intercept of $R2^*$ curve) is low.

19. Variance imaging and frequency analysis:[197, 198] This technique involves the collection of a time course series of echo planar images, then inspecting the series, in a voxel-wise manner, for noise characteristics. Large vessels seems to cause large MR signal intensity fluctuations at the cardiac and respiratory cycle rates, and are therefore identifiable. *Advantages*: It is relatively easy to implement. *Disadvantages*: The specificity of the technique is unreliable in that many regions other than large vessels (cerebral spinal fluid) can show large pulsatile effects. Also, Fourier analysis is performed best in conjunction with only the rapid sampling rate of EPI.

Improvements in functional spatial and temporal resolution are still being rapidly made at this stage. The maximum temporal and spatial resolution of fMRI can only be fully realized by the combination of: a) a high contrast to noise ratio, b) hemodynamic specificity, c) significant motion and artifact reduction, and d) well controlled and carefully executed experiments. Below is a summary of the issues in achieving high temporal and spatial resolution in fMRI.

9.3.4.2 Temporal Resolution

Two separate time scales are present and separately measurable: the time for the signal transition from one state to another; and the accuracy to which the location of the transition can be measured. Because the fMRI signal change arises from hemodynamic changes, the practical upper limit on functional temporal resolution is determined by the functional contrast to noise ratio and by the variation of the hemodynamic response latency in space and in time[173, 184, 199–201]. These variations may be due to differences in neuronal activation characteristics across tasks[174], but are more likely to be due to differences in vessel size[173], or to regional differences in the vascular transit rate. The latency of the hemodynamic response has been described as a shifting and smoothing transformation of the neuronal input[202]. While this smoothing creates a transition between activation states on the order of 5 to 8 sec, the accuracy in the measurement of the location of this transition can be much greater, and is limited primarily by variations in the hemodynamic response. The upper limit of temporal resolution discrimination has been empirically determined to be on the order of one second[203], or less[200].

The type of neuronal and/or hemodynamic information that may be obtained from signals elicited from brief stimuli paradigms may be qualitatively different from the information elicited by longer duration activation times. Transient activation durations (< 1 sec.) are detectable as MR signal changes which begin to increase 2 sec after the activation onset, and plateau

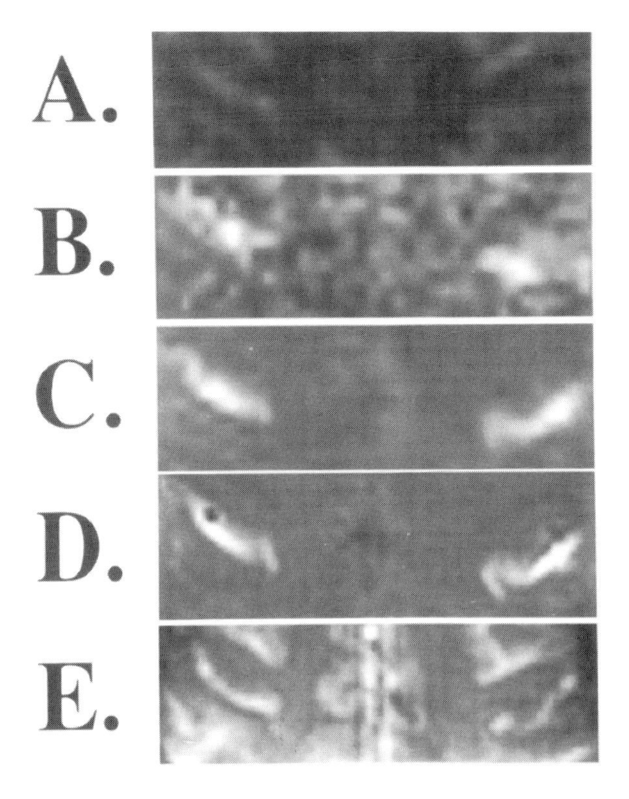

Figure 9.31. Magnification of selected images displayed in Figure 9.30 to emphasize the differences in the activation locations that appear with different hemodynamic sensitizations.
A. Baseline 64 × 64 perfusion image (magnification of 9.30 C).
B. 64 × 64 perfusion – only sensitive functional correlation image (magnification of 3. G).
C. 64 × 64 BOLD contrast functional correlation image (magnification of 9.30 E).
D. 96 × 96 BOLD contrast functional correlation image (magnification of 9.30 F).
E. 96 × 96 gradient-echo anatomical image (magnification of 9.30 B.) Dark lines in the image are likely to be due to deoxygenated veins (lower $T2^*$ and phase difference from other tissue in voxel, thereby causing dephasing).

at 3 to 4 sec. after activation[201, 204]. Figure 9.32 shows BOLD contrast dynamics related to activation durations lasting only 2 sec. Single shot gradient-echo EPI was performed using the same setup as described above. The FOV was 20 cm, and slice thickness was 5 mm. Matrix size was 96 × 96. A time course series of 1000 axial images (TR = 500 ms, TE = 40 ms) through the motor cortex was obtained during witch the subject performed bilateral finger tapping for 2 sec. followed by 18 sec. rest. This cycle was repeated for 500 sec. Figure 9.32a shows the time course from the motor cortex averaged across over time. This plot demonstrates that one limiting factor in upper temporal resolution is the standard deviation of the signal at each point. This variation may be due to the system noise of the hemodynamic variability over time. This plot was then used as the reference function for subsequent

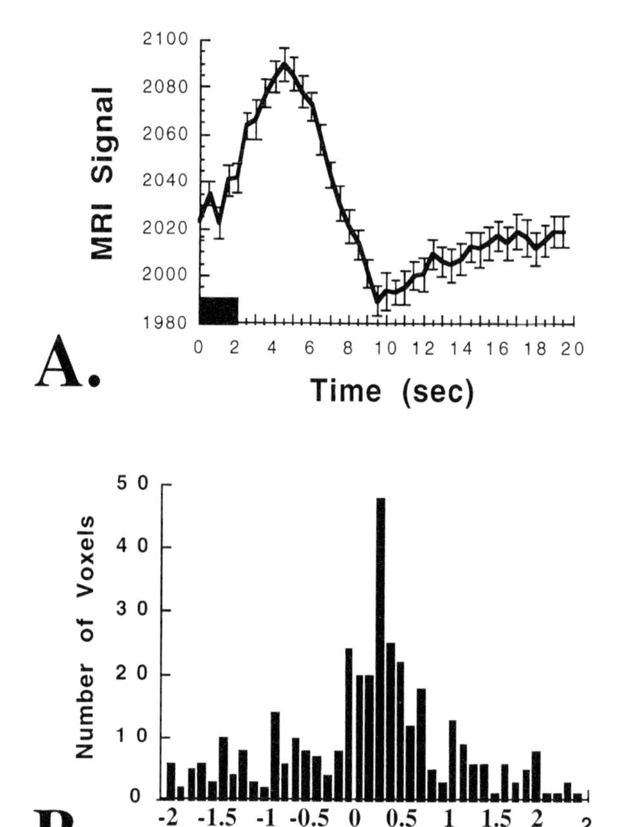

Figure 9.32. Demonstration of the limits of fMRI temporal resolution. Echo planar imaging was performed at 3 Tesla using a Bruker Biospec 3T/60 equipped with a local head gradient coil. An time course series of axial images (matrix size = 96 × 96, FOV = 20 cm, TE = 40 ms, TR = 500 ms, flip angle = 80°) through the motor cortex was obtained. Bilateral finger tapping was performed for 2 sec., alternating with 18 sec. rest. These figures demonstrate that the upper temporal resolution is determined by the variability of the signal change in time and space.
A. Time course of the signal elicited by tapping fingers for 2 sec. The standard deviation at each point was in the range of 1 to 2%. The standard deviation of the hemodynamic change, in time, is in the range of 200 ms to 500 ms.
B. Histogram of relative hemodynamic latencies of the hemodynamic response across active voxels in motor cortex.

correlation analysis. The histogram of the latencies, shown in Figure 9.32b demonstrate that a "spread" of 4 sec in latencies occurs over space. This spread is a second determining factor in of the upper temporal resolution. The areas that show the longest latency are likely to be "downstream" large draining vein effects.

For many types of investigations it may be desirable to use experimental paradigms similar to those used in event related potential recordings (ERP) or magneto – encephalography (MEG)[205], in which multiple runs of transient stimuli are averaged together. For this type of paradigm, requiring rapid sampling, EPI is optimal. As a side note, because of the brief collection time of EPI relative to typical *TR* values (e.g. 50 ms relative to about 1 sec), the between-image waiting time allows for performance of EEG in the scanner during the imaging session without electrical interference from MR pulse sequences[206].

9.3.4.3 Spatial Resolution

The upper limit on functional spatial resolution, similar to the limit on temporal resolution, is likely determined not by MRI resolution limits but by the hemodynamics through which neuronal activation is transduced. Evidence from *in vivo* high resolution optical imaging of the activation of ocular dominance columns[54–56] suggests that neuronal control of blood oxygenation occurs on a spatial scale of less than 0.5 mm. MR evidence suggests that the blood oxygenation increases that occur on brain activation are be more extensive than the actual activated regions[173, 180, 192, 207, 208]. In other words, it is possible that, while the local oxygenation may be regulated on a submillimeter scale, the subsequent changes in oxygenation may occur on a larger scale due to a "spill-over" effect. An effective counter-measure for the "spill-over" is mentioned in technique #8 (latency – tagging), which maintains a constant "spill – over" by always keeping stimuli "on" yet spatially modulated within a region – therefore discriminating subtle differences in activation within a large and less localized "umbrella" of increased oxygenation.

In general, to achieve the goal of high spatial resolution fMRI a high functional contrast to noise and reduced signal contribution from draining veins is neccssary. Greater hemodynamic specificity, accomplished by proper pulse sequence choice (selective to capillary effects), innovative activation protocol design (phase-tagging), or proper interpretation of signal change latency (latency mapping), may allow for greater functional spatial resolution. If the contribution to activation-induced signal changes from larger collecting veins or arteries can be easily identified and/or eliminated, then, not only will the confidence in brain activation localization increase, but also the upper limits of spatial resolution will be determined by scanner resolution and functional contrast to noise rather than variations in vessel architecture.

Currently, voxel volumes as low as 1.2 microliters have been obtained by functional FLASH techniques at 4T[169], and experiments specifically devoted to probing the upper limits of functional spatial resolution, using spiral scan techniques, have shown that fMRI can reveal activity localized to patches of cortex having a size of about 1.35 mm[177]. These studies and others using similar methods[175–177, 199, 209], have observed a close tracking of MR signal change along the calcarine fissure as the location of visual stimuli was varied.

The voxel dimensions typically used in single – shot EPI studies are in the range of 3 to 4 mm, in plane, and having 4 to 10 mm slice thicknesses. These dimensions are determined by practical limitations such as readout window length, sampling bandwidth, limits of dB/dt, SNR, and data storage capacity. Other ways to bypass the practical scanner limits in spatial resolution include partial k-space acquisition[158] and multi – shot mosaic or interleaved EPI[158, 210, 211]. In many fMRI situations, multishot EPI may be the optimum compromise between spatial resolution, SNR, and temporal resolution for fMRI.

9.3.4.4 Dynamic Range

While it is important not to interpret spatial differences in fMRI signal change magnitude as indications of differences in the degree of neuronal activation (because the signal is highly weighted by hemodynamic factors such as the distribution of blood volume across voxels), observation of differences in the fMRI signal change in the same regions but across incrementally modulated tasks is possible and may be a useful method for extracting more direct neuronal information from the fMRI time course series.

The first demonstration that fMRI response is not simply binary was made by Kwong et al.[58]. Both flow and oxygenation sensitized MR signal in V1 were measured as flicker rate was modulated. The signal behavior corresponded closely with that obtained with a previous PET study[47]. Other studies have revealed a responsivity in higher visual areas to contrast and flicker rate[178, 212]. In the primary motor cortex a linear signal dependence on finger tapping rate has been demonstrated[213]. In the primary auditory cortex, a sublinear dependence on syllable presentation rate has been demonstrated[214].

9.3.4.5 Sensitivity

Extraction of a 1% signal change, typical of fMRI, against a backdrop of motion, pulsation, and noise requires careful consideration of the variables which influence the signal detectability. These variables range from factors that increase signal, increase fMRI contrast, reduce physiologic noise, and reduce artifactual signal changes. Below is a list of some salient variables that are important to consider in relation to optimization of fMRI sensitivity.

1. **Averaging:** Averaging of sequentially obtained images increases the signal to noise by the square root of the number of images collected. A difficulty is that, if averaging is performed over too long of a period (over about 5 minutes) systematic artifacts (i.e. slow movement or drift) tend to outweigh the benefits obtained from averaging for that duration.
2. **Field strength:** As previously discussed, signal to noise and functional contrast increase with field strength. Difficulties such as increased shimming problems, increased physiologic fluctuations, and limitations on the possible RF coils used, also increase with field strength. It has yet to be determined if

gains in sensitivity and contrast obtained by increasing field strength cannot be achieved by other methods at lower fields, or if the gains in sensitivity and contrast outweigh the disadvantages of imaging at high field strengths.

3. Filtering: In most fMRI studies using EPI, the noise over time is dominated not by system noise but by physiologic fluctuations. These fluctuations are correspond with specific frequencies (i.e. heart and respiration rates). Filtering out of these frequencies can increase the functional contrast to noise ratio, or at least make the noise closer to Gaussian so that parametric statistical tests can be applied.

4. Gating: Gating is a technique with one serious drawback that has one potential solution. Gating involves triggering of the scanner to the heart beat so that an image is always collected at specific phase of the cardiac cycle. This is advantageous because a primary source of noise is collection of images at different phases of the cardiac cycle causing head misregistration (The brain moves with every heart beat) and pulsatile flow artifacts. Image collection at a single phase would eliminated this misregistration, thereby reducing the noise and potentially increasing the spatial resolution of fMRI (i.e. The brain would be imaged at a single position all of the time). The drawback to gating is that if the heart rate changes during the collection of images, the MR signal intensity also changes, depending on the tissue $T1$ and the average TR used. This generally causes very large fluctuations in the data – making gating relatively worthless in the context of fMRI. A technique has recently been developed to correct for the global fluctuations that occur with heart rate changes [215], therefore making gating a feasible option in fMRI. Gating would be especially useful for identifying activation in structures at the base of the brain since that is where pulsatile motion is greatest, where activation is most subtle, and where activated regions are the smallest – requiring the most consistent image to image registration.

5. Paradigm timing: The choice in fMRI timing is usually determined by the sluggishness of the hemodynamic response (It usually not useful to go much faster than an on – off cycle of 8 sec on and 8 sec off), the particular brain system that is being activated (cognitive tasks may have a more delayed response), and the predominant frequency power of the noise. As a rule of thumb, the goal is to maximize both the number of on – off cycles and amplitude of the cycle to maximize the power of post-processing techniques such as correlation analysis[216] to extract functional information. Generally, contrast to noise ratio is maximized and artifact is minimized by cycling the activation at the highest rate that the hemodynamics can keep up with and by having a time course series no longer than about 3 to 4 minutes long.

6. Post processing: Many approaches have been used to extract from fMRI data estimates of the significance, amplitude, and phase of the functional response, and there is still surprisingly little agreement on the appropriate techniques. If one knows exactly the shape and phase of the expected signal response, a matched filter (i.e. correlation) approach may be optimal. If the shape is unknown, use of a single expected response function, be it a boxcar function or a sine wave, may miss unique activation patterns. The challenge of accurately determining regions of significant activation from fMRI data is non-trivial and has yet to be solved. Some of the developments addressing this issue include: a) the development of accurate and robust motion correction[217, 218] and/or suppression methods, b) the determination of the noise distribution[197, 199, 219], c) the determination of the temporal[202] and spatial[220] correlation of activation-induced MR signal changes, and of baseline MR signal, d) the characterization or assessment of the temporal behavior or shape of activation-induced signal changes[199, 203, 214, 221, 222], and e) the characterization of how the above-mentioned factors vary in time, space[173, 184], across tasks[199, 221, 222], and with different pulse sequence parameters[58].

It is generally important to always inspect the data for motion, and not to assume too much about the expected response, yet, at the same time, use all of the current *a priori* information about hemodynamic responses and neuronal activation to extract meaningful information.

7. Pulse sequence: As described in detail above, pulse sequences that can be used for fMRI have a wide range of sensitivities – with gradient-echo sequences being the most sensitive and time efficient. Standard clinical multishot techniques (i.e. FLASH or GRASS) suffer from significantly more motion related noise than EPI techniques or spiral multi-shot techniques[223–225]. Also, application of navigator echoes[223, 226] or other types of image reconstruction – related post processing of multishot data can significantly reduce artifactual fluctuations.

8. RF coil choice: The tradeoff here is regarding spatial coverage vs. sensitivity. The smaller the coil used, the less brain tissue it couples to. This gives a higher signal to noise but much less brain coverage. Larger RF coils give more brain coverage but lower signal to noise. Where sensitivity is critical, a surface coil in a specific region may be desirable. Where whole brain imaging is desirable, a whole brain quadrature RF coil is optimal[227]. This coil is generally as close to the head as possible and couples only to brain region. It should noted here that typical whole head and neck coils used clinically are sub optimal for whole brain fMRI, since they couple also to the face and neck regions (only adding noise) and since they are generally not as close a possible to the head.

9. Voxel size: The signal to noise is directly proportional to voxel volume. Functional contrast to noise is optimized by matching the volume of the active region to the voxel volume. Since functional region sizes are not well characterized, and are likely to vary widely, the optimal voxel size is difficult to predict. Many have generally matched the voxel slice to the cortical thickness. Other groups have used a slightly thicker slice to increase brain coverage given a limitation in the number of slices obtainable. As described above, spatial resolution may actually be reduced with the use of smaller voxels if the contrast to noise is not high enough to detect more subtle capillary effects. In such a case of low contrast to noise, primarily downstream draining veins would be primarily detected. This phenomenon may explain the exclusive detection of large vessels by Lai and Haacke et al.[192, 208] using small voxels. Overall, small voxels are desirable as long the sensitivity remains high enough to detect a 1% signal changes.

9.3.4.6 Some Unknowns

While not directly related to the practical implementation of fMRI, some unexplained and controversial fMRI data can give an indication of possible directions that fMRI research and applications may take in the future. Listed are four "controversial" results accompanied by the hypotheses related to them.

1. Post–undershoot:[58, 228–230] After cessation of activation, the BOLD-weighted fMRI signal is commonly observed to undershoot the previous baseline signal intensity. The undershoot has been observed to last between 30 seconds and 2 minutes. The reasons for this are unclear. Two hypothesis have been suggested. The first is that on cessation of activation, neuronally – triggered flow returns to baseline but oxidative metabolic rate continues for several minutes, causing a reduction of signal (increased deoxyhemoglobin). The second hypothesis is that on cessation of neuronal stimulation, flow and oxygenation return to baseline levels but blood volume (possibly pooling in draining veins) takes longer to return to baseline levels, causing the signal to dip below baseline for a small amount of time. As a side note, the post activation undershoot is not observed using $T1$-weighted (flow – weighted) sequences[58].

2. Pre–undershoot:[168, 187] This phenomena is observed less frequently. Observations by Hennig et al. show a dip at 0.5 sec. Observations of Menon et al. show a dip at 2 sec. in agreement with reports of Grinvald et al.[55] using optical imaging. Menon et al. has put forth a hypothesis that is similar to that of Grinvald et al. – that on activation, an increase in oxidative metabolic rate occurs before a subsequent increase in flow. The observations of Hennig et al. not only differ in relative timing but also differ in the hypothesized origin. The signal is found to be only slightly $T2^*$ (oxygenation) related and primarily $T1$ related. The hypothesis is that changes in the ionic environment of the neurons caused by the influx of Na^+ may rapidly change the $T1$ of the tissue.

3. Long term effects:[228, 231, 232] The effect on sustained activation on fMRI signal intensity is controversial. Three studies with differing results have been reported. Hathout et al.[231] have suggested that local blood oxygenation returns from an initially elevated level to baseline after about 15 minutes of continuous stimulation. Frahm et al.[228] have observed a return of oxygenation sensitive MR signal to baseline after about 1 to 2 minutes of sustained activation, but has also observed sustained blood flow during the entire stimulation duration (233). Bandettini et al.[196, 201, 232–234] have demonstrated sustained flow and BOLD enhancement for entire stimulation durations. Stimulation durations were up to 20 minutes long. Possible explanations for these differences in results include differential effects of the particular stimuli on metabolic, hemodynamic, or neuronal changes or differential, and not fully understood pulse sequence sensitivities.

4. Noise correlation:[235, 236] This observation is that the noise in the fMRI data obtained during a resting state shows temporal correlation across regions that appear to be functionally connected (e.g., motor cortex). The predominant frequency that shows most correlation is in the 0.1 to 0.2 Hz. range. The origin of these suggests an oscillation in vascular tone that is synched across similar functional units in the brain. These findings may be clinically useful in determining vascular tone and/or diagnosing cerebrovascular pathologies.

9.3.5 Common fMRI Platforms

In an attempt to bring much of what has been mentioned together, this section describes some of the most commonly used platforms for fMRI. The three types of fMRI pulse sequences discussed are: EPI, conventional multi-shot imaging, and spiral scanning.

9.3.5.1 Echo Planar Imaging

Echo planar imaging (EPI), is an ultra fast MRI technique[83, 156–158], that has been and continues to be ubiquitous in the ongoing development and application of fMRI. In most of the growing number of centers that have EPI capability, it is the fMRI method of choice for most applications.

EPI has several drawbacks (low spatial resolution, high sensitivity to off – resonance effects, need for specialized hardware, potential for peripheral nerve stimulation, and need for specialized image reconstruction algorithms). The advantages of EPI (high temporal resolution, high flexibility for imaging several types of physiologic processes, high stability, low imaging duty cycle, low sensitivity to motion) still greatly outweigh the disadvantages for most purposes related to fMRI. Below is a brief description of some of these EPI characteristics.

Spatial resolution in single shot EPI is limited either by the area of k-space that can be sampled in approximately one $T2^*$ period or by the system bandwidth[237]. The area of k-space that can be covered can be limited by the velocity in k-space (gradient amplitude) or the acceleration in k-space (gradient slew rate) and is typically limited by both.

The requirement, with EPI, for strong and rapidly switching gradients is satisfied by: 1) increasing the gradient amplifier power or by using a speed-up circuit, 2) implementing resonant gradient technology, 3) reducing the inductance of the gradient coils such that they can be driven by conventional gradient amplifiers, or 4) increasing the field of view and/or lowering the resolution to match the speed at which standard gradient amplifiers can keep up.

The first strategy is probably among the least commonly used. The second strategy is likely to be the most common EPI technique as of yet. Both strategy 1 and 2 use whole body gradient coils, which allows performance of EPI for functional and/or kinematic studies on the heart, lungs, digestive system, kidneys, throat, joints, and muscles. In the context of fMRI, whole body gradients allow more accessibility for patients with mobility problems and for easy delivery of brain activation stimuli.

The third strategy is used primarily by several centers that have home built gradient coils (two examples are: National Institutes of Health[154], the Medical College of Wisconsin[189]) and marketed by Medical Advances, (using the coil design of E. C. Wong), Advanced NMR, and Siemens, among others. This strategy is implemented by using a gradient coil that is localized only to the head. The gradient fields are optimized for a region that usually covers the brain and/or the region of RF sensitivity.

Lastly, single – shot EPI can be carried out on a conventional imaging system without the use of local gradient coils (using the whole body gradient coil) by simply using a large FOV and/or a small image matrix size[238]. Functional MRI using EPI with voxel sizes of approximately 10 mm × 10 mm × 10 mm (approximately the resolution of a PET scanner) have been successfully performed on a standard GE 1.5T Signa system[198], with excellent results. This type of echo planar imaging capability exists on practically every clinical scanner in the world.

A major non-hardware related limitation on gradient slew rate is the biological threshold for neuronal stimulation due to time varying magnetic fields. At present, high performance gradient systems (either local gradient coils or high powered whole body systems) are capable of exceeding the FDA guidelines on gradient field slew rate (dB/dt). This is a large determinant of the upper limit on the resolution possible using single shot EPI to image humans.

The requirements for successful implementation of EPI for fMRI are not limited to hardware. In most cases, phase correction algorithms, applied during image reconstruction, are usually necessary to compensate for timing errors related to imperfections in the gradients, gradient-induced eddy currents, or static field inhomogeneities.

Because of the long sampling time and artifactual phase modulation, EPI is sensitive to two types of off-resonance related artifacts in EPI: signal dropout, and image distortion. Signal dropout is primarily due to intravoxel phase dispersion resulting from through plane variation of magnetic field. The problem of signal dropout in gradient-echo sequences can be reduced by reduction of the *TE*, reduction of the voxel volume, and/or by localized shimming. Also, this effect is greatly reduced in spin-echo EPI because the macroscopic off resonance effects are refocussed at the echo time.

Image distortion is caused by the an off resonance – related phase modulation that occurs during data acquisition. In EPI, this linear phase modulation creates primarily a linear distortion of the image in the phase encode direction. Several post-processing methods have been put forward for correcting image distortion in EPI[239, 240].

With the use of EPI, approximately 10 images may be obtained per second – giving the option to image the entire brain in under 2 seconds or to sample a smaller number of imaging planes to allow a more dense sampling of the time course. Another possibility in EPI is to sample less densely in space but to cover a large volume in a single shot. This technique is known as echo – volume imaging (EVI)[156, 241].

A practical but significant factor to be considered when performing fMRI with EPI is the rapidity with which large amounts of data are collected. This data may then go through several additional transformations (adding to the total required data storage capacity) before a functional image is created. If 10 slices having 64 × 64 resolution are acquired every 2 sec. (typical for multi-slice fMRI) then the data acquisition rate is approximately 2 MB per minute.

9.3.5.2 Conventional Multi-shot Imaging

High temporal resolution fMRI techniques developed for use with conventional gradients include multi-shot FLASH[59, 181, 208, 242–245], turbo-FLASH[246], low resolution EPI[150, 247], multi-shot or interleaved EPI[210, 211], echo-shifted flash[248, 249], keyhole imaging[250], and fast spin–echo[165].

Only a few centers have been able to successfully implement conventional multishot techniques in a routine and robust manner for fMRI[169, 181, 192, 242]. The advantage to multishot techniques is the ability to achieve relatively high in-plane spatial resolution, less sensitivity to off-resonance effects from poor shim, the availability of the technique on most clinical scanners. The disadvantages are: lower temporal resolution, increased noise due to non-repeated shot to shot misregistration of k-space lines[223–225] (from variable sampling of low frequency lines at different phases of the cardiac cycle), lower signal due to the need for short *TR* and low flip angles, reduced capability to perform multislice fMRI as rapidly as with EPI, less flexibility or “dead time,” (that comes with a long *TR* typically used for EPI) for other types of pulse sequence manipulations. More time – efficient and stable multishot techniques include fast spin echo[165] and spiral scan imaging[223–225].

9.3.5.3 Spiral Scanning

Of non-EPI techniques, multi-shot spiral-scan sequences, which involve traveling outward from the center of k-space in a spiral manner, and used in conjunction with a single point phase correction scheme have demonstrated the most temporal stability[224, 251]. Spiral scanning also involves oversampling at the center of k-space – where the acquisitions are intrinsically gradient moment nulled – providing less sensitivity to phase errors caused by brain, blood, or cerebral spinal fluid pulsations with the cardiac cycle.

Spiral scanning has been used for many fMRI applications[177, 209, 220, 252] and has demonstrated, when used in conjunction with a phase – tagging activation scheme, the highest functional resolution (1.35 mm)[177] to date. In studies where high spatial resolution is important or where EPI is unavailable, spiral scan appears to be the method of choice.

9.3.6 Applications

Most studies involving the development of fMRI from a contrast mechanism, pulse – sequence, and post – processing standpoint have used primary motor and visual cortex activation due to the easily elicited and robust signal changes. Listed below are some of the applications of fMRI that have gone beyond simple finger tapping or visual stimulation. The auditory cortex[174, 253], somatosensory cortex[254, 255] and cerebellum[256] have been studied. Detailed mapping of regions activated in the primary motor cortex[179, 213, 257–260] and visual cortex[175, 177, 199, 209, 261], have been performed as well. Activity elicited in the gustatory cortex has been mapped[262]. Other studies using fMRI have observed organizational differences related to handedness[263]. Activation changes during motor task learning have been observed in the primary motor cortex[264] and cerebellum[265].

Cognitive studies in normal subjects have included word generation[266–269], mental rehearsal of motor tasks and complex motor control[270, 271], visual processing[199, 272, 273], speech perception[174, 253], semantic processing[174, 274, 275], working memory[220], visual recall[276], and mental rotation[277].

Studies have also been performed involving specific pathologies. Changes in organization in the sensorimotor area after brain injury has been observed[259]. One study has demonstrated larger fMRI signal changes, on the average, in schizophrenic patients[278]. The ability to localize seizure activity has also been demonstrated by fMRI[279]. In addition, preliminary data demonstrating the effects of drugs on brain activation have been presented[280].

The immediate potential for clinical application is currently being explored. "Essential" areas of the sensory and motor cortex as well as language centers have been mapped using both fMRI and electrical stimulation techniques[258, 281, 282]. Activity foci observed across the two methods have shown a high spatial correlation, demonstrating the potential for fMRI to compliment or replace the invasive technique in the identification of cortical regions which should be avoided during surgery. In the context of presurgical mapping, fMRI has demonstrated the ability to reliably identify the hemisphere where language functions reside[174, 269, 274, 275], potentially complimenting or replacing the Wada test (hemisphere specific application of an anesthetic amobarbital) for language localization that is also currently used clinically prior to surgery[283].

Several review articles and chapters on fMRI techniques and applications are currently available[201, 205, 220, 272, 275, 284–291].

9.4 Acknowledgments

The authors would like to thank Eric C. Wong for his help in the text and Figure preparation. He also created the gradient coil on which all echo planar imaging was performed. This work was supported in part by grant MH51358 from the National Institutes of Health.

9.5 References

1. P. C. Lauterbur, *Nature* **242**, 190–191 (1973).
2. R. N. Bracewell, *The fourier transform and its applications* (McGraw-Hill, New York, 1965).
3. D. B. Twieg, *Medical Physics* **10**, 610–621 (1983).
4. R. Mezrich, *Radiology* **195**, 297–315 (1995).
5. S. Ljunggren, *J. Mag. Res.* **54**, 338–343 (1983).
6. J. Hennig, A. Nauerth, H. Friedburg, *Magn Reson Med* **3**, 823–833 (1986).
7. D. Chien, R. R. Edelman, *Magn. Reson. Quart* **7**, 31–56 (1991).
8. R. B. Buxton, C. R. Fisel, D. Chien, T. J. Brady, *J. Magn. Reson.* **83**, 576–585 (1989).
9. A. Haase, J. Frahm, D. Matthaei, W. Hanicke, K.-D. Merboldt, *J. Magn. Reson* **67**, 258–266 (1986).
10. D. J. Atkinson, D. Burstein, R. R. Edelman, *Radiology* **174**, 757–762 (1990).
11. E. O. Stejskal, J. E. Tanner, *J. Chem. Phys.* **42**, 288–292 (1965).
12. W. G. Bradley, V. Waluch, K.-S. Lai, E. J. Fernandez, C. Spalter, *AJR* **143**, 1167–1174 (1984).
13. G. K. VonShulthess, C. B. Higgins, *Radiology* **157**, 687–695 (1985).
14. P. E. Valk, et al., *AJR* **146** (1986).
15. R. R. Edelman, *AJR* **161**, 1–11 (1993).
16. R. Weissleder, et al., *AJR* **152**, 167–173 (1989).
17. K. M. Ludke, P. Roschmann, R. Tishler, *Magn. Reson. Imag.* **3**, 329–343 (1985).
18. K. M. Donahue, R. M. Weisskoff, D. Burstein, *J. Magn. Reson. Imaging.* **7**, 102–110 (1997).
19. P. E. Roland, *Brain Activation* (Wiley-Liss, Inc., New York, 1993).
20. P. S. Churchland, T. J. Sejnowski, *The Computational Brain* (MIT Press, Cambridge, 1992).
21. S. W. Kuffler, J. G. Nicholls, A. R. Martin, *From Neuron to Brain* (Sinauer Associates Inc., Sunderland, MA, 1984).
22. K. Krnjevic, in *Brain Work* D. H. Ingvar, N. A. Lassen, Eds. (Munksgaard, Copenhagen, 1975) pp. 65.
23. S. G. Hillyard, T. W. Picton, in *Handbook of Physiology, Section 1: Neurophysiology.* (American Physiological Society, New York, 1987) pp. 519.
24. R. Hari, *J. Clinical Neurophysiology* **8**, 157–169 (1991).
25. L. Kaufman, S. J. Williamson, *Annals of the New York Academy of Sciences* **388**, 197–213 (1982).
26. J. Prichard, et al., *Proc. Natl. Acad. Sci. USA* **88**, 5829–5831 (1991).
27. K.-D. Merboldt, H. Bruhn, W. Hanicke, T. Michaelis, J. Frahm, *Magn. Reson. Med.* **25**, 187–194 (1992).
28. P. T. Fox, M. E. Raichle, M. A. Mintun, C. Dence, *Science* **241**, 462- (1988).
29. M. E. Phelps, D. E. Kuhl, J. C. Mazziotta, *Science* **211**, 1445–1448 (1981).

30. J. C. Mazziotta, M. E. Phelps, in *Brain Imaging and Brain Function* L. Sokoloff, Ed. (Raven Press, New York, 1985) pp. 121.
31. J. L. Haxby, C. L. Grady, L. G. Ungerleider, B. Horowitz, *Neuropsychologia* **29**, 539–555 (1991).
32. C. S. Roy, C. S. Sherrington, *J. Physiol* **11**, 85–108 (1890).
33. F. Gotoh, K. Tanaka, in *Handbook of Clinical Neurology* P. J. Vinkin, G. W. Bruyn, H. L. Klawans, Eds. (Elsevier Science Publishing Co., Inc., New York, 1987) pp. 47.
34. W. Kushinsky, *Microcirculation* **2**, 357–378 (1982–1983).
35. D. W. Busija, D. D. Heistad, *Factors Involved in the Physiological Regulation of the Cerebral Circulation* (Springer Verlag, Berlin, 1984).
36. M. Ursino, *Critical Reviews in Biomedical Engineering* **18**, 255–288 (1991).
37. H. C. Lou, L. Edvinsson, E. T. MacKenzie, *Ann Neurol* **22**, 289–297 (1987).
38. W. Kuschinsky, *Arzneimittel-Forschung* **41**, 284–288 (1991).
39. Y. E. Moskalenko, G. B. Weinstein, I. T. Demchenko, Y. Y. Kislyakov, A. I. Krivchenko, *Biophysical Aspects of Cerebral Circulation* (Pergamon Press, Oxford, 1980).
40. C. Estrada, E. Mengual, C. Gonzalez, *J. Cereb. Blood Flow Metab.* **13**, 978–984 (1993).
41. U. Dirnagl, U. Lindauer, A.Villringer, *Neuroscience Letters* **149**, 43–46 (1993).
42. C. Iadecola, *TINS* **16**, 206–214 (1993).
43. G. Mchedlishvili, *Arterial Behavior and Blood Circulation in the Brain*. J. A. Bevan, Ed. (Plenum Press, New York, 1986).
44. D. H. Ingvar, in *Brain Work* D. H. Ingvar, N. A. Lassen, Eds. (Munksgaard, Copenhagen, 1975) pp. 397.
45. S. T. Grafton, R. P. Woods, J. C. Mazziotta, M. E. Phelps, *J. Neurophysiol* **66**, 735–743 (1991).
46. J. G. Colebatch, M.-P. Deiber, R. E. Passingham, K. J. Friston, R. S. J. Frackowiack, *J. Neurophysiol* **65**, 1392–1401 (1991).
47. P. T. Fox, M. E. Raichle, *Ann. Neurol* **17**, 303–305 (1985).
48. P. T. Fox, M. E. Raichle, *J. Neurophysiol* **51**, 1109–1120 (1991).
49. O. G. Cameron, J. G. Modell, R. D. Hichwa, B. W. Agranoff, R. A. Koeppe, *J. Cereb. Blood Flow Metab* **10**, 38–42 (1990).
50. C. A. Sandman, J. P. O'Halloran, R. Isenhart, *Science* **224**, 1355–1356 (1984).
51. J. W. Belliveau, et al., *Science* **254**, 716–719 (1991).
52. A. Villringer, J. Planck, C. Hock, L. Scheinkofer, U. Dirnagl, *Neuroscience Letters* **154**, 101–104 (1993).
53. P. T. Fox, M. E. Raichle, *Proc. Natl. Acad. Sci. USA* **83**, 1140–1144 (1986).
54. R. D. Frostig, E. E. Lieke, D. Y. Ts'o, A. Grinvald, *Proc. Natl. Acad. Sci. USA* **87**, 6082–6086 (1990).
55. A. Grinvald, R. D. Frostig, R. M. Siegel, E. Bratfeld, *Proc. Natl. Acad. Sci. USA* **88**, 11559–11563 (1991).
56. R. D. Frostig, in *Cerebral Cortex, vol. 10* A. Peters, K. S. Rockland, Eds. (Plenum Press, New York, 1994) pp. 331.
57. P. A. Bandettini, E. C. Wong, R. S. Hinks, R. S. Tikofsky, J. S. Hyde, *Magn. Reson. Med.* **25**, 390–397 (1992).
58. K. K. Kwong, et al., *Proc. Natl. Acad. Sci. USA.* **89**, 5675–5679 (1992).
59. S. Ogawa, et al., *Proc. Natl. Acad. Sci. USA.* **89**, 5951–5955 (1992).
60. J. Frahm, H. Bruhn, K.-D. Merboldt, W. Hanicke, D. Math, *JMRI* **2**, 501–505 (1992).
61. P. Mansfield, P. K. Grannell, *J. Phys. C., Solid State Phys*, L422–L426 (1973).
62. P. Mansfield, P. K. Grannell, *Phys. Rev. B* **12**, 3618–3634 (1975).
63. P. Mansfield, P. G. Morris, *NMR Imaging in Biomedicine* (Academic Press, New York, 1982).
64. P. G. Morris, *Nuclear Magnetic Resonance Imaging in Medicine and Biology* (Oxford University Press, Oxford, 1986).
65. W. A. Edelstein, P. A. Bottomly, H. R. Hart, L. S. Smith, *J. Comput. Assist. Tomogr* **7**, 391–401 (1983).
66. I. R. Young, M. Burl, B. M. Bydder, *J. Comput. Assist. Tomogr* **10**, 271–286 (1986).

67. R. A. Fox, P. W. Henson, *Med. Phys.* **13**, 635–643 (1986).
68. R. B. Buxton, R. R. Edelman, B. R. Rosen, G. L. Wismer, T. J. Brady, *J. Comput. Assist. Tomogr* **11**, 7–16 (1987).
69. F. W. Wehrli, et al., *J. Magn. Reson. Imag.* **2**, 3–16 (1984).
70. C. T. W. Moonen, P. C. M. vanZijl, J. A. Frank, D. LeBihan, E. D. Becker, *Science* **250**, 53–61 (1990).
71. V. J. Wedeen, et al., *Science* **230**, 946–948 (1988).
72. R. R. Edelman, H. P. Mattle, D. J. Atkinson, H. M. Hoogewood, *AJR* **154**, 937–946 (1990).
73. J. Listerud, *Magn. Reson. Quart.* **7**, 136–170 (1991).
74. H. Y. Carr, E. M. Purcell, *Phys. Rev.* **94**, 630–635 (1954).
75. D. LeBihan, R. Turner, C. T. Moonen, J. Pekar, *JMRI* **1**, 7–28 (1991).
76. D. LeBihan, et al., *Radiology* **168**, 497–505 (1988).
77. *Magn. Reson. Med* **19**, 209–333 (1991).
78. D. LeBihan, *Magn. Reson. Med.* **14**, 283–292 (1990).
79. J. A. Detre, J. S. Leigh, D. S. WIlliams, A. P. Koretsky, *Magn. Reson. Med.* **23**, 37–45 (1992).
80. D. LeBihan, *Invest Radiol* **27**, S6–S11 (1992).
81. J. A. Detre, et al., *NMR in Biomedicine* **7**, 75–82 (1994).
82. R. R. Edelman, B. Sievert, P. Wielopolski, J. Pearlman, S. Warach, *JMRI* **4(P), [Abstr.]**, 68 (1994).
83. R. Edelman, P. Wielopolski, F. Schmitt, *Radiology* **192**, 600–612 (1994).
84. R. R. Edelman, et al., *Magn. Reson. Med.* **31**, 233–238 (1994).
85. S. D. Wolff, R. S. Balaban, *Magn. Reson. Med.* **10**, 135–144 (1989).
86. R. S. Balaban, T. L. Ceckler, *Magn. Reson. Quart.*, 116–137 (1992).
87. T. R. Brown, B. M. Kincaid, K. Ugurbil, *Proc. Natl. Acad. Sci. USA* **79**, 3523–3526 (1982).
88. A. A. Maudsley, S. K. Hilal, W. H. Perman, H. E. Simon, *J. Magn. Reson. Med.* **52**, 147–151 (1983).
89. J. F. Schenck, *Annals of the New York Academy of Sciences* **649**, 285–301 (1992).
90. L. Pauling, C. D. Coryell, *Proc. Natl. Acad. Sci. USA* **22**, 210–216 (1936).
91. K. R. Thulborn, J. C. Waterton, P. M. Matthews, G. K. Radda, *Biochim. Biophys. Acta.* **714**, 265–270 (1982).
92. K. M. Brindle, F. F. Brown, I. D. Campbell, C. Grathwohl, P. W. Kuchell, *Biochem. J.* **180**, 37–44 (1979).
93. R. M. Weisskoff, S. Kiihne, *Magn. Reson. Med.* **24**, 375–383 (1992).
94. R. A. Brooks, G. D. Chiro, *Med. Phys* **14**, 903–913 (1987).
95. J. M. Gomori, R. J. Grossman, C. Yu-Ip, T. Asakura, *Journal of Computer Assisted Tomography* **11**, 684–690 (1987).
96. N. A. Matwiyoff, C. Gasparovic, R. Mazurchuk, G. Matwiyoff, *Magn. Reson. Imag.* **8**, 295–301 (1990).
97. L. A. Hayman, et al., *Magn. Reson. Imag.* **168**, 489–491 (1988).
98. P. A. Janick, D. B. Hackney, R. I. Grossman, T. Asakura, *AJNR* **12**, 891–897 (1991).
99. G. A. Wright, D. G. Nishimura, A. Macovski, *Magn. Reson. Med.* **17**, 126–140 (1991).
100. G. A. Wright, B. S. Hu, A. Macovski, *JMRI* **1**, 275–283 (1991).
101. K. R. Thulborn, T. J. Brady, *Magn. Reson. Quart.* **5**, 23–38 (1989).
102. B. E. Hoppel, et al., *Magn. Reson. Med* **30**, 715–723 (1993).
103. R. B. Lauffer, *Magn. Reson. Quart.* **6**, 65–84 (1990).
104. *Magn. Reson. Med.* **22**, 177–378 (1991).
105. N. Bloembergen, E. M. Purcell, R. V. Pound, *Phys. Rev.* **73**, 679 (1948).
106. B. R. Rosen, J. W. Belliveau, D. Chien, *Magn. Reson. Quart* **5**, 263–281 (1989).
107. A. Villringer, et al., *Magn. Reson. Med.* **6**, 164–174 (1988).
108. B. R. Rosen, J. W. Belliveau, J. M. Vevea, T. J. Brady, *Magn. Reson. Med.* **14**, 249–265 (1990).
109. K. M. Donahue, D. Burstein, W. J. Manning, M. L. Gray, *Magn. Reson. Med.* **32**, 66–76 (1994).

110. R. P. Kennan, J. Zhong, J. C. Gore, *Magn. Reson. Med.* **31**, 9–21 (1994).
111. C. R. Fisel, et al., *Magn. Reson. Med.* **17**, 336–347 (1991).
112. D. A. Yablonsky, E. M. Haacke, *Magn. Reson. Med* **32**, 749–763 (1994).
113. P. Gillis, S. H. Koenig, *Magn. Reson. Med.* **5**, 323–345 (1987).
114. P. A. Hardy, R. M. Henkleman, *Magn. Reson. Med.* **17**, 348–356 (1991).
115. P. A. Hardy, R. M. Henkleman, *Magn. Reson. Imag* **7**, 265–275 (1989).
116. S. C.-K. Chu, Y. Xu, J. A. Balschi, C. S. S. Jr, *Magn. Reson. Med* **13**, 239–262 (1990).
117. R. M. Weisskoff, B. J. Hoppel, B. R. Rosen, *JMRI* **2(P) [Abstr.]**, 77 (1992).
118. J. L. Boxerman, R. M. Weisskoff, B. E. Hoppel, B. R. Rosen, MR contrast due to microscopically heterogeneous magnetic susceptibility: cylindrical geometry, Proc., SMRM, 12th Annual Meeting, New York (1993).
119. W. B. Edmister, R. M. Weisskoff, Diffusion effects on *T*2 relaxation in microscopically inhomogeneous magnetic fields, Proc., SMRM, 12th Annual Meeting, New York (1993).
120. R. M. Weisskoff, C. S. Zuo, J. L. Boxerman, B. R. Rosen, *Magn. Reson. Med.* **31**, 601–610 (1994).
121. S. Ogawa, et al., *Biophysical J* **64**, 803–812 (1993).
122. E. C. Wong, P. A. Bandettini, A deterministic method for computer modelling of diffusion effects in MRI with application to BOLD contrast imaging, Proc., SMRM, 12th Annual Meeting, New York (1993).
123. C. S. Li, T. A. Frisk, M. B. Smith, Computer simulations of susceptibility effects: implications for lineshapes and frequency shifts in localized spectroscopy of the human head, Proc., SMRM, 12th Annual Meeting, New York (1993).
124. S. Posse, *Magn. Reson. Med* **25**, 12–29 (1992).
125. J. Lian, D. S. Williams, I. J. Lowe, *J. Magn. Reson.* **106**, 65–74 (1994).
126. T. J. Mosher, M. B. Smith, *Magn. Reson. Med.* **18**, 251–255 (1991).
127. J. Frahm, K.-D. Merboldt, W. Hanicke, *Magn. Reson. Med* **6**, 474–480 (1988).
128. I. J. Cox, et al., *J. Magn. Reson.* **70**, 163–168 (1986).
129. J. C. Ford, F. W. Wehrli, H.-W. Chung, *Magn. Reson. Med.* **30**, 373–379 (1993).
130. R. Bhagwandien, et al., *Magn. Reson. Imag* **10**, 299–313 (1992).
131. S. Majumdar, *Magn. Reson. Med.* **22**, 101–110 (1991).
132. S. Majumdar, J. C. Gore, *J. Magn. Reson.* **78**, 41–55 (1988).
133. D. T. Edmonds, M. R. Wormwald, *J. Magn. Reson.* **77**, 223–232 (1988).
134. F. W. Wehrli, J. C. Ford, M. Attie, H. Y. Kressel, F. S. Kaplan, *Radiology* **179**, 615–621 (1991).
135. S. Majumdar, D. Thomasson, A. Shimakawa, H. K. Genant, *Magn. Reson. Med.* **22**, 111–127 (1991).
136. T. A. Case, C. H. Durney, D. C. Ailion, A. G. Cutillo, A. H. Morris, *J. Magn. Reson.* **73**, 304–314 (1987).
137. S. Ogawa, T.-M. Lee, *Magn. Reson. Med* **16**, 9–18 (1990).
138. P. A. Bandettini, E. C. Wong, *International Journal of Imaging Systems and Technlogy* **6**, 134–152 (1995).
139. H. M. Duvernoy, S. Delon, J. L. Vannson, *Brain Research Bulletin* **7**, 519–579 (1981).
140. J. W. Belliveau, et al., *Magn. Reson. Med.* **14**, 538–546 (1990).
141. B. R. Rosen, et al., *Magn. Reson. Med.* **22**, 293–299 (1991).
142. D. S. Williams, J. A. Detre, J. S. Leigh, A. S. Koretsky, *Proc. Natl. Acad. Sci. USA* **89**, 212–216 (1992).
143. S.-G. Kim, *Magn. Reson. Med.* **34**, 293–301 (1995).
144. K. K. Kwong, et al., *Magn. Reson. Med.* **34**, 878–887 (1995).
145. E. C. Wong, R. B. Buxton, L. R. Frank, Quantitative imaging of perfusion using a single subtraction (QUIPSS), 2nd Int. Conf. of Func. Mapping of the Human Brain, Boston (1996).
146. E. C. Wong, R. B. Buxton, L. R. Frank, *Magn. Reson. Med* (submitted).
147. E. C. Wong, P. A. Bandettini, Two embedded techniques for simultaneous acquisition of flow and BOLD signals in functional MRI, Proc., ISMRM 4th Annual Meeting, New York (1996).

148. K. K. Kwong, *Magn. Reson. Quart.* **11**, 1–20 (1995).
149. M. K. Stehling, F. Schmitt, R. Ladebeck, *JMRI* **3**, 471–474 (1993).
150. A. M. Blamire, et al., *Proc. Natl. Acad. Sci. USA* **89**, 11069–11073 (1992).
151. S. Ogawa, T.-M. Lee, A. S. Nayak, P. Glynn, *Magn. Reson. Med.* **14**, 68–78 (1990).
152. S. Ogawa, T. M. Lee, A. R. Kay, D. W. Tank, *Proc. Natl. Acad. Sci. USA* **87**, 9868–9872 (1990).
153. R. Turner, D. LeBihan, C. T. W. Moonen, D. Despres, J. Frank, *Magn. Reson. Med.* **22**, 159–166 (1991).
154. R. Turner, et al., *Magn. Reson. Med.* **29**, 277–279 (1993).
155. P. Jezzard, et al., *NMR in Biomedicine* **7**, 35–44 (1994).
156. P. Mansfield, *J. Phys.* **C10**, L55–L58 (1977).
157. M. K. Stehling, R.Turner, P. Mansfield, *Science* **254**, 43–50 (1991).
158. M. S. Cohen, R. M. Weisskoff, *Magnetic Resonance Imaging* **9**, 1–37 (1991).
159. K. K. Kwong, D. A. Chesler, R. M. Weisskoff, B. R. Rosen, Perfusion MR imaging, Proc., SMR, 2nd Annual Meeting, San Francisco (1994).
160. D. A. Chesler, K. K. Kwong, *Int. J.ournal of Imag. Syst. and Tech.* **6**, 171–174 (1995).
161. R. B. Buxton, E. C. Wong, L. R. Frank, A quantitative model for epistar perfusion imaging, Proc., SMR, 2nd Annual Meeting, Nice (1995).
162. T. J. Brady, Future prospects for MR imaging, Proc., SMRM, 10th Annual Meeting, San Francisco (1991).
163. R. S. Menon, S. Ogawa, D. W. Tank, K. Ugurbil, *Magn. Reson. Med.* **30**, 380–386 (1993).
164. J. R. Baker, et al., Dynamic functional imaging of the complete human cortex using gradient-echo and asymmetric spin-echo echo-planar magnetic resonance imaging, Proc., SMRM, 12th Annual Meeting, New York (1993).
165. R. T. Constable, R. P. Kennan, A. Puce, G. McCarthy, J. C. Gore, *Magn. Reson. Med.* **31**, 686–690 (1994).
166. J. L. Boxerman, et al., *Magn. Reson. Med.* **34**, 4–10 (1995).
167. A. W. Song, E. C. Wong, S. G. Tan, J. S. Hyde, *Magn. Reson. Med.* **35**, 155–158 (1996).
168. R. S. Menon, S. Ogawa, J. P. Strupp, P. Anderson, K. Ugurbil, *Magn. Reson. Med.* **33**, 453–459 (1995).
169. K. Ugurbil, et al., *Magn. Reson. Quart.* **9**, 259–277 (1993).
170. P. A. Bandettini, et al., MRI of human brain activation at 0.5 T, 1.5 T, and 3 T: comparisons of $\Delta R2^*$ and functional contrast to noise ratio, Proc., SMR, 2nd Annual Meeting, San Francisco (1994).
171. A. Jesmanowicz, P. A. Bandettini, E. C. Wong, J. S. Hyde, Performance of fMRI using spin-echo and gradient-echo high resolution single-shotEPI at 3T, Proc., ISMRM 4th Annual Meeting, New York (1996).
172. P. A. Bandettini, E. C. Wong, *NMR in Biomedicine* (in press).
173. A. T. Lee, G. H. Glover, C. H. Meyer, *Magn. Reson. Med.* **33**, 745–754 (1995).
174. J. R. Binder, et al., *Arch. Neurol.* **52**, 593–601 (1995).
175. E. A. DeYoe, et al., *Proc. Natl. Acad. Sci.* **93**, 2382–2386 (1996).
176. M. I. Sereno, et al., *Science* **268**, 889–893 (1995).
177. S. A. Engel, et al., *Nature* **369, 370 [erratum]**, 525, 106 [erratum] (1994).
178. R. B. H. Tootell, et al., *The Journal of Neuroscience* **15**, 3215–3230 (1995).
179. S.-G. Kim, K. Hendrich, X. Hu, H. Merkle, K. Ugurbil, *NMR in Biomedicine* **7**, 69–74 (1994).
180. J. Frahm, K.-D. Merboldt, W. Hanicke, A. Kleinschmidt, H. Boecker, *NMR in Biomedicine* **7**, 45–53 (1994).
181. J. H. Duyn, C. T. W. Moonen, G. H. vanYperen, R. W. d. Boer, P. R. Luyten, *NMR in Biomedicine* **7**, 83–88 (1994).
182. J. H. Duyn, et al., *Int. J.ournal of Imag. Syst. and Tech.* **6**, 245–252 (1995).
183. J. H. Duyn, et al., *Magn. Reson. Med* **32**, 150–155 (1994).
184. A. T. Lee, C. H. Meyer, G. H. Glover, *JMRI* **3(P), [Abstr.]**, 59–60 (1993).
185. E. M. Haacke, S. Lai, D. A. Yablonski, W. Lin, *Int. J.ournal of Imag. Syst. and Tech.* **6**, 153–163 (1995).

186. H. Wen, et al., Phase and magnitude functional imaging of motor tasks and pain stimuli, Proc., SMRM, 12th Annual Meeting, New York (1993).
187. J. Hennig, C. Janz, O. Speck, T. Ernst, *Int. J.ournal of Imag. Syst. and Tech.* **6**, 203–208 (1995).
188. R. S. Menon, X. Hu, P. Anderson, K. Ugurbil, S. Ogawa, Cerebral oxy/deoxy hemoglobin changes during neural activation: MRI timecourse correlates to optical reflectance measurements, Proc., SMR, 2nd Annual Meeting, San Francisco (1994).
189. E. C. Wong, P. A. Bandettini, J. S. Hyde, Echo-planar imaging of the human brain using a three axis local gradient coil, Proc., SMRM, 11th Annual Meeting, Berlin (1992).
190. Z.-H. Cho, Y.-M. Ro, S.-T. Park, S.-C. Chung, *Magn. Reson. Med.* **36**, 1–5 (1996).
191. Z. H. Cho, Y. M. Ro, T. H. Lim, *Magn. Reson. Med.* **28**, 237–248 (1992).
192. E. M. Haacke, et al., *NMR in Biomedicine* **7**, 54–62 (1994).
193. P. A. Bandettini, E. C. Wong, A. Jesmanowicz, R. S. Hinks, J. S. Hyde, *NMR in Biomedicine* **7**, 12–19 (1994).
194. P. A. Bandettini, E. C. Wong, A. Jesmanowicz, R. S. Hinks, J. S. Hyde, Simultaneous mapping of activation–induced $\Delta R2^*$ and $\Delta R2$ in the human brain using a combined gradient-echo and spin-echo EPI pulse sequence, Proc., SMRM, 12th Annual Meeting, New York (1993).
195. P. A. Bandettini, et al., Simultaneous assessment of blood oxygenation and flow contributions to activation induced signal changes in the human brain, Proc., SMR, 2nd Annual Meeting, San Francisco (1994).
196. P. A. Bandettini, K. K. Kwong, E. C. Wong, R. B. Tootel, B. R. Rosen, Direct $R2^*$ measurements and flow insensitive $T2^*$ weighted studies indicate a sustained elevation of blood oxygenation during long term activation, Proc., ISMRM 4th Annual Meeting, New York (1996).
197. R. M. Weisskoff, et al., Power spectrum analysis of functionally-weighted MR data: what's in the noise?, Proc., SMRM, 12th Annual Meeting, New York (1993).
198. B. Biswal, P. A. Bandettini, A. Jesmanowicz, J. S. Hyde, Time-frequency analysis of functional EPI time-course series, Proc., SMRM, 12th Annual Meeting, New York (1993).
199. E. A. DeYoe, P. Bandettini, J. Neitz, D. Miller, P. Winans, *J. Neuroscience Methods* **54**, 171–187 (1994).
200. R. L. Savoy, et al., Pushing the temporal resolution of fMRI: studies of very brief visual stimuli, onset variablity and asynchrony, and stimulus-correlated changes in noise, Proc., SMR 3rd Annual Meeting, Nice (1995).
201. P. A. Bandettini, et al., in *Diffusion and Perfusion: Magnetic Resonance Imaging* D. LeBihan, Ed. (Raven Press, New York, 1995) pp. 335–349.
202. K. J. Friston, P. Jezzard, R. Turner, *Human Brain Mapping* **1**, 153–171 (1994).
203. J. R. Binder, et al., Analysis of phase differences in periodic functional MRI activation data, Proc., SMRM, 12th Annual Meeting, New York (1993).
204. R. L. Savoy, et al., Exploring the temporal bourdaries of fMRI: measuring responses to very brief visual stimuli, Book of Abstracts, Soc. for Neuroscience 24th Annual Meeting, Miami (1994).
205. W. W. Orrison, J. D. Lewine, J. A. Sanders, M. F. Hartshorne, *Functional Brain Imaging* (Mosby-Year Book, Inc., St. Louis, 1995).
206. J. R. Ives, S. Warach, F. Schmitt, R. R. Edelman, D. L. Schomer, *EEG and Clinical Neurophys* **87**, 417–420 (1993).
207. R. Turner, P. Jezzard, D. L. Bihan, A. Prinster, Contrast mechanisms and vessel size effects in BOLD contrast functional neuroimaging, Proc., SMRM, 12th Annual Meeting, New York (1993).
208. S. Lai, et al., *Magn. Reson. Med* **30**, 387–392 (1993).
209. W. Schneider, D. C. Noll, J. D. Cohen, *Nature* **365**, 150–153 (1993).
210. K. Butts, S. J. Riederer, R. L. Ehman, R. M. Thompson, C. R. Jack, *Magn. Reson. Med.* **31**, 67–72 (1994).
211. G. C. McKinnon, *Magn. Reson. Med.* **30**, 609–616 (1993).

212. E. A. DeYoe, P. W. Schmit, J. Neitz, Distinguishing cortical areas that are sensitive to task and stimulus variables with fMRI, Book of Abstracts, Soc. for Neuroscience 25'th Annual Meeting, San Diego (1995).
213. S. M. Rao, et al., *J. Cereb. Blood Flow and Metab.* **16**, 1250–1254 (1996).
214. J. R. Binder, et al., *Cognitive Brain Research* **2**, 31–38 (1994).
215. A. R. Guimaraes, J. R. Baker, R. M. Weisskoff, Cardiac-gated functional mri with $T1$ correction, Proc., SMR 3rd Annual Meeting, Nice (1995).
216. P. A. Bandettini, A. Jesmanowicz, E. C. Wong, J. S. Hyde, *Magn Reson. Med.* **30**, 161–173 (1993).
217. R. P. Woods, S. R. Cherry, J. C. Mazziotta, *J. Comp. Asst. Tomog.* **115**, 565–587 (1992).
218. K. J. Friston, et al., *Human Brain Mapping* (in press).
219. P. Jezzard, et al., An investigation of the contributions of physiological noise in human functional MRI studies at 1.5 Tesla and 4 Tesla, Proc., SMRM, 12th Annual Meeting, New York (1993).
220. J. D. Cohen, D. C. Noll, W. Schneider, *Behavior Research Methods, Instruments, & Computers* **25**, 101–113 (1993).
221. J. R. Binder, et al., Temporal characteristics of functional magnetic resonance signal changes in lateral frontal and auditory cortex, Proc., SMRM, 12th Annual Meeting, New York (1993).
222. E. A. DeYoe, J. Neitz, P. A. Bandettini, E. C. Wong, J. S. Hyde, Time course of event-related MR signal enhancement in visual and motor cortex, Proc., SMRM, 11th Annual Meeting, Berlin (1992).
223. D. C. Noll, *Int. J.ournal of Imag. Syst. and Tech.* **6**, 175–183 (1995).
224. D. C. Noll, J. D. Cohen, C. H. Meyer, W. Schneider, *JMRI* **5**, 49–56 (1995).
225. G. H. Glover, A. T. Lee, *Magn. Reson. Med.* **33**, 624–635 (1995).
226. X. Hu, S.-G. Kim, *Magn. Reson. Med.* **31**, 495–503 (1994).
227. E. C. Wong, E. Boskamp, J. S. Hyde, A volume optimized quadrature elliptical end-cap birdcage brain coil, Proc., SMRM, 11th Annual Meeting, Berlin (1992).
228. J. Frahm, G. Krüger, K.-D. Merboldt, A. Kleinschmidt, *Magn. Reson. Med.* **35**, 143–148 (1996).
229. S. Ogawa, T. M. Lee, *JMRI* **2(P)-WIP supplement, [Abstr.]**, S22 (1992).
230. C. E. Stern, K. K. Kwong, J. W. Belliveau, J. R. Baker, B. R. Rosen, MR tracking of physiological mechanisms underlying brain activity, Proc., SMRM, 11th Annual Meeting, Berlin (1992).
231. G. M. Hathout, et al., *JMRI* **4**, 537–543 (1994).
232. P. A. Bandettini, et al., *Human Brain Mapping* **5**, 93–109 (1997).
233. P. A. Bandettini, et al., FMRI demonstrates sustained blood oxygenation and flow enhancement during extended duration visual and motor cortex activation, Book of Abstracts, Soc. for Neuroscience 25th Annual Meeting, San Diego (1995).
234. P. A. Bandettini, et al., The functional dynamics of blood oxygen level dependent contrast in the motor cortex, Proc., SMRM, 12th Annual Meeting, New York (1993).
235. B. Biswal, F. Z. Yetkin, V. M. Haughton, J. S. Hyde, *Magn. Reson. Med.* **34**, 537–541 (1995).
236. B. Biswal, A. G. Hudetz, F. Z. Yetkin, V. M. Haughton, J. S. Hyde, *J. of Cereb. Blood Flow and Metab.* **17**, 301–308 (1996).
237. F. Farzaneh, S. J. Riederer, N. J. Pelc, *Magn. Reson. Med.* **14**, 123–139 (1990).
238. A. M. Blamire, R. G. Shulman, *Magnetic Resonance Imaging* **12**, 669–671 (1994).
239. P. Jezzard, R. S. Balaban, *Magn. Reson. Med.* **34**, 65–73 (1995).
240. R. M. Weisskoff, T. L. Davis, Correcting gross distortion on echo planar images, Proc., SMRM, 11th Annual Meeting, Berlin (1992).
241. A. W. Song, E. C. Wong, J. S. Hyde, *Magn. Reson. Med.* **32**, 668–671 (1994).
242. J. Frahm, K.-D. Merboldt, W. Hanicke, *Magn. Reson. Med.* **29**, 139–144 (1993).
243. A. Connelly, et al., *Radiology* **188**, 125–130 (1993).
244. Y. Cao, V. L. Towle, D. N. Levin, J. M. Balter, *JMRI* **3**, 869–875 (1993).
245. R. T. Constable, et al., *Magn. Reson. Imag* **11**, 451–459 (1994).

246. X. Hu, S.-G. Kim, *Magn. Reson. Med.* **30**, 512–517 (1993).
247. E. C. Wong, S. G. Tan, A comparison of signal to noise ratio and BOLD contrast between single voxel spectroscopy and echo-planar imaging, Proc., SMR, 2nd Annual Meeting, San Francisco (1994).
248. G. Liu, G. Sobering, A. W. Olson, P. v. Gelderen, C. T. Moonen, *Magn. Reson. Med.* **30**, 68–75 (1993).
249. C. T. Moonen, G. Liu, P. v. Gelderen, G. Sobering, *Magn. Reson. Med.* **26**, 184–189 (1992).
250. D. W. Shaw, et al., Reduced K-space (keyhole) functional imaging without contrast on a conventional clinical scanner, Proc., SMRM, 12th Annual Meeting, New York (1993).
251. G. H. Glover, A. T. Lee, C. H. Meyers, Motion artifacts in fMRI: comparison of 2DFT with PR and spiral scan methods, Proc., SMRM, 12th Annual Meeting, New York (1993).
252. J. D. Cohen, S. D. Forman, B. J. Casey, D. C. Noll, Spiral–scan imaging of dorsolateral prefrontal cortex during a working memory task, Proc., SMRM, 12th Annual Meeting, New York (1993).
253. J. R. Binder, et al., *Ann. Neurol.* **35**, 662–672 (1994).
254. K. Sakai, et al., *Magn. Reson. Med.* **33**, 736–743 (1995).
255. T. A. Hammeke, et al., *Neurosurgery* **35**, 677–681 (1994).
256. J. M. Ellerman, et al., *NMR in Biomedicine* **7**, 63–68 (1994).
257. S.-G. Kim, et al., *J. Neurophysiol* **69**, 297–302 (1993).
258. C. R. Jack, et al., *Radiology* **190**, 85–92 (1994).
259. Y. Cao, E. M. Vikingstad, P. R. Huttenlocher, V. L. Towle, D. N. Levin, *PNAS* **91**, 9612–9616 (1994).
260. S. M. Rao, et al., *Neurology* **45**, 919–924 (1995).
261. E. A. DeYoe, J. Neitz, D. Miller, J. Wieser, Functional magnetic resonance imaging (FMRI) of visual cortex in human subjects using a unique video graphics stimulator, Proc., SMRM, 12th Annual Meeting, New York (1993).
262. R. L. DeLaPaz, et al., Human gustatory cortex localization with functional MRI, Proc., SMR, 2nd Annual Meeting, San Francisco (1994).
263. S.-G. Kim, et al., *Science* **261**, 615–616 (1993).
264. P. Jezzard, et al., Practice makes perfect: a functional MRI study of long term motor cortex plasticity, Proc., SMR, 2nd Annual Meeting, San Francisco (1994).
265. J. M. Ellerman, et al., Cerebellar activation due to error detection/correction in a visuo-motor learning task: a functional magnetic resonance imaging study, Proc., SMR, 2nd Annual Meeting, San Francisco (1994).
266. G. McCarthy, A. M. Blamire, D. L. Rothman, R. Gruetter, R. G. Shulman, *Proc. Natl. Acad. Sci. USA* **90**, 4952–4956 (1993).
267. R. M. Hinke, et al., *Neuroreport* **4**, 675–678 (1993).
268. L. Rueckert, et al., *J. Neuroimaging* **4**, 67–70 (1994).
269. J. R. Binder, *Int. Journal of Imag. Syst. and Tech.* **6**, 280–288 (1995).
270. S. M. Rao, et al., *Neurology* **43**, 2311–2318 (1993).
271. S.-G. Kim, J. E. Jennings, J. P. Strupp, P. Anderson, K. Ugurbil, *Int. Journal of Imag. Syst. and Tech.* **6**, 271–279 (1995).
272. W. Schneider, B. J. Casey, D. Noll, *Human Brain Mapping* **1**, 117–133 (1994).
273. A. G. Sorensen, et al., Extrastriate activation in patients with visual field defects, Proc., SMRM, 12th Annual Meeting, New York (1993).
274. J. R. Binder, et al., *Neurology* **46**, 978–984 (1996).
275. J. R. Binder, S. M. Rao, in *Localization and Neuroimaging in Neuropsychology* A. Kertesz, Ed. (Academic Press, San Diego, 1994) pp. 185.
276. D. LeBihan, et al., *Proc. Natl. Acad. Sci. USA* **90**, 11802–11805 (1993).
277. M. S. Cohen, et al., *Brain* **119**, 89–100 (1996).
278. P. F. Renshaw, D. A. Yurgelun-Todd, B. M. Cohen, *Am. J. Psychiatry* **151**, 1493–1495 (1994).

279. G. D. Jackson, A. Connelley, J. H. Cross, I. Gordon, D. G. Gadian, *Neurology* **44**, 850–856 (1994).
280. F. Wenz, et al., Effects of neuroleptic drugs on signal intensity during motor cortex stimulation: functional MR-imaging performed with a standard 1.5 T clinical scanner, Proc., SMRM, 12th Annual Meeting, New York (1993).
281. Y. Cao, V. L. Towle, D. N. Levin, R. Grzeszczuk, J. F. Mullian, Conventional 1.5 T functional MRI localization of human hand senserimotor cortex with intraoperative electrophysiologic validation, Proc., SMRM, 12th Annual Meeting, New York (1993).
282. J. Pujol, et al., *J. Neurosurg* **84**, 7–13 (1996).
283. G. L. Morris, et al., *Epilepsia* **35**, 1194–1198 (1994).
284. R. Turner, P. Jezzard, in *Functional neuroimaging. Technical foundations.* R. W. Thatcher, M. Hallett, T. Zeffiro, E. R. John, M. Huerta, Eds. (Academic Press, San Diego, 1994) pp. 69–78.
285. R. G. Schulman, A. M. Blamire, D. L. Rothman, G. McCarthy, *Proc. Natl. Acad. Sci. USA* **90**, 3127–3133 (1993).
286. J. W. Prichard, B. R. Rosen, *Journal of Cerebral Blood Flow and Metabolism* **14**, 365–372 (1994).
287. M. E. Cohen, S. Y. Bookheimer, *TINS* **17**, 1994 (1994).
288. P. A. Bandettini, J. R. Binder, E. A. DeYoe, J. S. Hyde, in *Encyclopedia of Nuclear Magnetic Resonance* D. M. Grant, R. K. Harris, Eds. (John Wiley and Sons, Chichester, 1996), vol. 1, pp. 1051–1056.
289. P. A. Bandettini, et al., in *Diffusion and Perfusion Magnetic Resonance Imaging* D. LeBihan, Ed. (Raven Press, New York, 1995) pp. 351–362.
290. P. A. Bandettini, Ph.D. Dissertation, Medical College of Wisconsin (1995).
291. P. A. Bandettini, E. C. Wong, *Neurosurgery Clinics of North America* **8**, 345–371 (1997).

10 Imaging with Nonionizing Radiation

E. Russell Ritenour and Russell K. Hobbie

University of Minnesota

10.1 Introduction

There are a number of ways that non-ionizing radiation is used to make images of the body. The most familiar is a photograph using visible light. Photographs with ultraviolet or infrared light can also supply diagnostic information, though the former might be considered ionizing. Electric and magnetic fields that vary slowly with time also provide diagnostic information, though strictly speaking they are measured in the near field, not the radiation field. The most familiar such measurements are the electrocardiogram and electroencephalogram. Electric and magnetic source images represent a simple logical extension of these familiar tracings.

This chapter begins with a description of measurements of the electric potential on the body surface and of the magnetic field near the body. These measurements provide information about electrical activity in nerve and muscle cells, including fields that arise due to anisotropies in tissue conductivity. Measurements of fields due to ingested magnetic materials can provide information such as transit time in the gut. Electrical impedance tomography is an experimental technique for mapping conductivities within the body.

The chapter then turns to optical methods, which include measurement of infrared light emitted by the body and light from an external source diffusely reflected by tissue. These methods yield information such as blood oxygenation levels and have been investigated for possible applications in tumor characterization.

Ultrasound and magnetic resonance imaging also rely on non-ionizing radiation. However, these topics are discussed in other chapters.

10.2 Electric and Magnetic Surface Measurements

Electrical potential differences on the surface of the body and magnetic fields just outside the body arise from the same sources. This section reviews the basic physics of these sources and then discusses clinical measurements.

10.2.1 Origin of the Potential Difference and Magnetic Field

Electrical activity in a nerve or muscle cell generates electric and magnetic fields around the cell. The weak magnetic field from a group of cells can be measured outside the body. The electric field creates potential differences on the body surface. While quite sophisticated models exist for calculating these fields, our basic intuition for the problem is best developed using the simple cable model for the currents in and around a cell. (More detailed descriptions of the generation of the potential and the magnetic field can be found several places, such as in Hobbie[1] and in Wikswo.[2])

Imagine a long thin cell of radius a stretched along the x axis, with a voltage distribution $v_i(x, t)$ along the interior of the cell. The cell interior is an ohmic conducting medium with conductivity σ_i and resistance $r_i = 1/\pi a^2 \sigma_i$ per unit length. The current along the inside of the cell is $i_i = -(1/r_i)\partial v_i/\partial x$. Because charge is conserved, if i_i changes with position along the cell, the difference in current flows to the cell membrane. Part of this current charges or discharges the membrane capacitance (the displacement current), and the rest of it passes through the membrane (the conduction current). In the cable model this statement of charge conservation is

$$\frac{1}{2\pi a r_i}\frac{\partial^2 v_i}{\partial x^2} = c_m \frac{\partial(v_i - v_o)}{\partial t} + j_m, \tag{10.1}$$

where a is the radius of the cell, c_m is the membrane capacitance per unit area, j_m is the current per unit area flowing out through the membrane, and v_o is the potential at the outer surface of the membrane. This statement is independent of the mechanism producing the change in voltage along the cell. It depends only on conservation of charge and Ohm's law for the intracellular fluid. If the membrane current j_m is also described by Ohm's law, this equation leads to the classic model for electrotonus. A nonlinear model for j_m, such as the Hodgkin-Huxley model, leads to the propagation of an action potential.

10.2.1.1 The External Electric Field

Suppose that the cell is embedded in an infinite homogeneous medium with conductivity σ_o. After passing through the membrane, the conduction current j_m and displacement current $c_m\partial(v_i - v_o)/\partial t$ combine and flow in the external medium. The easiest way to visualize this flow is to use a piecewise-linear model to describe a depolarization wave that is traveling along the interior of the cell. A snapshot of the depolarization is shown in Figure 10.1. The abrupt changes of slope lead to δ functions for the second derivative, corresponding to a point source and sink of current in the external medium. The current along the cell between $x = 0$ and $x = x_2$ is $i_i = \Delta v_i \sigma_i \pi a^2/x_2$. The same current flows into the external medium from a source at x_2 and returns to a sink at the origin.

If the cell stretched along the x axis is very thin and does not appreciably change the isotropic homogeneity of the external medium, then a current source i_o at point x gives rise to a spherically symmetric current density j directed radially away from the source point and having magnitude $j = i_o/4\pi r^2$ at a distance r from the source. The potential a distance r is $v(r) = i_o/4\pi\sigma_o r$. If

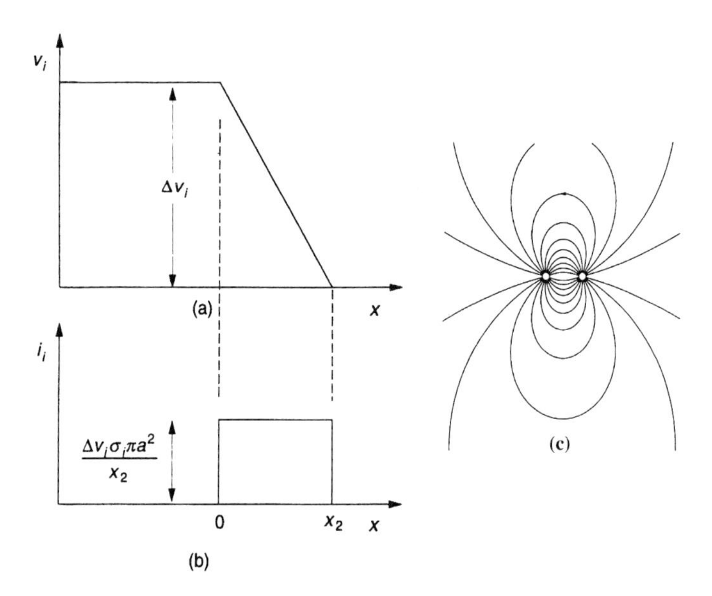

Figure 10.1. (a) The piecewise-linear approximation to the rising edge (depolarization) of an intracellular potential. (b) The current along the inside of the cell consistent with (a). (c) The electric field or current density lines outside the cell in an infinite homogeneous conductor.

a vector from the origin to the observation point has magnitude r_0 and makes an angle θ with the x axis, then the potential from both the source at $x = x_2$ and the sink at the origin is

$$v = \frac{\Delta v_i \pi a^2}{4\pi\sigma_o x_2}\left(\frac{1}{r_2} - \frac{1}{r_0}\right). \tag{10.2}$$

If the spacing of source and sink is small compared to the distance to the observation point, this is approximately

$$v = \frac{\Delta v_i \pi a^2 \sigma_i}{4\pi\sigma_o r^2}\cos\theta = \frac{p\cos\theta}{4\pi\sigma_o r^2}\,\frac{\boldsymbol{p}\cdot\boldsymbol{r}}{4\pi\sigma_o r^3}. \tag{10.3}$$

We have defined the *current dipole vector* $\boldsymbol{p}$ with magnitude $p = \pi a^2 \sigma_i \Delta v_i$. It points from the current sink to the current source. Its magnitude is equal to the product of the current and the distance between the source and sink. Field lines for the external electric field $\boldsymbol{E}$ and the current density $\boldsymbol{j}$, shown in Figure 10.1(c), are characteristic of a dipole.

A cell that depolarizes and then repolarizes can be modeled by adding a third current source at the trailing end of the pulse, $x = -x_1$ and increasing the strength of the sink, as shown in Figure 10.2. Far away the potential is

$$v = \left(\frac{2\pi a^2}{4\pi r^3}\right)\left(\frac{(x_1+x_2)\Delta v_i}{2}\right)\left(\frac{(3\cos^2\theta-1)}{2}\right). \tag{10.4}$$

We have represented the complete pulse by two current sources and one sink, or by two back-to-back dipoles. The angular and radial dependencies are those of a quadrupole source. The second term in parentheses is the area under the curve of $\boldsymbol{v}(x) - \boldsymbol{v}_{\text{resting}}$.

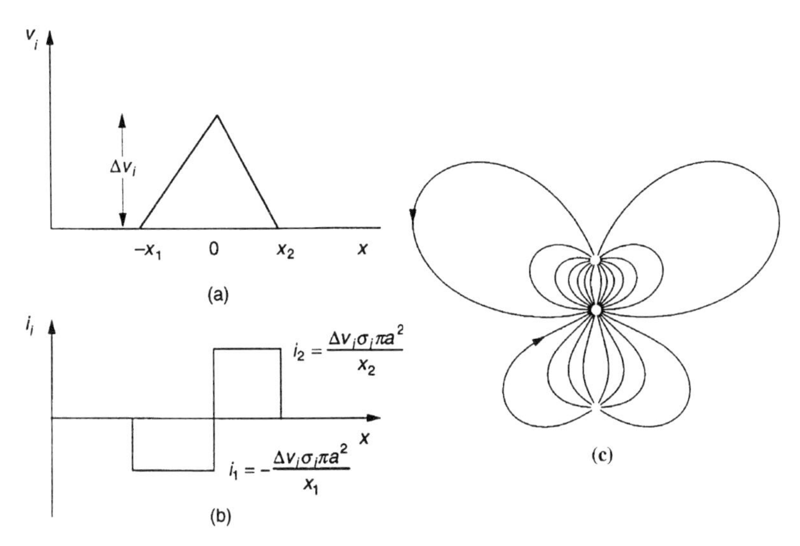

Figure 10.2. (a) A piecewise-linear approximation to an impulse of amplitude Δv_i. (b) The current along the cell consistent with the approximation in (a). (c) The electric field or current density lines outside the cell.

Equivalent results can be obtained without resorting to the piecewise-linear approximation. In general, using the approximation that the cell forms a line along the x axis, the potential is

$$v(\boldsymbol{R}) = \frac{\pi a^2 \sigma_i}{4\pi\sigma_o} \int \frac{\partial^2 v_i}{\partial x^2} \frac{1}{r}\, \mathrm{d}x, \tag{10.5}$$

where $\boldsymbol{R}$ is a vector from the origin to the observation point and r is the distance from point x to the observation point. This can be integrated numerically. Far from the cell, the first two terms of a multipole expansion of Eq. (10.5) give

$$v(\boldsymbol{R}) = \frac{\pi a^2 \sigma_i}{4\pi\sigma_o}\left(\frac{1}{R}\left[\frac{\partial v_i}{\partial x}\right]_{x_1}^{x_2} + \frac{\cos\theta}{R^2}[v_i(x_2) - v_i(x_1)] + \frac{3\cos^2\theta - 1}{2R^3} 2\int_{x_1}^{x_2} [v_i(x) - v_{\text{rest}}]\mathrm{d}x + \dots\right). \tag{10.6}$$

The first (monopole) term vanishes because x_1 and x_2 are chosen to be at places on the cell where the derivative vanishes. The second term is the dipole term seen in Eq. (10.3) and vanishes if we consider the entire pulse. The last term is the quadrupole term seen in Eq. (10.4).

10.2.1.2 The External Magnetic Field

We continue to adopt the model that the cell is embedded in an infinite, homogeneous conducting medium. The fact that the body actually has a surface will be addressed below. Current flows in three regions in this model: (i) along the cell, (ii) through the membrane, and (iii) in the infinite, homogeneous conducting medium outside the cell. The external current (iii) is the superposition of spherically symmetric radial currents from a series of point sources spread along the cell. One can show by symmetry that in this infinite, homogenous medium the spherically symmetric radial current from each source makes no contribution to the magnetic field. Also, the field from the current in the membrane (ii) is small because the membrane is so thin. Therefore, the magnetic field surrounding the cell is obtained by integrating the Biot-Savart law for the current (i) along the cell. Since the cell is stretched along the x axis, this is

$$\boldsymbol{B} = \frac{\mu_o}{4\pi} \int \frac{i_i(x)\hat{\boldsymbol{x}} \times \boldsymbol{r}}{r^3} \mathrm{d}x = -\frac{\mu_0 \pi a^2 \sigma_i}{4\pi} \int \frac{(\partial \upsilon_i / \partial x)(\hat{\boldsymbol{x}} \times \boldsymbol{r})}{r^3} \mathrm{d}x, \tag{10.7}$$

where $\boldsymbol{r}$ is the vector from x to the observation point. If there is no repolarization, but only depolarization that can be described by a current dipole $\boldsymbol{p}$ at the origin, integration gives

$$\boldsymbol{B} = \frac{\mu_o}{4\pi r^3} \boldsymbol{p} \times \boldsymbol{r}. \tag{10.8}$$

Note the similarity of Eq. (10.8) to Eq. (10.3). Both expressions are proportional to $\boldsymbol{p}$. However, the potential is a scalar proportional to $p \cos \theta$, while the magnetic field is a vector whose magnitude is proportional to $p \sin \theta$. The lines of $\boldsymbol{B}$ are circles, with the vector pointing in the direction of the fingers of the right hand when the thumb is in the direction of $\boldsymbol{p}$. (Barach[3] provides a good discussion of the different ways of calculating the magnetic field, which give different pictures of which currents generate the fields.)

Measurements are made on the surface of the body, either with electrodes that measure potential differences or with magnetometers that measure a component of the magnetic field. (We ignore for now the fact that the infinite homogenous conducting medium of our model has no surface; we assume that measurements on the surface will be the same as those predicted by this model.) Figure 10.3 shows a simple geometry in which measurements are made either in a plane perpendicular to $\boldsymbol{p}$ or a plane parallel to $\boldsymbol{p}$. The results of these measurements are shown in Figure 10.4. Two results are striking. First, complete measurements of either υ or $\boldsymbol{B}$ give the same information, because of the similarity of Eqs. (10.3) and (10.8). Second, some of the measurements show a positive peak and negative valley, with the zero crossing located directly above the dipole. The spacing between the peak and valley, Δ, is related to the depth of the dipole below the surface, d, by $d = \Delta/\sqrt{2}$.

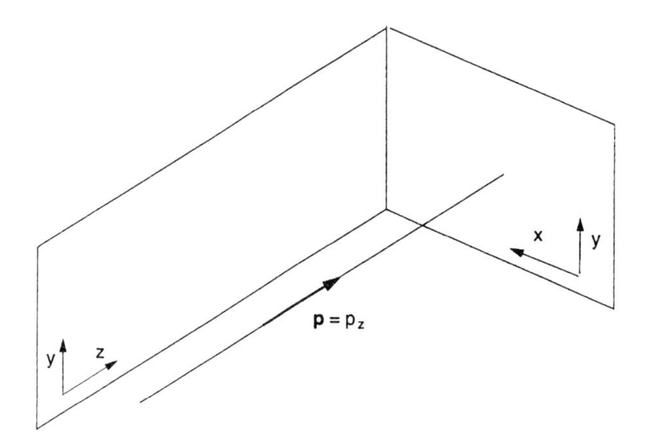

Figure 10.3. A cell is stretched along the x axis. A depolarization wavefront is described by a current dipole p. Measurements of $\boldsymbol{B}$ and v are made in the xy plane perpendicular to p and in the xz plane parallel to $\boldsymbol{p}$. The results are shown in Figure 10.4.

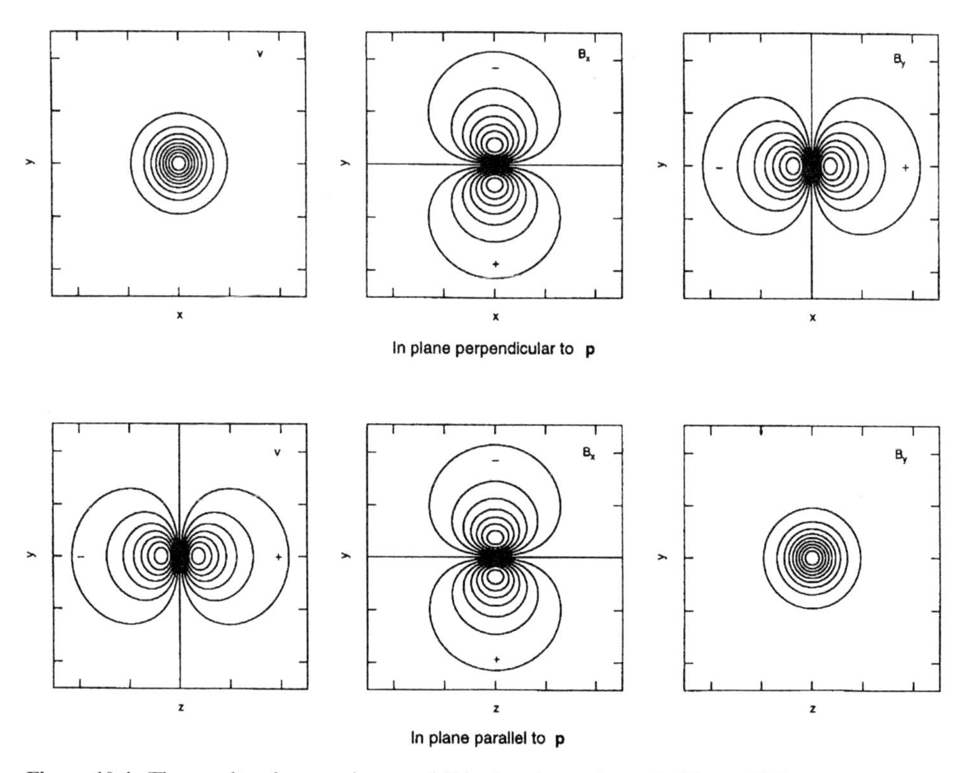

Figure 10.4. The results of measuring v and $\boldsymbol{B}$ in the planes shown in Figure 10.3.

When repolarization is also present, Eq. (10.8) can be expanded in powers of x/R. To express it in the simplest form, note that the lines of $\boldsymbol{B}$ are circles around the x axis and calculate B_z when $\boldsymbol{R}$ is in the xy plane.

$$B_z(\boldsymbol{R}) = B_z(R,\theta) = \frac{\mu_0 \pi a^2 \sigma_i}{4\pi R^2}\left[\sin\theta(\upsilon_i(x_1) - (\upsilon_i(x_2)) + \frac{3\sin\theta\cos\theta}{R}\int_{x_1}^{x_2}(\upsilon_i(x) - \upsilon_{\text{resting}})\mathrm{d}x\right] \tag{10.9}$$

Points x_1 and x_2 are selected to be where $\partial\upsilon_i/\partial x$ vanishes. The first term is analogous to Eq. (10.8). The second term is used for a complete impulse.

Figure 10.5 compares the potential and magnetic field along a line parallel to the x axis and one unit away. Both the depolarization wavefront and the complete pulse are located at the origin. The length of the complete pulse is 0.1 unit.

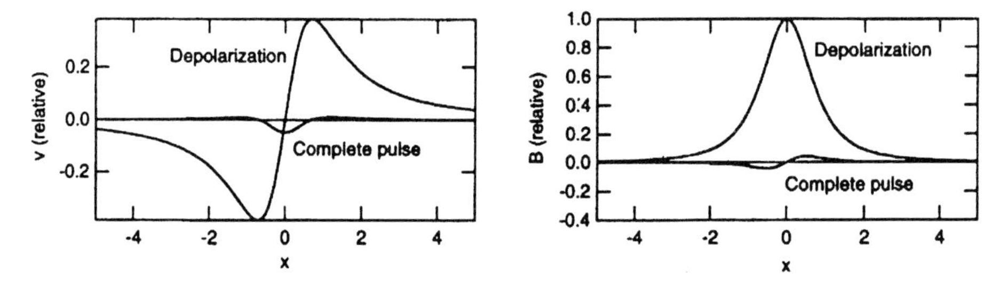

Figure 10.5. Comparison of the potential and the magnetic field parallel to the x axis and 1 unit away. The pulse has a length of 0.1 unit.

10.2.1.3 More Accurate Calculations

This model of a line source along the axis of an infinite homogeneous conductor, gives a simple physical picture of how the measured quantities arise. It can be improved in at least three respects. First, a more accurate calculation of the field and potential for this simple geometry can be made, taking into account the finite radius of the cell which generates the fields. Second, the fact that the body is not an infinite, homogenous, isotropic conducting medium can significantly alter both the potential and the magnetic field. Finally, real signals arise from a bundle of nerve or muscle cells in which the conductivity may be neither homogeneous nor isotropic.

10.2.1.4 More Accurate Models for the Cell

More accurate models that solve Laplace's equation inside and outside the cell subject to a given potential across the membrane have been studied by a number of authors. Trayanova, Henriquez and Plonsey[4] compared the line approximation presented above with more accurate calculations. They showed that for a single fiber, the length of the rising portion of the action potential is usually long compared to the fiber radius, in which case the line approximation is excellent. The line approximation is not as accurate when it is used to model a nerve bundle having a radius comparable to the length of the rise of the action potential.

10.2.1.5 Tissue Inhomogeneities and the Body Surface

Since the body is finite in extent and is surrounded by non-conducting air, return current that would have flowed outside the body flows inside. This distorts both the potential difference and the magnetic field. If there is a low-conductivity layer such as the skull just beneath the surface, the potential on the surface is attenuated. Because the magnetic permeability of tissue and air are nearly the same, there is no analogous attenuation of the magnetic field; the magnetic field is altered only by contributions from the return current that is no longer spherically symmetric and infinite in extent. Different organs have different conductivities. More sophisticated models include regions with different conductivity. Some, such as a spherical head, can be solved analytically. Most require numerical solutions of the differential equations for the potential or current density. Some models use piecewise-homogenous regions of different conductivity to represent various organs.

Such calculations for cardiography have been reviewed by Stroink, Lamothe, and Gardner.[5] References cited by Stroink[6] have shown that the effects of torso boundaries are as large for the magnetocardiogram as for the electrocardiogram. Differences in conductivity between blood and myocardium have the greatest effect on magnetic fields.[7]

A number of authors have discussed the distortions of the magnetic fields and the surface potentials due to conductivity differences at boundaries such as the skull.[8, 9, 10] The head is often modeled by four concentric spheres of varying conductivity representing brain, cerebrospinal fluid, bone, and scalp. Some volume conductor models adopt a more realistic shape including holes in the skull such as those for the eyes and at the base of the skull. The most realistic calculations assume a source, calculate the current throughout the head (from which the surface potential can be calculated), and then use the Biot-Savart law to calculate the magnetic field. Van den Broek, Zhou, and Peters[11] have investigated the accuracy of such calculations for the forward problem in a homogeneous spherical volume conductor where the result can be calculated exactly. Errors of about 3% were obtained with a mesh of about 1000 nodes subject to local refinement. The calculation required less than 30 minutes of a supercomputer time. Czapski, Ramon, Huntsman, Bardy, and Kim made a model study of the effect of volume currents on the MEG. The greatest effect was from blood.[12]

Results for the forward problem (given the source distribution, calculating the field or potential) are required in order to attack the inverse problem (determining the source distribution from the field or potential surface map). Unfortunately, the inverse problem is not well-posed and does not have a unique solution: an infinite number of different source arrangements can give the same external field or potential. The inverse solution is also quite susceptible to noise. Not only must tissue inhomogeneities[13, 14, 15] be taken into account, but various constraints must be imposed on the inverse solution.[16, 17] The inverse problem is currently under active investigation, both for the heart[18, 19, 20, 21, 22, 23] and for the head.[24, 25, 26]

10.2.1.6 Bundles of Cells: the Bidomain Model

The bidomain model assumes that a small volume element contains both an extracellular region and an intracellular region consisting of many interconnected cells. Each region can be described by Ohm's law. Except on the membrane separating the two regions, the divergence of the current density is zero. The bidomain region may be surrounded by another conducting region representing the rest of the body. The conductivities in the intracellular and extracellular regions of the bidomain are usually anisotropic and are described by conductivity tensors. A particularly clear calculation using the bidomain model has been presented by Roth.[27] Roth and Wikswo[28] have demonstrated that in a bidomain system where the conductivity is anisotropic, magnetic measurements can provide information about the conductivity tensor that is not available from measurements of the electric potential.

Calculations using the bidomain model show that when the conductivity tensor is anisotropic, the current dipole vector associated with the advancing wavefront need not be perpendicular to the wavefront. In particular, if the ratio of longitudinal to transverse conductivity is different in the extracellular and intracellular media, the depolarization wave front in a thick strand of cardiac muscle is curved, shaped like a meniscus. The action potential at the surface of the strand leads the action potential at the center.[29]

10.2.2 Measurements and Clinical Use

10.2.2.1 Cardiac: Electrocardiogram and Magnetocardiogram

The most common surface potential measurement is the electrocardiogram, based on 12 linear combinations of the potentials recorded by 9 surface electrodes. Larger numbers of sensors have been tried recently. Cardiac probes have up to 37 magnetic sensors and up to 117 surface electrodes. Stroink, Lamothe and Gardner[5] have reviewed electrocardiographic and magnetocardiographic mapping studies. Magnetic field maps have been used to locate the pre-excitation site in patients suffering from Wolff-Parkinson-White syndrome.

For research purposes, an array of SQUID magnetometers has been used 2.5 mm above a canine heart, scanning an area 23 by 23 mm. Octupolar currents were seen in anisotropic cardiac tissue.[30]

10.2.2.2 Brain: Electroencephalogram and Magnetoencephalogram

Magnetoencephalography instrumentation has been described by Hämäläinen *et al.*[31] These authors discuss various models for solving both the forward and inverse problem, as well as instrumentation, noise, and examples of neuromagnetic measurements. There has been a controversy about whether the EEG and MEG provide different information. The strengths and limitations of each have been reviewed by Wikswo, Gevins, and Williamson.[32] Up to 128 electrodes have been in an EEG array. The EEG is less expensive and can better record signals arising deep in the brain. The MEG is not influenced as much by the skull and costs about 25 times more than a clinical EEG system. The review also discusses magnetic source imaging: how the EEG or MEG data can be synthesized with data from PET, SPECT, and functional MRI studies.

The Wikswo, Gevins and Williamson review also discusses the use of surface Laplacian electrode arrays. With a rectangular array of electrodes, one calculates $v(i+1, j) + v(i-1, j) + v(i+1, j) + v(i-1, j) - 4v(i, j)$ as an approximation to $\partial^2 v/\partial^2 x + \partial^2 v/\partial y$. In analogy to the second spatial derivative in Eq. (10.1), the Laplacian shows current flowing from the cerebral cortex out

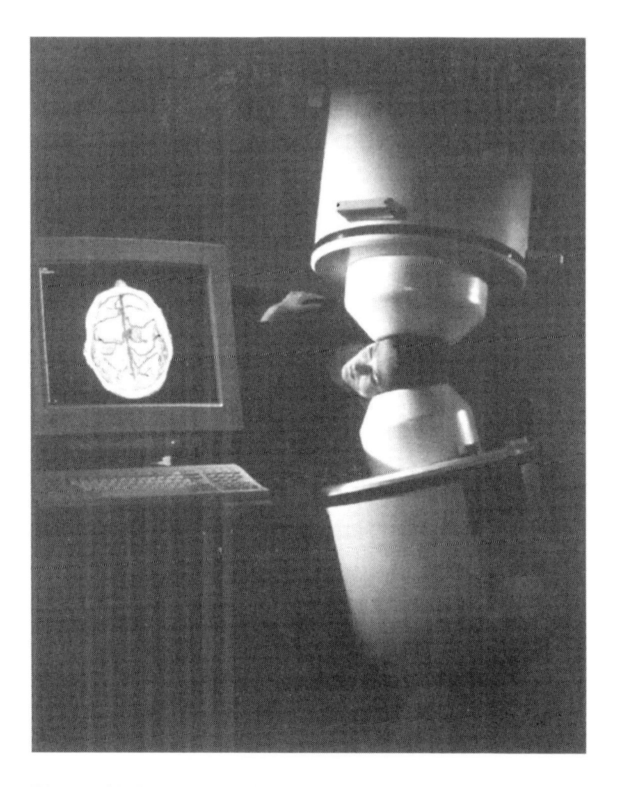

Figure 10.6. Superconducting quantum interference device (SQUID) detectors in a specially designed magnetometer measure the weak extracranial fields produced by the alpha rhythm of the brain. There are 37 detectors. From Biomagnetic Technologies, Inc. with permission.

through the skull and back into the interior. Laplacian maps[33] have also been used for the electrocardiogram.[34, 35]

The physiologic magnetic fields that can be measured at the surface of a patient are quite small. For example, the peak field at the surface of a patient during an epileptic seizure is about 10^{-12} T.[36] This requires the use of sensitive detectors based on superconducting quantum interference devices (Figure 10.6). Environmental sources such as power lines, building wiring, or the movement of nearby iron objects such as elevators, generate noise signals on the order of 10^{-7} to 10^{-6} T. Therefore, magnetically shielded rooms must usually be used, along with some method of subtracting fields from distant sources that vary slowly with distance across the subject (a gradiometer). Even with these measures, other sources of magnetic interference such as dental work, metallic implants, and even eyeliner and clothing, can produce serious errors.

Source localization based on a spherical head model is often unsatisfactory, particularly for components of $\boldsymbol{p}$ that are radial or deep in the head (Figure 10.7). Some groups have tried calculating the location of the source using models in which the conductivity distribution is based on the anatomy of that particular patient as determined by magnetic resonance imaging.[37]

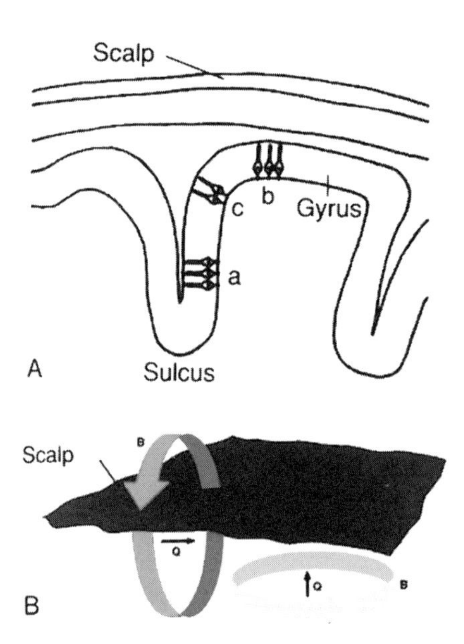

Figure 10.7. The convoluted surface of the brain forms gyri (convolutions on the surface of the brain) and sulci (fissures). The current dipoles in neuronal structures perpendicular to the walls of the sulci are parallel to the surface of the head and because of the right hand rule (see Figures 10.3 and 10.4) produce magnetic fields that are easily detected outside the skull. Current dipoles in the gyri are generally perpendicular to the surface of the skull and contribute little to the components of the magnetic field that are usually measured. From Gallen, et al. (Ref. 41). Used by permission.

Magnetic field measurements of brain responses evoked by auditory or visual stimuli are used extensively for research purposes. They are not reviewed here.

10.2.2.3 Magnetic Source Imaging

Data obtained from magnetoencephalography can be combined with images obtained by a technique such as computed tomography or magnetic resonance imaging, to match the location of the magnetic sources with anatomic landmarks (Figure 10.8). This combined modality is called magnetic source imaging (MSI). The magnetic sources are due to epilleptic seizures or somatosensory evoked responses. Resolution of 2–3 mm has been achieved.[38] MSI shows promise of localizing seizure foci before surgery and evaluating the position of tumors relative to important somatosensory or motor structures,

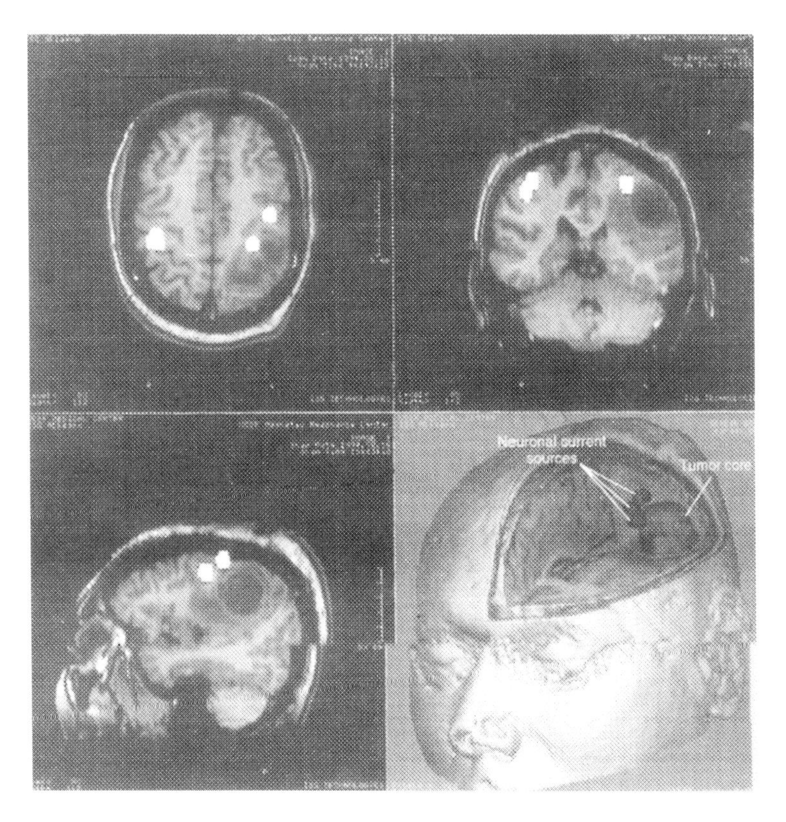

Figure 10.8. Preoperative localization using MSI as an aid to plan tumor resection to avoid unnecessary damage to critical areas of the cerebrum during surgery. Two-dimensional magnetic source images in axial (top left), coronal (top right) and saggital views (lower left). A three dimensional reconstruction of the composite MSI data set (lower right) corresponding to somatosensory stimulation of the central sulcus shows the location of neuronal current sources in relation to the tumor. From Roberts, et al. (Ref. 43.) Used by permission.

information needed to plan the path of surgical access.[39] In a retrospective survey[40] of seven neurosurgeons at two centers, six of the seven reviewers felt that MSI would be useful for functional localization before surgery, producing a net savings in terms of postoperative neurologic rehabilitation. Clinical uses of the MEG and MSI have been reviewed by Gallen, Hirschkoff and Buchanan.[41] Uses include brain mapping before surgery, either for tumors[40, 42] or to ablate the foci of epileptic seizures.[36, 43, 44, 45, 46]

10.2.2.4 Stomach and Gut

Intestinal ischemia is difficult to diagnose before irreversible damage has occurred. Because the small intestine exhibits a basic electrical rhythm, it has recently been possible to detect intestinal ischemia using a SQUID.[47, 48, 49] While these were rabbit studies, slow waves in humans have also been detected[50, 51] including pre- and post-prandial differences.[52] Electrical signals are also produced by the stomach.[53, 54, 55]

It is also possible to measure the transit of food through the stomach and gut by labeling the food with a harmless magnetic tracer.[56, 57, 58] If a ferromagnetic tracer is aligned by external coils, the signal is large enough so that it can be detected without using a SQUID.[59, 60]

10.2.3 Magnetic Susceptibility Measurements

The body contains paramagnetic and diamagnetic materials. Differences in susceptibility between the blood and myocardium and the lungs have been used to image the cardiac cycle in humans.[61, 62] Susceptibility measurements are also used to estimate the amount of iron load in the liver, as well as iron excretion rates.[63, 64, 65] Magnetic susceptibility tomography has been proposed by Sepulvéda, Thomas, and Wikswo.[66]

10.2.4 Electrical Impedance Tomography

Electrical impedance tomography (EIT) is the production of images representing the electrical impedance as a function of position in a two-dimensional slice through the body.[67] Typically, a set of electrodes (perhaps 16 or 32[68, 69]) are placed on the surface of the body. A current is sent between a pair of adjacent electrodes and the voltage developed between other pairs of adjacent electrodes is measured. A set of sequential measurements is made for all pairs of exciting electrodes. Various techniques for image reconstruction are used, often involving multiplication of the data by a matrix calculated from an assumed geometrical model. Supplements to *Physiological Measurement* (formerly *Clinical Physics and Physiological Measurement*) provide conference proceedings devoted to EIT.[70, 71, 72] EIT has been tried where large changes in conductivity either occur naturally or can be induced: the esophagus (swallowing boluses of different conductivity), gastric emptying, the heart, and the lung (tidal volume, emphysema, pulmonary edema). In vivo studies to date have been most positive for gastric emptying.[73, 74] Most reconstruction for measuring lung ventilation assumes that the conductivity of the chest changes. There is evidence that rib cage movement is also important.[75] One group has recently tried three-dimensional imaging, with 4 planes of 16 electrodes each.[76]

10.3 Optical Measurements

10.3.1 Theory

The probability that a photon interacts by a particular process in traversing matter is the product of the cross section for that type of interaction and the number of target entities per unit area in the path of the photon: $p = \sigma N_T$. If the photon travels a distance dz, $N_T = N_A \rho \mathrm{d}z/A$, and the attenuation coefficient μ is defined by

$$p = \mu \mathrm{d}z = (N_A \rho \sigma / A)\mathrm{d}z. \tag{10.10}$$

Both μ and σ are defined for each possible type of interaction or for the sum of all interactions.

X rays used for diagnostic images have typical values for μ of 0.2 cm^{-1} = 20 m^{-1}, and the image is formed by the beam that does not interact in passing through the subject. In that case z is the distance along the direction of the primary beam, and the unattenuated beam can be described by exponential decay (Beer's law). Visible or infrared light striking the body has much higher scattering and absorption cross sections. Typical values of the scattering coefficient might be up to 1000 times larger. Scattering cross sections are sometimes much larger than absorption cross sections. This leads to the multiple scattering characteristic of turbid media. If the medium absorbs light of certain colors but does not scatter very much, it will appear colored but transparent. If scattering is high it will appear colored and opaque.

The scattering angle θ is the direction of the scattered photon with respect to the incident beam. If the scattering cross section into solid angle dΩ depends on θ but not on the azimuthal scattering angle ϕ, the total scattering cross section is

$$\sigma_{\mathrm{scat}} = \int_0^{\pi} \frac{\mathrm{d}\sigma}{\mathrm{d}\Omega} 2\pi \sin\theta \, \mathrm{d}\,\theta.$$

The quantity $d\sigma/d\Omega$ is the differential cross section and is proportional to the phase function used in light scattering. A quantity often used to approximate the scattering is

$$g = \overline{\cos\theta} = \frac{\int_0^\pi \cos\theta (d\sigma/d\Omega) 2\pi \sin\theta \, d\theta}{\sigma_{scat}}. \tag{10.11}$$

If the scattering cross section is peaked at small angles, g is nearly 1. If the scattering is isotropic, $g = 0$. For backward scattering $g < 1$. The *reduced scattering coefficient*

$$\mu_s' = (1 - g)\mu_s \tag{10.12}$$

represents the loss of fluence in the forward direction. Typical values of the reduced scattering coefficient and the absorption coefficient are given in Table 10.1.

Table 10.1. Typical values of μ_s' and μ_a for red light at 633–635 nm.[77]

Tissue	μ_a (m^{-1})	$(1-g)\mu_s$ (m^{-1})
Brain (white matter)	160	200
Brain (grey matter)	160	720
Breast	30	1600
Lung	810	8100
Whole blood	130	610

When the body is illuminated with infrared, visible, or ultraviolet light for diagnosis or treatment, the high value of the cross section means that Beer's law cannot be used. The most accurate photon transport studies are done with "Monte Carlo" computer simulations, in which probabilistic calculations are used to follow a large number of photons as they repeatedly interact in the tissue. There are various approximations that are often useful. Photon dosimetry is reviewed by Star.[78] One of the approximations, the diffusion approximation, is described here.

If the photons have undergone enough scattering in a medium, all memory of their original direction is lost.[79] In that case the movement of the photons can be modeled by the diffusion equation. The concentration C of diffusing particles is described by Fick's second law:

$$\frac{\partial C}{\partial t} = D\nabla^2 C + Q.$$

The left hand side of the equation is the rate at which the number of particles per unit volume is increasing. The term $D\nabla^2 C$ is the net diffusive flow per unit volume into a small volume, the particle current being given by $j = -D\nabla C$. The last term is the rate of production or loss of particles within the volume by other processes, depending on whether Q is positive or negative.

In order to apply this to photons, the concentration must be the number of diffusing photons per unit volume. Any unscattered photons are not yet diffusing, but they may contribute to a source term, s. Photons are also being absorbed. They are traveling with a speed $c' = c/n$, where n is the index of refraction of the medium. In time dt they travel a distance $dx = c'dt$, and the probability that they are absorbed is $\mu_a dx = \mu_a c'dt$. Therefore the diffusion equation for photons is

$$\frac{\partial C}{\partial t} = D\nabla^2 C + s - \mu_a c'C. \tag{10.13}$$

In photon transfer, it is customary to make two changes in this equation. The first is to divide all terms by the speed of the photons in the medium, c'. The result is

$$\frac{1}{c'}\frac{\partial C}{\partial t} = D'\nabla^2 C - \mu_a C + \frac{s}{c'}, \tag{10.14}$$

where $D' = D/c'$ is referred to in the photon transfer literature as the "photon diffusion constant." It has dimensions of length. The second change writes the equation in terms of the photon fluence rate ϕ. (Two important quantities in radiation transfer are the *photon* or *particle fluence* and the *photon fluence rate*. The International Commission on Radiation Units (ICRU) defines the particle fluence as follows for any kind of particle, including photons: At the point of interest construct a small sphere of radius a. Let the number of particles striking the surface of the sphere during some time interval have an expectation value N. The particle fluence Φ is the ratio $N/\pi a^2$, where πa^2 is the area of a great circle of the sphere. The particle fluence rate is $\varphi = d\Phi/dt$.) The photon concentration is related to the photon fluence rate by $C = \varphi/c'$.) The photon diffusion equation becomes

$$\frac{1}{c'}\left(\frac{\partial \varphi}{\partial t}\right) = D'\nabla^2 \varphi - \mu_a \varphi + s. \tag{10.15}$$

This is the form that is usually found in the literature. The units of each term are photon m^{-3} s^{-1}. One can show that[80]

$$D' = \frac{1}{3[\mu_a + (1-g)\mu_s]} = \frac{1}{3[\mu_a + \mu_s']}. \tag{10.16}$$

10.3.1.1 Continuous Measurements

If the tissue is continuously irradiated with photons at a constant rate, the term containing the time derivative vanishes. If in addition we use a broad beam of photons so that we have a one-dimensional problem and we are far enough into the tissue so that the source term can be ignored, the model is

$$D'\frac{d^2\varphi}{dx^2} = \mu_a\varphi . \tag{10.17}$$

This has an exponential solution $\varphi = \varphi_0 e^{-\mu_{eff}x}$, where $\mu_{\text{eff}} = [3\mu_a(\mu_a + (1 - g)\mu_s)]^{1/2}$. Techniques have been suggested for determining the scattering and absorption coefficients from measurements.[81, 82, 83] It is interesting to see what these numbers mean. Using the values $\mu_s' = 1500$ and $\mu_a = 5\ \text{m}^{-1}$, the mean depth for the unattenuated beam is

$$\lambda_{\text{unatten}} = \frac{1}{\mu} = \frac{1}{\mu_a + \mu_s'} = \frac{1}{1505} = 0.66\ \text{mm}.$$

For the diffuse beam the mean depth is about ten times larger:

$$\lambda_{\text{diffuse}} = \frac{1}{\mu_{\text{eff}}} = \frac{1}{\sqrt{(3)(5)(1500)}} = 6.7\ \text{mm}.$$

10.3.1.2 Pulsed Measurements

Ultrashort light pulses from a laser are used to measure time-dependent diffusion, which allows determination of both μ_s' and μ_a. A very short (150 ps) pulse of light strikes a small region on the surface of the tissue. A detector placed on the surface and typically about 4 cm away records the arriving photons. Patterson *et.al.* [84] have shown that the reflected fluence rate at time t after a pulse is approximately

$$R(r, t) = (4\pi D'c't)^{-3/2} z_0 t^{-1} \exp(-\mu_a c't) \exp\left(-\frac{r^2 + z_0^2}{4D'c't}\right). \tag{10.18}$$

Here r is the lateral distance of the detector from the source along the surface of the skin, $c't$ is the total distance the photon has traveled before detection, and $z_0 = 1/[(1 - g)\mu_s]$ is the depth at which all the incident photons are assumed to scatter and become part of the diffuse photon pool. This curve fits the data fairly well and can be used to determine μ_a and $(1 - g)\mu_s$ (Figure 10.9). We can understand the various factors in Eq. (10.18). The last factor represents Gaussian spreading in two dimensions away from the z axis where the photons were injected. The next-to-last factor is the fraction of the photons in the pulse that have not been absorbed, $\exp(-\mu_a x)$, where $x = c't$ is the total distance the photons have traveled. The first factor is the normalization that reduces the amplitude of the Gaussian as it spreads.

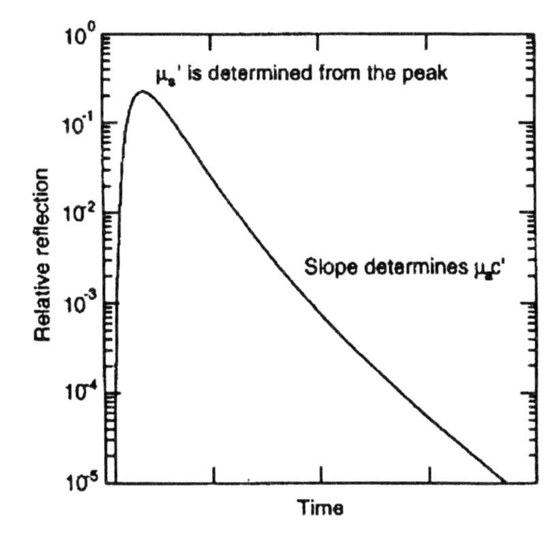

Figure 10.9. Measurement of the diffusely reflected light entering a detector a short distance from a pulsed laser source on the surface of the skin allows measurement of the photon scattering and attenuation coefficients.

A related technique is to apply a continuous laser beam whose amplitude is modulated at various frequencies between 50–800 MHz. The Fourier transform of Eq. (10.18) gives the change in amplitude and phase of the detected signal. Their variation with frequency can be used to determine μ_a and μ_s'.[85, 86]

10.3.1.3 Refinements to the Transport Model

The diffusion equation, Eq. (10.15), is an approximation, and the solution given, Eq. (10.18) requires some unrealistic assumptions about the boundary conditions at the surface of the medium ($z = 0$). Hielscher *et al.*[87] have recently compared experiment, Monte Carlo calculations, and solutions to the diffusion equation with three different boundary conditions. They found that Eq. (10.18) was the easiest to use but leads to errors in the estimates of the scattering coefficient that become worse when the detector and source are close together. Their Monte Carlo calculations and others[88] fit the data quite well.

10.3.2 Clinical Uses

Infrared radiation is used in two ways to image the body. The first technique is to measure infrared radiation emitted by the body. The second is to shine infrared radiation at the body and detect scattered radiation.

Significant emission of thermal radiation from human skin occurs in the range 4–30 μm, with a peak at 9 μm. The detectors used to measure it typically respond to wavelengths below 6 to 12 μm. Thermography began about 1957 with a report that skin temperature over a breast cancer was slightly elevated. There was hope that thermography would provide an inexpensive way to screen for breast cancer, but there have been too many technical problems. There is more variability in vascular patterns in normal breasts than was first realized, so that differences of temperature at corresponding points in each breast is not an accurate diagnostic criterion. The thermal environment in which the examination is done is extremely important. The sensitivity (ability to detect breast cancer) is too low to use thermography as a screening device. Thermography has also been proposed to detect and to diagnose various circulatory problems. Thermography is generally not accepted at this time,[89, 90] though it still has its proponents.

Infrared radiation from the tympanic membrane and ear canal is widely used to measure body temperature,[91] but it is not reviewed here.

In the second technique the subject is illuminated by an external source with wavelengths from 700–900 nm. If full illumination is used, a photograph can be taken if the camera is sensitive to the wavelengths used. The difference in absorption between oxygenated and non-oxygenated hemoglobin allows one to view veins lying within 2 or 3 mm of the skin. Either infrared film or a solid-state camera can be used.

Transillumination using visible and near-infrared light has been investigated for noninvasive characterization of structures in soft tissue such as breast.[92, 93, 94] Multiple scattering defeats the resolution of simple optical detection. Time-gated transillumination uses a pulsed source and fast optical detection circuits to accept only those photons that arrive at the detector without having undergone much multiple scattering. The signal from non-scattered photons is much weaker. Resolution is improved, but it is still much poorer than for X-ray imaging. A 5-cm thick tissue phantom ($\mu_s = 10\ \text{mm}^{-1}$, $g = 0.92$, $c = 0.225\ \text{mm ps}^{-1}$) with a high contrast (totally absorbing) insert at its center has a resolution that varies from 8 to 10 mm as the transit-time "gate" is varied from 100 to 600 ps (22 to 135 mm total light path).[95]

In another approach, time-resolved reflectance (TRR) is used to measure the large angle scattered and diffused component of photon fluence. In TRR, pairs of optical fibers are placed on the surface of the sample. The delay and temporal spread of the optical signal collected by the receiving fiber is used to solve the inverse problem, i.e., parameters are selected that best fit the measured scattering coefficient. Closed form expressions have been obtained for the diffusion approximation[84] given in Eq. (10.17), and for random walk models.[96] Both methods work well for long scattering times (large spacing between fibers) but tend to fail for short times.[97] Some success has been achieved at shorter times for a photon transport model that includes an empirical "time shift" parameter, strictly for the purpose of "curve fitting."[98] General TRR techniques have been employed in brain oxygen measurements[99] and as an aid in phototherapy,[100] the use of external photon sources to activate light-sensitive compounds for therapeutic purposes.

Near infrared spectroscopy illuminates a spot on the patient and measures scattered light nearby. Either pulsed or continuous techniques can be used. Recent papers include the determination of optical properties and blood oxygenation of tissue using continuous sources.[83] The scattering and absorption coefficients are such that measurements of brain oxygenation can be made through the skull.[101, 102, 103, 104] Some investigators find poor correlation (r^2 = 0.04) of changes in cerebral oxygenation measured by near infra red spectroscopy when compared to jugular venous bulb oximetry in patients with severe head injury.[105] Another group found much better correlation (r^2 = 0.69) in a group of children undergoing cardiac operations.[106]

These and other imaging methods have been reviewed recently.[107, 108] Coherent detection by optical heterodyning of the scattered light and the original signal is also used.[109]

Various optical spectroscopic methods are under investigation for characterizing tumors. Recent reviews by Bigio and Mourant[110] and by Andersson-Engles *et al.* [111] are available. Fluorescence spectroscopy has been proposed for the cervix, the skin, and with an endoscope for diagnosing tumors of the gastrointestinal tract. Both endogenous fluorescing molecules and drugs which concentrate in malignant tissues are being explored. Elastic-scattering spectroscopy is also being developed, using two probes as described above.

The scanning laser ophthalmoscope[112] scans the retina with a narrow beam of laser light. Reflected light can be recorded from the entire thickness of the retina or from layers in it. Other measuring techniques include fluorescein angiography (flow), and the amount of photopigment in the retina (densitometry).

10.4 References

1. R.K. Hobbie, *Intermediate Physics for Medicine and Biology*, 3rd. ed. (Springer-Verlag, New York, 1997).
2. P. Wikswo, in *Advances in Biomagnetism,* S.J. Williamson *et al.*, (Plenum, New York, 1989).
3. P. Barach, *J. Theor. Biol.* **125**, 187–191 (1987).
4. N. Trayanova, C.S. Henriquez, R. Plonsey, *IEEE Trans. Biomed Eng.* **37,** 22–35 (1990).
5. G. Stroink, M.J.R. Lamothe, M.J. Gardner, in *SQUID Sensors: Fundamentals, Fabrication and Applications,* H. Weinstock, Ed., (Kluewer, The Netherlands, 1996). NATO-ASI Series E: Applied Sciences. **329**, 413–444.
6. G. Stroink, in *Frontiers in Cardiovascular Imaging,* B.L. Zaret, L. Kaufman, A.S. Berson, R.A. Dunn, Eds., (Raven, New York, 1992).
7. P. Czapski, C. Ramon, L.L. Huntsman, G.H. Bardy, Y. Kim, *Phys. Med. Biol.* **41**, 1247–1263 (1996).
8. J-c. Huang, C. Nicholson, Y. Okada, *Biophys. J.* **57**, 1155–1166 (1990).
9. B.N. Cuffin, *IEEE Trans. Biomed. Eng.* **40**, 42–48 (1993).
10. B.N. Cuffin, *IEEE Trans. Biomed. Eng.* **43**, 299–303 (1996).
11. S.P. van den Broek, H. Zhou, M.J. Peters, *Med. & Biol. Eng. & Comput.* **34**, 21–26 (1996).
12. P. Czapski, C. Ramon, L.L. Huntsman, G.H. Bardy, Y. Kim, *IEEE Trans. Biomed. Eng.* **43**, 95–104 (1996).
13. W.J. Karlon, J.L. Lehr, S.R. Eisinberg, *IEEE Trans. Biomed. Eng.* **41**, 1010–1017 (1994).

14. H.A. Schlitt, L. Heller, R. Aaron, E. Best, D.M. Ranken, *IEEE Trans. Biomed. Eng.* **42**, 52–58 (1995).
15. B.N. Cuffin, *IEEE Trans. Biomed. Eng.* **42**, 68–71 (1995).
16. R. Srebro, *IEEE Trans. Biomed. Eng.* **41**, 997–1003 (1994).
17. A.C.K. Soong, Z.J. Koles, *IEEE Trans. Biomed. Eng.* **42**, 59–67 (1995).
18. R.D. Throne, L.G. Olson, *IEEE Trans. Biomed. Eng.* **42**, 1192–1200 (1995).
19. P.R. Johnston, R. Gulrajani, *IEEE Trans. Biomed. Eng.* **44**, 19–39 (1997).
20. R. Hren, X. Zhang, G. Stroink, *Med. & Biol. Eng. & Computing* **34**, 110–114 (1996).
21. J.C. Mosher, P.S. Lewis, R.M. Leahy, *IEEE Trans. Biomed. Eng.* **39**, 541–557 (1992).
22. H. Bruder, R. Killmann, W. Moshage, P. Weismüller, S. Achenbach, F. Bömmel, *Phys. Med. Biol.* **39**, 655–668 (1994).
23. A.S. Ferguson, G. Stroink, in *Biomagnetism: Fundamental Research and Clinical Applications*, C. Baumgartner, L. Deecke, G. Stroink, S.J. Williamson, Eds., (Proceedings 9th International Conference on Biomagnetism, Elsevier/Ios, 1995), pp. 641–646.
24. E. Menninghous, B. Lütkenhöner, S. L. Gonzalez, *IEEE Trans. Biomed. Eng.* **41**, 986–989 (1994).
25. R. Srebro, *IEEE Trans. Biomed. Eng.* **43**, 547–552 (1996).
26. C. Vaidyanathan, K.M. Buckley, *IEEE Trans. Biomed. Eng.* **44**, 94–97 (1997).
27. B.J. Roth, *Annals Biomed. Eng.* **16**, 609–637 (1988).
28. B.J. Roth, J.P. Wikswo, Jr., *Biophys. J.* **50**, 739–745 (1986).
29. B.J. Roth, *Circ. Res.* **68**, 172–173 (1991).
30. D.J. Staton, R.N. Friedman, J.P. Wikswo, Jr., *IEEE Trans. Appl. Superconductivity* **3**, 1934–1936 (1993).
31. M. Hämäläinen, R. Hari, R.J. Ilmoniemi, J. Knuutila, O.V. Lounasmaa, *Revs. Mod. Phys.* **65**, 413–497 (1993).
32. J.P. Wikswo, Jr., A. Gevins, S.J. Williamson, *Electroencephalography and Clinical Neurophysiology* **87**, 1–9 (1993).
33. T.F. Oostendorp, A. van Oosterom, *IEEE Trans. Biomed. Eng.* **43**, 394–405 (1996). See correction at **43**, 866 (1996).
34. P.R. Johnston, *IEEE Trans. Biomed. Eng.* **43**, 384–393.
35. B. He, R.J. Cohen, *IEEE Trans. Biomed. Eng.* **39**, 1179–1191 (1992); erratum in **41**, 410 (1994).
36. J.S. Ebersole, K.C. Squires, S.D. Eliashiv, J.R. Smith, in *Neuroimaging Clinics of North America: Functional Neuroimaging*, J. Kucharczyk, M.E. Mosely, T. Roberts, W.W. Orrison, Eds., **5,** 267–288 (1995).
37. J.S. George, C.J. Aine, J.C. Mosher, D.M. Schmidt, D.M. Ranken, H.A. Schlitt, C.C. Wood, J.D. Lewine, J.A. Sanders, J.W. Belliveau, *Journal of Clinical Neurophysiology* **12**, 406–431 (1995).
38. B.J. Schwartz, C.G. Gallen, S. Hampson, E. Hirschkoff, D. Sobel, K. Rieke, in *Biomagnetism: Clinical Aspects, Proceedings of the 8th International Conference on Biomagnetism, Munster, 19–24 August 1991,* M. Hoke, S.N. Erne, Y.C. Okada, G.L. Romani, Eds., (Excerpta Medica, International Congress Series 988, New York, 1992)., pp. 253–257.
39. C.G. Gallen, B.J. Schwartz, R.D. Bucholz, G. Malik, G.L. Barkley, J. Smith, H. Tung, B. Copeland, L. Bruno, S. Assam, E. Hirschkoff, F. Bloom, *J. Neurosurg.* **82**, 988–994 (1995).
40. J. Kucharczyk, J. Anson, E. Benzel, B. Copeland, G. Gerraf, A. Halliday, E. Marchand, H. Tung, *Acad. Radiol.* **3**, S131–S134. (1996).
41. C.G. Gallen, E.C. Hirschkoff, D.S. Buchanan, *Neuroimaging Clinics of North America.* **5**, 227–249 (1995).
42. T. Roberts, H. Rowley, J. Kucharczyk, *Neuroimaging Clinics of North America.* **5**, 251–266 (1995).
43. E. Knutsson, L. Gransberg, *Acta Neurochirurgica* – Supplementum. **64**, 74–78 (1995).
44. S.H. Chuang, H. Otsubo, P. Hwang, W.W. Orrison, Jr., J.D. Lewine, *Neuroimaging Clinics of North America.* **5**, 289–303 (1995).

45. M. Aung, D.F. Sobel, C.C. Gallen, E.C. Hirschkoff, *Neurosurgery* **37**, 1113–1121 (1995).
46. H. Stefan, P. Schüler, K. Abraham-Fuchs, S. Schneider, M. Gebhardt, U. Neubauer, C. Hummel, W.J. Huk, P. Thierauf, *Acta Neurol. Scand. Supplement* **152**, 83–88 (1994).
47. J. Golzarian, D.J. Staton, J.P. Wikswo, R.N. Friedman, W.O. Richards, *Amer. J. Surg.* **167**, 586–592 (1994).
48. W.O. Richards, C.L. Garrard, S.H. Allos, L.A. Bradshaw, D.J. Staton, J.P. Wikswo, Jr., *Annals Surg.* **221**, 696–705 (1995).
49. S.H. Allos, D.J. Staton, L.A. Bradshaw, S. Halter, J.P. Wikswo, Jr., W.O. Richards, *World J. Surg.* **21**, 173–178 (1997).
50. W.O. Richards, D. Staton, J. Golzarian, R.N. Friedman, J.P. Wikswo, Jr., in *Biomagnetism: Fundamental Research and Clinical Applications,* C. Baumgertner, *et al.*, Eds. (Elsevier Science, IOS Press, 1995), pp. 743–747.
51. W.O. Richards, L.A. Bradshaw, D.J. Staton, C.L. Garrard, F. Liu, S. Buchanan, J.P. Wikswo, Jr., *Digest. Dis. and Sci.* **41**, 2293–2301 (1996).
52. R.J. Petrie, G.K. Turnbull, G. Stroink, P. van Leeuwen, B. Brandts, S.J.O. Veldhuyzen van Zanten, *Can. J. Gastroenterol.* **10** (Suppl. A), S111 (1996) [Abstract].
53. B.O. Familoni, T.L. Abell, K.L. Bowes, *IEEE Trans. Biomed. Eng.* **42**, 647–657 (1995).
54. S. Comani, M. Basile, S. Casciardi, *et al.*, in *Biomagnetism: Clinical Aspects,* M. Hoke, *et al.*, Eds. (Elsevier Science, Amsterdam, 1992) 639–642.
55. T.L. Abell, J.R. Malagedala, *Dig. Dis. Sci.* **33**, 982–992 (1988).
56. S. Di Luzio, S. Comani, G.L. Romani, M. Basile, C. Del Gratta, V. Pizzella, *Nuovo Cimento* **11D**, 1853–1859 (1989).
57. M. Basile, M. Neri, A. Carriero, S. Casciardi, S. Comani, C. Del Gratta, L. Di Donato, S. Di Luzio, M.A. Macri, A. Pasquarelli, et al., *Digest. Dis. & Sci.* **37**, 1537–1543 (1992).
58. W. Weitschies, J. Wedemeyer, R. Stehr, L. Trahms, *IEEE Trans. Biomed. Eng.* **41**, 192–1955 (1994).
59. J.R. Miranda, O. Baffa, R.B. de Oliveira, N.M. Matsuda, *Med. Phys.* **19**, 445–448 (1992).
60. Y. Benmair, F. Dreyfus, B. Fischel, G. Frei, T. Gilat, *Gastroenterology* **73**, 1041–1045 (1977).
61. J.P. Wikswo, Jr., *Med. Phys.* **7**, 297–306 (1980).
62. J.P. Wikswo, Jr., J.E. Opfer, W.M. Fairbank, *Med. Phys.* **7**, 307–314 (1980).
63. G.M. Brittenham, D.E. Farrell, J.W. Harris, E.S. Feldman, E.H. Danish, W.A. Muir, J.H. Tripp, J.N. Brennon, E.M. Bellon, *Il Nuovo Cimento* **2D**, 567–581 (1983).
64. J.P. Kaltwasser, E. Werner, [Review] *Baillieres Clinical Haematology.* **2**, 363–389 (1989).
65. P. Nielsen, R. Fischer, R. Englehardt, P. Tondury, E.E. Gabbe, G.E. Janka, *Brit. J. Hematology* **91**, 827–833 (1995).
66. N.G. Sepulvéda, I.M. Thomas, J.P. Wikswo, Jr., *IEEE Trans. Magnetics* **30**, 5062–5069 (1994).
67. R.W.M. Smith, I.L. Freeston, B.H. Brown, *IEEE Trans. Biomed. Eng.* **42**, 133–140 (1995).
68. C.S. Koukourlis, G.A. Kyriacou, J.N. Sahalos, *IEEE Trans. Biomed. Eng.* **42**, 632–636 (1995).
69. P.M. Edic, G.J. Saulnier, J.C. Newell, D. Isaacson, *IEEE Trans. Biomed. Eng.* **42**, 849–859 (1995).
70. *Physiol. Meas.* **15**, Suppl. 2A, May 1994.
71. *Physiol. Meas.* **16**, Suppl. 3A, Aug. 1995.
72. *Physiol. Meas.* **17**, Suppl. 4A, Nov. 1996.
73. S. Nour, Y.F. Magnall, J.A. Dickson, A.G. Johnson, R.G. Pearse, *J. Ped. Gastroenterol. & Nutr.* **20**, 65–72 (1995).
74. H.C. Jongschaap, R. Wytch, J.M. Hutchison, V. Kulkarni, *European J. Radiol.* **18**, 165–174 (1994).

75. A. Adler, R. Guardo, Y. Berthiaume, *IEEE Trans. Biomed. Eng.* **43**, 414–420 (1996).
76. P. Metherall, D.C. Barber, R.H. Smallwood, B.H. Brown, *Nature* **380**, 509–512 (1996).
77. Adapted from L. Grossweiner. *The Science of Phototherapy*. (CRC Press, Boca Raton, 1994), Table 5.2.
78. W. M. Starr, *Phys. Med. Biol.* **42**, 763–787 (1997).
79. A photon has a random orientation after about $1/(1-g)$ scatters. This is about 5 scatters for typical biological tissues. See L. I. Grossweiner, *The science of Phototherapy*. (CRC Press, Boca Raton, 1994), p. 90.
80. See, for example, J. J. Duderstadt, and L. J. Hamilton. *Nuclear Reactor Analysis*. (Wiley, New York, 1976), pp. 133–136.
81. T.J. Farrell, M.S. Patterson, B. Wilson, *Med. Phys.* **19**, 879–888 (1992).
82. T.J. Farrell, B.C. Wilson, M.S. Patterson, *Phys. Med. Biol.* **37**, 2281–2286 (1992).
83. H. Liu, D.A. Boas, Y. Zhang, A.G. Yodh, B. Chance, *Phys. Med. Biol.* **40**, 1983–1993 (1995).
84. M.S. Patterson, B. Chance, B. C. Wilson, *Appl. Opt.* **28**, 2331–2336 (1989).
85. E.M. Sevick, B. Chance, J. Leigh, S. Nioka, M. Maris, *Analytical Biochem.* **195**, 330–351 (1991).
86. B.W. Pogue, M.S. Patterson, *Phys. Med. Biol.* **39**, 1157–1180 (1994).
87. A. Hielscher, S.L. Jacques, L. Wang, F.K. Tittel, *Phys. Med. Biol.* **40**, 1957–1975 (1995).
88. S.T. Flock, B.C. Wilson, M. S. Patterson, *IEEE Trans. Biomed. Eng.* **36**, 1169–1173 (1989).
89. P. Cotton, *JAMA* **267**, 1885–1887 (1992).
90. S. Blume, *Intl. J. Technology Assessment in Health Care* **9**, 335–345 (1993).
91. L.C. Rotello, L. Crawford, T.E. Terndrup, *Critical Care Medicine* **24**, 1501–1506 (1996).
92. J.C. Hebden, R. A. Kruger, *Med. Phys.* **17**, 351–356 (1991).
93. M.D. Duncan, R. Mahon, L.L. Tankersley, J. Reintjes, *Opt. Lett.* **16**, 1868–1870 (1991).
94. K. Suzuki, Y. Yamashita, K. Ohta, B. Chance, *Invest. Radiol.* **29**, 410–414 (1994).
95. V. Chernomordik, R. Nossal, A.H. Gandjbakhche, *Med. Phys.* **23**, 1857–1861 (1996).
96. A.H. Gandjbakhche, R. Nossal, R.F. Bonner, *Appl. Opt.* **32**, 504–516 (1993).
97. K.M. Yoo, F. Liu, R.R. Alfano, *Phys. Rev. Lett.* **64**, 2647–2650 (1990); *ibid.*, **65**, 2210–2211 (1990).
98. R. Cubeddu, A. Pifferi, P. Taroni, A. Torricelli, G. Valentini, *Med. Phys.* **23**, 1625–1633 (1996).
99. B. Chance, J.S. Leigh, H. Miyake, D.S. Smith, S. Nikoa, R. Greenfeld, M. Finander, K. Kaufmann, W. Levy, M. Young, P. Cohen, H. Yoshioka, R. Boretsky, *Proc. Natl. Acad. Sci. US.* **85**, 4971–4975 (1988).
100. M.S. Patterson, J.D. Moulton, B.C. Wilson, B. Chance, *Proc. Soc. Photo-Opt. Instrum. Eng.* **1203**, 62–75 (1990).
101. E. Okada, M. Firbank, D.T. Delpy, *Phys. Med. Biol.* **40**, 2093–2108 (1995).
102. M. Firbank, D. T. Delpy, *Phys. Med. Biol.* **39**, 1509–1513 (1994).
103. M. Firbank, S.R. Arridge, M. Schweiger, D.T. Delpy, *Phys. Med. Biol.* **41**, 767–783 (1996).
104. S.R. Arridge, M. Hiraoka, M. Schweiger, *Phys. Med. Biol.* **40**, 1539–1558 (1995).
105. S.B. Lewis, J.A. Myburgh, E.L. Thornton, P.L. Reilly, *Crit. Care Medicine* **24**, 1334–1338 (1996).
106. P.E. Daubeney, S.N. Pilkington, E. Janke, G.A. Charlton, D.C. Smith, S.A. Webber, *Annals Thoracic Surg.* **61**, 930–934 (1996).
107. J.C. Hebden, S.R. Arridge, D.P. Delpy, *Phys. Med. Biol.* **42**, 825–840 (1997).
108. S.R. Arrdige, J. C. Hebden, *Phys. Med. Biol.* **42**, 841–853 (1997).
109. K.P. Chan, B. Devaraj, M. Yamada, H. Inaba, *Phys. Med. Biol.* **42**, 855–867 (1997).
110. I.J. Bigio, J. R. Morant, *Phys. Med. Biol.* **42**, 803–814 (1997).
111. S. Andersson-Engels, C. a. Klintberg, K. Svanberg, S. Svanberg, *Phys. Med. Biol.* **42**, 815–824 (1997).
112. P.F. Sharp, A. Manivannan, *Phys. Med. Biol.* **42**, 951–966 (1997).

11 Image Quality

Marie Foley Kijewski

Harvard Medical School and Brigham and Women's Hospital

11.1 Introduction: Defining Image Quality

It is now generally accepted that a meaningful assessment of image quality must take into account the purpose for which the image is to be used[1]. Therefore, imaging systems should be evaluated on the basis of performance in well-defined tasks. Task-based criteria depend on the physical properties of the imaging system, such as resolution and noise, as well as the characteristics of the imaged distribution and the definition of the task.

11.1.1 General Principles

This chapter is based on linear systems theory. By a linear system we mean one in which the image of a combined object is equal to the superposition of the images of the constituents. Many medical imaging systems are linear (or linearizable), or nearly so. The deterministic properties of the imaging system are summarized by the system function, H; the expected output for an object, *f*, is given by:

$$\langle g(\vec{r})\rangle = \int H(\vec{r},\vec{r}')f(\vec{r}')\mathrm{d}\vec{r}'. \tag{11.1}$$

The expectation is over an ensemble of noisy images. If the system is shift-invariant, i.e., the input response is constant throughout the imaging volume, then Eq. (11.1) becomes:

$$\langle g(\vec{r})\rangle = \int H(\vec{r}-\vec{r}')f(\vec{r}')\mathrm{d}\vec{r}'. \tag{11.2}$$

Because in practice the output function is sampled by the system detectors, we can use vector notation. If we also assume that the noise is additive, we denote a single realization of the output data by:

$$\bar{g} = Hf + \bar{n}. \tag{11.3}$$

The measured data, $\bar{g}$, and the noise, $\bar{n}$, are always discrete. The object, *f*, may be discrete (in some computer simulations) or continuous; in the latter case, the system function represents a continuous-to-discrete transformation. For a thorough treatment of continuous-to-discrete transformations in medical imaging systems see Barrett and Gifford[2] or Barrett et al.[3] In some cases, such as radiographic imaging, the measured data themselves constitute the image; in others, such as computed tomography (CT), the image is formed by processing the measured data (see Section 11.1.4).

11.1.2 Definition of Imaging Tasks

Two broad categories of tasks are relevant to medical imaging[4]. One is classification – assignment of an image to one of two or more classes, e.g., normal and abnormal. Classification tasks are usually performed by humans, who make subjective judgments as to the status of the image. Although frequently there are several possible diagnostic classes, in this chapter we will consider only binary classification tasks. The other broad category of tasks involves estimation from the image of a quantity of interest, such as uptake of a radioisotope by an organ or rate of blood flow through a vessel. Estimation is almost always accomplished by objective procedures, i. e., calculation from image pixel values.

11.1.3 Physical Characteristics of Imaging Systems

In Section 11.2 below we describe methods of quantifying both deterministic and stochastic aspects of the imaging system. Deterministic properties are those that are constant over an ensemble of noisy images of the same object, while stochastic properties are those that vary over such an ensemble.

11.1.4 Stages of the Imaging Process

It is important to make a distinction between two stages of the imaging process: data acquisition and image processing (or reconstruction) and display. For some modalities the distinction is subtle. In film/screen radiography, for example, the acquired data are contained in the latent image on the X-ray film. Only development and placement on a viewbox are necessary to convert the film to an image display device. In the absence of equipment malfunction the effects of these operations on human observer performance are negligible. In SPECT imaging, on the other hand, the acquired data are projection images which cannot be readily interpreted by a human. Generation of a usable image requires correction for various physical effects (see Volume I, Chapter 3) as well as image reconstruction. Furthermore, display of a digital image, either on a monitor or on film, implies selection of a subset of the full dynamic range to be distributed across the available gray-scale levels by manipulation of display window and level. All of these can significantly alter the performance of the human (or model) observer.

The data acquisition stage sets fundamental limits on image quality – linear image processing can, at best, retain information, and almost always results in information loss. The performance of model observers, or estimation procedures, can usually be assessed using the acquired data, without the need for reconstruction and display. Human performance must be assessed using a displayed image. In some studies, model observers have been employed at the display stage to compare image reconstruction algorithms or display effects[5,6].

11.2 Physical Characteristics of Imaging Systems

11.2.1 Large-Area Transfer Function

The large-area transfer function describes the relationship between average (over a large area) imaging system input and average output. It incorporates any changes in units between the input and output. The system function, H, of Eq. (11.3) can be decomposed as follows:

$$H(\vec{r}) = K\,PSF(\vec{r}). \tag{11.4}$$

$PSF(\vec{r})$ is the system point spread function (see Section 11.2.2.1). K, the large-area transfer function, is a constant defined by:

$$K = \int H(\vec{r})\mathrm{d}\vec{r}, \tag{11.5}$$

where the integral is over all space. (For nonlinear imaging systems, the value of K varies with system input value; see Sharp et al.[1])

11.2.2 Spatial Resolution

Measures of spatial resolution quantify the blurring of the object by the imaging system. These can be analyzed using real-space functions, such as the point-spread function (PSF), line-spread function (LSF), and edge-response function (ERF), or using their Fourier-space counterparts, the optical transfer function (OTF) and modulation transfer function (MTF). For a comprehensive treatment of the analysis of resolution properties of medical imaging system, see Metz and Doi[7].

11.2.2.1 Spatial Domain Functions

A fundamental concept of linear systems is the point-spread function (PSF):

$$PSF(\vec{r}) = H(\vec{r})/K. \tag{11.6}$$

The PSF is the response of the system to a point input. From it, the response of the system to an extended object can be derived, for linear systems, using the principle of superposition. Any object can be decomposed into the sum of appropriately scaled and shifted points; the response of the system to N point inputs at locations, $\vec{r}_n$, $n = 1, N$ and amplitudes f_n; $n = 1, N$ will be:

$$I(\vec{r}) = \sum_{n=1}^{N} f_n PSF(\vec{r} - \vec{r}_n). \tag{11.7}$$

For a continuous object the image is its convolution[8] with the PSF:

$$I(\vec{r}) = \int f(\vec{r}')PSF(\vec{r} - \vec{r}')\mathrm{d}\vec{r}'. \tag{11.8}$$

The PSF is normalized so that its integral over all space is unity; this is consistent with the earlier definition of the large-area transfer characteristic as including all normalization factors. From a two-dimensional PSF, two related, one-dimensional functions can be derived. These are the line-spread function (LSF) and the edge-response function (ERF). Since it is easier to realize a line or edge than a point, these functions are often used in measurements[9,10]. The LSF is given by:

$$LSF(x) = \int_{-\infty}^{\infty} PSF(x, y)\mathrm{d}y \tag{11.9}$$

Eq. (11.9) assumes that the PSF is rotationally symmetric; in practice, it is often asymmetric and the LSF, consequently, is orientation-dependent. The ERF is derived from the LSF by:

$$ERF(x) = \int_{-\infty}^{x} LSF(x')\mathrm{d}x'. \tag{11.10}$$

For systems which are made up of several subcomponents, the system PSF is the convolution of the component PSF. For example, the overall PSF of a film screen system is the convolution of the PSF due to the non-zero width of the X-ray source, the PSF due to spreading of light in the fluorescent screen, and the PSF of the film:

$$PSF(\vec{r}) = PSF_g(\vec{r}) * PSF_s(\vec{r}) * PSF_f(\vec{r}) \tag{11.11}$$

The overall PSF is dominated by the broadest of the component PSF; therefore, improvements in the spatial resolution of the film, in this example, will have almost no effect on the system PSF.

11.2.2.2 Fourier Domain Functions

The optical transfer function (OTF) is the two-dimensional Fourier transform of the PSF. The OTF is, in general, a complex-valued function of spatial frequency. If, however, the PSF is symmetric (with respect to the distance origin in one dimension or angle in two dimensions), then it is real for all values of spatial frequency. The OTF corresponding to a rotationally symmetric two-dimensional PSF can be obtained by a Hankel transform[8]:

$$OTF(\nu) = 2\pi \int_{0}^{\infty} PSF(r)J_0(2\pi\nu r)r\mathrm{d}r, \tag{11.12}$$

where ν is radial spatial freqency and $J_0(\cdot)$ is a Bessel function of zero order[11]. The modulation-transfer function (MTF) is the modulus of the OTF. The OTF, and MTF, describe the ability of the imaging system to preserve information as a function of spatial frequency. The MTF can be interpreted in terms of the effects of the system on sinusoidal functions, the basis functions into which all objects can be decomposed. For a pure sinusoidal input, the MTF is the ratio of output peak amplitude to input peak amplitude as a function of frequency. The one-dimensional Fourier transform of the LSF yields the values along a spoke through the two-dimensional OTF, at the appropriate angle. Consistent with Eq. (11.11), the overall MTF of a system with multiple components is the product of the component MTF; because multiplication is more easily accomplished than convolution, it is usually more convenient to analyze imaging systems using frequency-space functions.

11.2.2.3 Examples

Example 11.2.2.3 (1): Gaussian point-spread functions
PSF of medical imaging systems can often be well-approximated by rotationally symmetric Gaussian functions, i.e.,

$$PSF(r) = \frac{1}{2\pi\sigma^2}\exp\left(\frac{-r^2}{2\sigma^2}\right). \qquad (11.13)$$

Here the PSF is shown as a function of a single coordinate, radius; note that it is normalized so that the integral under it is unity. From Eqs. (11.9), (11.10), and (11.12) we can obtain the LSF, ERF, and MTF (in this case, equal to the OTF):

$$LSF(x) = \frac{1}{\sqrt{2\pi}\sigma}\exp\left(-\frac{x^2}{2\sigma^2}\right), \qquad (11.14)$$

$$ERF(x) = \frac{1}{2}\left[1 - \mathrm{erf}\left(\frac{x}{\sqrt{2}\sigma}\right)\right], \qquad (11.15)$$

where erf(·) is the error function[11], and

$$MTF(\nu) = \exp\left(-2\pi^2\sigma^2\nu^2\right). \qquad (11.16)$$

Notc that these functions depend on a single parameter, σ, which completely characterizes the spatial resolution of systems for which these functions can be described by simple Gaussian forms. More commonly, the full-width-at-half-maximum (FWHM) of the PSF (or, equivalently, the LSF), which is equal to 2.355 σ, is given. System resolution is frequently specified in terms of FWHM even when the system PSF is not Gaussian in form; in this case, the single parameter does not completely characterize resolution. Figure 11.1 depicts these four functions for two systems with Gaussian PSF of different σ.

a)

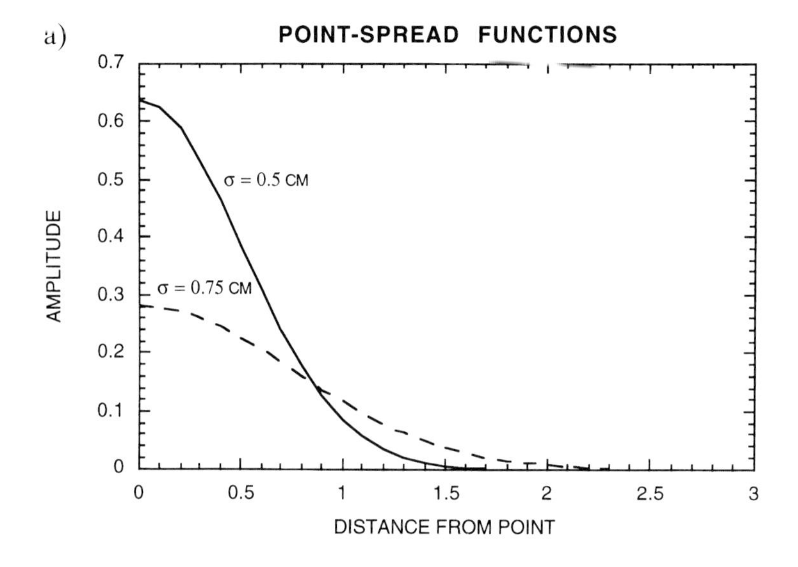

b)

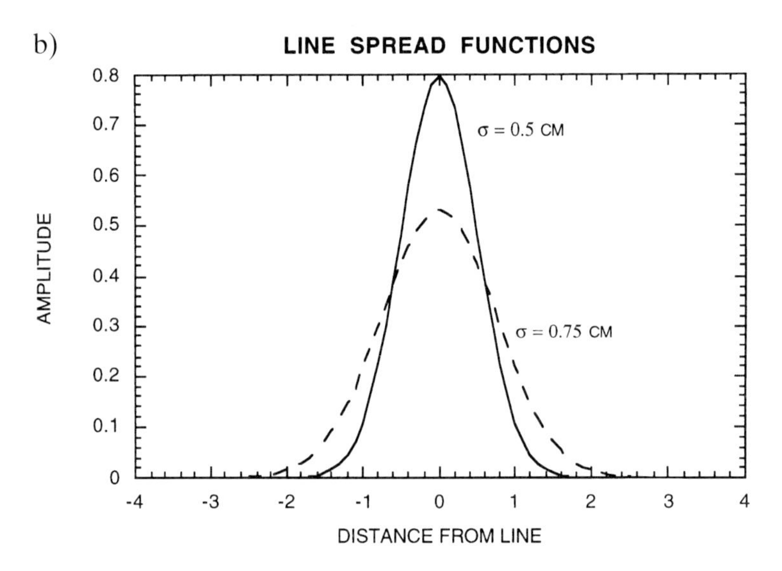

Figure 11.1. (a) Gaussian-shaped PSF, $\sigma = 0.5$ cm and 0.75 cm, and corresponding, (b) LSF.

c)

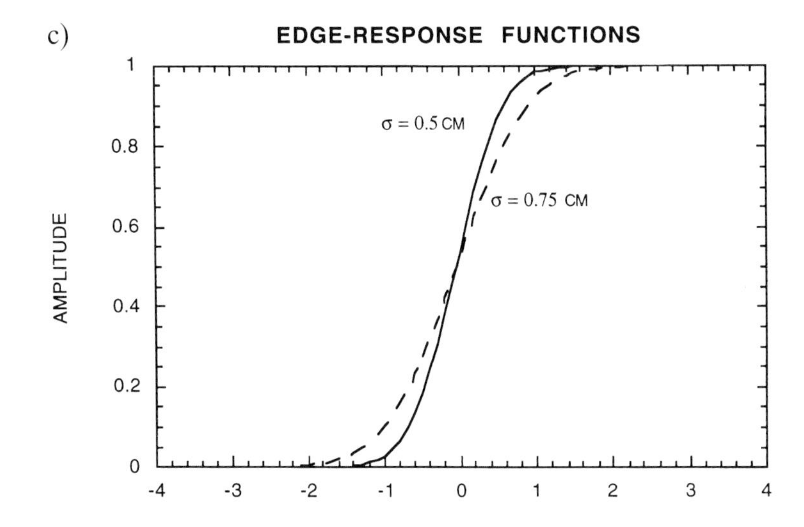

d)

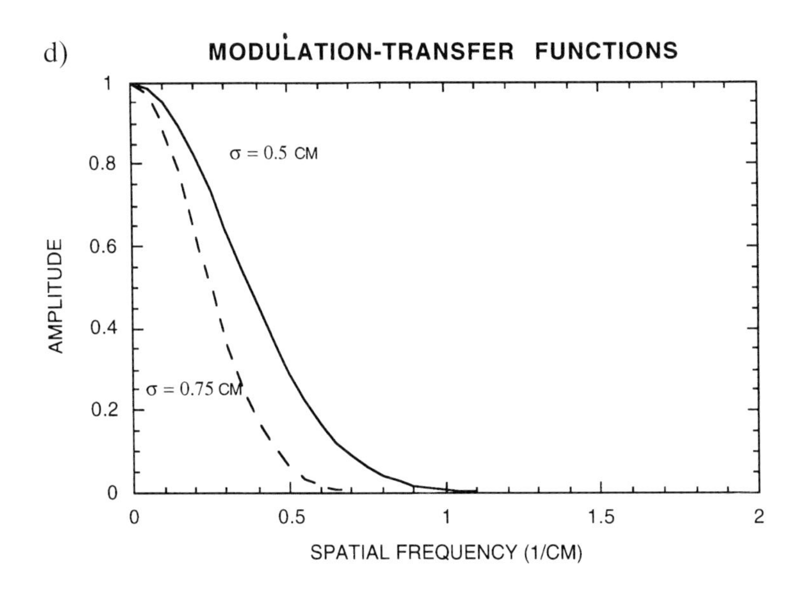

Figure 11.1. (c) ERF, and (d) MTF.

Note that the narrower spatial-frequency functions correspond to the wider MTF. A more compact PSF and, consequently, LSF, implies less blurring by the system and more faithful preservation of the higher frequency components of the object.

Example 11.2.2.3 (2): Long-tailed point-spread functions
Medical imaging systems are sometimes characterized by a PSF with a narrow central portion and long, low-amplitude tails. Two examples of systems with such a PSF are SPECT systems, where the long tails are due to detection of scattered photons, and image-intensifier-based fluoroscopy systems, where the long-range effects are due to veiling glare. These PSF can, in some cases, be modeled as a Gaussian function superimposed on an exponential function:

$$PSF(r) = \frac{\exp\left(-\dfrac{r^2}{2\sigma^2}\right) + b\exp(-ar)}{2\pi\left(\dfrac{b}{a^2} + \sigma^2\right)}, \tag{11.17}$$

where the terms in the denominator normalize the integrated PSF to unity. The corresponding OTF is given by:

$$OTF(f) = \frac{a^3 b}{\left(a^2 + 4\pi^2 f^2\right)^{3/2}\left(b + a^2\sigma^2\right)} + \frac{a^2\sigma^2 \exp\left(-2\pi^2\sigma^2 f^2\right)}{b + a^2\sigma^2}, \tag{11.18}$$

where f is radial spatial frequency. These functions are shown in Figure 11.2. The low-frequency peak in the OTF arises from the long tails of the PSF.

11.2.3 Stochastic Properties

The term 'noise' is sometimes used to encompass both artifacts and stochastic variation. In this chapter the term is restricted to random, rather than systematic, errors in the measured or processed data. Artifacts are discussed separately in Section 11.2.5.

Measures of noise are defined in terms of variation across an ensemble of images of the same object. Sources of noise in medical imaging systems include electronic shot noise, film grain noise, and photon noise. In a well-designed radiological imaging system, photon noise is dominant. Because the level of photon noise is directly proportional to patient radiation exposure, it is important to control other sources of noise so that the patient receives the maximum benefit from the potentially hazardous radiation exposure. In nuclear medicine imaging, the levels of noise are much higher than in other modalities. The amount of radioactivity that can be introduced into the

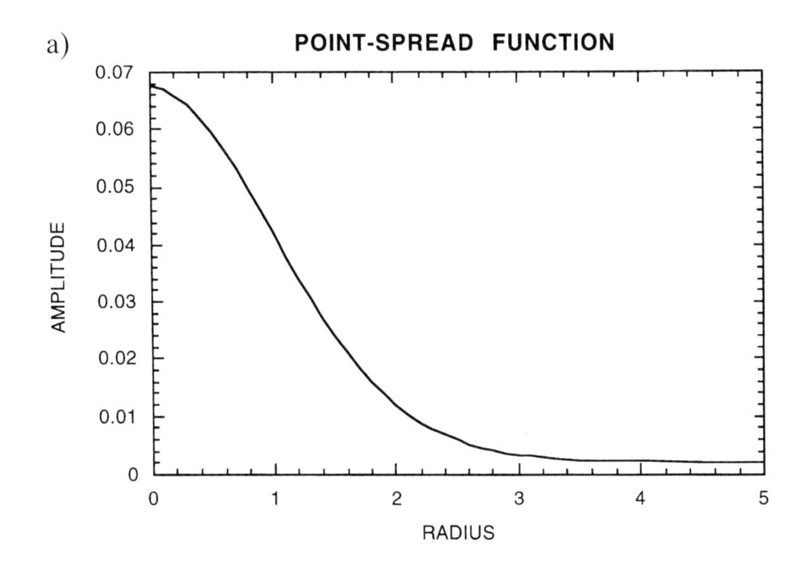

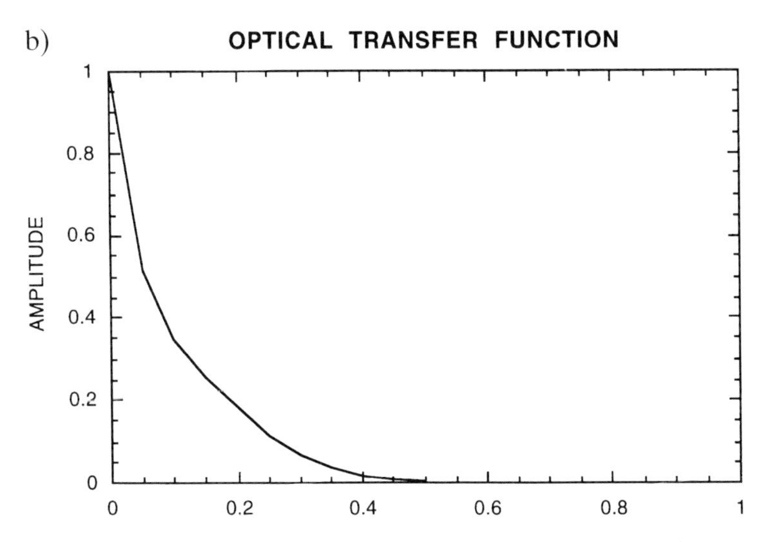

Figure 11.2. (a) Long-tailed PSF of Eq. (11.17); σ=1 cm, a=0.25 cm^{-1}, b–0.1, and (b) corresponding OTF.

patient is strictly limited for dosimetric reasons, since the patient is exposed not only during the imaging procedure, but for days or weeks afterwards while the isotope is decaying and being biologically eliminated.

For most medical imaging systems the noise is distributed according to either Gaussian or Poisson statistics[12]. Most of the discussion in this chapter assumes zero-mean Gaussian-distributed noise. Although the Poisson distribution, for all but very low-count conditions, is well-approximated by a

Gaussian distribution, there remains an important distinction between them. Gaussian noise is usually additive, i.e., the pixel variance does not depend on the expected value of the pixel, and often the variance is constant throughout the image. The variance of Poisson-distributed noise, on the other hand, is equal to the distribution mean, i.e., the expected number of counts in the pixel. Therefore, Poisson noise is nonstationary.

11.2.3.1 Pixel variance

Although this single-parameter measure of image noise is of limited value for image assessment, it is widely used, and is discussed here for completeness. The pixel variance is defined as the variance of the image value at a single pixel over an ensemble of noisy images. In practice, the ensemble averaging is often replaced by averaging over a nominally uniform area in a single image. This is equivalent to ensemble averaging only for white (uncorrelated) noise. The major deficiency of single-parameter measures, such as the pixel variance, is that they do not address the spatial correlations in noise which characterize many imaging systems.

11.2.3.2 Autocorrelation Function

In the following we assume a two-dimensional image; extension to other dimensions is straightforward. The noise autocorrelation function is defined by:

$$ACF\left(\vec{r},\vec{\Delta}\right) \equiv \left\langle \left(g(\vec{r}) - \bar{g}(\vec{r})\right)\left(g\left(\vec{r}+\vec{\Delta}\right) - \bar{g}\left(\vec{r}+\vec{\Delta}\right)\right)\right\rangle, \tag{11.19}$$

where $g(\vec{r})$ is the measured data at $\vec{r}$, $\bar{g}(\vec{r})$ is the ensemble mean of the measured data at $\vec{r}$, and the averaging is over an ensemble of noisy images of the same object. If the noise is stationary, then the autocorrelation function will be a function of the separation, $\vec{\Delta}$, only. The ACF for white noise is a delta function[8]. Positive correlations in the noise imply that if a given pixel value is randomly high, then the values of pixels surrounding it are likely to also be higher than their means. Negative noise correlations imply that the pixels surrounding a particular pixel are likely to deviate from their mean values in the opposite direction.

11.2.3.3 Noise Power Spectrum

The noise power (Weiner) spectrum is defined as the squared magnitude of the Fourier transform of an image containing only noise, averaged over an ensemble of images:

$$NPS(\vec{\nu}) = \lim_{X\to\infty} \frac{1}{X} \left\langle \left| \int_{-X/2}^{X/2} n(\vec{r}) \exp(-i2\pi\vec{\nu}\cdot\vec{r}) \mathrm{d}\vec{r} \right|^2 \right\rangle \tag{11.20}$$

Strictly speaking, the NPS is defined only for stationary noise, although some closely related functions have been used to describe the frequency-space characteristics of nonstationary noise[13,14]. For stationary noise, the NPS is the Fourier transform of the ACF. The integral under the NPS is equal to the pixel variance; the NPS can be viewed as a decomposition of the pixel variance into its spatial frequency components.

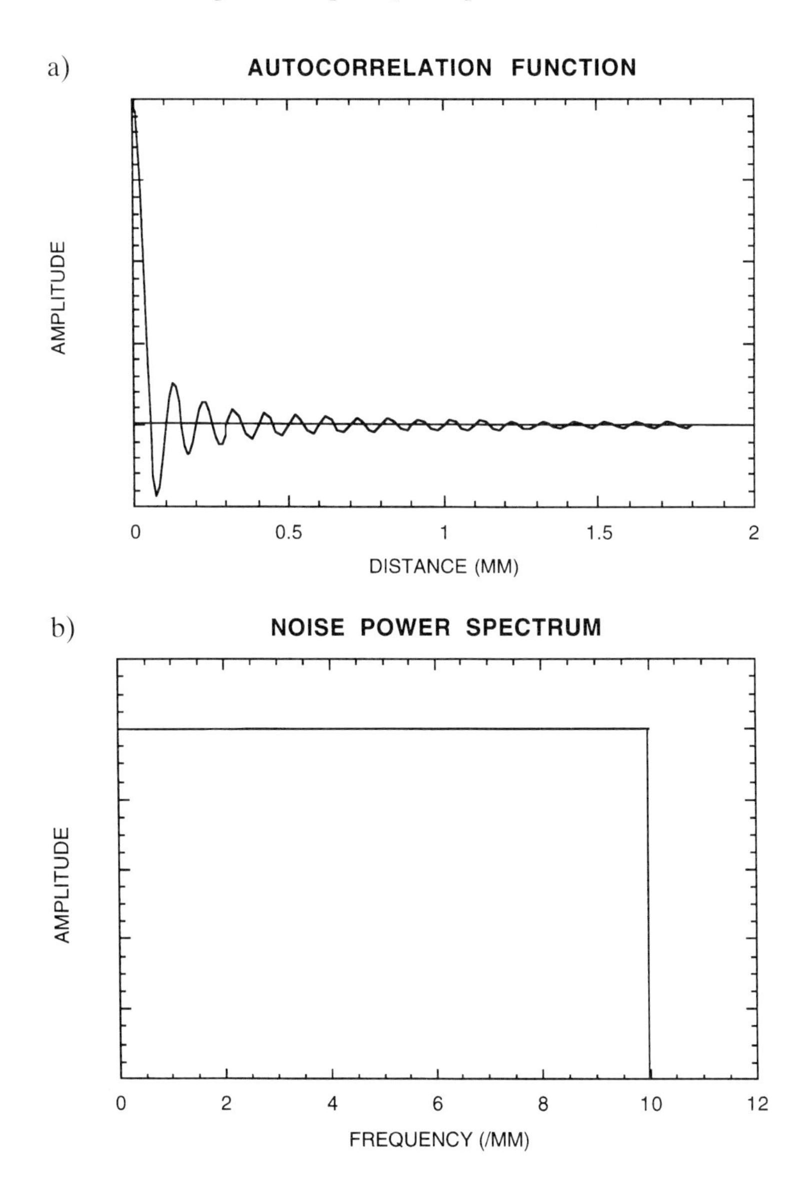

Figure 11.3. (a) ACF of white noise, ν_N=10 mm^{-1}, and (b) corresponding NPS.

11.2.3.4 Examples

Example 11.2.3.4 (1): Autocorrelation function/noise power spectrum pairs
In many medical imaging systems, such as radiography and planar scintigraphy systems, the noise is not apodized, or correlated, by the system PSF[12]. For white noise, the autocorrelation function is a delta function and the NPS is constant, up to the Nyquist frequency of the system. In Figure 11.3 are shown the (one-dimensional) ACF and NPS for a system with Nyquist frequency 10 mm^{-1}. The ACF can be obtained by Fourier transform of the NPS:

$$ACF(x) = 2\int_{0}^{\nu_N} \cos(2\pi\nu x)\mathrm{d}f = \frac{\sin(2\pi\nu_N x)}{\pi x}; \tag{11.21}$$

this function approaches a delta function as ν_N becomes large[8]. The noise at each pixel is independent of that at any other pixel.

In systems in which the noise is apodized by the system PSF, the noise is positively correlated. The (one-dimensional) ACF and NPS for a typical film/screen system are shown in Figure 11.4. There are two noise components, which can be seen in the NPS. The quantum noise, which is filtered by the (assumed Gaussian-shaped) PSF of the screen, is superimposed on the film-grain noise, which is not apodized by the screen PSF. The ACF consists of a low-frequency peak corresponding to the film-grain noise and a broader portion which represents the quantum noise. The ripples are a consequence of the abrupt cutoff of the NPS component due to film grain noise. The positive portion of the ACF represents the correlation distance, i.e., the region within which noise varies together. In this case, data points within about 0.3 mm of a given pixel tend to vary randomly in the same direction.

Images which are reconstructed from projections are characterized by negatively correlated noise (see example 11.2.3.4(2)); a typical (two-dimensional, rotationally symmetric) ACF/NPS pair are shown in Figure 11.5. A non-monotonically decreasing NPS implies negative correlations in the noise. In this example, pixels located 2–3 pixels from a given pixel tend to vary randomly in the opposite direction.

Example 11.2.3.4 (2): The noise-power spectrum of CT images
In this example we show how the image processing required for reconstruction of CT images from projections introduces noise correlations[15,16]. The effect of each operation on the noise power spectrum is demonstrated. The most common method of reconstruction from projections is the convolution-backprojection algorithm. First, the measured projections $proj(x,\phi)$, where x is distance along the projection and ϕ is projection angle, are convolved with a filter:

$$cproj(x,\phi) = \int proj(x',\phi)c(x-x')\mathrm{d}x', \tag{11.22}$$

where the integral is over the length of the projection; then, each convolved projection is backprojected onto the image array:

a)

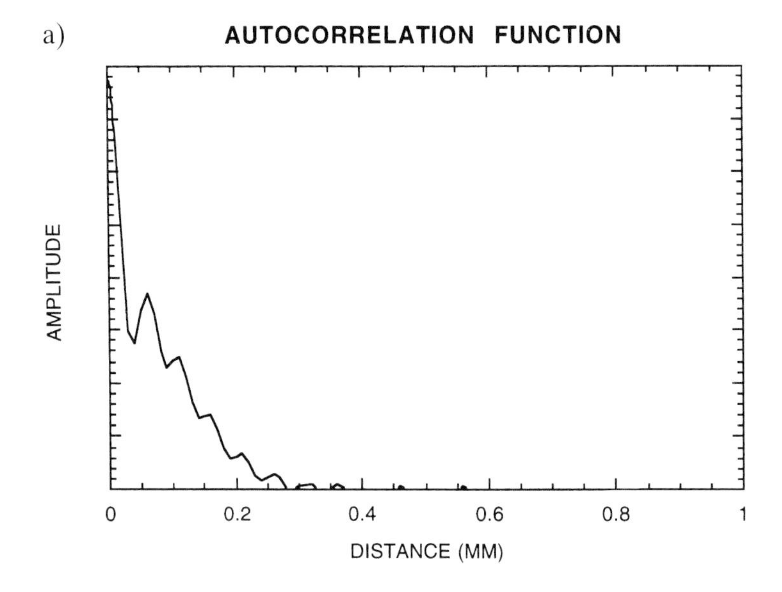

b)

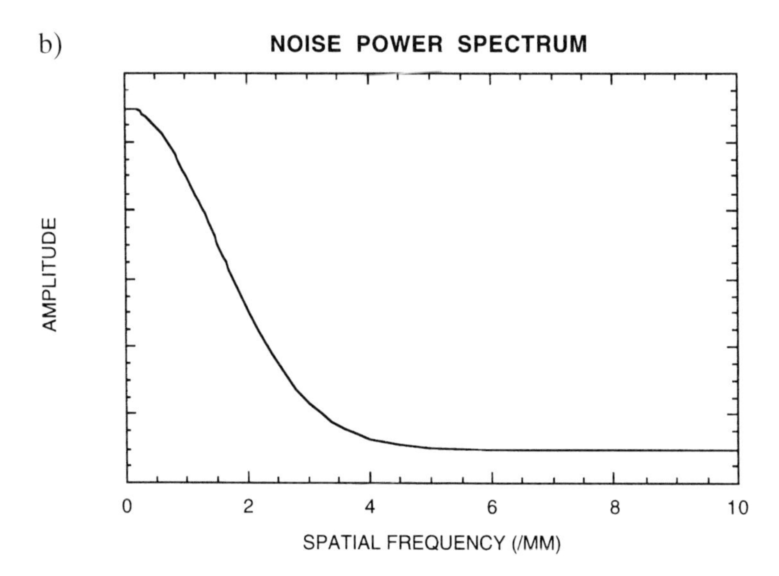

Figure 11.4. (a) ACF of positively correlated noise typical of film/screen systems, and (b) corresponding NPS.

$$I(r,\theta) = \frac{L}{\pi} \int_0^{\pi} cproj\left(r\cos(\theta - \phi)\right) \mathrm{d}\phi, \tag{11.23}$$

where L is the number of projections. We assume that the noise in the measured projections is white; this is a reasonably good assumption for many CT scanning conditions. The (one-dimensional) NPS of the projections is:

a)

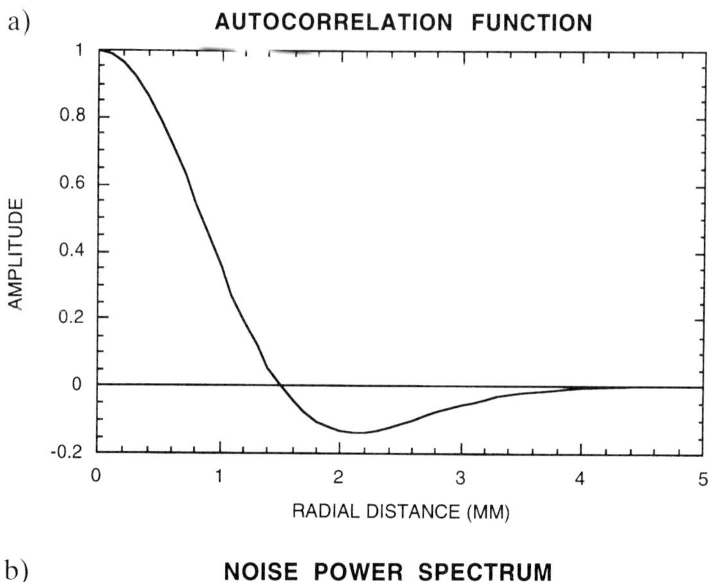

b)

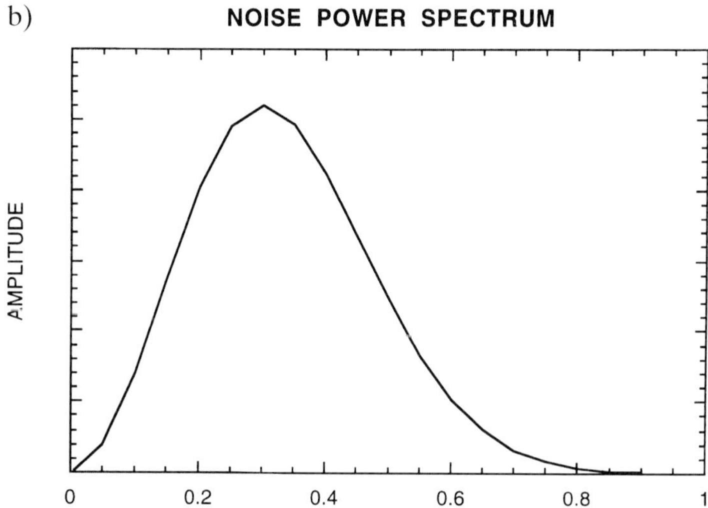

Figure 11.5. (a) ACF of negatively correlated noise typical of CT systems, and (b) corresponding NPS.

$$NPS(\nu) = \frac{\sigma^2}{P}\,\mathrm{rect}(a\nu), \tag{11.24}$$

where a is the projection sampling distance σ^2, is the variance of the projection values, P is the number of projection elements, and rect(...) is the rectangle function[8]; rect(x)=1 for $|x| < 1/2$, and 0 otherwise. Although the projections are discrete, their NPS is continuous. Unlike the signal, the noise is not apodized by the source-detector aperture. (Because the projections are sampled, the NPS is replicated at intervals of 1/a *ad infinitum* in frequency space. Only the main portion of the NPS, centered at the frequency origin, is considered here. For a treatment of the effects of noise aliasing on the NPS of CT images, see Kijewski and Judy[16].) Filtering the projections multiplies their power spectrum by the squared magnitude of the filter frequency response. The Fourier transform of the filter used in convolution-backprojection Eq. (11.22) is:

$$C(\nu) = |\nu| Q(\nu), \tag{11.25}$$

i.e., a ramp filter which is often apodized by a function $Q(\nu)$. The NPS of the filtered projection is:

$$NPS(\nu) = \frac{\sigma^2}{P} \nu^2 |Q(\nu)|^2 \operatorname{rect}(a\nu) \cdot \tag{11.26}$$

Because the filtered projections are discrete rather than continous, data are usually not available at the points along the projection at which they are required Eq. (11.23). Therefore, interpolation must be used to convert the sampled projections to a continuous function. The NPS of the interpolated, filtered projection is:

$$NPS(\nu) = \frac{\sigma^2}{P} |G(\nu)|^2 \nu^2 |Q(\nu)|^2 \operatorname{rect}(a\nu) \tag{11.27}$$

where $G(\nu)$ is the interpolating function. The NPS of the raw projections, the filtered projections, and the interpolated, filtered projections are shown in Figure 11.6 for an unapodized ramp filter, $Q(\nu) = 1$ and linear interpolation, $G(\nu) = a[\operatorname{sinc}(a\nu)]^2$, where $\operatorname{sinc}(x) = \dfrac{\sin(\pi x)}{\pi x}$.

a)

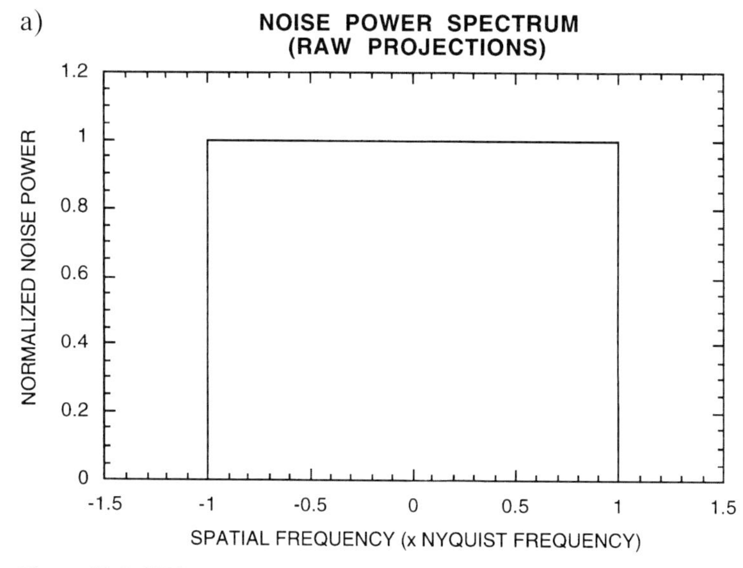

Figure 11.6. NPS at several stages of CT image acquisition and reconstruction: (a) NPS of raw projections.

b)

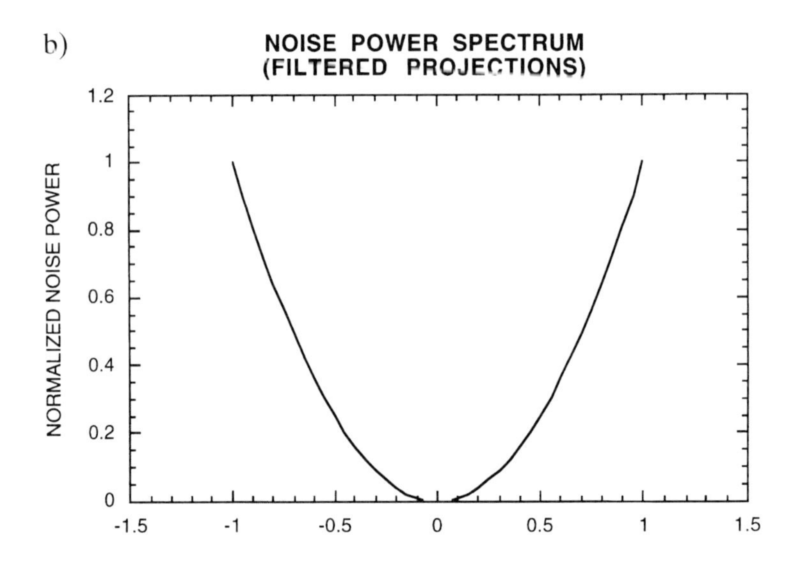

c)

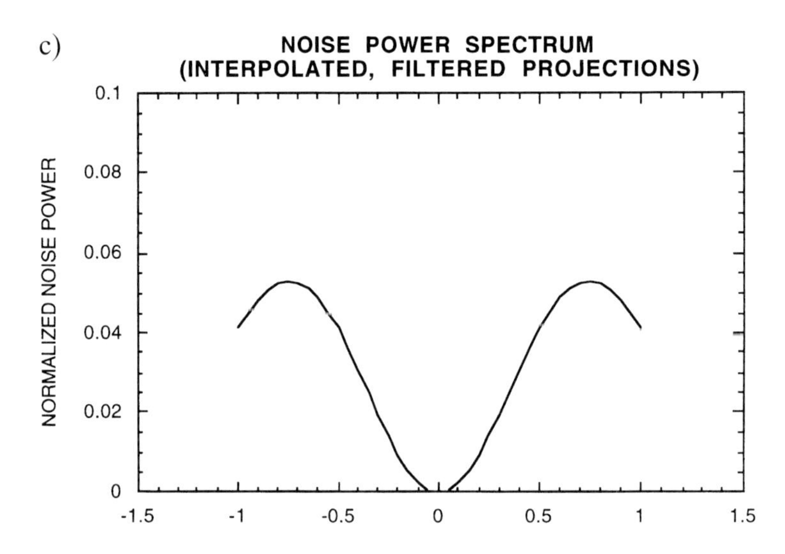

Figure 11.6. (b) NPS of ramp-filtered projections, and (c) NPS of interpolated, filtered projections.

By the projection-slice theorem[17], backprojecting a projection at a given angle superimposes its power spectrum along a spoke through the frequency-space origin at that angle. The spoke density is proportional to radial frequency; therefore, the (two-dimensional) image NPS is (neglecting normalization constants):

$$NPS(\nu_r) \propto \frac{\sigma^2}{P} \left| G(\nu_r) \right|^2 \nu_r \left| Q(\nu_r) \right|^2 \mathrm{rect}(a\nu_r). \tag{11.28}$$

(If sampling effects are considered, the image NPS is, in general, rotationally asymmetric. See Kijewski and Judy[16].) Figure 11.7 shows the (rotationally symmetric) NPS of the reconstructed image. The ramp-like spectrum at low frequency reflects a fundamental property of the data acquisition, while the high frequency rolloff is a result of image processing (smoothing or, in this case, interpolation)[18]. The implication of the shape of the low-frequency portion of the spectrum is that the noise in images reconstructed from projections is negatively correlated. The level of noise is lower at lower frequencies; this reflects the fact that the information is not measured equally at all frequencies. For example, each projection contains a measurement of the zero-frequency (mean) signal level.

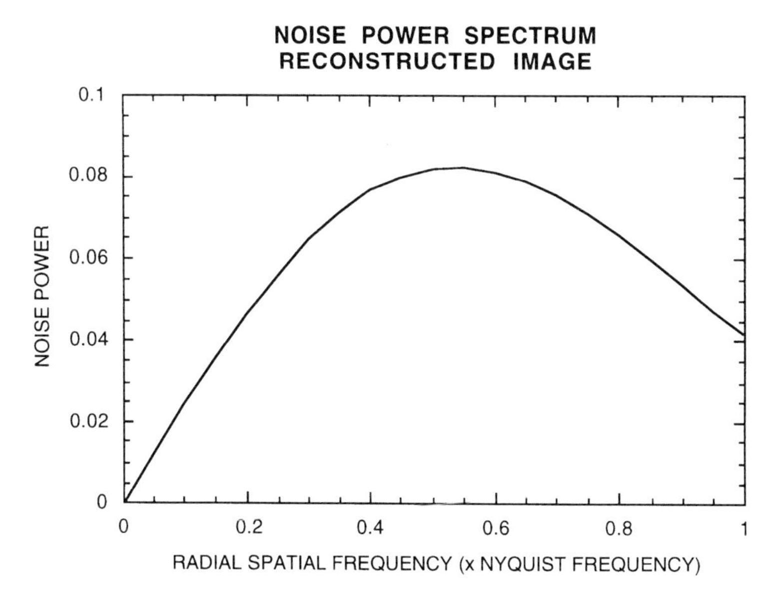

Figure 11.7. Two-dimensional, rotationally symmetric NPS of CT image.

11.2.4 Noise-Equivalent Quanta (NEQ)

The noise-equivalent quanta spectrum, $NEQ(\nu)$, summarizes the performance of the imaging system as a function of spatial frequency in units of photon density.

$$NEQ(\nu) \equiv \frac{K^2 MTF^2(\nu)}{N(\nu)}. \tag{11.29}$$

It is useful for specifying the efficiency with which the imaging system utilizes the information in the pattern of detected photons[18], since it can be interpreted as the number of input quanta needed to yield the observed output noise level for an ideal system. It is also invaluable in calculating the performance of ideal observers because it compactly represents the effects of the imaging system on their performance (see below). The value of $NEQ(\nu)$ is large for frequencies for which the MTF is large, or the noise-power spectrum is small. Barrett et al.[3] have shown that for certain imaging tasks, the NEQ concept can be extended to non-shift-invariant systems with nonstationary noise.

11.2.5 Artifacts

The term 'artifacts' refers to spurious elements of the image which are deterministic rather than stochastic. Artifacts cannot be analyzed as easily as can other aspects of image quality, because they are highly object-dependent. Furthermore, as Hanson has pointed out[19], some kinds of artifacts depend strongly on the relative positions of object and sampling grid. Despite these difficulties, there has been a great deal of progress in understanding the sources of artifacts in various imaging modalities, as well as in developing methods to diminish them. Most artifacts result from inadequate sampling[20, 21] (or, more generally, effects of the system 'null function'[4]) or inconsistencies between the physical aspects of data collection and the assumptions of the mathematical model used for image formation[22, 23, 24]. Barrett et al.[3] have recently proposed a powerful new formalism which may facilitate the analysis of artifacts which are due to the system null function.

11.3 Task-Dependent Measures of Image Quality: Detection and Discrimination

11.3.1 Ideal Observer Models

An ideal observer is defined as one that uses the image information optimally to yield the best possible performance in a specific task, given a particular imaging system. Models based on the ideal observer have been used extensively for binary detection and discrimination tasks as discussed in this section. The concept has more recently been extended to estimation tasks (see Section 11.4). The ideal observer is a mathematical construct which provides standards of performance to which the performance of real observers, such as the human in detection tasks, or clinical estimation procedures, can be compared. Ideal observer models are invaluable for design and optimization of imaging systems, since by their use multiple configurations can be rapidly evaluated. It is important to note that ideal observer models can be derived and evaluated without considering strategy, i.e., the method by which a decision in a binary task, or an estimate of a physical quantity, is extracted from the image. In some cases, for example, the matched filter described in Section 11.3.1.1, ideal observer models imply an optimal strategy. In other cases, such as the estimation task figures of merit discussed in Section 11.4.1.2, implementation procedures cannot be derived from the ideal observer models.

The ideal-observer approach was first applied to a category of binary classification tasks designated 'signal-known-exactly, background-known-exactly (SKE/BKE)'. These tasks require either detection of a lesion, whose size, shape, and location are known *a priori*, superimposed on a background, also specified exactly, or discrimination between two signals whose profiles are exactly known. In each case, there is no uncertainty about the expected image under the two hypotheses; the only question is whether or not the lesion is present, or which of the two possible signals the image represents. Because of concerns that SKE/BKE tasks may not be representative of clinical imaging tasks, methods have been developed to incorporate uncertainty into the lesion and background. Most of these models, discussed in Section 11.3.2, are suboptimal, or quasi-ideal. In Section 11.4.2, we show how estimation theory can be used to develop ideal-observer models of detection or discrimination which incorporate uncertainty about the lesion or background.

In the discussion below the term 'image' will be used to denote the measured data, whether it is acquired raw data or an actual reconstructed image. We will assume a linear, stationary imaging system with additive noise. For more general formulations see, e.g., Barrett et al.[3].

11.3.1.1 Signal-Detection Theory

Classical signal detection theory is applied to binary tasks, i.e., those for which there are two hypotheses, which we call H_1 and H_2. Under H_1, the object f_1 is present; under H_2, object f_2 is present. For detection tasks f_1 represents background only, while f_2 represents lesion plus background; for discrimination tasks, f_1 and f_2 represent the two alternative objects. Given an imaging system with large-area transfer characteristic constant K, point-spread function PSF, and additive, stationary, Gaussian-distributed white noise, the data under the two hypotheses would be given by $\mathbf{g_1}$ and $\mathbf{g_2}$, where:

$$\begin{aligned} g_{1i} &= K\,(PSF^{*}\,f_1)_i + n_i \\ g_{2i} &= K\,(PSF^{*}\,f_2)_i + n_i \end{aligned} \tag{11.30}$$

where n_i is zero-mean Gaussian-distributed noise of variance σ^2. Here g_{1i} represents the ith element of $\mathbf{g_1}$, $(PSF^{*}\,f_1)_i$ is the ith element of the convolution of object f_1 with the point-spread function, and other variables are defined similarly. For generality, a single index, i, is used to denote location within an array of any dimension; the index runs over all array dimensions. We denote the measured data by **g**, and we wish to determine the performance of the ideal observer in determining which of hypotheses H_1 and H_2 is more likely to be true. For Gaussian-distributed noise, the probability of the observed data given hypothesis H_k is:

$$p(\mathbf{g} \mid f_k) = \prod_{i=1}^{N} \frac{1}{\sqrt{2\pi}\sigma} \exp\left[-\frac{\left(g_i - (K\,PSF * f_k)_i\right)^2}{2\sigma^2}\right] \tag{11.31}$$

The ideal observer bases decisions on the value of the likelihood ratio, i.e., the probability that the data have arisen from f_2 rather than from f_1:

$$L = \frac{\prod_{i=1}^{N} \frac{1}{\sqrt{2\pi}\sigma} \exp\left[-\frac{\left(g_i - (K\,PSF * f_2)_i\right)^2}{2\sigma^2}\right]}{\prod_{i=1}^{N} \frac{1}{\sqrt{2\pi}\sigma} \exp\left[-\frac{\left(g_i - (K\,PSF * f_1)_i\right)^2}{2\sigma^2}\right]} \tag{11.32}$$

Decision performance is not altered by monotonic transformations; therefore, for convenience, we work with the logarithm of the likelihood ratio:

$$\log L = \sum_{i=1}^{N}\left[-\frac{\left(g_i - (K\,PSF * f_2)_i\right)^2}{2\sigma^2}\right] - \sum_{i=1}^{N}\left[-\frac{\left(g_i - (K\,PSF * f_1)_i\right)^2}{2\sigma^2}\right] \tag{11.33}$$

Expanding Eq. (11.33) and retaining only terms which depend on the data, we obtain the decision variable:

$$\xi = \frac{1}{\sigma^2}\sum_{i=1}^{N}\left(K\,PSF * (f_2 - f_1)\right)_i g_i \tag{11.34}$$

Implicit in Eq. (11.34) is the strategy for optimal detection (or discrimination): the measured data is multiplied point-by-point by the difference of the expected images under the two hypotheses, and the products summed over the image space. This ideal observer is known as the 'matched filter' for white noise; its counterpart for correlated noise is given in Eq. (11.41) below. Because ξ is formed by linear combination of Gaussian-distributed random variables, it is also Gaussian distributed. Each hypothesis will give rise to a Gaussian distribution in ξ (Figure 11.8).

The means and variances of the distributions under the two hypotheses can be calculated from Eq. (11.34):

$$\langle \xi \mid f_2 \rangle = \sum_{i=1}^{N} \frac{\left(K\,PSF * (f_2 - f_1)\right)_i \langle g_i \rangle}{\sigma^2} = \sum_{i=1}^{N} \frac{\left(K\,PSF * (f_2 - f_1)\right)_i \left(K\,PSF * f_2\right)_i}{\sigma^2} \tag{11.35}$$

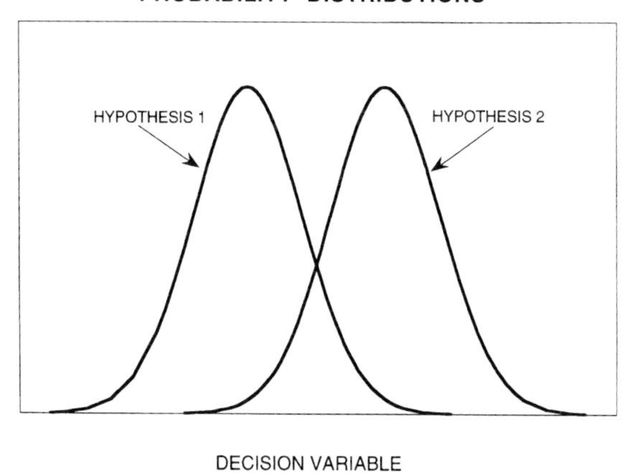

Figure 11.8. Distributions of decision variable under alternative hypotheses.

$$\langle \xi \mid f_1 \rangle = \sum_{i=1}^{N} \frac{(K\,PSF * (f_2 - f_1))_i (K\,PSF * f_1)_i}{\sigma^2} \tag{11.36}$$

$$\begin{aligned} \operatorname{var}(\xi \mid f_2) = \operatorname{var}(\xi \mid f_1) &= \sum_{i=1}^{N} \left(\frac{\partial \xi}{\partial g_i} \right)^2 \operatorname{var}(g_i) \\ &= \sum_{i=1}^{N} \left(\frac{(K\,PSF * (f_2 - f_1))_i}{\sigma^2} \right)^2 \sigma^2 = \frac{1}{\sigma^2} \sum_{i=1}^{N} (K\,PSF * (f_2 - f_1))_i^2 \end{aligned} \tag{11.37}$$

The square of the ideal-observer signal to noise ratio is defined as the squared difference of the means, divided by the average variance, i.e.:

$$SNR^2 = \frac{[\langle \xi \mid f_2 \rangle - \langle \xi \mid f_1 \rangle]^2}{\frac{1}{2}[\operatorname{var}(\xi \mid f_2) + \operatorname{var}(\xi \mid f_1)]} = \frac{1}{\sigma^2} \sum_{i=1}^{N} (K\,PSF * (f_2 - f_1))_i^2 \tag{11.38}$$

The decision variable can also be formulated in frequency space:

$$\xi = \frac{K}{N_0} \int [F_2(\nu) - F_1(\nu)] MTF(\nu) G(\nu) \mathrm{d}\nu \tag{11.39}$$

where N_0 is the value of the (constant) noise-power spectrum, F_2 and F_1 are the Fourier transforms of the alternative objects, $MTF(\nu)$ is the system modulation transfer function (see Section 11.2.2.2), and $G(\nu)$ is the Fourier transform of the data. If the noise is correlated, then the correlations must be removed by prewhitening, i.e., dividing the Fourier transform of the data by the square root of the noise-power spectrum. The matched filter must also be prewhitened, yielding:

$$\xi = K\int \frac{[F_2(\nu)-F_1(\nu)]MTF(\nu)G(\nu)}{N(\nu)}\mathrm{d}\nu, \tag{11.40}$$

the prewhitening matched filter. The corresponding signal-to-noise ratio is given by:

$$SNR_I^2 = K^2\int \frac{|F_2(\nu)-F_1(\nu)|^2\, MTF^2(\nu)}{N(\nu)}\mathrm{d}\nu \tag{11.41}$$

11.3.1.2 Spatial Frequency Dependence of Imaging Tasks

Eq. (11.41) can be reformulated to clarify the contributions of the imaging system and the task to the ideal-observer signal-to-noise ratio. The large-area transfer constant, K, was defined in Section 11.2.1 as the ratio of the mean output level to the mean input level. By incorporating these values into the integrand we can normalize the signal spectrum and the noise power spectrum, expressing them in relative terms:

$$SNR_I^2 = \int \frac{|(\Delta F(\nu))_{rel}|^2\, MTF^2(\nu)}{(N(\nu))_{rel}}\mathrm{d}\nu\cdot \tag{11.42}$$

Here the quantity $F_2(\nu) - F_1(\nu)$ has been replaced by $\Delta F(\nu)$. From Eq. (11.29) in Section 11.2.4 we recognize that:

$$SNR_I^2 = \int |(\Delta F(\nu))_{rel}|^2\, NEQ(\nu)\mathrm{d}\nu \tag{11.43}$$

From Eq. (11.43) it is clear that $NEQ(\nu)$, which summarizes the frequency response of the imaging system, is weighted by the signal spectrum $\Delta F(\nu)$, which represents the importance of various frequency components to a given imaging task.

11.3.1.3 Examples

Here we consider several binary tasks: detection, location discrimination, and discrimination of a single from a binary signal. These examples are one-dimensional; extension to higher dimensions is straightforward. This development follows that of Hanson[25].

The first task that we consider is detection of a Gaussian-shaped lesion superimposed on a zero-mean background. The two signals under the alternate hypotheses are:

$$f_1 = \exp\left(-x^2/\left(2\sigma^2\right)\right)$$
$$f_2 = 0 \tag{11.44}$$

These are shown in Figure 11.9a for $\sigma = 1$. The squared magnitude of the Fourier transform of the difference signal is (Figure 11.9b):

$$|\Delta F(\nu)|^2 = 2\pi \exp\left(-4\pi^2\sigma^2\nu^2\right). \tag{11.45}$$

The second task considered here is location discrimination, in which two identical signals, whose position differs by ε, must be discriminated. The alternate signals are:

$$f_1 = \exp\left(-x^2/\left(2\sigma^2\right)\right)$$
$$f_2 = \exp\left(-(x-\varepsilon)^2/\left(2\sigma^2\right)\right) \tag{11.46}$$

these are shown in Figure 11.9c for $\varepsilon = 2$. The squared magnitude of the Fourier transform of the difference signal is (Figure 11.9d):

$$|\Delta F(\nu)|^2 = 4\pi \exp\left(-4\pi^2\sigma^2\nu^2\right)\cos(2\pi\varepsilon\nu) \tag{11.47}$$

The third task is discrimination of a single Gaussian lesion from a pair of lesions of half its amplitude, separated by 2ε. The alternate signals are:

$$f_1 = \exp\left(-x^2/\left(2\sigma^2\right)\right)$$
$$f_2 = \left[\exp\left(-(x-\varepsilon)^2/\left(2\sigma^2\right)\right) + \exp\left(-(x+\varepsilon)^2/\left(2\sigma^2\right)\right)\right]/2; \tag{11.48}$$

these are shown in Figure 11.9e for $\varepsilon = 1.3$. The squared magnitude of the Fourier transform of the difference signal is (Figure 11.9f):

$$|\Delta F(\nu)|^2 = 2\pi \exp\left(-4\pi^2\sigma^2\nu^2\right)[1 - \cos(2\pi\varepsilon\nu)] \tag{11.49}$$

From Figure 11.9 it is clear that different tasks place different demands on the imaging system. For simple (SKE/BKE) detection, information at low frequencies is quite important; for the more complex tasks, low frequency information is less important, and zero-freqency information (i.e., the image mean) is not at all useful. Some implications of these task spectra are that for SKE/BKE detection, performance is optimized by a system which is characterized by low levels of noise; resolution is not important. For the more complex tasks, on the other hand, resolution is critically important in order to preserve the higher frequency components at which the alternative signals differ. (See also discussion of estimation tasks in Section 11.4)

a)

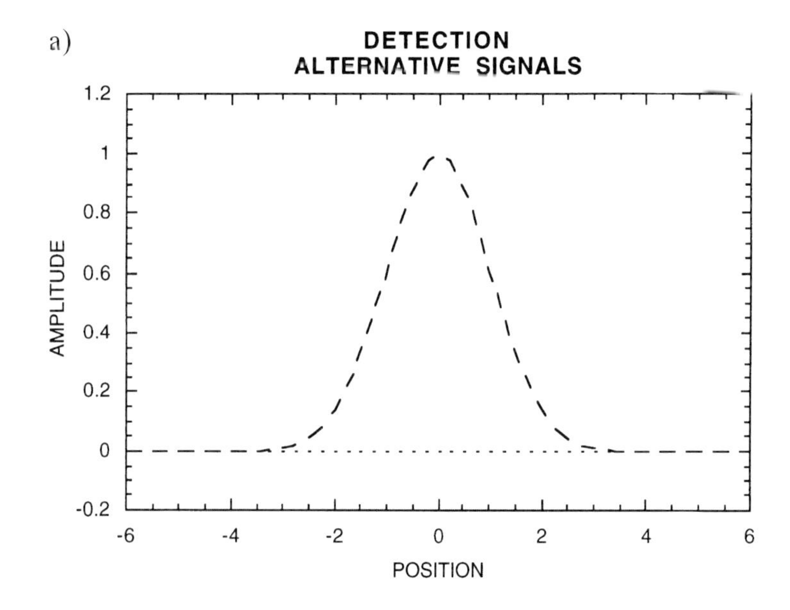

b)

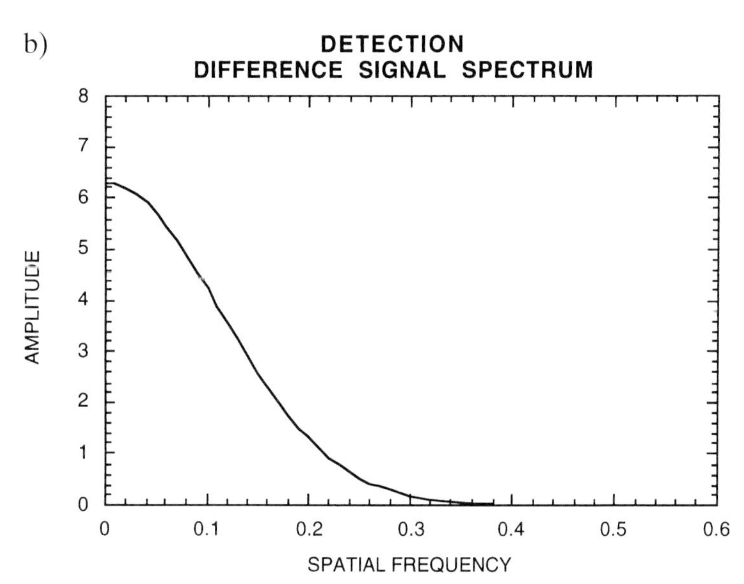

Figure 11.9. (a) Difference signal spectrum for detection of a Gaussian-shaped lesion and (b) its Fourier transform.

c)

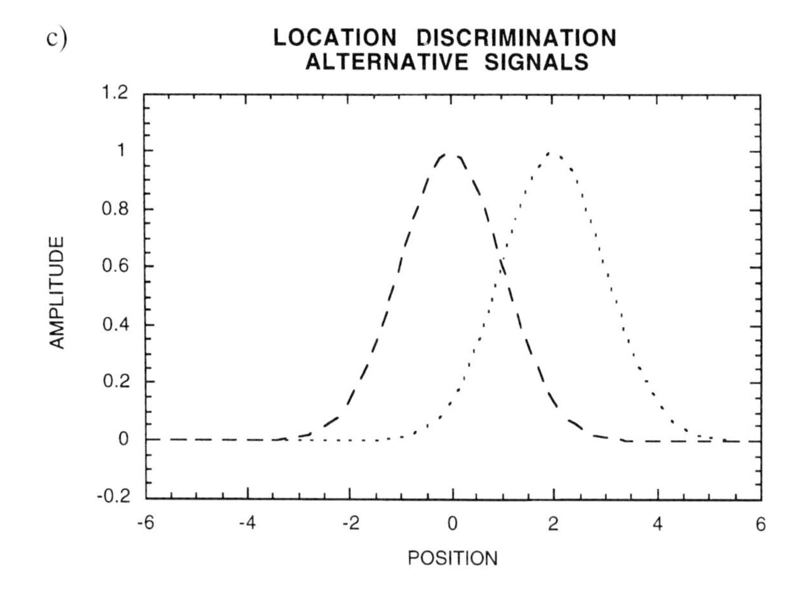

d)

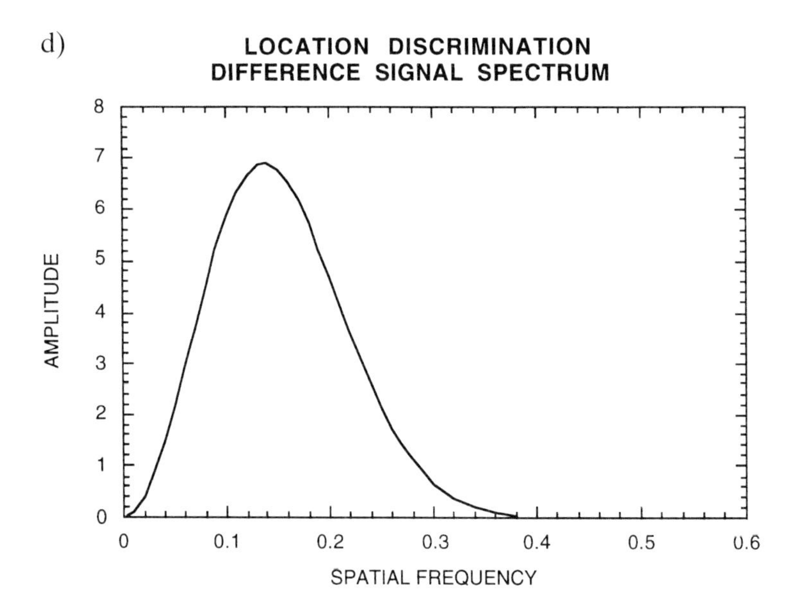

Figure 11.9. (c) Difference signal spectrum for location discrimination and (d) its Fourier transform.

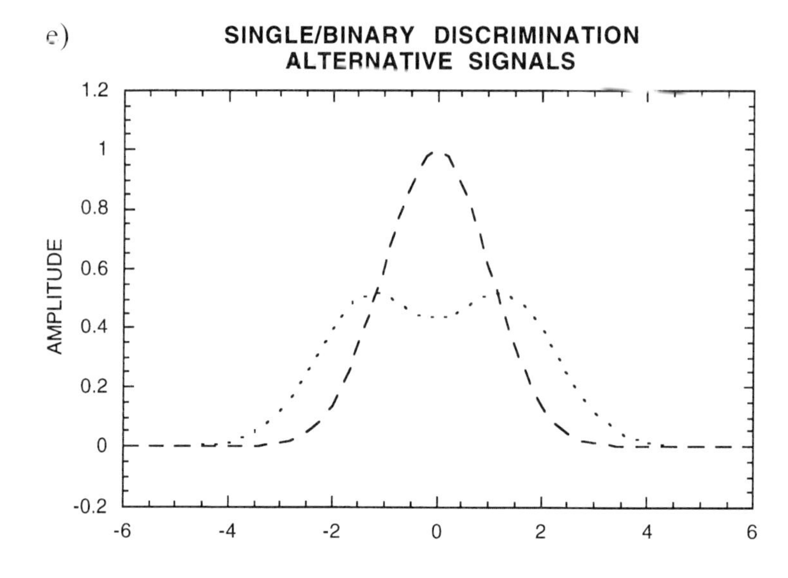

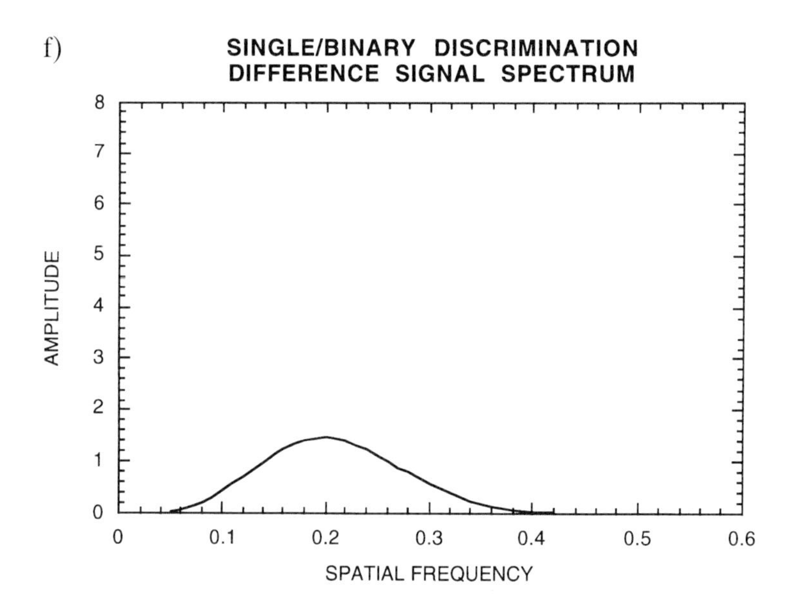

Figure 11.9. (e) Difference signal spectrum for discrimination between single and binary lesions and (f) its Fourier transform. Modified from Hanson[25].

11.3.2 Quasi-Ideal Observer Models

Quasi-ideal observer models have been formulated for two reasons. The first is that in some cases they have been found to be better predictors of human performance than are ideal-observer models. The nonprewhitening and channelized models, discussed in Section 11.3.2.1, were developed to emulate the human's apparent inability to use the information contained in the noise correlations. The second reason is related to the need to model more complex imaging tasks than the simple SKE/BKE tasks. In many cases ideal observer models are incapable of accommodating task complexity; even models that can do this (e.g., Section 11.4) involve nonlinear processing. Because it is believed that humans are not capable of such procedures, there is a need for linear, quasi-ideal observer models, such as the Fisher-Hotelling models described in Section 11.3.2.2.

11.3.2.1 Nonprewhitening and Channelized Models

The nonprewhitening matched filter (NPWMF) model, although otherwise optimal, is unable to account for correlations in noise (or, alternatively, to use the information implicit in the noise correlations). The SNR of the NPWMF model is given by:

$$SNR^2_{NPWMF} = \frac{K\left[\int \left|(\Delta F(\nu))_{rel}\right|^2 MTF^2(\nu)\mathrm{d}\nu\right]^2}{\int \left|(\Delta F(\nu))_{rel}\right|^2 MTF^2(\nu)N(\nu)\mathrm{d}\nu}. \tag{11.50}$$

It can be shown[18] that its performance is inferior to that of the ideal, prewhitening matched filter model of Eq. (11.42) in the presence of correlated noise. Visual perception studies have shown that for some tasks the nonprewhitening matched filter is a better predictor of human performance than is the ideal observer[26, 27]. For other tasks, however, it was less successful, and there were attempts to improve it by incorporating the frequency response of the human visual system[28, 29]. Burgess et al.[30] have recently shown that it is not possible to modify the NPWMF so that it predicts human performance in certain kinds of correlated noise. Current research on models of performance in correlated noise is directed toward channelized models[30].

The channelized quasi-ideal observer model was introduced by Myers and Barrett[31]. They showed that the predictions of the NPWMF model and the channelized model regarding detection in negatively correlated noise were indistinguishable, and argued that the channelized model was to be preferred, since it is consistent with the spatial frequency channels that are known to exist in the human visual system. The original model of Myers and Barrett incorporated a simple model of the spatial frequency channels; they were assumed to be nonoverlapping and rectangular in radial spatial frequency, with a cutoff frequency below which no information was transmitted, and channel widths of an octave in spatial frequency. A more recent version of the channelized model[30] allows the channels to extend to zero frequency, but multiplies the channel profiles by the frequency response of the eye, which goes to zero at zero frequency. Channel boundaries are not fixed, but are allowed to vary over an octave, and model predictions are averaged over boundary locations.

11.3.2.2 Fisher/Hotelling Observer

The SKE/BKE model is not a good description of most clinical classification tasks. The sizes, shapes, and locations of the features to be detected or discriminated are not known *a priori*, and the background is not uniform. When variability in the object and/or background is incorporated into the model, SNR_I is often uncalculable. In cases where it can be derived, it usually requires nonlinear operations of which the human may not be capable[32]. The Fisher-Hotelling observer, the optimum linear discriminator, was proposed for medical imaging purposes by Barrett[4,33]. It can be calculated for a given task if the statistics of the object and background, i.e., their means and covariance matrices across an ensemble of objects, are known for all image classes.

Two matrices are defined: S_1 and S_2. S_1, the interclass scatter matrix, measures the deviation of the group means from the overall mean:

$$S_1 = \sum_{j=1}^{J} P_j \left(\bar{f} - \bar{f}_j\right)\left(\bar{f} - \bar{f}_j\right)' \tag{11.51}$$

In Eq. (11.51), j indexes image class, P_j is the probability of class j, $\bar{f}$ is the mean image for class j, and $\bar{f}$ is the overall mean:

$$\bar{f} = \sum_{j=1}^{J} P_j \bar{f}_j \tag{11.52}$$

S_2, the intraclass scatter matrix, is the average covariance matrix:

$$S_2 = \sum_{j=1}^{J} P_j K_j. \tag{11.53}$$

It can be shown[4] that for binary tasks,

$$S_1 = P_1 P_2 \left(\bar{f}_2 - \bar{f}_1\right)\left(\bar{f}_2 - \bar{f}_1\right)'. \tag{11.54}$$

The decision variable for the Hotelling observer is:

$$\xi = \left(H\left(\bar{f}_2 - \bar{f}_1\right)\right)' S_2^{-1} H f. \tag{11.55}$$

The Hotelling trace, a measure of how well the classes can be separated, is defined as:

$$J = Tr\left(S_2^{-1} S_1\right); \tag{11.56}$$

this can also be viewed as the square of a SNR for the Fisher-Hotelling observer. If H is linear and shift-invariant, and S_2 is stationary, then the SNR can be expressed in Fourier space:

$$SNR_{Hot}^2 = K^2 \int \frac{\left|\bar{F}_2(\nu) - \bar{F}_1(\nu)\right|^2 MTF^2(\nu)}{W_g(\nu)} \mathrm{d}\nu, \tag{11.57}$$

where $W_g(\nu)$ is the total noise power spectrum, including contributions from object variability as well as photon statistics. SNR_{Hot} is analogous to SNR_I, the prewhitening matched filter, with the difference of two known signals replaced by the difference of the means of two signals known only statistically, and the concept of noise extended to include object variability.

It is evident from the above that classification performance increases as S_1, and therefore the difference in mean between the two classes, increases, and also as S_2, the intraclass scatter matrix, decreases. Values of J can be compared for various imaging systems. An absolute standard can be obtained by calculating the maximum possible value of J which would be obtained by a perfect imaging system, using the means and covariance matrices for the original objects.

The Fisher-Hotelling observer has successfully predicted human performance in some tasks incorporating object variability[34,35,36], and addition of a simple channel mechanism has extended the range of tasks for which it is a good model of human performance[37]. Current research suggests that a Fisher-Hotelling model, incorporating the more sophisticated frequency-channel model described above, is the most promising approach to prediction of human performance over a range of realistic classification tasks[30].

11.3.3 The Human Observer

For most purposes, medical images are interpreted by human observers. Therefore, human performance in clinically relevant tasks is the ultimate standard by which to evaluate imaging systems. The tasks for which human experiments are performed are almost always binary, i.e., there are two possible truth states, which we will call 'positive', or abnormal, or 'negative', or normal. The images used for the experiments may be obtained by computer simulation, by imaging of a simple or anatomically accurate phantom, or by collection of clinical images for which the truth has been established. In some cases, hybrid images, such as normal clinical images to which artificial lesions have been added, may be used[38].

11.3.3.1 General Principles

There are four possible outcomes when a human observer makes a binary judgement about an image: the image is correctly classified as abnormal (true positive decision), the image is correctly classified as normal (true negative), the image is incorrectly classified as abnormal (false positive), or the image is incorrectly classified as normal (false negative). In the medical literature, the efficacy of tests is often quantified by sensitivity, the true positive fraction, and specificity, one minus the false positive fraction. These measures are not entirely satisfactory, since they depend on the willingness of the observer to judge an image abnormal. As this criterion varies, both sensitivity and specificity change; each criterion is associated with a pair of sensitivity and specificity values. If these could be measured over the complete range of criteria, then human performance in a particular task with a particular imaging system would be completely characterized. This is accomplished by the receiver operating characteristic (ROC) curve (Figure 11.10), which plots the true-positive fraction (sensitivity) vs the false-positive fraction (1-specificity) as the criterion for a positive judgment varies. The ROC curve corresponding to chance performance – i.e., random judgments – is the dashed diagonal line. As the curves extend farther toward the upper left-hand corner, better classification performance is implied. Two commonly used summary measures of classification performance are the area under the ROC curve (ranging from 0.5 for chance performance to 1 for perfect performance) and d', a transformation of the ROC area.

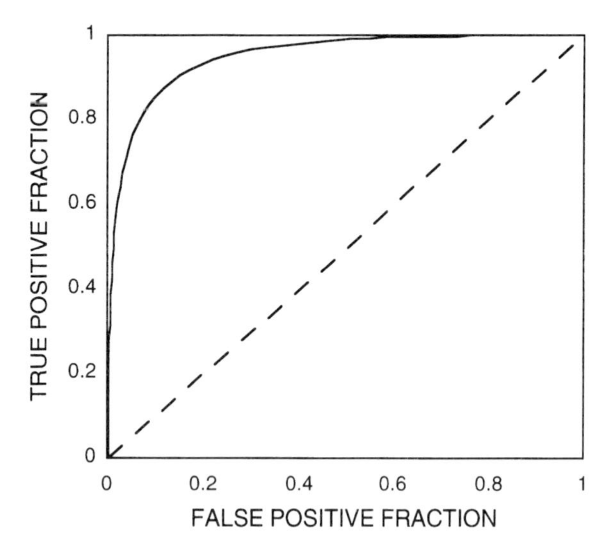

Figure 11.10. ROC curve

11.3.3.2 Rating-Method Experiments and ROC Analysis

To generate data for ROC studies using the rating method, humans are presented with a set of images which include both positive and negative cases, often in equal proportions. It is assumed that the observer looks at each image and forms a mental decision variable, or degree of certainty that the image is abnormal (positive). For the most commonly used ROC-fitting procedure[39], the values of the decision variable are assumed to be Gaussian-distributed for the negative and positive cases (Figure 11.11). For each image, the observer reports his degree of confidence that the image is abnormal. Often a five- or six-category confidence scale is used, although continuous confidence ratings can also be used[40]. In Figure 11.11, decision variable thresholds corresponding to a five-category confidence scale are illustrated. The categories and their confidence ranges are indicated below the abcissa. The probability that a given confidence rating will be reported, given a negative (or positive) case, is proportional to the area below that distribution between the decision variable thresholds defining that category. An ROC curve is fitted to the rating data (numbers of images per decision variable category for positive and negative images) using the binormal model discussed here. The fitting technique yields estimates of the means and standard deviations of the two categories, the values of the category boundaries, and the area under the ROC curve.

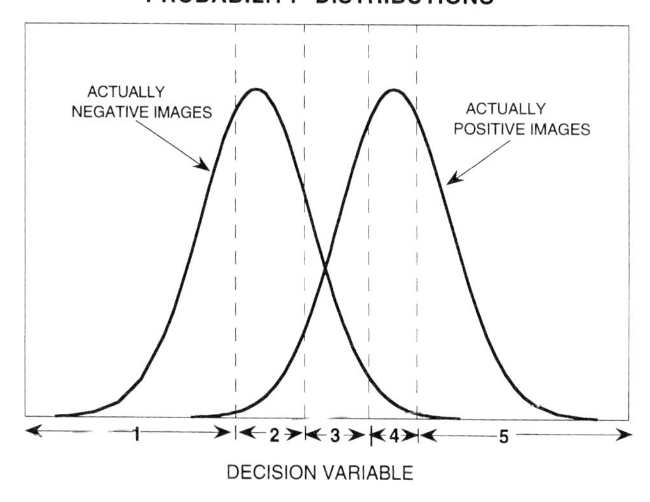

Figure 11.11. Distributions of decision variable under alternative hypotheses, with boundaries of a five-category rating scale superimposed.

11.3.3.3 Forced-Choice Experiments and Analysis

An alternate method of measuring human performance is by two-alternative forced-choice experiments. In this experimental paradigm, pairs of images are simultaneously presented to the observer. One of the two images is always positive, and one negative. The observer is required only to choose the one he believes is more likely to be positive; no confidence assessment is needed. Analysis of forced-choice data yields the area under the ROC curve, although the full ROC curve is not provided. The expected fraction of correct decisions equals the area under the ROC curve that would have been obtained by the rating data approach using the same image set. In many cases for which only the ROC area is needed, two-alternative forced-choice experiments are more efficient, since they are less demanding of observer time than are rating data experiments[41,42].

11.3.3.4 Comparing Model and Human Observers

Human observer performance can be directly compared to that of a model observer. A quantity analogous to d' can be obtained from the SNR for all model observers. The d' values for human and model observers are not expected to be equal in absolute terms; rather, a succesful model is one that predits changes in human performance as a result of changes in the imaging system. Human performance is, in almost all cases, lower than that of the ideal observer. The squared ratio of human to ideal (or quasi-ideal) performance is termed the 'statistical efficiency'[43], defined as:

$$F = \left[\frac{d'_H}{d'_I} \right]^2 \tag{11.58}$$

Efficiency values have been measured for a number of tasks; most of these values are near 50%[44]. It has been found that efficiencies are lower in correlated noise[27]. The determination of human efficiency in various tasks is an active area of research, as is the identification of sources of inefficiency. Several explanations for the inefficiency of the human have been proposed. First, imperfect *a priori* knowledge as to lesion size and/or location for a SKE/BKE task may contribute. Another plausible explanation is the presence of 'internal noise' in the human visual system. Another possible source, which would affect measurements by the rating data approach, but not the forced-choice approach, is instability in confidence thresholds.

11.4 Task-Dependent Measures of Image Quality: Estimation

11.4.1 Estimation Tasks

Estimation tasks are those which involve obtaining parameters from the measured data. Image reconstruction is sometimes viewed as an estimation task; the parameters are the image pixels, and estimation of the image itself is the goal. This approach, however, does not lead to task-dependent figures of merit, and therefore is not, strictly speaking, estimation in the sense we are using here. However, because such techniques, and figures of merit derived from them, are commonly used, they will be discussed here for completeness. Estimation of the image itself does not require (although it sometimes benefits from) *a priori* information. True estimation tasks require some *a priori* information. In some cases the object is seen as a model in one or more unknown parameters, which are to be estimated from the image.

11.4.1.1 Estimation of Image

In the presence of additive Gaussian noise, the optimal method of estimating the image components is by maximum likelihood techniques. This is equivalent to inverse filtering by the system function, assuming that it can be inverted. For planar imaging techniques, such as radiography, this approach is called 'deconvolution'. Attempts to compensate by deconvolution for blurring by the system function have the unwanted side effect of amplifying noise. For this reason, deconvolution is used only infrequently, and to a limited extent[45]. The system function is much more complex for systems in which the images must be reconstructed from projections. Because the system functions of most of these systems cannot be inverted, reconstruction is accomplished by either analytical or iterative means.

When estimation of the image itself is the goal, figures of merit measure fidelity. One such measure is the mean-square error, which is defined as the sum over the image of the squared differences between the estimated image values and the true values. Although this measure is in common use, it is not task-dependent.

11.4.1.2 Model-based Estimation Tasks

If the object can be modeled as a function of a small number of parameters, then there exist quite flexible approaches to evaluation or optimization of imaging systems even for nonlinear tasks. Mueller et al.[46] investigated the optimal tradeoff between collimator resolution and sensitivity using as a figure of merit the precision with which object parameters could be estimated by an unbiased, nonlinear maximum-likelihood procedure. They modeled the object as a disk of unknown activity concentration, size, and location, superimposed on a uniform (except for noise) background of unknown activity. They showed that for the estimation tasks considered (e.g., disk activity estimation with other parameters unknown, disk size estimation with other parameters unknown), performance was optimal at high resolution, even at the expense of large losses in sensitivity. This was consistent with the conclusions of Hanson[25], who analyzed complex binary tasks (see Section 11.3.1.3).

Although the analytical models are simplified for mathematical tractability, they can lead to general conclusions about optimization of imaging systems. It is hard to imagine, for example, that relationships between performance in estimating activity of a spherical lesion and system performance would change significantly for irregularly shaped lesions of similar size.

It is not necessary to specify an estimation procedure in order to determine the precision of an optimal unbiased (or biased) estimator. The performance of an optimal estimator in a given estimation task using a particular imaging system can be obtained by evaluating the Cramer-Rao lower bound(s) (CRB) on the variance with which the parameter(s) of interest can be estimated. Since the CRB represents the minimum variance for an unbiased (or, in a modified form, biased) estimator, its use as a figure of merit for assessing imaging systems is consistent with the 'ideal observer' approach used for binary classification tasks. In this way we can extend the ideal observer concept to estimation tasks; note, however, that there is no human judgment involved in these procedures.

We assume that the Fourier transform of the (two-dimensional) image is given by:

$$I(\vec{\nu}) = S(\vec{\nu};\vec{\theta}) + \eta(\vec{\nu}), \tag{11.59}$$

where

$$S(\vec{\nu};\vec{\theta}) = F(\vec{\nu};\vec{\theta})MTF(\vec{\nu}). \tag{11.60}$$

$F(\vec{\nu}\ \vec{\theta})$ is a function describing the Fourier tranform of the object in terms of the vector of unknown parameters, $\vec{\theta}$; the length of $\vec{\theta}$ is N, where N is the number of unknown parameters. $MTF(\vec{\nu})$ is the system modulation transfer function (see Section 11.2.2.2). The Cramer-Rao lower bounds on the variance of (unbiased) estimates of these parameters are the diagonal elements of the inverse of Fisher's information matrix[47]. The entries of Fisher's information matrix, **J**, are defined, for stationary additive white Gaussian noise, using the first derivatives of $S(\vec{\nu};\vec{\theta})$ with respect to the unknown parameters:

$$J_{ij} = \frac{1}{N_0} \int \frac{\partial S(\vec{\nu};\vec{\theta})}{\partial \theta_i} \frac{\partial S(\vec{\nu};\vec{\theta})}{\partial \theta_j} d\vec{\nu};\ i,j=1,\ldots N, \tag{11.61}$$

where $N_0 = \sigma_p^2 a^2$ is the height of the (constant) noise power spectrum, σ_p^2 is the pixel variance and a is the (square) pixel dimension. The integration is over all frequency space. For correlated noise, **J** is derived using a 'prewhitened' model, i.e., Eq. (11.60) is replaced by:

$$S(\vec{\nu};\vec{\theta}) = \frac{F(\vec{\nu};\vec{\theta})MTF(\vec{\nu})}{\sqrt{N(\vec{\nu})}}. \tag{11.62}$$

11.4.2 Relationship Between Estimation and Detection

As Hanson[25] has pointed out, signal detection is a special case of amplitude estimation; the task can be seen as estimation of the amplitude of the feature to be detected, where the two possible amplitudes are zero and A_f. The relationship can be clearly seen by constructing an SNR, based on estimation task metrics, and comparing it to the SNR_I for SKE/BKE detection of Eq. (11.41). For an optimal estimation procedure, the expected value of the estimates of A_f is equal to its true value, and the variance of the estimates is minimized. Therefore, the numerator of the SNR is A_f, and the denominator is the CRB on estimation of A_f. If the CRB on estimation of A_f is independent of its value, then we can assume that the estimation variance is the same under the two hypotheses, and:

$$SNR_{est}^2 = \frac{A_f^2}{CRB(A_f)}. \tag{11.63}$$

Since for this one-parameter estimation task,

$$CRB(A_f) = J_{11}^{-1} = \left[K^2 \int \frac{\left[\frac{\mathrm{d}F(\nu)}{\mathrm{d}A_f}\right]^2 MTF^2(\nu)}{N(\nu)} \mathrm{d}\nu \right]^{-1}, \tag{11.64}$$

and $F(\nu)$ is linear in A_f,

$$SNR_{est}^2 = K^2 \int \frac{|F(\nu)|^2 MTF^2(\nu)}{N(\nu)} \mathrm{d}\nu. \tag{11.65}$$

By comparison to Eq. (11.41) we see that SNR_{est} is equal to SNR_I for the SKE/BKE task. SNR_{est}, however, can be used for tasks involving uncertainty about the object, such as feature size or amplitude of the (assumed homogeneous) background (see Section 11.4.3). The CRB on the estimates of A_f will be greater in these multi-parameter estimation tasks and, consequently,

SNR_{est} will be reduced. This raises the possibility of nonlinear models of detection with uncertainty, based on estimation metrics; these models, however, have not yet been shown to be successful in predicting human performance.

Derivation of a quasi-ideal estimation model, analogous to the channelized ideal observer model discussed above (Section 11.3.2.1), is straightforward following the work of Myers and Barrett[31]. The entries of Fisher's information matrix are given by:

$$J_{ij} = \sum_k \frac{\int_{vl_k}^{vu_k} \frac{\partial S(\vec{v};\vec{\theta})}{\partial \theta_i} d\vec{v} \int_{vl_k}^{vu_k} \frac{\partial S(\vec{v};\vec{\theta})}{\partial \theta_j} d\vec{v}}{\int_{vl_k}^{vu_k} N(\vec{v}) d\vec{v}} \tag{11.66}$$

for k nonoverlapping channels which are rectangular in radial frequency. This model could, of course, be modified to incorporate the frequency-channel mechanism proposed by Burgess[30].

11.4.3 Examples

1 Estimation of feature activity concentration with unknown size

We model the (two-dimensional) object as a Gaussian-shaped feature centered on a circular background, and assume that the MTF is Gaussian in form and the noise is white, additive and Gaussian-distributed. The Fourier transform of the expected image is:

$$S(\nu) = \left\{ 2\pi\sigma_g^2 A_g \exp\left(-2\pi^2\sigma_g^2\nu^2\right) + \frac{A_b r_b J_1(2\pi r_b \nu)}{\nu} \right\} \exp\left(-2\pi^2\sigma_m^2\nu^2\right) \tag{11.67}$$

where ν is radial frequency, A_g and A_b are the activity concentrations of the feature and the background, σ_g is the size parameter of the Gaussian-shaped feature, r_b is the radius of the background, and σ_m is the parameter of the Gaussian MTF. The parameter of interest is the activity concentration, A_g, of the feature; the other unknown parameter is σ_g. Because there are two unknown parameters, Fisher's information matrix is of order 2×2; its entries are obtained by Eq. (11.61):

$$J_{11} = \frac{1}{N_0} \int \left[\frac{\partial S(\vec{v};\vec{\theta})}{\partial A_d} \right]^2 d\vec{v} = \frac{\pi\sigma_g^4}{N_0\left(\sigma_g^2 + \sigma_m^2\right)} \tag{11.68}$$

$$J_{22} = \frac{1}{N_0} \int \left[\frac{\partial S(\vec{v};\vec{\theta})}{\partial \sigma_g} \right]^2 d\vec{v} = \frac{2\pi A_g^2 \sigma_g^4 \left(\sigma_g^4 + 2\sigma_g^2\sigma_m^2 + 2\sigma_m^4\right)}{N_0\left(\sigma_g^2 + \sigma_m^2\right)^3} \tag{11.69}$$

$$J_{12} = J_{21} = \frac{1}{N_0} \int \frac{\partial S(\vec{\nu};\vec{\theta})}{\partial A_g} \frac{\partial S(\vec{\nu};\vec{\theta})}{\partial \sigma_g} \mathrm{d}\vec{\nu} = \frac{\pi A_g \sigma_g^3 (\sigma_g^2 + 2\sigma_m^2)}{N_0 (\sigma_g^2 + \sigma_m^2)^2} \tag{11.70}$$

The CRB on the variance with which A_g can be estimated by an unbiased procedure is given by:

$$[\mathbf{J}^{-1}]_{11} = \frac{2N_0(\sigma_g^2 + \sigma_m^2)(\sigma_g^4 + 2\sigma_g^2\sigma_m^2 + 2\sigma_m^4)}{\pi \sigma_g^8}. \tag{11.71}$$

We can construct a SNR by:

$$SNR^2 = \frac{\langle \hat{A}_g \rangle^2}{\mathrm{var}(\hat{A}_g)}; \tag{11.72}$$

for an optimal estimation procedure, the expected value of the estimates of A_g is equal to its true value, and the variance of the estimates is minimized. Therefore,

$$SNR_I^2 = \frac{A_g^2}{[\mathbf{J}^{-1}]_{11}} = \frac{\pi \sigma_g^8 A_g^2}{2N_0(\sigma_g^2 + \sigma_m^2)(\sigma_g^4 + 2\sigma_g^2\sigma_m^2 + 2\sigma_m^4)}. \tag{11.73}$$

If the feature size is known, then the CRB on the variance of $\hat{A}_g$ is given by:

$$J_{11}^{-1} = \frac{N_0(\sigma_g^2 + \sigma_m^2)}{\pi \sigma_g^4} \tag{11.74}$$

and

$$SNR_I^2 = \frac{A_g^2}{J_{11}^{-1}} = \frac{\pi \sigma_g^4 A_g^2}{N_0(\sigma_g^2 + \sigma_m^2)} \tag{11.75}$$

2 Estimation of Disk Activity Concentration with Disk Size and Background Activity Concentration Unknown

We model the object as in Example 1, above. For this task, the parameter of interest is the activity concentration, A_g, of the feature, and there are two other unknown parameters: σ_g and A_b. In this case Fisher's information matrix is of order 3 × 3. The four entries in the upper left-hand corner are as given in Eqs. (11.68, 11.69 and 11.70), and the additional entries are:

$$J_{33} = \frac{1}{N_0} \int \left[\frac{\partial S(\vec{\nu};\vec{\theta})}{\partial A_b} \right]^2 \mathrm{d}\vec{\nu} = \frac{\pi r_b^2}{N_0} \left[1 - \exp\left(-\frac{r_b^2}{2\sigma_m^2} \right) \left(I_0\left(\frac{r_b^2}{2\sigma_m^2} \right) - I_1\left(\frac{r_b^2}{2\sigma_m^2} \right) \right) \right] \tag{11.76}$$

$$J_{13} = J_{31} = \frac{1}{N_0} \int \frac{\partial S(\vec{v};\vec{\theta})}{\partial A_g} \frac{\partial S(\vec{v};\vec{\theta})}{\partial A_b} \mathrm{d}\vec{v} = \frac{2\pi\sigma_g^2}{N_0}\left[1-\exp\left(-\frac{r_b^2}{2(\sigma_g^2+2\sigma_m^2)}\right)\right] \quad (11.77)$$

$$J_{23} = J_{32} = \frac{1}{N_0} \int \frac{\partial S(\vec{v};\vec{\theta})}{\partial \sigma_g} \frac{\partial S(\vec{v};\vec{\theta})}{\partial A_b} \mathrm{d}\vec{v}$$

$$= \frac{4\pi A_g \sigma_g}{N_0}\left[1-\exp\left(-\frac{r_b^2}{2(\sigma_g^2+2\sigma_m^2)}\right)\right] - \frac{2\pi A_g r_b^2 \sigma_g^3 \exp\left(-\frac{r_b^2}{2(\sigma_g^2+2\sigma_m^2)}\right)}{N_0(\sigma_g^2+2\sigma_m^2)^2} \quad (11.78)$$

The expression for the SNR for amplitude estimation with unknown size and background is too complex to give here. Its dependence on σ_m is shown for two values of r_b in Figure 11.12, along with the SNR of Eqs. (11.73) and (11.75). Note that the more complex tasks are more demanding in terms of resolution (which improves toward the left of the figure).

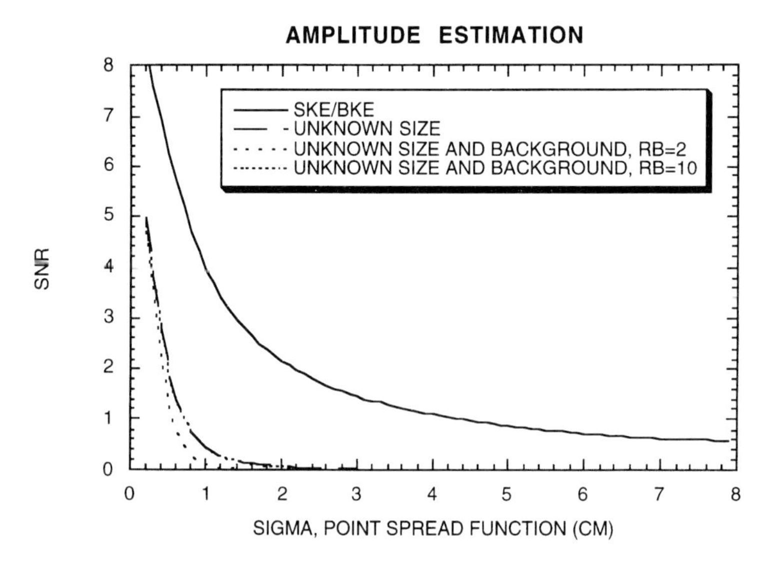

Figure 11.12. SNR for amplitude estimation as a function of PSF size. Three estimation tasks are shown: SKE/BKE amplitude estimation (Eq. 11.75), amplitude estimation with size uncertainty (Eq. 11.73), and amplitude estimation with size uncertainty and background amplitude uncertainty for two background diameters (2 cm and 10 cm).

11.5 References

1. Sharp P, Barber D C, Brown D G, et al, Medical imaging – the assessment of image quality. Report 54, (1996) International Commission on Radiation Units and Measurements.
2. Barrett H H, Gifford H C, Cone-beam tomography with discrete data sets. *Phys. Med. Biol.* **39** 451–476 (1994).
3. Barrett H H, Denny J L, Wagner R F, Myers K J, Objective assessment of image quality. II. Fisher information, Fourier crosstalk, and figures of merit for task performance. *J. Opt. Soc. Am. A* **12** 834–852 (1995).
4. Barrett H H, Objective assessment of image quality: effects of quantum noise and object variability. *J. Opt. Soc. Am. A* **7** 1266–1278 (1990).
5. Judy P F, Swensson R G, Szulc M, Lesion detection and signal-to-noise ratio in CT images. *Med. Phys.* **8** 13–23 (1981).
6. Judy P F, Swensson R G, Display thresholding of images and observer detection performance. *J. Opt. Soc. Am. A* **4** 954–965 (1987).
7. Metz C E, Doi K, Transfer function analysis of radiographic imaging systems. *Phys. Med. Biol.* **24** 1079–1106 (1979).
8. Bracewell R N, *The Fourier Transform and Its Applications* (New York: Mc Graw Hill) (1978).
9. Metz C E, Strubler K A, Rossmann K, Choice of line spread function sampling distance for computing the MTF of radiographic screen-film systems. *Phys. Med. Biol.* **17** 638–647 (1972).
10. Judy P F, The line spread function and modulation transfer function of a computed tomographic scanner. *Med. Phys.* **3** 233–236 (1976).
11. Spanier J, Oldham K B, *An Atlas of Functions* (Washington: Hemisphere Publishing Corporation) (1987).
12. Barrett H H, Swindell W, *Radiological Imaging: The Theory of Image Formation, Detection, and Processing* (New York: Academic Press) (1981).
13. Metz C E, A mathematical investigation of radioisotope scan image processing. Ph.D. Thesis, University of Pennsylvania (1969).
14. Moore S C, Kijewski M F, Mueller S P, Holman B L, SPECT image noise power: effects of nonstationary projection noise and attenuation compensation. *J. Nucl. Med.* **29** 1704–1709 (1988).
15. Riederer S J, Pelc N J, Chesler D A, The noise power spectrum in computed X-ray tomography. *Phys. Med. Biol.* **23** 446–454 (1978).
16. Kijewski M F, Judy P F, The noise-power spectrum of CT images. *Phys. Med. Biol.* **32** 565–575 (1987).
17. Deans S R, *The Radon Transform and Some of its Applications* (New York: John Wiley) (1983).
18. Wagner R F, Brown D G, Unified SNR analysis of medical imaging systems. *Phys. Med. Biol* . **30** 498–518 (1985).
19. Hanson K M, Method of evaluating image-recovery algorithms based on task performance. *J. Opt. Soc. Am. A* **7** 1294–1304 (1990).
20. Joseph P M, Schulz R A, View sampling requirements in fan beam computed tomography. *Med. Phys.* **7** 692–702 (1980).
21. Giger M L, Doi K, Investigation of basic imaging properties in digital radiography. 3. Effect of pixel size on SNR and threshold contrast. *Med. Phys.* **12** 201–208 (1985).
22. Glover G H, Pelc N J, Nonlinear partial volume artifacts in X-ray computed tomography. *Med. Phys.* **7** 238–248 (1980).
23. Glover G H, Pelc N J, An algorithm for the reduction of metal clip artifacts in CT reconstructions. *Med. Phys.* **8** 799–807 (1981).
24. Joseph P M, Spital R D, The exponential edge-gradient effect in X-ray computed tomography. *Phys Med Biol* **26** 473–487 (1981).

25. Hanson K M, Variations in task and the ideal observer. *Proc. Soc. Photo-Opt. Instr. Eng.* **419** 60–67 (1983).
26. Burgess A E, Wagner R F, Jennings R J, Barlow H B, Efficiency of human visual signal discrimination. *Science* **214** 93–94 (1981).
27. Myers K J, Barrett H H, Borgstrom M C, et al, Effect of noise correlations on detectability of disk signals in medical imaging. *J. Opt. Soc. Am. A* **2** 1752–1759 (1985).
28. Loo L D, Doi K, Metz C E, A comparison of physical image quality indices and observer performance in the radiographic detection of nylon beads. *Phys. Med. Biol.* **29** 837–856 (1984).
29. Burgess A E, Statistically defined backgrounds: performance of a modified nonprewhitening matched filter model. *J. Opt. Soc. Am. A* **11** 1237–1242 (1994).
30. Burgess A E, Li X, Abbey C K, Visual signal detectability with two noise components: anomalous masking effects. *J. Opt. Soc. Am. A* **14** 2420–2442 (1997).
31. Myers K J, Barrett H H, Addition of a channel mechanism to the ideal-observer model. *J. Opt. Soc. Am. A* **4** 2447–2457 (1987).
32. Wagner R F, Myers K J, Brown D G, et al, Higher-order tasks: Human vs. machine performance. *Proc. Soc. Photo-Opt. Instr. Eng.* **1090** 183–194 (1989).
33. Barrett H H, Smith W E, Myers K J, et al, Quantifying the performance of imaging systems. *Proc. Soc. Photo-Opt. Instr. Eng.* **535** 65–69 (1985).
34. Fiete R D, Barrett H H, Smith W E, Myers K J, Hotelling trace criterion and its correlation with human-observer performance. *J. Opt. Soc. Am. A* **4** 945–953 (1987).
35. Fiete R D, Barrett H H, Cargill E B, et al, Psychophysical validation of the Hotelling trace criterion as a metric for system performance. *Proc. Soc. Photo-Opt. Instr. Eng.* **767** 298–305 (1987).
36. Rolland J P, Barrett H H, Effect of random background inhomogeneity on observer detection performance. *J. Opt. Soc. Am. A* **9** 649–658 (1992).
37. Yao J, Barrett H H, Predicting human performance by a channelized Hotelling observer model. *Proc. Soc. Photo-Opt. Instr. Eng.* **1768** 161–168 (1992).
38. Wester D, Judy P F, Polger M, et al, Influence of visual distractors on detectability of liver nodules on contrast-enhanced spiral computed tomography scans. *Acad. Radiol.* **4** 335–342 (1997).
39. Metz C E, Kronman H B, Statistical significance tests for binormal ROC curves. *J. Math. Psych.* **22** 218–243 (1980).
40. Metz C E, Herman B A, Shen J-H, Maximum-likelihood estimation of ROC curves from continuously-distributed data. *Statistics in Medicine* **17** 1033–1053 (1998).
41. Green D M, Swets J A, *Signal Detection Theory and Psychophysics* (New York: Wiley) (1966).
42. Burgess A E, Comparison of receiver operating characteristic and forced choice observer performance measurement methods. *Med. Phys.* **22** 643–655 (1995).
43. Tanner W P, Birdsall T G, Definitions of d' and h as psychophysical measures. *J. Acoust. Soc. Am.* **30** 922–928 (1958).
44. Burgess A E, Jennings R J, Wagner R F, Statistical efficiency: a measure of human visual signal-detection performance. *J. Appl. Photogr. Eng.* **8** 76–78 (1982).
45. King M A, Schwinger R B, Doherty P W, Penney B C, Two-dimensional filtering of SPECT images using the Metz and Wiener filters. *J. Nucl. Med.* **25** 1234–1240 (1984).
46. Mueller S P, Kijewski M F, Moore S C, Holman B L, Maximum-likelihood estimation – a mathematical model for quantitation in nuclear medicine. *J. Nucl. Med.* **31** 1693–1701 (1990).
47. Van Trees H, *Detection, Estimation and Modulation Theory* (New York: Wiley) (1968).

12 A Glimpse Into the Future: Diagnostic Applications

William R. Hendee, Ph.D.

Medical College of Wisconsin

12.1 Introduction

The preceding chapters of this volume reveal the extraordinary contributions of radiation-based imaging technologies to the elucidation of human anatomy and physiology and to the detection and diagnosis of disease and injury. Before the discovery of X rays in 1895, physicians had few tools beyond their own human senses to identify the causes of patient complaints and disorders. The present century has witnessed a remarkable expansion in noninvasive tools to probe the internal structure and function of patients and to characterize the causes of human illness and injury. This expansion has led to a phenomenal growth in the knowledge of human anatomy, physiology, biochemistry, and metabolism, and in the understanding of cellular, molecular and genetic causes underlying human disorders and disease. The ingenuity, creativity and dedication of scientists and physicians working worldwide are what have made this growth in knowledge and understanding possible.

12.2 Frontiers in Medical Imaging

Medical science is an exciting and challenging career path. Many young people with exceptional backgrounds in mathematics, the physical and biological sciences, engineering and medicine are entering the discipline. This influx of bright, energetic and dedicated young people helps to ensure the continued evolution of diagnostic technologies to improve human health and alleviate human pain and suffering. Advances are being made today in every aspect of medical science, including the innovation frontiers listed in Table 12.1.

Table 12.1. Innovation Frontiers in Medical Science

Cellular Biology	Molecular Biology
Computational Biology	Cybernetics
Genetics	Imaging
Neural Networks	Information Systems
Nanotechnologies Applications	Microelectronics Applications

These frontiers, and the rate of advances in each of them, serve to solidify the foundation for further development of imaging technologies applied to medical challenges.

As revealed in the preceding chapters of this volume, many energy sources have been employed to produce medical images. These sources are summarized in Table 12.2.

Table 12.2. Energy Sources Employed in Medical Imaging

Visible and UV Light	Infrared Radiation
Microwaves	Ultrasound
X Rays	Gamma Rays
Annihilation Photons	Applied Voltages
Electric Fields	Magnetic Fields

The different energy sources provide a spectrum of approaches to exploitation of tissue properties in the formation of medical images. Some of the relevant tissue properties are described in Table 12.3.

Table 12.3. Tissue Properties Exploited in the Production of Medical Images

Physical Density	Electron Density
Atomic Number	Pharmaceutical Localization
Proton Density	Relaxation Constants
Velocity of Propogation	Electrical Conductivity
Blood Flow Patterns	Neural Activation

The energy sources and tissue properties depicted in Tables 12.2 and 12.3, together with other possibilities not yet identified, offer a matrix of opportunities for continued development of approaches to medical imaging. This matrix suggests that the scientific potential remains strong for continued development of radiation-based imaging technologies as an innovation frontier in diagnostic medicine. Major prospects for advances in medical imaging in the near future include those listed in Table 12.4.

Table 12.4. Major Prospects for Near-Term Medical Imaging Advances

Functional MRI	Digital X-ray Imaging
MR Spectroscopy	Image Superposition
Image-Guided Therapy	3D Visualization
Teleimaging	Image Networking
ComputerAided Diagnosis	Low-Cost Ultrasound

These prospects reflect the trends in applications of medical imaging depicted in Table 12.5.

Table 12.5. Trends in Medical Imaging

From	To
Anatomic	Physiologic and Biochemical
Static	Dynamic
Qualitative	Quantitative
Analog	Digital
Nonspecific Agents	Tissue-Targeted Agents
Detection/Diagnosis	Diagnosis/Therapy

12.3 Challenges to Medical Imaging

Medical imaging technologies are a mainstay of modern diagnostic medicine. In the United States alone, approximately 300 million imaging examinations and procedures are conducted each year, at an estimated cost of $20 billion. This cost contributes about 3.5 percent to the total bill for healthcare services in the country, with high-technology procedures such as MRI, computed tomography and image-guided therapy accounting for about half of the cost. Although the direct cost of medical imaging is only a small fraction of the total healthcare bill, imaging technologies are an easily-identifiable target for cost-reduction measures proposed by advocates of market-driven healthcare strategies. In 1987 Schwartz[1] suggested that:

"Long-term control of the rate of increase in expenditures thus requires that we curb the development and diffusion of clinically useful technology."

The development and diffusion of medical technologies are influenced by a variety of factors, including those described in Table 12.6.

Table 12.6. Factors Influencing the Development and Diffusion of Medical Technologies

Improvement in Quality of Care*	Improvement in Access to Care*
Reduction in Cost of Care*	Acceptance by Physicians and Patients*
and	
Case Reports in the Scientific Literature	
Career Advancement by Technology Proponents	
Marketing Advantages by Owners of Technologies	
Use of Technologies to Attract Patients	
Desire to be Portrayed on the Leading Edge of Medicine	
Financial Gain by Those Developing and Employing Technologies	
Pressure from Referring Physicians and Patients	
Protection of Imaging "Turf"	

* Criteria identified by the National Academy of Sciences' Institute of Medicine as legitimate measures for assessment of the added value of medical technologies[2]

With the exception of criteria identified by the Institute of Medicine as valid measures of a technology's worth, the factors described in Table 12.6 may be less-than-satisfactory causes of the diffusion of a technology within the arena of clinical medicine.

12.4 Control of Medical Technology Diffusion

Since the mid-1970s, several efforts have been directed in the United States to control the rate of diffusion and use of expensive medical technologies, including high-profile imaging technologies such as computed tomography and MRI. The earliest efforts began with establishment of state-run Health Systems Agencies and Utilization Review Panels to review applications by healthcare providers to acquire and use new medical technologies, and to expand healthcare facilities. This "front-end" approach added several layers of bureaucracy to the acquisition and use of medical technologies, but did little to suppress the diffusion of new technologies or slow the rate of increase in healthcare costs. After several years the approach was widely recognized as failing to achieve its intended purpose.

In the mid-1980s a "back-end" approach was developed to control the diffusion of medical technologies. This approach was initiated with passage of the Tax Equity and Fiscal Responsibility Act (TEFRA) of 1983. This act led directly to establishment of Diagnosis-Related Groups (DRGs) by the Health Care Financing Administration (HCFA) to cap the reimbursement of hospital (Part A) and physician (Part B) charges paid by Medicare. Since reimbursement by other third-party payers often keys off payments by HCFA, DRGs promised to have an impact extending far beyond reimbursement for Medicare patients. Although DRGs provided an initial reduction in the rate of increase in healthcare costs, the effect was temporary, and costs resumed their earlier rate of increase after a hiatus of a couple of years.

In the early 1990s healthcare services in the United States entered a new era described by Relman[3] as the Era of Accountability. This era is often characterized by the terms "Managed Care" and "Healthcare Reform." These terms are euphimisms because they mask the real transformation of healthcare services over the past few years from a protected environment into one governed by market forces. Managed care was pioneered many years ago by health maintenance organizations (HMOs) which provide healthcare services to a group of people ("covered lives") for a contractural fee set in advance. Under the traditional approach to payment for healthcare services (fee for services rendered), technologies such as medical imaging were viewed as "profit centers" by healthcare institutions because each use of a technology yielded a payment to both the institution and the physician. Under the new approach of "capitated care" (i.e., fixed-cost contracts for healthcare services to groups of individuals), medical technologies were immediately transformed into "cost centers," because each use creates a cost for which there is no added reimbursement. This transformation has had a chilling effect on the development and acquisition of new medical technologies, including imaging technologies employing radiation. The ultimate response to this effect by developers and users of technologies will greatly influence the future rate of growth of medical technologies in support of improvements in patient care.

12.5 Accountability in Health Care

Accountability in health care means much more than controlling healthcare costs. The overarching concept of accountability is depicted in Figure 12.1.

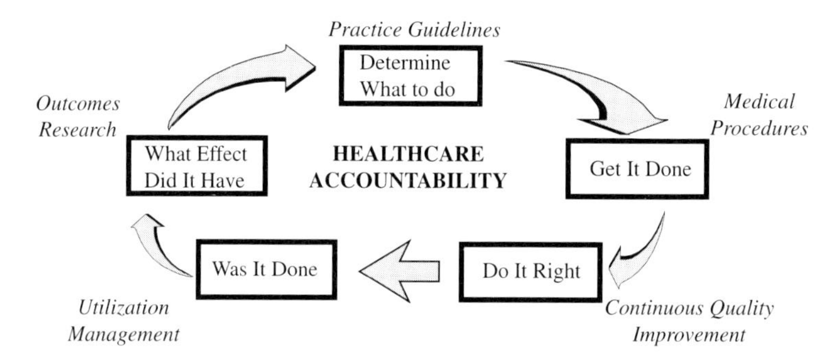

Figure 12.1. Schematic of Healthcare Accountability (courtesy of D. Eddy)

In Figure 12.1, outcomes research refers to the evaluation of medical procedures to determine their impact on the well-being of patients. Typical outcome measures related to patient welfare include those described by the Institute of Medicine[2] and listed in Table 12.7.

Table 12.7. Representative Outcome Measures for Healthcare Services

Clinical Status *physiologic* *cognitive*	Feeling of Well-Being *mental* *emotional*
Feeling of Vitality Reduced Morbidity Return to Work	Functional Capacity Reduced Mortality Absence of Rehospitalization
Quality-Adjusted Life Years Intermediate Outcome Measures *biochemical* *metabolic* *physiologic* *service needs*	

The outcome measures in Table 12.7 work reasonably well for therapeutic procedures in which a causal pathway is apparant between a procedure and one or more measurable health outcomes. For most diagnostic procedures, however, including those employing medical technologies such as imaging methods, the causal relationship may be difficult to identify. These technologies have a number of dimensions, as demonstrated in Table 12.8. However, the correlation between their use and some measurable health outcome is usually impossible to demonstrate, often because the use and the outcome are separated by one or more therapeutic procedures, or because the diagnostic method yields negative findings, for which no therapeutic procedure is indicated.

Table 12.8. Dimensions of Diagnostic Technologies (after Fineberg et al[4])

Technical Capacity:	capability, safety, reliability
Diagnostic Accuracy:	correct information
Diagnostic Impact	contribution to correct diagnosis
Therapeutic Impact	effect on patient management
Patient Outcomes:	reduced mortality, reduced morbidity, improved well-being, etc

In responding to the demand for healthcare accountability, persons employing a diagnostic technology face the challenge of identifying the particular dimension in Table 12.8 that presents the most appropriate target for determining the technology's value. For many diagnostic technologies, the degree of therapeutic impact may be the most appropriate target. That is, the technology makes a positive contribution to patient welfare if, in an adequate percentage of cases, it influences whether the patient is treated and the way the patient is managed through the course of treatment.

12.6 Healthcare Changes and Strategies

In the United States, the delivery of health care is undergoing a massive transformation. At present, this transformation is focused on controlling the rate of increase in healthcare costs. Changes occuring as a result of this transformation include those depicted in Table 12.9.

Table 12.9. Changes in Healthcare Delivery (after Brody[5])

Reduced Emphasis	Enhanced Emphasis
Inpatient Capacity *reduced beds* *closed hospitals*	Utilization Control *gatekeepers* *technology protocols*
Patient Care Activity	Healthcare Outcomes
Individual Patients *entrepreneurial practices* *patient choice*	Patient Populations *group contracts* *covered lives*
Fee-For Service	Managed Care
Episodic Care	Disease and Injury Prevention
Specialty Care	Primary Care
Healthcare as Cottage Industry	Merger of Healthcare Resources
Individual Responsibility	Team Approach to Health Care

These changes are occurring today across the country. Some communities are far along in accommodating the changes, while others are just beginning to feel their impact. In these communities, patients, providers and payers are experiencing a major transformation in the way health care is offered. Even where the transformation in healthcare delivery is in its final stages, many issues remain unanswered. These issues, some of which are denoted in Table 12.10, provide a target for future efforts to improve the quality and cost-effectiveness of healthcare services.

Table 12.10. Remaining Issues in Healthcare Delivery (after Brody[5])

Reconnect Patients and Providers *identify common values* *form provider-patient teams*	Improve Healthcare Management *quality* *cost*
Measure Quality	Employ Utilization Review
Assess Outcomes *individual patients* *patient populations*	Improve Information Systems *electronic patient records* *patient data repositories*
Assess Medical Technologies	Address Long-Term Healthcare Needs
Improve Rural Health Care	Address Urban Healthcare Needs

Present and future issues arising from the transformation in healthcare delivery yield several challenges for medical imaging. Some of these challenges are listed in Table 12.11.

Table 12.11. Future Challenges for Medical Imaging

Improve Imaging Applications	Medical Decision-Making
image-guided therapy	*improve understanding*
image-monitored therapy	*develop imaging protocols/guidelines*
Assess Imaging Technologies	Emphasize Imaging as Information
quality and cost	Manufacturer/Purchaser Risk Sharing
cost savings	Horizontal and Vertical Integration

These challenges, and others that will evolve from continued developments in medical technology and healthcare delivery, present many opportunities for scientists, physicians and technologists. Health care is changing dramatically in the United States and throughout the developed world. A time of change is always a time of opportunity for those willing to seize it. As Harry Lime (Orson Welles) said in the movie The Third Man:

"In Italy for 30 years under the Borgias they had warfare, terror, murder, bloodshed. They produced Michelangelo, Leonardo de Vinci and the Renaissance. In Switzerland they had brotherly love, 500 years of democracy and peace. And what did that produce? The cuckoo clock."

12.7 References

1. W.B. Schwartz, *JAMA* **257**: 220 (1987).
2. Institute of Medicine, *Telemedicine: A Guide to Assessing Telecommunications in Health Care*, M.J. Field, Ed. (National Academy Press, Washington, DC, 1996).
3. A.S. Relman, *N. Engl. J. Med.* **319**, 1220 (1989).
4. H.V. Fineberg, in *Assessing Medical Technologies,* Institute of Medicine (National Academy Press, Washington, DC, 1985), pp. 176-210.
5. W. Brody, *keynote address* of the Sixth Annual Event of the American Institute of Medical and Biological Engineering, Washington, DC, March 1-4, 1997.